Quantum Theory of Materials

This accessible new text introduces the theoretical concepts and tools essential for graduate-level courses on the physics of materials in condensed matter physics, physical chemistry, materials science and engineering, and chemical engineering.

Topics covered range from fundamentals such as crystal periodicity and symmetry, and derivation of single-particle equations, to modern additions including graphene, two-dimensional solids, carbon nanotubes, topological states, and Hall physics. Advanced topics such as phonon interactions with phonons, photons, and electrons, and magnetism, are presented in an accessible way, and a set of appendices reviewing crucial fundamental physics and mathematical tools makes this text suitable for students from a range of backgrounds.

Students will benefit from the emphasis on translating theory into practice, with worked examples explaining experimental observations, applications illustrating how theoretical concepts can be applied to real research problems, and 242 informative full-color illustrations. End-of-chapter problems are included for homework and self-study, with solutions and lecture slides for instructors available online.

Efthimios Kaxiras is the John Hasbrouck Van Vleck Professor of Pure and Applied Physics at Harvard University. He holds joint appointments in the Department of Physics and the School of Engineering and Applied Sciences, and is an affiliate of the Department of Chemistry and Chemical Biology. He is the Founding Director of the Institute for Applied Computational Science, a Fellow of the American Physical Society, and a Chartered Physicist and Fellow of the Institute of Physics, London.

John D. Joannopoulos is the Francis Wright Davis Professor of Physics at MIT, where he is Director of the Institute for Soldier Nanotechnologies. He is a member of the National Academy of Sciences and American Academy of Arts and Sciences, a Fellow of the American Association for the Advancement of Science, a Fellow of the American Physical Society, and a Fellow of the World Technology Network. His awards include the MIT School of Science Graduate Teaching Award (1991), the William Buechner Teaching Prize of the MIT Department of Physics (1996), the David Adler Award (1997) and Aneesur Rahman Prize (2015) of the American Physical Society, and the Max Born Award of the Optical Society of America.

Quantum Theory of Materials

EFTHIMIOS KAXIRAS

Harvard University

JOHN D. JOANNOPOULOS

Massachusetts Institute of Technology

CAMBRIDGE
UNIVERSITY PRESS

CAMBRIDGE
UNIVERSITY PRESS

University Printing House, Cambridge CB2 8BS, United Kingdom

One Liberty Plaza, 20th Floor, New York, NY 10006, USA

477 Williamstown Road, Port Melbourne, VIC 3207, Australia

314–321, 3rd Floor, Plot 3, Splendor Forum, Jasola District Centre, New Delhi – 110025, India

79 Anson Road, #06-04/06, Singapore 079906

Cambridge University Press is part of the University of Cambridge.

It furthers the University's mission by disseminating knowledge in the pursuit of education, learning, and research at the highest international levels of excellence.

www.cambridge.org
Information on this title: www.cambridge.org/9780521117111
DOI: 10.1017/9781139030809

First published 2019

Printed in Singapore by Markono Print Media Pte Ltd

A catalogue record for this publication is available from the British Library.

ISBN 978-0-521-11711-1 Hardback

Additional resources for this publication at www.cambridge.org/9780521117111

Contents

List of Figures

List of Tables

Preface

Why do various materials behave the way they do? For instance, what makes a material behave like a good insulator, instead of being a good conductor or a semiconductor? What determines the strength of a material? How can we account for the color of different solids? Questions like these have attracted curious minds for centuries. Materials, after all, are of central importance to humanity: they define the stage of civilization, as in "Stone Age," "Bronze Age," "Iron Age," and the current "Silicon Age." The scientific study of the properties of materials in the last two centuries has produced a body of knowledge referred to as the "physics of materials" that goes a long way toward explaining and even *predicting* their properties from first-principles theoretical concepts. Our book aims to present these concepts in a concise and accessible manner.

The book emerged as the result of many years of teaching this subject at Harvard and MIT. The intended audience is graduate or advanced undergraduate students in physics, applied physics, materials science, chemistry, and related engineering and applied science fields. There are classic textbooks on the subject, the venerable work by N. W. Ashcroft and N. D. Mermin, *Solid State Physics*, being a standard example; there are also numerous more recent works, for instance *Fundamentals of Condensed Matter Physics* by M. L. Cohen and S. G. Louie, a work of great depth and clarity, and the delightfully intuitive *Physics of Solids* by E. N. Economou. We mention most of these books as suggestions for further reading at the end of each chapter, as appropriate. Taken together, these sources quite nicely cover all important aspects of the subject. The present work aims to fill a gap in the literature, by providing a single book that covers all the essential topics, including recent advances, at a level that can be accessible to a wider audience than the typical graduate student in condensed matter physics. This is what prompted us to use the word "materials" (rather than "solids" or "condensed matter") in the title of the book. Consistent with this aim, we have included topics beyond the standard fare, like elasticity theory and group theory, that hopefully cover the needs, and address the interests, of this wider community of readers.

To facilitate accessibility, we have intentionally kept the mathematical formalism at the simplest possible level, for example, avoiding second quantization notation except when it proved absolutely necessary (the discussion of the BCS model for superconductivity, Chapter 8). Instead, we tried to emphasize physical concepts and supply all the information needed to motivate how they translate into specific expressions that relate physical quantities to experimental measurements.

The book concentrates on theoretical concepts and tools, developed during the last few decades to understand the properties of materials. As such, we did not undertake an extensive survey of experimental data. Rather, we compare the results of the theoretical models to key experimental findings throughout the book. We also give examples of how the theory can be applied to explain what experiment observes, as well as several compilations of experimental data to capture the range of behavior encountered in various types of materials.

The book can be used to teach a one-semester graduate-level course (approximately 40 hours of lecture time) on the physics of materials. For an audience with strong physics and math background and some previous exposure to solid-state physics, this can be accomplished by devoting an introductory lecture to Chapter 1, and covering the contents of Chapters 2–7 thoroughly. Topics from Chapters 8, 9, and 10 can then be covered as time permits and the instructor's interest dictates. An alternative approach, emphasizing more the applications of the theory and aimed at an audience with no prior exposure to solid-state physics, is to cover thoroughly Chapters 1 and 2, skip Chapter 3, cover Sections 4.1–4.7, Chapters 5 and 6, Sections 7.1–7.5, Sections 8.3 and 8.4, and selected topics from Chapters 9 and 10 as time permits.

Many examples and applications are carefully worked out in the text, illustrating how the theoretical concepts and tools can be applied to simple and more sophisticated models. Not all of these need to be presented in lectures; in fact, the reason for giving their detailed solutions was to make it possible for the student to follow them on their own, reserving lecture time for discussions of key ideas and derivations. We have also included several problems at the end of each chapter and we strongly encourage the interested student to work through them in detail, as this is the only meaningful way of mastering the subject.

Finally, we have included an extensive set of appendices, covering basic mathematical tools and elements from classical electrodynamics, quantum mechanics, and thermodynamics and statistical mechanics. The purpose of these appendices is to serve as a reference for material that students may have seen in a different context or at a different level, so that they can easily refresh their memory of it, or become familiar with the level required for understanding the discussion in the main text, without having to search a different source.

Acknowledgments

When trying to explain something, it is often difficult to keep track of how your own understanding of the ideas was formed. Acknowledging all the sources of inspiration and insight can therefore be an impossible task, but some of these sources definitely stand out.

We wish to thank our teacher and mentor Marvin Cohen as an especially important influence on our thinking and understanding of physics. EK wishes to express a deep debt of gratitude to his teachers at various stages of his career, John Joannopoulos, Kosal Pandey, and Lefteris Economou, all of whom served as role models and shaped his thinking. Many colleagues have played an equally important role, including Bert Halperin, David Nelson, Nihat Berker, and Stratos Manousakis.

The students who patiently followed several iterations of the manuscript, helping to improve it in ways big and small, are the ultimate reason for putting a vast collection of hand-written notes into a coherent text – we thank them sincerely, and hope they benefitted from the experience. Some of them put an extraordinary amount of energy and care into writing solutions to the problems at the end of the chapters, for which we are thankful. The most recent and careful compilation was made by Cedric Flamant, who also corrected some problem statements and several inaccuracies in the text – he deserves special thanks. Daniel Larson read Chapter 9 very carefully, and provided much useful feedback and suggestions for improving the presentation. We also thank Eugene Mele for many useful comments on the physics of topological states.

Last but not least, we wish to acknowledge the essential, or more appropriately, existential role that our partners Eleni and Kyriaki played in this endeavor; they have put up with innumerable hours of distraction while we were working out some idea to include in the book. Without their support and patience, the book would simply not exist; we gratefully dedicate it to them.

1 From Atoms to Solids

Materials exhibit an extremely wide range of properties, which is what makes them so useful and indispensable to humankind. The extremely wide range of the properties of materials is surprising, because most of them are made up from a relatively small subset of the elements in the Periodic Table: about 20 or 30 elements, out of more than 100 total, are encountered in most common materials. Moreover, most materials contain only very few of these elements, from one to half a dozen or so. Despite this relative simplicity in composition, materials exhibit a huge variety of properties over ranges that differ by many orders of magnitude. It is quite extraordinary that even among materials composed of single elements, physical properties can differ by many orders of magnitude.

One example is the ability of materials to conduct electricity. What is actually measured in experiments is the resistivity, that is, the difficulty with which electrical current passes through a material. Some typical single-element metallic solids (like Ag, Cu, Al) have room-temperature resistivities of 1–5 $\mu\Omega$·cm, while some metallic alloys (like nichrome) have resistivities of 10^2 $\mu\Omega$·cm. All these materials are considered good conductors of electrical current. Certain single-element solids (like Si and Ge) have room-temperature resistivities much higher than good conductors, for instance, 2.3×10^{11} $\mu\Omega$·cm for Si, and they are considered semiconductors. Finally, certain common materials like wood (with a rather complex structure and chemical composition) or quartz (with a rather simple structure and composed of two elements, Si and O) have room-temperature resistivities of 10^{16}–10^{19} $\mu\Omega$·cm (for wood) to 10^{25} $\mu\Omega$·cm (for quartz). These solids are considered insulators. The range of electrical resistivities covers an impressive 25 orders of magnitude. Even for two materials that are composed of the same element, carbon, their resistivity can differ by many orders of magnitude: graphitic carbon has resistivity 3.5×10^3 $\mu\Omega$·cm and is considered a semimetal, while in the diamond form, carbon has a much higher resistivity of $\sim 10^{22}$ $\mu\Omega$·cm, the difference being entirely due to the different crystal structure of the two forms, as shown in Fig. 1.1.

Another example has to do with the mechanical properties of materials. Solids are classified as ductile when they yield plastically when stressed, or brittle when they do not yield easily, but instead break when stressed. A useful measure of this behavior is the yield stress σ_Y, which is the stress up to which the solid behaves as a linear elastic medium when stressed, that is, it returns to its original state when the external stress is removed. Yield stresses in materials, measured in units of MPa, range from 40 in Al, a rather soft and ductile metal, to 5×10^4 in diamond, the hardest material, a brittle insulator. The yield stresses of common steels range from 200 to 2000 MPa. Again we see an impressive range of more than three orders of magnitude in how a material responds to an external agent, in this case a mechanical stress.

Different forms of carbon-based solids. **Left**: A diamond in raw form, in the shape of an ideal octahedron. The structure of the diamond crystal offers many equivalent planes along which it can be cleaved, exposing a shape of very high symmetry in polished ("cut") form. **Right**: Graphite [*source:* Rob Lavinsky, iRocks.com CC-BY-SA-3.0, via Wikimedia Commons].

Naively, one might expect that the origin of these widely different properties is related to great differences in the concentration of atoms, and correspondingly that of electrons. This is far from the truth. Concentrations of atoms in a solid range from 10^{22} cm^{-3} in Cs, a representative alkali metal, to 17×10^{22} cm^{-3} in C, a representative covalently bonded solid. Anywhere from one to a dozen electrons per atom participate actively in determining the properties of solids. These considerations give a range of atomic concentrations of roughly 20, and of available electron concentrations[1] of roughly 100. These ranges are nowhere close to the ranges of yield stresses and electrical resistivities mentioned above. Rather, the variation of the properties of solids has to do with the specific ways in which the available electrons of the constituent atoms interact when these atoms are brought together at distances of a few angstroms (1 Å $= 10^{-10}$ m $= 10^{-1}$ nm). Typical distances between nearest-neighbor atoms in solids range from 1.5 to 3 Å. The way in which the available electrons interact determines the atomic structure, and this in turn determines all the other properties of the solid, including mechanical, electrical, optical, thermal, and magnetic properties.

1.1 Electronic Structure of Atoms

We invoked above, on several occasions, the term "available electrons" for specific types of atoms. This is not a very precise term, but was intended to convey the message that not all of the electrons of an atom are necessarily involved in determining the properties of a solid in which this atom exists. Some electrons eagerly participate in the formation of the solid, by being shared among several atoms and thus serving as the "glue" that holds those atoms together, while other electrons are very tightly bound to the nuclei of the constituent

[1] The highest concentration of atoms does not correspond to the highest number of available electrons per atom.

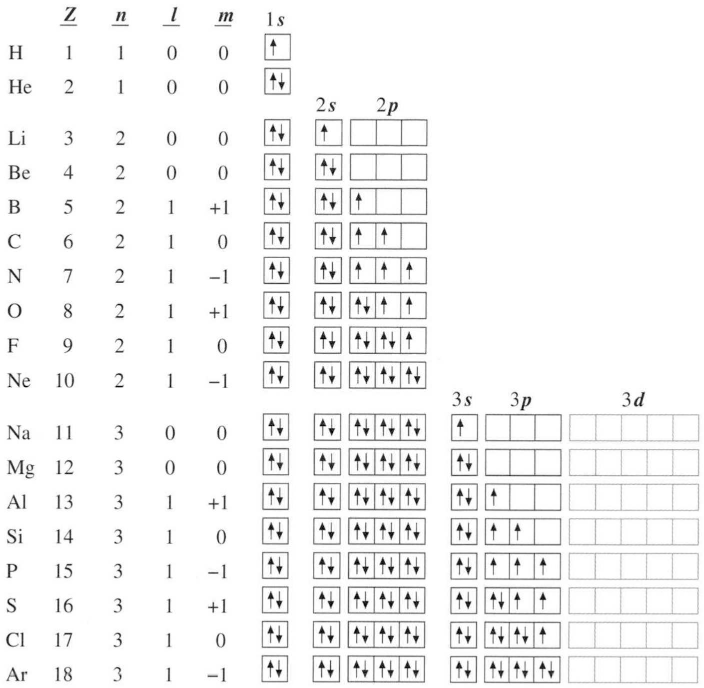

	Z	n	l	m
H	1	1	0	0
He	2	1	0	0
Li	3	2	0	0
Be	4	2	0	0
B	5	2	1	+1
C	6	2	1	0
N	7	2	1	−1
O	8	2	1	+1
F	9	2	1	0
Ne	10	2	1	−1
Na	11	3	0	0
Mg	12	3	0	0
Al	13	3	1	+1
Si	14	3	1	0
P	15	3	1	−1
S	16	3	1	+1
Cl	17	3	1	0
Ar	18	3	1	−1

Fig. 1.2 Filling of electronic shells for the elements in the Periodic Table with atomic numbers $Z = 1$–18: s shells ($l = 0, m = 0$) are represented by one red box, p shells ($l = 1, m = 0, \pm 1$) by three blue boxes, and d shells ($l = 2, m = 0, \pm 1, \pm 2$) by five green boxes. Each box can contain up to two electrons of opposite spin (up and down arrows). For each element, we show the values of the principal (n) and angular momentum (l) quantum numbers, as well as the value of the magnetic (m) quantum number of the last electron.

atoms, almost in exactly the same way as if the atoms were in complete isolation from each other; the latter electrons play essentially no role in determining the properties of the solid. This is an important distinction, which deserves further discussion.

The solution of the Schrödinger equation for the isolated atom produces a number of states in which electrons can exist. The total number of electrons in the neutral atom, equal to the total number of protons in its nucleus, called the **atomic number** Z, determines how many of the allowed states are occupied. The occupation of the allowed states leads to a certain number of shells being completely full (the ones with the lowest energy and wavefunctions that are closest to the nucleus), while some shells can be partially filled (the ones with higher energy and wavefunctions that extend farther from the nucleus).[2] A schematic representation of the filling of electronic shells for atoms with $Z = 1$–36 is given in Figs 1.2 and 1.3, where it becomes obvious which shells are completely filled and which are partially filled by electrons for each element. A continuation of this sequence of filling of electronic shells leads to the Periodic Table, presented in the next section.

The periodic filling of the electronic shells is actually quite evident in some basic properties of atoms, like their physical size or their tendency to attract electrons, referred

---

[2] A more detailed discussion is given in Appendix C.

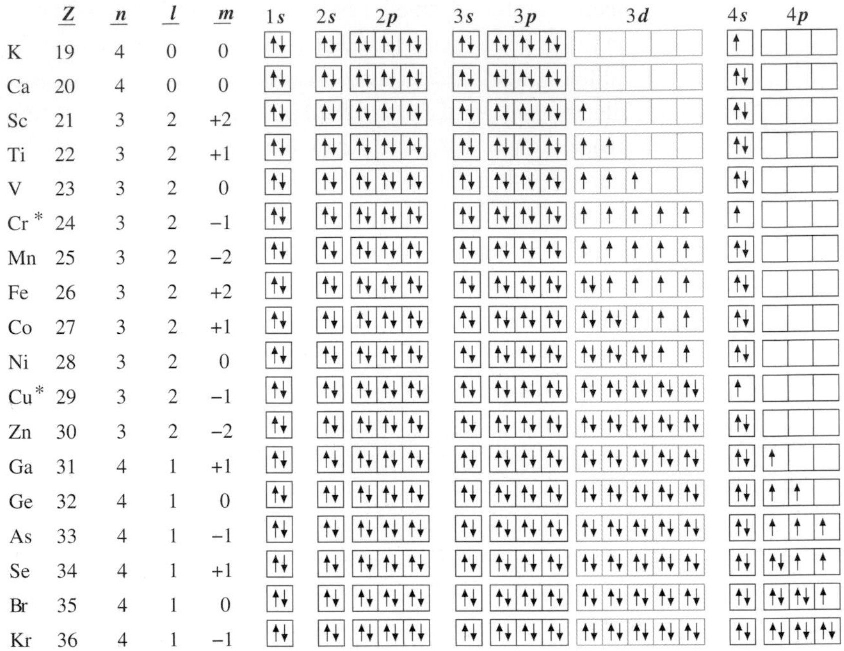

	Z	n	l	m	1s	2s	2p	3s	3p	3d	4s	4p
K	19	4	0	0	↑↓	↑↓	↑↓ ↑↓ ↑↓	↑↓	↑↓ ↑↓ ↑↓		↑	
Ca	20	4	0	0	↑↓	↑↓	↑↓ ↑↓ ↑↓	↑↓	↑↓ ↑↓ ↑↓		↑↓	
Sc	21	3	2	+2	↑↓	↑↓	↑↓ ↑↓ ↑↓	↑↓	↑↓ ↑↓ ↑↓	↑	↑↓	
Ti	22	3	2	+1	↑↓	↑↓	↑↓ ↑↓ ↑↓	↑↓	↑↓ ↑↓ ↑↓	↑ ↑	↑↓	
V	23	3	2	0	↑↓	↑↓	↑↓ ↑↓ ↑↓	↑↓	↑↓ ↑↓ ↑↓	↑ ↑ ↑	↑↓	
Cr*	24	3	2	−1	↑↓	↑↓	↑↓ ↑↓ ↑↓	↑↓	↑↓ ↑↓ ↑↓	↑ ↑ ↑ ↑ ↑	↑	
Mn	25	3	2	−2	↑↓	↑↓	↑↓ ↑↓ ↑↓	↑↓	↑↓ ↑↓ ↑↓	↑ ↑ ↑ ↑ ↑	↑↓	
Fe	26	3	2	+2	↑↓	↑↓	↑↓ ↑↓ ↑↓	↑↓	↑↓ ↑↓ ↑↓	↑↓ ↑ ↑ ↑ ↑	↑↓	
Co	27	3	2	+1	↑↓	↑↓	↑↓ ↑↓ ↑↓	↑↓	↑↓ ↑↓ ↑↓	↑↓ ↑↓ ↑ ↑ ↑	↑↓	
Ni	28	3	2	0	↑↓	↑↓	↑↓ ↑↓ ↑↓	↑↓	↑↓ ↑↓ ↑↓	↑↓ ↑↓ ↑↓ ↑ ↑	↑↓	
Cu*	29	3	2	−1	↑↓	↑↓	↑↓ ↑↓ ↑↓	↑↓	↑↓ ↑↓ ↑↓	↑↓ ↑↓ ↑↓ ↑↓ ↑↓	↑	
Zn	30	3	2	−2	↑↓	↑↓	↑↓ ↑↓ ↑↓	↑↓	↑↓ ↑↓ ↑↓	↑↓ ↑↓ ↑↓ ↑↓ ↑↓	↑↓	
Ga	31	4	1	+1	↑↓	↑↓	↑↓ ↑↓ ↑↓	↑↓	↑↓ ↑↓ ↑↓	↑↓ ↑↓ ↑↓ ↑↓ ↑↓	↑↓	↑
Ge	32	4	1	0	↑↓	↑↓	↑↓ ↑↓ ↑↓	↑↓	↑↓ ↑↓ ↑↓	↑↓ ↑↓ ↑↓ ↑↓ ↑↓	↑↓	↑ ↑
As	33	4	1	−1	↑↓	↑↓	↑↓ ↑↓ ↑↓	↑↓	↑↓ ↑↓ ↑↓	↑↓ ↑↓ ↑↓ ↑↓ ↑↓	↑↓	↑ ↑ ↑
Se	34	4	1	+1	↑↓	↑↓	↑↓ ↑↓ ↑↓	↑↓	↑↓ ↑↓ ↑↓	↑↓ ↑↓ ↑↓ ↑↓ ↑↓	↑↓	↑↓ ↑ ↑
Br	35	4	1	0	↑↓	↑↓	↑↓ ↑↓ ↑↓	↑↓	↑↓ ↑↓ ↑↓	↑↓ ↑↓ ↑↓ ↑↓ ↑↓	↑↓	↑↓ ↑↓ ↑
Kr	36	4	1	−1	↑↓	↑↓	↑↓ ↑↓ ↑↓	↑↓	↑↓ ↑↓ ↑↓	↑↓ ↑↓ ↑↓ ↑↓ ↑↓	↑↓	↑↓ ↑↓ ↑↓

Fig. 1.3   Filling of electronic shells for elements with atomic numbers $Z = 19$–36. The notation is the same as in Fig. 1.2. Elements marked by an asterisk have electronic levels filled out of the regular order: for example, Cu has a $3d^{10}4s^1$ shell rather than a $3d^9 4s^2$ shell.

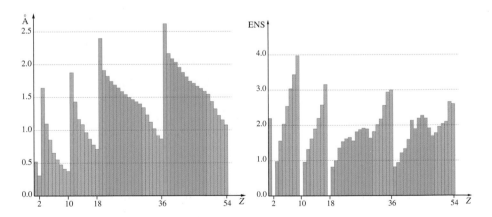

Fig. 1.4   Periodic behavior of the properties of the elements: atomic radii (Å) and electronegativity scale (ENS), for elements with atomic numbers Z from 1 to 54. The values corresponding to closed electronic shells, $Z = 2, 10, 18, 36, 54$, that is, the elements He, Ne, Ar, Kr, Xe, are indicated. The color coding corresponds to the different types of electronic shells that are being gradually filled: red for the s shells, blue for the p shells, and green for the d shells.

to as the "electronegativity scale" (ENS), as shown in Fig. 1.4. Interestingly, as the number of protons in the nucleus increases (Z increases), the electron wavefunctions become more compact and the atom shrinks in size, while the levels for a particular (n, l) set are being gradually filled by electrons. After a shell is full, as is the case for the elements He ($Z = 2$), Ne ($Z = 10$), Ar ($Z = 18$), and Kr ($Z = 36$), the so-called noble gases, the size is a local minimum; there is an abrupt increase in size for the elements with one more proton, when the next electronic shell starts being filled. Similarly, the ENS exhibits a regular pattern, that is, with steadily increasing values within each partially filled shell and the highest value corresponding to those elements that are just one electron short of having a completely filled electronic shell, like F ($Z = 9$), Cl ($Z = 17$), Br ($Z = 35$).

1.2 Forming Bonds Between Atoms

The electrons in the outermost, partially filled shells of the isolated atom are the ones that interact strongly with similar electrons in neighboring atoms in the solid; these are called **valence electrons**, while the remaining electrons in the completely filled shells of the atom are called **core electrons**. This is shown schematically in Fig. 1.5. Core electrons are essentially unaffected when the atom is surrounded by its neighbors in the solid. For most practical purposes it is quite reasonable to neglect the presence of the core electrons as far as the solid is concerned, and consider how the valence electrons behave. The core electrons and the nucleus form the **ion**, which has a positive net charge equal in magnitude to the number of valence electrons. The description of the solid can then be based on the behavior of ions and the accompanying valence electrons.

The classification of electrons in the atom into core and valence is very useful when we address the issue of how atoms come together to form a solid. This is illustrated next in two representative and very broad cases. The first concerns the type of bonding found in metallic solids, the second the type of bonding common in insulating or semiconducting materials, called covalent bonding.

Fig. 1.5　Schematic representation of the core and valence electrons in an atom, shown as the red and blue clouds, respectively.

1.2.1 The Essence of Metallic Bonding: The Free-Electron Model

The simplest possible arrangement of the valence electrons is to have them stripped apart from the ions and distributed uniformly through the solid. This is illustrated in Fig. 1.6. The bonding comes from the fact that the electrons are now shared by all the ions and act as a glue between them. The total energy E_{tot} of the system is then lower than the energy of isolated, neutral ions. The difference is defined as the "binding energy," E_b, of the solid. This quantity is characteristic of the stability of the solid. To make it a number independent of the size of the solid, we define it as the energy per atom gained by putting all the atoms together so they can share their valence electrons. For a system composed of N atoms, each with energy E_0 when isolated from all the others, we have:

$$E_b = \frac{1}{N}(E_{\text{tot}} - NE_0) \tag{1.1}$$

The general behavior of the binding energy is shown in Fig. 1.6: when the atoms are far apart, the binding energy is zero. The valence electrons are bound to the ions, occupying states with energy lower than zero; in the simple example shown in Fig. 1.6, this energy is labeled ϵ_0 for the case of the isolated atoms. When the atoms are brought together, the attractive potential due to the presence of many ions leads to lower values for the energy of the bound valence electrons (blue line). However, there is also a Coulomb repulsion E_c between the positive ions (green line). The competition between the two terms leads to a binding energy per atom E_b that is negative, that is, overall the system gains energy by having all the electrons shared among all the ions. When the distance between the

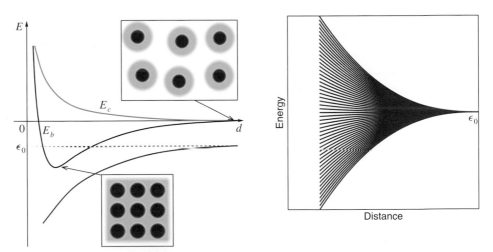

Fig. 1.6 Illustration of metallic bonding between many atoms. **Left**: The binding energy E_b of the solid as a function of the distance between the ions; E_c is the Coulomb repulsion between the ions and ϵ_0 is the energy of the valence electrons in the isolated neutral atom. The insets indicate schematically the charge distribution, with red circles representing the ions and blue clouds representing the valence electrons which end up being uniformly distributed between the ions in the solid.
Right: The corresponding energy-level diagram. The original N degenerate levels of the isolated individual atoms (at large distances) form a distribution of states in the solid that span a wide energy range, as the distance between atoms decreases.

ions becomes too small, the Coulomb repulsion between ions becomes too strong and the binding energy rises and eventually becomes positive, at which point the solid is not stable any longer. There is an optimal distance between the ions at which the binding energy attains its lowest value, which corresponds to the optimal structure of the solid.

An interesting issue is the behavior of the electron energy levels as the distance between the ions decreases. As shown in the right panel of Fig. 1.6, these energy levels are degenerate in the case of isolated atoms. Because electrons are fermions, they cannot all reside on the same energy level when the atoms are very close to each other. As a result, when the atoms get closer together the electron energy levels break the degeneracy and form a distribution over a range of values. The closer the distance between ions, the wider this energy range is, as the distance between ions approaches that of interatomic distances in the solid. In the limit of a very large number of atoms (as is the case for typical solids), the difference between consecutive energy values becomes infinitesimally small and the electron energy levels form a continuum distribution.

To put these notions on a more quantitative basis, we will use a very simple model in which the ionic potential is a uniformly distributed positive background. This frees us from needing to consider the detailed behavior of the ions. Moreover, we will neglect any kind of interaction between the electrons (coming from their Coulomb repulsion or from exchange and correlation effects). This is referred to as the "free-electron" or "jellium" model. Although evidently oversimplified, this model is quite useful in describing many of the properties of metallic solids, at least in a qualitative way.

In this model, the electronic states must also reflect this symmetry of the potential which is uniform, so they must be plane waves:

$$\psi_i(\mathbf{r}) = \frac{1}{\sqrt{V}} e^{i\mathbf{k}_i \cdot \mathbf{r}} \tag{1.2}$$

where V is the volume of the solid and $\mathbf{k}_i$ the wave-vector which characterizes state ψ_i. Since the wave-vectors suffice to characterize the single-particle states, we will use those as the only index, that is, $\psi_i \rightarrow \psi_{\mathbf{k}}$. Plane waves are actually a very convenient and useful basis for expressing various physical quantities. In particular, they allow the use of Fourier transform techniques, which simplify the calculations. In the following we will be using relations implied by the Fourier transform method, which are discussed in Appendix A.

We identify the state of such an electron by its wave-vector $\mathbf{k}$, which takes values from 0 up to a maximum magnitude k_F, called the "Fermi momentum." In three-dimensional (3D) cartesian coordinates the wave-vector is written as $\mathbf{k} = k_x\hat{\mathbf{x}} + k_y\hat{\mathbf{y}} + k_z\hat{\mathbf{z}}$ and the energy of the state with wave-vector $\mathbf{k}$ is:

$$\epsilon_{\mathbf{k}} = \frac{\hbar^2 \mathbf{k}^2}{2m_e} \tag{1.3}$$

We consider that the solid is a rectangular object with sides L_x, L_y, L_z, with $V = L_x L_y L_z$ its total volume. For a typical solid, the values of L_x, L_y, L_z are very large on the scale of atomic units, of order 10^8–10^9 a_0, where a_0 is the Bohr radius ($a_0 = 0.529177$ Å, with 1 Å $= 10^{-10}$ m). From Fourier analysis, we can take dk_x, dk_y, dk_z to be given by:

$$dk_x dk_y dk_z = \frac{(2\pi)^3}{L_x L_y L_z} \implies \frac{d\mathbf{k}}{(2\pi)^3} = \frac{1}{V} \implies \lim_{L_x, L_y, L_z \to \infty} \sum_{\mathbf{k}} = V \int \frac{d\mathbf{k}}{(2\pi)^3} \tag{1.4}$$

as discussed in Appendix A, Eq. (A.46). For each free-electron state characterized by $\mathbf{k}$ there is a multiplicity factor of 2 coming from the spin, which allows each electron to exist in two possible states, spin up or down, for a given value of its energy. Therefore, the total number of states available to the system is:

$$2 \sum_{|\mathbf{k}| \leq k_F} = 2V \int_0^{k_F} \frac{d\mathbf{k}}{(2\pi)^3} = \frac{k_F^3}{3\pi^2} V = N \tag{1.5}$$

which we have set equal to the total number of electrons N in the solid. This leads to the following relation between the density $n = N/V$ and the Fermi momentum k_F:

$$n = \frac{k_F^3}{3\pi^2} \tag{1.6}$$

which in turn gives for the energy of the highest occupied state ϵ_F, called the "Fermi energy":

$$\epsilon_F = \frac{\hbar^2 k_F^2}{2m_e} = \frac{\hbar^2 (3\pi^2 n)^{2/3}}{2m_e} \tag{1.7}$$

These results help us make some sense of the spectrum of energy levels in the solid: the energy of electronic states is now given by Eq. (1.3), which is a continuous set of values, from 0 up to ϵ_F. This is to be compared to the set of N individual atomic levels, when the atoms are isolated, as illustrated in Fig. 1.6. Moreover, if we assume that each atom has one valence electron in the atomic state, we will need only $N/2$ free-electron states to accommodate all these electrons in the solid due to spin degeneracy, as worked out above, in Eq. (1.5). The binding or cohesive energy in this case comes from the much more delocalized nature of the states in the solid as compared to those in individual isolated atoms.

Example 1.1   1D model of atoms in a box

To illustrate these ideas, we consider a simple example of six "atoms" in one dimension, each represented by a box of size a, as shown in Fig. 1.7. Consistent with the free-electron model, we will take the electron in such an atom to have a sinusoidal wavefunction $\psi_0(x)$ and energy ϵ_0 given by:

$$\psi_0(x) = \sqrt{\frac{2}{a}} \sin\left(\frac{\pi x}{a}\right), \quad \epsilon_0 = \frac{\hbar^2 \pi^2}{2m_e a^2}$$

corresponding to the largest possible wavelength of $\lambda_0 = 2a$ within the box of the isolated atom. If we condense these six atoms into a one-dimensional (1D) "solid" of size $L = 6a$, the lowest energy level in the solid will have energy $\epsilon_1 = \epsilon_0/36 = 0.028\epsilon_0$. Assuming that each atom has one valence electron, all the electrons in this 1D solid can be accommodated by the three lowest states with

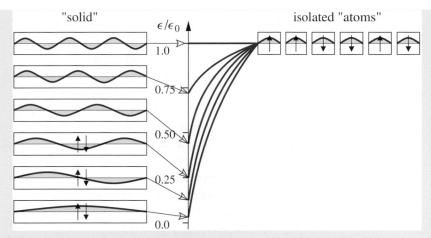

Fig. 1.7 A simple 1D model of the energy levels associated with the free-electron model for a system of six "atoms." The red sinusoidal curves represent the single-particle wavefunctions. In each atom, contained in a small box of size a, the lowest free-electron state within the box has wavelength $\lambda = 2a$ and energy $\epsilon_0 = \hbar^2\pi^2/2m_e a^2$. For six atoms forming a "solid" of size $L = 6a$, the six lowest-energy states and their corresponding energies in units of ϵ_0 are shown on the vertical axis.

energies $\epsilon_1, \epsilon_2 = 0.111\epsilon_0, \epsilon_3 = 0.25\epsilon_0$, giving a much lower total energy for the condensed system of $E_{\text{tot}} = 0.778\epsilon_0$ compared to the six isolated atoms, and a binding energy per atom of $E_b = -0.870\epsilon_0$. This component, namely the electronic kinetic energy contribution to the total energy, is the dominant component within our simple model. The actual total energy of the solid must include also the Coulomb repulsion of the electrons, as well as the electron–ion attraction, but these terms cancel each other for uniform and equal densities of the two oppositely charged sets of particles.

Generalizing this simple model, and considering only the kinetic energy of the electrons in the free-electron model, including a factor of 2 for the spin degeneracy, we find that the total energy of this system is given by:

$$E^{\text{kin}} = 2 \sum_{|\mathbf{k}| \leq k_F} \epsilon_{\mathbf{k}} = 2V \int_0^{k_F} \frac{d\mathbf{k}}{(2\pi)^3} \frac{\hbar^2 k^2}{2m_e} = \frac{V}{\pi^2} \frac{\hbar^2 k_F^5}{10 m_e} = \frac{3}{5} N \epsilon_F \tag{1.8}$$

This quantity is evidently positive and only depends on the density of electrons in the solid, n. The attraction of electrons to the positive charge of the ions provides the bonding for the atoms in the solid, but we have eliminated these terms from the free-electron model. In thinking about the properties of metallic solids, it is often useful to express equations in terms of another quantity, called r_s, which is defined as the radius of the sphere whose volume corresponds to the average volume per electron in the solid:

$$\frac{4\pi}{3} r_s^3 = \frac{V}{N} = n^{-1} = \frac{3\pi^2}{k_F^3} \tag{1.9}$$

and r_s is typically measured in atomic units. This gives the following expression for k_F:

$$k_F = \frac{(9\pi/4)^{1/3}}{r_s} \Longrightarrow k_F a_0 = \frac{(9\pi/4)^{1/3}}{(r_s/a_0)} \tag{1.10}$$

where the last expression contains the dimensionless combinations of variables $k_F a_0$ and r_s/a_0. In actual metals (r_s/a_0) varies between 2 and 6.

It is helpful to introduce here the units of energy that are typically involved in the formation of bonds in solids. The rydberg (Ry) is the natural unit for energies in atoms:

$$\frac{\hbar^2}{2m_e a_0^2} = \frac{e^2}{2a_0} = 1 \text{ Ry} \tag{1.11}$$

Another useful unit is the electron-volt (eV), the potential energy gained by an electron when it moves across a voltage difference of 1 V. The rydberg and the electron-volt are related by:

$$1 \text{ Ry} = 13.6058 \text{ eV} \tag{1.12}$$

Typical binding energies per atom, for metallic solids, are in the range of a few electron-volts.

1.2.2 The Essence of Covalent Bonding

We consider next what can happen when two atoms of the same type come together to form a diatomic molecule. For simplicity, we will assume that each atom has only one valence electron outside its core, in a state with energy ϵ_1, as indicated in Fig. 1.8; this corresponds,

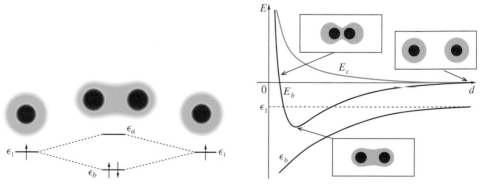

Fig. 1.8 Schematic representation of the formation of a bond between two atoms with one valence electron each. **Left**: When separated, the atoms each have a valence state of energy ϵ_1 with one electron in it; when they are brought together, new states are formed, the bonding state with energy $\epsilon_b < \epsilon_1$ and the anti-bonding state with energy $\epsilon_a > \epsilon_1$; the two available valence electrons occupy the lower-energy bonding state. **Right**: The binding energy E_b per atom (black line) of a diatomic molecule as a function of the distance d between the ions composing it. The energy of the bonding state ϵ_b is shown as a red line; for large distances it approaches the value ϵ_1 of the electron state in the isolated atom. The Coulomb repulsion between the two cores (E_c) is shown as a green line.

for example, to an element like Li ($Z = 3$).[3] When two such atoms come closer together, the presence of the two ions in close proximity creates new levels for the valence electrons, labeled ϵ_b (for bonding, the lower-energy state) and ϵ_a (for anti-bonding, the higher-energy state). The core electrons are unaffected, but the valence electrons can take advantage of the new states, by occupying the lower-energy one which can accommodate both of them. The wavefunction of the new state with lower energy is different from the wavefunction of the valence state in the isolated atom, and has more weight between the two ions. This creates a bond between the two ions, which share both of the valence electrons. The net result is a very stable unit consisting of two ions and the two shared valence electrons, that is, a diatomic molecule. The stability is the result of the lower energy of the entire system, compared to the energy of two isolated atoms, because each of the valence electrons now occupies a lower-energy state, $\epsilon_a < \epsilon_1$. As a result of the rearrangement of the charge, there is a strong electrostatic attraction between the two shared valence electrons lying near the center of the molecule and the two positively charged ions. The higher-energy state is called an anti-bonding state because its wavefunction has different character: it has a node in the region between the two ions and more weight in the regions away from the molecule center. If this state were to be occupied by electrons, it would lead to a weakening of the bond, because of the unfavorable distribution of electrons relative to the positions of the ions (this is the reason it is called an anti-bonding state).

As in the case of the metallic solid, there is an optimal distance between the two atoms, which corresponds to maximum bonding by sharing of the valence electrons. At distances larger than this optimal distance, the bond is weaker because there is not enough concentration of electrons between the two ions. At distances shorter than the optimal one, there is not enough space between the ions to accommodate the valence electrons (which also feel a mutually repulsive force), and the valence electrons are displaced from the inter-ionic region, which also weakens the bond. When the two ions come very close together the net interaction is quite repulsive, because the valence electrons are not able to shield the two ionic cores. The binding energy E_b has a minimum at the optimal distance between the two atoms in the diatomic molecule ($N = 2$); it approaches 0 for very large distances and it starts rising from the minimum for distances smaller than the optimal distance and eventually becomes positive, which corresponds to a net repulsion between the ions, as illustrated in Fig. 1.8.

Example 1.2   1D model of covalent bond in a two-atom molecule
We wish to demonstrate in a simple 1D example that symmetric and antisymmetric combinations of single-particle orbitals give rise to bonding and anti-bonding states. Though evidently oversimplified, this example is relevant to the hydrogen molecule which is discussed at length in Chapter 4. We begin with an atom consisting of an ion of charge $+e$ and a single valence electron: the electron–ion interaction potential is $\mathcal{V}(x) = -e^2/|x|$, arising from the ion which is situated at

---

[3] The case of the H atom and the H_2 molecule is also covered by this discussion, the only difference being that H lacks a core, since it only has one electron and its core is the bare proton.

$x = 0$; we will take the normalized wavefunction for the ground state of the electron to be:

$$\phi(x) = \frac{1}{\sqrt{a}} e^{-|x|/a} \qquad (1.13)$$

where a is a constant. We next consider two such atoms, the first ion at $x = -b/2$ and the second at $x = +b/2$, with b the distance between them (also referred to as the "bond length"). From the two wavefunctions associated with the electrons in each originally isolated atom:

$$\phi_1(x) = \frac{1}{\sqrt{a}} e^{-|x-b/2|/a}, \quad \phi_2(x) = \frac{1}{\sqrt{a}} e^{-|x+b/2|/a} \qquad (1.14)$$

we construct the symmetric ($+$) and antisymmetric ($-$) combinations:

$$\psi^{(\pm)}(x) = \frac{1}{\sqrt{\mathcal{N}^{(\pm)}}} [\phi_1(x) \pm \phi_2(x)], \quad \mathcal{N}^{(\pm)} = 2\left[1 \pm e^{-b/a}\left(1 + \frac{b}{a}\right)\right] \qquad (1.15)$$

where $\mathcal{N}^{(\pm)}$ are the normalization factors.

The difference between the probability of the electron being in the symmetric or the antisymmetric state rather than in the average of the states $\phi_1(x)$ and $\phi_2(x)$, is given by:

$$\delta n^{(\pm)}(x) = \pm \frac{1}{\mathcal{N}^{(\pm)}} \left[2\phi_1(x)\phi_2(x) - e^{-b/a}\left(1 + \frac{b}{a}\right)\left(|\phi_1(x)|^2 + |\phi_2(x)|^2\right)\right] \qquad (1.16)$$

A plot of the probabilities $|\psi^{(\pm)}(x)|^2$ and the differences $\delta n^{(\pm)}(x)$ is given in Fig. 1.9 for the choice $b = 2.5a$. Using this plot, we can justify the bonding character of state $\psi^{(+)}(x)$ and the anti-bonding character of state $\psi^{(-)}(x)$, taking into account the enhanced Coulomb attraction between the electron and the two ions in the region

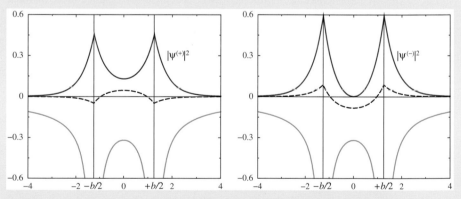

Fig. 1.9 Symmetric ($+$), left panel, and antisymmetric ($-$), right panel, linear combinations of single-particle orbitals: the probability densities $|\psi^{(\pm)}(x)|^2$ are shown by the red line for the ($+$) and the blue line for the ($-$) and the differences $\delta n^{(\pm)}(x)$ between each of those and the average occupation of states $\phi_1(x), \phi_2(x)$ by black dashed lines; the green lines show the attractive potential of the two ions located at $\pm b/2$ (positions indicated by thin vertical lines). In this example $b = 2.5a$ and x is given in units of a.

$-b/2 < x < +b/2$. In the case of $\psi^{(+)}$ more electronic charge is accumulated in the region between the two ions (the dashed black line is mostly positive), contributing to enhanced bonding. In the case of $\psi^{(-)}$, electronic charge is depleted from the region between the two ions (the black dashed curve is mostly negative), which is detrimental to the bonding between the ions. The charge is of course conserved, so the amount of electronic charge accumulated (or depleted) between the two ions is compensated by corresponding changes in the amount of charge near the position of each ion and in the regions farther to the left of the left ion or to the right of the right ion.

To make this argument a bit more quantitative, we define the change in potential energy for the combined system versus the two isolated ions:

$$\Delta \mathcal{V} \equiv \langle \psi^{(\pm)} | [\mathcal{V}_1(x) + \mathcal{V}_2(x)] | \psi^{(\pm)} \rangle - \frac{1}{2} [\langle \phi_1 | \mathcal{V}_1(x) | \phi_1 \rangle + \langle \phi_2 | \mathcal{V}_2(x) | \phi_2 \rangle] \quad (1.17)$$

where we have defined the potentials associated with the two ions, $\mathcal{V}_1(x), \mathcal{V}_2(x)$:

$$\mathcal{V}_1(x) = \frac{-e^2}{|x - b/2|}, \quad \mathcal{V}_2(x) = \frac{-e^2}{|x + b/2|}$$

A common approximation is to take the symmetric and antisymmetric combinations to be defined as:

$$\psi^{(\pm)}(x) = \frac{1}{\sqrt{2}} [\phi_1(x) \pm \phi_2(x)]$$

that is, $\mathcal{N}^{(\pm)} = 2$, which is reasonable in the limit $b \gg a$. In this limit, we can show (either analytically or by numerical integration) that the change in potential energy $\Delta \mathcal{V}$ is always negative for $\psi^{(+)}$ and always positive for $\psi^{(-)}$; moreover, the kinetic energy is *reduced* for $\psi^{(+)}$ and *increased* for $\psi^{(-)}$. All these considerations support the picture of $\psi^{(+)}$ corresponding to a bonding state and $\psi^{(-)}$ to an anti-bonding state.

The preceding discussion can be generalized to the case of many atoms and more than one valence state per atom, especially when these states have pronounced lobes pointing in different directions in space. An example of this situation is shown in Fig. 1.10: here, for the purposes of illustration, we consider a hypothetical atom with two valence states, each having pronounced lobes which point in orthogonal directions in space. As in the case of the diatomic molecule discussed above, new levels are created when two such atoms are in close proximity, with energy lower than (bonding states) or higher than (anti-bonding states) the energy of the original states in the isolated atom. The bonding and anti-bonding states from a pair of atoms are separated by an energy difference, to which we refer as an "energy gap." The presence of more ions creates a sequence of bonding states and a corresponding sequence of anti-bonding states, separated by a gap in energy. Assuming that the isolated atoms have one electron per valence state available in their neutral configuration, the bonding states will be fully occupied by the valence electrons and the anti-bonding states will be empty. This leads to a lower-energy, stable structure for the assembly of atoms relative to the isolated atoms, since the occupation of bonding states by valence electrons reduces their energy. As the number of atoms in the system increases, this pattern will evolve into a large set of bonding states and anti-bonding states, with the former occupied

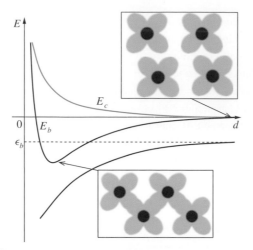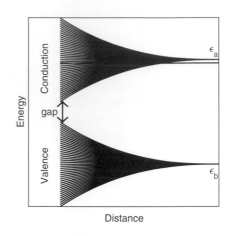

Fig. 1.10 Illustration of covalent bonding between many atoms: **Left**: The binding energy E_b of the solid as a function of the distance between the ions; E_c is the Coulomb repulsion between the ions and ϵ_b is the energy of the bonding combination of orbitals of the isolated neutral atoms. The insets indicate schematically the charge distribution, with red circles representing the ions and blue clouds representing the valence electrons leading to directed covalent bonds in which the electronic charge is distributed predominantly between nearest-neighbor pairs of ions. **Right**: The corresponding energy-level diagram. The original N degenerate levels of the isolated individual atoms (at large distances) form bonding and anti-bonding combinations (with energies ϵ_b, ϵ_a) which evolve into a distribution of states in the solid, each of which covers a wide energy range as the distance decreases. In this distribution, valence and conduction states are separated by an energy gap.

by valence electrons and the latter empty, but always separated by a gap in the energy due to their very different nature. This is the essence of a covalently bonded solid, also discussed at length in following chapters. Figure 1.10 illustrates the behavior of the binding energy per atom (left panel) and the distribution of electronic energy levels (right panel) as a function of the distance between the ions. The main differences between this situation and the case of metallic bonding is the non-uniform distribution of electron charge in space, which forms the directed covalent bonds between ions, and the presence of an energy gap in the distribution of energy levels, separating the filled bonding (also referred to as "valence") states from the empty anti-bonding (also referred to as "conduction") states.

1.2.3 Other Types of Bonding in Solids

The metallic and covalent bonding that we have described so far are the dominant types that produce stable crystals. There are other types of bonding, generally weaker than the metallic and covalent bonding, that are also encountered in solids. One type of bonding, that can actually be quite strong and is responsible for the stability of many solids encountered in everyday life, is *ionic* bonding. In this case, there are two (or sometimes more) kinds of ions in the solid, one that is positively charged having lost some of its valence electrons and one that is negatively charged having gained some electrons due to its higher electronegativity. This type of bonding is encountered, for example, in sodium chloride (NaCl), common salt. Another type of generally weaker but quite common bonding is the so-called *van der Waals* bonding, due mostly to dipolar interactions between neutral atoms or small atomic units. Yet another type of bonding, encountered in solids that contain hydrogen atoms and certain highly electronegative elements (like oxygen and nitrogen), is called *hydrogen* bonding, and has to do with the fact that the hydrogen atom has no core electrons and its core essentially consists of a proton which is very small on the atomic scale. In many solids, especially those composed of several kinds of atoms, the cohesion is due to a combination of different types of bonding, occasionally including all of the types mentioned. We discuss these other types of bonding in solids and their combinations in more detail below, when we deal with specific examples of solids where they are encountered.

1.3 The Architecture of Crystals

The atomic structure of materials is typically discussed in terms of the properties of crystals. There is good reason for this. A crystal, from a mathematical point of view, is a perfectly periodic arrangement of atoms in space which extends to infinite size; in practice, crystals are large enough when measured in units of interatomic distances that the mathematical notion of the crystal can be adopted as a very good approximation. Although no real material is a perfect crystal, in a wide range of materials the atoms are organized in a fairly regular manner, approximating the order inherent in the mathematical definition of the crystal. In some cases, the similarity is quite spectacular, as in precious stones that can be cut in ways which reveal the underlying symmetry of the atomic arrangement. In other cases the similarity is not so obvious but still exists in very deep ways, as in the cases of chocolate bars, ice, candles, and many other objects familiar in everyday life. In many cases, the overall shape of a material is due to the atomic structure of the crystal, as we will show through some examples in this chapter.

The structure of crystals can be understood by a close look at the properties of the atoms from which they are composed. We can identify several broad categories of atoms, depending on the nature of the valence electrons that participate actively in the formation of the solid. Accordingly, we will discuss below the crystal structure of various solids based on the properties of the valence electronic states of the constituent atoms.

We are only concerned here with the basic features of the crystal structures that the various atoms form, such as their coordination number (number of nearest neighbors) and the nature of the bonds they form, without paying close attention to the details; these will come later. Finally, we will only concern ourselves with low-temperature structures, which correspond to the lowest-energy static configuration; dynamical effects, which can produce a different structure at higher temperatures, will not be considered.[4] We begin our discussion with those solids formed by atoms of one element only, called elemental solids, and then proceed to more complicated structures involving several types of atoms.

Some basic properties of the elemental solids are collected in the Periodic Table. There is no information for the elements H and He (atomic numbers 1 and 2) because these elements

---

[4] J. Friedel, *J. de Phys. (Paris)* **35**, L-59 (1974).

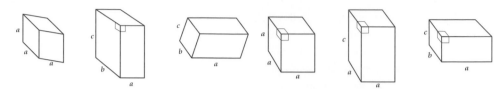

Fig. 1.11 Shapes of the unit cells in some lattices that appear in the Periodic Table. The corners drawn in thin black lines indicate right angles between edges. The lattices are discussed in more detail in Chapter 3 (see Section 3.3).

do not form solids under normal conditions: H forms a molecular solid at sufficiently low temperature or high pressure, while He remains liquid at temperatures very close to absolute zero. For the rest of the elements, we list:

- The crystal structure of the most common phase. The acronyms for the crystal structures that appear in the Periodic Table stand for: BCC = body centered cubic, FCC = face centered cubic, HCP = hexagonal close packed, GRA = graphite, TET = tetragonal, DIA = diamond, CUB = cubic, MCL = monoclinic, ORH = orthorhombic, RHL = rhombohedral, TRI = triclinic. Selected shapes of the corresponding crystal unit cells are shown in Fig. 1.11.
- For crystals which can be characterized by a single structural parameter, the lattice constant, the value of this parameter is given in units of angstroms. Such crystals are, for example, FCC, BCC, HCP. For other crystal types, more structural parameters are required to fully specify the unit of the crystal.
- The melting temperature (K). We note here that typical values of the cohesive energy of solids are in the range of a few electron-volts (1 eV = 11,604 K), which means that the melting temperature is only a small fraction of the cohesive energy, typically a few percent.
- The atomic concentration of the most common crystal phase in 10^{22} cm^{-3}.
- The electrical resistivity (in units of $\mu\Omega\cdot$cm); for most elemental solids the resistivity is of order 1–100 in these units, except for some good insulators which have resistivities 10^3, 10^6, or 10^9 times higher.

1.3.1 Atoms with No Valence Electrons

The first category consists of those elements which have no valence electrons. These are the atoms with all their electronic shells completely filled, which are chemically inert and are referred to as the "noble elements," He, Ne, Ar, Kr, Xe. When these atoms are brought together to form solids, they interact very weakly. Their outer electrons are not disturbed much, since they are essentially core electrons. The weak attractive interaction between such atoms is the result of slight polarization of the electronic wavefunction in one atom due to the presence of other atoms around it. This interaction is referred to as a "fluctuating dipole" or van der Waals interaction. Since the interaction is weak, the solids are not very stable and they have very low melting temperatures, well below room temperature. The main concern of the atoms in forming such solids is to have as many neighbors as possible,

Fig. 1.12 Illustration of the face centered cubic (FCC) crystal structure. **Left**: The conventional cubic cell of the FCC crystal, shown in thin red lines, with atoms at the corners of the cube and at the centers of the cube faces. **Right**: Several planes (outlined by thin colored lines) of close-packed spheres in the FCC crystal, and the six nearest neighbors of one atom indicated by yellow arrows.

in order to maximize the cohesion, since all interactions are attractive. The crystal structure that corresponds to this type of atom is one of the close-packing geometries, that is, atomic arrangements in space which allow the closest packing of hard spheres. The particular crystal structure that noble-element atoms assume in solid form is FCC. Each atom has 12 equidistant nearest neighbors in this crystal, which lie on different equivalent planes. For example, an atom at the center of a particular cube face has four neighbors at the corners of the same cube face and eight more neighbors, four each on a plane parallel to, and on either side of, this face. Alternatively, atoms can be viewed as having six neighbors on a plane and six more, three each on a plane parallel to and on either side of it. These structural features are shown in Fig. 1.12.

Thus, in the simplest case, atoms that have no valence electrons at all behave like hard spheres, which attract each other with weak forces, but are not deformed. They form weakly bonded solids in the FCC crystal structure, in which the attractive interactions are optimized by maximizing the number of nearest neighbors in a close-packing arrangement. The only exception to this rule is He, in which the attractive interaction between atoms is so weak that it is overwhelmed by the zero-point motion of the atoms. Except when external pressure is applied to enhance this attractive interaction, He remains a liquid. This is an indication that in some cases it will prove unavoidable to treat the nuclei as quantum particles (see also the discussion below about hydrogen).

The other close-packing arrangement of hard spheres is the hexagonal structure (HCP), with 12 neighbors which are separated into two groups of six atoms each: the first group forms a planar six-member ring surrounding an atom at the center, while the second group consists of two equilateral triangles, one above and one below the six-member ring, with the central atom situated above or below the geometrical center of each equilateral triangle, as shown in Fig. 1.13. The HCP structure bears a certain relation to FCC: we can view both structures as planes of spheres closely packed in two dimensions, which gives a hexagonal lattice; for close packing in three dimensions the successive planes must be situated so that a sphere in one plane sits directly above the center of a triangle formed by three spheres in the previous plane. There are two ways to form such a stacking of hexagonal close-packed planes: $\ldots ABCABC \ldots$ and $\ldots ABABAB \ldots$, where A, B, C represent the three possible

PERIODIC

I–A s^1	II–A s^2
H [1] Hydrogen	**He** [2] Helium
Li [3] Lithium BCC 3.51 453.7 4.70 9.4	**Be** [4] Beryllium HCP 2.29 1560 12.1 3.3
Na [11] Sodium BCC 4.29 370.9 2.65 4.75	**Mg** [12] Magnesium HCP 3.21 923 4.30 4.46

Legend:

Symbol → **Li** [3]
Name → Lithium
Crystal structure → BCC 3.51 ← Lattice constant (Å)
Melting point → 453.7 (K)
Atomic concentration → 4.70 ($10^{22}/\text{cm}^3$)
Resistivity → 9.4 ($\mu\Omega$cm)

III–B $s^2 d^1$	IV–B $s^2 d^2$	V–B $s^2 d^3$	VI–B $s^2 d^4$	VII–B $s^2 d^5$	VIII $s^2 d^6$	VIII $s^2 d^7$
K [19] Potassium BCC 5.33 336.5 1.40 21.6	**Ca** [20] Calcium FCC 5.59 1115 2.30 3.7					
Sc [21] Scandium HCP 3.31 1814 4.27 51	**Ti** [22] Titanium HCP 2.95 1941 5.66 47.8	**V** [23] Vanadium BCC 3.03 2183 7.22 24.8	**Cr*** [24] Chromium BCC 2.91 2944 8.33 12.9	**Mn** [25] Manganese CUB 1519 8.18 139	**Fe** [26] Iron BCC 2.87 1811 8.50 9.71	**Co** [27] Cobalt HCP 2.51 1768 8.97 6.34
Rb [37] Rubidium BCC 5.59 312.5 1.15 12.1	**Sr** [38] Strontium FCC 6.08 1050 1.78 22.8	**Y** [39] Yttrium HCP 3.65 1799 3.02 60	**Zr** [40] Zirconium HCP 3.23 2128 4.29 41.4	**Nb*** [41] Niobium BCC 3.30 2750 5.56 15.2	**Mo*** [42] Molybdenum BCC 3.15 2896 6.42 5.17	**Tc** [43] Technetium HCP 2.74 2430 7.04 14	**Ru*** [44] Ruthenium HCP 2.71 2607 7.36 7.2	**Rh*** [45] Rhodium FCC 3.80 2237 7.26 4.5
Cs [55] Cesium BCC 6.14 301.6 0.91 20	**Ba** [56] Barium BCC 5.03 1000 1.60 50	**La** [57] Lanthanum HCP 3.77 1193 2.70 80	**Hf** [72] Hafnium HCP 3.20 2506 4.52 35.1	**Ta** [73] Tantalum BCC 3.30 3290 5.55 13.5	**W** [74] Wolframium BCC 3.17 3695 6.30 5.6	**Re** [75] Rhenium HCP 2.76 3459 6.80 19.3	**Os** [76] Osmium HCP 2.73 3306 7.14 8.1	**Ir** [77] Iridium FCC 3.84 2739 7.06 5.1

$s^2 d^0 f^2$	$s^2 d^0 f^3$	$s^2 d^0 f^4$	$s^2 d^0 f^5$	$s^2 d^0 f^6$	$s^2 d^0 f^7$
Ce [58] Cerium HCP 3.62 1068 2.91 85.4	**Pr** [59] Praseodymium HCP 3.67 1208 2.92 68	**Nd** [60] Neodymium HCP 3.66 1297 2.93 64.3	**Pm** [61] Promethium	**Sm** [62] Samarium TRI 1345 3.03 105	**Eu** [63] Europium BCC 4.58 1099 3.04 91

TABLE

	III–A s^2p^1	IV–A s^2p^2	V–A s^2p^3	VI–A s^2p^4	VII–A s^2p^5	Noble s^2p^6
	B [5] Boron RHL 2349 13.0 4×10^6	**C** [6] Carbon GRA 2.46 3800 17.6 3.5×10^3	**N** [7] Nitrogen HCP 3.86 63	**O** [8] Oxygen MCL 54.8	**F** [9] Fluorine MCL 53.5	**Ne** [10] Neon FCC 4.43 24.6 4.36
	Al [13] Aluminum FCC 4.05 933 6.02 2.67	**Si** [14] Silicon DIA 5.43 1687 5.00 230×10^9	**P** [15] Phosphorus TCL 317	**S** [16] Sulfur ORH 388	**Cl** [17] Chlorine ORH 172	**Ar** [18] Argon FCC 5.26 83.8 2.66

VIII s^2d^8	I–B s^2d^9	II–B s^2d^{10}						
Ni [28] Nickel FCC 3.52 1728 9.14 6.84	**Cu*** [29] Copper FCC 3.61 1357 8.45 1.67	**Zn** [30] Zinc HCP 2.66 693 6.55 5.92	**Ga** [31] Gallium ORH 303 5.10	**Ge** [32] Germanium DIA 5.66 1211 4.42 47×10^6	**As** [33] Arsenic TRI 1090 4.65 12×10^6	**Se** [34] Selenium MCL 494 3.67	**Br** [35] Bromine ORH 266 2.36	**Kr** [36] Krypton FCC 5.71 116 2.17
Pd* [46] Palladium FCC 3.89 1828 6.80 9.93	**Ag*** [47] Silver FCC 4.09 1235 5.85 1.63	**Cd*** [48] Cadmium HCP 2.98 594 4.64 6.83	**In** [49] Indium TET 430 3.83 8.37	**Sn** [50] Tin TET 505 2.91 11	**Sb** [51] Antimony TRI 904 3.31 39	**Te** [52] Tellurium TRI 723 2.94 160×10^3	**I** [53] Iodine ORH 387 2.36	**Xe** [54] Xenon FCC 6.20 161 1.64
Pt* [78] Platinum FCC 3.92 2041 6.80 9.85	**Au*** [79] Gold FCC 4.08 1337 7.14 2.12	**Hg** [80] Mercury RHL 234 7.06 96	**Tl** [81] Thallium HCP 3.46 577 3.50 18	**Pb** [82] Lead FCC 4.95 601 3.30 20.6	**Bi** [83] Bismuth MCL 544 2.82 107	**Po** [84] Polonium	**At** [85] Astatine	**Rn** [86] Radon

$s^2d^1f^7$	$s^2d^1f^8$	$s^2d^0f^{10}$	$s^2d^0f^{11}$	$s^2d^0f^{12}$	$s^2d^0f^{13}$	$s^2d^0f^{14}$	$s^2d^1f^{14}$
Gd* [64] Gadolinium HCP 3.64 1585 3.02 131	**Tb*** [65] Terbium HCP 3.60 1629 3.22 114.5	**Dy** [66] Dysprosium HCP 3.59 1680 3.17 92.6	**Ho** [67] Holmium HCP 3.58 1734 3.22 81.4	**Er** [68] Erbium HCP 3.56 1770 3.26 86.0	**Tm** [69] Thulium HCP 3.54 1818 3.32 67.6	**Yb** [70] Ytterbium FCC 5.48 1097 3.02 25.1	**Lu** [71] Lutetium HCP 3.50 1925 3.39 58.2

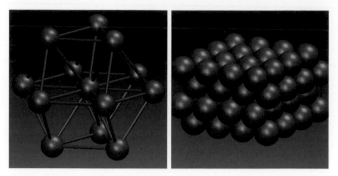

Fig. 1.13 Illustration of the hexagonal close packed (HCP) crystal structure. **Left**: One atom and its 12 neighbors in the HCP crystal; the size of the spheres representing atoms is chosen so as to make the neighbors and their distances apparent. **Right**: The stacking of close-packed planes of atoms in the HCP crystal.

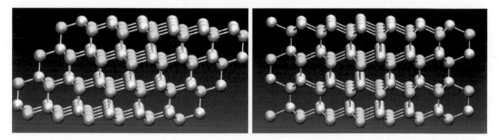

Fig. 1.14 Illustration of two interpenetrating FCC (left) or HCP (right) lattices; the two original lattices are denoted by sets of white and yellow spheres. These structures correspond to the diamond lattice if the two atom types are the same or the zincblende lattice if they are different (left), and to the wurtzite lattice (right). The lattices are viewed from the side, with the vertical direction corresponding to the direction along which close-packed planes of the FCC or HCP lattices would be stacked.

relative positions of spheres in successive planes according to the rules of close packing. The first sequence corresponds to the FCC lattice, the second to the HCP lattice.

An interesting variation of the close-packing theme of the FCC and HCP lattices is the following: consider two interpenetrating such lattices, that is, two FCC or two HCP lattices, arranged so that in the resulting crystal the atoms in each sublattice have as nearest equidistant neighbors atoms belonging to the other sublattice. These arrangements give rise to the diamond lattice or the zincblende lattice (when the two original lattices are FCC) and to the wurtzite lattice (when the two original lattices are HCP). This is illustrated in Fig. 1.14. Interestingly, in both cases each atom finds itself at the center of a tetrahedron with exactly four nearest neighbors. Since the nearest neighbors are exactly the same, these two types of lattices differ only in the relative positions of second (or farther) neighbors. It should be evident that the combination of two close-packed lattices cannot produce another close-packed lattice. Consequently, the diamond, zincblende, and wurtzite lattices are encountered in covalent or ionic structures in which fourfold coordination is preferred. For example: tetravalent group IV elements like C, Si, Ge form the diamond lattice; combinations of two different group IV elements or complementary elements (like group

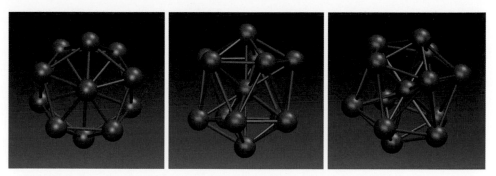

Fig. 1.15 Three different views of the regular icosahedron: the left view is along one of the fivefold symmetry axes; in the middle view, a fivefold axis is along the vertical direction; the right view is along a random orientation.

III–group V, group II–group VI, group I–group VII) form the zincblende lattice; certain combinations of group III–group V elements form the wurtzite lattice. These structures are discussed in more detail below. A variation of the wurtzite lattice is also encountered in ice and is due to hydrogen bonding.

Yet another version of the close-packing arrangement is the icosahedral structure. In this case an atom has again 12 equidistant neighbors, which are at the apexes of an icosahedron. The icosahedron is one of the Platonic solids in which all the faces are perfect planar shapes; in the case of the icosahedron, the faces are 20 equilateral triangles. The icosahedron has 12 apexes arranged in fivefold symmetric rings,[5] as shown in Fig. 1.15. In fact, it turns out that the icosahedral arrangement is optimal for close packing of a small numbers of atoms, but it is not possible to fill 3D space in a periodic fashion with icosahedral symmetry. This fact is a simple geometrical consequence (see also Chapter 3 on crystal symmetries). Based on this observation it was thought that crystals with perfect fivefold (or tenfold) symmetry could not exist, unless defects were introduced to allow for deviations from the perfect symmetry.[6] The discovery of solids that exhibited fivefold or tenfold symmetry in their diffraction patterns, in the mid-1980s,[7] caused quite a sensation. These solids were named "quasicrystals," and their study created a new exciting subfield in condensed matter physics.

1.3.2 Atoms with *s* Valence Electrons

The second category consists of atoms that have only *s* valence electrons. These are Li, Na, K, Rb, and Cs (the alkalis) with one valence electron, and Be, Mg, Ca, Sr, and Ba with two valence electrons. The wavefunctions of valence electrons of all these elements extend far from the nucleus. In solids, the valence electron wavefunctions at one site

---

[5] An *n*-fold symmetry means that rotation by $2\pi/n$ around an axis leaves the structure invariant.
[6] F. C. Frank, *Proc. Roy. Soc. A* **215**, 43 (1952); F. C. Frank and J. S. Kasper, *Acta Cryst.* **11**, 184 (1958); *Acta Cryst.* **12**, 483 (1959).
[7] D. Shechtman, I. Blech, D. Gratias, and J. W. Cahn, *Phys. Rev. Lett.* **53**, 1951 (1984). D. Shechtman was awarded the 2011 Nobel Prize in Chemistry for this discovery.

have significant overlap with those at the nearest-neighbor sites. Since the s states are spherically symmetric, the wavefunctions of valence electrons do not exhibit any particular preference for orientation of the nearest neighbors in space. For the atoms with one and two s valence electrons, a simplified picture consists of all the valence electrons overlapping strongly, and thus being shared by all the atoms in the solid forming a "sea" of negative charge. The nuclei with their core electrons form ions which are immersed in this sea of valence electrons. The ions have charge $+1$ for the alkalis and $+2$ for the atoms with two s valence electrons. The resulting crystal structure is the one which optimizes the electrostatic repulsion of the positively charged ions with their attraction by the sea of electrons. The actual structures are BCC for all the alkalis, and FCC or HCP for the two-s-valence electron atoms, except Ba, which prefers the BCC structure. In the BCC structure each atom has eight equidistant nearest neighbors, as shown in Fig. 1.16, which is the second highest number of nearest neighbors in a simple crystalline structure, after FCC and HCP. In Ba, which is the heaviest of the column II-A elements, the filled-shell electrons start having wavefunctions that extend far from the nucleus, simply because they correspond to large principal quantum numbers, and thus are no longer unaffected by the presence of other nearby atoms; the interactions between such electronic states lead to the BCC structure being more stable than the FCC structure for this element.

One point deserves further clarification: we mentioned that the valence electrons have significant overlap with the electrons in neighboring atoms, and thus they are shared by all atoms in the solid, forming a sea of electrons. It may seem somewhat puzzling that we can jump from one statement – the overlap of electron orbitals in nearby atoms – to the other – the sharing of valence electrons by all atoms in the solid. The physical symmetry which allows us to make this jump is the periodicity of the crystalline lattice. This symmetry is the main feature of the external potential that the valence electrons feel in the bulk of a crystal: they are subjected to a periodic potential in space, in all three dimensions, which for all practical purposes extends to infinity – an idealized situation we discussed earlier. The electronic wavefunctions must transform according to the symmetry operations of the hamiltonian, as we will discuss in detail in later chapters; for the purposes of the

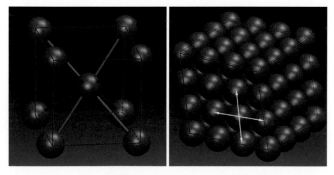

Fig. 1.16   Illustration of the body centered cubic (BCC) crystal structure. **Left**: The conventional cubic cell (shown in thin red lines) with one atom and its eight neighbors in the BCC crystal. **Right**: The stacking of close-packing planes, indicated by thin colored lines, in the BCC crystal; the four nearest neighbors of an atom on such a plane are shown by yellow arrows.

present discussion, this implies that the wavefunctions themselves must be periodic, up to a phase. The mathematical formulation of this statement is called Bloch's theorem. A periodic wavefunction implies that if two atoms in the crystal share an electronic state due to overlap between atomic orbitals, then all equivalent atoms of the crystal share the same state equally, that is, the electronic state is delocalized over the entire solid. This behavior is central to the physics of solids, and represents a feature that is qualitatively different from what happens in atoms and molecules, where electronic states are localized (except in certain large molecules that possess symmetries akin to lattice periodicity).

1.3.3 Atoms with s and p Valence Electrons

The next level of complexity in crystal structure arises from atoms that have both s and p valence electrons. The individual p states are not spherically symmetric, so they can form linear combinations with the s states that have directional character: a single p state has two lobes of opposite sign pointing in diametrically opposite directions. The s and p states, illustrated in Fig. 1.17, can then serve as the new basis for representing electron wavefunctions and their overlap with neighboring wavefunctions of the same type can lead to interesting ways for arranging the atoms into a stable crystalline lattice (see Appendix C on the character of atomic orbitals).

In the following we will use the symbols $s(\mathbf{r})$, $p_l(\mathbf{r})$ ($l = x, y, z$) and $d_m(\mathbf{r})$ ($m = xy, yz, zx, x^2 - y^2, 3z^2 - r^2$) to denote atomic orbitals as they would exist in an isolated atom, which are functions of $\mathbf{r}$. When they are related to an atom A at position $\mathbf{R}_A$, these

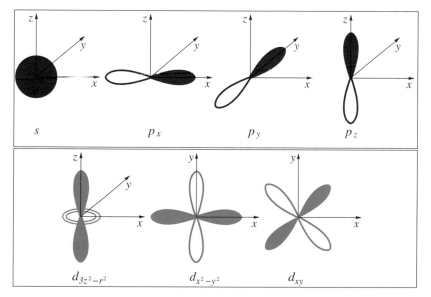

Fig. 1.17 Representation of the character of s, p, d atomic orbitals. The lobes of positive sign in the p_x, p_y, p_z and $d_{3z^2-r^2}, d_{x^2-y^2}, d_{xy}$ orbitals are shown shaded. The d_{yz}, d_{zx} orbitals are similar to the d_{xy} orbital, but lie on the y, z and z, x planes.

become functions of $\mathbf{r} - \mathbf{R}_A$ (the origin is translated by $-\mathbf{R}_A$ so that the new orbital is centered at the position of the nucleus) and these will be denoted as $s^A(\mathbf{r}) = s(\mathbf{r} - \mathbf{R}_A)$, $p_l^A(\mathbf{r}) = p_l(\mathbf{r} - \mathbf{R}_A)$, and $d_m^A(\mathbf{r}) = d_m(\mathbf{r} - \mathbf{R}_A)$. We will use the symbols $\phi_i^A(\mathbf{r})$ $(i = 1, 2, ...)$ for states that are linear combinations of the atomic orbitals s^A, p_l^A, d_m^A at site A, and the symbols $\psi^n(\mathbf{r})$ $(n = a, b)$ for combinations of the $\phi_i^X(\mathbf{r})$s $(X = A, B, ...; i = 1, 2, ...)$ which are appropriate for the description of electronic states in the crystal.

Our first example concerns a two-dimensional (2D) crystal of carbon atoms, which, when stacked together with many other such planes, forms graphite. In this case, the first three orbitals, $\phi_1^A, \phi_2^A, \phi_3^A$, belonging to a carbon atom labeled A, point along three directions on the x, y plane separated by $120°$, while the last one, ϕ_4^A, points in a direction perpendicular to the x, y plane, as shown in Fig. 1.18 and defined in Table 1.1. If the orbitals s^A, p_i^A $(i = x, y, z)$ have energies ϵ_s and ϵ_p, then the states ϕ_k $(k = 1, 2, 3)$ have energy $(\epsilon_s + 2\epsilon_p)/3$ (see Problem 5); these states, since they are composed of one s and two p atomic orbitals, are called sp^2 orbitals. Imagine now a second identical atom, which we label B, with the linear combinations shown in Fig. 1.18 and defined in Table 1.1 as ϕ_i^B $(i = 1 - 4)$. The orbitals $\phi_1^B, \phi_2^B, \phi_3^B$ also point along three directions on the x, y plane separated by $120°$, but in the opposite sense (rotated by $180°$) from those of atom A. For example, ϕ_1^A points along the $-\hat{\mathbf{x}}$ direction, while ϕ_1^B points along the $+\hat{\mathbf{x}}$ direction. Now imagine that we place atoms A and B next to each other along the x-axis, first atom A and to its right atom B, at a distance a. We arrange the distance so that there is significant overlap between orbitals ϕ_1^A and ϕ_1^B, which are pointing toward each other, thereby maximizing the interaction between these two orbitals. Let us assume that in the neutral isolated state of the atom we can occupy each of these orbitals by one electron; note that this is *not* the ground state of the atom. We can form two linear combinations, $\psi_1^b = (\phi_1^A + \phi_1^B)/\sqrt{2}$ and $\psi_1^a = (\phi_1^A - \phi_1^B)/\sqrt{2}$, of which the first maximizes the overlap and the second has a node at the midpoint between atoms A and B. As usual, we expect the symmetric linear combination of single-particle orbitals (called bonding state) to have lower energy than the antisymmetric one (called anti-bonding state) in the system of the two atoms; this is a general feature of how combinations of single-particle orbitals behave (see Problem 4), as was discussed in our earlier heuristic model of two atoms with one valence electron each (see Fig. 1.8 and related discussion). The exact energy of the bonding and anti-bonding states will depend on the overlap of the orbitals ϕ_1^A, ϕ_1^B. We can place two electrons, one from each atomic orbital, in the symmetric linear combination because of their spin degree of freedom; this is based on the assumption that the spin wavefunction of the two electrons is antisymmetric (a spin singlet), so that the total wavefunction, the product of the spatial and spin parts, is antisymmetric upon exchange of their coordinates, as it should be due to their fermionic nature. Through this exercise we have managed to lower the energy of the system, since the energy of ψ^b is lower than the energy of ϕ_1^A or ϕ_1^B. This is the essence of the chemical bond between two atoms, which in this case is called a covalent σ bond, as already mentioned in relation to the simple model of Fig. 1.10.

Imagine next that we repeat this exercise: we take another atom with the same linear combinations of orbitals as B, which we will call B_1, and place it in the direction of the vector $(\hat{\mathbf{x}} - \sqrt{3}\hat{\mathbf{y}})/2$ relative to the position of atom A, and at the same distance a as atom B from A. Due to our choice of orbitals, ϕ_2^A and $\phi_2^{B_1}$ will be pointing toward each other. We

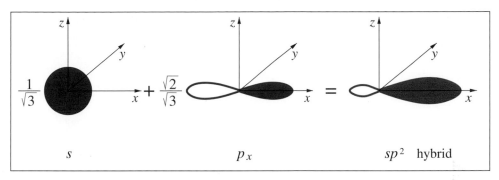

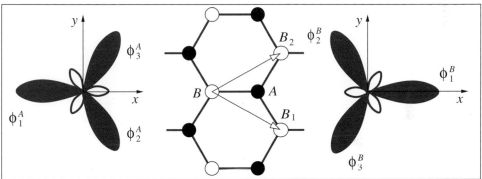

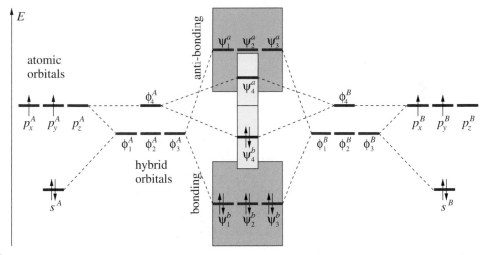

Fig. 1.18   Illustration of covalent bonding in a single plane of graphite, called "graphene." **Top**: Example of the formation of hybrid sp^2 orbitals from linear combinations of s and p orbitals, with relative weights 1 and 2, respectively: this is the hybrid orbital labeled ϕ_1^B in Table 1.1. **Middle**: The sp^2 linear combinations of s and p atomic orbitals corresponding to the atoms labeled A and B, as defined in Table 1.1. **Bottom**: The energy-level diagram for the s, p atomic states, their sp^2 linear combinations (ϕ_i^As and ϕ_i^Bs), and the bonding (ψ_i^bs) and anti-bonding (ψ_i^as) states (up–down arrows indicate electron spins). The shaded regions indicate the broadening of the individual energy levels over a range of values into the valence and conduction bands.

Table 1.1 The linear combinations of s and p atomic orbitals that form the sp^2 hybrids for the two atoms, labeled A and B, per unit cell in the graphite structure.

$$\phi_1^A = \frac{1}{\sqrt{3}}s^A - \frac{\sqrt{2}}{\sqrt{3}}p_x^A \qquad \phi_1^B = \frac{1}{\sqrt{3}}s^B + \frac{\sqrt{2}}{\sqrt{3}}p_x^B$$

$$\phi_2^A = \frac{1}{\sqrt{3}}s^A + \frac{1}{\sqrt{6}}p_x^A - \frac{1}{\sqrt{2}}p_y^A \qquad \phi_2^B = \frac{1}{\sqrt{3}}s^B - \frac{1}{\sqrt{6}}p_x^B + \frac{1}{\sqrt{2}}p_y^B$$

$$\phi_3^A = \frac{1}{\sqrt{3}}s^A + \frac{1}{\sqrt{6}}p_x^A + \frac{1}{\sqrt{2}}p_y^A \qquad \phi_3^B = \frac{1}{\sqrt{3}}s^B - \frac{1}{\sqrt{6}}p_x^B - \frac{1}{\sqrt{2}}p_y^B$$

$$\phi_4^A = p_z^A \qquad\qquad\qquad \phi_4^B = p_z^B$$

can form symmetric and antisymmetric combinations from them, occupy the symmetric (lower-energy) one with two electrons as before, and create a second σ bond between atoms A and B_1. Finally, we repeat this procedure with a third atom B_2 placed along the direction of the vector $(\hat{\mathbf{x}}+\sqrt{3}\hat{\mathbf{y}})/2$ relative to the position of atom A, and at the same distance a as the previous two neighbors. Through the same procedure we can form a third σ bond between atoms A and B_2, by forming the symmetric and antisymmetric linear combinations of the orbitals ϕ_3^A and $\phi_3^{B_2}$. Now, as far as atom A is concerned, its three neighbors are exactly equivalent, so we consider the vectors that connect them as the repeat vectors at which equivalent atoms in the crystal should exist. If we place atoms of type B at all the possible integer multiples of these vectors, we form a lattice. To complete the lattice we have to place atoms of type A also at all the possible integer multiples of the same vectors, relative to the position of the original atom A. The resulting lattice is called the honeycomb lattice. Each atom of type A is surrounded by three atoms of type B and vice versa, as illustrated in Fig. 1.18. Though this example may seem oversimplified, it actually corresponds to the structure of graphite, one of the most stable crystalline solids. In graphite, planes of C atoms in the honeycomb lattice are placed on top of each other to form a 3D solid, as illustrated in Fig. 1.19, but the interaction between planes is rather weak (similar to the van der Waals interaction). An indication of this weak bonding between planes compared to the in-plane bonds is that the distance between nearest-neighbor atoms on a plane is 1.42 Å, whereas the distance between successive planes is 3.35 Å, a factor of 2.36 larger.

What about the orbitals p_z (or ϕ_4), which so far have not been used? If each atom had only three valence electrons, then these orbitals would be left empty since they have higher energy than the orbitals ϕ_1, ϕ_2, ϕ_3, which are linear combinations of s and p orbitals (the original s atomic orbitals have lower energy than p). In the case of C, each atom has four valence electrons so there is one electron left per atom when all the σ bonds have been formed. These electrons remain in the p_z orbitals which are perpendicular to the x, y plane and thus parallel to each other. Symmetric and antisymmetric combinations of neighboring p_z^A and p_z^B orbitals can also be formed (the states ψ_4^b, ψ_4^a respectively), and the energy can be lowered by occupying the symmetric combination. In this case the overlap between neighboring p_z orbitals is significantly smaller and the corresponding gain in energy significantly less than in σ bonds. This is referred to as a π bond, which is

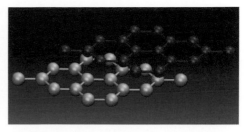

Fig. 1.19 Illustration of atomic arrangement in graphite in a top view (left) and a perspective view (right): atoms form planes with every atom on a plane having three neighbors bonded by strong covalent bonds, and the planes are stacked on top of each other. The atoms on a single plane are shown in the same color but are colored differently in successive planes (blue and red), which are shifted relative to each other. The next plane in the sequence would be exactly above the first (blue) one, along the direction perpendicular to the plane, repeating the pattern to very large distances. Bonding between the planes is relatively weak, due to van der Waals interactions, and is not shown in this representation.

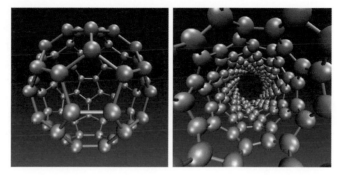

Fig. 1.20 The structure of the C_{60} molecule or "buckyball" (left), consisting of pentagonal and hexagonal rings of C atoms, and of carbon nanotubes (right), consisting of hexagonal rings of C atoms as in a plane of graphite, but wrapped into a cylinder (for more details on these and similar structures, see Chapter 5).

generally weaker than a σ bond. Carbon is a special case, in which the π bonds are almost as strong as the σ bonds.

An intriguing variation of this theme is a structure that contains pentagonal rings as well as the regular hexagons of the honeycomb lattice, while maintaining the threefold coordination and bonding of the graphitic plane. The presence of pentagons introduces curvature in the structure, and the right combination of pentagonal and hexagonal rings produces the almost perfect sphere, shown in Fig. 1.20. This structure actually exists in nature! It was discovered in 1985 and has had a great impact on carbon chemistry and physics – its discoverers, R. F. Curl, H. W. Kroto, and R. E. Smalley, received the 1996 Nobel Prize for Chemistry. Many more interesting variations of this structure have also been produced, including "onions" – spheres within spheres – and "tubes" – cylindrical arrangements of threefold-coordinated carbon atoms. The tubes in particular seem promising for applications in technologically and biologically relevant systems. These structures have been nicknamed after Buckminster Fuller, an American scientist

Table 1.2 The linear combinations of s and p atomic orbitals that form the sp^3 hybrids for the two atoms, labeled A and B, per unit cell in the diamond structure.

$$\phi_1^A = \frac{1}{2}\left[s^A - p_x^A - p_y^A - p_z^A\right] \qquad \phi_1^B = \frac{1}{2}\left[s^B + p_x^B + p_y^B + p_z^B\right]$$

$$\phi_2^A = \frac{1}{2}\left[s^A + p_x^A - p_y^A + p_z^A\right] \qquad \phi_2^B = \frac{1}{2}\left[s^B - p_x^B + p_y^A - p_z^B\right]$$

$$\phi_3^A = \frac{1}{2}\left[s^A + p_x^A + p_y^A - p_z^A\right] \qquad \phi_3^B = \frac{1}{2}\left[s^B - p_x^B - p_y^B + p_z^B\right]$$

$$\phi_4^A = \frac{1}{2}\left[s^A - p_x^A + p_y^A + p_z^A\right] \qquad \phi_4^B = \frac{1}{2}\left[s^B + p_x^B - p_y^B - p_z^B\right]$$

and practical inventor of the early twentieth century, who designed architectural domes based on pentagons and hexagons; their nicknames are buckminsterfullerene or buckyball for C_{60}, bucky-onions, and bucky-tubes.

There is a different way of forming bonds between C atoms: consider the following linear combinations of the s and p atomic orbitals for atom A, defined in Table 1.2 as ϕ_i^A ($i = 1 - 4$). The energy of these states, which are degenerate, is equal to $(\epsilon_s + 3\epsilon_p)/4$, where ϵ_s, ϵ_p are the energies of the original s and p atomic orbitals (see Problem 5); the new states, which are composed of one s and three p orbitals, are called sp^3 orbitals. These orbitals point along the directions from the center to the corners of a regular tetrahedron, as illustrated in Fig. 1.21. We can now imagine placing atoms B, B_1, B_2, B_3 at the corners of the tetrahedron, with which we associate linear combinations of s and p orbitals just like those for atom A, but having all the signs of the p orbitals reversed, defined in Table 1.2 as ϕ_i^B ($i = 1 - 4$). Then we will have a situation where the ϕ orbitals on neighboring A and B atoms will be pointing toward each other, and we can form symmetric and antisymmetric combinations of those, ψ^b, ψ^a respectively, to create four σ bonds around atom A. The exact energy of the ψ orbitals will depend on the overlap between the ϕ^A and ϕ^B orbitals; for sufficiently strong overlap, we can expect the energy of the ψ^b states to be lower than the original s atomic orbitals and those of the ψ^a states to be higher than the original p atomic orbitals, as shown schematically in Fig. 1.21. The vectors connecting the equivalent B, B_1, B_2, B_3 atoms define the repeat vectors at which atoms must be placed to form an infinite crystal. By placing both A-type and B-type atoms at all the possible integer multiples of these vectors, we create the diamond lattice shown in Fig. 1.22. This is the other stable form of bulk C; an experimental image of this structure taken with a transmission electron microscope (TEM) is shown in Fig. 1.23. Since C has four valence electrons and each atom at the center of a tetrahedron forms four σ bonds with its neighbors, all electrons are taken up by the bonding states. This results in a very stable and strong 3D crystal. Surprisingly, graphite has a somewhat lower internal energy than diamond, that is, the thermodynamically stable solid form of carbon is the soft, black, cheap graphite rather than the very strong, brilliant, expensive diamond crystal (which means that diamonds are definitely *not* forever)!

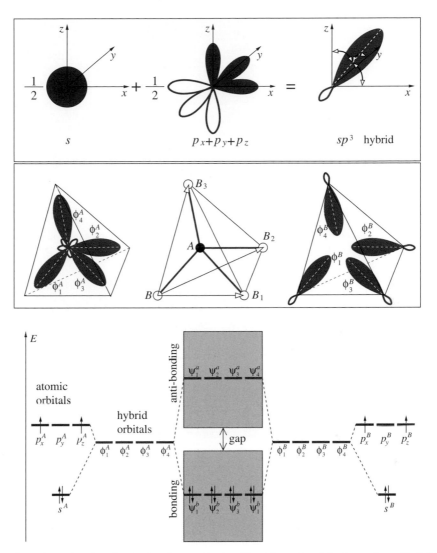

Fig. 1.21 Illustration of covalent bonding in the diamond lattice. **Top**: Example of the formation of hybrid sp^3 orbitals from linear combinations of one s and three p orbitals: this is the hybrid orbital labeled ϕ_1^B in Table 1.2. **Middle**: The sp^3 linear combinations of s and p atomic orbitals corresponding to the atoms labeled A and B, as defined in Table 1.2. **Bottom**: The energy-level diagram for the s, p atomic states, their sp^3 linear combinations (ϕ_i^As and ϕ_i^Bs), and the bonding (ψ_i^bs) and anti-bonding (ψ_i^as) states (up–down arrows indicate electron spins). The shaded regions indicate the broadening of the individual energy levels over a range of values into the valence and conduction bands.

The diamond lattice, with four neighbors per atom, is relatively open compared to the close-packed lattices. Its stability comes from the very strong covalent bonds formed between the atoms. Two other elements with four valence s and p electrons, namely Si and Ge, also crystallize in the diamond, but not the graphite, lattice. There are two more elements with four valence s and p electrons in the Periodic Table, Sn and Pb. Sn forms crystal structures that are distorted variants of the diamond lattice, since its σ bonds are

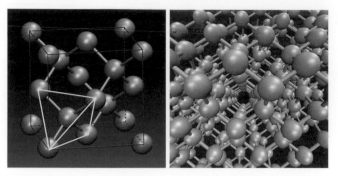

Fig. 1.22   The diamond crystal. **Left**: The conventional cube, outlined in red lines, with C atoms at the cube corners and face centers, forming a regular FCC lattice. A second FCC lattice is formed by C atoms at the centers of tetrahedra defined by the atoms of the first FCC lattice – one such tetrahedron is outlined in yellow lines. The two interpenetrating FCC lattices are displaced relative to each other by 1/4 of the main diagonal of the cube. **Right**: Perspective view of larger portion of the crystal.

Fig. 1.23   The structure of the diamond lattice as revealed by the transmission electron microscope (TEM) [adapted from F. Hosokawa, H. Sawada, Y. Kondo, K. Takayanagi, and K. Suenaga, "Development of Cs and Cc correctors for transmission electron microscopy," *Microscopy* **62**, 2341 (2013)]. The figure on the upper right shows an intensity scan along the vertical dashed line in the main figure, with peaks corresponding to the atomic positions separated by 89 pm (0.89 Å). The lower-left corner of the TEM image contains a simulated image of the structure. The figure on the lower left shows a ball-and-stick representation of the crystal oriented along the same crystallographic direction.

not as strong as those of the other group-IV-A elements, and it can gain some energy by increasing the number of neighbors (from four to six) at the expense of perfect tetrahedral σ bonds. Pb, on the other hand, behaves more like a metal, preferring to optimize the number of neighbors and form the FCC crystal. Interestingly, elements with only three valence s and p electrons, like B, Al, Ga, In, and Tl, do not form the graphite structure, as alluded to above. They instead form more complex structures in which they try to optimize bonding given their relatively small number of valence electrons per atom. Some examples: the common structural unit for B is the icosahedron, shown in Fig. 1.15, and such units are close packed to form the solid; Al forms the FCC crystal and is the representative metal with s and p electrons and a close-packed structure; Ga forms quite complicated crystal structures with six or seven near neighbors (not all of them at the same distance); In forms a distorted version of the cubic close packing in which the 12 neighbors are split into a

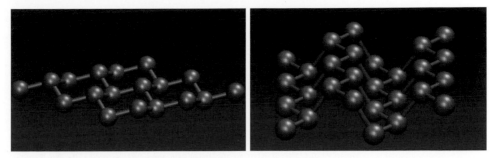

Fig. 1.24 Typical layer structures that group-V elements form, with all atoms threefold coordinated as in a graphitic plane, but with bond angles between nearest neighbors being close to 90°, representative of *p*-orbital bonds. The planes are stacked on top of each other to form the 3D solids. **Left**: A layer of blue phosphorus. **Right**: A layer of black phosphorus (also referred to as "phosphorene").

group of four and another group of eight equidistant atoms. None of these structures can easily be described in terms of the notions introduced above to handle *s* and *p* valence electrons, demonstrating the limitations of this simple approach.

Of the other elements in the Periodic Table with *s* and *p* valence electrons, those with five electrons, N, P, As, Sb, and Bi, tend to form complex structures where atoms have three σ bonds to their neighbors but not in a planar configuration. A characteristic pattern is one in which the three *p* valence electrons participate in covalent bonding while the two *s* electrons form a filled state which does not contribute much to the cohesion of the solid; this filled state is called the "lone pair" state. If the covalent bonds were composed of purely *p* orbitals, the bond angles between nearest neighbors would be 90°; instead, the covalent bonds in these structures are a mixture of *s* and *p* orbitals with predominant *p* character, and the bond angles are somewhere between 120° (sp^2 bonding) and 90° (pure *p* bonding), as illustrated in Fig. 1.24. As an example, two different structures of solid P are represented by this kind of local bonding arrangement. In one structure, blue phosphorus, the covalent bonds are arranged in puckered hexagons which form planes. In another structure, black phosphorus, the atoms form bonded chains along one direction, displaced in the other two directions so that the resulting arrangement is a layer with the chains slightly above and below the average position in the direction perpendicular to the plane of the layer. In both structures, the layers are stacked on top of each other to form the solid. The interaction between layers is much weaker than that between atoms in a single layer: an indication of this difference in bonding is the fact that the distance between nearest neighbors in a layer of blue phosphorus is 2.17 Å, while the closest distance between atoms on successive layers is 3.87 Å, almost a factor of 2 larger. The structures of As, Sb, and Bi follow the same general pattern with threefold bonded atoms, but in those solids there exist additional covalent bonds between atoms in different layers so that the structure is not as clearly layered as in the case for P. An exception to this general tendency is nitrogen, the lightest element with five valence electrons, which forms a crystal composed of nitrogen molecules since the N_2 unit is particularly stable.

The elements with six *s* and *p* valence electrons, O, S, Se, Te, and Po, tend to form molecular-like ring or chain structures with two nearest neighbors per atom, which are then packed to form 3D crystals. These rings or chains are puckered and form bonds at angles that try to satisfy bonding requirements analogous to those described for the solids with four s and p valence electrons. Since these elements have a valence of 6, they tend to keep four of their electrons in one filled s and one filled p orbital and form covalent bonds to two neighbors with their other two p orbitals. This picture is somewhat oversimplified, since significant hybridization takes place between s and p orbitals that participate in bonding, so that the preferred angles between the bonding orbitals are not 90°, as pure p bonding would imply, but range between 102° and 108°. Typical distances between nearest-neighbor atoms in the rings or chains are 2.06 Å for S, 2.32 Å for Se, and 2.86 Å for Te, while typical distances between atoms in successive units are 3.50 Å for S, 3.46 Å for Se, and 3.74 Å for Te; that is, the ratio of distances between atoms within a bonding unit and across bonding units is 1.7 for S, 1.5 for Se, and 1.3 for Te. An exception to this general tendency is oxygen, the lightest element with six valence electrons, which forms a crystal composed of oxygen molecules; the O_2 unit is particularly stable. The theme of diatomic molecules as the basic unit of the crystal, already mentioned for nitrogen and oxygen, is common in elements with seven s and p valence electrons also: chlorine, bromine, and iodine form solids by close packing of diatomic molecules.

1.3.4   Atoms with s and d Valence Electrons

This category includes all the atoms in the middle columns of the Periodic Table, that is, columns I-B to VII-B and VIII. The d orbitals in atoms have directional nature like the p orbitals. However, since there are five d orbitals it is difficult to construct linear combinations with s orbitals that would neatly point toward nearest neighbors in 3D space and produce a crystal with simple σ bonds. Moreover, the d valence orbitals typically lie lower in energy than the s valence orbitals and therefore do not participate as much in bonding (see for example the discussion about Ag in Chapter 5). Note that d orbitals *can* form strong covalent bonds by combining with p orbitals of other elements, as we discuss in a subsequent section. Thus, elements with s and d valence electrons tend to form solids where the s electrons are shared among all atoms in the lattice, just like elements with one or two s valence electrons. This is particularly true for the elements Cu, Ag, and Au (called the "noble elements" due to their relatively chemically inert nature), which have electronic shells of the type $d^{10}s^1$, that is, a filled d-shell and one s electron outside it; the d states, however, are close in energy to the s state, and there is significant interaction between them. The predominantly d-electron elements form space-filling close-packed crystals, of the FCC, HCP, and BCC types. There are very few exceptions to this general tendency, namely Mn which forms a very complex structure with a cubic lattice and a very large number of atoms (58) in the unit cell, and Hg which forms a low-symmetry rhombohedral structure. Even those structures, however, are slight variations of the basic close-packing structures already mentioned. For instance, the Mn structure, in which atoms have from 12 to 16 neighbors, is a slight variation of the BCC structure. The crystals formed by most of these elements typically have metallic character.

1.3.5 Atoms with s, d, and f Valence Electrons

The same general trends are found in the rare earth elements, which are grouped in the lanthanides (atomic numbers 58–71) and the actinides (atomic numbers 90 and beyond). Of those we discuss briefly the lanthanides as the more common of the rare earths that are found in solids. In these elements the f-shell is gradually filled as the atomic number increases, starting with an occupation of two electrons in Ce and completing the shell with 14 electrons in Lu. The f orbitals have directional character, which is even more complicated than that of d orbitals. The solids formed by these elements are typically close-packed structures like FCC and HCP, with a couple of exceptions (Sm which has rhombohedral structure and Eu which has BCC structure). They are metallic solids with high atomic densities. More interesting are structures formed by these elements and other elements of the Periodic Table, in which the complex character of the f orbitals can be exploited in combination with orbitals from neighboring atoms to form strong bonds. Alternatively, these elements are used as dopants in complicated crystals, where they donate some of their electrons to states formed by other atoms. One such example is discussed in the following sections.

1.3.6 Solids with Two Types of Atoms

Some of the most interesting and useful solids involve two types of atoms, which in some ways are complementary. One example that comes immediately to mind are solids composed of atoms in the first (group-I-A) and seventh (group-VII-A) columns of the Periodic Table, which have one and seven valence electrons, respectively. Solids composed of such elements are referred to as "alkali halides." It is natural to expect that the atom with one valence electron will lose it to the more electronegative atom with seven valence electrons, which then acquires a closed electronic shell, completing the s and p levels. This of course leads to one positively and one negatively charged ion, which are repeated periodically in space to form a lattice. The easiest way to arrange such atoms is at the corners of a cube, with alternating corners occupied by atoms of the opposite type. This arrangement results in the sodium chloride (NaCl) or rocksalt structure, one of the most common crystals. Many combinations of group-I-A and group-VII-A atoms form this kind of crystal. In this case each ion has six nearest neighbors of the opposite type. A different way to arrange the ions is to have one ion at the center of a cube formed by ions of the opposite type. This arrangement forms two interpenetrating cubic lattices and is known as the cesium chloride (CsCl) structure. In this case each ion has eight nearest neighbors of the opposite type. Several combinations of group-I-A and group-VII-A atoms crystallize in this structure. Since in all these structures the group-I-A atoms lose their s valence electron to the group-VII-A atoms, this type of crystal is representative of ionic bonding. Both of these ionic structures are shown in Fig. 1.25.

Another way of achieving a stable lattice composed of two kinds of ions with opposite sign is to place them in the two interpenetrating FCC sublattices of the diamond lattice, described earlier. In this case each ion has four nearest neighbors of the opposite type, as

Fig. 1.25 Illustration of cubic crystals formed by ionic solids. **Left**: The rocksalt, NaCl structure, in which the ions of each type form an FCC lattice with each ion surrounded by six neighbors of the opposite type; the two interpenetrating FCC lattices are displaced relative to each other by 1/2 of the main diagonal of the cube. **Middle**: The CsCl structure, in which the ions of each type form a simple cubic lattice with each ion surrounded by eight neighbors of the opposite type. The two interpenetrating cubic lattices are displaced relative to each other by 1/2 of the main diagonal of the cube. **Right**: NaCl crystals exhibiting the characteristic cubic shape dictated by the atomic-scale structure of the crystal [*source:* Wikimedia CC BY-SA 3.0, https://commons.wikimedia.org/w/index.php?curid=227730].

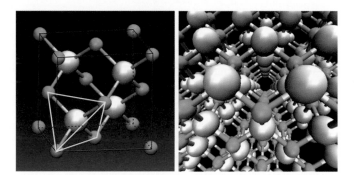

Fig. 1.26 The zincblende lattice in which every atom is surrounded by four neighbors of the opposite type. **Left**: The conventional cubic cell is outlined in thin red lines and a tetrahedron of green atoms is outlined in yellow lines. **Right**: Perspective view of a larger portion of the zincblende crystal.

shown in Fig. 1.26. Many combinations of atoms in the I-B column of the Periodic Table and group-VII-B atoms crystallize in this structure, which is called the zincblende structure from the German term for ZnS, the representative solid with this lattice.

The elements in the I-B column have a filled d-shell (10 electrons) and one extra s valence electron, so it is natural to expect them to behave in some ways similarly to the alkali metals. However, the small number of neighbors in this structure, as opposed to the rocksalt and cesium chloride structures, suggest that the cohesion of these solids cannot be attributed to simple ionic bonding alone. In fact, this becomes more pronounced when atoms from the second (group-II-B) and sixth (group-VI-A) columns of the Periodic Table form the zincblende structure (ZnS itself is the prime example). In this case we would have to assume that the group-II atoms lose their two electrons to the group-VI atoms, but since the electronegativity difference is not as great between these two types of elements as between group-I-A and group-VII-A elements, something more than ionic bonding must be involved. Indeed, the crystals of group-II and group-VI atoms in the zincblende structure are good examples of mixed ionic and covalent bonding. This trend extends to one more class of solids: group-III-A and group-V-A atoms also form zincblende crystals, like AlP, GaAs, InSb, etc. In this case, the bonding is even more tilted toward covalent character, similar to the case of group-IV atoms which form the diamond lattice. Finally, there are combinations of two group-IV atoms that form the zincblende structure; some interesting examples are SiC and GeSi alloys.

A variation on this theme is a class of solids composed of Si and O. In these solids, each Si atom has four O neighbors and is situated at the center of a tetrahedron, while each O atom has two Si neighbors, as illustrated in Fig. 1.27. In this manner the valence of both Si and O is perfectly satisfied, so that the structure can be thought of as covalently bonded. Due to the large electronegativity of O, the covalent bonds are polarized to a large extent, so that the two types of atoms can be considered partially ionized. This results again in a mixture of covalent and ionic bonding. The tetrahedra of Si–O atoms are very stable units and the relative positions of atoms in a tetrahedron are essentially fixed. The position of these tetrahedra relative to each other, however, can be changed with little cost in energy, because this type of structural distortion involves only a slight bending of bond angles, without changing bond lengths. This freedom in relative tetrahedron orientation produces a very wide variety of solids based on this structural unit, including amorphous structures, like the common glass illustrated in Fig. 1.27, and structures with many open spaces in them, like the zeolites.

Yet another category of solids composed of two types of atoms are the so-called transition-metal chalcogenides. In these materials, the transition-metal atom contributes to the bonding with its partially filled d states (typical elements are Sn, Fe, Mo, Ta, W) and the chalcogen atom contributes to the bonding with its p electrons (typical elements are S, Se, Te). Some materials involve a one-to-one ratio of the metal to the chalcogen atoms (the transition-metal monochalcogenides, like SnS, FeSe), while in others the ratio is one-to-two (the transition-metal dichalcogenides, like MoS_2, WSe_2). What is

Fig. 1.27   Example of an ionic-covalent amorphous structure, the silicon-dioxide glass. **Left**: Illustration of an amorphous network of SiO_2, with every Si atom (yellow spheres) having four O neighbors and every O atom (red spheres) having two Si neighbors. **Right**: An enlargement of a small part of the amorphous SiO_2 network, in which the local bonding between Si and O atoms is clearly seen.

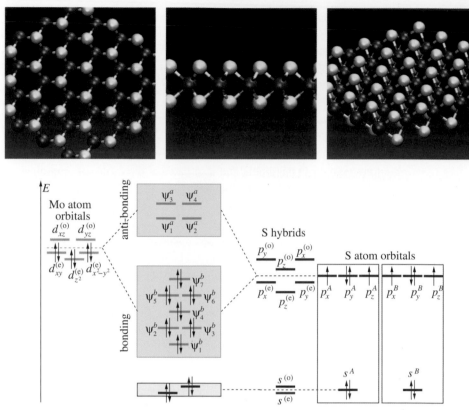

Fig. 1.28 Illustration of the atomic and electronic structure of a layered transition-metal dichalcogenide crystal, molybdenum disulfide. **Top row**: A single layer of the MoS_2 solid, with yellow spheres representing the S atoms and brown spheres representing the Mo atoms: top, side, and perspective views. **Bottom row**: Schematic representation of the energy levels of atomic orbitals of Mo and S atoms, the symmetric S hybrid orbitals, and their combination into bonding and anti-bonding bands.

interesting about these solids is that they consist of layers within which the atoms are strongly bonded, while the interlayer bonding is much weaker, of van der Waals type. A typical example, molybdenum disulfide (MoS_2), is shown in Fig. 1.28. In this case, the corresponding energy-level diagram of atomic-like states and their bonding and anti-bonding combinations produces a band gap when the bands are filled by the available valence electrons. Many of these materials behave like semiconductors, with band gaps in the range of 2–3 eV. For this layer, which has a mirror-plane symmetry on the plane of the metal atoms, the states of the chalcogen atoms that participate in the bonding within one layer are symmetric (even and odd) linear combinations of the atomic p orbitals. The bonding between these hybrid p orbitals and the d orbitals of the metal atom involves both covalent and ionic character.

The 3D crystal is formed by stacking of such layers. The layered nature of these materials makes them very good lubricants. The possibility of extracting a single layer of one material, and combining it with one or more layers of another similar material, has opened up new exciting avenues for manipulating the properties of solids to achieve interesting electronic and optical responses. The hope of employing these structures in the fabrication of novel electronic and optical devices has sparked intensive research into their properties. Equally exciting is the prospect of exploring the physics of electrons confined in these 2D layered heterostructures, including quantum Hall effects, topologically protected edge states that can carry spin currents, and superconductivity; all of these topics are under intense investigation at the time of writing.

1.3.7 Hydrogen: A Special One-s-Valence-Electron Atom

So far we have left H out of the discussion. This is because H is a special case: it has no core electrons. Its interaction with the other elements, as well as between H atoms is unusual, because when H tries to share its one valence s electron with other atoms, what is left is a bare proton rather than a nucleus shielded partially by core electrons. The proton is an ion much smaller in size than the other ions produced by stripping the valence electrons from atoms: its size is 10^{-15} m, five orders of magnitude smaller than typical ions which have a size of order 1 Å. It also has the smallest mass, which gives it a special character: in all other cases (except He) we can consider the ions as classical particles, due to their large mass, while in the case of hydrogen, its light mass implies a large zero-point motion which makes it necessary to take into account the quantum nature of the proton's motion. Yet another difference between hydrogen and all other elements is the fact that its s valence electrons are very strongly bound to the nucleus: the ionization energy is 13.6 eV, whereas typical ionization energies of valence electrons in other elements are in the range 1–2 eV. Due to its special character, H forms a special type of bond called a "hydrogen bond." This is encountered in many structures composed of molecules that contain H atoms, such as organic molecules and water.

The solid in which hydrogen bonding plays the most crucial role is ice. Ice forms many complex phases.[8] In its ordinary phase called Ih, the H_2O molecules are placed so that the O atoms occupy the sites of a wurtzite lattice (see Fig. 1.14), while the H atoms are along lines that join O atoms.[9] There are two H atoms attached to each O atom by short covalent bonds (of length 1.00 Å), while the distance between O atoms is 2.75 Å. There is one H atom along each line joining two O atoms. The bond between a H atom and an O atom to which it is *not* covalently bonded is called a hydrogen bond, and in this system, has length 1.75 Å; it is these hydrogen bonds that give stability to the crystal. This is illustrated in Fig. 1.29. The hydrogen bond is much weaker than the covalent bond between H and O in the water molecule: the energy of the hydrogen bond is 0.3 eV, while that of the covalent H–O bond is 4.8 eV. There are many ways of arranging the H atoms within these constraints for a fixed lattice of O atoms, giving rise to a large configurational entropy. Other forms of ice have different lattices, but this motif of local bonding is common.

Within the atomic orbital picture discussed earlier for solids with s and p electrons, we can construct a simple argument to rationalize hydrogen bonding in the case of ice. The O atom has six valence electrons in its s and p-shells and therefore needs two more

---

[8] N. H. Fletcher, *The Chemical Physics of Ice* (Cambridge University Press, Cambridge, 1970).
[9] J. D. Bernal and R. H. Fowler, *J. Chem. Phys.* **1**, 515 (1933).

Fig. 1.29   Illustration of hydrogen bonding between water molecules in ice: the O atom (red larger circles) is at the center of a tetrahedron formed by other O atoms, and the H atoms (white smaller circles) are along the directions joining the center to the corners of the tetrahedron. The O–H covalent bond distance is $a = 1.00$ Å (thicker black lines), while the H–O hydrogen bond distance is $b = 1.75$ Å (denoted by dashed thinner lines). This is the structure of Ih ice in which the O atoms sit at the sites of a wurtzite lattice (cf. Fig. 1.14) and the H atoms are along the lines joining O atoms; there is one H atom along each such line, and two H atoms bonded by short covalent bonds to each O atom. On the right, a picture of a snowflake exhibiting the characteristic sixfold symmetry of the underlying ice crystal.

electrons to complete its electronic structure. The two H atoms that are attached to it to form the water molecule provide these two extra electrons, at the cost of an anisotropic bonding arrangement (a completed electronic shell should be isotropic, as in the case of Ne, which has two more electrons than O). The cores of the H atoms (the protons), having lost their electrons to O, experience a Coulomb repulsion. The most favorable structure for the molecule which optimizes this repulsion would be to place the two H atoms in diametrically opposite positions relative to the O atom, but this would involve only one p orbital of the O atom to which both H atoms would bond. This is an unfavorable situation as far as formation of covalent bonds is concerned, because it is not possible to form two covalent bonds with only one p orbital and two electrons from the O atom. A compromise between the desire to form strong covalent bonds and the repulsion between the H cores is the formation of four sp^3 hybrids from the orbitals of the O atom, two of which form covalent bonds with the H atoms, while the other two are filled by two electrons each. This produces a tetrahedral structure with two lobes which have more positive charge (the two sp^3 orbitals to which the H atoms are bonded) than the other two lobes (the two sp^3 orbitals which are occupied by two electrons each) which have more negative charge. It is natural to expect that bringing similar molecular units together would produce some attraction between the lobes of opposite charge in neighboring units. This is precisely the arrangement of molecules in the structure of ice discussed above and shown in Fig. 1.29. This rationalization, however, is somewhat misleading as it suggests that the hydrogen bond, corresponding to the attraction between oppositely charged lobes of the H_2O tetrahedrons, is essentially ionic. In fact, the hydrogen bond has significant covalent character as well: the two types of orbitals pointing toward each other form bonding (symmetric) and anti-bonding (antisymmetric) combinations, leading to covalent bonds between them. This point of view was originally suggested by L. Pauling[10] and has remained controversial until recently, when sophisticated scattering experiments and quantum-mechanical calculations provided convincing evidence in its support.[11]

The solid phases of pure hydrogen are also unusual. At low pressure and temperature, H is expected to form a crystal composed of H_2 molecules in which every molecule behaves almost like an inert unit, with very weak interactions to the other molecules. At higher pressure, H is supposed to form an atomic solid when the molecules have approached each other enough so that their electronic distributions are forced to overlap strongly.[12] However, the conditions of pressure and temperature at which this transition occurs, and the structure of the ensuing atomic solid, are still a subject of active research.[13] The latest estimates are that it takes more than 5 Mbar of pressure to form the atomic H solid, which can only be reached under very special conditions in the laboratory. There is considerable debate about what the crystal structure at this pressure should be, and although the BCC structure seems to be the most likely phase by analogy to all other alkalis, this has not been unambiguously proven to date.

1.3.8 Solids with More Than Two Types of Atoms

If we allow several types of atoms to participate in the formation of a crystal, many more possibilities open up. There are indeed many solids with complex composition, but the types of bonding that occur in these situations are variants of the types we have already discussed: metallic, covalent, ionic, van der Waals, and hydrogen bonding. In many situations, several of these types of bonding are present simultaneously.

One interesting example of such complex structures is the class of ceramic materials in which high-temperature superconductivity (HTSC) was observed in the mid-1980s (this discovery, by J. G. Bednorz and K. A. Müller, was recognized by the 1987 Nobel Prize for Physics). In these materials, strong covalent bonding between Cu and O forms 1D or 2D structures where the basic building block is oxygen octahedra; rare earth atoms are then placed at hollow positions of these backbond structures and become partially ionized, giving rise to mixed ionic and covalent bonding (see for example Fig. 1.30).

The motif of oxygen octahedra with a metal atom at the center to which the O atoms are covalently bonded, supplemented by atoms which are easily ionized, is also the basis for a class of structures called "perovskites." The chemical formula of perovskites is ABO_3, where A is the easily ionized element and B the element which is bonded to the oxygens. The basic unit is shown in Fig. 1.30: bonding in the xy plane is accomplished through the overlap between the p_x and p_y orbitals of the first (O_1) and second (O_2) oxygen atoms,

[10] L. Pauling, *J. Am. Chem. Soc.* **57**, 2680 (1935).
[11] E. D. Isaacs, A. Shukla, P. M. Platzman, *et al.*, *Phys. Rev. Lett.* **82**, 600 (1999).
[12] E. Wigner and H. B. Huntington, *J. Chem. Phys.* **3**, 764 (1935); I. F. Silvera, *Rev. Mod. Phys.* **52**, 393 (1980); H. K. Mao and R. J. Hemley, *Am. Sci.* **80**, 234 (1992).
[13] R. P. Dias and I. F. Silvera, Science, **355**, 715 (2017).

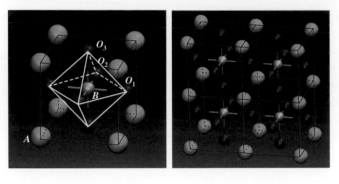

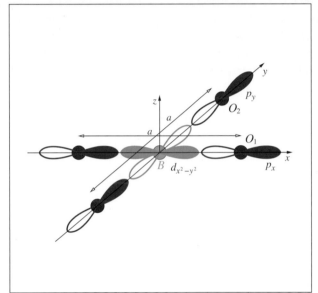 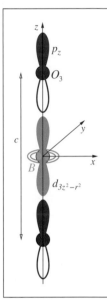

Fig. 1.30 Illustration of the structure and bonding in perovskite crystals. **Top, left**: A Cu atom (cyan sphere, labeled B) surrounded by six O atoms (red spheres, three of them labeled O_1, O_2, O_3), which form an octahedron (shown in yellow); the Cu–O atoms are bonded by strong covalent bonds. The green spheres at the cube corners (labeled A) represent ionized metal atoms. **Top, right**: Corner-sharing O octahedra, forming a 2D square lattice; when joined at the remaining apexes with other octahedra they form a 3D lattice. **Bottom**: The p_x, p_y, p_z orbitals of the three O atoms (labeled O_1, O_2, O_3 in red) and the $d_{x^2-y^2}, d_{3z^2-r^2}$ orbitals of the Cu atoms (labeled B in cyan) that participate in the formation of covalent bonds in the octahedron are shown schematically.

respectively, and the $d_{x^2-y^2}$ orbital of B; bonding along the z-axis is accomplished through the overlap between the p_z orbital of the third (O_3) oxygen atom and the $d_{3z^2-r^2}$ orbital of B (see Fig. 1.17 for the nature of these p and d orbitals). The A atoms provide the necessary number of electrons to satisfy all the covalent bonds. Thus, the overall bonding involves both strong covalent character between B and O, as well as ionic character between the B–O units and the A atoms. The complexity of the structure gives rise to several interesting properties, such as ferroelectricity, that is, the ability of the solid to acquire and maintain an internal dipole moment. The dipole moment is associated with a displacement of the

B atom away from the center of the octahedron, which breaks the symmetry of the cubic lattice. These solids have very intriguing behavior: when external pressure is applied on them it tends to change the shape of the unit cell of the crystal and therefore produces an electrical response since it affects the internal dipole moment; conversely, an external electric field can also affect the internal dipole moment and the solid changes its shape to accommodate it. This coupling of mechanical and electrical responses is very useful for practical applications, such as sensors and actuators and non-volatile memories. The solids that exhibit this behavior are called piezoelectrics; some examples are $CaTiO_3$ (calcium titanate), $PbTiO_3$ (lead titanate), $BaTiO_3$ (barium titanate), and $PbZrO_3$ (lead zirconate).

Another example of complex solids is the class of crystals formed by fullerene clusters and alkali metals: there is strong covalent bonding between C atoms in each fullerene cluster, weak van der Waals bonding between the fullerenes, and ionic bonding between the alkali atoms and the fullerene units. The clusters act just like the group-VII atoms in ionic solids, taking up the electrons of the alkali atoms and becoming ionized. It is intriguing that these solids also exhibit superconductivity at relatively high temperatures!

1.4 Bonding in Solids

In our discussion on the formation of solids from atoms we encountered five general types of bonding in solids:

1. *Van der Waals bonding*, which is formed by atoms that do not have valence electrons available for sharing (like the noble elements) and is rather weak; the solids produced in this way are not particularly stable.
2. *Metallic bonding*, which is formed when electrons are shared by all the atoms in the solid, producing a uniform "sea" of negative charge; the solids produced in this way are the usual metals.
3. *Covalent bonding*, which is formed when electrons in well-defined directional orbitals, which can be thought of as linear combinations of the original atomic orbitals, have strong overlap with similar orbitals in neighboring atoms; the solids produced in this way are semiconductors or insulators.
4. *Ionic bonding*, which is formed when two different types of atoms are combined, one that prefers to lose some of its valence electrons and become a positive ion, and one that prefers to grab electrons from other atoms and become a negative ion. Combinations of such elements are I–VII, II–VI, and III–V. In the first case bonding is purely ionic, in the other two there is a degree of covalent bonding present.
5. *Hydrogen bonding*, which is formed when H is present, due to its lack of core electrons, light mass, and high ionization energy.

For some of these cases, it is possible to estimate the strength of bonding without involving a detailed description of the electronic behavior. Specifically, for van der Waals bonding and for purely ionic bonding it is sufficient to assume simple classical models. For van der Waals bonding, one assumes that there is an attractive potential between the atoms which behaves like $\sim r^{-6}$ with distance r between atoms (this behavior can actually be derived from perturbation theory, see Problem 8). The potential must become repulsive at very short range, as the electronic densities of the two atoms start overlapping, but electrons have no incentive to form bonding states (as was the case in covalent bonding) since all electronic shells are already filled. For convenience the repulsive part is taken to be proportional to r^{-12}, which gives the famous Lennard–Jones 6–12 potential:

$$\mathcal{V}_{\mathrm{LJ}}(r) = 4\epsilon \left[\left(\frac{a}{r} \right)^{12} - \left(\frac{a}{r} \right)^{6} \right] \tag{1.18}$$

with ϵ and a constants that determine the energy and length scales. These have been determined for the different elements by referring to the thermodynamic properties of the noble gases; the values of these parameters for the usual noble gas elements are shown in Table 1.3.

Use of this potential can then provide a quantitative measure of cohesion in these solids. One measure of the strength of these potentials is the vibrational frequency that would correspond to a harmonic oscillator potential with the same curvature at the minimum; this is indicative of the stiffness of the bond between atoms. In Table 1.3 we present the frequencies corresponding to the Lennard–Jones potentials of the common noble gas elements (see following discussion and Table 1.4 for the relation between this frequency

Table 1.3 Parameters for the Lennard–Jones potential for noble gases (for original sources see Ashcroft and Mermin, mentioned in Further Reading). For the calculation of $\hbar\omega$ using the Lennard–Jones parameters see the following discussion and Table 1.4.

	Ne	Ar	Kr	Xe
ϵ (meV)	3.1	10.4	14.0	20.0
a (Å)	2.74	3.40	3.65	3.98
$\hbar\omega$ (meV)	2.213	2.310	1.722	1.510

Table 1.4 Comparison of the three effective potentials, Lennard–Jones $\mathcal{V}_{\mathrm{LJ}}(r)$, Morse $\mathcal{V}_{\mathrm{M}}(r)$, and harmonic oscillator $\mathcal{V}_{\mathrm{HO}}(r)$. The relations between the parameters that ensure the three potentials have the same minimum and curvature at the minimum are also given (the parameters of the Morse and harmonic oscillator potentials are expressed in terms of the Lennard–Jones parameters).

	$\mathcal{V}_{\mathrm{LJ}}(r)$	$\mathcal{V}_{\mathrm{M}}(r)$	$\mathcal{V}_{\mathrm{HO}}(r)$
Potential	$4\epsilon \left[\left(\frac{a}{r} \right)^{12} - \left(\frac{a}{r} \right)^{6} \right]$	$\epsilon \left[e^{-2(r-r_0)/b} - 2e^{-(r-r_0)/b} \right]$	$-\epsilon + \frac{1}{2}m\omega^2(r-r_0)^2$
$\mathcal{V}_{min}$	$-\epsilon$	$-\epsilon$	$-\epsilon$
r_{min}	$(2^{\frac{1}{6}})a$	r_0	r_0
$\mathcal{V}''(r_{min})$	$(72/2^{\frac{1}{3}})(\epsilon/a^2)$	$2(\epsilon/b^2)$	$m\omega^2$
Relations		$r_0 = (2^{\frac{1}{6}})a$	$r_0 = (2^{\frac{1}{6}})a$
		$b = (2^{\frac{1}{6}}/6)a$	$\omega = (432^{\frac{1}{3}})\sqrt{\epsilon/ma^2}$

and the Lennard–Jones potential parameters). The vibrational frequency ω is related to the spring constant κ of the harmonic oscillator by the expression

$$\omega \sim \left(\frac{\kappa}{m}\right)^{1/2}$$

where m is the mass of the oscillating particle. For comparison, the vibrational frequency of the H_2 molecule, the simplest type of covalent bond between two atoms, is about 500 meV, a factor of $\omega_H/\omega_{Ne} \sim 200$ larger, while the masses of the Ne and H atoms give a factor of $\sqrt{m_{Ne}/m_H} \sim 3$; we conclude from this comparison that the spring constant for Ne–Ne interactions must be about four orders of magnitude smaller than that of the H–H bond, that is, the Lennard–Jones potentials for the noble gases correspond to very soft bonds indeed!

A potential of similar nature, also used to describe effective interactions between atoms, is the Morse potential:

$$\mathcal{V}_M(r) = \epsilon \left[e^{-2(r-r_0)/b} - 2e^{-(r-r_0)/b} \right] \tag{1.19}$$

where again ϵ and b are the constants that determine the energy and length scales and r_0 is the position of the minimum energy. It is instructive to compare these two potentials to the harmonic oscillator potential which has the same minimum and curvature, given by:

$$\mathcal{V}_{HO}(r) = -\epsilon + \frac{1}{2}m\omega^2(r - r_0)^2 \tag{1.20}$$

with ω the frequency, m the mass of the particle in the potential, and r_0 the position of the minimum.

The definitions of the three potentials are such that they all have the same value of the energy at their minimum, namely $-\epsilon$. The relations between the values of the other parameters which ensure that the minimum in the energy occurs at the same value of r and that the curvature at the minimum is the same are given in Table 1.4; a plot of the three potentials with these parameters is given in Fig. 1.31. The harmonic oscillator potential is what we would expect near the equilibrium of any normal interaction potential. The other two potentials extend the range far from the minimum; both potentials have a much sharper increase of the energy for distances shorter than the equilibrium value, and a much weaker increase of the energy for distances larger than the equilibrium value, relative to the harmonic oscillator. The overall behavior of the two potentials is quite similar. One advantage of the Morse potential is that it can be solved exactly, by analogy to the harmonic oscillator potential (see Appendix C). This allows a comparison between the energy levels associated with this potential and the corresponding energy levels of the harmonic oscillator; the latter are given by:

$$E_n^{HO} = \left(n + \frac{1}{2}\right)\hbar\omega \tag{1.21}$$

whereas those of the Morse potential are given by:

$$E_n^M = \left(n + \frac{1}{2}\right)\hbar\omega \left[1 - \frac{\hbar\omega}{4\epsilon}\left(n + \frac{1}{2}\right)\right] \tag{1.22}$$

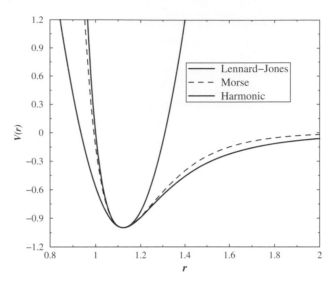

Fig. 1.31 The three effective potentials discussed in the text, Lennard–Jones [Eq. (1.18)], Morse [Eq. (1.19)], and harmonic oscillator [Eq. (1.20)], with the same minimum and curvature at the minimum. The energy is in units of ϵ and the distance in units of a, the two parameters of the Lennard–Jones potential.

for the parameters defined in Table 1.4. We thus see that the spacing of levels in the Morse potential is smaller than in the corresponding harmonic oscillator, and that it becomes progressively smaller as the index of the levels increases. This is expected from the behavior of the potential mentioned above, and in particular from its asymptotic approach to zero for large distances. Since the Lennard–Jones potential has an overall shape similar to the Morse potential, we expect its energy levels to behave in the same manner.

For purely ionic bonding, one assumes that what keeps the crystal together is the attractive interaction between the positively and negatively charged ions, again in a purely classical picture. For the ionic solids with rocksalt, cesium chloride, and zincblende lattices discussed already, it is possible to calculate the cohesive energy which only depends on the ionic charges, the crystal structure, and the distance between ions. This is called the Madelung energy. The main difficulty in evaluating this energy is that the summation converges very slowly, because the interaction potential (Coulomb) is long range. In fact, formally this sum does not converge, and any simple way of summing successive terms gives results that depend on the choice of terms. The formal way to treat periodic structures using the concept of reciprocal space, which we will develop in detail in Chapter 2, makes the calculation of the Madelung energy through the Ewald summation trick much more efficient. For an example of the difficulties of calculating the Madelung energy in simple cases, see Problem 2. We will revisit this issue in Chapter 4 in much more detail, once we have developed all the methodology for handling the ionic energy contribution accurately and efficiently (see Section 4.10).

The other types of bonding, metallic, covalent, and mixed bonding, are much more difficult to describe quantitatively. For metallic bonding, even if we think of the electrons as a uniform sea, we need to know the energy of this uniform "liquid" of fermions, which is not a trivial matter. In addition to the electronic contributions, we have to consider the energy of the positive ions that exist in the uniform negative background of the electron sea. This is another Madelung sum, which converges very slowly. As far as covalent bonding is concerned, although the approach we used by combining atomic orbitals is conceptually simple, much more information is required to render it a realistic tool for calculations, and the electron interactions again come into play in an important way. This will also be discussed in detail in subsequent chapters.

The descriptions that we mentioned for the metallic and covalent solids are also referred to by more technical terms. The metallic sea of electrons paradigm is referred to as the "jellium" model in the extreme case when the ions (atoms stripped of their valence electrons) are considered to form a uniform positive background; in this limit the crystal itself does not play an important role, other than it provides the background for forming the electronic sea. The description of the covalent bonding paradigm is referred to as the linear combination of atomic orbitals (LCAO) approach, since it relies on the use of a basis of atomic orbitals in linear combinations that make the bonding arrangement transparent, as was explained above for the graphite and diamond lattices. We will revisit these notions in more detail in the following chapters.

Further Reading

1. *Solid State Physics*, N. W. Ashcroft and N. D. Mermin (Saunders College Publishing, Philadelphia, PA, 1976). This is a comprehensive and indispensable source on the physics of solids; it provides an inspired coverage of most topics that had been the focus of research up to its publication.

2. *Introduction to Solid State Theory*, O. Madelung (Springer-Verlag, Berlin, 1981). This book represents a balanced introduction to the theoretical formalism needed for the study of solids at an advanced level; it covers both the single-particle and the many-body pictures.

3. *Basic Notions of Condensed Matter Physics*, P. W. Anderson (Benjamin-Cummings Publishing, Menlo Park, CA, 1984).

4. *Quantum Theory of Solids*, C. Kittel (Wiley, New York, 1963).

5. *Solid State Theory*, W. A. Harrison (McGraw-Hill, New York, 1970).

6. *Principles of the Theory of Solids*, J. M. Ziman (Cambridge University Press, Cambridge, 1972).

7. *The Nature of the Chemical Bond and the Structure of Molecules and Solids*, L. Pauling (Cornell University Press, Ithaca, NY, 1960). This is a classic treatment of the nature of bonding between atoms; it discusses extensively bonding in molecules but there is also a rich variety of topics relevant to the bonding in solids.

8. *Crystal Structures*, R. W. G. Wyckoff (Wiley, New York, 1963). This is a very useful compilation of all the structures of elemental solids and a wide variety of common compounds.

9. *Physics of Solids*, E. N. Economou (Springer-Verlag, Berlin, 2010). This is a careful and intuitive introduction to the physics of solids with many useful examples.
10. *Condensed Matter Physics*, M. P. Marder (Wiley, Hoboken, NJ, 2010). This is a comprehensive and modern treatment of the physics of solids.

Problems

1. An important consideration in the formation of crystals is the so-called packing ratio or filling fraction. For each of the elemental cubic crystals (simple cubic, FCC, BCC), calculate the packing ratio, that is, the percentage of volume occupied by the atoms modeled as touching hard spheres. For the NaCl and CsCl structures, assuming that each type of ion is represented by a hard sphere and the nearest-neighbor spheres touch, calculate the packing ratio as a function of the ratio of the two ionic radii and find its extrema and the asymptotic values (for the ratio approaching infinity or zero, that is, with one of the two ions being negligible in size relative to the other one). Using values of the atomic radii for the different elements, estimate the filling fraction in the actual solids.

2. The three ionic lattices, rocksalt, cesium chloride, and zincblende, discussed in this chapter are called bipartite lattices, because they include two equivalent sites per unit cell which can be occupied by the different ions so that each ion type is completely surrounded by the other. Describe the corresponding bipartite lattices in two dimensions. Are they all different from each other? Try to obtain the Madelung energy for one of them, and show how the calculation is sensitive to the way in which the infinite sum is truncated.

3. Consider the single-particle hamiltonian:

$$\hat{\mathcal{H}}^{\mathrm{sp}} = \frac{\mathbf{p}^2}{2m} + \mathcal{V}(\mathbf{r})$$

where $\mathbf{p}$ is the momentum operator

$$\mathbf{p} = -i\hbar\nabla_{\mathbf{r}}$$

m the mass of the particle, and $\mathcal{V}(\mathbf{r})$ the potential energy. Show that the commutator of the hamiltonian with the potential energy term, defined as

$$[\hat{\mathcal{H}}^{\mathrm{sp}}, \mathcal{V}(\mathbf{r})] = \hat{\mathcal{H}}^{\mathrm{sp}}\mathcal{V}(\mathbf{r}) - \mathcal{V}(\mathbf{r})\hat{\mathcal{H}}^{\mathrm{sp}}$$

is not zero, except when the potential is a constant $\mathcal{V}(\mathbf{r}) = \mathcal{V}_0$. Based on this result, provide an argument to the effect that the energy eigenfunctions can be simultaneous eigenfunctions of the momentum operator only for free particles.

4. In the example of bond formation in a 1D molecule, using the definitions of the wavefunctions for the isolated atoms, Eq. (1.14), and the bonding and anti-bonding states, Eq. (1.15), show that:

(a) The difference in the probability of an electron being in the bonding or anti-bonding states, rather than in the average of the two isolated-atom wavefunctions, is given by the expression in Eq. (1.16).

(b) The change in the potential energy for the combined system versus the two isolated ions, given by Eq. (1.17), is negative for the $\psi^{(+)}$ state and positive for the $\psi^{(-)}$ state (take $b = 2.5a$ and perform a numerical integration).

(c) The change in the kinetic energy for the combined system versus the two isolated ions is negative for the $\psi^{(+)}$ state and positive for the $\psi^{(-)}$ state (take $b = 2.5a$ and perform a numerical integration).

5.   Derive the value of the energy of sp^2 and sp^3 hybrids in terms of the values of the energy of the s and p orbitals, ϵ_s and ϵ_p. What are the assumptions that you need to make in order for these expressions to be valid?

6.   Produce an energy-level diagram for the orbitals involved in the formation of the covalent bonds in the water molecule. Provide an argument of how different combinations of orbitals than the ones discussed in the text would not produce as favorable a covalent bond between H and O. Describe how the different orbitals combine to form hydrogen bonds in the solid structure of ice.

7.   Consider a simple excitation of the ground state of the free-electron system, consisting of taking an electron from a state with momentum $\mathbf{k}_1$ and putting it in a state with momentum $\mathbf{k}_2$; since the ground state of the system consists of filled single-particle states with momentum up to the Fermi momentum k_F, we must have $|\mathbf{k}_1| \leq k_F$ and $|\mathbf{k}_2| > k_F$. Removing the electron from state $\mathbf{k}_1$ leaves a "hole" in the Fermi sphere, so this excitation is described as an "electron–hole pair." Discuss the relationship between the total excitation energy and the total momentum of the electron–hole pair; show a graph of this relationship in terms of reduced variables, that is, the excitation energy and momentum in units of the Fermi energy ϵ_F and the Fermi momentum k_F. (At this point we are not concerned with the nature of the physical process that can create such an excitation and with how momentum is conserved in this process.)

8.   In order to derive the attractive part of the Lennard–Jones potential, we consider two atoms with Z electrons each and filled electronic shells. In the ground state, the atoms will have spherical electronic charge distributions and, when sufficiently far from each other, they will not interact. When they are brought closer together, the two electronic charge distributions will be polarized because each will feel the effect of the ions and electrons of the other. We are assuming that the two atoms are still far enough from each other so that their electronic charge distributions do not overlap, and therefore we can neglect exchange of electrons between them. Thus, it is the polarization that gives rise to an attractive potential; for this reason this interaction is sometimes also referred to as the "fluctuating dipole interaction."

To model the polarization effect, we consider the interaction potential between the two neutral atoms:

$$V_{\text{int}} = \frac{Z^2 e^2}{|\mathbf{R}_1 - \mathbf{R}_2|} - \sum_i \frac{Ze^2}{|\mathbf{r}_i^{(1)} - \mathbf{R}_2|} - \sum_j \frac{Ze^2}{|\mathbf{r}_j^{(2)} - \mathbf{R}_1|} + \sum_{ij} \frac{e^2}{|\mathbf{r}_i^{(1)} - \mathbf{r}_j^{(2)}|}$$

where $\mathbf{R}_1, \mathbf{R}_2$ are the positions of the two nuclei and $\mathbf{r}_i^{(1)}, \mathbf{r}_j^{(2)}$ are the sets of electronic coordinates associated with each nucleus. In the above equation, the first term is the repulsion between the two nuclei, the second term is the attraction of the electrons of the first atom to the nucleus of the second, the third term is the attraction of the electrons of the second atom to the nucleus of the first, and the last term is the repulsion between the two sets of electrons in the two different atoms. The summations over i, j run from 1 to Z.

From second-order perturbation theory, the energy change due to this interaction is given by:

$$\Delta E = \langle \Psi_0^{(1)} \Psi_0^{(2)} | V_{\text{int}} | \Psi_0^{(1)} \Psi_0^{(2)} \rangle \tag{1.23}$$
$$+ \sum_{nm} \frac{1}{E_0 - E_{nm}} \left| \langle \Psi_0^{(1)} \Psi_0^{(2)} | V_{\text{int}} | \Psi_n^{(1)} \Psi_m^{(2)} \rangle \right|^2$$

where $\Psi_0^{(1)}, \Psi_0^{(2)}$ are the ground-state many-body wavefunctions of the two atoms, $\Psi_n^{(1)}, \Psi_m^{(2)}$ are their excited states, and E_0, E_{nm} are the corresponding energies of the two-atom system in their unperturbed states.

We define the electronic charge density associated with the ground state of each atom through:

$$n_0^{(I)}(\mathbf{r}) = Z \int \left| \Psi_0^{(I)}(\mathbf{r}, \mathbf{r}_2, \mathbf{r}_3, ..., \mathbf{r}_Z) \right|^2 d\mathbf{r}_2 d\mathbf{r}_3 \cdots d\mathbf{r}_Z \tag{1.24}$$

$$= \sum_{i=1}^{Z} \int \delta(\mathbf{r} - \mathbf{r}_i) \left| \Psi_0^{(I)}(\mathbf{r}_1, \mathbf{r}_2, ..., \mathbf{r}_Z) \right|^2 d\mathbf{r}_1 d\mathbf{r}_2 \cdots d\mathbf{r}_Z$$

with $I = 1, 2$ (the expression for the density $n(\mathbf{r})$ in terms of the many-body wavefunction $|\Psi\rangle$ is discussed in detail in Appendix C). Show that the first-order term in ΔE corresponds to the electrostatic interaction energy between the charge density distributions $n_0^{(1)}(\mathbf{r}), n_0^{(2)}(\mathbf{r})$. Assuming that there is no overlap between these two charge densities, show that this term vanishes (the two charge densities in the unperturbed ground state are spherically symmetric).

The wavefunctions involved in the second-order term in ΔE will be negligible, unless the electronic coordinates associated with each atom are within the range of non-vanishing charge density. This implies that the distances $|\mathbf{r}_i^{(1)} - \mathbf{R}_1|$ and $|\mathbf{r}_j^{(2)} - \mathbf{R}_2|$ should be small compared to the distance between the atoms $|\mathbf{R}_2 - \mathbf{R}_1|$, which defines the distance at which interactions between the two charge densities become negligible. Show that expanding the interaction potential in the small quantities $|r_i^{(1)} - R_1|/|R_2 - R_1|$ and $|r_j^{(2)} - R_2|/|R_2 - R_1|$ gives, to lowest order:

$$-\frac{e^2}{|R_2 - R_1|} \sum_{ij} 3 \frac{(r_i^{(1)} - R_1) \cdot (R_2 - R_1)}{(R_2 - R_1)^2} \cdot \frac{(r_j^{(2)} - R_2) \cdot (R_2 - R_1)}{(R_2 - R_1)^2}$$

$$+ \frac{e^2}{|R_2 - R_1|} \sum_{ij} \frac{(r_i^{(1)} - R_1) \cdot (r_j^{(2)} - R_2)}{(R_2 - R_1)^2} \tag{1.25}$$

Using this expression, show that the leading-order term in the energy difference ΔE behaves like $|R_2 - R_1|^{-6}$ and is negative. This establishes the origin of the attractive term in the Lennard–Jones potential.

9.   We wish to determine the eigenvalues of the Morse potential, Eq. (1.19). One method is to consider an expansion in powers of $(r - r_0)$ near the minimum and relate it to the harmonic oscillator potential with higher-order terms. Specifically, the potential

$$V(r) = \frac{1}{2} m\omega^2 (r - r_0)^2 - \alpha(r - r_0)^3 + \beta(r - r_0)^4 \tag{1.26}$$

has eigenvalues

$$\epsilon_n = (n + \frac{1}{2})\hbar\omega \left[1 - \gamma(n + \frac{1}{2}) \right]$$

where

$$\gamma = \frac{3}{2\hbar\omega} \left(\frac{\hbar}{m\omega} \right)^2 \left[\frac{5}{2} \frac{\alpha^2}{m\omega^2} - \beta \right]$$

First, check to what extent the expansion (1.26) with up to fourth-order terms in $(r - r_0)$ is a good representation of the Morse potential; what are the values of α and β in terms of the parameters of the Morse potential? Use this approach to show that the eigenvalues of the Morse potential are given by Eq. (1.22).

10. An important simple model that demonstrates some of the properties of electron states in infinite-periodic solids is the so-called Kronig–Penney model.[14] In this model, a particle of mass m experiences a 1D periodic potential with period a:

$$V(x) = 0, \quad 0 < x < (a - l)$$
$$= V_0, \quad (a - l) < x < a$$
$$V(x + a) = V(x)$$

where we will take $V_0 > 0$. The wavefunction $\psi(x)$ obeys the Schrödinger equation

$$\left[-\frac{\hbar^2}{2m} \frac{d^2}{dx^2} + V(x) \right] \psi(x) = \epsilon \psi(x)$$

---

[14] R. de L. Kronig and W. G. Penney, *Proc. Roy. Soc. (London), A* **130**, 499 (1931).

(a) Choose the following expression for the particle wavefunction:

$$\psi(x) = e^{ikx} u(x)$$

and show that the function $u(x)$ must obey the equation

$$\frac{d^2 u(x)}{dx^2} + 2ik\frac{du(x)}{dx} - \left[k^2 - \frac{2m\epsilon}{\hbar^2} + \frac{2mV(x)}{\hbar^2}\right] u(x) = 0$$

Assuming that $u(x)$ is finite for $x \to \pm\infty$, the variable k must be real so that the wavefunction $\psi(x)$ is finite for all x.

(b) We first examine the case $\epsilon > V_0 > 0$. Consider two solutions $u_1(x), u_2(x)$ for the ranges $0 < x < (a-l)$ and $(a-l) < x < a$, respectively, which obey the equations

$$\frac{d^2 u_1(x)}{dx^2} + 2ik\frac{du_1(x)}{dx} - [k^2 - \kappa^2]u_1(x) = 0, \quad 0 < x < (a-l)$$

$$\frac{d^2 u_2(x)}{dx^2} + 2ik\frac{du_2(x)}{dx} - [k^2 - \lambda^2]u_2(x) = 0, \quad (a-l) < x < a$$

where we have defined the quantities

$$\kappa = \sqrt{\frac{2m\epsilon}{\hbar^2}}, \quad \lambda = \sqrt{\frac{2m(\epsilon - V_0)}{\hbar^2}}$$

which are both real for $\epsilon > 0$ and $(\epsilon - V_0) > 0$. Show that the solutions to these equations can be written as:

$$u_1(x) = c_1 e^{i(\kappa - k)x} + d_1 e^{-i(\kappa + k)x}, \quad 0 < x < (a-l)$$

$$u_2(x) = c_2 e^{i(\lambda - k)x} + d_2 e^{-i(\lambda + k)x}, \quad (a-l) < x < a$$

By matching the values of these solutions and of their first derivatives at $x = 0$ and $x = a - l$, find a system of four equations for the four unknowns, c_1, d_1, c_2, d_2. Show that requiring this system to have a non-trivial solution leads to the following condition:

$$-\frac{\kappa^2 + \lambda^2}{2\kappa\lambda}\sin(\kappa(a-l))\sin(\lambda l) + \cos(\kappa(a-l))\cos(\lambda l) = \cos(ka)$$

Next, show that with the definition

$$\tan(\theta) = -\left(\frac{\kappa^2 + \lambda^2}{2\kappa\lambda}\right)\tan(\lambda l)$$

the above condition can be written as:

$$\left[1 + \frac{(\kappa^2 - \lambda^2)^2}{4\kappa^2\lambda^2}\sin^2(\lambda l)\right]^{1/2} \cos(\kappa(a-l) - \theta) = \cos(ka)$$

Show that this last equation admits real solutions for k only in certain intervals of ϵ; determine these intervals of ϵ and the corresponding values of k. Plot the values of ϵ as a function of k and interpret the physical meaning of these solutions. How does the solution depend on the ratio ϵ/V_0? How does it depend on the ratio l/a?

(c) Repeat the above problem for the case $V_0 > \epsilon > 0$. Discuss the differences between the two cases.

2   Electrons in Crystals: Translational Periodicity

In the previous chapter we explored the structure of solids from a general viewpoint, emphasizing the crystalline nature of most common materials. We also described two simple, yet powerful, ways of understanding how sharing of electrons between the atoms that constitute the solid can explain the nature of bonds responsible for the cohesion and stability of the solid: these were the free-electron model, appropriate for metallic bonding, and the formation of bonding and anti-bonding combinations of atomic orbitals that can capture the essence of covalent bonding. Other types of bonding that we encounter in solids can be thought of as combinations of these two simple concepts. In this chapter we will explore the constraints that the periodic arrangement of atoms in a crystal impose on the behavior of electrons. The essence of these constraints is captured by Bloch's theorem, a cornerstone of the mathematical description of electronic behavior in crystals. We will then illustrate through a series of simple examples how these constraints are manifest in the behavior of electrons, using again as conceptual models the two extreme cases: free electrons and covalent bonding as captured by the linear combinations of atomic-like orbitals.

In exploring all these concepts, we are making the assumption that electrons in a solid can be viewed essentially as independent single particles moving in some effective potential. This assumption needs to be properly justified, because each electron in a solid interacts with all the ions by an *attractive* Coulomb force, as well as with all the other electrons by a *repulsive* Coulomb force, of the same range and strength. From the fact that electrons are indistinguishable particles and have to obey the Pauli exclusion principle, and from their strong Coulomb repulsion, it is far from clear that they can be treated as independent particles; in fact, there are many phenomena that suggest a highly correlated nature in the behavior of electrons. In Chapter 4 we return to the issue of how and when the single-particle picture of electrons can be justified, and the approximations and limitations inherent in this picture.

2.1   Translational Periodicity: Bloch States

A crystal consists of atoms that form a regular pattern in space. In order to describe this regularity in atomic-scale structure mathematically, we need to define where exactly each atom is located and what its environment looks like. The environment of an atom in a crystal is defined by the number and type of neighbors the atom has at any given direction and distance. Two atoms, at a distance $\mathbf{R}$ apart, are considered equivalent within the crystal when they are atoms of the same element *and* their environment in any given direction in space and at any given distance from their positions is exactly the same (this statement makes sense only for an infinite crystal, an implicit assumption throughout the discussion presented in this chapter). According to this description, we need to be able to specify two things in order to fully define the crystal:

 (i) all the vectors $\mathbf{R}$ which connect equivalent atoms in the crystal, and
(ii) a basic unit of atoms, called the primitive unit cell (PUC), which contains all the distinct, inequivalent (not connected by an $\mathbf{R}$ vector) atoms.

In 3D space, a crystal can be described fully through the definition of:

 (i) the primitive lattice vectors $\mathbf{a}_1, \mathbf{a}_2, \mathbf{a}_3$, which are the three shortest, linearly independent vectors that connect equivalent atoms, forming the PUC, and
(ii) the positions of atoms inside the PUC, called the "basis."

The lattice vectors $\mathbf{R}$ are constructed through all the possible combinations of primitive lattice vectors, multiplied by integers:

$$\mathbf{R} = n_1 \mathbf{a}_1 + n_2 \mathbf{a}_2 + n_3 \mathbf{a}_3, \quad n_1, n_2, n_3 : \text{integers} \tag{2.1}$$

We give some examples of the primitive lattice vectors for representative simple crystals in Table 2.1 and Figs 2.1 and 2.2.

Table 2.1 Examples of 2D and 3D crystals. The vectors $\mathbf{a}_1, \mathbf{a}_2, \mathbf{a}_3$ define the primitive unit cell. The position of one atom in the PUC is always assumed to be at the origin; when there are two atoms in the PUC, the position of the second atom $\mathbf{t}_2$ is given with respect to the origin. All vectors are given in cartesian coordinates and in terms of the standard lattice parameter a, the side of the conventional cube or parallelepiped. For the HCP lattice, a second parameter is required, namely the c/a ratio. d_{NN} is the distance between nearest neighbors in terms of the lattice constant a. These crystals are illustrated in Figs 2.1 and 2.2.

Lattice	$\mathbf{a}_1$	$\mathbf{a}_2$	$\mathbf{a}_3$	$\mathbf{t}_2$	c/a	d_{NN}
Square	$(a, 0, 0)$	$(0, a, 0)$				a
Hexagonal	$\left(\frac{\sqrt{3}a}{2}, \frac{a}{2}, 0\right)$	$\left(\frac{\sqrt{3}a}{2}, -\frac{a}{2}, 0\right)$				a
Honeycomb	$\left(\frac{\sqrt{3}a}{2}, \frac{a}{2}, 0\right)$	$\left(\frac{\sqrt{3}a}{2}, -\frac{a}{2}, 0\right)$		$\left(\frac{a}{\sqrt{3}}, 0, 0\right)$		$\frac{a}{\sqrt{3}}$
Simple cubic	$(a, 0, 0)$	$(0, a, 0)$	$(0, 0, a)$			a
BCC	$\left(\frac{a}{2}, -\frac{a}{2}, -\frac{a}{2}\right)$	$\left(\frac{a}{2}, \frac{a}{2}, -\frac{a}{2}\right)$	$\left(\frac{a}{2}, \frac{a}{2}, \frac{a}{2}\right)$			$\frac{a\sqrt{3}}{2}$
FCC	$\left(\frac{a}{2}, \frac{a}{2}, 0\right)$	$\left(\frac{a}{2}, 0, \frac{a}{2}\right)$	$\left(0, \frac{a}{2}, \frac{a}{2}\right)$			$\frac{a}{\sqrt{2}}$
Diamond	$\left(\frac{a}{2}, \frac{a}{2}, 0\right)$	$\left(\frac{a}{2}, 0, \frac{a}{2}\right)$	$\left(0, \frac{a}{2}, \frac{a}{2}\right)$	$\left(\frac{a}{4}, \frac{a}{4}, \frac{a}{4}\right)$		$\frac{a}{4\sqrt{3}}$
HCP	$\left(\frac{\sqrt{3}a}{2}, \frac{a}{2}, 0\right)$	$\left(\frac{\sqrt{3}a}{2}, -\frac{a}{2}, 0\right)$	$(0, 0, c)$	$\left(\frac{2a}{\sqrt{3}}, 0, \frac{c}{2}\right)$	$\sqrt{\frac{8}{3}}$	$\frac{a}{\sqrt{3}}$

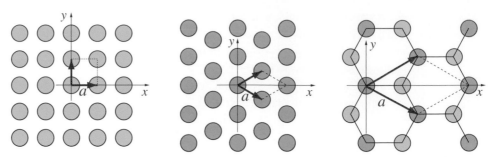

Fig. 2.1   The 2D crystals defined in Table 2.1: square, hexagonal, and honeycomb. The lattice vectors are indicated by red arrows and the primitive unit cells are outlined by dashed lines.

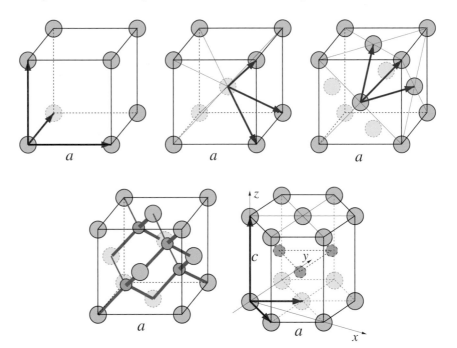

Fig. 2.2   The 3D crystals defined in Table 2.1. **Top row, left to right**: Simple cubic (SC), body centered cubic (BCC), face centered cubic (FCC), with a the size of the cube side (in the latter two structures, thinner colored lines join the atoms at the cube corners to those at the center of the cube in BCC and the centers of the cube faces in FCC). **Bottom row**: Diamond and hexagonal close-packed (HCP). In all cases the lattice vectors are indicated by arrows (the lattice vectors for the diamond lattice are identical to those for the FCC lattice). For the diamond and HCP lattices two different symbols, blue and red circles, are used to denote the two atoms in the unit cell.

The volume of the PUC, enclosed by the three primitive lattice vectors, is given by:

$$V_{\text{PUC}} =| \mathbf{a}_1 \cdot (\mathbf{a}_2 \times \mathbf{a}_3) | \tag{2.2}$$

This is a useful definition: we need to determine all relevant real-space functions for $\mathbf{r}$ *only within* the PUC since, due to the periodicity of the crystal, these functions must have the same value at an equivalent point of any other unit cell related to the PUC by a translation

R; this is referred to as the "translational periodicity" of the crystal. There can be one or many atoms inside the PUC, and the origin of the coordinate system can be located at any position in space; for convenience, it is often chosen to be the position of one of the atoms in the PUC.

From the requirements of translational periodicity, the lattice vectors **R** connect not just all equivalent *atoms* in the crystal, but all equivalent *points* in space. A set of points connected by the **R** vectors is referred to as the "Bravais lattice." The fact that equivalent points in space are connected by the Bravais lattice vectors **R** implies that the potential experienced by electrons in the crystal has the following periodicity:

$$\mathcal{V}(\mathbf{r} + \mathbf{R}) = \mathcal{V}(\mathbf{r}) \tag{2.3}$$

To analyze the consequences of translational periodicity, we will employ Fourier transforms which give a convenient representation of quantities of interest. The periodicity of the ionic potential gives, when we take its Fourier transform:

$$\mathcal{V}(\mathbf{r}) = \int \mathcal{V}(\mathbf{q}) e^{i\mathbf{q}\cdot\mathbf{r}} \, d\mathbf{q} = \mathcal{V}(\mathbf{r} + \mathbf{R}) = \int \mathcal{V}(\mathbf{q}) e^{i\mathbf{q}\cdot(\mathbf{r}+\mathbf{R})} \, d\mathbf{q}$$

which shows, by comparing the two integrals, that only special values of the vectors **q** produce non-vanishing components $\mathcal{V}(\mathbf{q})$, namely those for which

$$e^{i\mathbf{q}\cdot\mathbf{R}} = 1 \tag{2.4}$$

We denote these special values of **q** as a new set of vectors **G**. Since the product of **q** vectors with **r** vectors must be a dimensionless number (the argument of the complex exponential in the above expressions), we refer to the **q** vectors, and to the space they span, as the "reciprocal" vectors and space. The foundation for describing the behavior of electrons in a crystal is the reciprocal lattice, the set of vectors **G** which is the "inverse" of the real Bravais lattice vectors **R**, in the sense that they satisfy Eq. (2.4).

By analogy to the procedure we followed in real space, we can define the reciprocal-space primitive lattice vectors as:

$$\mathbf{b}_1 = \frac{2\pi(\mathbf{a}_2 \times \mathbf{a}_3)}{\mathbf{a}_1 \cdot (\mathbf{a}_2 \times \mathbf{a}_3)}, \quad \mathbf{b}_2 = \frac{2\pi(\mathbf{a}_3 \times \mathbf{a}_1)}{\mathbf{a}_2 \cdot (\mathbf{a}_3 \times \mathbf{a}_1)}, \quad \mathbf{b}_3 = \frac{2\pi(\mathbf{a}_1 \times \mathbf{a}_2)}{\mathbf{a}_3 \cdot (\mathbf{a}_1 \times \mathbf{a}_2)} \tag{2.5}$$

which satisfy the relation

$$\mathbf{a}_i \cdot \mathbf{b}_j = 2\pi \delta_{ij} \tag{2.6}$$

The vectors $\mathbf{b}_i, i = 1, 2, 3$, define a cell in reciprocal space which also has useful consequences, as we describe below. The volume of that cell in reciprocal space is given by:

$$| \mathbf{b}_1 \cdot (\mathbf{b}_2 \times \mathbf{b}_3) | = \frac{(2\pi)^3}{| \mathbf{a}_1 \cdot (\mathbf{a}_2 \times \mathbf{a}_3) |} = \frac{(2\pi)^3}{V_{\text{PUC}}} \tag{2.7}$$

As for the case of real-space Bravais lattice vectors, defined in Eq. (2.1), we construct the **G** vectors which connect all equivalent points in reciprocal space, through:

$$\mathbf{G} = m_1\mathbf{b}_1 + m_2\mathbf{b}_2 + m_3\mathbf{b}_3, \quad m_1, m_2, m_3 : \text{integers} \tag{2.8}$$

By construction, the dot product of any $\mathbf{R}$ vector with any $\mathbf{G}$ vector gives:

$$\mathbf{R} \cdot \mathbf{G} = 2\pi l, \quad l = n_1 m_1 + n_2 m_2 + n_3 m_3 \tag{2.9}$$

where l is always an integer, and therefore:

$$e^{i\mathbf{G} \cdot \mathbf{R}} = 1 \tag{2.10}$$

which is the desired relation between all $\mathbf{R}$ and $\mathbf{G}$ vectors defined by Eqs (2.1) and (2.8). Generalizing the translational periodicity property of the potential, we conclude that any function which has the periodicity of the Bravais lattice, $f(\mathbf{r} + \mathbf{R}) = f(\mathbf{r})$, can be written as:

$$f(\mathbf{r}) = \sum_{\mathbf{G}} e^{i\mathbf{G} \cdot \mathbf{r}} f(\mathbf{G}) \tag{2.11}$$

with $f(\mathbf{G})$ the Fourier transform (FT) components; as mentioned above, $f(\mathbf{r})$ need only be studied for $\mathbf{r}$ within the PUC. This statement applied to the electron wavefunctions is known as "Bloch's theorem."

Bloch's Theorem – 1 *Consider a single-particle hamiltonian of the form:*

$$\hat{\mathcal{H}}^{\mathrm{sp}} = -\frac{\hbar^2 \nabla_{\mathbf{r}}^2}{2m_e} + V(\mathbf{r}), \quad V(\mathbf{r} + \mathbf{R}) = V(\mathbf{r}) \tag{2.12}$$

that is, one in which the potential $V(\mathbf{r})$ possesses the translational periodicity for all $\mathbf{R}$ of a Bravais lattice. The single-particle eigenfunctions can be chosen to be labeled by a vector $\mathbf{k}$ and be of the form:

$$\psi_{\mathbf{k}}(\mathbf{r} + \mathbf{R}) = e^{i\mathbf{k} \cdot \mathbf{R}} \psi_{\mathbf{k}}(\mathbf{r}) \tag{2.13}$$

The states $\psi_{\mathbf{k}}(\mathbf{r})$ are referred to as "Bloch states." At this point $\mathbf{k}$ is just a subscript index for identifying the wavefunctions. We will explore later the physical meaning of the vector $\mathbf{k}$. A different formulation of Bloch's theorem is the following.

Bloch's Theorem – 2 *The single-particle eigenfunctions of the hamiltonian of Eq. (2.12) must have the form:*

$$\psi_{\mathbf{k}}(\mathbf{r}) = \frac{e^{i\mathbf{k} \cdot \mathbf{r}}}{\sqrt{V}} u_{\mathbf{k}}(\mathbf{r}), \quad u_{\mathbf{k}}(\mathbf{r} + \mathbf{R}) = u_{\mathbf{k}}(\mathbf{r}) \tag{2.14}$$

that is, the wavefunctions $\psi_{\mathbf{k}}(\mathbf{r})$ can be expressed as the product of the phase factor $\exp(i\mathbf{k} \cdot \mathbf{r})$ multiplied by the functions $u_{\mathbf{k}}(\mathbf{r})$, which have the full translational periodicity of the Bravais lattice.

Proof　We will first prove that the two formulations of Bloch's theorem are equivalent. For any wavefunction $\psi_{\mathbf{k}}(\mathbf{r})$ that can be put in the form of Eq. (2.14), the relation of Eq. (2.13) must obviously hold. Conversely, if Eq. (2.13) holds, we can factor out of $\psi_{\mathbf{k}}(\mathbf{r})$ the phase factor $\exp(i\mathbf{k} \cdot \mathbf{r})$, in which case the remainder

$$u_{\mathbf{k}}(\mathbf{r}) = \sqrt{V} e^{-i\mathbf{k} \cdot \mathbf{r}} \psi_{\mathbf{k}}(\mathbf{r})$$

must have the translational periodicity of the Bravais lattice by virtue of Eq. (2.13).

A convenient way to prove Bloch's theorem is through the definition of translation operators, whose eigenvalues and eigenfunctions can easily be determined. We define the translation operator $\hat{\mathcal{R}}_{\mathbf{R}}$ which acts on any function $f(\mathbf{r})$ and changes its argument by a lattice vector $-\mathbf{R}$:

$$\hat{\mathcal{R}}_{\mathbf{R}} f(\mathbf{r}) \equiv f(\mathbf{r} - \mathbf{R}) \tag{2.15}$$

This operation effectively translates the value of f at $\mathbf{r}$ to the position $\mathbf{r} + \mathbf{R}$. The operator $\hat{\mathcal{R}}_{\mathbf{R}}$ commutes with the hamiltonian $\hat{\mathcal{H}}^{\text{sp}}$: it obviously commutes with the kinetic energy operator and it leaves the potential energy unaffected, since this potential has the translational periodicity of the Bravais lattice. Consequently, we can choose all eigenfunctions of $\hat{\mathcal{H}}^{\text{sp}}$ to be simultaneous eigenfunctions of $\hat{\mathcal{R}}_{\mathbf{R}}$:

$$\hat{\mathcal{H}}^{\text{sp}} \psi_{\mathbf{k}}(\mathbf{r}) = \epsilon_{\mathbf{k}} \psi_{\mathbf{k}}(\mathbf{r}) \quad \text{and} \quad \hat{\mathcal{R}}_{\mathbf{R}} \psi_{\mathbf{k}}(\mathbf{r}) = \lambda_{\mathbf{k}}(\mathbf{R}) \psi_{\mathbf{k}}(\mathbf{r}) \tag{2.16}$$

with $\lambda_{\mathbf{k}}(\mathbf{R})$ the eigenvalue corresponding to the operator $\hat{\mathcal{R}}_{\mathbf{R}}$ and eigenfunction $\psi_{\mathbf{k}}(\mathbf{r})$. Our goal is to determine the eigenfunctions of $\hat{\mathcal{R}}_{\mathbf{R}}$ so that we can use them as the basis to express the eigenfunctions of $\hat{\mathcal{H}}^{\text{sp}}$. To this end, we will first determine the eigenvalues of $\hat{\mathcal{R}}_{\mathbf{R}}$. We notice that:

$$\hat{\mathcal{R}}_{\mathbf{R}} \hat{\mathcal{R}}_{\mathbf{R}'} = \hat{\mathcal{R}}_{\mathbf{R}'} \hat{\mathcal{R}}_{\mathbf{R}} = \hat{\mathcal{R}}_{\mathbf{R}+\mathbf{R}'} \Rightarrow \lambda_{\mathbf{k}}(\mathbf{R} + \mathbf{R}') = \lambda_{\mathbf{k}}(\mathbf{R}) \lambda_{\mathbf{k}}(\mathbf{R}') \tag{2.17}$$

Considering $\lambda_{\mathbf{k}}(\mathbf{R})$ as a function of $\mathbf{R}$, we conclude that it must be an exponential in $\mathbf{R}$, which is the only function that satisfies the above relation. Without loss of generality, we define:

$$\lambda_{\mathbf{k}}(\mathbf{a}_j) = e^{-i\kappa_j} \quad (j = 1, 2, 3) \tag{2.18}$$

where κ_j is an unspecified complex number, so that $\lambda_{\mathbf{k}}(\mathbf{a}_j)$ can take any complex value. By virtue of Eq. (2.6), the definition of $\lambda_{\mathbf{k}}(\mathbf{a}_j)$ produces, for the general eigenvalue $\lambda_{\mathbf{k}}(\mathbf{R})$:

$$\lambda_{\mathbf{k}}(\mathbf{R}) = e^{-i\mathbf{k} \cdot \mathbf{R}}, \quad \text{where} \quad \mathbf{k} = \kappa_1 \mathbf{b}_1 + \kappa_2 \mathbf{b}_2 + \kappa_3 \mathbf{b}_3 \tag{2.19}$$

with the index $\mathbf{k}$, introduced earlier to label the wavefunctions, expressed in terms of the reciprocal lattice vectors $\mathbf{b}_j$ and the complex constants κ_j. Having established that the eigenvalues of the operator $\hat{\mathcal{R}}_{\mathbf{R}}$ are $\lambda_{\mathbf{k}}(\mathbf{R}) = \exp(-i\mathbf{k} \cdot \mathbf{R})$, we find by inspection that the eigenfunctions of this operator are $\exp[i(\mathbf{k} + \mathbf{G}) \cdot \mathbf{r}]$, since

$$\hat{\mathcal{R}}_{\mathbf{R}} e^{i(\mathbf{k}+\mathbf{G}) \cdot \mathbf{r}} = e^{i(\mathbf{k}+\mathbf{G}) \cdot (\mathbf{r}-\mathbf{R})} = e^{-i\mathbf{k} \cdot \mathbf{R}} e^{i(\mathbf{k}+\mathbf{G}) \cdot \mathbf{r}} = \lambda_{\mathbf{k}}(\mathbf{R}) e^{i(\mathbf{k}+\mathbf{G}) \cdot \mathbf{r}} \tag{2.20}$$

because $\exp(-i\mathbf{G} \cdot \mathbf{R}) = 1$. Then we can write the eigenfunctions of $\hat{\mathcal{H}}^{\text{sp}}$ as an expansion over all eigenfunctions of $\hat{\mathcal{R}}_{\mathbf{R}}$ corresponding to the same eigenvalue of $\hat{\mathcal{R}}_{\mathbf{R}}$:

$$\psi_{\mathbf{k}}(\mathbf{r}) = \frac{1}{\sqrt{V}} \sum_{\mathbf{G}} \alpha_{\mathbf{k}}(\mathbf{G}) e^{i(\mathbf{k}+\mathbf{G}) \cdot \mathbf{r}} = \frac{e^{i\mathbf{k} \cdot \mathbf{r}}}{\sqrt{V}} u_{\mathbf{k}}(\mathbf{r}), \quad u_{\mathbf{k}}(\mathbf{r}) = \sum_{\mathbf{G}} \alpha_{\mathbf{k}}(\mathbf{G}) e^{i\mathbf{G} \cdot \mathbf{r}} \tag{2.21}$$

which proves Bloch's theorem, since $u_{\mathbf{k}}(\mathbf{r}+\mathbf{R}) = u_{\mathbf{k}}(\mathbf{r})$ for $u_{\mathbf{k}}(\mathbf{r})$ defined in Eq. (2.21). ∎

When the Bloch form of the wavefunction is inserted in the single-particle Schrödinger equation, we obtain the equation for $u_\mathbf{k}(\mathbf{r})$:

$$\left[\frac{1}{2m_e} \left(-i\hbar\nabla_\mathbf{r} + \hbar\mathbf{k} \right)^2 + \mathcal{V}(\mathbf{r}) \right] u_\mathbf{k}(\mathbf{r}) = \epsilon_\mathbf{k} u_\mathbf{k}(\mathbf{r}) \tag{2.22}$$

Solving this last equation determines $u_\mathbf{k}(\mathbf{r})$, which with the factor $\exp(i\mathbf{k} \cdot \mathbf{r})$ makes up the solution to the original single-particle equation. The great advantage is that we only need to solve this equation for $\mathbf{r}$ within a PUC of the crystal, since $u_\mathbf{k}(\mathbf{r} + \mathbf{R}) = u_\mathbf{k}(\mathbf{r})$, where $\mathbf{R}$ is any vector connecting equivalent Bravais lattice points.

The result of Eq. (2.22) can also be thought of as equivalent to changing the momentum operator in the hamiltonian $\hat{\mathcal{H}}^{\mathrm{sp}}(\hat{\mathbf{p}}, \mathbf{r})$ by $+\hbar\mathbf{k}$, when dealing with the states $u_\mathbf{k}$ instead of the states $\psi_\mathbf{k}$:

$$\hat{\mathcal{H}}^{\mathrm{sp}}(\hat{\mathbf{p}}, \mathbf{r})\psi_\mathbf{k}(\mathbf{r}) = \epsilon_\mathbf{k}\psi_\mathbf{k}(\mathbf{r}) \Rightarrow \hat{\mathcal{H}}^{\mathrm{sp}}(\hat{\mathbf{p}} + \hbar\mathbf{k}, \mathbf{r})u_\mathbf{k}(\mathbf{r}) = \epsilon_\mathbf{k} u_\mathbf{k}(\mathbf{r}) \tag{2.23}$$

A useful relation between the two forms of the single-particle hamiltonian is obtained by multiplying the first expression from the left by $\sqrt{V}\exp(-i\mathbf{k} \cdot \mathbf{r})$:

$$\sqrt{V}e^{-i\mathbf{k}\cdot\mathbf{r}}\hat{\mathcal{H}}^{\mathrm{sp}}(\hat{\mathbf{p}}, \mathbf{r})\psi_\mathbf{k}(\mathbf{r}) = e^{-i\mathbf{k}\cdot\mathbf{r}}\epsilon_\mathbf{k} e^{i\mathbf{k}\cdot\mathbf{r}}u_\mathbf{k}(\mathbf{r}) = \epsilon_\mathbf{k} u_\mathbf{k}(\mathbf{r}) = \hat{\mathcal{H}}^{\mathrm{sp}}(\hat{\mathbf{p}} + \hbar\mathbf{k}, \mathbf{r})u_\mathbf{k}(\mathbf{r})$$

and comparing the first and last terms. We conclude that:

$$e^{-i\mathbf{k}\cdot\mathbf{r}}\hat{\mathcal{H}}^{\mathrm{sp}}(\hat{\mathbf{p}}, \mathbf{r})e^{i\mathbf{k}\cdot\mathbf{r}} = \hat{\mathcal{H}}^{\mathrm{sp}}(\hat{\mathbf{p}} + \hbar\mathbf{k}, \mathbf{r}) \tag{2.24}$$

This last expression will prove useful in describing the motion of crystal electrons under the influence of an external electric field.

Before we proceed to take further advantage of the relations derived so far, we introduce another important feature of the single-particle eigenvalues and eigenfunctions. Specifically, the discrete boundary condition obeyed by the functions $u_\mathbf{k}(\mathbf{r})$ due to translational periodicity, $u_\mathbf{k}(\mathbf{r} + \mathbf{R}) = u_\mathbf{k}(\mathbf{r})$, Eq. (2.14), with $\mathbf{R}$ a Bravais lattice vector, implies that these functions, and therefore the wavefunctions $\psi_\mathbf{k}(\mathbf{r})$ and the corresponding eigenvalues $\epsilon_\mathbf{k}$ as well, form a set of discrete states for each value of $\mathbf{k}$, which requires the introduction of an additional index n to label them:

$$\psi_\mathbf{k}^{(n)}(\mathbf{r}), \quad \epsilon_\mathbf{k}^{(n)}, \quad n = 1, 2, 3, \ldots$$

If we now consider the index $\mathbf{k}$ to be a real vector which takes continuous values in reciprocal space, as we prove explicitly in the next section, we might expect that the eigenvalues belonging to the same value of n are a smooth and continuous function of $\mathbf{k}$, in the manner illustrated in Fig. 2.3. This is indeed the case, as will become evident through the discussion in the following sections. These smooth and continuous energy eigenvalues are called "energy bands" and are one of the key concepts in the description of the physics of crystals. A plot of the energy bands is called the "band structure."

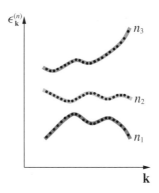

Fig. 2.3 Energy bands in crystals: the black dots represent the discrete energy eigenvalues $\epsilon_{\mathbf{k}}^{(n)}$ at each value of the index $\mathbf{k}$ for different values of the index, n_1, n_2, n_3. The colored lines are guides to the eye, connecting the eigenvalues that correspond to the same value of n.

2.2 Reciprocal Space: Brillouin Zones

We next develop the formal framework for classifying the energy eigenvalues and corresponding single-particle wavefunctions, and for reducing the region of reciprocal space which must be explored to obtain a full solution of the problem. To this end, we prove first that the wave-vector $\mathbf{k}$ is a real quantity, we then define the notions of Brillouin zones and Bragg planes, and we use these concepts to discuss the filling of states in simple examples of crystals.

2.2.1 Nature of Wave-Vector $\mathbf{k}$

In the previous section we introduced $\mathbf{k} = \kappa_1 \mathbf{b}_1 + \kappa_2 \mathbf{b}_2 + \kappa_3 \mathbf{b}_3$ as a convenient index to label the wavefunctions. Here we will show that this index actually has physical meaning. Consider that the crystal is composed of N_j unit cells in the direction of vector $\mathbf{a}_j$ ($j = 1, 2, 3$), where we think of the values of N_j as macroscopically large. $N = N_1 N_2 N_3$ is equal to the total number of unit cells in the crystal (of order Avogadro's number, 6.023×10^{23}). We need to specify some boundary conditions for the single-particle states within this crystal. One convenient choice is *periodic* boundary conditions, also known as the Born–von Karman boundary conditions:

$$\psi_{\mathbf{k}}^{(n)}(\mathbf{r}) = \psi_{\mathbf{k}}^{(n)}(\mathbf{r} + N_j \mathbf{a}_j) \tag{2.25}$$

Bloch's theorem, however, tells us that:

$$\psi_{\mathbf{k}}^{(n)}(\mathbf{r} + N_j \mathbf{a}_j) = e^{i\mathbf{k}\cdot(N_j \mathbf{a}_j)} \psi_{\mathbf{k}}^{(n)}(\mathbf{r})$$

Thus, to satisfy both conditions, we must have:

$$e^{i\mathbf{k}\cdot(N_j \mathbf{a}_j)} = 1 \Rightarrow e^{i2\pi \kappa_j N_j} = 1 \Rightarrow \kappa_j = \frac{n_j}{N_j} \tag{2.26}$$

where n_j is any integer. This shows two important things:

(1) The vector $\mathbf{k}$ is real because the parameters κ_j are real. Since $\mathbf{k}$ is defined in terms of the reciprocal lattice vectors $\mathbf{b}_j$, it can be thought of as a wave-vector: $\exp(i\mathbf{k} \cdot \mathbf{r})$ represents a plane wave of wave-vector $\mathbf{k}$. The physical meaning of this result is that the wavefunction does not decay within the crystal but rather extends throughout the crystal like a wave modified by the periodic function $u_\mathbf{k}(\mathbf{r})$. This fact was first introduced in Chapter 1; here we develop its mathematical expression.

(2) The number of distinct values that $\mathbf{k}$ may take is $N = N_1 N_2 N_3$, because n_j can take N_j inequivalent values that satisfy Eq. (2.26), which can be any N_j consecutive integer values. Values of n_j beyond this range are equivalent to values within this range, because they correspond to adding integer multiples of $2\pi i$ to the argument of the exponential in Eq. (2.26). Values of $\mathbf{k}$ that differ by a reciprocal lattice vector $\mathbf{G}$ are equivalent, since adding a vector $\mathbf{G}$ to $\mathbf{k}$ corresponds to a difference of an integer multiple of $2\pi i$ in the argument of the exponential in Eq. (2.26). This statement is valid even in the limit when $N_j \to \infty$, that is, in the case of an infinite crystal when the values of $\mathbf{k}$ become continuous.

We calculate next the differential volume change in $\mathbf{k}$, which is:

$$\Delta^3 \mathbf{k} = \Delta\mathbf{k}_1 \cdot (\Delta\mathbf{k}_2 \times \Delta\mathbf{k}_3) = \frac{\Delta n_1 \mathbf{b}_1}{N_1} \cdot \left(\frac{\Delta n_2 \mathbf{b}_2}{N_2} \times \frac{\Delta n_3 \mathbf{b}_3}{N_3} \right) \Rightarrow |d\mathbf{k}| = \frac{(2\pi)^3}{N V_{\text{PUC}}} \quad (2.27)$$

where we have used $\Delta n_j = 1$ and $N = N_1 N_2 N_3$; we have also made use of Eq. (2.7) for the volume of the basic cell in reciprocal space. For an infinite crystal $N \to \infty$, so that the spacing of $\mathbf{k}$ values becomes infinitesimal and $\mathbf{k}$ becomes a continuous variable. In this limit, the product $N|d\mathbf{k}|$ takes a finite value:

$$N|d\mathbf{k}| = \frac{(2\pi)^3}{V_{\text{PUC}}} \quad (2.28)$$

which turns out to be a key quantity in reciprocal space, as discussed in more detail below.

2.2.2 Brillouin Zones and Bragg Planes

Statement (2) in the previous subsection has important consequences: it restricts the inequivalent values of $\mathbf{k}$ to a volume in reciprocal space, which is the analog of the PUC in real space, so that one needs to solve:

$$\hat{\mathcal{H}}^{\text{sp}}(\hat{\mathbf{p}} + \hbar\mathbf{k})u_\mathbf{k}^{(n)}(\mathbf{r}) = \epsilon_\mathbf{k}^{(n)} u_\mathbf{k}^{(n)}(\mathbf{r})$$

only within this volume in reciprocal space, known as the first Brillouin zone (BZ in the following). By convention, we choose the first BZ to correspond to the following N_j consecutive values of the index n_j:

$$n_j = -\frac{N_j}{2} + 1, ..., 0, ..., \frac{N_j}{2} \quad (j = 1, 2, 3) \quad (2.29)$$

where we assume N_j to be an even integer (since we are interested in the limit $N_j \to \infty$ this assumption does not impose any significant restrictions). This implies that the

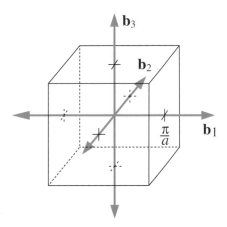

Fig. 2.4   Illustration of boundaries in reciprocal space that bisect the reciprocal lattice vectors $\pm\mathbf{b}_1, \pm\mathbf{b}_2, \pm\mathbf{b}_3$ (the three sets of green arrows) for a cubic lattice. The small black crosses indicate where the planes bisect the reciprocal lattice vectors.

corresponding values of the vector $\mathbf{k}$ in each direction will range from $-\pi/a$ to $+\pi/a$, for the simplest case of a cubic lattice with reciprocal lattice vectors $\mathbf{b}_1 = (2\pi/a)\hat{x}, \mathbf{b}_2 = (2\pi/a)\hat{y}, \mathbf{b}_3 = (2\pi/a)\hat{z}$. We can visualize this situation as the volume in reciprocal space contained between the set of planes that bisect the six reciprocal lattice vectors $\pm\mathbf{b}_1, \pm\mathbf{b}_2, \pm\mathbf{b}_3$, as illustrated in Fig. 2.4.

To generalize the concept of the BZ to reciprocal lattices that are more complex, we first introduce the notion of Bragg planes. Consider a plane wave of incident radiation and wave-vector $\mathbf{q}$, which is scattered by the planes of atoms in a crystal to a wave-vector $\mathbf{q}'$. For elastic scattering, $|\mathbf{q}| = |\mathbf{q}'|$. As the schematic representation of Fig. 2.5 shows, the difference in paths along incident and reflected radiation from two consecutive planes is:

$$d\cos\theta + d\cos\theta' = \mathbf{d}\cdot\hat{\mathbf{q}} - \mathbf{d}\cdot\hat{\mathbf{q}}' \tag{2.30}$$

with $\hat{\mathbf{q}}$ the unit vector along $\mathbf{q}$ ($\hat{\mathbf{q}} = \mathbf{q}/|\mathbf{q}|$) and $d = |\mathbf{d}|$, $\mathbf{d}$ being a vector that connects equivalent lattice points. For constructive interference between the two reflected waves, this difference must be equal to $l\lambda$, where l is an integer and λ is the wavelength. Using $\mathbf{q} = (2\pi/\lambda)\hat{\mathbf{q}}$, we obtain the condition for constructive interference:

$$\mathbf{R}\cdot(\mathbf{q} - \mathbf{q}') = 2\pi l \Rightarrow \mathbf{q} - \mathbf{q}' = \mathbf{G} \tag{2.31}$$

where we have made use of two facts: first, that $\mathbf{d} = \mathbf{R}$ since $\mathbf{d}$ represents a distance between equivalent lattice points in neighboring atomic planes; second, that the reciprocal lattice vectors are defined through the relation $\mathbf{G}\cdot\mathbf{R} = 2\pi l$, as shown in Eq. (2.9).

From the above equation we find $\mathbf{q}' = \mathbf{q} - \mathbf{G}$. By squaring both sides of this equation and using the fact that for elastic scattering $|\mathbf{q}| = |\mathbf{q}'|$, we obtain:

$$\mathbf{q}\cdot\hat{\mathbf{G}} = \frac{1}{2}|\mathbf{G}| \tag{2.32}$$

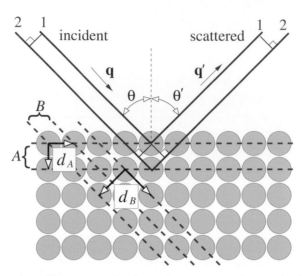

Fig. 2.5 Schematic representation of Bragg scattering from atoms on successive atomic planes. Two families of parallel atomic planes are identified by sets of parallel blue lines (labeled A, B), together with the pairs of lattice vectors (black arrows, labeled d_A, d_B) that are perpendicular to the planes and join equivalent atoms along the planes; notice that the closer the planes are spaced, the longer is the corresponding perpendicular lattice vector. The difference in path for two rays (shown as red lines, labeled "1" and "2") which are scattered from the family of planes labeled A is the part contained between the two perpendicular projections of the point where the first ray touches the atom on the first crystal plane onto the second ray.

This is the definition of the Bragg plane: it is formed by the tips of all the vectors $\mathbf{q}$ which satisfy Eq. (2.32) for a given $\mathbf{G}$. This relation determines all vectors $\mathbf{q}$ that lead to constructive interference of the reflected rays. Since the angle of incidence and magnitude of the wave-vector $\mathbf{q}$ can be varied arbitrarily, Eq. (2.32) serves to identify all the families of planes that can reflect radiation constructively. Therefore, by scanning the values of the angle of incidence and the magnitude of $\mathbf{q}$, we can determine all the $\mathbf{G}$ vectors and from those all the $\mathbf{R}$ vectors, that is, the Bravais lattice of the crystal.

Now consider the origin of reciprocal space and around this point all other points that can be reached without crossing a Bragg plane. This corresponds to the first BZ. The condition (2.32) means that the projection of $\mathbf{q}$ on $\mathbf{G}$ is equal to half the length of $\mathbf{G}$, indicating that the tip of the vector $\mathbf{q}$ must lie on a plane perpendicular to $\mathbf{G}$ that passes through its midpoint. This gives a convenient recipe for defining the first BZ: draw all reciprocal lattice vectors $\mathbf{G}$ and the planes that are perpendicular to them at their mid-points, which by the above arguments are identified as the Bragg planes; the volume enclosed by the first such set of Bragg planes around the origin is the first BZ. It also provides a convenient definition for the second, third, ..., BZs: the second BZ is the volume enclosed between the first set of Bragg planes and the second set of Bragg planes, going outward from the origin. A more rigorous definition is that the first BZ is the set of points that can be reached from the origin without crossing any Bragg planes; the second BZ is the set of points that can be reached from the origin by crossing only one Bragg plane, excluding the points in the first BZ, etc.

It turns out that every BZ has the same volume, given by:

$$V_{BZ} = |\mathbf{b}_1 \cdot (\mathbf{b}_2 \times \mathbf{b}_3)| = \frac{(2\pi)^3}{|\mathbf{a}_1 \cdot (\mathbf{a}_2 \times \mathbf{a}_3)|} = \frac{(2\pi)^3}{V_{PUC}} \tag{2.33}$$

which is explicitly shown for the example of the 2D square lattice below. By comparing the result of Eq. (2.33) with Eq. (2.28), we conclude that in each BZ there are N distinct values of $\mathbf{k}$, where N is the total number of PUCs in the crystal. This is a very useful observation: if there are n electrons in the PUC (that is, nN electrons in the crystal), then we need exactly $nN/2$ different $\psi_{\mathbf{k}}^{(n)}(\mathbf{r})$ states to accommodate them, taking into account spin degeneracy (two electrons with opposite spins can coexist in state $\psi_{\mathbf{k}}^{(n)}(\mathbf{r})$). Since the first BZ contains N distinct values of $\mathbf{k}$, it can accommodate up to $2N$ electrons. Similarly, each subsequent BZ can accommodate $2N$ electrons because it has the same volume in reciprocal space. For n electrons per unit cell, we need to fill completely the states that occupy a volume in $\mathbf{k}$-space equivalent to $n/2$ BZs. Which states will be filled is determined by their energy: in order to minimize the total energy of the system, the lowest-energy states must be occupied first. The Fermi level is defined as the value of the energy below which all single-particle states are occupied.

Example 2.1 Brillouin zones of a 2D square lattice

For a concrete example of these ideas, let us consider a square lattice with lattice constant a. The explicit construction of the first three BZs for this case, by bisecting the corresponding reciprocal space vectors, is shown in Fig. 2.6. The reconstructed version of the first six Brillouin zones, obtained by displacing the different parts by reciprocal space vectors, is shown in Fig. 2.7, which proves by construction that each BZ has the same volume.

To demonstrate the filling of the BZs, we consider that there is one atom per primitive unit cell in this model. We shall assume that this atom has Z electrons so that there are Z/a^2 electrons per unit cell, that is, a total of NZ electrons in the crystal of volume Na^2, where N is the number of unit cells. We shall also assume, for simplicity, that states are equally occupied in all directions of $\mathbf{k}$ for the same value of $|\mathbf{k}|$, up to the highest value needed to accommodate all the electrons, which we defined as the Fermi momentum k_F (this is actually equivalent to assuming free-electron bands, a model discussed in Chapter 1). Therefore, the Fermi momentum can be obtained by integrating over all $\mathbf{k}$-vectors until we have enough states to accommodate all the electrons. Taking into account a factor of 2 for spin, the total number of states we need in reciprocal space to accommodate all the electrons of the crystal is given by:

$$\sum_{\mathbf{k},|\mathbf{k}|<k_F} 2 \rightarrow \frac{2}{(2\pi)^2}(Na^2)\int_{|\mathbf{k}|<k_F} d\mathbf{k} = NZ \Rightarrow \frac{2}{(2\pi)^2}\int_{|\mathbf{k}|<k_F} d\mathbf{k} = \frac{Z}{a^2} \tag{2.34}$$

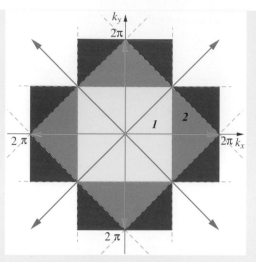

Fig. 2.6

Illustration of the construction of Brillouin zones in a 2D crystal with $\mathbf{a}_1 = \hat{\mathbf{x}}, \mathbf{a}_2 = \hat{\mathbf{y}}$. The first two sets of reciprocal lattice vectors, $\mathbf{G} = \pm 2\pi\hat{\mathbf{x}}, \pm 2\pi\hat{\mathbf{y}}$, light-green arrows, and $\mathbf{G} = 2\pi(\pm\hat{\mathbf{x}}\pm\hat{\mathbf{y}})$, dark-green arrows, are shown, along with the Bragg planes that bisect them, as dashed lines of the same color as the arrows. The first BZ, shown in yellow and labeled 1, is the central square; the second BZ, shown in pink and labeled 2, is composed of the four triangles around the central square; the third BZ, shown in red and labeled 3, is composed of the eight smaller triangles around the second BZ.

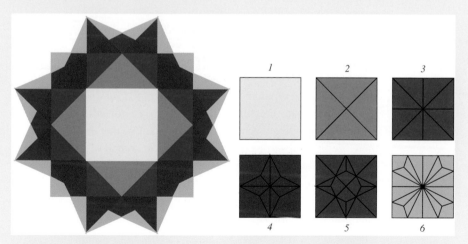

Fig. 2.7

The first six Brillouin zones of the 2D square lattice, coded in color, and their reconstructed version through translations by reciprocal lattice vectors, showing explicitly that they have exactly the same volume.

where we have used $\mathbf{dk} = (2\pi)^2/(Na^2)$ for the 2D square lattice, by analogy to the general result for the 3D lattice $\mathbf{dk} = (2\pi)^3/(NV_{\text{PUC}})$, see Eq. (2.27). Using spherical coordinates for the integration over $\mathbf{k}$, we obtain:

$$\frac{2}{(2\pi)^2} 2\pi \int_0^{k_F} k\mathrm{d}k = \frac{1}{2\pi}k_F^2 = \frac{Z}{a^2} \Rightarrow k_F = \left(\frac{2\pi}{a}\right)\left(\frac{Z}{2\pi}\right)^{1/2} \quad (2.35)$$

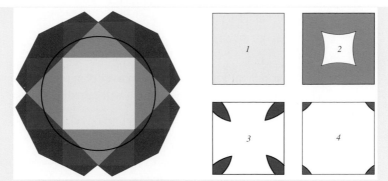

Fig. 2.8 The shape of occupied portions of the various Brillouin zones for the 2D square lattice with $Z = 4$ electrons per unit cell. The black circle represents the Fermi surface: this corresponds to the first BZ being full, the second BZ having a large hole at the center, and the third and fourth BZs with electron pockets at the corners.

and from this we can obtain the Fermi energy ϵ_F:

$$\epsilon_F = \frac{\hbar^2 k_F^2}{2m_e} = \frac{\hbar^2}{2m_e} \left(\frac{2\pi}{a} \right)^2 \frac{Z}{2\pi} \tag{2.36}$$

The value of k_F determines the so-called "Fermi surface" in reciprocal space, which contains all the occupied states for the electrons; the Fermi surface in the present case is a sphere in **k**-space.

   This Fermi sphere corresponds to a certain number of BZs with all states occupied by electrons, and a certain number of partially occupied BZs with interestingly shaped regions for the occupied portions. The number of full BZs and the shapes of occupied regions in partially filled BZs depend on k_F through Z; an example for $Z = 4$ is shown in Fig. 2.8. In this case, the first BZ is fully occupied, the second BZ is mostly occupied but has some empty portion (missing electrons or "holes") at the center, the third BZ has small sections near the corners which are occupied (called "electron pockets"), and the fourth BZ has even smaller electron pockets at the corners.

2.2.3 Periodicity in Reciprocal Space

The usefulness of BZs is that they play an analogous role in reciprocal space as the PUCs do in real space. We saw above that due to crystal periodicity, we only need to solve the single-particle equations inside the PUC. We also saw that values of **k** are equivalent if one adds to them any vector **G**. Thus, we only need to solve the single-particle equations for values of **k** within the first BZ, or within any single BZ: points in other BZs are related by **G** vectors, which make them equivalent. To prove this, suppose we have calculated the complete set of wavefunctions at some value of **k**, given by the usual expression:

$$\psi_{\mathbf{k}}^{(n)}(\mathbf{r}) = \frac{e^{i\mathbf{k}\cdot\mathbf{r}}}{\sqrt{V}} \sum_{\mathbf{G}'} \alpha_{\mathbf{k}}^{(n)}(\mathbf{G}') e^{i\mathbf{G}'\cdot\mathbf{r}} \Rightarrow \hat{\mathcal{H}}^{sp} \psi_{\mathbf{k}}^{(n)}(\mathbf{r}) = \epsilon_{\mathbf{k}}^{(n)} \psi_{\mathbf{k}}^{(n)}(\mathbf{r}) \tag{2.37}$$

We rewrite the sum as follows:

$$\psi_{\mathbf{k}}^{(n)}(\mathbf{r}) = \frac{1}{\sqrt{V}}\left[\cdots + e^{i\mathbf{k}\cdot\mathbf{r}}\alpha_{\mathbf{k}}^{(n)}(\mathbf{G}')e^{i\mathbf{G}'\cdot\mathbf{r}} + \cdots\right] \tag{2.38}$$

where $\mathbf{G}'$ is the symbol for the infinite sum over reciprocal space vectors. A wavefunction with index $\mathbf{k} + \mathbf{G}$, using the same notation, will be expressed as:

$$\psi_{\mathbf{k}+\mathbf{G}}^{(m)}(\mathbf{r}) = \frac{1}{\sqrt{V}}\left[\cdots + e^{i(\mathbf{k}+\mathbf{G})\cdot\mathbf{r}}\alpha_{\mathbf{k}+\mathbf{G}}^{(m)}(\mathbf{G}'')e^{i\mathbf{G}''\cdot\mathbf{r}} + \cdots\right] \tag{2.39}$$

with the symbol for the infinite sum over reciprocal space vectors being $\mathbf{G}''$. Let us now consider the term of this sum for which $\mathbf{G}'' = \mathbf{G}' - \mathbf{G}$:

$$e^{i(\mathbf{k}+\mathbf{G})\cdot\mathbf{r}}\alpha_{\mathbf{k}+\mathbf{G}}^{(m)}(\mathbf{G}' - \mathbf{G})e^{i(\mathbf{G}'-\mathbf{G})\cdot\mathbf{r}} = e^{i\mathbf{k}\cdot\mathbf{r}}\alpha_{\mathbf{k}+\mathbf{G}}^{(m)}(\mathbf{G}' - \mathbf{G})e^{i\mathbf{G}'\cdot\mathbf{r}} \tag{2.40}$$

but this term contains exactly the same two exponentials that appear in the expansion of the wavefunction $\psi_{\mathbf{k}}^{(n)}(\mathbf{r})$, only with a different coefficient, that is, with the coefficient $\alpha_{\mathbf{k}+\mathbf{G}}^{(m)}(\mathbf{G}' - \mathbf{G})$, instead of $\alpha_{\mathbf{k}}^{(n)}(\mathbf{G}')$. Since we have already calculated the entire set of wavefunctions that correspond to the index $\mathbf{k}$, the wavefunction $\psi_{\mathbf{k}+\mathbf{G}}^{(m)}(\mathbf{r})$, which has the same exponentials term by term as $\psi_{\mathbf{k}}^{(n)}(\mathbf{r})$, must be one of the set of those wavefunctions with index $\mathbf{k}$, that is, for some value of the band index n we must have:

$$\alpha_{\mathbf{k}+\mathbf{G}}^{(m)}(\mathbf{G}' - \mathbf{G}) = \alpha_{\mathbf{k}}^{(n)}(\mathbf{G}') \Rightarrow \psi_{\mathbf{k}+\mathbf{G}}^{(m)}(\mathbf{r}) = \psi_{\mathbf{k}}^{(n)}(\mathbf{r}) \Rightarrow \epsilon_{\mathbf{k}+\mathbf{G}}^{(m)} = \epsilon_{\mathbf{k}}^{(n)} \tag{2.41}$$

In other words, the set of wavefunctions and eigenvalues at the index $\mathbf{k} + \mathbf{G}$ is the same as the set of wavefunctions and eigenvalues at the index $\mathbf{k}$, possibly with some reordering of the band index values. This allows us to do the calculation of the wavefunctions and corresponding energy eigenvalues only for those values of $\mathbf{k}$ that are not related by a reciprocal lattice vector $\mathbf{G}$, that is, for $\mathbf{k}$ within a single (by convention the first) BZ. Keeping only the first BZ is referred to as the "reduced zone" scheme, while keeping all BZs is referred to as the "extended zone" scheme. The eigenvalues and eigenfunctions in the two schemes are related by relabeling of the superscript band indices. The above relations also imply:

$$u_{\mathbf{k}+\mathbf{G}}^{(m)}(\mathbf{r}) = \sqrt{V}e^{-i(\mathbf{G}+\mathbf{k})\cdot\mathbf{r}}\psi_{\mathbf{k}+\mathbf{G}}^{(m)}(\mathbf{r}) = \sqrt{V}e^{-i\mathbf{G}\cdot\mathbf{r}}e^{-i\mathbf{k}\cdot\mathbf{r}}\psi_{\mathbf{k}}^{(n)}(\mathbf{r}) = e^{-i\mathbf{G}\cdot\mathbf{r}}u_{\mathbf{k}}^{(n)}(\mathbf{r}) \tag{2.42}$$

a relation that is referred to as "periodic gauge," which will prove useful in later chapters. Referring back to the discussion of filling the energy band in a crystal with $n/2$ electrons per PUC, in the extended zone scheme we need to occupy states that correspond to the lowest energy band and take up the equivalent of $n/2$ BZs. In the reduced zone scheme we need to occupy a number of states that corresponds to a total of $n/2$ bands per $\mathbf{k}$-point inside the first BZ.

2.2.4 Symmetries Beyond Translational Periodicity

A crystal generally has several other types of symmetries, beyond translational periodicity. This has consequences for the behavior of the electronic eigenvalues and eigenfunctions. We shall cover this topic in Chapter 3, but we briefly introduce here some concepts that will significantly facilitate our discussion of the band structure of simple, representative examples. This will also enable us to present those examples in a form compatible with the literature where the symmetries of each crystal are taken into account in displaying the corresponding band structure.

Taking into account symmetries of the crystal beyond translational periodicity allows us to restrict the solutions of the single-particle Schrödinger equation to a small subsection of the first BZ, referred to as the "irreducible Brillouin zone" (IBZ). The mathematical formulation of the problem, which justifies this restriction and provides a full account of the consequences, is based on group theory. Here we give an intuitive argument and illustrate it in two cases that we will encounter in following examples. The basic idea is that the symmetries, other than translational periodicity, that leave a crystal invariant involve rotations around an axis, reflections on a plane, inversion, and combinations of these operations. When these operations are applied in real space they map the crystal back to itself, thus they should be reflected in how electrons behave. Symmetries that leave the crystal invariant in real space also apply to reciprocal space, with a slight modification in certain cases; these details will be dealt with in Chapter 3. The invariance of the reciprocal space under these operations produces the effect that certain points in reciprocal space are related by symmetry to other points. The energy eigenvalue and corresponding wavefunction at reciprocal space points related by a symmetry operation are also related: specifically, the energy eigenvalues are equal, and the wavefunctions are either equal or differ by a well-defined phase factor. Accordingly, we only need to calculate the energy eigenvalues and eigenfunctions of the hamiltonian for those points that are *not* related by symmetry. This set of points constitutes the IBZ.

These concepts are similar to the consequences of translational periodicity: this type of symmetry allowed us to restrict the solutions to the first BZ, since solutions in other parts of reciprocal space are related to those by the expressions derived in the previous subsection. The main difference is that, in the case of translational periodicity, there is an infinite number of translations in real space, $\{\mathbf{R}\}$, the Bravais lattice vectors; this gave rise to the infinite set of lattice vectors $\{\mathbf{G}\}$ in reciprocal space, through which we were able to fold the infinite reciprocal space into the first BZ. In the case of symmetries beyond translational periodicity, the number of symmetry operations is finite, and therefore the IBZ is a *finite* portion of the full BZ; the size of this portion in inversely proportional to the number of symmetries that the crystal possesses. We will demonstrate these ideas next in two illustrative examples.

Example 2.2 Symmetries of the Brillouin zone

As the first example to show the symmetries of the BZ, we choose the square lattice in two dimensions, as shown in Fig. 2.1. Assuming a crystal with one atom per unit cell in this lattice, it is easy to see that the symmetries involve rotation by $n(2\pi/4)$ around an axis perpendicular to the 2D plane, where $n = 1, 2, 3$, that is, rotations by 90° referred to as C_4, 180° referred to as C_4^2 or C_2, and 270° referred to as C_4^3. There are also four mirror planes, which are perpendicular to the plane of the crystal. These are: σ_x and σ_y, passing through the x and y axes; σ_1 and σ_3, passing through

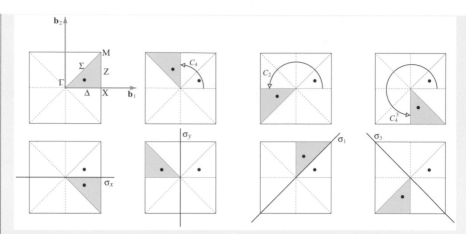

Fig. 2.9

Symmetries of the 2D square in reciprocal space: the first BZ is shown by the black square, and the primitive lattice vectors are indicated as $\mathbf{b}_1, \mathbf{b}_2$ (green arrows). The symmetries include the three rotations C_4, C_2, C_4^3, and the four mirror planes $\sigma_x, \sigma_y, \sigma_1, \sigma_2$. The IBZ is shown shaded in yellow; each symmetry operation is described by how it maps the IBZ triangle in the first square (upper-left corner) to other equivalent triangles in the first BZ.

the diagonals of the square. All these symmetries apply to the reciprocal space, as shown in Fig. 2.9. Using these symmetries, we can map an arbitrary point in the shaded triangle of the first square to other, equivalent points in the remaining seven triangles by applying the symmetry operations. We can also think of this mapping in the opposite sense: the points in any of the seven non-shaded triangles are mapped by the symmetry operations to equivalent points within the shaded triangle. Thus, we only need to solve the single-particle Schrödinger equation for all values of $\mathbf{k}$ in the shaded triangle, which is the IBZ for this case, and corresponds to 1/8th of the first BZ. The lines that form the boundary of this triangle are of particular interest, because they have higher symmetry than points in the interior of the triangle; the highest-symmetry points are the center of the BZ, denoted as Γ, the center of the side, denoted as X, and the corner point, denoted as M. The points along the line segments connecting the three high-symmetry points Γ–X–M are denoted as Δ, Z, Σ. Typically, we calculate the solutions of the Schrödinger equation for all values of $\mathbf{k}$ along these lines, which give a good representation of the solutions over the entire IBZ; solutions in the triangle interior are smooth interpolations of those at the boundary.

Another commonly encountered case is the 2D hexagonal lattice and its variations, like the honeycomb lattice, shown in Fig. 2.1, that we will encounter in Example 2.7. Assuming again one atom per unit cell, it is easy to see that the 2D hexagonal lattice has the following symmetries: rotations by $n(2\pi/3)$ around an axis perpendicular to the 2D plane, where $n = 1, 2$, that is, rotations by 120° referred to as C_3, and 240° referred to as C_3^2; reflections on three mirror planes perpendicular to the plane of the lattice, that pass through the lines connecting the centers of opposite sides of the hexagon, labeled $\sigma_1, \sigma_2, \sigma_3$. In addition, there is inversion symmetry, that is, mapping of a point $\mathbf{r}$ to the point $-\mathbf{r}$, and the combination of this operation with each

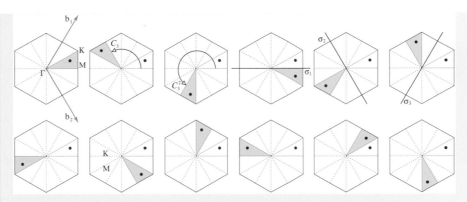

Fig. 2.10

Symmetries of the 2D hexagon in reciprocal space: the first BZ is shown by the black hexagon, and the primitive lattice vectors are indicated as $\mathbf{b}_1, \mathbf{b}_2$ (green arrows). The symmetries include: the two rotations C_3, C_3^2 and the three mirror-plane reflections $\sigma_1, \sigma_2, \sigma_3$, shown in the first row; inversion and its combination with each of the other five operations (two rotations and three mirror-plane reflections), shown in the second row. The IBZ is shown shaded in yellow; each symmetry operation is described by how it maps the IBZ triangle in the first hexagon (upper-left corner) to other equivalent triangles in the first BZ.

of the five other operations (two rotations and three mirror-plane reflections). Other symmetry operations could also be considered, like rotations around the vertical axis by multiples of $2\pi/6$, or mirror planes perpendicular to the plane of the crystal that pass through opposite corners of the hexagon, but these end up being identical to one of the operations mentioned already, and therefore they do no add any useful information. What is essential is to have a complete set of symmetry operations, defined as the "point group" of the crystal, a notion we discuss in detail in Chapter 3. As before, we can use these operations in reciprocal space to identify the IBZ. This is shown in Fig. 2.10, with the IBZ shown by the shaded triangle in the first hexagon. An arbitrary point (marked by a dot) in this triangle is mapped to equivalent points in the other 11 triangles of equal size within the first BZ. As in the previous example, this mapping implies that we only need to solve the problem for points within the IBZ, which corresponds to 1/12th of the first BZ. The high-symmetry points are the center of the BZ (Γ), the center of the hexagon side (M), and the corner of the hexagon (K).

2.3 The Free-Electron and Nearly Free-Electron Models

To illustrate the concepts introduced above, we consider a simple model, in which the external potential experienced by the electrons is very weak. In the limit where this potential is vanishingly small, $\mathcal{V}(\mathbf{r}) \approx 0$, we think of the electrons as being "free," hence the name of the model, the "free-electron" model. The only aspect of the potential which influences the behavior of the electrons is the crystal periodicity, which is behind all the concepts discussed above. For this system of electrons the single-particle energy is given by the kinetic energy term

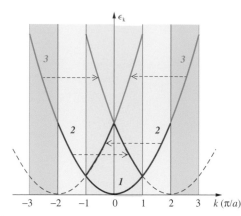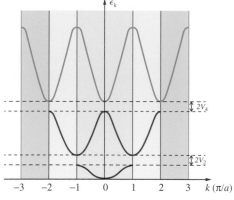

Fig. 2.11 Energy bands of the free-electron and nearly free-electron models in one dimension. **Left**: The 1D band structure of free electrons illustrating the reduced zone scheme. The first band within the first BZ ($-\pi/a < k < \pi/a$, shown shaded red) is colored red; the second band is colored blue and comes from shifting the values of ϵ_k by a reciprocal lattice vector $\pm(2\pi/a)$ (as indicated by the black arrows) from the second BZ ($-2\pi/a < k < -\pi/a$ and $\pi/a < k < 2\pi/a$, shown shaded blue), or alternatively, from the energy curves centered at $k = \pm 2\pi/a$; similarly for the third band, which is colored green. Note that the same bands are generated within the first BZ by the original parabola displaced by $\pm 2\pi/a$ (dashed blue curves). **Right**: The 1D band structure of electrons in a weak potential (nearly free electrons) in the reduced and extended zone schemes, with splitting of the energy levels at BZ boundaries (Bragg planes): $\mathcal{V}_2 = |\mathcal{V}(2\pi/a)|$, and $\mathcal{V}_4 = |\mathcal{V}(4\pi/a)|$.

$$\epsilon_\mathbf{k} = \frac{\hbar^2 \mathbf{k}^2}{2m_e}$$

since the potential energy term is negligible. To take a simple example, in one dimension the lattice constant of the crystal is $\mathbf{a} = a\hat{\mathbf{x}}$ and the reciprocal lattice constant is $\mathbf{b} = (2\pi/a)\hat{\mathbf{x}}$, with reciprocal lattice vectors given by $\mathbf{G}_m = m\mathbf{b}$, m : an integer. The corresponding band structure in the reduced and the extended zone schemes for the free-electron model is shown in Fig. 2.11. Since the energy is quadratic in $|\mathbf{k}|$, the bands have a simple form and there are no gaps at the BZ boundaries. For $n = 2$ electrons per PUC in this 1D model, the Fermi level must be such that the first band in the first BZ is completely full. In the free-electron case this corresponds to the value of the energy at the first BZ boundary. The case of electrons in a weak periodic potential gives rise to a more interesting situation, discussed in detail next, with a "band gap" between occupied and unoccupied states.

A simple and useful generalization of the free-electron model is to consider that the crystal potential is not exactly zero but very weak; we refer to this as the "nearly free-electron" model. Using the Fourier expansion of the potential and the Bloch states:

$$\mathcal{V}(\mathbf{r}) = \sum_{\mathbf{G}''} \mathcal{V}(\mathbf{G}'') e^{i\mathbf{G}''\cdot\mathbf{r}}, \quad \psi_\mathbf{k}^{(n)}(\mathbf{r}) = \frac{e^{i\mathbf{k}\cdot\mathbf{r}}}{\sqrt{V}} \sum_{\mathbf{G}'} \alpha_\mathbf{k}(\mathbf{G}') e^{i\mathbf{G}'\cdot\mathbf{r}}$$

in the single-particle Schrödinger equation we obtain the following equation:

$$\sum_{\mathbf{G}'} \left[\frac{\hbar^2}{2m_e}(\mathbf{k}+\mathbf{G}')^2 - \epsilon_\mathbf{k} + \sum_{\mathbf{G}''} \mathcal{V}(\mathbf{G}'') e^{i\mathbf{G}''\cdot\mathbf{r}} \right] \alpha_\mathbf{k}(\mathbf{G}') e^{i(\mathbf{k}+\mathbf{G}')\cdot\mathbf{r}} = 0 \qquad (2.43)$$

Multiplying by $\exp[-i(\mathbf{G} + \mathbf{k}) \cdot \mathbf{r}]$ and integrating over $\mathbf{r}$ gives:

$$\left[\frac{\hbar^2}{2m_e}(\mathbf{k} + \mathbf{G})^2 - \epsilon_{\mathbf{k}} \right] \alpha_{\mathbf{k}}(\mathbf{G}) + \sum_{\mathbf{G}'} \mathcal{V}(\mathbf{G} - \mathbf{G}')\alpha_{\mathbf{k}}(\mathbf{G}') = 0 \qquad (2.44)$$

where we have used the relation

$$\int_{\mathrm{PUC}} e^{i\mathbf{q} \cdot \mathbf{r}} \, d\mathbf{r} = V_{\mathrm{PUC}}\delta(\mathbf{q})$$

which holds for any reciprocal-space vector $\mathbf{q}$, and in this case is used to reduce the summations over $\mathbf{G}''$ and $\mathbf{G}'$ that appear in Eq. (2.43) to a single term. This is a linear system of equations in the unknowns $\alpha_{\mathbf{k}}(\mathbf{G})$, which can be solved to determine the values of these unknowns and hence find the eigenfunctions $\psi_{\mathbf{k}}^{(n)}(\mathbf{r})$. We can always assume $\mathcal{V}(\mathbf{G} = 0) = 0$, since this is the average value of the potential (a constant), and we can set it equal to zero because if it had any other value it would simply shift the whole spectrum of eigenvalues by this constant amount, which does not affect the properties of the system. If the potential is very weak, then we should have $\mathcal{V}(\mathbf{G}) \approx 0$ for all the $\mathbf{G}$ vectors, which means that the wavefunction cannot have any components $\alpha_{\mathbf{k}}(\mathbf{G})$ for $\mathbf{G} \neq 0$, since these components can only arise from corresponding features in the potential. In this case we take $\alpha_{\mathbf{k}}(0) = 1$ and obtain:

$$\psi_{\mathbf{k}}^{(n)}(\mathbf{r}) = \frac{1}{\sqrt{V}}e^{i\mathbf{k} \cdot \mathbf{r}}, \quad \epsilon_{\mathbf{k}}^{(n)} = \frac{\hbar^2 k^2}{2m_e} \qquad (2.45)$$

as we would expect for free electrons.

Now suppose that all Fourier components of the potential are negligible except for one, $\mathcal{V}(\mathbf{G}_0)$, which is small but not negligible, and consequently all coefficients $\alpha_{\mathbf{k}}(\mathbf{G})$ are negligible, except for $\alpha_{\mathbf{k}}(\mathbf{G}_0)$. Then Eq. (2.44) reduces to:

$$\alpha_{\mathbf{k}}(\mathbf{G}_0) = \frac{\mathcal{V}(\mathbf{G}_0)}{\hbar^2 \left[k^2 - (\mathbf{k} + \mathbf{G}_0)^2 \right]/2m_e} \, \alpha_{\mathbf{k}}(0) \qquad (2.46)$$

where we have used the zeroth-order approximation for $\epsilon_{\mathbf{k}} = (\hbar^2/2m_e)k^2$ and $\alpha_{\mathbf{k}}(0)$ is a constant of order unity. Given that $\mathcal{V}(\mathbf{G}_0)$ is itself small, $\alpha_{\mathbf{k}}(\mathbf{G}_0)$ is indeed very small as long as the denominator is finite. The only chance for the coefficient $\alpha_{\mathbf{k}}(\mathbf{G}_0)$ to be large is if the denominator is vanishingly small, which happens for

$$(\mathbf{k} + \mathbf{G}_0)^2 = k^2 \Rightarrow \mathbf{k} \cdot \hat{\mathbf{G}}_0 = -\frac{1}{2}|\mathbf{G}_0| \qquad (2.47)$$

and this is the condition for Bragg planes! In this case, in order to obtain the correct solution we have to consider both $\alpha_{\mathbf{k}}(0)$ and $\alpha_{\mathbf{k}}(\mathbf{G}_0)$ simultaneously. Since all $\mathcal{V}(\mathbf{G}) = 0$ except for $\mathbf{G}_0$, we obtain the following linear system of equations:

$$\left[\frac{\hbar^2 k^2}{2m_e} - \epsilon_{\mathbf{k}} \right] \alpha_{\mathbf{k}}(0) + \mathcal{V}^*(\mathbf{G}_0)\alpha_{\mathbf{k}}(\mathbf{G}_0) = 0$$

$$\left[\frac{\hbar^2 (\mathbf{k} + \mathbf{G}_0)^2}{2m_e} - \epsilon_{\mathbf{k}} \right] \alpha_{\mathbf{k}}(\mathbf{G}_0) + \mathcal{V}(\mathbf{G}_0)\alpha_{\mathbf{k}}(0) = 0$$

where we have used $\mathcal{V}(-\mathbf{G}) = \mathcal{V}^*(\mathbf{G})$. Solving this system we obtain:

$$\epsilon_{\mathbf{k}}^{(1)} = \frac{\hbar^2 \mathbf{k}^2}{2m_e} + |\mathcal{V}(\mathbf{G}_0)|, \quad \epsilon_{\mathbf{k}}^{(2)} = \frac{\hbar^2 \mathbf{k}^2}{2m_e} - |\mathcal{V}(\mathbf{G}_0)|$$

for the two possible solutions. Thus, at Bragg planes, that is, at the boundaries of the BZ, the energy of the free electrons is modified by the terms $\pm|\mathcal{V}(\mathbf{G})|$ for the non-vanishing components of $\mathcal{V}(\mathbf{G})$. This is illustrated for the 1D case in Fig. 2.11.

For $n = 2$ electrons per PUC, there will be a gap between the highest energy of occupied states (the top of the first band) and the lowest energy of unoccupied states (the bottom of the second band); this gap, denoted by $2\mathcal{V}_2$ in Fig. 2.11, is referred to as the "band gap." Given the above definition of the Fermi level, its position could be anywhere within the band gap. A more detailed examination of the problem reveals that actually the Fermi level is at the middle of the gap (see Chapter 5).

The discussion of the free-electron and nearly free-electron bands can easily be extended to crystals in higher dimensions, by simply following the same procedure as above. For example, the free-electron bands for the FCC lattice (a 3D crystal) are shown in Fig. 2.12. This is a useful example, because in certain solids the bands look surprisingly similar to free-electron bands, indicating that the behavior in these crystals is quite close to that of the free-electron model, which provides the basis for simple intuitive arguments about the physical properties of these materials. Aluminum is a good case in point (discussed in more detail in Chapter 5). In fact, it is remarkable how close to free-electron bands the actual bands of Al look, as is clearly evident in Fig. 2.12. The deviation from the ideal

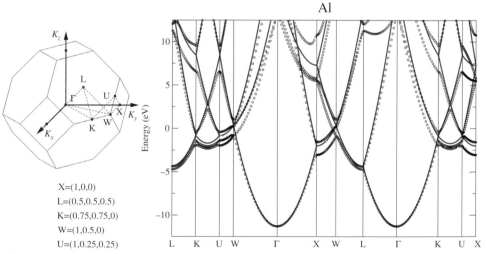

Fig. 2.12 Free electron bands in the 3D FCC lattice and comparison to the bands of the aluminum crystal. **Left**: Brillouin zone for the FCC lattice, with the special **k**-points X, L, K, W, U identified and expressed in units of $2\pi/a$, where a is the lattice constant; Γ is the center of the BZ. **Right**: The band structure for the 3D free-electron model in an FCC crystal, in red lines. The black dots are the band structure of Al in the FCC crystal structure, as obtained from realistic, first-principles calculations (see Chapter 5). The zero of the energy scale is set at the Fermi level.

free-electron bands is mostly near the boundaries of the BZ, like at the X, L, and K points, as we would expect from the nearly free-electron model.

Example 2.3   Free-electron bands of a 2D square lattice

To illustrate the power of the concepts developed so far, we take up again the free-electron model in a representative 2D lattice. Suppose that we want to plot the energy levels as a function of the **k**-vector, as we did in the 1D case: what do these look like, and what does the occupation of states in the various BZs mean? The answer to these questions will help us visualize and understand the behavior of the physical system. First, we realize that the **k**-vectors are 2D so we would need a 3D plot to plot the energy eigenvalues, two axes for the k_x, k_y components and one for the energy. Moreover, in the extended zone scheme each component of **k** will span different ranges of values in different BZs, making the plot very complicated. Using the reduced zone scheme is certainly helpful, since each BZ will be represented by a different "energy surface" as a function of the reciprocal space vector **k** which spans values within the first BZ.

Taking into account symmetries of the crystal beyond translational periodicity further simplifies the problem by allowing us to focus on the irreducible part of the BZ. Even the use of the remaining symmetries of the crystal leaves us with the task of calculating and displaying the values of the energy bands through a portion of the first BZ, which is rather demanding in more than one dimension. To circumvent some of the difficulties of this task, especially in what concerns the display of the bands, what is typically done is that the bands are obtained only along certain lines of the BZ, connecting points of high symmetry. Since both $\epsilon_\mathbf{k}$ and $\psi_\mathbf{k}^{(n)}(\mathbf{r})$ are smooth functions of **k**, obtaining them along the high-symmetry lines of the first BZ essentially provides a full description of the physics.

For the particular model that we are considering here, we will calculate the bands along the lines joining the center of reciprocal space, denoted by Γ, to the center of one side of the square which corresponds to the first BZ, denoted by X, and to the corner of this square, denoted by M. A point along the Γ–X line is denoted by Δ, a point along the X–M line is denoted by Z, and a point along the M–Γ line is denoted by Σ. The lines Γ–X, X–M, and M–Γ form a triangle, which is actually the irreducible part of the first BZ. The energy for the first nine bands along the high-symmetry lines is shown in Fig. 2.13. In this simple model, the energy of single-particle levels is given by:

$$\epsilon_\mathbf{k}^{(n)} = \frac{\hbar^2}{2m_e}|\mathbf{k}|^2$$

that is, it depends only on the magnitude of the wave-vector **k**. The bands are obtained by scanning the values of the wave-vector **k** along the various directions in the first, second, third, etc. BZ, and then folding them within the first BZ. To illustrate the origin of the bands along the high-symmetry directions in the BZ, we show in

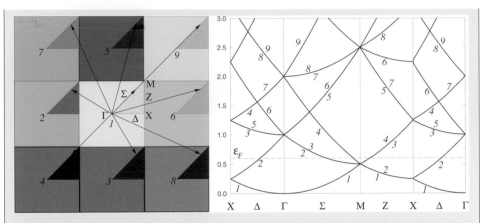

Fig. 2.13 The band structure for the 2D free-electron model along the high-symmetry lines in the IBZ, for the first nine bands coming from the first nine BZs, which are shown on the left. The position of the Fermi level ϵ_F for $Z = 4$ electrons per unit cell is indicated as a dashed line. The energy scale is in units of $(\hbar^2/2m_e)(2\pi/a)^2$.

Fig. 2.13 the wave-vectors that correspond to equivalent Σ points in the first nine BZs, along the Γ–M direction. All these points differ by reciprocal lattice vectors, so in the reduced zone scheme they are mapped to the same Σ point in the first BZ. This makes it evident why the bands labeled 2 and 3 are degenerate along the Γ–M line: the wave-vectors corresponding to these points in the BZs with labels 2 and 3 have equal magnitude. For the same reason, the pairs of bands labeled 5 and 6, or 7 and 8, are also degenerate along the Γ–M line. Analogous degeneracies are found along the Γ–X and M–X lines, but they involve different pairs of bands. The labeling of the bands is of course arbitrary and serves only to identify electronic states with different energy at the same wave-vector **k**; the only requirement is that the label be kept consistent for the various parts of the BZ, as is done in Fig. 2.13. In this case, it is convenient to choose band labels that make the connection to the various BZs transparent.

Using this band structure, we can now interpret the filling of the different BZs in Fig. 2.8. The first band is completely full, being entirely below the Fermi level, and so is the first BZ. The second band is full along the direction M–X, but only partially full in the directions Γ–X and Γ–M, giving rise to the almost full second BZ with an empty region in the middle, which corresponds to the Fermi surface shape depicted. The third band has only a small portion occupied, in the M–X and M–Γ directions, and the third BZ has occupied slivers near the four corners. Finally, the fourth band has a tiny portion occupied around the point M, corresponding to the so-called "pockets" of occupied states at the corners. As this discussion exemplifies, the behavior of the energy bands along the high-symmetry directions of the BZ provides essentially a complete picture of the eigenvalues of the single-particle hamiltonian throughout reciprocal space in the reduced zone scheme.

2.4 Effective Mass, "$\mathbf{k} \cdot \mathbf{p}$" Perturbation Theory

We next introduce a useful approximation that facilitates the description of energy bands near special points of the BZ. We begin by noting that if we know the solutions of the single-particle hamiltonian at some value of $\mathbf{k}_0$, $\psi_{\mathbf{k}_0}^{(n)}(\mathbf{r}), \epsilon_{\mathbf{k}_0}^{(n)}$, for all values of the band index n, then we can obtain the energy at nearby points that differ from $\mathbf{k}_0$ by a small vector $\mathbf{k}$ using second-order perturbation theory (see Problem 3). The result is:

$$\epsilon_{\mathbf{k}_0+\mathbf{k}}^{(n)} = \epsilon_{\mathbf{k}_0}^{(n)} + \frac{\hbar}{m_e}\mathbf{k} \cdot \mathbf{p}^{(nn)}(\mathbf{k}_0) + \frac{\hbar^2 k^2}{2m_e} + \frac{\hbar^2}{m_e^2}\sum_{n' \neq n}\frac{|\mathbf{k} \cdot \mathbf{p}^{(nn')}(\mathbf{k}_0)|^2}{\epsilon_{\mathbf{k}_0}^{(n)} - \epsilon_{\mathbf{k}_0}^{(n')}} \qquad (2.48)$$

where the quantities $\mathbf{p}^{(nn')}(\mathbf{k}_0)$ are defined as:

$$\mathbf{p}^{(nn')}(\mathbf{k}_0) = \frac{\hbar}{i}\langle \psi_{\mathbf{k}_0}^{(n')} | \nabla_{\mathbf{r}} | \psi_{\mathbf{k}_0}^{(n)} \rangle \qquad (2.49)$$

Because of the appearance of terms $\mathbf{k} \cdot \mathbf{p}^{(nn')}(\mathbf{k}_0)$ in the above expressions, this approach is known as "$\mathbf{k} \cdot \mathbf{p}$" perturbation theory. The quantities defined in Eq. (2.49) are elements of a two-index matrix (n and n'); the diagonal matrix elements are simply the expectation value of the momentum operator in state $\psi_{\mathbf{k}_0}^{(n)}(\mathbf{r})$. We can also calculate the same quantity from the Taylor expansion of $\epsilon_{\mathbf{k}}^{(n)}$, viewed as a function of $\mathbf{k}$:

$$\left[\nabla_{\mathbf{k}}\epsilon_{\mathbf{k}}^{(n)}\right]_{\mathbf{k}_0} = \lim_{\mathbf{k} \to 0}\frac{\partial \epsilon_{\mathbf{k}_0+\mathbf{k}}^{(n)}}{\partial \mathbf{k}} = \frac{\hbar}{m_e}\mathbf{p}^{(nn)}(\mathbf{k}_0) \qquad (2.50)$$

which shows that the gradient of $\epsilon_{\mathbf{k}}^{(n)}$ with respect to $\mathbf{k}$ (multiplied by the factor $m_e/\hbar$) gives the expectation value of the momentum for the crystal states.

Let us consider the second derivative of $\epsilon_{\mathbf{k}}^{(n)}$ with respect to components of the vector $\mathbf{k}$, denoted by k_i, k_j, evaluated at $\mathbf{k}_0$:

$$\left[\frac{1}{\hbar^2}\frac{\partial^2 \epsilon_{\mathbf{k}}^{(n)}}{\partial k_i \partial k_j}\right]_{\mathbf{k}_0} = \lim_{\mathbf{k} \to 0}\frac{1}{\hbar^2}\frac{\partial^2 \epsilon_{\mathbf{k}_0+\mathbf{k}}^{(n)}}{\partial k_i \partial k_j} \Rightarrow$$

$$\frac{1}{\overline{m}_{ij}^{(n)}(\mathbf{k}_0)} \equiv \frac{1}{m_e}\delta_{ij} + \frac{1}{m_e^2}\sum_{n' \neq n}\frac{p_i^{(nn')}(\mathbf{k}_0)p_j^{(n'n)}(\mathbf{k}_0) + p_j^{(nn')}(\mathbf{k}_0)p_i^{(n'n)}(\mathbf{k}_0)}{\epsilon_{\mathbf{k}_0}^{(n)} - \epsilon_{\mathbf{k}_0}^{(n')}} \qquad (2.51)$$

The dimensions of this expression are 1/mass. This can then be identified directly as the inverse effective mass of the single-particle states, which is no longer a simple scalar quantity but a second-rank tensor. As this expression demonstrates, the effective mass of a crystal electron depends on the wave-vector $\mathbf{k}_0$ and band index n of its wavefunction, as well as on the wavefunctions and energies of all other crystal electrons with the same $\mathbf{k}_0$-vector. Since the effective mass involves complicated dependence on the direction of the $\mathbf{k}_0$-vector and the momenta and energies of many states, it can have different magnitude and even different signs along different crystallographic directions! This is referred to as the "quasiparticle" nature of electrons, a notion to which we return in Chapter 4.

2.5 The Tight-Binding Approximation

The simplest method for calculating band structures, both conceptually and computationally, is the so-called tight-binding approximation (TBA), also referred to as linear combination of atomic orbitals (LCAO). This approach is also referred to as the Slater–Koster method, from the authors of the paper that originally developed it.[1] The latter term is actually used in a wider sense, as we will explain below. The basic assumption in the TBA is that we can use orbitals that are very similar to atomic states (that is, wavefunctions tightly bound to the atoms, hence the term "tight-binding") as a basis for expanding the crystal wavefunctions. We will deal with the general theory of the TBA first and then illustrate how it is applied through several examples.

We consider a set of atomic wavefunctions

$$\phi_l(\mathbf{r} - \mathbf{t}_i) \tag{2.52}$$

where $\mathbf{t}_i$ is the position of the atom with label i in the PUC, and $\phi_l(\mathbf{r})$ is one of the atomic orbitals associated with this atom. The index l can take the usual values for an atom, that is, the angular momentum character $s, p, d, \ldots$ The state $\phi_l(\mathbf{r} - \mathbf{t}_i)$ is centered at the position of the atom with index i. It is assumed that we need as many orbitals as the number of valence states in the atom (this is referred to as the "minimal basis"). In effect, we are assuming that we can solve the electronic problem at the level of individual atoms to obtain the states $\phi_l(\mathbf{r})$, and then use the same states (in appropriate linear combinations) to represent the electronic states in the crystal.

Our first task is to construct states which can be used as the basis for expansion of the crystal wavefunctions. These states must obey Bloch's theorem, and we call them $\chi_{\mathbf{k}li}(\mathbf{r})$:

$$\chi_{\mathbf{k}li}(\mathbf{r}) = \frac{1}{\sqrt{N}} \sum_{\mathbf{R}'} e^{i\mathbf{k}\cdot\mathbf{R}'} \phi_l(\mathbf{r} - \mathbf{t}_i - \mathbf{R}') \tag{2.53}$$

with the summation running over all the N unit cells in the crystal (the vectors $\mathbf{R}'$), for a given pair of indices i (used to denote the position $\mathbf{t}_i$ of the atom in the PUC) and l (used for the type of orbital). We first verify that these states have Bloch character:

$$\chi_{\mathbf{k}li}(\mathbf{r} + \mathbf{R}) = \frac{1}{\sqrt{N}} \sum_{\mathbf{R}'} e^{i\mathbf{k}\cdot(\mathbf{R}'-\mathbf{R})} e^{i\mathbf{k}\cdot\mathbf{R}} \phi_l((\mathbf{r} + \mathbf{R}) - \mathbf{t}_i - \mathbf{R}')$$

$$= e^{i\mathbf{k}\cdot\mathbf{R}} \frac{1}{\sqrt{N}} \sum_{\mathbf{R}'} e^{i\mathbf{k}\cdot(\mathbf{R}'-\mathbf{R})} \phi_l(\mathbf{r} - \mathbf{t}_i - (\mathbf{R}' - \mathbf{R}))$$

$$= e^{i\mathbf{k}\cdot\mathbf{R}} \frac{1}{\sqrt{N}} \sum_{\mathbf{R}''} e^{i\mathbf{k}\cdot\mathbf{R}''} \phi_l(\mathbf{r} - \mathbf{t}_i - \mathbf{R}'') = e^{i\mathbf{k}\cdot\mathbf{R}} \chi_{\mathbf{k}li}(\mathbf{r}) \tag{2.54}$$

[1] J. C. Slater and G. F. Koster, *Phys. Rev.* **94**, 1498 (1954).

that is, Bloch's theorem is satisfied for our choice of $\chi_{\mathbf{k}li}(\mathbf{r})$, with the obvious definition $\mathbf{R}'' = \mathbf{R}' - \mathbf{R}$, which is another lattice vector. Now we can expand the crystal single-particle eigenstates in this basis:

$$\psi_{\mathbf{k}}^{(n)}(\mathbf{r}) = \sum_{l,i} c_{\mathbf{k}li}^{(n)} \chi_{\mathbf{k}li}(\mathbf{r}) \tag{2.55}$$

and all that remains to do is determine the coefficients $c_{\mathbf{k}li}^{(n)}$, assuming that the $\psi_{\mathbf{k}}^{(n)}(\mathbf{r})$ are solutions to the appropriate single-particle equation:

$$\hat{\mathcal{H}}^{\text{sp}}\psi_{\mathbf{k}}^{(n)}(\mathbf{r}) = \epsilon_{\mathbf{k}}\psi_{\mathbf{k}}^{(n)}(\mathbf{r}) \Rightarrow \sum_{l,i}\left[\langle\chi_{\mathbf{k}mj}|\hat{\mathcal{H}}^{\text{sp}}|\chi_{\mathbf{k}li}\rangle - \epsilon_{\mathbf{k}}^{(n)}\langle\chi_{\mathbf{k}mj}|\chi_{\mathbf{k}li}\rangle\right]c_{\mathbf{k}li}^{(n)} = 0 \tag{2.56}$$

In the above equation we only need to consider matrix elements of states with the same $\mathbf{k}$ index, because

$$\langle\psi_{\mathbf{k}}^{(n)}|\psi_{\mathbf{k}'}^{(n')}\rangle \sim \delta(\mathbf{k} - \mathbf{k}') \tag{2.57}$$

where we are restricting the values of $\mathbf{k}, \mathbf{k}'$ to the IBZ. In Eq. (2.56) we have a secular equation of size equal to the total number of atomic orbitals in the PUC: the sum is over the number of different types of atoms and the number of orbitals associated with each type of atom. This is exactly the number of solutions (bands) that we can expect at each $\mathbf{k}$-point. In order to solve this linear system we need to be able to evaluate the following integrals:

$$\begin{aligned}
\langle\chi_{\mathbf{k}mj}|\chi_{\mathbf{k}li}\rangle &= \frac{1}{N}\sum_{\mathbf{R}',\mathbf{R}''} e^{i\mathbf{k}\cdot(\mathbf{R}'-\mathbf{R}'')}\langle\phi_m(\mathbf{r}-\mathbf{t}_j-\mathbf{R}'')|\phi_l(\mathbf{r}-\mathbf{t}_i-\mathbf{R}')\rangle \\
&= \frac{1}{N}\sum_{\mathbf{R},\mathbf{R}'} e^{i\mathbf{k}\cdot\mathbf{R}}\langle\phi_m(\mathbf{r}-\mathbf{t}_j)|\phi_l(\mathbf{r}-\mathbf{t}_i-\mathbf{R})\rangle \\
&= \sum_{\mathbf{R}} e^{i\mathbf{k}\cdot\mathbf{R}}\langle\phi_m(\mathbf{r}-\mathbf{t}_j)|\phi_l(\mathbf{r}-\mathbf{t}_i-\mathbf{R})\rangle
\end{aligned} \tag{2.58}$$

where we have used the obvious definition $\mathbf{R} = \mathbf{R}' - \mathbf{R}''$, and we eliminated one of the sums over the lattice vectors with the factor $1/N$, since in the last line of Eq. (2.58) there is no explicit dependence on $\mathbf{R}'$. We call the brackets in the last expression the "overlap matrix elements" between atomic states. In a similar fashion we obtain:

$$\langle\chi_{\mathbf{k}mj}|\hat{\mathcal{H}}^{\text{sp}}|\chi_{\mathbf{k}li}\rangle = \sum_{\mathbf{R}} e^{i\mathbf{k}\cdot\mathbf{R}}\langle\phi_m(\mathbf{r}-\mathbf{t}_j)|\hat{\mathcal{H}}^{\text{sp}}|\phi_l(\mathbf{r}-\mathbf{t}_i-\mathbf{R})\rangle \tag{2.59}$$

and we call the brackets on the right-hand side of Eq. (2.59) the "hamiltonian matrix elements" between atomic states.

At this point we introduce an important approximation: in the spirit of the TBA, we take the overlap matrix elements in Eq. (2.58) to be non-zero only for the same orbitals on the same atom, that is, only for $m = l, j = i, \mathbf{R} = 0$, which is expressed by the relation:

$$\langle\phi_m(\mathbf{r}-\mathbf{t}_j)|\phi_l(\mathbf{r}-\mathbf{t}_i-\mathbf{R})\rangle = \delta_{lm}\delta_{ij}\delta(\mathbf{R}) \tag{2.60}$$

This is referred to as an "orthogonal basis," since any overlap between different orbitals on the same atom or orbitals on different atoms is taken to be zero. Strictly speaking, if the overlap matrix elements between the $\phi_m(\mathbf{r})$ orbitals were exactly zero, there would be no interactions between nearest neighbors. Nevertheless, this is both a practical approximation and a reasonable one. It is practical in the sense that it simplifies the problem of solving the single-particle Schrödinger equation considerably, since we do not need to worry about the overlap between neighboring orbitals which complicates the matrix equations to be solved; we will revisit this issue in a following section to discuss how to relax this constraint. Moreover, the approximation makes physical sense, since the orbitals on different atoms can be thought of as *essentially* orthogonal; it is their interaction through the hamiltonian matrix elements that matters most, rather than their overlap. Turning to the hamiltonian matrix elements defined in Eq. (2.59), we will assume them to be non-zero only if the orbitals are on the same atom, that is, for $j = i$, $\mathbf{R} = 0$, which are referred to as the "on-site energies":

$$\langle \phi_m(\mathbf{r} - \mathbf{t}_j)|\hat{\mathcal{H}}^{\text{sp}}|\phi_l(\mathbf{r} - \mathbf{t}_i - \mathbf{R})\rangle = \delta_{lm}\delta_{ij}\delta(\mathbf{R})\epsilon_l \tag{2.61}$$

or if the orbitals are on different atoms but situated at nearest-neighbor sites, denoted in general as $\mathbf{d}_{nn}$:

$$\langle \phi_m(\mathbf{r} - \mathbf{t}_j)|\hat{\mathcal{H}}^{\text{sp}}|\phi_l(\mathbf{r} - \mathbf{t}_i - \mathbf{R})\rangle = \delta((\mathbf{t}_j - \mathbf{t}_i - \mathbf{R}) - \mathbf{d}_{nn})t_{lm,ij} \tag{2.62}$$

The $t_{lm,ij}$ are also referred to as "hopping" matrix elements. When the nearest neighbors are in the same unit cell, $\mathbf{R}$ can be zero, when they are across unit cells, $\mathbf{R}$ can be one of the primitive lattice vectors. The equations that define the TBA model with an orthogonal basis are summarized below.

Equations of tight-binding approximation model with orthogonal basis

Bloch basis	$\chi_{\mathbf{k}li}(\mathbf{r}) = \dfrac{1}{\sqrt{N}} \sum_{\mathbf{R}} e^{i\mathbf{k}\cdot\mathbf{R}} \phi_l(\mathbf{r} - \mathbf{t}_i - \mathbf{R})$
crystal states	$\psi_{\mathbf{k}}^{(n)}(\mathbf{r}) = \sum_{l,i} c_{\mathbf{k}li}^{(n)} \chi_{\mathbf{k}li}(\mathbf{r})$
secular equation	$\sum_{l,i} \left[\langle \chi_{\mathbf{k}mj}|\hat{\mathcal{H}}^{\text{sp}}|\chi_{\mathbf{k}li}\rangle - \epsilon_{\mathbf{k}}^{(n)} \langle \chi_{\mathbf{k}mj}|\chi_{\mathbf{k}li}\rangle \right] c_{\mathbf{k}li}^{(n)} = 0$
orthogonal basis	$\langle \phi_m(\mathbf{r} - \mathbf{t}_j)|\phi_l(\mathbf{r} - \mathbf{t}_i - \mathbf{R})\rangle = \delta_{lm}\delta_{ij}\delta(\mathbf{R})$
on-site terms	$\langle \phi_m(\mathbf{r} - \mathbf{t}_j)|\hat{\mathcal{H}}^{\text{sp}}|\phi_l(\mathbf{r} - \mathbf{t}_i - \mathbf{R})\rangle = \delta_{lm}\delta_{ij}\delta(\mathbf{R})\epsilon_l$
hopping terms	$\langle \phi_m(\mathbf{r} - \mathbf{t}_j)|\hat{\mathcal{H}}^{\text{sp}}|\phi_l(\mathbf{r} - \mathbf{t}_i - \mathbf{R})\rangle = \delta((\mathbf{t}_j - \mathbf{t}_i - \mathbf{R}) - \mathbf{d}_{nn})t_{lm,ij}$

The first three equations are general, based on the atomic orbitals $\phi_l(\mathbf{r} - \mathbf{t}_i - \mathbf{R})$ of type l centered at an atom situated at the position $\mathbf{t}_i$ of the unit cell identified by the lattice vector **R**. The last three correspond to an orthogonal basis of orbitals and nearest-neighbor interactions only, which define the on-site and hopping matrix elements of the hamiltonian.

Even with the drastic approximation of an orthogonal basis, we still need to calculate the values of the matrix elements that we have kept. The parametrization of the hamiltonian matrix in an effort to produce a method with quantitative capabilities has a long history, starting with the work of Harrison (see Further Reading), and continues to be an active area of research. In principle, these matrix elements can be calculated using one of the single-particle hamiltonians discussed in Chapter 4 (this approach is being actively pursued as a means of performing fast and reliable electronic structure calculations).[2] However, it is often more convenient to consider these matrix elements as parameters, which are fitted to reproduce certain properties and can then be used to calculate other properties of the solid.[3] We illustrate these concepts through two simple examples, the first concerning a 1D lattice, the second a 2D lattice of atoms.

Example 2.4   1D linear chain with s or p orbitals

We consider first the simplest possible case, a linear periodic chain of atoms. Our system has only one type of atom and only one orbital associated with each atom. The first task is to construct the basis for the crystal wavefunctions using the atomic wavefunctions, as was done for the general case in Eq. (2.53). We notice that because of the simplicity of the model, there are no summations over the indices l (there is only one type of orbital for each atom) and i (there is only one atom per unit cell). We keep the index l to identify different types of orbitals in our simple model. Therefore, the basis for the crystal wavefunctions in this case will be simply:

$$\chi_{kl}(x) = \sum_{n=-\infty}^{\infty} e^{ikx}\phi_l(x - na) \tag{2.63}$$

where we have further simplified the notation since we are dealing with a 1D example, with the position vector **r** set equal to the position x on the 1D axis and the reciprocal-space vector **k** set equal to k, while the lattice vectors **R** are given by na, with a the lattice constant and n an integer. We will consider atomic wavefunctions $\phi_l(x)$ which have either s-like or p-like character. The real parts of the wavefunction $\chi_{kl}(x)$, $l = s, p$, for a few values of k are shown in Fig. 2.14.

With these states, we can now attempt to calculate the band structure for this model. The TBA with an orthogonal basis and nearest-neighbor interactions implies that the overlap matrix elements are non-zero only for orbitals $\phi_l(x)$ on the same atom, that is:

$$\langle \phi_l(x) | \phi_l(x - na) \rangle = \delta_{n0} \tag{2.64}$$

---

[2] P. Blaudeck, Th. Frauenheim, D. Porezag, G. Seifert, and E. Fromm, *J. Phys.: Cond. Mat. C* **4**, 6389 (1992).

[3] R. E. Cohen, M. J. Mehl, and D. A. Papaconstantopoulos, *Phys. Rev. B* **50**, 14694 (1994); M. J. Mehl and D. A. Papaconstantopoulos, *Phys. Rev. B* **54**, 4519 (1996).

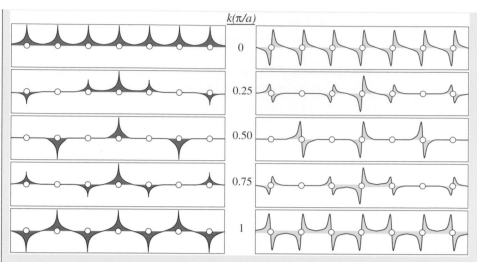

$k(\pi/a)$

Fig. 2.14
Bloch states in the tight-binding approximation for a 1D, linear chain model: Real parts of the crystal wavefunctions $\chi_{kl}(x)$ for $k = 0, 0.25, 0.5, 0.75,$ and 1 (in units of π/a). **Left**: s-like state (red curves). **Right**: p-like state (blue curves). The black circles represent the atoms in the 1D chain.

Similarly, nearest-neighbor interactions require that the hamiltonian matrix elements are non-zero only for orbitals that are on the same or neighboring atoms. If the orbitals are on the same atom, then we define the hamiltonian matrix element to be:

$$\langle \phi_l(x) | \hat{\mathcal{H}}^{\mathrm{sp}} | \phi_l(x - na) \rangle = \epsilon_l \delta_{n0} \qquad (2.65)$$

while if they are on neighboring atoms, that is $n = \pm 1$, we define the hamiltonian matrix element to be:

$$\langle \phi_l(x) | \hat{\mathcal{H}}^{\mathrm{sp}} | \phi_l(x - na) \rangle = t_l \delta_{n\pm1} \qquad (2.66)$$

where ϵ_l is the on-site hamiltonian matrix element and t_l is the hopping matrix element. We expect this interaction between orbitals on neighboring atoms to contribute to the cohesion of the solid, which implies that $t_s < 0$ for s-like orbitals and $t_p > 0$ for p-like orbitals, as we explain in more detail below.

We are now ready to use the $\chi_{kl}(x)$ functions as the basis to construct crystal wavefunctions and with these calculate the single-particle energy eigenvalues, that is, the band structure of the model. The crystal wavefunctions are obtained from the general expression Eq. (2.55):

$$\psi_k(x) = c_k \chi_{kl}(x) \qquad (2.67)$$

where only the index k has survived due to the simplicity of the model (the index l simply denotes the character of the atomic orbitals). Inserting these wavefunctions in the secular equation, Eq. (2.56), we find that we have to solve a 1×1 matrix, because we have only one orbital per atom and one atom per unit cell. With the above definitions of the hamiltonian matrix elements between the atomic orbitals ϕ_l, we obtain:

$$[\langle \chi_{kl}(x)|\hat{\mathcal{H}}^{sp}|\chi_{kl}(x)\rangle - \epsilon_k \langle \chi_{kl}(x)|\chi_{kl}(x)\rangle]c_k = 0 \Rightarrow$$

$$\sum_n e^{ikna} \langle \phi_l(x)|\hat{\mathcal{H}}^{sp}|\phi_l(x-na)\rangle = \epsilon_k \sum_n e^{ikna} \langle \phi_l(x)|\phi_l(x-na)\rangle \Rightarrow$$

$$\sum_n e^{ikna}[\epsilon_l \delta_{n0} + t_l \delta_{n\pm1}] = \epsilon_k \sum_n e^{ikna} \delta_{n0}$$

The solution to the last equation is straightforward, giving the energy band for this simple model:

$$\text{1D chain:} \quad \epsilon_k = \epsilon_l + 2t_l \cos(ka) \tag{2.68}$$

The behavior of the energy in the first BZ of the model, that is, for $-\pi/a \leq k \leq \pi/a$, is shown in Fig. 2.15 for the s and p orbitals. Since the coefficient c_k is undefined by the secular equation, we can take it to be unity, in which case the crystal wavefunctions $\psi_k(x)$ are the same as the basis functions $\chi_{kl}(x)$, which we have already discussed above (see Fig. 2.14).

We elaborate briefly on the sign of the hopping matrix elements and the dispersion of the bands. It is assumed that the single-particle hamiltonian is spherically symmetric. The s orbitals are spherically symmetric and have everywhere the same sign, so that the overlap between s orbitals situated at nearest-neighbor sites is positive (we are concerned here with the sign implied only by the angular momentum character of the wavefunction). In order to produce an attractive interaction between these orbitals, the hopping matrix element must be negative:

$$t_s = \int \phi_s^*(x)\hat{\mathcal{H}}^{sp}(x)\phi_s(x-a)dx < 0$$

We conclude that the negative sign of this matrix element is due to the hamiltonian, since the product of the wavefunctions is positive. On the other hand, the p orbitals have a positive and a negative lobe (see Appendix C); consequently, the overlap between p orbitals situated at nearest-neighbor sites and oriented in the same sense as required by translational periodicity is negative, because the positive lobe of one is closest to the negative lobe of the next. Therefore, in order to produce an attractive interaction between these orbitals, and since the hamiltonian is the same as in the previous case, the hopping matrix element must be positive:

$$t_p \equiv \int \phi_p^*(x)\hat{\mathcal{H}}^{sp}(x)\phi_p(x-a)dx > 0$$

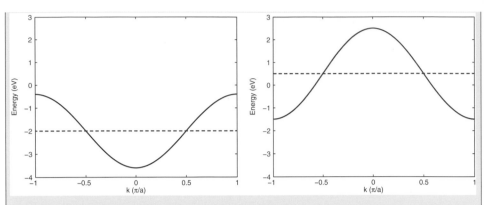

Fig. 2.15 Single-particle energy eigenvalues $\epsilon_k^{(s)}$ and $\epsilon_k^{(p)}$ in the first BZ $(-\pi/a \leq k \leq \pi/a)$, for the 1D infinite chain model, with one atom per unit cell and one orbital per atom, s only (left panel, red curves) or p only (right panel, blue curves), in the TBA with nearest-neighbor interactions. The horizontal dashed lines are at ϵ_s (red) and at ϵ_p (blue). The values of the parameters for this example are: $\epsilon_s = -2$ (red dashed line), $\epsilon_p = 0.5$ (blue dashed line), $t_s = -0.8$, $t_p = 1.0$ (eV).

Thus, the band structure for a 1D model with one s-like orbital per unit cell will have a maximum at $k = \pm\pi/a$ and a minimum at $k = 0$, while that of the p-like orbital will have the positions of the extrema reversed, as shown in Fig. 2.15. Moreover, we expect that in general there will be a larger overlap between the neighboring p orbitals than between the s orbitals, due to the directed lobes of the former, and therefore $|t_p| > |t_s|$, leading to larger dispersion for the p bands.

The generalization of the model to a 2D square lattice with one s-like orbital and one atom per unit cell is straightforward; the energy eigenvalues are given by:

$$\text{2D square}: \quad \epsilon_{\mathbf{k}} = \epsilon_s + 2t_s[\cos(k_x a) + \cos(k_y a)] \tag{2.69}$$

with the 2D reciprocal-space vector defined as $\mathbf{k} = k_x\hat{\mathbf{x}} + k_y\hat{\mathbf{y}}$. Similarly, the generalization to the 3D cubic lattice with one s-like orbital and one atom per unit cell leads to the energy eigenvalues:

$$\text{3D cube}: \quad \epsilon_{\mathbf{k}} = \epsilon_s + 2t_s[\cos(k_x a) + \cos(k_y a) + \cos(k_z a)] \tag{2.70}$$

where $\mathbf{k} = k_x\hat{\mathbf{x}} + k_y\hat{\mathbf{y}} + k_z\hat{\mathbf{z}}$ is the 3D reciprocal-space vector. From these expressions, we can immediately deduce that for this simple model the band width of the energy eigenvalues is given by:

$$W = 4dt_s = 2zt_s \tag{2.71}$$

where d is the dimensionality of the model ($d = 1, 2, 3$ in the above examples), or equivalently, z is the number of nearest neighbors ($z = 2, 4, 6$ in the above examples). The case of p orbitals is less straightforward, because of different types of hamiltonian matrix elements between such orbitals when they are arranged along a line pointing toward each other or parallel to each other, as we discuss in more detail in the next section.

Finally, we consider a 1D linear chain with atoms that have both an s and a p orbital. In this case, we will have a 2×2 hamiltonian that must be diagonalized, because we will also need to account for the hopping matrix element between the s orbital on one site and the p orbital on adjacent sites; we call this hopping matrix element t_{sp}. The hamiltonian now takes the form:

$$\mathcal{H}_{\mathbf{k}} = \begin{pmatrix} \epsilon_s + 2t_s \cos(ka) & t_{sp}\left[e^{-ika} - e^{ika}\right] \\ t_{sp}\left[e^{ika} - e^{-ika}\right] & \epsilon_p + 2t_p \cos(ka) \end{pmatrix} \tag{2.72}$$

which can easily be diagonalized to give, for the two eigenvalues:

$$\epsilon_k^{(\pm)} = \left(\frac{\epsilon_s + \epsilon_p}{2}\right) + (t_s - t_p)\cos(ka)$$

$$\pm \left\{\left[\left(\frac{\epsilon_s - \epsilon_p}{2}\right) + (t_s - t_p)\cos(ka)\right]^2 + 4t_{sp}^2 \sin^2(ka)\right\}^{1/2}$$

This solution introduces qualitatively different behavior, namely a gap opens between the lower-energy band and the higher-energy band, as shown in Fig. 2.16. Note that each band is now a *hybrid* between the s-like and p-like states, with s or p character that varies with the value of k. The size and position of the minimum gap depends on the value of t_{sp}. For small values of t_{sp}, the model is a small perturbation of the non-interacting s and p orbitals, so the gap is small and appears near the point where the s and p bands of the non-interacting case intersect; for larger values of t_{sp} the gap moves to the boundary of the BZ, $k = \pm \pi/a$.

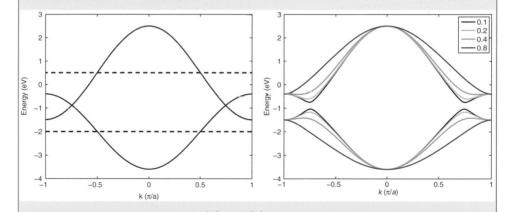

Fig. 2.16

Single-particle energy eigenvalues $\epsilon_k^{(+)}$ and $\epsilon_k^{(-)}$ in the first BZ ($-\pi/a \leq k \leq \pi/a$), for the 1D infinite chain model, with one atom per unit cell that has one s orbital and one p orbital, in the TBA with nearest-neighbor interactions. The horizontal dashed lines are at ϵ_s (red) and at ϵ_p (blue). The values of the parameters for this example are: $\epsilon_s = -2$ (horizontal solid red line), $\epsilon_p = 0.5$ (horizontal solid blue line), $t_s = -0.8$, $t_p = 1.0$ (eV). When $t_{sp} = 0$, the two bands intersect and they are the superposition of the two single-orbital, non-interacting cases, as shown in the left panel (see Fig. 2.15). For $t_{sp} \neq 0$ a gap opens up, as shown in the right panel for various values of this parameter.

Example 2.5   2D square lattice with *s* and *p* orbitals

We next consider a slightly more complex case, the 2D square lattice with one atom per unit cell. We assume that there are four atomic orbitals per atom, one *s*-type and three *p*-type (p_x, p_y, p_z). We work again within the orthogonal basis of orbitals and nearest-neighbor interactions only, as described by the equations listed in the pink-shaded box on p. 78. The overlap matrix elements in this case (see Table 2.2) are:

$$\langle \phi_m(\mathbf{r}) | \phi_l(\mathbf{r} - \mathbf{R}) \rangle = \delta_{lm} \delta(\mathbf{R}) \Rightarrow$$

$$\langle \chi_{\mathbf{k}m} | \chi_{\mathbf{k}l} \rangle = \sum_{\mathbf{R}} e^{i\mathbf{k}\cdot\mathbf{R}} \langle \phi_m(\mathbf{r}) | \phi_l(\mathbf{r} - \mathbf{R}) \rangle = \sum_{\mathbf{R}} e^{i\mathbf{k}\cdot\mathbf{R}} \delta_{lm} \delta(\mathbf{R})$$

$$= \delta_{lm} \tag{2.73}$$

while the hamiltonian matrix elements are:

$$\langle \phi_m(\mathbf{r}) | \hat{\mathcal{H}}^{\mathrm{sp}} | \phi_l(\mathbf{r} - \mathbf{R}) \rangle \neq 0 \ \text{ only for } \ \left[\mathbf{R} = \pm a\hat{x}, \pm a\hat{y}, 0 \right] \Rightarrow$$

$$\langle \chi_{\mathbf{k}m} | \hat{\mathcal{H}}^{\mathrm{sp}} | \chi_{\mathbf{k}l} \rangle = \sum_{\mathbf{R}} e^{i\mathbf{k}\cdot\mathbf{R}} \langle \phi_m(\mathbf{r}) | \hat{\mathcal{H}}^{\mathrm{sp}} | \phi_l(\mathbf{r} - \mathbf{R}) \rangle$$

$$\neq 0 \ \text{ only for } \ \left[\mathbf{R} = \pm a\hat{x}, \pm a\hat{y}, 0 \right] \tag{2.74}$$

A number of different on-site and hopping matrix elements are generated from all possible combinations of $\phi_m(\mathbf{r})$ and $\phi_l(\mathbf{r})$ in Eq. (2.74), which we define as follows:

$$\epsilon_s = \langle \phi_s(\mathbf{r}) | \hat{\mathcal{H}}^{\mathrm{sp}} | \phi_s(\mathbf{r}) \rangle$$

$$\epsilon_p = \langle \phi_{p_x}(\mathbf{r}) | \hat{\mathcal{H}}^{\mathrm{sp}} | \phi_{p_x}(\mathbf{r}) \rangle = \langle \phi_{p_y}(\mathbf{r}) | \hat{\mathcal{H}}^{\mathrm{sp}} | \phi_{p_y}(\mathbf{r}) \rangle = \langle \phi_{p_z}(\mathbf{r}) | \hat{\mathcal{H}}^{\mathrm{sp}} | \phi_{p_z}(\mathbf{r}) \rangle$$

$$t_{ss} = \langle \phi_s(\mathbf{r}) | \hat{\mathcal{H}}^{\mathrm{sp}} | \phi_s(\mathbf{r} \pm a\hat{x}) \rangle = \langle \phi_s(\mathbf{r}) | \hat{\mathcal{H}}^{\mathrm{sp}} | \phi_s(\mathbf{r} \pm a\hat{y}) \rangle$$

$$t_{sp} = \langle \phi_s(\mathbf{r}) | \hat{\mathcal{H}}^{\mathrm{sp}} | \phi_{p_x}(\mathbf{r} - a\hat{x}) \rangle = -\langle \phi_s(\mathbf{r}) | \hat{\mathcal{H}}^{\mathrm{sp}} | \phi_{p_x}(\mathbf{r} + a\hat{x}) \rangle$$

$$t_{sp} = \langle \phi_s(\mathbf{r}) | \hat{\mathcal{H}}^{\mathrm{sp}} | \phi_{p_y}(\mathbf{r} - a\hat{y}) \rangle = -\langle \phi_s(\mathbf{r}) | \hat{\mathcal{H}}^{\mathrm{sp}} | \phi_{p_y}(\mathbf{r} + a\hat{y}) \rangle$$

$$t_{pp\sigma} = \langle \phi_{p_x}(\mathbf{r}) | \hat{\mathcal{H}}^{\mathrm{sp}} | \phi_{p_x}(\mathbf{r} \pm a\hat{x}) \rangle = \langle \phi_{p_y}(\mathbf{r}) | \hat{\mathcal{H}}^{\mathrm{sp}} | \phi_{p_y}(\mathbf{r} \pm a\hat{y}) \rangle$$

$$t_{pp\pi} = \langle \phi_{p_y}(\mathbf{r}) | \hat{\mathcal{H}}^{\mathrm{sp}} | \phi_{p_y}(\mathbf{r} \pm a\hat{x}) \rangle = \langle \phi_{p_x}(\mathbf{r}) | \hat{\mathcal{H}}^{\mathrm{sp}} | \phi_{p_x}(\mathbf{r} \pm a\hat{y}) \rangle$$

$$= \langle \phi_{p_z}(\mathbf{r}) | \hat{\mathcal{H}}^{\mathrm{sp}} | \phi_{p_z}(\mathbf{r} \pm a\hat{x}) \rangle = \langle \phi_{p_z}(\mathbf{r}) | \hat{\mathcal{H}}^{\mathrm{sp}} | \phi_{p_z}(\mathbf{r} \pm a\hat{y}) \rangle \tag{2.75}$$

The hopping matrix elements are shown schematically in Fig. 2.17. By the symmetry of the atomic orbitals we can deduce:

$$\langle \phi_s(\mathbf{r}) | \hat{\mathcal{H}}^{\mathrm{sp}} | \phi_{p_\alpha}(\mathbf{r}) \rangle = 0 \quad (\alpha = x, y, z)$$

$$\langle \phi_s(\mathbf{r}) | \hat{\mathcal{H}}^{\mathrm{sp}} | \phi_{p_\alpha}(\mathbf{r} \pm a\hat{x}) \rangle = 0 \quad (\alpha = y, z)$$

$$\langle \phi_{p_\alpha}(\mathbf{r}) | \hat{\mathcal{H}}^{\mathrm{sp}} | \phi_{p_\beta}(\mathbf{r} \pm a\hat{x}) \rangle = 0 \quad (\alpha, \beta = x, y, z; \alpha \neq \beta)$$

$$\langle \phi_{p_\alpha}(\mathbf{r}) | \hat{\mathcal{H}}^{\mathrm{sp}} | \phi_{p_\beta}(\mathbf{r}) \rangle = 0 \quad (\alpha, \beta = x, y, z; \alpha \neq \beta) \tag{2.76}$$

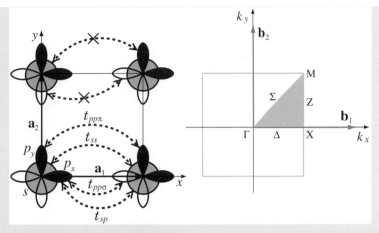

Fig. 2.17

Left: The 2D square lattice with atoms that possess s orbitals (red circles) and p orbitals (blue lobes, only p_x and p_y on the x, y plane are shown). The unit cell is outlined by the lattice vectors, $\mathbf{a}_1$ and $\mathbf{a}_2$, and the black lines. The thick-dashed colored lines with double-headed arrows indicate the non-vanishing hopping matrix elements $t_{ss}, t_{sp}, t_{pp\sigma}, t_{pp\pi}$ and the vanishing ones due to symmetry. **Right**: The corresponding BZ outlined in yellow with the irreducible part shaded.

as can be seen in the diagrams of Fig. 2.17, with the single-particle hamiltonian $\hat{\mathcal{H}}^{\mathrm{sp}}$ assumed to contain only spherically symmetric terms.

Having defined all these matrix elements, we can calculate the matrix elements between crystal states that enter in the secular equation; we find for our example crystal:

$$\langle \chi_{\mathbf{k}s}(\mathbf{r})|\hat{\mathcal{H}}^{\mathrm{sp}}|\chi_{\mathbf{k}s}(\mathbf{r})\rangle = \langle \phi_s(\mathbf{r})|\hat{\mathcal{H}}^{\mathrm{sp}}|\phi_s(\mathbf{r})\rangle +$$

$$\langle \phi_s(\mathbf{r})|\hat{\mathcal{H}}^{\mathrm{sp}}|\phi_s(\mathbf{r} - a\hat{\mathbf{x}})\rangle \mathrm{e}^{\mathrm{i}\mathbf{k}\cdot a\hat{\mathbf{x}}} +$$

$$\langle \phi_s(\mathbf{r})|\hat{\mathcal{H}}^{\mathrm{sp}}|\phi_s(\mathbf{r} + a\hat{\mathbf{x}})\rangle \mathrm{e}^{-\mathrm{i}\mathbf{k}\cdot a\hat{\mathbf{x}}} +$$

$$\langle \phi_s(\mathbf{r})|\hat{\mathcal{H}}^{\mathrm{sp}}|\phi_s(\mathbf{r} - a\hat{\mathbf{y}})\rangle \mathrm{e}^{\mathrm{i}\mathbf{k}\cdot a\hat{\mathbf{y}}} +$$

$$\langle \phi_s(\mathbf{r})|\hat{\mathcal{H}}^{\mathrm{sp}}|\phi_s(\mathbf{r} + a\hat{\mathbf{y}})\rangle \mathrm{e}^{-\mathrm{i}\mathbf{k}\cdot a\hat{\mathbf{y}}}$$

$$= \epsilon_s + 2\, t_{ss}\left[\cos(k_x a) + \cos(k_y a)\right] \qquad (2.77)$$

and similarly for the rest of the matrix elements:

$$\langle \chi_{\mathbf{k}s}(\mathbf{r})|\hat{\mathcal{H}}^{\mathrm{sp}}|\chi_{\mathbf{k}p_x}(\mathbf{r})\rangle = \mathrm{i}\, 2\, t_{sp}\sin(k_x a)$$

$$\langle \chi_{\mathbf{k}s}(\mathbf{r})|\hat{\mathcal{H}}^{\mathrm{sp}}|\chi_{\mathbf{k}p_y}(\mathbf{r})\rangle = \mathrm{i}\, 2\, t_{sp}\sin(k_y a)$$

$$\langle \chi_{\mathbf{k}p_z}(\mathbf{r})|\hat{\mathcal{H}}^{\mathrm{sp}}|\chi_{\mathbf{k}p_z}(\mathbf{r})\rangle = \epsilon_p + 2\, t_{pp\pi}\left[\cos(k_x a) + \cos(k_y a)\right]$$

$$\langle \chi_{\mathbf{k}p_x}(\mathbf{r})|\hat{\mathcal{H}}^{\mathrm{sp}}|\chi_{\mathbf{k}p_x}(\mathbf{r})\rangle = \epsilon_p + 2\, t_{pp\sigma}\cos(k_x a) + 2 t_{pp\pi}\cos(k_y a)$$

$$\langle \chi_{\mathbf{k}p_y}(\mathbf{r})|\hat{\mathcal{H}}^{\mathrm{sp}}|\chi_{\mathbf{k}p_y}(\mathbf{r})\rangle = \epsilon_p + 2\, t_{pp\pi}\cos(k_x a) + 2 t_{pp\sigma}\cos(k_y a) \qquad (2.78)$$

With these we can now construct the hamiltonian matrix for each value of $\mathbf{k}$, and obtain the eigenvalues and eigenfunctions by diagonalizing the secular equation.

For a quantitative discussion of the energy bands we will concentrate on certain portions of BZ, which correspond to high-symmetry points or directions in the IBZ. Using the results we derived for the IBZ for the high-symmetry points for this lattice, we conclude that we need to calculate the band structure along $\Gamma - \Delta - X - Z - M - \Sigma - \Gamma$. We find that at $\Gamma \to 0$, the matrix is already diagonal and the eigenvalues are given by:

$$\epsilon_\Gamma^{(1)} = \epsilon_s + 4t_{ss}, \quad \epsilon_\Gamma^{(2)} = \epsilon_p + 4t_{pp\pi}, \quad \epsilon_\Gamma^{(3)} = \epsilon_\Gamma^{(4)} = \epsilon_p + 2t_{pp\pi} + 2t_{pp\sigma} \qquad (2.79)$$

The same is true for the point $M \to \frac{\pi}{a}(\hat{\mathbf{x}} + \hat{\mathbf{y}})$, where we get:

$$\epsilon_M^{(1)} = \epsilon_M^{(3)} = \epsilon_p - 2t_{pp\pi} - 2t_{pp\sigma}, \quad \epsilon_M^{(2)} = \epsilon_p - 4t_{pp\pi}, \quad \epsilon_M^{(4)} = \epsilon_s - 4t_{ss} \qquad (2.80)$$

Finally, at the point $X \to \frac{\pi}{a}\hat{\mathbf{x}}$ we have another diagonal matrix with eigenvalues:

$$\epsilon_X^{(1)} = \epsilon_p + 2t_{pp\pi} - 2t_{pp\sigma}, \quad \epsilon_X^{(2)} = \epsilon_p, \quad \epsilon_X^{(3)} = \epsilon_s, \quad \epsilon_X^{(4)} = \epsilon_p - 2t_{pp\pi} + 2t_{pp\sigma} \qquad (2.81)$$

We have chosen the labels of those energy levels to match the band labels as displayed in Fig. 2.18(a). Notice that there are doubly degenerate states at Γ and at M, dictated by symmetry, that is, by the values of $\mathbf{k}$ at those points and the form of the hopping matrix elements within the nearest-neighbor approximation. For the three other high-symmetry points, Δ, Z, Σ, we obtain matrices of the type:

$$\begin{bmatrix} A_\mathbf{k} & B_\mathbf{k} & 0 & 0 \\ B_\mathbf{k}^* & C_\mathbf{k} & 0 & 0 \\ 0 & 0 & D_\mathbf{k} & 0 \\ 0 & 0 & 0 & E_\mathbf{k} \end{bmatrix} \qquad (2.82)$$

The matrices for Δ and Z can be put in this form straightforwardly, while the matrix for Σ requires a change of basis in order to be brought into this form, namely:

$$\chi_{\mathbf{k}1}(\mathbf{r}) = \frac{1}{\sqrt{2}} \left[\chi_{\mathbf{k}p_x}(\mathbf{r}) + \chi_{\mathbf{k}p_y}(\mathbf{r}) \right]$$

$$\chi_{\mathbf{k}2}(\mathbf{r}) = \frac{1}{\sqrt{2}} \left[\chi_{\mathbf{k}p_x}(\mathbf{r}) - \chi_{\mathbf{k}p_y}(\mathbf{r}) \right] \qquad (2.83)$$

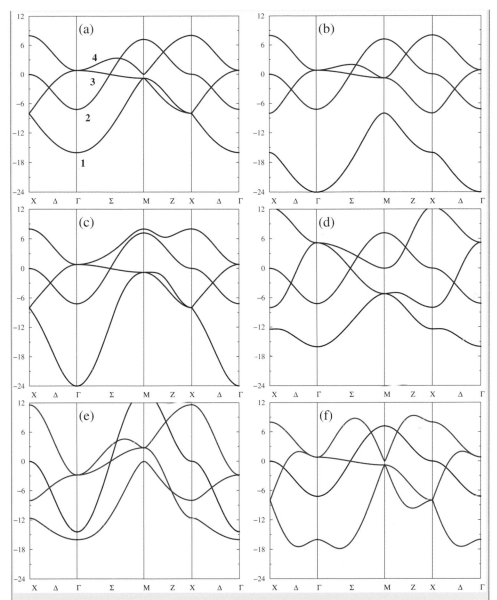

Fig. 2.18 The TBA band structure of the 2D square lattice with parameters from Table 2.3. Bands of *s* character are shown as red lines, *p* character as blue, and mixed character as purple.

with the other two functions, $\chi_{\mathbf{k}s}(\mathbf{r})$ and $\chi_{\mathbf{k}p_z}(\mathbf{r})$, the same as before. The different high-symmetry $\mathbf{k}$-points result in the matrix elements tabulated in Table 2.2. These matrices are then easily solved for the eigenvalues, giving:

$$\epsilon_{\mathbf{k}}^{(1,2)} = \frac{1}{2}\left[(A_{\mathbf{k}} + C_{\mathbf{k}}) \pm \sqrt{(A_{\mathbf{k}} - C_{\mathbf{k}})^2 + 4|B_{\mathbf{k}}|^2}\right], \quad \epsilon_{\mathbf{k}}^{(3)} = D_{\mathbf{k}}, \quad \epsilon_{\mathbf{k}}^{(4)} = E_{\mathbf{k}} \quad (2.84)$$

Table 2.2 Matrix elements for the 2D square lattice with s, p_x, p_y, p_z orbitals at the high-symmetry points Δ, Z, Σ. In all cases $0 < k < \pi/a$.

$\mathbf{k}$	$\Delta \to k\,\hat{\mathbf{x}}$	$Z \to (\pi/a)\,\hat{\mathbf{x}} + k\,\hat{\mathbf{y}}$	$\Sigma \to k\,\hat{\mathbf{x}} + k\,\hat{\mathbf{y}}$
$A_{\mathbf{k}}$	$\epsilon_s + 2\,t_{ss}(\cos(ka)+1)$	$\epsilon_s + 2\,t_{ss}(\cos(ka)-1)$	$\epsilon_s + 4\,t_{ss}\cos(ka)$
$B_{\mathbf{k}}$	$\mathrm{i}\,2\,t_{sp}\sin(ka)$	$\mathrm{i}\,2\,t_{sp}\sin(ka)$	$\mathrm{i}\,\sqrt{8}\,t_{sp}\sin(ka)$
$C_{\mathbf{k}}$	$\epsilon_p + 2\,t_{pp\sigma}\cos(ka) + 2\,t_{pp\pi}$	$\epsilon_p + 2\,t_{pp\sigma}\cos(ka) - 2\,t_{pp\pi}$	$\epsilon_p + 2(t_{pp\sigma} + t_{pp\pi})\cos(ka)$
$D_{\mathbf{k}}$	$\epsilon_p + 2\,t_{pp\sigma} + 2\,t_{pp\pi}\cos(ka)$	$\epsilon_p + 2\,t_{pp\pi}\cos(ka) - 2\,t_{pp\sigma}$	$\epsilon_p + 2(t_{pp\sigma} + t_{pp\pi})\cos(ka)$
$E_{\mathbf{k}}$	$\epsilon_p + 2\,t_{pp\pi}(\cos(ka)+1)$	$\epsilon_p + 2\,t_{pp\pi}(\cos(ka)-1)$	$\epsilon_p + 4\,t_{pp\pi}\cos(ka)$

We have then obtained the eigenvalues for all the high-symmetry points in the IBZ. All that remains to be done is to determine the numerical values of the hamiltonian matrix elements.

In principle, one can imagine calculating the values of the hamiltonian matrix elements using one of the single-particle hamiltonians discussed in Chapter 4. There is a question as to what exactly the appropriate atomic basis functions $\phi_l(\mathbf{r})$ should be. States associated with free atoms are not a good choice, because in the solid the corresponding single-particle states are more compressed due to the presence of other electrons nearby. One possibility then is to solve for atomic-like states in fictitious atoms where the single-particle wavefunctions are compressed, by imposing for instance a constraining potential (typically a harmonic well) in addition to the Coulomb potential of the nucleus.

Alternatively, one can try to guess the values of the hamiltonian matrix so that they reproduce some important features of the band structure, which can be determined independently from experiment. Let us try to predict at least the sign and relative magnitude of the hamiltonian matrix elements, in an attempt to guess a set of reasonable values. First, the diagonal matrix elements ϵ_s, ϵ_p should have a difference approximately equal to the energy difference of the corresponding eigenvalues in the free atom. Notice that if we think of the atomic-like functions $\phi_l(\mathbf{r})$ as corresponding to compressed wavefunctions, then the corresponding eigenvalues ϵ_l are not identical to those of the free atom, but we could expect the compression of eigenfunctions to have similar effects on the different eigenvalues. Since the energy scale is arbitrary, we can choose ϵ_p to be the zero of energy and ϵ_s to be lower in energy by approximately the energy difference of the corresponding free-atom states. The choice $\epsilon_s = -8$ eV is representative of this energy difference for several second-row elements in the Periodic Table.

The matrix element t_{ss} represents the interaction of two $\phi_s(\mathbf{r})$ states at a distance a, the lattice constant of our model crystal. We expect this interaction to be attractive, that is, to contribute to the cohesion of the solid. Therefore, by analogy to our earlier analysis for the 1D model, we expect t_{ss} to be negative. The choice $t_{ss} = -2$ eV for this interaction would be consistent with our choice of the difference between ϵ_s and ϵ_p. Similarly, we expect the interaction of two p states to be attractive in general. In

Table 2.3 Values of the on-site and hopping matrix elements for the band structure of the 2D square lattice with an orthogonal s and p basis and nearest-neighbor interactions. ϵ_p is taken zero in all cases. All values (eV); (a)–(f) refer to parts in Fig. 2.18.

	(a)	(b)	(c)	(d)	(e)	(f)
ϵ_s	−8.0	−16.0	−8.0	−8.0	−8.0	−8.0
t_{ss}	−2.0	−2.0	−4.0	−2.0	−2.0	−2.0
$t_{pp\sigma}$	+2.2	+2.2	+2.2	+4.4	+2.2	+2.2
$t_{pp\pi}$	−1.8	−1.8	−1.8	−1.8	−3.6	−1.8
t_{sp}	−2.1	−2.1	−2.1	−2.1	−2.1	−4.2

the case of $t_{pp\sigma}$ we are assuming the neighboring $\phi_{p_x}(\mathbf{r})$ states to be oriented along the x-axis in the same sense, that is, with positive lobes pointing in the positive direction as required by translational periodicity. This implies that the negative lobe of the state to the right is closest to the positive lobe of the state to the left, so that the overlap between the two states will be negative. Because of this negative overlap, $t_{pp\sigma}$ should be positive so that the net effect is an attractive interaction, by analogy to what we discussed earlier for the 1D model. We expect this matrix element to be roughly of the same magnitude as t_{ss} and a little larger in magnitude, to reflect the larger overlap between the directed lobes of p states. A reasonable choice is $t_{pp\sigma} = +2.2$ eV. In the case of $t_{pp\pi}$, the two p states are parallel to each other at a distance a, so we expect the attractive interaction to be a little weaker than in the previous case, when the orbitals were pointing toward each other. A reasonable choice is $t_{pp\pi} = -1.8$ eV. Finally, we define t_{sp} to be the matrix element with $\phi_{p_x}(\mathbf{r})$ to the left of $\phi_s(\mathbf{r})$, so that the positive lobe of the p orbital is closer to the s orbital and their overlap is positive. As a consequence of this definition, this matrix element, which also contributes to attraction, must be negative; we expect its magnitude to be somewhere between the t_{ss} and $t_{pp\sigma}$ matrix elements. A reasonable choice is $t_{sp} = -2.1$ eV. With these choices, the model yields the band structure shown in Fig. 2.18(a). Notice that in addition to the doubly degenerate states at Γ and M, which are expected from symmetry, there is also a doubly degenerate state at X; this is purely accidental, due to our choice of parameters, as the following discussion also illustrates.

In order to elucidate the influence of the various matrix elements on the band structure, we also show in Fig. 2.18 a number of other choices for their values. To keep the comparisons simple, in each of the other choices we increase one of the matrix elements by a factor of 2 relative to its value in the original set and keep all other values the same; the values for each case are given explicitly in Table 2.3. The corresponding Fig. 2.18(b)–(f) provides insights into the origin of the bands. To facilitate the comparison we label the bands 1 to 4, according to their order in energy near Γ.

Comparing Fig. 2.18(a) and (b) we conclude that band 1 arises from interaction of the s orbitals in neighboring atoms: a decrease of the corresponding eigenvalue ϵ_s from −8 to −16 eV splits this band off from the rest, by lowering its energy throughout the BZ by 8 eV, without affecting the other three bands, except for some minor changes in the neighborhood of M, where bands 1 and 3 were originally degenerate. Since in plot (b) band 1 has split from the rest, now bands 3 and 4 have become degenerate at M, because there must be a doubly degenerate eigenvalue at M independent of the values of the parameters, as we found in Eq. (2.80). An increase of the magnitude of t_{ss} by a factor of 2, which leads to the band structure of plot (c), has as a major effect the increase of the dispersion of band 1; this confirms that band 1 is primarily due to the interaction between s orbitals. There are also some changes in band 4, which at M depends on the value of t_{ss}, as found in Eq. (2.80). Increasing the magnitude of $t_{pp\sigma}$ by a factor of 2 affects significantly bands 3 and 4, somewhat less band 1, and not at all band 2, as seen from the comparison between plots (a) and (d). This indicates that bands 3 and 4 are essentially related to σ interactions between the p_x and p_y orbitals on neighboring atoms. This is also supported by plot (e), in which increasing the magnitude of $t_{pp\pi}$ by a factor of 2 has as a major effect the dramatic increase of the dispersion of band 2; this leads to the conclusion that band 2 arises from π-bonding interactions between p_z orbitals. The other bands are also affected by this change in the value of $t_{pp\pi}$, because they contain π-bonding interactions between p_x and p_y orbitals, but the effect is not as dramatic since in the other bands there are also contributions from σ-bonding interactions, which lessen the importance of the $t_{pp\pi}$ matrix element. Finally, increasing the magnitude of t_{sp} by a factor of 2 affects all bands except band 2, as seen in plot (f); this is because all other bands except band 2 involve orbitals s and p interacting through σ bonds.

Two other features of the band structure are also worth mentioning: First, that bands 1 and 3 in Fig. 2.18(a) and (b) are nearly parallel to each other throughout the BZ. This is an accident related to our choice of parameters for these two plots, as the other four plots prove. This type of behavior has important consequences for the optical properties, as discussed in Chapter 6, particularly when the lower band is occupied (it lies entirely below the Fermi level) and the upper band is empty (it lies entirely above the Fermi level). The second interesting feature is that the lowest band is parabolic near Γ, in all plots of Fig. 2.18 except for (f). The parabolic nature of the lowest band near the minimum is also a feature of the simple 1D model, as well as of the free-electron model discussed previously. In all these cases, the lowest band near the minimum has essentially pure s character, and its dispersion is dictated by the periodicity of the lattice rather than interaction with other bands. Only for the choice of parameters in plot (f) is the parabolic behavior near the minimum altered; in this case the interaction between s and p orbitals (t_{sp}) is much larger than the interaction between s orbitals, so that the nature of the band near the minimum is not pure s any longer but involves also the p states. This last situation is unusual. Far more common is the behavior exemplified by plots (a)–(d), where the nature of the lowest band is clearly associated with the atomic orbitals with the lowest energy. This is demonstrated in more realistic examples later in this chapter.

Example 2.6   2D square lattice with two-atom basis

We consider next an example which involves two atoms per unit cell and exhibits richer physics than the models discussed so far. We will assume that these atoms are different, and the relevant atomic orbitals are of $d_{x^2-y^2}$ character for one atom type and of p_x, p_y for the other atom type. We will also take the lattice to be a square 2D lattice with each type of atom surrounded by four nearest neighbors of the other type, as illustrated in Fig. 2.19, with real-space lattice vectors $\mathbf{a}_1, \mathbf{a}_2$ and corresponding reciprocal-space vectors $\mathbf{b}_1, \mathbf{b}_2$ given by:

$$\mathbf{a}_1 = \frac{a}{\sqrt{2}}(\hat{\mathbf{x}} - \hat{\mathbf{y}}), \quad \mathbf{a}_2 = \frac{a}{\sqrt{2}}(\hat{\mathbf{x}} + \hat{\mathbf{y}}); \quad \mathbf{b}_1 = \frac{\sqrt{2}\pi}{a}(\hat{\mathbf{x}} - \hat{\mathbf{y}}), \quad \mathbf{b}_2 = \frac{\sqrt{2}\pi}{a}(\hat{\mathbf{x}} + \hat{\mathbf{y}})$$

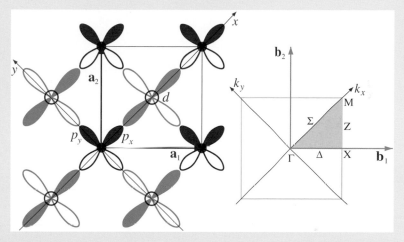

Fig. 2.19

Left: The square lattice with a two-atom basis, represented by the two different symbols, larger open black circles and smaller filled black circles. The lattice constant is a in the two directions; the unit cell is outlined by the two lattice vectors $\mathbf{a}_1, \mathbf{a}_2$ and the black lines. The larger atoms have one d orbital (of $d_{x^2-y^2}$ character, green lobes) while the smaller atoms have two p orbitals (p_x and p_y, blue lobes). **Right**: The corresponding first BZ outlined in yellow line, with the irreducible part shaded, along with the reciprocal lattice vectors $\mathbf{b}_1, \mathbf{b}_2$, and the special $\mathbf{k}$-points Γ, M, X.

In this model, for each of the p orbitals, the diagonal matrix elements of the hamiltonian in the TBA are given by:

$$\langle \chi_{\mathbf{k}p_\alpha}(\mathbf{r}) | \hat{\mathcal{H}}^{\mathrm{sp}} | \chi_{\mathbf{k}p_\alpha}(\mathbf{r}) \rangle = \langle \phi_{p_\alpha}(\mathbf{r}) | \hat{\mathcal{H}}^{\mathrm{sp}} | \phi_{p_\alpha}(\mathbf{r}) \rangle +$$

$$\langle \phi_{p_\alpha}(\mathbf{r}) | \hat{\mathcal{H}}^{\mathrm{sp}} | \phi_{p_\alpha}(\mathbf{r} - \mathbf{a}_1) \rangle \mathrm{e}^{\mathrm{i}\mathbf{k}\cdot\mathbf{a}_1} +$$

$$\langle \phi_{p_\alpha}(\mathbf{r}) | \hat{\mathcal{H}}^{\mathrm{sp}} | \phi_{p_\alpha}(\mathbf{r} + \mathbf{a}_1) \rangle \mathrm{e}^{-\mathrm{i}\mathbf{k}\cdot\mathbf{a}_1} +$$

$$\langle \phi_{p_\alpha}(\mathbf{r}) | \hat{\mathcal{H}}^{\mathrm{sp}} | \phi_{p_\alpha}(\mathbf{r} - \mathbf{a}_2) \rangle \mathrm{e}^{\mathrm{i}\mathbf{k}\cdot\mathbf{a}_2} +$$

$$\langle \phi_{p_\alpha}(\mathbf{r}) | \hat{\mathcal{H}}^{\mathrm{sp}} | \phi_{p_\alpha}(\mathbf{r} + \mathbf{a}_2) \rangle \mathrm{e}^{-\mathrm{i}\mathbf{k}\cdot\mathbf{a}_2}$$

$$= \epsilon_p + 2\, t_{pp} \left[\cos((k_x - k_y)a/\sqrt{2}) + \cos((k_x + k_y)a/\sqrt{2}) \right]$$

where the subscript p_α in wavefunctions in the hamiltonian matrix elements stands for either p_x or p_y, and similarly the diagonal matrix element for the d orbitals is given by:

$$\langle \chi_{\mathbf{k}d}(\mathbf{r})|\hat{\mathcal{H}}^{\mathrm{sp}}|\chi_{\mathbf{k}d}(\mathbf{r})\rangle = \epsilon_d + 2\, t_{dd}\left[\cos((k_x - k_y)a/\sqrt{2}) + \cos((k_x + k_y)a/\sqrt{2})\right]$$

Finally, the non-vanishing hopping matrix elements between p_x, p_y and d orbitals at the two neighboring sites are given by:

$$\langle \chi_{\mathbf{k}p_x}(\mathbf{r})|\hat{\mathcal{H}}^{\mathrm{sp}}|\chi_{\mathbf{k}d}(\mathbf{r})\rangle = \langle \phi_{p_x}(\mathbf{r})|\hat{\mathcal{H}}^{\mathrm{sp}}|\phi_d(\mathbf{r})\rangle +$$

$$\langle \phi_{p_x}(\mathbf{r})|\hat{\mathcal{H}}^{\mathrm{sp}}|\phi_d(\mathbf{r} - (\mathbf{a}_1 + \mathbf{a}_2))\rangle e^{i\mathbf{k}\cdot(\mathbf{a}_1 + \mathbf{a}_2)}$$

$$= t_{pd}\left[1 - e^{i\mathbf{k}\cdot(\mathbf{a}_1 + \mathbf{a}_2)}\right] = -t_{pd}\, e^{ik_x\, a/\sqrt{2}}\, 2i\sin(k_x\, a/\sqrt{2})$$

$$\langle \chi_{\mathbf{k}p_y}(\mathbf{r})|\hat{\mathcal{H}}^{\mathrm{sp}}|\chi_{\mathbf{k}d}(\mathbf{r})\rangle = \langle \phi_{p_y}(\mathbf{r})|\hat{\mathcal{H}}^{\mathrm{sp}}|\phi_d(\mathbf{r} - \mathbf{a}_1)\rangle e^{i\mathbf{k}\cdot\mathbf{a}_1} +$$

$$\langle \phi_{p_y}(\mathbf{r})|\hat{\mathcal{H}}^{\mathrm{sp}}|\phi_d(\mathbf{r} - \mathbf{a}_2)\rangle e^{i\mathbf{k}\cdot\mathbf{a}_2}$$

$$= -t_{pd}\left[e^{i\mathbf{k}\cdot\mathbf{a}_1} - e^{i\mathbf{k}\cdot\mathbf{a}_2}\right] = t_{pd}\, e^{ik_x\, a/\sqrt{2}}\, 2i\sin(k_y\, a/\sqrt{2})$$

with all other matrix elements being zero by symmetry. Diagonalizing the resulting 3×3 hamiltonian matrix for $\mathbf{k}$ values along the $\Gamma - M - X - \Gamma$ directions in the IBZ gives the bands shown in Fig. 2.20. For these plots, we define:

$$\bar{\epsilon} = \frac{\epsilon_p + \epsilon_d}{2}, \quad \mathcal{V}_0 = \epsilon_d - \epsilon_p$$

and use for the first case $\mathcal{V}_0 = 3, t_{dd} = 0.10$ and for the second case $\mathcal{V}_0 = 2, t_{dd} = 0.25$, while for both cases we take $\bar{\epsilon} = -0.5, t_{pp} = -0.25, t_{pd} = 2$ (all in units of eV).

We discuss next the two new interesting features that this model introduces, which are captured by the behavior of the bands in the two cases shown in Fig. 2.20. We will assume that there are four valence electrons per unit cell, associated with the two types of atoms in each unit cell (for example, three from the atom with the p_x, p_y orbitals and one from the atom with the d orbital). Then the Fermi level should be such that two bands are fully occupied throughout the first BZ. In the first case, defined by the choice of parameters that produce a band gap, this means that the Fermi level could be anywhere in the range of values shown by the gray-shaded area in the corresponding band structure of Fig. 2.20 (left panel). This gray-shaded area is the "band gap," separating occupied ("valence") from unoccupied ("conduction") bands. This is the situation that describes semiconductors or insulators, that is, the presence of a gap in the single-particle energy spectrum that separates the occupied from the unoccupied manifold of states. For a pure semiconductor or insulator, the Fermi level actually falls in the middle of the band gap. The presence of a band gap requires at least two different orbitals per unit cell. In the case illustrated here, the band gap is called "direct" because the maximum value of the valence bands and the minimum value of the conduction bands occur at the same point in the BZ (here at the M point, see Fig. 2.20).

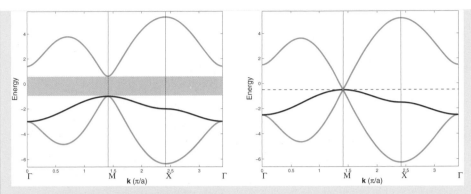

Fig. 2.20

Bands of the two-atom unit cell with p_x, p_y orbitals in one of the atoms and a $d_{x^2-y^2}$ orbital in the other atom, in the 2D square lattice: **Left**: Parameters that give a direct band gap (shaded region). **Right**: Parameters that give no band gap and a Dirac point (linear dispersion of bands) at the M point. Cyan bands come from hybridization between p and d orbitals, while the blue ones come from p-like bands.

A different interesting scenario is illustrated by the second choice of parameters, that produce no gap in the band structure shown in Fig. 2.20 (right panel). The Fermi level for the same number of electrons per unit cell is denoted by a horizontal dashed line. What is interesting here is that near the M point of the BZ, the dispersion of some bands is *linear*, instead of the usual quadratic form we have encountered so far. This is surprising, since we have found that the curvature of the bands is related to the effective mass of crystal electrons, see Eq. (2.51). If the bands near the band extremum at M are linear, that is, the curvature is zero, then their effective mass is not a finite quantity, and the implied energy vs. crystal-momentum relation is:

$$\epsilon_{\mathbf{k}} = v_{\mathrm{F}} \, \hbar |\mathbf{k}| \qquad (2.85)$$

with v_{F} the Fermi velocity; this is the relation that characterizes the behavior of relativistic massless particles, like photons! Particles that exhibit a linear relation of the energy with momentum are called "Dirac particles," from the relativistic version of the wave equation that applies to photons. When the linear behavior of the bands near a critical point extends in all directions of reciprocal space, it is referred to as the "Dirac cone." The slope of the Dirac cone gives the velocity of electrons in these states with linear dispersion, equivalent to the velocity of light for photons; in the above expression, the Fermi velocity v_{F} plays the role of the speed of light. This situation is actually realized in 2D solids like graphene, which is discussed in the following example; in that case the linear relation of the energy with momentum holds in every direction of reciprocal space, which forms a true Dirac cone. In the case of the square lattice discussed above, the energy is a linear function of momentum in one direction and quadratic in the perpendicular direction, as shown in Fig. 2.21; this is described as "semi-Dirac" particles.

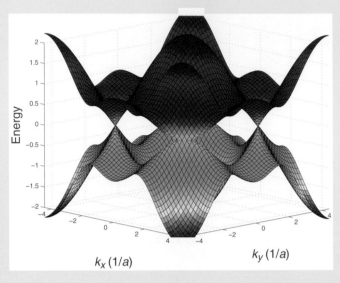

Fig. 2.21 The landscape of the bands in the 2D square lattice with a two-atom basis which have a d orbital on one site and p_x, p_y orbitals on the other; the energy is on the vertical axis and the values of k_x, k_y on the horizontal plane. Only the p–d hybrid bands are shown (cyan-colored in Fig. 2.20), shaded green/red (negative values) and blue/purple (positive values). The semi-Dirac cone is evident, with linear dispersion in one $\mathbf{k}$-space direction and quadratic dispersion in the direction orthogonal to it.

Example 2.7   2D honeycomb lattice with two-atom basis: graphene

Our next example consists of the 2D honeycomb lattice, defined by the lattice vectors $\mathbf{a}_1, \mathbf{a}_2$ and reciprocal lattice vectors $\mathbf{b}_1, \mathbf{b}_2$:

$$\mathbf{a}_{1,2} = \frac{a}{2}\left(\sqrt{3}\hat{x} \pm \hat{y}\right), \ \mathbf{b}_{1,2} = \frac{2\pi}{\sqrt{3}a}\left(\hat{x} \pm \sqrt{3}\hat{y}\right)$$

and the two atoms at positions $\mathbf{t}_1 = 0$, $\mathbf{t}_2 = (\mathbf{a}_1 + \mathbf{a}_2)/3 = (a/\sqrt{3})\hat{x}$, as shown in Fig. 2.22. This is the structure of an actual 2D material, graphene, a single sheet of C atoms, each bonded to its three nearest neighbors in a honeycomb lattice. This important 2D material was mentioned already in Chapter 1, and we will return to it in following chapters.

We will assume that there is one orbital per atom, of p_z character, which we label $\phi_{p_1}(\mathbf{r})$ and $\phi_{p_2}(\mathbf{r})$, respectively, for the two basis atoms, and take the on-site energy to be

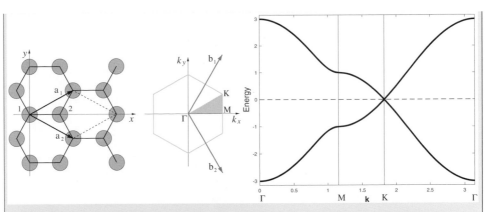

Fig. 2.22
The graphene lattice and π bands. **Left**: The real-space structure of the 2D honeycomb lattice with a two-atom basis and the primitive unit vectors $\mathbf{a}_1$, $\mathbf{a}_2$ (red arrows). The dashed lines indicate a unit cell, with the two basis atoms labeled "1" and "2." **Middle**: The reciprocal-space unit vectors $\mathbf{b}_1$, $\mathbf{b}_2$ (green arrows) and first Brillouin zone (yellow hexagon) of the 2D honeycomb lattice; the high-symmetry points K (corners of the hexagon) and M (centers of the hexagon edges) are identified. **Right**: Band structure of the 2D honeycomb lattice with one p_z orbital per atom, with $\epsilon_p = 0$ and $t_\pi = 1$.

ϵ_p and the nearest-neighbor hopping matrix element between these orbitals to be given by t_π, a shorthand notation for the $t_{pp\pi}$ type of interaction mentioned earlier for parallel p orbitals. With the restrictions of an orthogonal basis and nearest-neighbor interactions, the diagonal matrix elements of the single-particle hamiltonian are equal to ϵ_p and the off-diagonal ones are:

$$\langle \chi_{\mathbf{k}p_1}(\mathbf{r})|\hat{\mathcal{H}}^{\mathrm{sp}}|\chi_{\mathbf{k}p_2}(\mathbf{r})\rangle = \langle \phi_{p_1}(\mathbf{r})|\hat{\mathcal{H}}^{\mathrm{sp}}|\phi_{p_2}(\mathbf{r})\rangle +$$

$$\langle \phi_{p_1}(\mathbf{r})|\hat{\mathcal{H}}^{\mathrm{sp}}|\phi_{p_2}(\mathbf{r}+\mathbf{a}_1)\rangle e^{-i\mathbf{k}\cdot\mathbf{a}_1} +$$

$$\langle \phi_{p_1}(\mathbf{r})|\hat{\mathcal{H}}^{\mathrm{sp}}|\phi_{p_2}(\mathbf{r}+\mathbf{a}_2)\rangle e^{-i\mathbf{k}\cdot\mathbf{a}_2}$$

$$= t_\pi \left[1 + e^{-i\mathbf{k}\cdot\mathbf{a}_1} + e^{-i\mathbf{k}\cdot\mathbf{a}_2} \right]$$

This leads to a simple 2×2 hamiltonian matrix:

$$\mathcal{H}_{\mathbf{k}} = \begin{pmatrix} \epsilon_p & t_\pi f_{\mathbf{k}}^* \\ t_\pi f_{\mathbf{k}} & \epsilon_p \end{pmatrix}, \quad f_{\mathbf{k}} \equiv \left[1 + e^{i\mathbf{k}\cdot\mathbf{a}_1} + e^{i\mathbf{k}\cdot\mathbf{a}_2} \right] \tag{2.86}$$

which can easily be diagonalized to give as eigenvalues:

$$\epsilon_{\mathbf{k}}^{(\pm)} = \epsilon_p \pm t_\pi \left[1 + 4\cos\left(\frac{\sqrt{3}a}{2}k_x\right)\cos\left(\frac{a}{2}k_y\right) + 4\cos^2\left(\frac{a}{2}k_y\right) \right]^{1/2} \tag{2.87}$$

We plot these two eigenvalues along the high-symmetry lines of the BZ, from the center Γ to the point M $= (2\pi/\sqrt{3}a)\hat{x}$ at the center of one of the BZ edges, to K $= (2\pi/3a)(\sqrt{3}\hat{x} + \hat{y})$, the corner of the hexagonal BZ, and back to Γ, as shown in Fig. 2.22; by symmetry, there are six equivalent K points and six equivalent M points.

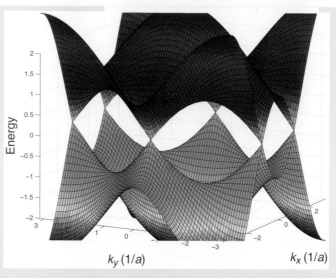

Fig. 2.23 The landscape of the π bands in the 2D honeycomb lattice with a two-atom basis and a p_z-like orbital at each site (a simplified version of graphene); the six equivalent Dirac cones are evident.

For simplicity, since the value of ϵ_p just shifts the spectrum of energies uniformly, we might as well choose $\epsilon_p = 0$; we give the bands in this plot for $t_\pi = 1$, that is, the energy scale is in units of the hopping matrix element. It is evident from this plot that the energy of the bands around K is linear in $|\mathbf{k}|$, at least in the directions K–Γ and K–M. If we plot the energy of the bands as a function of the vector $\mathbf{k}$, we find that the energy is indeed linear in $|\mathbf{k}|$ in all directions away from K. This corresponds to a true Dirac cone with its apex at K, as shown in Fig. 2.23. Since the six corners of the hexagonal BZ are equivalent by symmetry, there is a Dirac cone associated with each one of these points.

2.5.1 Generalizations of the TBA

The case we discussed above is the simplest version of the TBA, with only orthogonal basis functions and nearest-neighbor interactions, as defined in Eqs (2.73) and (2.74), respectively. We also encountered matrix elements in which the p wavefunctions are either parallel or point toward one another along the line that separates them; the case when they are perpendicular results in zero matrix elements by symmetry, see Fig. 2.17. It is easy to generalize all this to a more flexible model, as we discuss next. A comprehensive treatment of the tight-binding method and its application to elemental solids is given in the book by D. A. Papaconstantopoulos (see Further Reading). It is also worth mentioning that the TBA methods are increasingly employed to calculate the total energy of a solid. This practice is motivated by the desire to have a reasonably fast method for total energy and force calculations while maintaining the flexibility of a quantum-mechanical treatment as opposed to resorting to effective interatomic potentials.

Fig. 2.24 **Left**: Two p orbitals at arbitrary directions θ_1, θ_2 relative to the line that joins their centers. **Right**: An s orbital and a p orbital at an angle θ relative to the line that joins their centers.

Arbitrary Orientation of Orbitals

First, it is straightforward to include configurations in which the p orbitals are not just parallel or lie on the line that joins the atomic positions. We can consider each p orbital as composed of a linear combination of two p orbitals, one lying along the line that joins the atomic positions, the other perpendicular to it. This then leads to the general description of the interaction between two p-type orbitals oriented in random directions θ_1 and θ_2 relative to the line that joins the atomic positions where they are centered, as shown in Fig. 2.24:

$$\phi_{p_1}(\mathbf{r}) = \phi_{p_{1x}}(\mathbf{r})\cos\theta_1 + \phi_{p_{1y}}(\mathbf{r})\sin\theta_1$$

$$\phi_{p_2}(\mathbf{r}) = \phi_{p_{2x}}(\mathbf{r})\cos\theta_2 + \phi_{p_{2y}}(\mathbf{r})\sin\theta_2$$

$$\langle\phi_{p_1}|\hat{\mathcal{H}}^{\mathrm{sp}}|\phi_{p_2}\rangle = \langle\phi_{p_{1x}}|\hat{\mathcal{H}}^{\mathrm{sp}}|\phi_{p_{2x}}\rangle\cos\theta_1\cos\theta_2 + \langle\phi_{p_{1y}}|\hat{\mathcal{H}}^{\mathrm{sp}}|\phi_{p_{2y}}\rangle\sin\theta_1\sin\theta_2$$

$$= t_{pp\sigma}\cos\theta_1\cos\theta_2 + t_{pp\pi}\sin\theta_1\sin\theta_2 \tag{2.88}$$

where the line joining the atomic centers is taken to be the x-axis, the direction perpendicular to it the y-axis, and from symmetry we have $\langle\phi_{p_{1x}}|\hat{\mathcal{H}}^{\mathrm{sp}}|\phi_{p_{2y}}\rangle = 0$ and $\langle\phi_{p_{1y}}|\hat{\mathcal{H}}^{\mathrm{sp}}|\phi_{p_{2x}}\rangle = 0$. The matrix elements between a p and an s orbital with arbitrary orientation relative to the line joining their centers is handled by the same procedure, leading to:

$$\phi_p(\mathbf{r}) = \phi_{p_x}(\mathbf{r})\cos\theta + \phi_{p_y}(\mathbf{r})\sin\theta$$

$$\langle\phi_p|\hat{\mathcal{H}}^{\mathrm{sp}}|s\rangle = \langle\phi_{p_x}|\hat{\mathcal{H}}^{\mathrm{sp}}|s\rangle\cos\theta + \langle\phi_{p_y}|\hat{\mathcal{H}}^{\mathrm{sp}}|s\rangle\sin\theta$$

$$= t_{sp}\cos\theta \tag{2.89}$$

for the relative orientation of the p and s orbitals shown in Fig. 2.24.

Non-orthogonal Basis

Another generalization is to consider a basis of orbitals that are not orthogonal, that is, to relax the constraint discussed earlier, that orbitals on nearest-neighbor atoms have no overlap. In this more general case we will have:

$$\langle\phi_m(\mathbf{r} - \mathbf{R}' - \mathbf{t}_j)|\phi_l(\mathbf{r} - \mathbf{R} - \mathbf{t}_i)\rangle = S_{\mu\nu} \tag{2.90}$$

where we use the indices μ, ν to denote all three indices associated with each atomic orbital, that is, $\nu \rightarrow (li\mathbf{R})$ and $\mu \rightarrow (mj\mathbf{R}')$. This new matrix is no longer diagonal, $S_{\mu\nu} \neq \delta_{ml}\delta_{ji}\delta(\mathbf{R} - \mathbf{R}')$, as we had assumed earlier, Eq. (2.73). We now need to solve the general secular equation [Eq. (2.56)] with the general definitions of the hamiltonian

[Eq. (2.59)] and overlap [Eq. (2.58)] matrix elements. A common approximation is to take:

$$S_{\mu v} = S_{mj,li}(r_{ij})f(r_{ij}) \tag{2.91}$$

$$r_{ij} = |(\mathbf{R} + \mathbf{t}_j) - (\mathbf{R}' + \mathbf{t}_i)|$$

with r_{ij} the distance between the atomic sites where the two orbitals are centered and $f(r)$ a function which is equal to unity for a given range of values and falls smoothly to zero beyond this range, $f(r) = 0$ for $r > r_c$; this is known as the cutoff function. In this case, consistency requires that we adopt a similar expression for the hamiltonian matrix elements:

$$t_{\mu v} = f(r_{ij})t_{mj,li}(r_{ij}) \tag{2.92}$$

which are cut off for $r > r_c$ by the same function $f(r)$. The larger r_c is, the more matrix elements we will need to calculate (or fit), and the approximation becomes more computationally demanding.

Multi-center Integrals

The formulation of the TBA up to this point assumed that the TBA hamiltonian matrix elements depend only on two single-particle wavefunctions centered at two different atomic sites. For example, we assumed that the hamiltonian matrix elements depend only on the relative distance and orientation of the two atomic-like orbitals between which we calculate the expectation value of the single-particle hamiltonian. This is referred to as the two-center approximation, but it is obviously another implicit approximation, on top of restricting the basis to the atomic-like wavefunctions. In fact, it is plausible that in the environment of the solid, the presence of other electrons nearby will affect the interaction of any two given atomic-like wavefunctions. In principle, we should consider all such interactions. An example of such terms is a three-center matrix element of the hamiltonian in which one orbital is centered at some atomic site, a second orbital at a different atomic site, and a term in the hamiltonian (the ionic potential) includes the position of a third atomic site. One way of taking into account these types of interactions is to make the hamiltonian matrix elements environment dependent. In this case, the value of a two-center hamiltonian matrix element, involving explicitly the positions of only two atoms, will depend on the position of all other atoms around it and on the type of atomic orbitals on these other atoms. To accomplish this, we need to introduce more parameters to allow for the flexibility of having several possible environments around each two-center matrix element, making the approach much more complicated. The increase in realistic representation of physical systems is always accompanied by an increase in complexity and computational cost.

Excited-State Orbitals in Basis

Finally, we can consider our basis as consisting not only of the valence states of the atoms, but including unoccupied (excited) atomic states. This is referred to as going beyond the minimal basis. The advantages of this generalization are obvious, since including more basis functions always gives a better approximation (by the variational principle). This, however, presents certain difficulties: the excited states tend to be more diffuse in space, with tails extending farther away from the atomic core. This implies that the overlap between such states will not fall off fast with distance between their centers, and it will be difficult to truncate the non-orthogonal overlap matrix and the hamiltonian matrix at a reasonable distance. To avoid this problem, we perform the following operations. We orthogonalize first the states in the minimal basis. This is accomplished by diagonalizing the non-orthogonal overlap matrix and using as the new basis the linear combination of states that corresponds to the eigenvectors of the non-orthogonal overlap matrix. Next, we orthogonalize the excited states to the states in the orthogonal minimal basis, and finally we orthogonalize the new excited states among themselves. Each orthogonalization involves the diagonalization of the corresponding overlap matrix. The advantage of this procedure is that with each diagonalization, the energy of the new states is raised (since they are orthogonal to all previous states), and the overlap between them is reduced. In this way we create a basis that gives rise to a hamiltonian which can be truncated at a reasonable cutoff distance. Nevertheless, the increase in variational freedom that comes with the inclusion of excited states increases the computational complexity, since we will have a larger basis and a correspondingly larger number of matrix elements that we need to calculate or obtain from fitting to known results.

This extension is suggestive of a more general approach: we can use an arbitrary set of functions centered at atomic sites to express the hamiltonian matrix elements. A popular set is composed of normalized gaussian functions (see Appendix A) multiplied by the appropriate spherical harmonics to resemble a set of atomic-like orbitals. We can then calculate the hamiltonian and overlap matrix elements using this set of functions and diagonalize the resulting secular equation to obtain the desired eigenvalues and eigenfunctions. To the extent that these functions represent accurately all possible electronic states (if the original set does not satisfy this requirement we can simply add more basis functions), we can then consider that we have a variationally correct description of the system. This is the more general LCAO method. It is customary in this case to use an explicit form for the single-particle hamiltonian and calculate the hamiltonian and overlap matrix elements exactly, either analytically, if the choice of basis functions permits it, or numerically. The number of basis functions is no longer determined by the number of valence states in the constituent atoms, but rather by variational requirements.

2.6 General Band-Structure Methods

Since the early days of solid-state theory, a number of approaches have been introduced to solve the single-particle hamiltonian and obtain the eigenvalues (band structure) and eigenfunctions. These methods were the foundation on which modern approaches for electronic structure calculations have been developed. We review the basic ideas of these methods next.

Cellular or linearized muffin-tin orbital (LMTO) method: This approach, originally developed by Wigner and Seitz,[4] considers the solid as made up of cells (the Wigner–Seitz or WS cells), which are the real-space analog of the Brillouin zones; a simple example is shown in Fig. 2.25 for a 2D hexagonal lattice. In each cell, the potential felt by the electrons is the Coulomb potential of the atomic nucleus, which is spherically symmetric around the nucleus. However, the boundaries are those of the WS cell, whose shape is dictated by the crystal. Due to the Bloch character of wavefunctions, the following boundary conditions must be obeyed at the boundary of the WS cell denoted by $\mathbf{r}_b$:

$$\psi_{\mathbf{k}}(\mathbf{r}_b) = e^{-i\mathbf{k}\cdot\mathbf{R}}\psi_{\mathbf{k}}(\mathbf{r}_b + \mathbf{R})$$

$$\hat{\mathbf{n}}(\mathbf{r}_b) \cdot \nabla\psi_{\mathbf{k}}(\mathbf{r}_b) = -e^{-i\mathbf{k}\cdot\mathbf{R}}\hat{\mathbf{n}}(\mathbf{r}_b + \mathbf{R}) \cdot \nabla\psi_{\mathbf{k}}(\mathbf{r}_b + \mathbf{R}) \qquad (2.93)$$

where $\hat{\mathbf{n}}(\mathbf{r}_b)$ is the vector normal to the surface of the WS cell at $\mathbf{r}_b$; the minus sign on the right-hand side of the second equation is a consequence of the convention that the surface normal vector always points outward from the volume enclosed by the surface. Since the potential inside the WS cell is assumed to be spherical, we can use the standard expansion in spherical harmonics $Y_{lm}(\hat{\mathbf{r}})$ and radial wavefunctions $\phi_{\mathbf{k}l}(r)$ (for details see Appendix C) which obey the following equation:

$$\left[-\frac{\hbar^2}{2m_e}\left(\frac{d^2}{dr^2} + \frac{2}{r}\frac{d}{dr}\right) + V(r) + \frac{l(l+1)}{r^2} \right]\phi_{\mathbf{k}}(r) = \epsilon_{\mathbf{k}}\phi_{\mathbf{k}}(r) \qquad (2.94)$$

where the dependence of the radial wavefunction on $\mathbf{k}$ enters through the eigenvalue $\epsilon_{\mathbf{k}}$. In terms of these functions, the crystal wavefunctions become:

$$\psi_{\mathbf{k}}(\mathbf{r}) = \sum_{lm}\alpha_{\mathbf{k}lm}Y_{lm}(\hat{\mathbf{r}})\phi_{\mathbf{k}l}(r) \qquad (2.95)$$

Taking matrix elements of the hamiltonian between such states creates a secular equation which can be solved to produce the desired eigenvalues. Since the potential cannot be truly spherical throughout the WS cell, it is reasonable to consider it to be spherical within a sphere which lies entirely within the WS, and to be zero outside that sphere. This gives rise to a potential that looks like a muffin-tin, hence the name of the method. This method is used for calculations of the band structure of complex solids. The basic assumption of the method is that a spherical potential around the nuclei is a reasonable approximation to the true potential experienced by the electrons in the solid.

Augmented plane wave (APW) method: This method, introduced by J. C. Slater,[5] consists of expanding the wavefunctions in plane waves in the regions between the atomic spheres, and in functions with spherical symmetry within the spheres. Then the two expressions must be matched at the sphere boundary so that the wavefunctions and their first and second derivatives are continuous. For core states, the wavefunctions are essentially unchanged within the spheres. It is only valence states that have significant weight in the regions outside the atomic spheres.

[4] E. Wigner and F. Seitz, *Phys. Rev.* **43**, 804 (1933).
[5] J. C. Slater, *Phys. Rev.* **51**, 846 (1937).

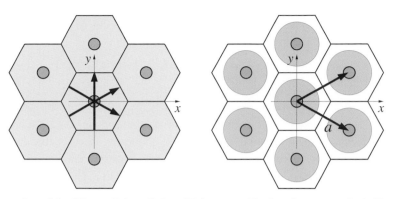

Fig. 2.25   **Left**: Illustration of the Wigner–Seitz cell for a 2D hexagonal lattice, shown as a shaded hexagon around each atom (blue circle). The periodicity (red arrows) is captured as boundary conditions on the wavefunctions at the edges of the WS cell. **Right**: Illustration of the muffin-tin potential, in which a region (green-shaded area) around each atom represents the atomic-like spherical potential and the rest of the space is at a constant potential (white). The red arrows are the primitive lattice vectors of the lattice.

Both the LMTO and the APW methods treat all the electrons in the solid, that is, valence as well as core electrons. Accordingly, they are referred to as "all-electron" methods. The two methods share the basic concept of separating space in the solid into the spherical regions around the nuclei and the interstitial regions between these spheres. In the APW method, the spheres are touching while in the LMTO method, they are overlapping. In many cases, especially in crystal structures other than the close-packed ones (FCC, HCP, BCC), this separation of space leads to inaccuracies, which can be corrected by elaborate extensions described as "full potential" treatment. There is also an all-electron electronic structure method based on multiple scattering theory known as the Korringa–Kohn–Rostocker (KKR) method. A detailed exposition of these methods falls beyond the scope of this book; they are discussed in specialized articles or books (see for example the book by D. Singh),[6] often accompanied by descriptions of computer codes which are necessary for their application. In the remainder of this section we will examine other band structure methods in which the underlying concept is a separation between the electronic core and valence states. This separation makes it possible to treat a larger number of valence states, with a relatively small sacrifice in accuracy. The advantage is that structures with many more atoms in the unit cell can then be studied efficiently.

Orthogonalized plane wave (OPW) method: This method, due to C. Herring,[7] is an elaboration on the APW approach. The basis states $|\phi_{\mathbf{k}}^{(v)}\rangle$ for expanding the valence wavefunctions are written at the outset as a combination of plane waves and core states:

$$|\phi_{\mathbf{k}}^{(v)}\rangle = |\mathbf{k}\rangle + \sum_{c'} \beta_{\mathbf{k}}^{(c')} |\psi_{\mathbf{k}}^{(c')}\rangle \qquad (2.96)$$

[6] D. Singh, *Planewaves, Pseudopotentials and the LAPW Method* (Kluwer Academic, Boston, MA, 1994).
[7] C. Herring, *Phys. Rev.* **57**, 1169 (1940).

where $|\mathbf{k}\rangle$ represents a plane wave of wave-vector $\mathbf{k}$ and $|\psi_{\mathbf{k}}^{(c)}(\mathbf{r})\rangle$ are Bloch states formed out of core states:

$$|\psi_{\mathbf{k}}^{(c)}\rangle = \sum_{\mathbf{R}} e^{i\mathbf{k}\cdot\mathbf{R}}|\phi_{\mathbf{R}}^{(c)}\rangle \tag{2.97}$$

with $|\phi_{\mathbf{R}}^{(c)}\rangle$ a core state centered at $\mathbf{R}$. In essence, we are assuming that the core states $\phi^{(c)}(\mathbf{r})$ can easily be obtained at the level of individual atoms, and that they do not change when the atoms are in the crystal environment, a reasonable approximation. We can then use those core states in a Bloch sum to construct the appropriate electronic states in the crystal that correspond to core electrons. For the valence states, we created a basis which, in addition to core states, includes plane waves. With the choice of the parameters

$$\beta_{\mathbf{k}}^{(c)} = -\langle \psi_{\mathbf{k}}^{(c)}|\mathbf{k}\rangle$$

we make sure that the basis functions $\phi_{\mathbf{k}}^{(v)}(\mathbf{r})$ are orthogonal to the core Bloch states:

$$\langle \phi_{\mathbf{k}}^{(v)}|\psi_{\mathbf{k}}^{(c)}\rangle = \langle \mathbf{k}|\psi_{\mathbf{k}}^{(c)}\rangle - \sum_{c'}\langle \mathbf{k}|\psi_{\mathbf{k}}^{(c')}\rangle\langle \psi_{\mathbf{k}}^{(c')}|\psi_{\mathbf{k}}^{(c)}\rangle = 0 \tag{2.98}$$

where we have used $\langle \psi_{\mathbf{k}}^{(c')}|\psi_{\mathbf{k}}^{(c)}\rangle = \delta_{cc'}$ to reduce the sum to a single term. We can then use these states as the basis for the expansion of the true valence states:

$$|\psi_{\mathbf{k}}^{(v)}\rangle = \sum_{\mathbf{G}} \alpha_{\mathbf{k}}(\mathbf{G})|\phi_{\mathbf{k}+\mathbf{G}}^{(v)}\rangle \tag{2.99}$$

Taking matrix elements between such states produces a secular equation which can be diagonalized to obtain the eigenvalues of the energy.

Pseudopotential plane wave (PPW) method: We can manipulate the expression in the last equation to obtain something quite interesting. First notice that:

$$|\psi_{\mathbf{k}+\mathbf{G}}^{(c)}\rangle = \sum_{\mathbf{R}} e^{i(\mathbf{k}+\mathbf{G})\cdot\mathbf{R}}|\phi_{\mathbf{R}}^{(c)}\rangle = |\psi_{\mathbf{k}}^{(c)}\rangle \tag{2.100}$$

and with this, we obtain:

$$|\psi_{\mathbf{k}}^{(v)}\rangle = \sum_{\mathbf{G}} \alpha_{\mathbf{k}}(\mathbf{G})\left[|\mathbf{k}+\mathbf{G}\rangle - \sum_{c}\langle \psi_{\mathbf{k}+\mathbf{G}}^{(c)}|\mathbf{k}+\mathbf{G}\rangle|\psi_{\mathbf{k}+\mathbf{G}}^{(c)}\rangle\right]$$

$$= \left(\sum_{\mathbf{G}} \alpha_{\mathbf{k}}(\mathbf{G})|\mathbf{k}+\mathbf{G}\rangle\right) - \sum_{c}\langle \psi_{\mathbf{k}}^{(c)}|\left(\sum_{\mathbf{G}} \alpha_{\mathbf{k}}(\mathbf{G})|\mathbf{k}+\mathbf{G}\rangle\right)|\psi_{\mathbf{k}}^{(c)}\rangle$$

which, with the definition

$$|\tilde{\phi}_{\mathbf{k}}^{(v)}\rangle = \sum_{\mathbf{G}} \alpha_{\mathbf{k}}(\mathbf{G})|\mathbf{k}+\mathbf{G}\rangle \rightarrow \tilde{\phi}_{\mathbf{k}}^{(v)}(\mathbf{r}) = \frac{e^{i\mathbf{k}\cdot\mathbf{r}}}{\sqrt{V}}\sum_{\mathbf{G}} \alpha_{\mathbf{k}}(\mathbf{G})e^{i\mathbf{G}\cdot\mathbf{r}} \tag{2.101}$$

can be rewritten as:

$$|\psi_{\mathbf{k}}^{(v)}\rangle = |\tilde{\phi}_{\mathbf{k}}^{(v)}\rangle - \sum_{c}\langle \psi_{\mathbf{k}}^{(c)}|\tilde{\phi}_{\mathbf{k}}^{(v)}\rangle|\psi_{\mathbf{k}}^{(c)}\rangle \tag{2.102}$$

This is an interesting expression, because it tells us that we can express the true valence wavefunctions $\psi_{\mathbf{k}}^{(v)}$ in terms of the auxiliary functions $\tilde{\phi}_{\mathbf{k}}^{(v)}$ from which the overlap with the core states has been subtracted out. These are known as "pseudo-wavefunctions" and they obey a single-particle Schrödinger equation with a modified potential, known as the "pseudopotential," first introduced by Phillips and Kleinman[8] (see also Chapter 4, Section 4.9, for a more extensive discussion). Notice that the pseudo-wavefunction $\tilde{\phi}_{\mathbf{k}}^{(v)}(\mathbf{r})$ is given as an expansion over plane waves only, Eq. (2.101), while it involves exactly the same coefficients $\alpha_{\mathbf{k}}(\mathbf{G})$ as the true valence wavefunction $\psi_{\mathbf{k}}^{(v)}(\mathbf{r})$. This implies that finding these coefficients by solving the appropriate equation obeyed by $\tilde{\phi}_{\mathbf{k}}^{(v)}(\mathbf{r})$ is equivalent to solving the full problem for the original valence wavefunction; the advantage of dealing with $\tilde{\phi}_{\mathbf{k}}^{(v)}$ rather than with $\psi_{\mathbf{k}}^{(v)}$ is that the former is quite simple, since it only involves plane waves. The concepts of the pseudo-wavefunction and the pseudopotential are very important for realistic calculations in solids, and we will return to them in Chapter 4, where a more thorough justification of these ideas and their applications will be presented.

2.6.1 Crystal Pseudopotentials

Having developed pseudopotentials for individual atoms, we return to the discussion of valence electron states in crystals. The crystal potential that the pseudo-wavefunctions experience is given by:

$$\tilde{\mathcal{V}}^{\text{xtl}}(\mathbf{r}) = \sum_{\mathbf{R},i} \tilde{\mathcal{V}}^{\text{ion}}(\mathbf{r} - \mathbf{t}_i - \mathbf{R}) \tag{2.103}$$

where $\tilde{\mathcal{V}}^{\text{ion}}(\mathbf{r} - \mathbf{t}_i)$ is the pseudopotential of a particular atom in the unit cell, at position $\mathbf{t}_i$. We can expand the crystal pseudopotential in the plane-wave basis of the reciprocal lattice vectors:

$$\tilde{\mathcal{V}}^{\text{xtl}}(\mathbf{r}) = \sum_{\mathbf{G}} \tilde{\mathcal{V}}^{\text{xtl}}(\mathbf{G}) e^{i\mathbf{G}\cdot\mathbf{r}} \tag{2.104}$$

It turns out that the pseudopotential is much smoother than the true Coulomb potential of the ions, and therefore we expect its Fourier components $\tilde{\mathcal{V}}^{\text{xtl}}(\mathbf{G})$ to fall fast with the magnitude of $\mathbf{G}$. To simplify the situation, we will assume that we are dealing with a solid that has several atoms of the same type in each unit cell; this is easily generalized to the case of several types of atoms. Using the expression from above in terms of the atomic pseudopotentials, the Fourier components take the form (with NV_{PUC} the volume of the crystal):

$$\tilde{\mathcal{V}}^{\text{xtl}}(\mathbf{G}) = \frac{1}{NV_{\text{PUC}}} \int \tilde{\mathcal{V}}^{\text{xtl}}(\mathbf{r}) e^{-i\mathbf{G}\cdot\mathbf{r}} \, d\mathbf{r} = \sum_{\mathbf{R},i} \frac{1}{NV_{\text{PUC}}} \int \tilde{\mathcal{V}}^{\text{ion}}(\mathbf{r} - \mathbf{t}_i - \mathbf{R}) e^{-i\mathbf{G}\cdot\mathbf{r}} \, d\mathbf{r}$$

$$= \frac{1}{N} \sum_{\mathbf{R},i} e^{i\mathbf{G}\cdot\mathbf{t}_i} \left[\frac{1}{V_{\text{PUC}}} \int \tilde{\mathcal{V}}^{\text{ion}}(\mathbf{r}) e^{-i\mathbf{G}\cdot\mathbf{r}} \, d\mathbf{r} \right] = \tilde{\mathcal{V}}^{\text{ion}}(\mathbf{G}) \sum_{i} e^{i\mathbf{G}\cdot\mathbf{t}_i} \tag{2.105}$$

[8] J. C. Phillips and L. Kleinman, *Phys. Rev.* **116**, 287 (1959).

where we have eliminated a factor of N (the number of PUCs in the crystal) with a summation $\sum_{\mathbf{R}}$, since the summand at the end does not involve an explicit dependence on $\mathbf{R}$. We have also defined the Fourier transform of the atomic pseudopotential $\tilde{\mathcal{V}}^{\mathrm{at}}(\mathbf{G})$ as the content of the square brackets in the next-to-last expression in the above equation. The sum appearing in the last step of this equation:

$$S(\mathbf{G}) = \sum_i e^{i\mathbf{G}\cdot\mathbf{t}_i} \tag{2.106}$$

is called the "structure factor." Depending on the positions of atoms in the unit cell, this summation can vanish for several values of the vector $\mathbf{G}$. This means that the values of the crystal pseudopotential for these values of $\mathbf{G}$ are not needed for a band-structure calculation.

From the above analysis, we conclude that relatively few Fourier components of the pseudopotential survive, since those corresponding to large $|\mathbf{G}|$ are negligible because of the smoothness of the pseudopotential, whereas among those with small $|\mathbf{G}|$, several may be eliminated due to vanishing values of the structure factor $S(\mathbf{G})$. The idea then is to use a basis of plane waves $\exp(i\mathbf{G}\cdot\mathbf{r})$ to expand both the pseudo-wavefunctions and the pseudopotential, which will lead to a secular equation with a relatively small number of non-vanishing elements. Solving this secular equation produces the eigenvalues (band structure) and eigenfunctions for a given system.

To put these arguments in quantitative form, consider the single-particle equations which involve a pseudopotential (we neglect for the moment all the electron interaction terms, which are anyway isotropic; the full problem is considered in more detail in Chapter 4):

$$\left[-\frac{\hbar^2}{2m_e}\nabla^2 + \tilde{\mathcal{V}}^{\mathrm{xtl}}(\mathbf{r}) \right] \tilde{\phi}_{\mathbf{k}}^{(n)}(\mathbf{r}) = \epsilon_{\mathbf{k}}^{(n)} \tilde{\phi}_{\mathbf{k}}^{(n)}(\mathbf{r}) \tag{2.107}$$

These must be solved by considering the expansion for the pseudo-wavefunction in terms of plane waves, Eq. (2.101). Taking matrix elements of the hamiltonian with respect to plane-wave states, we arrive at the following secular equation:

$$\sum_{\mathbf{G}'} \hat{\mathcal{H}}_{\mathbf{k}}^{\mathrm{sp}}(\mathbf{G},\mathbf{G}')\alpha_{\mathbf{k}}^{(n)}(\mathbf{G}') = \epsilon_{\mathbf{k}}^{(n)}\alpha_{\mathbf{k}}^{(n)}(\mathbf{G}) \tag{2.108}$$

where the matrix elements of the single-particle hamiltonian $\hat{\mathcal{H}}^{\mathrm{sp}}$, which contains the ionic pseudopotential $\tilde{\mathcal{V}}^{\mathrm{ion}}$, are given by:

$$\hat{\mathcal{H}}_{\mathbf{k}}^{\mathrm{sp}}(\mathbf{G},\mathbf{G}') = \frac{\hbar^2(\mathbf{k}+\mathbf{G})^2}{2m_e}\delta(\mathbf{G}-\mathbf{G}') + \tilde{\mathcal{V}}^{\mathrm{ion}}(\mathbf{G}-\mathbf{G}')S(\mathbf{G}-\mathbf{G}') \tag{2.109}$$

Diagonalization of the hamiltonian matrix gives the eigenvalues of the energy $\epsilon_{\mathbf{k}}^{(n)}$ and corresponding eigenfunctions $\tilde{\phi}_{\mathbf{k}}^{(n)}(\mathbf{r})$. Obviously, Fourier components of $\tilde{\mathcal{V}}^{\mathrm{ion}}(\mathbf{r})$ which are multiplied by vanishing values of the structure factor $S(\mathbf{G})$ will not be of use in the above equation.

2.7 Localized Wannier Functions

Up to this point we have emphasized the description of the single-particle states for electrons in terms of plane-wave-like functions $\psi_{\mathbf{k}}^{(n)}(\mathbf{r})$, or "band states," which are delocalized. Often, it is desirable to use localized states to describe the electrons, especially when we want to determine the effect of localized perturbations on the properties of the solid. These localized states are referred to as "Wannier functions." There is a one-to-one correspondence between such states and the delocalized wave-like states with which we are familiar. We derive these relations here and comment on the usefulness of the localized states. We denote as $\phi_{\mathbf{R}}^{(n)}(\mathbf{r})$ the Wannier function that is localized in the unit cell defined by the Bravais lattice vector $\mathbf{R}$. We define the Wannier functions as follows:

$$\phi_{\mathbf{R}}^{(n)}(\mathbf{r}) = \frac{1}{\sqrt{N}} \sum_{\mathbf{k}} e^{-i\mathbf{k}\cdot\mathbf{R}} \psi_{\mathbf{k}}^{(n)}(\mathbf{r}) \tag{2.110}$$

We can obtain the band states from the Wannier functions by a similar linear transformation, namely:

$$\psi_{\mathbf{k}}^{(n)}(\mathbf{r}) = \frac{1}{\sqrt{N}} \sum_{\mathbf{R}'} e^{i\mathbf{k}\cdot\mathbf{R}'} \phi_{\mathbf{R}'}^{(n)}(\mathbf{r}) \tag{2.111}$$

To show that the two transformations are equivalent, we begin with the last expression and multiply both sides by $e^{-i\mathbf{k}\cdot\mathbf{R}}$ and sum over all $\mathbf{k}$:

$$\sum_{\mathbf{k}} e^{-i\mathbf{k}\cdot\mathbf{R}} \psi_{\mathbf{k}}^{(n)}(\mathbf{r}) = \frac{1}{\sqrt{N}} \sum_{\mathbf{k}} e^{-i\mathbf{k}\cdot\mathbf{R}} \sum_{\mathbf{R}'} \phi_{\mathbf{R}'}^{(n)}(\mathbf{r}) e^{i\mathbf{k}\cdot\mathbf{R}'} = \frac{1}{\sqrt{N}} \sum_{\mathbf{k}} \sum_{\mathbf{R}'} e^{-i\mathbf{k}\cdot(\mathbf{R}-\mathbf{R}')} \phi_{\mathbf{R}'}^{(n)}(\mathbf{r})$$

In the last expression, the sum of the complex exponentials over all values of $\mathbf{k}$ gives a δ-function in the lattice vectors:

$$\frac{1}{N} \sum_{\mathbf{k}} e^{-i\mathbf{k}\cdot(\mathbf{R}-\mathbf{R}')} = \delta(\mathbf{R} - \mathbf{R}')$$

which, with the summation over $\mathbf{R}'$, produces the desired relation, that is, Eq. (2.110). Similarly, we can show straightforwardly that the Wannier functions are orthonormal:

$$\langle \phi_{\mathbf{R}'}^{(m)} | \phi_{\mathbf{R}}^{(n)} \rangle = \frac{1}{N} \sum_{\mathbf{k}\mathbf{k}'} e^{-i\mathbf{k}\cdot\mathbf{R}} e^{-i\mathbf{k}'\cdot\mathbf{R}'} \langle \psi_{\mathbf{k}'}^{(m)} | \psi_{\mathbf{k}}^{(n)} \rangle$$

$$= \frac{1}{N} \sum_{\mathbf{k}\mathbf{k}'} e^{-i\mathbf{k}\cdot\mathbf{R}} e^{i\mathbf{k}'\cdot\mathbf{R}'} \delta_{\mathbf{k}\mathbf{k}'} \delta_{mn} = \delta(\mathbf{R} - \mathbf{R}') \delta_{mn}$$

where we have used the orthonormality of the band states, $\langle \psi_{\mathbf{k}'}^{(m)} | \psi_{\mathbf{k}}^{(n)} \rangle = \delta_{\mathbf{k}\mathbf{k}'} \delta_{mn}$.

We discuss next some interesting aspects of the Wannier functions. The first one has to do with the degree of localization of Wannier functions. Using our familiar expression for the Bloch function:

$$\psi_{\mathbf{k}}^{(n)}(\mathbf{r}) = u_{\mathbf{k}}^{(n)}(\mathbf{r}) e^{i\mathbf{k}\cdot\mathbf{r}} \Rightarrow \phi_{\mathbf{R}}^{(n)}(\mathbf{r}) = \frac{1}{\sqrt{N}} \sum_{\mathbf{k}} e^{-i\mathbf{k}\cdot\mathbf{R}} u_{\mathbf{k}}^{(n)}(\mathbf{r}) e^{i\mathbf{k}\cdot\mathbf{r}}$$

and taking $u_{\mathbf{k}}^{(n)}(\mathbf{r}) \approx u_0^{(n)}(\mathbf{r})$ to be approximately independent of $\mathbf{k}$:

$$\phi_{\mathbf{R}}^{(n)}(\mathbf{r}) = \frac{u_0^{(n)}(\mathbf{r})}{\sqrt{N}} \sum_{\mathbf{k}} e^{-i\mathbf{k}\cdot\mathbf{R}} e^{i\mathbf{k}\cdot\mathbf{r}} \propto \int_{\mathbf{k}\in BZ} e^{-i\mathbf{k}\cdot(\mathbf{R}-\mathbf{r})} \, d\mathbf{k}$$

with the integration over $\mathbf{k}$ in the first BZ, which has a typical radius of $k_{\max} \sim \pi/a$, where a is the lattice constant. This last integral then, assuming spherically symmetric BZ, gives:

$$\int_{\mathbf{k}\in BZ} e^{-i\mathbf{k}\cdot(\mathbf{R}-\mathbf{r})} d\mathbf{k} \propto \int_0^{k_{\max}} k^2 \frac{e^{-ik|\mathbf{R}-\mathbf{r}|} - e^{ik|\mathbf{R}-\mathbf{r}|}}{-ik|\mathbf{R}-\mathbf{r}|} \, dk \propto \frac{1}{|\mathbf{R}-\mathbf{r}|^2}$$

where we have neglected numerical factors to obtain the qualitative dependence of the integral on $|\mathbf{R} - \mathbf{r}|$. The last expression shows that the Wannier function $\phi_{\mathbf{r}}^{(n)}(\mathbf{r})$ is reasonably localized around the position $\mathbf{R}$ of the unit cell. A much better degree of localization can be obtained by taking advantage of the fact that the Bloch functions can include an arbitrary phase $\varphi_n(\mathbf{k})$:

$$\psi_{\mathbf{k}}^{(n)}(\mathbf{r}) = e^{i\varphi_n(\mathbf{k})} u_{\mathbf{k}}^{(n)}(\mathbf{r}) e^{i\mathbf{k}\cdot\mathbf{r}}$$

This freedom in the definition of the Bloch function can be exploited to produce the so-called "maximally localized Wannier functions," as originally proposed by N. Marzari and D. Vanderbilt.[9] These functions have found wide use in understanding the results of electronic structure calculations in a more physically transparent manner.

The second aspect is the usefulness of the Wannier functions in dealing with localized perturbations of the crystal lattice. Suppose we have the hamiltonian $\hat{\mathcal{H}}^{\mathrm{sp}}$ that includes the part $\hat{\mathcal{H}}_0^{\mathrm{sp}}$, which is the familiar ideal crystal and a weak, localized, time-dependent perturbation potential:

$$\hat{\mathcal{H}}^{\mathrm{sp}} = \hat{\mathcal{H}}_0^{\mathrm{sp}} + \mathcal{V}(\mathbf{r}, t)$$

We can then use the Wannier functions of the unperturbed hamiltonian, $\phi_{\mathbf{R}}^{(n)}(\mathbf{r})$, as the basis to express the solution of the time-dependent Schrödinger equation for the full hamiltonian:

$$\hat{\mathcal{H}}^{\mathrm{sp}}\psi(\mathbf{r}, t) = i\hbar \frac{\partial}{\partial t}\psi(\mathbf{r}, t), \quad \psi(\mathbf{r}, t) = \sum_{n,\mathbf{R}} c_{\mathbf{R}}^{(n)}(t)\phi_{\mathbf{R}}^{(n)}(\mathbf{r}) \tag{2.112}$$

Inserting the expression for $\psi(\mathbf{r}, t)$ in the time-dependent Schrödinger equation and multiplying both sides by $\phi_{\mathbf{R}'}^{*(m)}(\mathbf{r})$ and integrating over $\mathbf{r}$ we obtain:

$$\sum_{n,\mathbf{R}} c_{\mathbf{R}}^{(n)}(t)\langle\phi_{\mathbf{R}'}^{(m)}|\hat{\mathcal{H}}_0^{\mathrm{sp}} + \mathcal{V}(\mathbf{r}, t)|\phi_{\mathbf{R}}^{(n)}\rangle = i\hbar \sum_{n,\mathbf{R}} \frac{\partial c_{\mathbf{R}}^{(n)}(t)}{\partial t}\langle\phi_{\mathbf{R}'}^{(m)}|\phi_{\mathbf{R}}^{(n)}\rangle = i\hbar\frac{\partial c_{\mathbf{R}'}^{(m)}(t)}{\partial t}$$

where we have taken advantage of the orthogonality of Wannier functions, namely:

$$\langle\phi_{\mathbf{R}'}^{(m)}|\phi_{\mathbf{R}}^{(n)}\rangle = \delta(\mathbf{R} - \mathbf{R}')\delta_{mn}$$

---

[9] N. Marzari and D. Vanderbilt, *Phys. Rev. B*, **56**, 12847 (1997).

Using the Bloch eigenfunctions of $\hat{\mathcal{H}}_0^{sp}$, $\psi_{\mathbf{k}}^{(n)}(\mathbf{r})$, their eigenvalues $\epsilon_{\mathbf{k}}^{(n)}$, and their relation to the Wannier functions, $\phi_{\mathbf{R}}^{(n)}(\mathbf{r})$, in the above equation we arrive at:

$$\hat{\mathcal{H}}_0^{sp}\phi_{\mathbf{R}}^{(n)}(\mathbf{r}) = \frac{1}{\sqrt{N}}\sum_{\mathbf{k}} e^{-i\mathbf{k}\cdot\mathbf{R}}\epsilon_{\mathbf{k}}^{(n)}\psi_{\mathbf{k}}^{(n)}(\mathbf{r}) \Rightarrow \langle\phi_{\mathbf{R}'}^{(m)}|\hat{\mathcal{H}}_0^{sp}|\phi_{\mathbf{R}}^{(n)}\rangle = \frac{1}{N}\sum_{\mathbf{k}} e^{-i\mathbf{k}\cdot(\mathbf{R}-\mathbf{R}')}\epsilon_{\mathbf{k}}^{(n)}\delta_{mn}$$

The earlier equation for the time derivative of $c_{\mathbf{R}'}^{(m)}(t)$ then becomes:

$$i\hbar\frac{\partial c_{\mathbf{R}'}^{(m)}(t)}{\partial t} = \sum_{n,\mathbf{R}} c_{\mathbf{R}}^{(n)}(t)\left[\delta_{mn}\epsilon_{\mathbf{R}-\mathbf{R}'}^{(n)} + V_{\mathbf{R}\mathbf{R}'}^{(nm)}(t)\right] \tag{2.113}$$

with the definitions

$$\epsilon_{\mathbf{R}-\mathbf{R}'}^{(n)} = \frac{1}{N}\sum_{\mathbf{k}}\epsilon_{\mathbf{k}}^{(n)}e^{-i\mathbf{k}\cdot(\mathbf{R}-\mathbf{R}')}, \quad V_{\mathbf{R}\mathbf{R}'}^{(nm)}(t) = \langle\phi_{\mathbf{R}'}^{(m)}|V(\mathbf{r},t)|\phi_{\mathbf{R}}^{(n)}\rangle$$

Equation (2.113) for the time evolution of the expansion coefficients is a useful expression for a perturbing potential $V(\mathbf{r},t)$ that is well localized, because then very few matrix elements $V_{\mathbf{R}\mathbf{R}'}^{(nm)}(t)$ survive, since the Wannier functions are also well localized. The resulting system of a few coupled equations for the time evolution of the coefficients $c_{\mathbf{R}}^{(n)}(t)$ can easily be solved. If instead we had used the Bloch states of the unperturbed hamiltonian to express the wavefunction $\psi(\mathbf{r},t)$, we would need to employ a very large number of coupled equations to solve for the time evolution of the system, since a localized perturbation like $V(\mathbf{r},t)$ would require a large number of Fourier components, the matrix elements $\langle\psi_{\mathbf{k}'}^{(m)}|V(\mathbf{r},t)|\psi_{\mathbf{k}}^{(n)}\rangle$, for its accurate representation.

2.8 Density of States

An important feature of the electronic structure of a solid is the density of electronic states as a function of the energy. This is a key quantity in describing many of the electrical and optical properties of solids. This is the quantity measured experimentally in a number of situations, like in tunneling experiments or in photoemission and inverse photoemission experiments.

The density of states, $g(\epsilon)$, for energies in the range $[\epsilon, \epsilon+d\epsilon]$ is defined as the sum over all states labeled by n with energy ϵ_n in that range, per volume V of the system:

$$g(\epsilon)d\epsilon = \frac{1}{V}\sum_{n,\ \epsilon_n\in[\epsilon,\epsilon+d\epsilon]} 1 \Rightarrow g(\epsilon) = \frac{1}{V}\sum_{n}\delta(\epsilon - \epsilon_n) \tag{2.114}$$

where the last expression is obtained by taking the limit $d\epsilon \to 0$, in which case the unit under the sum is replaced by a δ-function in the energy ϵ, centered at the value ϵ_n. The units of the density of states $g(\epsilon)$ are 1/[Energy][Volume]. The index of the energy states n can take the usual expressions, for example, a set of discrete values for a finite system, the momentum $\mathbf{p} = \hbar\mathbf{k}$ for a system of free particles with eigenstates $\epsilon_{\mathbf{k}}$, or the crystal

where $\theta(\epsilon)$ is the Heavyside step-function and therefore its derivative is a δ-function (see Appendix A):

$$\frac{\partial f^{(M)}(\epsilon - eV_b)}{\partial V_b} = -e \frac{\partial f^{(M)}(\epsilon')}{\partial \epsilon'} = e\delta(\epsilon' - \epsilon_F)$$

where we have introduced the auxiliary variable $\epsilon' = \epsilon - eV_b$. Using this result in the expression for the differential conductance we find:

$$\left[\frac{dI}{dV_b}\right]_{T=0} = e|\mathcal{T}|^2 g^{(S)}(\epsilon_F + eV_b) g^{(M)}(\epsilon_F) \tag{2.118}$$

which shows that by scanning the voltage V_b, one samples the DOS of the semiconductor. Thus, the measured differential conductance will reflect all the features of the semiconductor DOS, including the gap.

2.8.1 Free-Electron Density of States

To illustrate the calculation of the DOS for real systems, we consider first the free-electron model in three dimensions. Since the states are characterized by their wave-vector $\mathbf{k}$, we simply need to add up all states with energy in the interval of interest. Taking into account the usual factor of 2 for spin degeneracy and applying the general expression of Eq. (2.114), we get:

$$g_{FE}^{(3D)}(\epsilon)d\epsilon = \frac{1}{V}\sum_{\mathbf{k},\epsilon_{\mathbf{k}}\in[\epsilon,\epsilon+d\epsilon]} 2 = \frac{2}{(2\pi)^3}\int_{\epsilon_{\mathbf{k}}\in[\epsilon,\epsilon+d\epsilon]}d\mathbf{k} = \frac{1}{\pi^2}k^2 dk \tag{2.119}$$

where we used spherical coordinates in $\mathbf{k}$-space to obtain the last result as well as the fact that in the free-electron model the energy does not depend on the angular orientation of the wave-vector:

$$\epsilon_{\mathbf{k}} = \frac{\hbar^2|\mathbf{k}|^2}{2m_e} \implies k\,dk = \frac{m_e}{\hbar^2}d\epsilon, \quad k = \left(\frac{2m_e\epsilon}{\hbar^2}\right)^{1/2} \tag{2.120}$$

for $\epsilon_{\mathbf{k}} \in [\epsilon, \epsilon + d\epsilon]$. These relations give, for the DOS in this simple model in three dimensions:

$$g_{FE}^{(3D)}(\epsilon) = \frac{1}{2\pi^2}\left(\frac{2m_e}{\hbar^2}\right)^{3/2}\epsilon^{1/2} \tag{2.121}$$

To facilitate comparison, we define first a length scale a (which could be, for example, the lattice constant of a crystal), and in terms of this an energy scale $\tilde{\epsilon}$:

$$\tilde{\epsilon} = \frac{\hbar^2}{2m_e a^2} \tag{2.122}$$

in terms of which the DOS in three dimensions becomes:

$$g_{FE}^{(3D)}(\epsilon) = \frac{1}{2\pi^2}\frac{1}{\tilde{\epsilon}a^3}\left(\frac{\epsilon}{\tilde{\epsilon}}\right)^{1/2} \tag{2.123}$$

which has the dimensions of number of states per energy per volume.

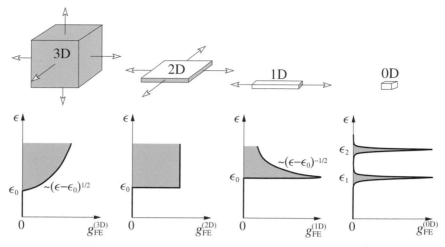

Fig. 2.28   Illustration of the behavior of the DOS as a function of the energy ϵ, in the free-electron model in three, two, one, and zero dimensions; the shapes in gray indicate schematically a 3D, 2D, 1D, and 0D (finite size) solid (the arrows show the directions in which the solid in each case extends to macroscopic size).

By analogous arguments we find that in two and one dimensions the corresponding densities of states are given by:

$$g_{FE}^{(2D)}(\epsilon) = \frac{1}{\pi}\left(\frac{m_e}{\hbar^2}\right) = \frac{1}{2\pi}\frac{1}{\tilde{\epsilon}a^2}\left(\frac{\epsilon}{\tilde{\epsilon}}\right)^0 \tag{2.124}$$

which has the dimensions of number of states per energy per area, and

$$g_{FE}^{(1D)}(\epsilon) = \frac{1}{\pi}\left(\frac{2m_e}{\hbar^2}\right)^{1/2}\epsilon^{-1/2} = \frac{1}{\pi}\frac{1}{\tilde{\epsilon}a}\left(\frac{\epsilon}{\tilde{\epsilon}}\right)^{-1/2} \tag{2.125}$$

which has the dimensions of number of states per energy per length. Finally, for a structure which is finite in all dimensions, that can be considered a 0D structure, the energy spectrum consists of discrete energy levels and the corresponding DOS consists of a series of δ-functions at the values of the energy levels, with dimensions of number of states per energy. As is evident from these expressions, the dependence of the DOS on energy is a very strong function of dimensionality, as illustrated in Fig. 2.28.

2.8.2 Local Density of States

For systems where the electrons are localized, as opposed to the uniform density of the free-electron model, an important concept is the "localized density of states" or "projected density of states" (PDOS). This concept makes it possible to resolve the total density into components related to specific local orbitals in the system. This becomes intuitively clear in cases where the basis for expanding electronic states consists of localized orbitals, hence the term "localized density of states," as in the case of the tight-binding approximation or the Wannier functions. In such situations, suppose that the state $|\psi_n\rangle$ can be expressed in the basis of the orthonormal localized orbitals $|\phi_i\rangle$:

$$|\psi_n\rangle = \sum_i c_i^{(n)} |\phi_i\rangle, \quad \langle\phi_j|\phi_i\rangle = \delta_{ij} \Rightarrow \langle\psi_n|\psi_n\rangle = \sum_i |c_i^{(n)}|^2 = 1$$

Inserting the last expression in the definition of the density of states $g(\epsilon)$ we get:

$$g(\epsilon) = \frac{1}{V} \sum_n \delta(\epsilon - \epsilon_n) \sum_i |c_i^{(n)}|^2 = \sum_i \left[\frac{1}{V} \sum_n |c_i^{(n)}|^2 \delta(\epsilon - \epsilon_n) \right] = \sum_i g_i(\epsilon)$$

where we have defined the PDOS, $g_i(\epsilon)$, corresponding to the orbital $|\phi_i\rangle$, as:

$$g_i(\epsilon) = \frac{1}{V} \sum_n |c_i^{(n)}|^2 \delta(\epsilon - \epsilon_n) \tag{2.126}$$

This quantity gives the density of states at the value ϵ that is related to the specific orbital $|\phi_i\rangle$ in the basis.

2.8.3 Crystal DOS: Van Hove Singularities

In a crystal we need to use the full band-structure calculation for $\epsilon_{\mathbf{k}}^{(n)}$ with n the band index and $\mathbf{k}$ the reciprocal-space vector in the first BZ, in the expression for the density of states, which takes the form:

$$g(\epsilon) = \frac{1}{V} \sum_{n,\mathbf{k}} 2\delta(\epsilon - \epsilon_{\mathbf{k}}^{(n)}) = \frac{2}{(2\pi)^3} \sum_n \int \delta(\epsilon - \epsilon_{\mathbf{k}}^{(n)}) d\mathbf{k}$$

$$= \frac{2}{(2\pi)^3} \sum_n \int_{\epsilon_{\mathbf{k}}^{(n)} = \epsilon} \frac{1}{|\nabla_{\mathbf{k}} \epsilon_{\mathbf{k}}^{(n)}|} dS_{\mathbf{k}} \tag{2.127}$$

where the last integral is over a surface in $\mathbf{k}$-space on which $\epsilon_{\mathbf{k}}^{(n)}$ is constant and equal to ϵ. The last step in the above equation can be proven by using the properties of the δ-function, see Appendix A. In this final expression, the roots of the denominator are of first order and therefore contribute a finite quantity to the integration over a smooth 2D surface represented by $S_{\mathbf{k}}$; these roots introduce sharp features in the function $g(\epsilon)$, which are called "van Hove singularities." For the values $\mathbf{k}_0$ where $\nabla_{\mathbf{k}} \epsilon_{\mathbf{k}} = 0$, we can expand the energy in a Taylor expansion (from here on we consider the contribution of a single band and drop the band index for simplicity):

$$[\nabla_{\mathbf{k}} \epsilon_{\mathbf{k}}]_{\mathbf{k}=\mathbf{k}_0} = 0 \Rightarrow \epsilon_{\mathbf{k}} = \epsilon_{\mathbf{k}_0} + \sum_{i=1}^{d} \alpha_i (k_i - k_{0,i})^2 \tag{2.128}$$

The expansion is over as many principal axes as the dimensionality of our system: in three dimensions ($d = 3$) there are three principal axes, characterized by the symbols $\alpha_i, i = 1, 2, 3$. Depending on the signs of these coefficients, the extremum can be a minimum (0 negative coefficients, referred to as "type 0 critical point"), two types of saddle point ("type 1 and 2 critical points"), or a maximum ("type 3 critical point"). There is a useful theorem that tells us exactly how many critical points of each type we can expect.

Table 2.4 Symbols (l, M_l), multiplicity $= d! / l! (d - l)!$, type, and characteristic behavior of the coefficients α_i, $i = 1, 2, 3$ along the principal axes for critical points in $d = 3$ dimensions.

l	Symbol	Multiplicity	Type	Coefficients
0	M_0	1	minimum	$\alpha_1, \alpha_2, \alpha_3 > 0$
1	M_1	3	saddle point	$\alpha_1, \alpha_2 > 0, \alpha_3 < 0$
2	M_2	3	saddle point	$\alpha_1 > 0, \alpha_2, \alpha_3 < 0$
3	M_3	1	maximum	$\alpha_1, \alpha_2, \alpha_3 < 0$

Theorem   *Given a function of d variables periodic in all of them, there are $d! / l! (d - l)!$ critical points of type l, where l is the number of negative coefficients in the Taylor expansion of the energy, Eq. (2.128).*

With the help of this theorem, we can obtain the number of each type of critical point in $d = 3$ dimensions, as given in Table 2.4.

Next we want to extract the behavior of the DOS explicitly near each type of critical point. Let us first consider the critical point of type 0, M_0, in which case $\alpha_i > 0, i = 1, 2, 3$. In order to perform the **k**-space integrals involved in the DOS we first make the following changes of variables: in the neighborhood of the critical point at $\mathbf{k}_0$

$$\epsilon_{\mathbf{k}} - \epsilon_{\mathbf{k}_0} = \alpha_1 k_1^2 + \alpha_2 k_2^2 + \alpha_3 k_3^2 \tag{2.129}$$

where **k** is measured relative to $\mathbf{k}_0$. We can choose the principal axes so that α_1, α_2 have the same sign; we can always do this since there are at least two coefficients with the same sign (see Table 2.4). We rescale these axes so that $\alpha_1 = \alpha_2 = \beta$ after the scaling and introduce cylindrical coordinates for the rescaled variables k_1, k_2:

$$q^2 = k_1^2 + k_2^2, \quad \theta = \tan^{-1}(k_2/k_1) \tag{2.130}$$

With these changes, the DOS function takes the form:

$$g(\epsilon) = \frac{\lambda}{(2\pi)^3} \int \delta(\epsilon - \epsilon_{\mathbf{k}_0} - \beta q^2 - \alpha_3 k_3^2) \, d\mathbf{k} \tag{2.131}$$

where the factor λ comes from rescaling the principal axes 1 and 2. We can rescale the variables q, k_3 so that their coefficients become unity, to obtain:

$$g(\epsilon) = \frac{\lambda}{(2\pi)^2 \beta \alpha_3^{1/2}} \int \delta(\epsilon - \epsilon_{\mathbf{k}_0} - q^2 - k_3^2) \, q \, dq \, dk_3 \tag{2.132}$$

Now we can consider the expression in the argument of the δ-function as a function of k_3:

$$f(k_3) = \epsilon - \epsilon_{\mathbf{k}_0} - q^2 - k_3^2 \Rightarrow f'(k_3) = -2k_3 \tag{2.133}$$

and we can integrate over k_3 with the help of the expression for δ-function integration Eq. (A.61) (derived in Appendix A), which gives:

$$g(\epsilon) = \lambda_0 \int_0^Q \frac{1}{(\epsilon - \epsilon_{\mathbf{k}_0} - q^2)^{1/2}} \, q \, dq \tag{2.134}$$

where the factor λ_0 embodies all the constants in front of the integral from rescaling and integration over k_3, and we have dropped primes from the remaining variable of integration for simplicity. The upper limit of integration Q for the variable q is determined by the condition

$$k_3^2 = \epsilon - \epsilon_{\mathbf{k}_0} - q^2 \geq 0 \Rightarrow Q = (\epsilon - \epsilon_{\mathbf{k}_0})^{1/2} \qquad (2.135)$$

and with this we can now perform the final integration to obtain:

$$g(\epsilon) = -\lambda_0 \left[(\epsilon - \epsilon_{\mathbf{k}_0} - q^2)^{1/2} \right]_0^{(\epsilon - \epsilon_{\mathbf{k}_0})^{1/2}} = \lambda_0 (\epsilon - \epsilon_{\mathbf{k}_0})^{1/2} \qquad (2.136)$$

This result holds for $\epsilon > \epsilon_{\mathbf{k}_0}$. For $\epsilon < \epsilon_{\mathbf{k}_0}$, the δ-function cannot be satisfied for the case we are investigating, with $\alpha_i > 0, i = 1, 2, 3$, so the DOS must be 0.

By an exactly analogous calculation, we find that for the maximum M_3, with $\alpha_i < 0, i = 1, 2, 3$, the DOS behaves like:

$$g(\epsilon) = \lambda_3 (\epsilon_{\mathbf{k}_0} - \epsilon)^{1/2} \qquad (2.137)$$

for $\epsilon < \epsilon_{\mathbf{k}_0}$ and it is 0 for $\epsilon > \epsilon_{\mathbf{k}_0}$.

For the other two cases, M_1, M_2, we can perform a similar analysis. We outline briefly the calculation for M_1: in this case, we have after rescaling $\alpha_1 = \alpha_2 = \beta > 0$ and $\alpha_3 < 0$, which leads to:

$$g(\epsilon) = \frac{\lambda}{(2\pi)^2 \beta \alpha_3^{1/2}} \int \delta(\epsilon - \epsilon_{\mathbf{k}_0} - q^2 + k_3^2) \, q \, dq \, dk_3$$

$$\rightarrow \lambda_1 \int_{Q_1}^{Q_2} \frac{1}{(q^2 + \epsilon_{\mathbf{k}_0} - \epsilon)^{1/2}} \, q \, dq \qquad (2.138)$$

and we need to specify the limits of the last integral from the requirement that $q^2 + \epsilon_{\mathbf{k}_0} - \epsilon \geq 0$. There are two possible situations:

(i) For $\epsilon < \epsilon_{\mathbf{k}_0}$ the condition $q^2 + \epsilon_{\mathbf{k}_0} - \epsilon \geq 0$ is always satisfied, so that the lower limit of q is $Q_1 = 0$, and the upper limit is any positive value $Q_2 = Q > 0$. Then the DOS becomes:

$$g(\epsilon) = \lambda_1 \left[(q^2 + \epsilon_{\mathbf{k}_0} - \epsilon)^{1/2} \right]_0^Q = \lambda_1 (Q^2 + \epsilon_{\mathbf{k}_0} - \epsilon)^{1/2} - \lambda_1 (\epsilon_{\mathbf{k}_0} - \epsilon)^{1/2} \quad (2.139)$$

For $\epsilon \rightarrow \epsilon_{\mathbf{k}_0}$, expanding in powers of the small quantity $(\epsilon_{\mathbf{k}_0} - \epsilon)$ gives:

$$g(\epsilon) = \lambda_1 Q - \lambda_1 (\epsilon_{\mathbf{k}_0} - \epsilon)^{1/2} + \mathcal{O}(\epsilon_{\mathbf{k}_0} - \epsilon) \qquad (2.140)$$

(ii) For $\epsilon > \epsilon_{\mathbf{k}_0}$ the condition $q^2 + \epsilon_{\mathbf{k}_0} - \epsilon \geq 0$ is satisfied for a lower limit $Q_1 = (\epsilon - \epsilon_{\mathbf{k}_0})^{1/2}$ and an upper limit being any positive value $Q_2 = Q > (\epsilon - \epsilon_{\mathbf{k}_0})^{1/2} > 0$. Then the DOS becomes:

$$g(\epsilon) = \lambda_1 \left[(q^2 + \epsilon_{\mathbf{k}_0} - \epsilon)^{1/2} \right]_{(\epsilon - \epsilon_{\mathbf{k}_0})^{1/2}}^Q = \lambda_1 (Q^2 + \epsilon_{\mathbf{k}_0} - \epsilon)^{1/2} \qquad (2.141)$$

For $\epsilon \rightarrow \epsilon_{\mathbf{k}_0}$, expanding in powers of the small quantity $(\epsilon_{\mathbf{k}_0} - \epsilon)$ gives:

$$g(\epsilon) = \lambda_1 Q + \frac{\lambda_1}{2Q} (\epsilon_{\mathbf{k}_0} - \epsilon) + \mathcal{O}[(\epsilon_{\mathbf{k}_0} - \epsilon)^2] \qquad (2.142)$$

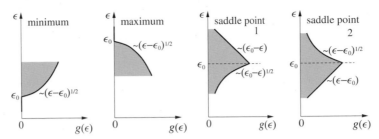

Fig. 2.29 The behavior of DOS, $g(\epsilon)$, near critical points of different type in three dimensions.

By an exactly analogous calculation, we find that for the other saddle point M_2, with $\alpha_1, \alpha_2 < 0, \alpha_3 > 0$, the DOS behaves like:

$$g(\epsilon) = \lambda_2 Q + \frac{\lambda_2}{2Q}(\epsilon - \epsilon_{\mathbf{k}_0}) + \mathcal{O}[(\epsilon - \epsilon_{\mathbf{k}_0})^2] \tag{2.143}$$

for $\epsilon < \epsilon_{\mathbf{k}_0}$ and

$$g(\epsilon) = \lambda_2 Q - \lambda_2 (\epsilon - \epsilon_{\mathbf{k}_0})^{1/2} + \mathcal{O}(\epsilon - \epsilon_{\mathbf{k}_0}) \tag{2.144}$$

for $\epsilon > \epsilon_{\mathbf{k}_0}$. The behavior of the DOS for all the critical points is summarized graphically in Fig. 2.29. We should caution the reader that very detailed calculations are required to resolve these critical points. In particular, methods must be developed that allow the inclusion of eigenvalues at a very large number of sampling points in reciprocal space, usually by interpolation between points at which electronic eigenvalues are actually calculated.[11]

Further Reading

1. *Wave Propagation in Periodic Structures*, L. Brillouin (McGraw-Hill, New York, 1946). This is a classic book on wave and periodic structures, introducing the notion that has become known as "Brillouin zones," with excellent illustrative examples.

2. *Electronic Structure and the Properties of Solids*, W. A. Harrison (W.H. Freeman, San Francisco, 1980). This book contains a general discussion of the properties of solids based on the tight-binding approximation.

3. *Physics of Solids*, E. N. Economou (Springer-Verlag, Berlin, 2010). This book contains a careful discussion of the physics of the tight-binding approximation and its many applications.

4. *Handbook of the Band Structure of Elemental Solids*, D. A. Papaconstantopoulos (Plenum Press, New York, 1989). This is a comprehensive account of the tight-binding approximation and its application to the electronic structure of elemental solids.

[11] O. Jepsen and O. K. Andersen, *Solid State Comm.* **9**, 1763 (1971); G. Lehmann and M. Taut, *Phys. Stat. Solidi* **54**, 469 (1972).

Problems

1.   (a) Calculate the reciprocal lattice vectors for the crystal lattices given in Table 2.1.

     (b) Determine the lattice vectors, the positions of atoms in the primitive unit cell, and the reciprocal lattice vectors, for the NaCl, CsCl, and zincblende crystal structures discussed in Chapter 1.

2.   (a) Show that the single-particle equation obeyed by $u_{\mathbf{k}}(\mathbf{r})$, the part of the wavefunction which has full translational symmetry, is Eq. (2.22).

     (b) Show that the coefficients $\alpha_{\mathbf{k}}(\mathbf{G})$ in the Fourier expansion of $u_{\mathbf{k}}^{(n)}(\mathbf{r})$ obey Eq. (2.44).

3.   We want to derive the expression for the energy at wave-vector $\mathbf{k}_0 + \mathbf{k}$ in terms of the energy and wavefunctions at wave-vector $\mathbf{k}_0$, as given in Eq. (2.48). The simplest way to do this is to work with the part of the wavefunctions that are fully symmetric in the lattice periodicity, $u_{\mathbf{k}}^{(n)}(\mathbf{r})$, and the corresponding hamiltonian for which they are solutions, $\hat{\mathcal{H}}^{\mathrm{sp}}(\hat{\mathbf{p}} + \hbar\mathbf{k})$ [see Eq. (2.22)]. Consider as the unperturbed hamiltonian the one corresponding to $\mathbf{k}_0$, $\hat{\mathcal{H}}_0^{\mathrm{sp}} = \hat{\mathcal{H}}^{\mathrm{sp}}(\hat{\mathbf{p}} + \hbar\mathbf{k}_0)$, with the known solutions $u_{\mathbf{k}_0}^{(n)}(\mathbf{r}), \epsilon_{\mathbf{k}_0}^{(n)}$, and then consider the hamiltonian at $(\mathbf{k}_0 + \mathbf{k})$: the latter has extra terms relative to $\hat{\mathcal{H}}_0^{\mathrm{sp}}$ which can be identified as the perturbation $\hat{\mathcal{H}}_1^{\mathrm{sp}}$. We assume that there are no band degeneracies ($\epsilon_{\mathbf{k}_0}^{(n)} \neq \epsilon_{\mathbf{k}_0}^{(n')}$ for $n \neq n'$), so that we can use non-degenerate perturbation theory (see Appendix C). From the result for the energy at $\mathbf{k}_0 + \mathbf{k}$, derive the expression for the inverse effective mass, given in Eq. (2.51).

4.   For the 2D square lattice with lattice constant a, with one atom per unit cell, in the free-electron approximation, find the Fermi momentum and Fermi energy if there is one valence electron per atom. Show the Fermi sphere and draw carefully the occupied portions of the various BZs.

5.   (a) Determine the first Brillouin zone of the FCC lattice. In order to do this, find the first set of reciprocal space vectors with the lowest magnitude, draw these vectors and the planes that are perpendicular to them at their mid-points. What shape does the envelope of these planes produce in reciprocal space? Then, find the set of reciprocal-space vectors that have the next higher magnitude, and the planes that bisect these vectors. Describe the intersection of one of these planes with the corresponding planes of the first set. This should be sufficient to fully determine by symmetry the first BZ. Compare your result to what is shown in Fig. 2.12.

     (b) Draw the first few free-electron bands for the 3D FCC lattice along the high-symmetry lines shown in Fig. 2.12, in units of $(\hbar^2/2m_e)(2\pi/a)^2$; compare your results to those shown in Fig. 2.12.

     (c) For an FCC crystal with one atom per unit cell, which has $Z = 3$ valence electrons, estimate the position of the Fermi level on the band-structure plot and describe the features of the resulting 3D Fermi sphere qualitatively.

6. Prove that the orthogonality relation, Eq. (2.57), holds in general (not only in the context of the tight-binding approximation).

7. The relationship between the band width and the number of nearest neighbors, Eq. (2.71), was derived for the simple chain, square, and cubic lattices in one, two, and three dimensions, using the simplest tight-binding model with one atom per unit cell and one s-like orbital per atom. For these lattices, the number of neighbors z is always $2d$, where d is the dimensionality. Consider the same simple tight-binding model for the close-packed lattices in two and three dimensions, that is, the simple 2D hexagonal lattice and the 3D FCC lattice, and derive the corresponding relation between the band width and the number of nearest neighbors.

8. Prove the Bloch statement for the many-body wavefunction, that is, show that the many-body wavefunction will have Bloch-like symmetry:

$$\Psi_{\mathbf{K}}(\mathbf{r}_1 + \mathbf{R}, \mathbf{r}_2 + \mathbf{R}, ...) = e^{i\mathbf{K}\cdot\mathbf{R}}\Psi_{\mathbf{K}}(\mathbf{r}_1, \mathbf{r}_2, ...) \qquad (2.145)$$

Then relate the value of the wave-vector $\mathbf{K}$ to the single-particle wave-vectors $\mathbf{k}_i, i = 1, \ldots, N$, when the many-body wavefunction is written as a product of single-particle states.

9. For a Bloch state defined as:

$$\phi_{\mathbf{k},s}^{(n)}(\mathbf{r}) = \frac{e^{i\mathbf{k}\cdot\mathbf{r}}}{\sqrt{V}}u_{\mathbf{k}}^{(n)}(\mathbf{r})\lambda_s$$

where λ_s is a two-component space-independent spinor, find the hamiltonian in the weakly relativistic limit [see Appendix C, Eq. (C.80)] that applies to the space-dependent part of the wavefunction $u_{\mathbf{k}}^{(n)}(\mathbf{r})$.

10. Derive the density of states for the free-electron model in two dimensions, Eq. (2.124) and one dimension, Eq. (2.125).

11. Derive the type, multiplicity, and behavior of the DOS at critical points in one and two dimensions.

12. Derive the behavior of the DOS near the Dirac point in one, two, and three dimensions, where the energy is given by Eq. (2.85).

13. For the simple model of the graphene lattice, which consists of one p_z-like orbital per atom in the 2D honeycomb lattice, described in Section 2.5:

    (a) Show that the energy of the two bands is given by Eq. (2.87).
    (b) Show that at the corner of the Brillouin zone (K point) this expression leads to Dirac behavior, Eq. (2.85).
    (c) Calculate numerically the DOS near the critical points Γ, M, and K and show that it has the expected behavior.

14. Consider the following 2D model: the lattice is a square of side a with lattice vectors $\mathbf{a}_1 = a\hat{x}$ and $\mathbf{a}_2 = a\hat{y}$; there are three atoms per unit cell, one Cu atom and two O atoms at distances $\mathbf{a}_1/2$ and $\mathbf{a}_2/2$, as illustrated in Fig. 1.30. We will assume that the atomic orbitals associated with these atoms are orthogonal.

   (a) Although there are five d orbitals associated with the Cu atom and three p orbitals associated with each O atom, only one of the Cu orbitals (the $d_{x^2-y^2}$ one) and two of the O orbitals (the p_x one on the O atom at $\mathbf{a}_1/2$ and the p_y one on the O atom at $\mathbf{a}_2/2$) are relevant, because of the geometry; the remaining Cu and O orbitals do not interact with their neighbors. Explain why this is a reasonable approximation, and in particular why the Cu–$d_{3z^2-r^2}$ orbitals do not interact with their nearest-neighbor O–p_z orbitals.

   (b) Define the hamiltonian matrix elements between the relevant Cu–d and O–p nearest-neighbor orbitals and take the on-site energies to be ϵ_d and ϵ_p, with $\epsilon_p < \epsilon_d$. Use these matrix elements to calculate the band structure for this model.

   (c) Discuss the position of the Fermi level for the case where there is one electron in each O orbital and one or two electrons in the Cu orbital.

   Historical note: Even though this model may appear artificial, it has been used extensively to describe the basic electronic structure of the copper oxide–rare earth materials, which are high-temperature superconductors.[12]

---

[12] L. F. Mattheis, *Phys. Rev. Lett.* **58**, 1028 (1987).

3 Symmetries Beyond Translational Periodicity

In the previous chapter we discussed in detail the effects of lattice periodicity on the single-particle wavefunctions and the energy eigenvalues. We also touched on the notion that a crystal can have symmetries beyond the translational periodicity, such as rotations around axes, reflections on planes, and combinations of these operations with translations by vectors that are *not* lattice vectors, called "non-primitive" translations. All these symmetry operations are useful in calculating and analyzing the physical properties of a crystal. There are two basic advantages to using the symmetry operations of a crystal in describing its properties. First, the volume in reciprocal space for which solutions need to be calculated is further reduced, usually to a small fraction of the first Brillouin zone, called the "irreducible" part; for example, in the FCC crystals with one atom per unit cell, the irreducible part is 1/48 of the full BZ. Second, certain selection rules and compatibility relations are dictated by symmetry alone, leading to a deeper understanding of the physical properties of the crystal as well as to simpler ways of calculating these properties in the context of the single-particle picture; for example, using symmetry arguments it is possible to identify the allowed optical transitions in a crystal, which involve excitation or de-excitation of electrons by absorption or emission of photons, thereby elucidating its optical properties.

In this chapter we address the issue of crystal symmetries beyond translational periodicity in much more detail. To achieve this, we must make an investment in the mathematical ideas of group theory. We restrict ourselves to the bare minimum necessary to explore the effects of symmetry on the electronic properties of the crystal. In several cases, we will find it useful to state and prove theorems to the extent that this exercise provides insight into how the theory actually works. When we give proofs, we shall mark their end by the symbol ■, as is commonly done in mathematical texts; this may be useful in case the reader wishes to skip the proofs and continue with the exposition of the theory.

3.1 Time-Reversal Symmetry for Spinless Fermions

In addition to the symmetries a crystal has in real space, there is also another type of symmetry in reciprocal space which allows us to connect the properties of states with wave-vector $\mathbf{k}$ to those with wave-vector $-\mathbf{k}$. This is called "inversion symmetry" and holds irrespective of whether the real-space crystal possesses the corresponding symmetry or not. In real space, inversion symmetry implies $\mathbf{r} \rightarrow -\mathbf{r}$, that is, the crystal remains invariant if we move every point in space to its inverse with respect to some properly chosen origin. The corresponding operation in reciprocal space is $\mathbf{k} \rightarrow -\mathbf{k}$, that is, the properties of the crystal are invariant if we replace every value of the wave-vector with its opposite. The inversion symmetry in reciprocal space always holds, even if the crystal does not have inversion in real space. This remarkable property is a consequence of time-reversal symmetry, that is, the fact that the hamiltonian of a solid does not change if we change the sign of the time variable, $t \rightarrow -t$. Evidently, when a term in the hamiltonian breaks time-reversal symmetry (e.g. an external magnetic field), the inversion in reciprocal space is also broken. We discuss this symmetry first, and then turn our attention to the description of all other crystal symmetries beyond periodicity.

Kramers' Theorem   *For spinless fermions, since the hamiltonian is real, that is, the system is time-reversal invariant, we must have for any state characterized by the wave-vector* $\mathbf{k}$:

$$\epsilon_{\mathbf{k}}^{(n)} = \epsilon_{-\mathbf{k}}^{(m)} \tag{3.1}$$

Proof   To prove Kramers' theorem, we take the complex conjugate of the single-particle Schrödinger equation

$$\hat{\mathcal{H}}^{\mathrm{sp}} \psi_{\mathbf{k}}^{(n)}(\mathbf{r}) = \epsilon_{\mathbf{k}}^{(n)} \psi_{\mathbf{k}}^{(n)}(\mathbf{r}) \Rightarrow \hat{\mathcal{H}}^{\mathrm{sp}} \psi_{\mathbf{k}}^{*(n)}(\mathbf{r}) = \epsilon_{\mathbf{k}}^{(n)} \psi_{\mathbf{k}}^{*(n)}(\mathbf{r}) \tag{3.2}$$

that is, the wavefunctions $\psi_{\mathbf{k}}^{(n)}(\mathbf{r})$ and $\psi_{\mathbf{k}}^{*(n)}(\mathbf{r})$ have the same (real) eigenvalue $\epsilon_{\mathbf{k}}^{(n)}$. But we can identify $\psi_{\mathbf{k}}^{*(n)}(\mathbf{r})$ with $\psi_{-\mathbf{k}}^{(m)}(\mathbf{r})$, because:

$$\psi_{-\mathbf{k}}^{(m)}(\mathbf{r}) = \frac{e^{-i\mathbf{k}\cdot\mathbf{r}}}{\sqrt{V}} \sum_{\mathbf{G}} \alpha_{-\mathbf{k}}^{(m)}(\mathbf{G}) e^{i\mathbf{G}\cdot\mathbf{r}} \quad \text{and} \quad \psi_{\mathbf{k}}^{*(n)}(\mathbf{r}) = \frac{e^{-i\mathbf{k}\cdot\mathbf{r}}}{\sqrt{V}} \sum_{-\mathbf{G}} \alpha_{\mathbf{k}}^{*(n)}(-\mathbf{G}) e^{i\mathbf{G}\cdot\mathbf{r}} \tag{3.3}$$

where in the second sum we have used $-\mathbf{G}$ as the symbol for the summation over the infinite set of reciprocal lattice vectors. In the above expressions, the two sums contain exactly the same exponentials, term by term, with the same overall wave-vector $-\mathbf{k}$, as determined by the exponential outside the sum. Therefore, the second wavefunction must be one of the set of wavefunctions corresponding to the wave-vector $-\mathbf{k}$, that is:

$$\alpha_{\mathbf{k}}^{*(n)}(-\mathbf{G}) = \alpha_{-\mathbf{k}}^{(m)}(\mathbf{G}) \Rightarrow \psi_{\mathbf{k}}^{*(n)}(\mathbf{r}) = \psi_{-\mathbf{k}}^{(m)}(\mathbf{r}) \Rightarrow \epsilon_{-\mathbf{k}}^{(m)} = \epsilon_{\mathbf{k}}^{(n)} \tag{3.4}$$

as desired.   ∎

The identification of the two wavefunctions may require a change in the band labels, but this is inconsequential since those labels are purely for keeping track of all the solutions at each wave-vector value and have no other physical meaning. This theorem is relevant to electrons in crystals when each state characterized by $\mathbf{k}$ is occupied by both spin-up and spin-down particles, in which case the spin does not play a major role, other than doubling the occupancy of the $\mathbf{k}$-labeled states. For systems with equal numbers of up and down spins, Kramers' theorem amounts to inversion symmetry in reciprocal space.

A more detailed analysis which takes into account spin states explicitly reveals that for spin-1/2 particles, without spin–orbit coupling, Kramers' theorem becomes:

$$\epsilon_{-\mathbf{k},\downarrow}^{(m)} = \epsilon_{\mathbf{k},\uparrow}^{(n)}, \quad \psi_{-\mathbf{k},\downarrow}^{(m)}(\mathbf{r}) = i\sigma_y \psi_{\mathbf{k},\uparrow}^{*(n)}(\mathbf{r}) \tag{3.5}$$

where σ_y is a Pauli matrix (see Appendix C). This will be discussed in more detail in Chapter 9.

3.2 Crystal Groups: Definitions

Taking full advantage of the crystal symmetries requires the use of group theory. This very interesting branch of mathematics is particularly well suited to reduce the amount of work by effectively using group representations in the description of single-particle eigenfunctions. Here, we will develop some of the basic concepts of group theory including group representations, and we will employ them in simple illustrative examples.

We first define the notion of a group: it is a finite or infinite set of operations, which satisfy the following four conditions.

1. **Closure**: if P and Q are elements of the group, so is $R = PQ$.
2. **Associativity**: if P, Q, R are members of a group, then $(PQ)R = P(QR)$.
3. **Unit element**: there is one element E of the group for which $EP = PE = P$, where P is any other element of the group.
4. **Inverse element**: for every element P of the group, there exists another element called its inverse P^{-1}, for which $PP^{-1} = P^{-1}P = E$.

A **subgroup** is a subset of operations of a group which form a group by themselves (they satisfy the above four conditions). Two groups are called **isomorphic** when there is a one-to-one correspondence between their elements $P_1, P_2, ..., P_n$ and $Q_1, Q_2, ..., Q_n$, such that

$$P_1 P_2 = P_3 \Rightarrow Q_1 Q_2 = Q_3 \tag{3.6}$$

Finally, we define the notion of a **class**. A subset of elements Q_j of the group $\mathcal{G}$ constitutes a class $\mathcal{C}$ if, for all elements Q_i of the group, the following relation holds:

$$S_i \, Q_j \, S_i^{-1} = Q_k \tag{3.7}$$

that is, when we multiply an element of the class Q_j by any other group element S_i (on the left) and by its inverse S_i^{-1} (on the right), the result is an element Q_k which also belongs to the class. In practical terms, we choose an element of the group Q_j and go through all the other elements of the group S_i, multiplying Q_j by S_i and S_i^{-1}, to produce the entire class $\mathcal{C}$ to which Q_j belongs. Implicit in this definition is the fact that a true class cannot be broken into subclasses. We will give examples of groups and classes below, for specific crystals; the notion of a class plays an important role in group representations, which we will discuss in a later section.

For a crystal, we can define the following groups.

(a) $\mathcal{S}$: the space group, which contains all symmetry operations that leave the crystal invariant.
(b) $\mathcal{R}$: the translation group, which contains all the translations that leave the crystal invariant, that is, all the Bravais lattice vectors **R**. $\mathcal{R}$ is a subgroup of $\mathcal{S}$.

(c) $\mathcal{Q}$: the point group, which contains all space-group operations with the translational part set equal to 0. $\mathcal{Q}$ is not necessarily a subgroup of $\mathcal{S}$, because when the translational part is set to 0, some operations may no longer leave the crystal invariant if they involved a translation by a vector not equal to any of the Bravais lattice vectors.

The space group can contain the following types of operations.

- Lattice translations (all the Bravais lattice vectors **R**), which form $\mathcal{R}$.
- Proper rotations by 0 degrees (which corresponds to the identity or unit element, $\{E|0\}$) and by the special angles $\frac{\pi}{3}, \frac{\pi}{2}, \frac{2\pi}{3}, \pi, \frac{4\pi}{3}, \frac{3\pi}{2}, \frac{5\pi}{3}$ around an axis.
- Improper rotations: inversion (denoted by I), or reflection on a plane (denoted by σ).
- Screw axes, which involve a proper rotation and a translation by $\mathbf{t} \neq \mathbf{R}$, parallel to the rotation axis.
- Glide planes, which involve a reflection about a plane and a translation by $\mathbf{t} \neq \mathbf{R}$ along that plane.

Some general remarks on group-theory applications to crystals are in order. We saw that $\mathcal{Q}$ consists of all space-group operators with the translational part set equal to 0. Therefore, $\mathcal{Q} \times \mathcal{R} \neq \mathcal{S}$ in general, because several of the group symmetries may involve non-lattice vector translations $\mathbf{t} \neq \mathbf{R}$. The point group leaves the Bravais lattice invariant but not necessarily the crystal invariant (recall that the crystal is defined by the Bravais lattice and the atomic basis in each unit cell). If the crystal has glide planes and screw rotations, that is, symmetry operations that involve $\{U|\mathbf{t}\}$ with $\mathbf{t} \neq 0$ and $\{E|\mathbf{t}\} \notin \mathcal{R}$, then $\mathcal{Q}$ does not leave the crystal invariant. So why do we bother to consider the point group $\mathcal{Q}$ at all? As we shall see in the next section, even if $\mathcal{Q}$ does *not* leave the crystal invariant, it *does* leave the energy spectrum $\epsilon_{\mathbf{k}}^{(n)}$ invariant! Groups that include such symmetry operations are referred to as **non-symmorphic** groups. A **symmorphic** group is one in which a proper choice of the origin of coordinates eliminates all non-lattice translations, in which case $\mathcal{Q}$ is a true subgroup of $\mathcal{S}$. Finally, all symmetry operators in the space group commute with the hamiltonian, since they leave the crystal, and therefore the external potential, invariant.

Example 3.1   Honeycomb lattice with basis of two *different* atoms: C_{3v} group
This model is described by the real-space lattice vectors $\mathbf{a}_1 = (a/2)[\sqrt{3}\hat{x} + \hat{y}], \mathbf{a}_2 = (a/2)[\sqrt{3}\hat{x} - \hat{y}]$ and reciprocal-space vectors $\mathbf{b}_1 = (2\pi/\sqrt{3}a)[\hat{x} + \sqrt{3}\hat{y}], \mathbf{b}_2 = (2\pi/\sqrt{3}a)[\hat{x} - \sqrt{3}\hat{y}]$, with the positions of the two atoms within the PUC given by $\mathbf{t}_1 = 0, \mathbf{t}_2 = (\mathbf{a}_1 + \mathbf{a}_2)/3$.

As illustrated in Fig. 3.1, the symmetries of this model are:

- the identity E
- rotation by $2\pi/3$ around an axis perpendicular to the plane, C_3
- rotation by $4\pi/3$ around an axis perpendicular to the plane, C_3^2
- reflection on the x-axis, σ_1
- reflection on the axis at $\theta = 2\pi/3$ from the x-axis, σ_2
- reflection on the axis at $\theta = 4\pi/3$ from the x-axis, σ_3.

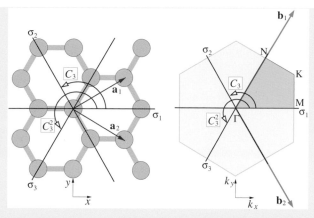

Fig. 3.1

Left: The symmetry operations of the 2D honeycomb lattice with a basis of two *different* atoms (circles colored red and blue). These operations form the C_{3v} group and are labeled $\sigma_1, \sigma_2, \sigma_3$ for the reflections and C_3, C_3^2 for the rotations, which are indicated by the curves with arrows. The red arrows labeled $\mathbf{a}_1, \mathbf{a}_2$ show the primitive lattice constants. **Right**: The irreducible Brillouin zone (yellow rhombus) for the 2D honeycomb lattice with C_{3v} symmetry; the special points are labeled Γ, M, K, N. The green arrows labeled $\mathbf{b}_1, \mathbf{b}_2$ show the reciprocal-space lattice constants.

Table 3.1 Symmetry operations and group multiplication table for symmetries of the C_{3v} group: each entry in the table is the result of taking the product of two operations, with the operation on the left of the row acting *first* and the operation at the top of the column acting *second*. The horizontal and vertical lines in the multiplication table separate the group into the three classes $\mathcal{C}_1, \mathcal{C}_2, \mathcal{C}_3$. We have used the definitions: $u_1 = [x - \sqrt{3}y]/2, u_2 = [x + \sqrt{3}y]/2,$ $v_1 = [\sqrt{3}x - y]/2, v_2 = [\sqrt{3}x + y]/2.$

U	$\{U\|0\}\,(x,y)$	$\hat{O}_{\{U\|0\}}\,(x,y)$	$\mathcal{C}_1$	$\mathcal{C}_2$			$\mathcal{C}_3$	
E	$(\ \ x,\ \ y)$	$(\ \ x,\ \ y)$	E	C_3	C_3^2	σ_1	σ_2	σ_3
C_3	$(-u_1, -v_2)$	$(-u_2,\ \ v_1)$	C_3	C_3^2	E	σ_2	σ_3	σ_1
C_3^2	$(-u_2,\ \ v_1)$	$(-u_1, -v_2)$	C_3^2	E	C_3	σ_3	σ_1	σ_2
σ_1	$(\ \ x, -y)$	$(\ \ x, -y)$	σ_1	σ_3	σ_2	E	C_3^2	C_3
σ_2	$(\ \ u_1, -v_2)$	$(\ \ u_1, -v_2)$	σ_2	σ_1	σ_3	C_3	E	C_3^2
σ_3	$(\ \ u_2,\ \ v_1)$	$(\ \ u_2,\ \ v_1)$	σ_3	σ_2	σ_1	C_3^2	C_3	E

These symmetries constitute the point group $\mathcal{Q}$ for this physical system, called C_{3v}. We assume that the origin is at the position of one of the two different atoms, so that the group is symmorphic. This point group has the group multiplication table given in Table 3.1: an entry in this table is the result of multiplying the element at the top of its column with the element at the left of its row. The definition of each element in terms of its action on an arbitrary vector (x, y) on the plane is also given in the left half of Table 3.1. This group multiplication table can be used to prove that all the group properties are satisfied. It is also very useful in identifying the classes in the group:

the element E is in a class by itself, as is always the case for any group; the elements C_3, C_3^2 form a second class; the elements $\sigma_1, \sigma_2, \sigma_3$ form a third class. The separation of the group into these three classes can easily be verified by applying the definition in Eq. (3.7).

3.3 Symmetries of 3D Crystals

In three dimensions there are 14 different types of Bravais lattices, 32 different types of point groups, and a total of 230 different space groups, of which 73 are symmorphic and 157 are non-symmorphic. The 14 Bravais lattices are grouped into six crystal systems called triclinic, monoclinic, orthorhombic, tetragonal, hexagonal, and cubic, in order of increasing symmetry. The hexagonal system is often split into two parts, called the trigonal and hexagonal subsystems. The definitions of these systems in terms of the relations between the sides of the unit cell (a, b, c) and the angles between them (with $\alpha = $ the angle between sides a, c; $\beta = $ the angle between sides b, c; and $\gamma = $ the angle between sides a, b) are given in Table 3.2. Figure 3.2 shows the conventional unit cells for the 14 Bravais lattices. Each corner of the cell is occupied by sites that are equivalent by translational symmetry. When there are no other sites equivalent to the corners, the conventional cell is the same as the primitive cell for this lattice and it is designated P. When the cell has an equivalent site at its geometric center (a body centered cell) it is designated I for the implied inversion symmetry; the primitive unit cell in this case is one-half the conventional cell. When the cell has an equivalent site at the center of one face it is designated C (by translational symmetry it must also have an equivalent site at the opposite face); the primitive unit cell in this case is one-half the conventional cell. When the cell has an equivalent site at the center of each face (a face centered cell) it is designated F; the

Table 3.2 The seven crystal systems (first column) and the associated 32 point groups (last column) for crystals in three dimensions. The relations between the cell sides (a, b, c) and cell angles (α, β, γ) are shown and the corresponding lattices are labeled P = primitive, I = body centered, C = side centered, F = face centered, R = rhombohedral.

System	Cell sides	Cell angles	Lattices	Point groups
Triclinic	$a \neq b \neq c$	$\alpha \neq \beta \neq \gamma$	P	C_1, C_i
Monoclinic	$a \neq b \neq c$	$\alpha = \beta = \frac{\pi}{2} \neq \gamma$	P, C	C_2, C_s, C_{2h}
Orthorhombic	$a \neq b \neq c$	$\alpha = \beta = \gamma = \frac{\pi}{2}$	P, I, C, F	D_2, D_{2h}, C_{2v}
Tetragonal	$a = b \neq c$	$\alpha = \beta = \gamma = \frac{\pi}{2}$	P, I	$C_4, S_4, C_{4h}, D_4,$ C_{4v}, D_{2d}, D_{4h}
Trigonal	$a = b = c$	$\alpha = \beta = \gamma \neq \frac{\pi}{2}$	P	$C_3, C_{3i}, C_{3v}, D_3, D_{3d}$
Hexagonal	$a = b \neq c$	$\alpha = \beta = \frac{\pi}{2}, \gamma = \frac{2\pi}{3}$	P, R	$C_6, C_{3h}, C_{6h}, C_{6v},$ D_{3h}, D_6, D_{6h}
Cubic	$a = b = c$	$\alpha = \beta = \gamma = \frac{\pi}{2}$	P, I, F	T, T_h, O, T_d, O_h

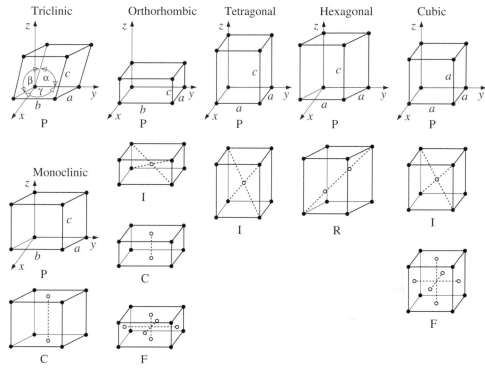

Fig. 3.2 The conventional unit cells of the 14 Bravais lattices in the six systems in three dimensions; the trigonal system unit cell is included in the conventional cell of the hexagonal system. The dots indicate equivalent sites in the unit cell: those at the corners are equivalent by translational symmetry, the rest indicate the presence of additional symmetries implying a primitive unit cell smaller than the conventional one. Dashed lines within the cells connect equivalent points for easier visualization. The cells are labeled P, I, C, F, R, according to the convention of Table 3.2. For the cells labeled P in each system, we also indicate the cartesian coordinate axes, labeled x, y, z, and the length of the unit cell sides, as a, b, c; in the triclinic cell, we also show the cell angles α, β, γ between the cell sides.

primitive unit cell in this case is one-quarter the conventional cell. Finally, in the hexagonal system there exists a lattice which has two more equivalent sites inside the conventional cell, at height $c/3$ and $2c/3$ along the main diagonal (see Fig. 3.2), and it is designated R for rhombohedral; the primitive unit cell in this case is one-third the conventional cell.

A number of possible point groups are associated with each of the 14 Bravais lattices, depending on the symmetry of the basis. There are 32 point groups in all, denoted by the following names:

- C for "cyclic," when there is a single axis of rotation; a number subscript m indicates the m-fold symmetry around this axis (m can be 2, 3, 4, 6).
- D for "dihedral," when there are twofold axes at right angles to another axis.
- T for "tetrahedral," when there are four sets of rotation axes of threefold symmetry, as in a tetrahedron (see Fig. 3.3).
- O for "octahedral," when there are fourfold rotation axes combined with perpendicular twofold rotation axes, as in an octahedron (see Fig. 3.3).

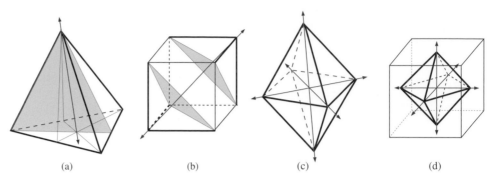

(a)   (b)   (c)   (d)

Fig. 3.3 Illustration of symmetry axes, shown as double-headed red arrows, of the tetrahedron and the octahedron. (a) A threefold rotation axis of the *regular* tetrahedron passes through one of its corners and the geometric center of the equilateral triangle directly across it; there are four such axes in the regular tetrahedron, each passing through one of its corners. One symmetry plane, passing through one of the tetrahedron edges and the center of the opposite edge, is highlighted in pink. (b) Two tetrahedra (in this case *not* regular ones) with a common axis of threefold rotational symmetry in the cube are identified by thicker red lines; the triangular bases of these tetrahedra, which lie entirely within the cube, are highlighted in pink. There are four pairs of such tetrahedra in the cube, whose threefold rotational symmetry axes are the main diagonals of the cube. (c) Three different symmetry axes of an octahedron; when the edges of the horizontal square are not of the same length as the other edges, the vertical axis corresponds to fourfold rotation, while the two horizontal axes correspond to twofold rotation. When all edges are of equal length, all three axes correspond to fourfold rotational symmetry. (d) One such octahedron, with all edges equal and three fourfold rotation axes, shown in thicker lines, is inscribed in the cube.

In all these cases the existence of additional mirror-plane symmetries is denoted by a second subscript which can be h for "horizontal," v for "vertical," or d for "diagonal" planes relative to the rotation axes. Inversion symmetry is denoted by the letter subscript i. In the case of a onefold rotation axis, when inversion symmetry is present the notation C_i is adopted (instead of what would normally be called C_{1i}), while when a mirror-plane symmetry is present the notation C_s is used (instead of what would normally be called C_{1h}). Finally, the group generated by the symmetry operation of $\pi/2$ rotation followed by reflection on a vertical plane, the so-called "roto-reflection" group, is denoted by S_4 (this is different from C_{4v}, in which the $\pi/2$ rotation and the reflection on a vertical plane are independently symmetry operations). This set of conventions for describing the 32 crystallographic groups is referred to as the Schoenflies notation; the names of the groups in this notation are given in Table 3.2. There is a somewhat more rational set of conventions for naming crystallographic point groups described in the *International Tables for Crystallography*. This scheme is more complicated; we refer the interested reader to those tables for further details.

There exists a useful way of visualizing the symmetry operations of the various point groups, referred to as "stereograms." These consist of 2D projections of a point on the surface of a sphere and its images generated by acting on the sphere with the various symmetry operations of the point group. In order to illustrate the action of the symmetry operations, the initial point is chosen as some arbitrary point on the sphere, that is, it does not belong to any special axis or plane of symmetry. The convention for the equivalent

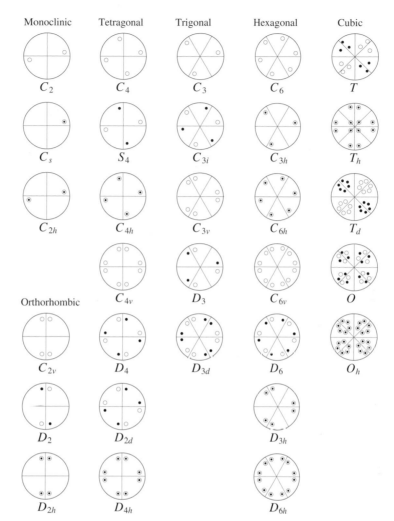

Fig. 3.4 Stereograms for 30 point groups in three dimensions. The lines within the circles are visual aids, which in several cases correspond to reflection planes. The trigonal and hexagonal subsystems are shown separately, and the triclinic system (groups C_1 and C_i) is not shown at all since it has trivial representations.

points on the sphere is that those in the northern hemisphere are drawn as open circles, while those in the southern hemisphere are shown as solid dots and the sphere is viewed from the northern pole. In Fig. 3.4 we show the stereograms for 30 point groups belonging to six crystal systems; the stereograms for the triclinic system (point groups C_1 and C_i) are not shown because they are trivial: they contain only one point each, since these groups do not have any symmetry operations other than the trivial ones (identity and inversion).

As an illustration, we discuss briefly the symmetries of the cube, which has the highest degree of symmetry compatible with 3D periodicity. When only twofold, threefold, and fourfold rotations are allowed, as required by 3D periodicity (see Problem 1), there are 24

Table 3.3 The 24 symmetry operations of the cube that involve rotations by π (the classes $\mathcal{C}_5$ and $\mathcal{C}_2$), $\pi/3$ (the class $\mathcal{C}_3$), $\pi/2$ (the class $\mathcal{C}_4$), and 2π (the identity E, which forms a class $\mathcal{C}_1$ by itself). 24 more symmetry operations can be generated by combining each of the rotations with inversion I, which corresponds to $(-x, -y, -z)$, that is, a change of all signs.

$\mathcal{C}_1$	E	x	y	z		$U_{x/4}$	x	z	$-y$
	$U_{x/2}$	x	$-y$	$-z$		$\overline{U}_{x/4}$	x	$-z$	y
$\mathcal{C}_5$	$U_{y/2}$	$-x$	y	$-z$	$\mathcal{C}_4$	$U_{y/4}$	$-z$	y	x
	$U_{z/2}$	$-x$	$-y$	z		$\overline{U}_{y/4}$	z	y	$-x$
	U_1	y	z	x		$U_{z/4}$	y	$-x$	z
	$\overline{U}_1$	z	x	y		$\overline{U}_{z/4}$	$-y$	x	z
	U_2	z	$-x$	$-y$		$U_{\bar{x}/2}$	$-x$	z	y
$\mathcal{C}_3$	$\overline{U}_2$	$-y$	$-z$	x		$U'_{\bar{x}/2}$	$-x$	$-z$	$-y$
	U_3	$-z$	x	$-y$	$\mathcal{C}_2$	$U_{\bar{y}/2}$	z	$-y$	x
	$\overline{U}_3$	y	$-z$	$-x$		$U'_{\bar{y}/2}$	$-z$	$-y$	$-x$
	U_4	$-y$	z	$-x$		$U_{\bar{z}/2}$	y	x	$-z$
	$\overline{U}_4$	$-z$	$-x$	y		$U'_{\bar{z}/2}$	$-y$	$-x$	$-z$

images of an arbitrary point in space (x, y, z): these are given in Table 3.3. The associated 24 operations are separated into five classes:

1. The identity E.
2. Twofold (π) rotations around the axes x, y, z denoted as a class by $\mathcal{C}_4^2$ and as individual operations by $U_{v/2}$ ($v = x, y, z$).
3. Fourfold ($2\pi/4$) rotations around the axes x, y, z denoted as a class by $\mathcal{C}_4$ and as individual operations by $U_{v/4}$ ($v = x, y, z$) for counter-clockwise or by $\overline{U}_{v/4}$ for clockwise rotation.
4. Threefold ($2\pi/3$) rotations around the main diagonals of the cube denoted as a class by $\mathcal{C}_3$ and as individual operations by U_n ($n = 1, 2, 3, 4$) for counter-clockwise or by $\overline{U}_n$ for clockwise rotation.
5. Twofold (π) rotations around axes that are perpendicular to the x, y, z axes and bisect the in-plane angles between the cartesian axes, denoted as a class by $\mathcal{C}_2$ and as individual operations by $U_{\bar{v}/2}$ ($v = x, y, z$) or $U'_{\bar{v}/2}$ with the subscript indicating the cartesian axis perpendicular to the rotation axis.

These symmetry operations are represented schematically in Fig. 3.5. When inversion is added to these operations, the total number of images of the arbitrary point becomes 48, since inversion can be combined with any rotation to produce a new operation. It is easy to rationalize this result: there are 48 different ways of rearranging the three cartesian components of the arbitrary point, including changes of sign. The first coordinate can be any of six possibilities ($\pm x, \pm y, \pm z$), the second coordinate can be any of four possibilities, and the last coordinate any of two possibilities.

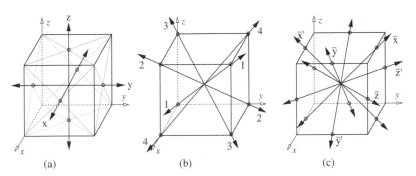

Fig. 3.5   Symmetry operations of the cube. (a) The rotation axes for the C_4 (fourfold rotations) and C_5 (twofold rotations) classes, for rotations around axes that pass through the centers of opposite faces of the cube. (b) The C_3 class consisting of threefold rotation axes which correspond to the main diagonals of the cube (arrows labeled 1–4). (b) The C_2 class consisting of twofold rotation axes that are perpendicular to the cartesian axes, and bisect the angles between them. All rotation axes are shown as double-headed arrows and the points at which they intersect the cube [face centers in (a), corners in (b), edge centers in (c)] are indicated by small red circles.

If these 48 operations are applied with respect to the center of the cube, then the cube remains invariant. Therefore, in a crystal with a cubic Bravais lattice and a basis that does not break the cubic symmetry (such as a single atom per unit cell referred to as simple cubic, or an FCC or a BCC lattice), all 48 symmetry operations of the cube will leave the crystal invariant and therefore all these operations form the point group of the crystal. This is the largest point group compatible with 3D periodicity. There are other, larger point groups in three dimensions but they are not compatible with 3D periodicity. For example, the icosahedral group has a total of 120 operations, but its fivefold symmetry is not compatible with translational periodicity in three dimensions.

3.4 Symmetries of the Band Structure

To apply symmetry operations on various functions, we introduce the operators corresponding to the symmetries. These will be symbolized by $\{U|\mathbf{t}\}$, where U corresponds to a proper or improper rotation and $\mathbf{t}$ to a translation. These operators act on real-space vectors according to the rule

$$\{U|\mathbf{t}\}\mathbf{r} \equiv U\mathbf{r} + \mathbf{t} \tag{3.8}$$

In terms of these operators, the translations can be described as $\{E|\mathbf{R}\}$ (where E is the identity, or unit element), while the pure rotations (proper or improper) are described as $\{U|0\}$. The first set corresponds to the translation group $\mathcal{R}$, the second to the point group $\mathcal{Q}$. We note that the proper or improper rotations leave the Bravais lattice invariant, that is:

$$\{U|0\}\mathbf{R} = U\mathbf{R} = \mathbf{R}', \quad \mathbf{R}, \mathbf{R}' \in \mathcal{R} \tag{3.9}$$

For any two symmetry operators belonging to the space group:

$$\{U_1|\mathbf{t}_1\}\{U_2|\mathbf{t}_2\}\mathbf{r} = \{U_1|\mathbf{t}_1\}[U_2\mathbf{r} + \mathbf{t}_2] = U_1 U_2 \mathbf{r} + U_1 \mathbf{t}_2 + \mathbf{t}_1 \tag{3.10}$$

which means that the operator $\{U_1 U_2|U_1\mathbf{t}_2 + \mathbf{t}_1\}$ corresponds to an element of the space group. Using this rule for multiplication of operators, we can easily show that the inverse of $\{U|\mathbf{t}\}$ is:

$$\{U|\mathbf{t}\}^{-1} = \{U^{-1}| - U^{-1}\mathbf{t}\} \tag{3.11}$$

The validity of this can be verified by applying $\{U^{-1}| - U^{-1}\mathbf{t}\}\{U|\mathbf{t}\}$ on $\mathbf{r}$, whereupon we get back $\mathbf{r}$. We can use the expression for the inverse of $\{U|\mathbf{t}\}$ to prove the following relation for the inverse of a product of operators in terms of their inverses:

$$\{U_2|\mathbf{t}_2\}^{-1}\{U_1|\mathbf{t}_1\}^{-1} = [\{U_1|\mathbf{t}_1\}\{U_2|\mathbf{t}_2\}]^{-1} \tag{3.12}$$

which can easily be proved from the definition of the inverse element in Eq. (3.11).

In the following we will need to apply symmetry operations to functions of the space variable $\mathbf{r}$, so we define a new set of operators whose effect is to change $\mathbf{r}$:

$$\hat{\mathcal{O}}_{\{U|\mathbf{t}\}} f(\mathbf{r}) \equiv f(\{U|\mathbf{t}\}^{-1}\mathbf{r}) \tag{3.13}$$

In this definition of $\hat{\mathcal{O}}_{\{U|\mathbf{t}\}}$, the action is always on the vector $\mathbf{r}$ itself.

Theorem   *The group formed by the operators $\hat{\mathcal{O}}_{\{U|\mathbf{t}\}}$ is isomorphic to the group of operators* $\{U|\mathbf{t}\}$.

Proof   We note that

$$\hat{\mathcal{O}}_{\{U_1|\mathbf{t}_1\}} \hat{\mathcal{O}}_{\{U_2|\mathbf{t}_2\}} f(\mathbf{r}) = \hat{\mathcal{O}}_{\{U_1|\mathbf{t}_1\}} f(\{U_2|\mathbf{t}_2\}^{-1}\mathbf{r}) = \hat{\mathcal{O}}_{\{U_1|\mathbf{t}_1\}} g(\mathbf{r})$$
$$= g(\{U_1|\mathbf{t}_1\}^{-1}\mathbf{r}) = g(\mathbf{r}')$$

where we have introduced the definitions

$$g(\mathbf{r}) = f(\{U_2|\mathbf{t}_2\}^{-1}\mathbf{r}) \quad \text{and} \quad \mathbf{r}' = \{U_1|\mathbf{t}_1\}^{-1}\mathbf{r}$$

to facilitate the proof. Using these definitions, we have

$$g(\mathbf{r}') = f(\{U_2|\mathbf{t}_2\}^{-1}\mathbf{r}') = f(\{U_2|\mathbf{t}_2\}^{-1}\{U_1|\mathbf{t}_1\}^{-1}\mathbf{r})$$
$$= f([\{U_1|\mathbf{t}_1\}\{U_2|\mathbf{t}_2\}]^{-1}\mathbf{r}) = \hat{\mathcal{O}}_{\{U_1|\mathbf{t}_1\}\{U_2|\mathbf{t}_2\}} f(\mathbf{r})$$
$$\Rightarrow \hat{\mathcal{O}}_{\{U_1|\mathbf{t}_1\}} \hat{\mathcal{O}}_{\{U_2|\mathbf{t}_2\}} = \hat{\mathcal{O}}_{\{U_1|\mathbf{t}_1\}\{U_2|\mathbf{t}_2\}} \tag{3.14}$$

The last equality in Eq. (3.14) proves the isomorphism between the groups $\{U|\mathbf{t}\}$ and $\hat{\mathcal{O}}_{\{U|\mathbf{t}\}}$, since their elements satisfy the condition of Eq. (3.6) in an obvious correspondence between elements of the two groups.   ∎

Having defined the basic formalism for taking advantage of the crystal symmetries, we will now apply it to simplify the description of the eigenfunctions and eigenvalues of the single-particle hamiltonian.

Theorem *For any element of the space group $\{U|t\} \in \mathcal{S}$, when we apply it to an eigenstate of the hamiltonian $\psi_{\mathbf{k}}^{(n)}(\mathbf{r})$ with eigenvalue $\epsilon_{\mathbf{k}}^{(n)}$, we obtain a new state which we will denote as $\tilde{\psi}_{\mathbf{q}}^{(n)}(\mathbf{r})$ with eigenvalue $\tilde{\epsilon}_{\mathbf{q}}^{(n)}$ such that:*

$$\tilde{\epsilon}_{\mathbf{q}}^{(n)} = \epsilon_{U\mathbf{k}}^{(n)} = \epsilon_{\mathbf{k}}^{(n)} \quad \text{and} \quad \tilde{\psi}_{\mathbf{q}}^{(n)}(\mathbf{r}) = \hat{O}_{\{U|t\}} \psi_{\mathbf{k}}^{(n)}(\mathbf{r}) = \tilde{\psi}_{U\mathbf{k}}^{(n)}(\mathbf{r}) = \psi_{\mathbf{k}}^{(n)}(U^{-1}\mathbf{r} - U^{-1}t)$$

$$(3.15)$$

Proof We need to prove first, that $\psi_{\mathbf{k}}^{(n)}(\mathbf{r})$ and $\tilde{\psi}_{\mathbf{q}}^{(n)}(\mathbf{r})$ are eigenstates of the hamiltonian with the same eigenvalue and second, that $\tilde{\psi}_{\mathbf{q}}^{(n)}(\mathbf{r})$ is a Bloch state of wave-vector $\mathbf{q} = U\mathbf{k}$. We first note that the symmetry operators of the space group commute with the hamiltonian, by their definition (they leave the crystal, hence the total potential invariant):

$$\hat{O}_{\{U|t\}} \left[\hat{\mathcal{H}}^{\mathrm{sp}} \psi_{\mathbf{k}}^{(n)}(\mathbf{r}) \right] = \hat{O}_{\{U|t\}} \left[\epsilon_{\mathbf{k}}^{(n)} \psi_{\mathbf{k}}^{(n)}(\mathbf{r}) \right] \Rightarrow \hat{\mathcal{H}}^{\mathrm{sp}} \left[\hat{O}_{\{U|t\}} \psi_{\mathbf{k}}^{(n)}(\mathbf{r}) \right]$$

$$= \epsilon_{\mathbf{k}}^{(n)} \left[\hat{O}_{\{U|t\}} \psi_{\mathbf{k}}^{(n)}(\mathbf{r}) \right]$$

which shows that $\psi_{\mathbf{k}}^{(n)}(\mathbf{r})$ and $\tilde{\psi}_{\mathbf{q}}^{(n)}(\mathbf{r})$ are eigenfunctions with the same eigenvalue, proving the first relation in Eq. (3.15). To prove the second relation, we apply the operator $\hat{O}_{\{E|-\mathbf{R}\}}$ to the function $\tilde{\psi}_{\mathbf{q}}^{(n)}(\mathbf{r})$: this is simply a translation operator (E is the identity), and therefore changes the argument of a function by $+\mathbf{R}$. We expect to find that this function is then changed by a phase factor $\exp[iU\mathbf{k} \cdot \mathbf{R}]$, if indeed it is a Bloch state of wave-vector $U\mathbf{k}$. To prove this we note:

$$\{U|t\}^{-1}\{E| - \mathbf{R}\}\{U|t\} = \{U|t\}^{-1}\{U|t - \mathbf{R}\} = \{U^{-1}| - U^{-1}t\}\{U|t - \mathbf{R}\}$$

$$= \{E|U^{-1}t - U^{-1}\mathbf{R} - U^{-1}t\} = \{E| - U^{-1}\mathbf{R}\}$$

and multiplying the initial and final parts of this equation by $\{U|t\}$ from the left, we get

$$\{E| - \mathbf{R}\}\{U|t\} = \{U|t\}\{E| - U^{-1}\mathbf{R}\} \Rightarrow \hat{O}_{\{E|-\mathbf{R}\}\{U|t\}} = \hat{O}_{\{U|t\}\{E|-U^{-1}\mathbf{R}\}}$$

due to isomorphism of the groups $\{U|t\}$ and $\hat{O}_{\{U|t\}}$. When we apply the operators in this last equation to $\psi_{\mathbf{k}}^{(n)}(\mathbf{r})$, we find:

$$\hat{O}_{\{E|-\mathbf{R}\}}\hat{O}_{\{U|t\}} \psi_{\mathbf{k}}^{(n)}(\mathbf{r}) = \hat{O}_{\{U|t\}}\hat{O}_{\{E|-U^{-1}\mathbf{R}\}} \psi_{\mathbf{k}}^{(n)}(\mathbf{r}) \Rightarrow$$

$$\hat{O}_{\{E|-\mathbf{R}\}}\tilde{\psi}_{\mathbf{q}}^{(n)}(\mathbf{r}) = \hat{O}_{\{U|t\}} \psi_{\mathbf{k}}^{(n)}(\mathbf{r} + U^{-1}\mathbf{R})$$

Using the fact that $\psi_{\mathbf{k}}^{(n)}(\mathbf{r})$ is a Bloch state, we obtain:

$$\psi_{\mathbf{k}}^{(n)}(\mathbf{r} + U^{-1}\mathbf{R}) = e^{i\mathbf{k} \cdot U^{-1}\mathbf{R}} e^{i\mathbf{k}\cdot\mathbf{r}} u^{(n)}(\mathbf{r} + U^{-1}\mathbf{R}) = e^{i\mathbf{k} \cdot U^{-1}\mathbf{R}} e^{i\mathbf{k}\cdot\mathbf{r}} u^{(n)}(\mathbf{r} + \mathbf{R}')$$

$$= e^{iU\mathbf{k}\cdot\mathbf{R}} \psi_{\mathbf{k}}^{(n)}(\mathbf{r})$$

where we have used the fact that $\mathbf{k} \cdot U^{-1}\mathbf{R} = U\mathbf{k} \cdot \mathbf{R}$, that is, a dot product of two vectors does not change if we rotate both vectors by U. We have also taken advantage of the fact that the operation $\{U|0\}^{-1}$ on $\mathbf{R}$ produces another Bravais lattice vector $\mathbf{R}'$, Eq. (3.9), and that $u^{(n)}(\mathbf{r})$ is the part of the Bloch state that has the full periodicity of the Bravais lattice, $u^{(n)}(\mathbf{r} + \mathbf{R}') = u^{(n)}(\mathbf{r})$. When the last result is substituted in the previous equation, we get:

$$\hat{O}_{\{E|-\mathbf{R}\}}\tilde{\psi}_{\mathbf{q}}^{(n)}(\mathbf{r}) = e^{iU\mathbf{k}\cdot\mathbf{R}}\tilde{\psi}_{\mathbf{q}}^{(n)}(\mathbf{r})$$

which proves that $\tilde{\psi}_{\mathbf{q}}^{(n)}(\mathbf{r})$, defined in Eq. (3.15), is indeed a Bloch state of wave-vector $\mathbf{q} = U\mathbf{k}$. Therefore, the corresponding eigenvalue must be $\epsilon_{U\mathbf{k}}^{(n)}$, which is equal to $\epsilon_{\mathbf{k}}^{(n)}$ since we have already proven that the two Bloch states $\psi_{\mathbf{k}}^{(n)}(\mathbf{r})$ and $\tilde{\psi}_{\mathbf{q}}^{(n)}(\mathbf{r})$ have the same eigenvalue. Thus, the energy spectrum $\epsilon_{\mathbf{k}}^{(n)}$ has the full symmetry of the point group $\mathcal{Q} = [\{U|0\}]$. Furthermore, we have shown that the two Bloch states at $\mathbf{k}$ and $U\mathbf{k}$ are related by:

$$\tilde{\psi}_{U\mathbf{k}}^{(n)}(\mathbf{r}) = \hat{O}_{\{U|\mathbf{t}\}}\psi_{\mathbf{k}}^{(n)}(\mathbf{r}) = \psi_{\mathbf{k}}^{(n)}(\{U|\mathbf{t}\}^{-1}\mathbf{r}) = \psi_{\mathbf{k}}^{(n)}(U^{-1}\mathbf{r} - U^{-1}\mathbf{t})$$

which completes our proof of the relations in Eq. (3.15).   ∎

The meaning of the theorem we just proved is that from the states and their eigenvalues at the wave-vector value $\mathbf{k}$ we can obtain the states and corresponding eigenvalues at the wave-vector values $U\mathbf{k}$, related to the original value by the elements of the point group $\mathcal{Q} = \{\{U|0\}\}$. The set of vectors $U\mathbf{k}$ for $\{U|0\} \in \mathcal{Q}$ is called the star of $\mathbf{k}$. Another way to connect the eigenvalues and eigenstates at $\mathbf{k}$ to those at $U\mathbf{k}$ is through the hamiltonian that applies to $u_{\mathbf{k}}^{(n)}(\mathbf{r})$, the part of the Bloch wavefunction that has the full translational symmetry of the Bravais lattice:

$$\left[\frac{(\mathbf{p} + \hbar\mathbf{k})^2}{2m_e} + \mathcal{V}(\mathbf{r})\right]u_{\mathbf{k}}^{(n)}(\mathbf{r}) = \hat{\mathcal{H}}_{\mathbf{k}}^{sp}u_{\mathbf{k}}^{(n)}(\mathbf{r}) = \epsilon_{\mathbf{k}}^{(n)}u_{\mathbf{k}}^{(n)}(\mathbf{r}) \qquad (3.16)$$

with $\mathcal{V}(\mathbf{r})$ the single-particle potential that has the full symmetry of the crystal. We first note that since $\{U|\mathbf{t}\} \in \mathcal{S}$, $\hat{O}_{\{U|\mathbf{t}\}}$ leaves the potential invariant: $\hat{O}_{\{U|\mathbf{t}\}}\mathcal{V}(\mathbf{r}) = \mathcal{V}(U^{-1}\mathbf{r} - U^{-1}\mathbf{t}) = \mathcal{V}(\mathbf{r})$. We will find the following three relations useful:

$$\text{(i)} \quad \hat{O}_{\{U|\mathbf{t}\}}\mathbf{p} = U^{-1}\mathbf{p}$$

because $\mathbf{p} = -i\hbar\nabla_{\mathbf{r}}$, which means that the translation by $\mathbf{t}$ does not affect $\mathbf{p}$;

$$\text{(ii)} \quad \mathbf{k} \cdot (U^{-1}\mathbf{p}) = (U\mathbf{k}) \cdot \mathbf{p}$$

because when both vectors $\mathbf{k}$ and $U^{-1}\mathbf{p}$ are rotated by U, their dot product does not change;

$$\text{(iii)} \quad (U\mathbf{a})^2 = (U^{-1}\mathbf{a})^2 = \mathbf{a}^2$$

since the magnitude of a vector does not change when it is rotated by any proper or improper rotation operator U (or by its inverse U^{-1}). We apply the operator $\hat{O}_{\{U|\mathbf{t}\}}$ to $\hat{\mathcal{H}}_{\mathbf{k}}^{sp}f(\mathbf{r})$ where $f(\mathbf{r})$ is an arbitrary function, and use the invariability of the potential term and the relations (i), (ii), (iii) derived above to find:

$$\hat{O}_{\{U|\mathbf{t}\}}\left[\hat{\mathcal{H}}_{\mathbf{k}}^{sp}f(\mathbf{r})\right] = \left[\frac{\mathbf{p}^2}{2m_e} + \frac{\hbar\mathbf{k} \cdot (U^{-1}\mathbf{p})}{m_e} + \frac{\hbar^2\mathbf{k}^2}{2m_e} + \mathcal{V}(\mathbf{r})\right]f(U^{-1}\mathbf{r} - \mathbf{t})$$

$$= \left[\frac{\mathbf{p}^2}{2m_e} + \frac{\hbar(U\mathbf{k}) \cdot \mathbf{p}}{m_e} + \frac{\hbar^2(U\mathbf{k})^2}{2m_e} + \mathcal{V}(\mathbf{r})\right]\left[\hat{O}_{\{U|\mathbf{t}\}}f(\mathbf{r})\right]$$

$$\Rightarrow \hat{O}_{\{U|\mathbf{t}\}}\hat{\mathcal{H}}_{\mathbf{k}}^{sp} = \hat{\mathcal{H}}_{U\mathbf{k}}^{sp}\hat{O}_{\{U|\mathbf{t}\}} \qquad (3.17)$$

This result shows that $\hat{O}_{\{U|\mathbf{t}\}}$ does not commute with $\hat{\mathcal{H}}_{\mathbf{k}}^{sp}$, because it changes the value of $\mathbf{k}$ to $U\mathbf{k}$; we can use this to obtain the same relations between the eigenvalues and eigenfunctions at the two values, $\mathbf{k}$ and $U\mathbf{k}$, as those that we derived earlier, Eq. (3.15) – we leave the proof of this to the reader (see Problem 5) We note that for the group of operations $\{U|0\}$ that leave $\mathbf{k}$ invariant up to a reciprocal space vector $\mathbf{G}$, $U\mathbf{k} = \mathbf{k} + \mathbf{G}$, the hamiltonian $\hat{\mathcal{H}}_{\mathbf{k}}^{\mathrm{sp}}$, defined in Eq. (3.16), is also left invariant.

The statements we have proved above on the symmetries of the band structure offer a way to achieve great savings in computation, since we need only solve the single-particle equations in a small portion of the Brillouin zone which can then be used to obtain all the remaining solutions. This portion of the Brillouin zone, which can be "unfolded" by the symmetry operations of the point group to give us all the solutions in the entire Brillouin zone, is called the irreducible Brillouin zone (IBZ).

Example 3.2   Square lattice with one atom per unit cell: C_{4v} group

The symmetries for this model, described by the real-space lattice vectors $\mathbf{a}_1 = a\hat{\mathbf{x}}$, $\mathbf{a}_2 = a\hat{\mathbf{y}}$ and reciprocal-space vectors $\mathbf{b}_1 = (2\pi/a)\hat{\mathbf{x}}$, $\mathbf{b}_2 = (2\pi/a)\hat{\mathbf{y}}$ are:

- The identity E.
- Rotation by $(2\pi)/4$ around an axis perpendicular to the plane, C_4.
- Rotation by $(2\pi)/2$ around an axis perpendicular to the plane, C_2.
- Rotation by $3(2\pi)/4$ around an axis perpendicular to the plane, C_4^3.
- Reflection on the x-axis, σ_x.
- Reflection on the y-axis, σ_x.
- Reflection on the axis at $\theta = \pi/4$ from the x-axis, σ_1.
- Reflection on the axis at $\theta = 3\pi/4$ from the x-axis, σ_3.

These symmetries constitute the point group $\mathcal{Q}$ for this physical system. We assume that the origin is at the position of the atom, so that the group is symmorphic. This point group has the group multiplication table given in Table 3.4: an entry in this table is the result of multiplying the element at the top of its column by the element at the left of its row. The definition of each element in terms of its action on an arbitrary vector (x, y) on the plane is also given in the left half of Table 3.4. This group multiplication table can be used to prove that all the group properties are satisfied. It is also very useful in identifying the classes in the group: the element E is in a class by itself, as is always the case for any group; the elements C_4, C_4^3 form a second class; the element C_2 is in a class by itself; the elements σ_x, σ_y form a fourth class; the elements σ_1, σ_3 form a fifth class. The separation of the group into these three classes can easily be verified by applying the definition in Eq. (3.7).

For this system the point group is a subgroup of the space group, because it is a symmorphic group and therefore the space group $\mathcal{S}$ is given by $\mathcal{S} = \mathcal{Q} \times \mathcal{S}$, where $\mathcal{S}$ is the translation group. Using these symmetries we deduce that the symmetries of the point group give an IBZ which is 1/8 that of the first BZ, shown in Fig. 3.6.

There are a few high-symmetry points in this IBZ (in (k_x, k_y) notation):

(1) $\Gamma = (0, 0)$, which has the full symmetry of the point group.
(2) $M = (1, 1)(\pi/a)$, which also has the full symmetry of the point group.

Table 3.4 Group multiplication table for symmetries of the 2D square lattice: each entry in the table is the result of taking the product of two operations, with the operation on the left of the row acting *first* and the operation at the top of the column acting *second*. The horizontal and vertical lines in the multiplication table separate the group into the five classes, $\mathcal{C}_1, \mathcal{C}_2, \mathcal{C}_3, \mathcal{C}_4, \mathcal{C}_5$.

U	$\{U\|0\}\,(x,y)$	$\hat{\mathcal{O}}_{\{U\|0\}}\,(x,y)$	$\mathcal{C}_1$	$\mathcal{C}_2$	$\mathcal{C}_3$	$\mathcal{C}_4$		$\mathcal{C}_5$	
E	$(\ x,\ \ y)$	$(\ x,\ \ y)$	E	C_2	C_4	C_4^3	σ_x	σ_y	σ_1	σ_3
C_2	$(-x,-y)$	$(-x,-y)$	C_2	E	C_4^3	C_4	σ_y	σ_x	σ_3	σ_1
C_4	$(-y,\ \ x)$	$(\ y,-x)$	C_4	C_4^3	C_2	E	σ_3	σ_1	σ_x	σ_y
C_4^3	$(\ y,-x)$	$(-y,\ \ x)$	C_4^3	C_4	E	C_2	σ_1	σ_3	σ_y	σ_x
σ_x	$(\ x,-y)$	$(\ x,-y)$	σ_x	σ_y	σ_1	σ_3	E	C_2	C_4	C_4^3
σ_y	$(-x,\ \ y)$	$(-x,\ \ y)$	σ_y	σ_x	σ_3	σ_1	C_2	E	C_4^3	C_4
σ_1	$(\ y,\ \ x)$	$(\ y,\ \ x)$	σ_1	σ_3	σ_y	σ_x	C_4^3	C_4	E	C_2
σ_3	$(-y,-x)$	$(-y,-x)$	σ_3	σ_1	σ_x	σ_y	C_4	C_4^3	C_2	E

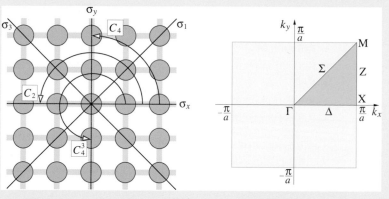

Fig. 3.6

Left: The symmetry operations of the 2D square lattice; the thick straight lines indicate the reflections (labeled $\sigma_x, \sigma_y, \sigma_1, \sigma_3$) and the curves with arrows the rotations (labeled C_4, C_2, C_4^3). These operations form the C_{4v} group. **Right**: The IBZ (yellow triangle) for the 2D square lattice with C_{4v} symmetry; the special points are labeled $\Gamma, \Sigma, \Delta, M, Z, X$.

(3) $X = (1,0)(\pi/a)$, which has the symmetries $E, C_2, \sigma_x, \sigma_y$.
(4) $\Delta = (k,0)(\pi/a), 0 < k < 1$, which has the symmetrics E, σ_x.
(5) $\Sigma = (k,k)(\pi/a), 0 < k < 1$, which has the symmetries E, σ_1.
(6) $Z = (1,k)(\pi/a), 0 < k < 1$, which has the symmetries E, σ_y.

Any point inside the IBZ that is not one of the high-symmetry points mentioned above does not have any symmetry, other than the identity E.

To summarize the discussion so far, we have found that the symmetry properties of the energy spectrum and the single-particle wavefunctions simplify the task of obtaining the band structure considerably, by requiring that we solve the single-particle Schrödinger equations in the IBZ only. Furthermore, both the energy and the wavefunctions are smooth functions of **k**, which in the limit of an infinite crystal is a continuous variable, so we can understand the behavior of the energy bands by considering their values along a few high-symmetry directions: the rest interpolate in a smooth fashion between these directions.

One important point to keep in mind is that Kramers' theorem always applies, which for systems with equal numbers of spin-up and spin-down electrons implies inversion symmetry in *reciprocal space*. Thus, even if the crystal does not have inversion symmetry as one of the point group elements, we can always use inversion symmetry, in addition to all the other point group symmetries imposed by the lattice, to reduce the size of the irreducible portion of the BZ. In many cases, another symmetry operation is actually equivalent to inversion symmetry, so we do not have to add it explicitly in the list of symmetry operations when deriving the size of the IBZ.

In the following, since we will be dealing with the point group symmetries that apply to the properties in reciprocal space, we will denote the arbitrary proper or improper rotation that belongs to this group simply by U.

3.5 Application: Special k-Points

Another useful application of group theory arguments is in the simplification of reciprocal-space integrations. Very often we need to calculate quantities of the type

$$\bar{g} = \frac{1}{N} \sum_{\mathbf{k} \in \mathrm{BZ}} g(\mathbf{k}) = \frac{V_{\mathrm{PUC}}}{(2\pi)^3} \int g(\mathbf{k}) \, \mathrm{d}\mathbf{k} \tag{3.18}$$

where $\mathbf{k} \in \mathrm{BZ}$ stands for all values of the wave-vector $\mathbf{k}$ inside a Brillouin zone (it is convenient to assume that we are dealing with the reduced zone scheme and the first BZ). For example, when calculating the electronic number density $n(\mathbf{r})$ in the single-particle picture we have to calculate the quantities

$$n_{\mathbf{k}}(\mathbf{r}) = \sum_{j, \epsilon_{\mathbf{k}}^{(j)} < \epsilon_{\mathrm{F}}} |\psi_{\mathbf{k}}^{(j)}(\mathbf{r})|^2, \quad n(\mathbf{r}) = \sum_{\mathbf{k}} n_{\mathbf{k}}(\mathbf{r}) = \frac{V_{\mathrm{PUC}}}{(2\pi)^3} \int n_{\mathbf{k}}(\mathbf{r}) \, \mathrm{d}\mathbf{k} \tag{3.19}$$

with the single-particle wavefunctions $\psi_{\mathbf{k}}^{(j)}(\mathbf{r})$ normalized to unity in the PUC. We can take advantage of the crystal symmetry to reduce the summation over **k** values inside the IBZ. In doing this, we need to keep track of the multiplicity of each point in the IBZ, that is, the number of equivalent points to which a particular **k**-point in the IBZ gets mapped when the different symmetry operations are applied to it. While this simplifies our task significantly, there is an even greater simplification: with the use of a very few **k** values we can obtain an excellent approximation to the sum in Eq. (3.18). These are called special **k**-points[1] and we discuss them next (we will follow closely the derivation given by J. D. Joannopoulos and M. L. Cohen, see footnote 1).

---

[1] A. Baldereschi, *Phys. Rev. B* **7**, 5212 (1973); D. J. Chadi and M. L. Cohen, *Phys. Rev. B* **7**, 692 (1973); J. D. Joannopoulos and M. L. Cohen, *J. Phys. C: Solid State Phys.* **6**, 1572 (1973); H. J. Monkhorst and J. D. Pack, *Phys. Rev. B* **13**, 5188 (1976).

We begin by defining the function

$$f(\mathbf{k}) = \frac{1}{\mathcal{N}} \sum_{U \in \mathcal{Q}} g(U\mathbf{k}) \tag{3.20}$$

where $\mathcal{Q}$ is the point group, U is an operation in the point group, and $\mathcal{N}$ is the total number of operations in $\mathcal{Q}$. Then we will have $f(\mathbf{k}) = f(U\mathbf{k})$, since applying U to any operation in $\mathcal{Q}$ gives another operation in $\mathcal{Q}$ due to closure, and the definition of $f(\mathbf{k})$ includes already a summation over all the operations in $\mathcal{Q}$. We can also deduce that the sum of $f(\mathbf{k})$ over all $\mathbf{k}$-points in the BZ is equal to the sum of $g(\mathbf{k})$, because:

$$\sum_{\mathbf{k} \in \mathrm{BZ}} f(\mathbf{k}) = \sum_{\mathbf{k} \in \mathrm{BZ}} \frac{1}{\mathcal{N}} \sum_{U \in \mathcal{Q}} g(U\mathbf{k}) = \frac{1}{\mathcal{N}} \sum_{U \in \mathcal{Q}} \sum_{U\mathbf{k} \in \mathrm{BZ}} g(U\mathbf{k}) \tag{3.21}$$

where we have used the fact that summation over $\mathbf{k} \in \mathrm{BZ}$ is the same as summation over $U\mathbf{k} \in \mathrm{BZ}$. Now, summation over $U\mathbf{k} \in \mathrm{BZ}$ of $g(U\mathbf{k})$ is the same for all U, so that doing this summation for all $U \in \mathcal{Q}$ gives $\mathcal{N}$ times the same result:

$$\frac{1}{\mathcal{N}} \sum_{U \in \mathcal{Q}} \left[\sum_{U\mathbf{k} \in \mathrm{BZ}} g(U\mathbf{k}) \right] = \sum_{U\mathbf{k} \in \mathrm{BZ}} g(U\mathbf{k}) = \sum_{\mathbf{k} \in \mathrm{BZ}} g(\mathbf{k}) \tag{3.22}$$

Combining the results of the last two equations, we obtain the desired result for $\bar{f}, \bar{g}$:

$$\bar{f} = \frac{1}{N} \sum_{\mathbf{k} \in \mathrm{BZ}} f(\mathbf{k}) = \frac{1}{N} \sum_{\mathbf{k} \in \mathrm{BZ}} g(\mathbf{k}) = \bar{g} \tag{3.23}$$

We can expand the function $f(\mathbf{k})$ in a Fourier expansion with coefficients $\tilde{f}(\mathbf{R})$ as follows:

$$f(\mathbf{k}) = \sum_{\mathbf{R}} \tilde{f}(\mathbf{R}) e^{-i\mathbf{k}\cdot\mathbf{R}} \rightarrow \tilde{f}(\mathbf{R}) = \frac{1}{N} \sum_{\mathbf{k} \in \mathrm{BZ}} f(\mathbf{k}) e^{i\mathbf{k}\cdot\mathbf{R}} \tag{3.24}$$

which gives for the sum $\bar{f}$:

$$\bar{f} = \frac{1}{N} \sum_{\mathbf{R}} \sum_{\mathbf{k} \in \mathrm{BZ}} \tilde{f}(\mathbf{R}) e^{-i\mathbf{k}\cdot\mathbf{R}} = \frac{1}{N} \sum_{\mathbf{R}} \tilde{f}(\mathbf{R}) \sum_{\mathbf{k} \in \mathrm{BZ}} e^{-i\mathbf{k}\cdot\mathbf{R}} \tag{3.25}$$

Now we can use the δ-function relations that result from summing the complex exponential $\exp[\pm i\mathbf{k}\cdot\mathbf{R}]$ over real-space lattice vectors or reciprocal-space vectors within the BZ, which are proven in Appendix A, to simplify Eq. (3.25). With these relations, $\bar{f}$ takes the form

$$\bar{f} = \sum_{\mathbf{R}} \tilde{f}(\mathbf{R}) \delta(\mathbf{R}) = \tilde{f}(0) = \bar{g} = \frac{1}{N} \sum_{\mathbf{k} \in \mathrm{BZ}} g(\mathbf{k}) \tag{3.26}$$

so that we have found the sum of $g(\mathbf{k})$ in the BZ to be equal to the Fourier coefficients $\tilde{f}(\mathbf{R})$ of $f(\mathbf{k})$ evaluated at $\mathbf{R} = 0$. But we also have:

$$f(\mathbf{k}) = \sum_{\mathbf{R}} \tilde{f}(\mathbf{R}) e^{-i\mathbf{k}\cdot\mathbf{R}} = \tilde{f}(0) + \sum_{\mathbf{R} \neq 0} \tilde{f}(\mathbf{R}) e^{-i\mathbf{k}\cdot\mathbf{R}} \tag{3.27}$$

We would like to find a value of $\mathbf{k}_0$ that makes the second term on the right side of this equation vanishingly small, because then:

$$\sum_{\mathbf{R}\neq 0}\tilde{f}(\mathbf{R})e^{-i\mathbf{k}_0\cdot\mathbf{R}} \approx 0 \Rightarrow \bar{g} = \tilde{f}(0) \approx f(\mathbf{k}_0) \tag{3.28}$$

The value $\mathbf{k}_0$ is a special point which allows us to approximate the sum of $g(\mathbf{k})$ over the entire BZ by calculating a single value of $f(\mathbf{k})$!

In practice, it is not possible to make the second term in Eq. (3.27) identically zero, so we must find a reasonable way of determining values of the special $\mathbf{k}$-point. To this end we notice that:

$$\tilde{f}(U\mathbf{R}) = \frac{1}{N}\sum_{\mathbf{k}\in BZ} f(\mathbf{k})e^{i\mathbf{k}\cdot(U\mathbf{R})} = \frac{1}{N}\sum_{\mathbf{k}\in BZ} f(\mathbf{k})e^{i(U^{-1}\mathbf{k})\cdot\mathbf{R}}$$

$$= \frac{1}{N}\sum_{U^{-1}\mathbf{k}\in BZ} f(U^{-1}\mathbf{k})e^{i(U^{-1}\mathbf{k})\cdot\mathbf{R}} = \tilde{f}(\mathbf{R}) \tag{3.29}$$

where we took advantage of the fact that $f(\mathbf{k}) = f(U\mathbf{k})$ for any operation $U \in \mathcal{Q}$; we have also used the expression for the inverse Fourier transform $\tilde{f}(\mathbf{R})$ from Eq. (3.24) (see also the discussion of Fourier transforms in Appendix A). Then, for all $\mathbf{R}$ of the same magnitude which are connected by U operations, we will have $\tilde{f}(\mathbf{R}) = \tilde{f}(|\mathbf{R}|)$ and consequently we can break the sum over $\mathbf{R}$ values into a sum over $|\mathbf{R}|$ and a sum over $U \in \mathcal{Q}$, which connect $\mathbf{R}$s of the same magnitude:

$$f(\mathbf{k}_0) = \tilde{f}(0) + \sum_{|\mathbf{R}|}\tilde{f}(|\mathbf{R}|)\left[\sum_{U\in\mathcal{Q}} e^{-i\mathbf{k}_0\cdot(U\mathbf{R})}\right]_{|\mathbf{R}|} \tag{3.30}$$

where the summation inside the square brackets is done at constant $|\mathbf{R}|$. These sums are called "shells of $\mathbf{R}$," and their definition depends on the Bravais lattice. This last equation gives us a practical means of determining $\mathbf{k}_0$: it must make as many shells of $\mathbf{R}$ vanish as possible. Typically, $\tilde{f}(|\mathbf{R}|)$ falls fast as its argument increases, so it is only necessary to make sure that the first few shells of $\mathbf{R}$ vanish for a given $\mathbf{k}_0$, to make it a reasonable candidate for a special $\mathbf{k}$-point. A generalization of this is to consider a set of special $\mathbf{k}$-points for which the first few shells of $\mathbf{R}$ vanish, in which case we need to evaluate $f(\mathbf{k})$ at all of these points to obtain a good approximation for $\bar{g}$.

3.6 Group Representations

We turn our attention to another aspect of group theory, group representations, which will give us the ability to understand essential features of solids that depend on the symmetry of the crystal. One such example are selection rules for transitions between energy eigenvalues by absorption or emission of light quanta, which are involved in the optical properties of solids. The theory of group representations is quite elaborate. Our goal here is not to give a comprehensive treatment of the mathematics of group representations;

rather, we aim to provide a solid background in the theory through a discussion of relevant definitions and theorems, and then to provide detailed examples of their applications.

Definition 1 – Representation *A representation of a group is a set of square, non-singular matrices denoted as* $\underline{\mathbf{D}}(U_i)$ *associated with the elements* U_i *of a group* $\mathcal{G}$, *which obey the same multiplication rules as the group elements, that is:*

$$U_i\, U_j = U_k \Rightarrow \underline{\mathbf{D}}(U_i) \cdot \underline{\mathbf{D}}(U_j) = \underline{\mathbf{D}}(U_k) \tag{3.31}$$

The size (order) of the square matrices is called the **dimension** of the representation. The representation matrix that corresponds to the identity element of the group, E, is a square matrix with all diagonal elements equal to 1 and all off-diagonal elements equal to 0, that is, the unit square matrix of a given size:

$$\underline{\mathbf{D}}(E) = \begin{bmatrix} 1 & 0 & \cdots & 0 \\ 0 & 1 & \cdots & 0 \\ & & \cdot & \\ & & \cdot & \\ & & \cdot & \\ 0 & 0 & \cdots & 1 \end{bmatrix} \tag{3.32}$$

A **faithful** representation is one in which all matrices $\underline{\mathbf{D}}(U_i)$ are different from each other; an **unfaithful** representation is one in which some matrices associated with different group elements $U_i \neq U_j$ are the same, $\underline{\mathbf{D}}(U_i) = \underline{\mathbf{D}}(U_j)$. The **identical** representation is one in which all matrices are equal to the identity matrix, that is, $\underline{\mathbf{D}}(U_i) = \underline{\mathbf{D}}(E) = 1, \forall i$. Evidently, the identical representation is unfaithful while a faithful representation is **isomorphic** to the group $\mathcal{G}$.

Theorem 1 – Similarity transformation *Given a representation* $\underline{\mathbf{D}}$ *of a group* $\mathcal{G}$, *a set of matrices* $\underline{\mathbf{D}}'$ *of the same order, defined through*

$$\underline{\mathbf{D}}' = \underline{\mathbf{S}} \cdot \underline{\mathbf{D}} \cdot \underline{\mathbf{S}}^{-1} \tag{3.33}$$

where $\underline{\mathbf{S}}$ *is a non-singular square matrix of the same order and* $\underline{\mathbf{S}}^{-1}$ *is its inverse, is also a representation of* $\mathcal{G}$. *The transformation between the representations* $\underline{\mathbf{D}}$ *and* $\underline{\mathbf{D}}'$ *is called a* **similarity transformation**.

Corollary *Given a representation* $\underline{\mathbf{D}}$ *of* $\mathcal{G}$, *we can generate an infinite number of other representations of the same dimension by simply multiplying the matrices of* $\underline{\mathbf{D}}$ *by an arbitrary non-singular matrix* $\underline{\mathbf{S}}$ *and its inverse, in the order defined in Eq. (3.33); all the representations generated by similarity transformations have the same trace.*

Proof   From the definition of the similarity transformation, we have for the matrix elements D'_{ij} of the matrix $\underline{\mathbf{D}}'$:

$$D'_{ij} = \sum_{kl} S_{ik}\, D_{kl}\, S^{-1}_{kj}$$

and of course, for the non-singular matrix $\underline{S}$:

$$\sum_k S_{ik}\, S_{kj}^{-1} = \sum_k S_{ik}^{-1}\, S_{kj} = \delta_{ij}$$

Using these relations, we can calculate the trace of the matrix $\underline{D}'$:

$$\sum_i D'_{ii} = \sum_i \sum_{kl} S_{ik}\, D_{kl}\, S_{li}^{-1} = \sum_{kl} \left(\sum_i S_{ik}\, S_{li}^{-1} \right) D_{kl} = \sum_{kl} \delta_{kl} D_{kl} = \sum_k D_{kk}$$

with the last expression being the trace of the matrix $\underline{D}$. ■

Definition 2 – Equivalent representation *An* **equivalent** *representation $\underline{D}'$ to a given representation $\underline{D}$ is a representation which can be obtained from $\underline{D}$ by a similarity transformation. An* **inequivalent** *representation $\underline{C}$ to a given representation $\underline{D}$ is one for which no non-singular matrix $\underline{S}$ can be found to relate it to $\underline{D}$ by a similarity transformation:*

$$\underline{C} \neq \underline{S} \cdot \underline{D} \cdot \underline{S}^{-1} \Rightarrow \underline{C}, \underline{D} : \text{ inequivalent}$$

Definition 3 – Unitary representation *A* **unitary** *representation is one for which the following relation holds:*

$$\underline{D}(U_k) \cdot \underline{D}^{\dagger}(U_k) = 1, \ \forall \ k, \ D_{ij}^{\dagger} = D_{ji}^{*} \tag{3.34}$$

where $\underline{D}^{\dagger}$ is the **conjugate transpose** *of the matrix $\underline{D}$.*

Theorem 2 – Unitarity *Any representation is equivalent to a unitary representation.* In other words, given an arbitrary representation which is not unitary, we can always find a similarity transformation to transform it into a unitary one. As a consequence, we can assume that we are always working with unitary representations.

Definition 4 – Reducible representation *A representation is called* **reducible** *if it can be put into a block-diagonal form by a similarity transformation, that is, in the form:*

$$\underline{D}(U) = \begin{bmatrix} \underline{D}^{(1)}(U) & 0 & \cdots & 0 \\ 0 & \underline{D}^{(2)}(U) & \cdots & 0 \\ & & \ddots & \\ & & & \\ 0 & 0 & \cdots & \underline{D}^{(n)}(U) \end{bmatrix} \tag{3.35}$$

where $\underline{D}^{(\alpha)}(U), \alpha = 1, 2, \ldots, n$, are square matrices of order smaller than the order of $\underline{D}(U)$ (not necessarily the same for the different values of the index α), and the sets of matrices $\underline{D}^{(\alpha)}(U)$ are representations of the group $\mathcal{G}$ for each value of α. In the above expression, the 0s represent rectangular matrices with zero matrix elements. An **irreducible** *representation is one for which it is impossible to put all the matrices into a block-diagonal form by the same similarity transformation.*

Theorem 3 – Schur's lemma *Any matrix that commutes with all the matrices of an irreducible representation must be a constant matrix, that is, a matrix which is a multiple by a factor of c of the identity matrix, where c is a complex number.*

Example 3.3   Representations of the C_{3v} group

To illustrate these ideas, we give different representations of the C_{3v} group discussed previously. The first one, $\underline{\mathbf{D}}''(U)$, is a *faithful* but *not unitary* representation:

$$\underline{\mathbf{D}}''(E) = \begin{bmatrix} 1 & 0 & 0 \\ 0 & 1 & 0 \\ 0 & 0 & 1 \end{bmatrix}, \quad \underline{\mathbf{D}}''(C_3) = \begin{bmatrix} 0 & 0 & \frac{1}{3} \\ 2 & 0 & 0 \\ 0 & \frac{3}{2} & 0 \end{bmatrix}, \quad \underline{\mathbf{D}}''(C_3^2) = \begin{bmatrix} 0 & \frac{1}{2} & 0 \\ 0 & 0 & \frac{2}{3} \\ 3 & 0 & 0 \end{bmatrix}$$

$$\underline{\mathbf{D}}''(\sigma_1) = \begin{bmatrix} 1 & 0 & 0 \\ 0 & 0 & \frac{2}{3} \\ 0 & \frac{3}{2} & 0 \end{bmatrix}, \quad \underline{\mathbf{D}}''(\sigma_2) = \begin{bmatrix} 0 & 0 & \frac{1}{3} \\ 0 & 1 & 0 \\ 3 & 0 & 0 \end{bmatrix}, \quad \underline{\mathbf{D}}''(\sigma_3) = \begin{bmatrix} 0 & \frac{1}{2} & 0 \\ 2 & 0 & 0 \\ 0 & 0 & 1 \end{bmatrix}$$

The validity of this representation can easily be checked by comparing the products of various matrix pairs with the expected result from the group multiplication table: for instance, $\underline{\mathbf{D}}''(C_3) \cdot \underline{\mathbf{D}}''(C_3^2) = \underline{\mathbf{D}}''(E)$, just like $C_3 C_3^2 = E$, or $\underline{\mathbf{D}}''(C_3) \cdot \underline{\mathbf{D}}''(\sigma_1) = \underline{\mathbf{D}}''(\sigma_3)$, just like $C_3 \sigma_1 = \sigma_3$, etc. This representation is faithful because all matrices are different from each other, and is not unitary because $\underline{\mathbf{D}}''(U) \cdot \underline{\mathbf{D}}''^{\dagger}(U) \neq 1$, as can easily be checked for any member U of the group, other than the identity E. We can transform this representation into a unitary one using the following similarity transformation $\underline{\mathbf{D}}'(U) = \underline{\mathbf{S}}' \cdot \underline{\mathbf{D}}''(U) \cdot \underline{\mathbf{S}}'^{-1}$, with the matrix $\underline{\mathbf{S}}$:

$$\underline{\mathbf{S}}' = \begin{bmatrix} 1 & 0 & 0 \\ 0 & 2 & 0 \\ 0 & 0 & 3 \end{bmatrix}, \quad \underline{\mathbf{S}}'^{-1} = \begin{bmatrix} 1 & 0 & 0 \\ 0 & \frac{1}{2} & 0 \\ 0 & 0 & \frac{1}{3} \end{bmatrix}$$

The resulting faithful, unitary representation $\underline{\mathbf{D}}'(U)$ is:

$$\underline{\mathbf{D}}'(E) = \begin{bmatrix} 1 & 0 & 0 \\ 0 & 1 & 0 \\ 0 & 0 & 1 \end{bmatrix}, \quad \underline{\mathbf{D}}'(C_3) = \begin{bmatrix} 0 & 0 & 1 \\ 1 & 0 & 0 \\ 0 & 1 & 0 \end{bmatrix}, \quad \underline{\mathbf{D}}'(C_3^2) = \begin{bmatrix} 0 & 1 & 0 \\ 0 & 0 & 1 \\ 1 & 0 & 0 \end{bmatrix}$$

$$\underline{\mathbf{D}}'(\sigma_1) = \begin{bmatrix} 1 & 0 & 0 \\ 0 & 0 & 1 \\ 0 & 1 & 0 \end{bmatrix}, \quad \underline{\mathbf{D}}'(\sigma_2) = \begin{bmatrix} 0 & 0 & 1 \\ 0 & 1 & 0 \\ 1 & 0 & 0 \end{bmatrix}, \quad \underline{\mathbf{D}}'(\sigma_3) = \begin{bmatrix} 0 & 1 & 0 \\ 1 & 0 & 0 \\ 0 & 0 & 1 \end{bmatrix}$$

To check if this is a reducible or irreducible representation, we look for a matrix that is *not* a constant matrix, which commutes with all matrices of $\underline{\mathbf{D}}'(U)$. It is easy to verify that the following matrix:

$$\underline{\mathbf{M}} = \begin{bmatrix} 1 & 1 & 1 \\ 1 & 1 & 1 \\ 1 & 1 & 1 \end{bmatrix}$$

commutes with all matrices of $\underline{\mathbf{D}}'(U)$, which means that this representation is *reducible* (Schur's lemma, Theorem 3). We can further transform the representation by another similarity transformation $\underline{\mathbf{D}}(U) = \underline{\mathbf{S}} \cdot \underline{\mathbf{D}}'(U) \cdot \underline{\mathbf{S}}^{-1}$, with $\underline{\mathbf{S}}$ given by:

$$\underline{\mathbf{S}} = \frac{1}{\sqrt{6}} \begin{bmatrix} \sqrt{2} & \sqrt{2} & \sqrt{2} \\ 0 & \sqrt{3} & -\sqrt{3} \\ -2 & 1 & 1 \end{bmatrix}, \quad \underline{\mathbf{S}}^{-1} = \frac{1}{\sqrt{6}} \begin{bmatrix} \sqrt{2} & 0 & -2 \\ \sqrt{2} & \sqrt{3} & 1 \\ \sqrt{2} & -\sqrt{3} & 1 \end{bmatrix}$$

to produce the following block-diagonal representation:

$$\underline{\mathbf{D}}(E) = \begin{bmatrix} 1 & 0 & 0 \\ 0 & 1 & 0 \\ 0 & 0 & 1 \end{bmatrix}, \quad \underline{\mathbf{D}}(C_3) = \begin{bmatrix} 1 & 0 & 0 \\ 0 & -\frac{1}{2} & -\frac{\sqrt{3}}{2} \\ 0 & \frac{\sqrt{3}}{2} & -\frac{1}{2} \end{bmatrix}, \quad \underline{\mathbf{D}}(C_3^2) = \begin{bmatrix} 1 & 0 & 0 \\ 0 & -\frac{1}{2} & \frac{\sqrt{3}}{2} \\ 0 & -\frac{\sqrt{3}}{2} & -\frac{1}{2} \end{bmatrix}$$

$$\underline{\mathbf{D}}(\sigma_1) = \begin{bmatrix} 1 & 0 & 0 \\ 0 & -1 & 0 \\ 0 & 0 & 1 \end{bmatrix}, \quad \underline{\mathbf{D}}(\sigma_2) = \begin{bmatrix} 1 & 0 & 0 \\ 0 & \frac{1}{2} & \frac{\sqrt{3}}{2} \\ 0 & \frac{\sqrt{3}}{2} & -\frac{1}{2} \end{bmatrix}, \quad \underline{\mathbf{D}}(\sigma_3) = \begin{bmatrix} 1 & 0 & 0 \\ 0 & \frac{1}{2} & -\frac{\sqrt{3}}{2} \\ 0 & -\frac{\sqrt{3}}{2} & -\frac{1}{2} \end{bmatrix}$$

which consists of 1×1 submatrices on the upper-left corner and 2×2 submatrices on the lower-right corner, with the remaining matrix elements equal to zero. The upper-left 1×1 submatrices of $\underline{\mathbf{D}}(U)$ form a 1D irreducible representation (the identical representation) and the lower-right 2×2 submatrices form a 2D irreducible representation of C_{3v}.

Theorem 4 – Number of IIRs *The number of inequivalent irreducible representations (IIRs) for a group is equal to the number of classes in the group.*

Theorem 5 – Grand orthogonality *The following relationship holds:*

$$\sum_{n=1}^{\mathcal{N}} D_{ij}^{*(\alpha)}(U_n) \, D_{kl}^{(\beta)}(U_n) = \frac{\mathcal{N}}{\eta_\alpha} \delta_{\alpha\beta} \, \delta_{ik} \, \delta_{jl} \tag{3.36}$$

where $\mathcal{N}$ is the number of elements in group $\mathcal{G}$, the summation runs through the $\mathcal{N}$ elements U_n of the group, $\underline{\mathbf{D}}^{(\alpha)}$ and $\underline{\mathbf{D}}^{(\beta)}$ are two irreducible representations of the group with matrix elements $D_{ij}^{(\alpha)}$ and $D_{kl}^{(\beta)}$, respectively, and η_α is the order of $\underline{\mathbf{D}}^{(\alpha)}$.

This theorem is very helpful because it generates a number of constraints on the representations. For instance, for a given irreducible representation labeled α, there are η_α^2 independent pairs of numbers (i, j), when each of the integers i and j take the values $1, 2, \ldots, \eta_\alpha$; these pairs identify all the matrices of the representation $\underline{\mathbf{D}}^{(\alpha)}$. Applying the above theorem for $\beta = \alpha, k = i, l = j$, we can then generate η_α^2 equations corresponding to the independent values of the pair (i, j), each of them reading:

$$\sum_{n=1}^{\mathcal{N}} |D_{ij}^{(\alpha)}(U_n)|^2 = \frac{\mathcal{N}}{\eta_\alpha} \tag{3.37}$$

Corollary *A direct consequence of the orthogonality theorem is that, for any group, summing over all the inequivalent, irreducible representations gives:*

$$\sum_{\alpha} \eta_{\alpha}^2 = \mathcal{N} \tag{3.38}$$

We will prove this relation below, after we state a few more useful theorems.

Definition 5 – Characters *Given a representation $\underline{\mathbf{D}}^{(\alpha)}$ with matrix elements $D_{ij}^{(\alpha)}$, we define the **character** of each group element U, denoted by $\chi^{(\alpha)}(U)$, as the trace (sum of diagonal matrix elements) of the matrix corresponding to U:*

$$\chi^{(\alpha)}(U) \equiv \mathrm{Tr}\left[\underline{\mathbf{D}}^{(\alpha)}(U)\right] = \sum_{i} D_{ii}^{(\alpha)}(U) \tag{3.39}$$

All elements in a class have the same character. This is easily shown from the following relations. First, we can show that for a product of two matrices $\underline{\mathbf{A}}$ and $\underline{\mathbf{B}}$, the trace of $\underline{\mathbf{A}} \cdot \underline{\mathbf{B}}$ is equal to the trace of $\underline{\mathbf{B}} \cdot \underline{\mathbf{A}}$:

$$\mathrm{Tr}[\underline{\mathbf{A}} \cdot \underline{\mathbf{B}}] = \sum_{i}[AB]_{ii} = \sum_{i}\left[\sum_{j} A_{ij} B_{ji}\right] = \sum_{j}\left[\sum_{i} B_{ji} A_{ij}\right] = \sum_{j}[BA]_{jj} = \mathrm{Tr}[\underline{\mathbf{B}} \cdot \underline{\mathbf{A}}]$$

From the class definition, if two elements of the group U and W belong to a class, then there is some other group element S which relates them through $U = S\,W\,S^{-1}$, and

$$\mathrm{Tr}[\underline{\mathbf{D}}(U)] = \mathrm{Tr}[\underline{\mathbf{D}}(S) \cdot \underline{\mathbf{D}}(W) \cdot \underline{\mathbf{D}}(S^{-1})] = \mathrm{Tr}[(\underline{\mathbf{D}}(S) \cdot \underline{\mathbf{D}}(W)) \cdot \underline{\mathbf{D}}(S^{-1})]$$
$$= \mathrm{Tr}[\underline{\mathbf{D}}(S^{-1}) \cdot (\underline{\mathbf{D}}(S) \cdot \underline{\mathbf{D}}(W))] = \mathrm{Tr}[\underline{\mathbf{D}}(W)]$$

where we first associated the matrix $\underline{\mathbf{D}}(S) \cdot \underline{\mathbf{D}}(W)$ with $\underline{\mathbf{A}}$ and the matrix $\underline{\mathbf{D}}(S^{-1})$ with $\underline{\mathbf{B}}$ in the earlier expression for the trace of products, then used the property of the traces of products of matrices to change the order of $\underline{\mathbf{A}} \cdot \underline{\mathbf{B}}$ in the product to $\underline{\mathbf{B}} \cdot \underline{\mathbf{A}}$, and finally we used the fact that $S^{-1}S = 1 \Rightarrow \underline{\mathbf{D}}(S) \cdot \underline{\mathbf{D}}(S^{-1}) = 1$ to obtain the desired result.

Theorem 6 – Character orthogonality *The following relation holds between the characters of two inequivalent, irreducible representations with labels α and β:*

$$\sum_{n=1}^{\mathcal{N}} \chi^{*(\alpha)}(U_n)\, \chi^{(\beta)}(U_n) = \mathcal{N}\, \delta_{\alpha\beta} \tag{3.40}$$

Since, as we showed already, all elements of the same class have the same character, if we split the sum over all group elements U into a sum over the classes C_i, with the class labeled i containing N_i elements, then the above theorem becomes:

$$\sum_{i} \chi^{*(\alpha)}(C_i)\, \chi^{(\beta)}(C_j)\, N_i = \mathcal{N}\, \delta_{\alpha\beta} \tag{3.41}$$

Now, using the definition of Eq. (3.35) we can decompose the arbitrary representation $\underline{\mathbf{D}}$, by a similarity transformation, as a block sum:

$$\underline{\mathbf{D}} = \sum_{\alpha} \mu_{\alpha} \underline{\mathbf{D}}^{(\alpha)} \tag{3.42}$$

with the summation index α running over the number of IIRs, that is, the arbitrary representation contains μ_α copies of the αth IIR. From this we obtain the characters:

$$\chi(U) = \sum_\alpha \mu_\alpha \, \chi^{(\alpha)}(U) \Rightarrow \sum_{n=1}^{\mathcal{N}} \chi^{*(\beta)}(U_n) \, \chi(U_n)$$

$$= \sum_\alpha \sum_{n=1}^{\mathcal{N}} \chi^{*(\beta)}(U_n)\mu_\alpha \, \chi^{(\alpha)}(U_n) = \mu_\beta \, \mathcal{N}$$

where, in the last step, we have used Theorem 6, Eq. (3.40). This leads to:

$$\mu_\beta = \frac{1}{\mathcal{N}} \sum_{n=1}^{\mathcal{N}} \chi^{*(\beta)}(U_n) \, \chi(U_n)$$

that is, we can find how many times a particular IIR is contained within an arbitrary representation by multiplying the characters of the representation $\chi(U_n)$ by the characters of that particular IIR, $\chi^{*(\beta)}(U_n)$, and summing over all elements of the group.

Theorem 7 – Character projection *Summing over the number of IIRs labeled by α, the product of characters belonging to classes C_i and C_j gives:*

$$\sum_\alpha \chi^{*(\alpha)}(C_i)\chi^{(\alpha)}(C_j) = \frac{\mathcal{N}}{N_i}\delta_{ij} \tag{3.43}$$

with $\mathcal{N}$ the number of group elements and N_i the number of elements in C_i.

Definition 6 – Inner product of classes *If we multiply the elements in class C_i by those in class C_j (any possible such product), we obtain the elements of other classes C_k with the multiplicities v_{ijk}:*

$$C_i \cdot C_j = \sum_k v_{ijk} \, C_k \tag{3.44}$$

Theorem 8 – Character summation *Using the multiplicities v_{ijk} defined in Eq. (3.44), the following relation holds between the characters of class C_i and class C_j:*

$$N_i\chi^{(\alpha)}(C_i) \, N_j\chi^{(\alpha)}(C_j) = \eta_\alpha \sum_k v_{ijk} \, N_k\chi^{(\alpha)}(C_k) \tag{3.45}$$

We can now prove the corollary of Eq. (3.38): using Theorem 7 for class C_i and applying it for the identity class $C_1 = \{E\}$, with $N_1 = 1$, its only element being the identity matrix of order η_α, we find:

$$\sum_\alpha \chi^{*(\alpha)}(C_i)\chi^{(\alpha)}(C_i) = \frac{\mathcal{N}}{N_i} \quad \text{and} \quad \chi^{(\alpha)}(E) = \eta_\alpha \Rightarrow \sum_\alpha \eta_\alpha^2 = \mathcal{N}$$

The rules for constructing a **character table** for each group are summarized below.

Rules for constructing character tables

1. Number of inequivalent irreducible representations = number of classes.
2. With η_α the order of the matrices in the αth IIR, and $\mathcal{N}$ the size of the group:

$$\sum_\alpha \eta_\alpha^2 = \mathcal{N}$$

3. The rows of the character table are orthogonal vectors, with the number of elements N_i in class $\mathcal{C}_i$ as weights:

$$\sum_{n=1}^{\mathcal{N}} \chi^{*(\alpha)}(U_n)\, \chi^{(\beta)}(U_n) = \mathcal{N}\, \delta_{\alpha\beta} \ \text{ or } \ \sum_i N_i \chi^{*(\alpha)}(\mathcal{C}_i)\, \chi^{(\beta)}(\mathcal{C}_i) = \mathcal{N}\, \delta_{\alpha\beta}$$

Taking the αth representation to be the identical and the βth any other IIR:

$$\sum_{n=1}^{\mathcal{N}} \chi^{(\beta)}(U_n) = 0 \ \text{ or } \ \sum_i N_i \chi^{(\beta)}(\mathcal{C}_i) = 0$$

4. The columns of the character table are orthogonal vectors:

$$\sum_\alpha \chi^{*(\alpha)}(\mathcal{C}_i)\chi^{(\alpha)}(\mathcal{C}_i) = \frac{\mathcal{N}}{N_i}\delta_{ij}$$

5. The characters of the αth IIR, with v_{ijk} the multiplicities, obey:

$$N_i\chi^{(\alpha)}(\mathcal{C}_i)\, N_j\chi^{(\alpha)}(\mathcal{C}_j) = \eta_\alpha \sum_k v_{ijk}\, N_k\chi^{(\alpha)}(\mathcal{C}_k), \ \ \mathcal{C}_i \cdot \mathcal{C}_j = \sum_k v_{ijk}\, \mathcal{C}_k$$

Example 3.4   Character table of the C_{3v} group

We illustrate these concepts by calculating the character table of the C_{3v} group. This group has three classes, $\mathcal{C}_1 = \{E\}$, $\mathcal{C}_2 = \{C_3, C_3^2\}$, and $\mathcal{C}_3 = \{\sigma_1, \sigma_2, \sigma_3\}$. From the group multiplication table we obtain the class multiplication table shown in Table 3.5 . This gives the following multiplicities:

$$v_{111} = v_{122} = v_{133} = v_{212} = v_{222} = v_{313} = 1$$
$$v_{221} = v_{233} = v_{323} = 2 \ \text{ and } \ v_{331} = v_{332} = 3$$

with all other multiplicites being 0. We also know from Rule 1 that there are three representations, the same as the number of classes. Assume that the orders of these representations are η_1, η_2, η_3, then, from Rule 2, we have:

$$\eta_1^2 + \eta_2^2 + \eta_3^2 = 6 \Rightarrow \eta_1 = 1, \ \eta_2 = 1, \ \eta_3 = 2$$

because these are the only three positive integers that satisfy Rule 2. We call the three representations $\Gamma_1, \Gamma_2, \Gamma_3$ and we choose Γ_1 to be the identical representation,

Table 3.5 Class multiplication table for the C_{3v} group.

E	C_3	C_3^2	σ_1	σ_2	σ_3
C_3	C_3^2	E	σ_2	σ_3	σ_1
C_3^2	E	C_3	σ_3	σ_1	σ_2
σ_1	σ_3	σ_2	E	C_3^2	C_3
σ_2	σ_1	σ_3	C_3	E	C_3^2
σ_3	σ_2	σ_1	C_3^2	C_3	E

$\Rightarrow$

	$\mathcal{C}_1$	$\mathcal{C}_2$	$\mathcal{C}_3$
$\mathcal{C}_1$	$\mathcal{C}_1$	$\mathcal{C}_2$	$\mathcal{C}_3$
$\mathcal{C}_2$	$\mathcal{C}_2$	$2\mathcal{C}_1 + \mathcal{C}_2$	$2\mathcal{C}_3$
$\mathcal{C}_3$	$\mathcal{C}_3$	$2\mathcal{C}_3$	$3\mathcal{C}_1 + 3\mathcal{C}_2$

whose characters are all equal to 1. Γ_2 is the other 1D representation ($\eta_2 = 1$) and Γ_3 is the 2D representation ($\eta_3 = 2$). The character table must then be that shown in Table 3.6, since for the first class, which consists of E, the matrices are the identity matrix of order 1 (for Γ_1 and Γ_2) and of order 2 (for Γ_3). The quantities x, y, u, v will be specified by application of the remaining rules. From Rule 3, we have:

$$1 + 2|x|^2 + 3|y|^2 = 6, \quad \text{for } \alpha = \beta = 2 \quad \text{and} \quad 1 + 2x + 3y = 0, \quad \text{for } \beta = 2$$

while from Rule 5, using the multiplicities v_{ijk} calculated already, we find for $i = j = 2$ and $\alpha = 2$:

$$4x^2 = 2 + 2x \Rightarrow x = 1 \text{ or } -\frac{1}{2}$$

Of these two possible values of x, only $x = 1$ is compatible with the previous relations we derived, and this value yields $y = -1$. From Rule 4 we find:

$$(i,j) = (1,2) \rightarrow 1 + x + 2u = 0 \Rightarrow u = -1$$

and $\quad (i,j) = (1,3) \rightarrow 1 + y + 2v = 0 \Rightarrow v = 0$

which completes the solution of the character table. With these characters, we can now figure out which IIR, and how many times each, is contained in the reducible representation $\underline{D}(U)$ of the previous example. From the actual matrices of this representation, we find that the characters are $\chi(\mathcal{C}_1) = 3, \chi(\mathcal{C}_2) = 0, \chi(\mathcal{C}_3) = 1$, and with these values we obtain:

$$\mu_1 = \frac{1}{6}\sum_{i=1}^{3}\chi^{*(1)}(\mathcal{C}_i)\chi(\mathcal{C}_i)N_i = 1, \quad \mu_2 = \frac{1}{6}\sum_{i=1}^{3}\chi^{*(2)}(\mathcal{C}_i)\chi(\mathcal{C}_i)N_i = 0, \quad \text{and}$$

$$\mu_3 = \frac{1}{6}\sum_{i=1}^{3}\chi^{*(3)}(\mathcal{C}_i)\chi(\mathcal{C}_i)N_i = 1$$

that is, the 3D reducible representation $\underline{D}(U)$ contains once the 1D irreducible representation Γ_1 (the 1×1 submatrices on the upper-left corner) and once the 2D irreducible representation Γ_3 (the 2×2 submatrices on the lower-right corner).

Table 3.6 Character table for the irreducible representations of the C_{3v} group.

C_i	C_1	C_2	C_3
N_i	1	2	3
Γ_1	1	1	1
Γ_2	1	x	y
Γ_3	2	u	v

Now consider a K-fold degenerate eigenvalue ϵ_k of the hamiltonian, which corresponds to K linearly independent solutions $\psi_k(\mathbf{r}), k = 1, \ldots, K$. For any symmetry operation of the hamiltonian, we will have:

$$\left[\hat{O}_U, \hat{\mathcal{H}}^{\mathrm{sp}}\right] = 0, \ \hat{\mathcal{H}}^{\mathrm{sp}}\psi_k(\mathbf{r}) = \epsilon_k\psi_k(\mathbf{r}) \Rightarrow \hat{\mathcal{H}}^{\mathrm{sp}}\left[\hat{O}_U\psi_k(\mathbf{r})\right] = \hat{O}_U\left[\hat{\mathcal{H}}^{\mathrm{sp}}\psi_k(\mathbf{r})\right]$$

$$= \epsilon_k\left[\hat{O}_U\psi_k(\mathbf{r})\right]$$

that is, the states $\psi_k(\mathbf{r})$ and $\left[\hat{O}_U\psi_k(\mathbf{r})\right]$ are eigenfunctions with the same eigenvalue.

Corollary   *The set $\psi_k(\mathbf{r})$ forms a complete subspace of states, that is, the state $[\hat{O}_U\psi_k(\mathbf{r})]$ for any operation corresponding to an element U of the symmetry group of the hamiltonian must be a linear combination of the states $\{\psi_k(\mathbf{r})\}$:*

$$\left[\hat{O}_U\psi_k(\mathbf{r})\right] = \sum_{j=1}^{K} \psi_j(\mathbf{r})D_{jk}(U) \tag{3.46}$$

where $D_{jk}(U)$ are complex numbers.

Theorem 9 *The matrices $\underline{\mathbf{D}}(U)$ defined by the relations of Eq. (3.46) form a representation of the symmetry group of the hamiltonian $\hat{\mathcal{H}}^{\mathrm{sp}}$.*

Proof   All we need to show is that the matrices $\underline{\mathbf{D}}(U)$ obey the same multiplication rules as the group operators $\hat{O}_U$. We consider the row vector $\boldsymbol{\Psi}$ of all the states $\psi_k(\mathbf{r}), k = 1, \ldots, K$:

$$\boldsymbol{\Psi}(\mathbf{r}) = (\psi_1(\mathbf{r}), \psi_2(\mathbf{r}), \ldots, \psi_K(\mathbf{r}))$$

From our definition of $\underline{\mathbf{D}}(U)$, we have:

$$\hat{O}_U\boldsymbol{\Psi}(\mathbf{r}) = \boldsymbol{\Psi}(\mathbf{r}) \cdot \underline{\mathbf{D}}(U)$$

and from this, taking the product of two operators corresponding to U, W:

$$\hat{O}_W\hat{O}_U\boldsymbol{\Psi}(\mathbf{r}) = \hat{O}_W\boldsymbol{\Psi}(\mathbf{r}) \cdot \underline{\mathbf{D}}(U) = \boldsymbol{\Psi}(\mathbf{r}) \cdot \underline{\mathbf{D}}(W) \cdot \underline{\mathbf{D}}(U)$$

and by definition we will also have:

$$\hat{O}_{WU}\boldsymbol{\Psi}(\mathbf{r}) = \boldsymbol{\Psi}(\mathbf{r}) \cdot \underline{\mathbf{D}}(WU)$$

but we also know that:

$$\hat{O}_W \hat{O}_U \Psi(\mathbf{r}) = \hat{O}_{WU} \Psi(\mathbf{r}) \Rightarrow \Psi(\mathbf{r}) \cdot \underline{\mathbf{D}}(W) \cdot \underline{\mathbf{D}}(U) = \Psi(\mathbf{r}) \cdot \underline{\mathbf{D}}(WU) \Rightarrow \underline{\mathbf{D}}(W) \cdot \underline{\mathbf{D}}(U) = \underline{\mathbf{D}}(WU)$$

which completes the proof.   ∎

Theorem 10 *All representations of the symmetry group of the hamiltonian associated with a particular eigenvalue of the hamiltonian are equivalent.*

Proof Suppose we have two representations $\underline{\mathbf{D}}$ and $\underline{\mathbf{D}}'$ defined by:

$$\hat{O}_U \Psi(\mathbf{r}) = \Psi(\mathbf{r}) \cdot \underline{\mathbf{D}}(U), \quad \hat{O}_U \Phi(\mathbf{r}) = \Phi(\mathbf{r}) \cdot \underline{\mathbf{D}}'(U)$$

Since we are dealing with the same eigenvalue, each state in the set $\Phi(\mathbf{r})$ must be a linear combination of the states in the set $\Psi(\mathbf{r})$, that is, the two sets are related by:

$$\Phi(\mathbf{r}) = \Psi(\mathbf{r}) \cdot \underline{\mathbf{S}} \tag{3.47}$$

where $\underline{\mathbf{S}}$ is the matrix of the coefficients relating the two sets of states. Then we will have:

$$\hat{O}_U \Phi(\mathbf{r}) = \hat{O}_U \Psi(\mathbf{r}) \cdot \underline{\mathbf{S}} = \Psi(\mathbf{r}) \cdot \underline{\mathbf{D}}(U) \cdot \underline{\mathbf{S}} \text{ and } \hat{O}_U \Phi(\mathbf{r}) = \Phi(\mathbf{r}) \cdot \underline{\mathbf{D}}'(U)$$

$$= \Psi(\mathbf{r}) \cdot \underline{\mathbf{S}} \cdot \underline{\mathbf{D}}'(U)$$

From these two relations we obtain, multiplying both from the left by $\underline{\mathbf{S}}^{-1}$:

$$\Psi(\mathbf{r}) \cdot \underline{\mathbf{D}}(U) = \Psi(\mathbf{r}) \cdot \underline{\mathbf{S}} \cdot \underline{\mathbf{D}}'(U) \cdot \underline{\mathbf{S}}^{-1} \Rightarrow \underline{\mathbf{D}}(U) = \underline{\mathbf{S}} \cdot \underline{\mathbf{D}}'(U) \cdot \underline{\mathbf{S}}^{-1}$$

that is, the two representations $\underline{\mathbf{D}}(U)$ and $\underline{\mathbf{D}}'(U)$ are equivalent because they are related by a similarity transformation.   ∎

Definition 7 – Irreducible space *We define the space spanned by the set of states $\Psi(\mathbf{r})$ to be an* **irreducible space** *if the representation $\underline{\mathbf{D}}(U)$ is irreducible; otherwise, the space is called a* **reducible space**.

Definition 8 – Normal degeneracy *We call the degeneracy of the eigenvalue ϵ_k* **normal degeneracy** *if the set of states generated by $\{\hat{O}_U \Psi(\mathbf{r})\}$ spans the same vector space as $\Psi(\mathbf{r})$, for all choices of $\Psi(\mathbf{r})$; otherwise, it is called* **accidental degeneracy**.

Theorem 11 *If the degeneracy of an eigenvalue is not normal, then the representation generated by the states corresponding to this eigenvalue is reducible.*

Definition 9 – Partner functions *A set of functions $\psi_k^{(\alpha)}(\mathbf{r}), k = 1, \ldots, K$, denoted collectively as $\Psi^{(\alpha)}(\mathbf{r})$, are called* **partner functions** *if they transform into each other as:*

$$\hat{O}_U \Psi^{(\alpha)}(\mathbf{r}) = \Psi^{(\alpha)}(\mathbf{r}) \cdot \underline{\mathbf{D}}^{(\alpha)}(U)$$

where $\underline{\mathbf{D}}^{(\alpha)}(U)$ is a set of unitary, irreducible representation matrices. $\psi_k^{(\alpha)}(\mathbf{r}), k = 1, \ldots, K$, transform into each other like the eigenfunctions corresponding to an eigenvalue ϵ_k with normal degeneracy n, but are not necessarily the eigenfunctions themselves.

Theorem 12 *Any function $\psi(\mathbf{r})$ can be expanded in terms of components $\{\psi_k^{(\alpha)}(\mathbf{r})\}$ which are partner functions of the αth irreducible representation of the symmetry group.*

Definition 10 – Projection operator *A* **projection operator** *is an object which, when applied to a function, projects out the part that has some special property.*

Theorem 13 *The projection operator for component i of the αth irreducible representation is:*

$$\hat{P}_k^{(\alpha)} = \frac{\eta_\alpha}{\mathcal{N}} \sum_{n=1}^{\mathcal{N}} D_{kk}^{*(\alpha)}(U_n) \hat{O}_{U_n}$$

Proof Take the component j or the βth irreducible representation and apply $\hat{P}_k^{(\alpha)}$ to it:

$$\hat{P}_k^{(\alpha)} \psi_l^{(\beta)}(\mathbf{r}) = \frac{\eta_\alpha}{\mathcal{N}} \sum_{n=1}^{\mathcal{N}} D_{kk}^{*(\alpha)}(U_n) \hat{O}_{U_n} \psi_l^{(\beta)}(\mathbf{r}) = \frac{\eta_\alpha}{\mathcal{N}} \sum_{n=1}^{\mathcal{N}} D_{kk}^{*(\alpha)}(U_n) \sum_j \psi_j^{(\beta)}(\mathbf{r}) D_{jl}^{(\beta)}(U_n)$$

$$= \sum_j \psi_j^{(\beta)}(\mathbf{r}) \frac{\eta_\alpha}{\mathcal{N}} \sum_{n=1}^{\mathcal{N}} D_{kk}^{*(\alpha)}(U_n) D_{jl}^{(\beta)}(U_n) = \sum_j \psi_j^{(\beta)}(\mathbf{r}) \frac{\eta_\alpha}{\mathcal{N}} \delta_{\alpha\beta} \delta_{kj} \delta_{kl} = \psi_k^{(\beta)}(\mathbf{r}) \delta_{kl} \delta_{\alpha\beta}$$

If an arbitrary function $\psi(\mathbf{r})(\mathbf{r})$ is expressed in terms of the complete basis set of functions that correspond to all the components labeled by l of all the IIRs labeled by β, then:

$$\psi(\mathbf{r}) = \sum_\beta \sum_{l=1}^K c_l^{(\beta)} \psi_l^{(\beta)}(\mathbf{r}) \Rightarrow \hat{P}_k^{(\alpha)} \psi(\mathbf{r}) = \sum_\beta \sum_{l=1}^K c_l^{(\beta)} \hat{P}_k^{(\alpha)} \psi_l^{(\beta)}(\mathbf{r}) = c_k^{(\alpha)} \psi_k^{(\alpha)}(\mathbf{r})$$

which proves the desired property of the projection operator $\hat{P}_k^{(\alpha)}$. ∎

Theorem 14 *The operator*

$$\hat{\mathcal{P}}^{(\alpha)} = \sum_{k=1}^K \hat{P}_k^{(\alpha)} = \sum_{k=1}^K \frac{\eta_\alpha}{\mathcal{N}} \sum_{n=1}^{\mathcal{N}} D_{kk}^*(U_n) \hat{O}_{U_n} = \frac{\eta_\alpha}{\mathcal{N}} \sum_{n=1}^{\mathcal{N}} \chi^{*(\alpha)} \hat{O}_{U_n} \qquad (3.48)$$

projects out of an arbitrary function the part that transforms according to the αth irreducible representation.

We summarize below the rules for constructing irreducible representations.

Rules for constructing irreducible representations

1. Using the character table, construct the projection operator defined in Eq. (3.48).
2. Pick any function $\psi(\mathbf{r})$ and project out the part $\psi^{(\alpha)}(\mathbf{r})$ that transforms like the αth IIR by applying the operator $\hat{\mathcal{P}}^{(\alpha)}$ to it.
3. Using $\psi^{(\alpha)}(\mathbf{r})$ construct the functions $\hat{O}_U \psi^{(\alpha)}(\mathbf{r})$ by applying all the symmetry operators of the group.
4. Choose among all these functions the linearly independent ones and orthogonalize them to form the set $\boldsymbol{\Psi}_o^{(\alpha)}(\mathbf{r}) = \{\psi_k^{(\alpha)}(\mathbf{r})\}$, $k = 1, \ldots, K$, with the number of linearly independent functions $K \leq \mathcal{N}$: the size of the group.
5. This set forms a basis for determining the irreducible representation $\underline{\mathbf{D}}^{(\alpha)}$:

$$\hat{O}_{U_n} \psi_k^{(\alpha)}(\mathbf{r}) = \sum_{l=1}^K \psi_l^{(\alpha)}(\mathbf{r}) D_{lk}^{(\alpha)}(U_n) \ : \ \text{defines} \ D_{lk}^{(\alpha)}(U_n), \ n = 1, \ldots, \mathcal{N}$$

Example 3.5   Irreducible representations of the C_{4v} group

To illustrate these points, we give an example of how all the irreducible representations of the C_{4v} group can be constructed. This group has five classes, $\mathcal{C}_1 = \{E\}, \mathcal{C}_2 = \{C_4, C_4^3\}, \mathcal{C}_3 = \{C_2\}, \mathcal{C}_4 = \{\sigma_x, \sigma_y\}$, and $\mathcal{C}_5 = \{\sigma_1, \sigma_3\}$, and therefore five IIRs, which we call $\Gamma_1, \Gamma_2, \Gamma_3, \Gamma_4, \Gamma_5$. The size of the group is $\mathcal{N} = 8$ and the order of the different representations must be $\eta_1 = \eta_2 = \eta_3 = \eta_4 = 1$ and $\eta_5 = 2$, which are the only positive integers that satisfy Rule 2 for constructing the character table. Following the rest of the rules for constructing the character table, we find the results shown in Table 3.6.

For the 1D representations, $\Gamma_i, i = 1-4$, the characters are the matrices themselves. We only need to calculate explicitly the matrices for the 2D representation Γ_5. From Rule 1 we construct the projection operator

$$\hat{\mathcal{P}}^{(5)} = \frac{2}{8}\left(2\hat{O}_E - 2\hat{O}_{C_2}\right) = \frac{1}{2}\left(\hat{O}_E - \hat{O}_{C_2}\right)$$

and we choose the trial function $\psi(\mathbf{r}) = x$ from which we project out the part that transforms like the Γ_5 representation by applying $\hat{\mathcal{P}}^{(5)}$ to it (Rule 2):

$$\psi^{(5)}(\mathbf{r}) = \hat{\mathcal{P}}^{(5)}\psi(\mathbf{r}) = \frac{1}{2}\left(\hat{O}_E - \hat{O}_{C_2}\right)\psi(\mathbf{r}) = \frac{1}{2}\left(\psi([E]^{-1}\mathbf{r}) - \psi([C_2]^{-1}\mathbf{r})\right)$$

where we have followed the standard procedure for applying an operator $\hat{O}_U$ on a function of $\mathbf{r}$, that is, we applied U^{-1} to the argument of the function. Since we have chosen the function to be $\psi(\mathbf{r}) = \psi(x\hat{x} + y\hat{y}) = x$, the above operations give:

$$[E]^{-1}\mathbf{r} = E\mathbf{r} = \mathbf{r}, \quad [C_2]^{-1}\mathbf{r} = C_2\mathbf{r} = -\mathbf{r} \Rightarrow$$

and with these, we finally find:

$$\psi^{(5)}(\mathbf{r}) = \frac{1}{2}\left(\psi(\mathbf{r}) - \psi(-\mathbf{r})\right) = \frac{1}{2}(x - (-x)) = x$$

so the projection of the component that transforms like Γ_5 is the function itself. We next apply the symmetry operators of the group to $\psi^{(5)}(\mathbf{r})$ (Rule 3), following the same process as above:

$$\hat{O}_E\psi^{(5)}(\mathbf{r}) = x, \quad \hat{O}_{C_4}\psi^{(5)}(\mathbf{r}) = y, \quad \hat{O}_{C_2}\psi^{(5)}(\mathbf{r}) = -x, \quad \hat{O}_{C_4^3}\psi^{(5)}(\mathbf{r}) = -y$$

$$\hat{O}_{\sigma_x}\psi^{(5)}(\mathbf{r}) = x, \quad \hat{O}_{\sigma_y}\psi^{(5)}(\mathbf{r}) = -x, \quad \hat{O}_{\sigma_1}\psi^{(5)}(\mathbf{r}) = y, \quad \hat{O}_{\sigma_2}\psi^{(5)}(\mathbf{r}) = -y$$

Table 3.7 Character table for the irreducible representations of the C_{4v} group.

C_i	C_1	C_2	C_3	C_4	C_5
N_i	1	2	1	2	2
Γ_1	1	1	1	1	1
Γ_2	1	1	1	-1	-1
Γ_3	1	-1	1	-1	1
Γ_4	1	-1	1	1	-1
Γ_5	2	0	-2	0	0

so the partner functions generated by all the group operations on $\psi^{(5)}(\mathbf{r})$ are $x, -x, y,$ $-y$. The linearly independent among those are x, y, which are already orthogonal, so the vector is (Rule 4) $\boldsymbol{\Psi}^{(5)} = (x, y)$. We then determine the matrices of the representation by requiring that they give the same result when multiplied by the vector $\boldsymbol{\Psi}^{(5)}$ as when the corresponding operator is applied to this vector (Rule 5). For example, for the matrices corresponding to C_4, C_4^3, C_2 we find:

$$\hat{O}_{C_4}(x, y) = (y, -x) = (x, y) \cdot \underline{\mathbf{D}}^{(5)}(C_4) \Rightarrow \underline{\mathbf{D}}^{(5)}(C_4) = \begin{bmatrix} 0 & -1 \\ 1 & 0 \end{bmatrix}$$

$$\hat{O}_{C_4^3}(x, y) = (-y, x) = (x, y) \cdot \underline{\mathbf{D}}^{(5)}(C_4^3) \Rightarrow \underline{\mathbf{D}}^{(5)}(C_4^3) = \begin{bmatrix} 0 & 1 \\ -1 & 0 \end{bmatrix}$$

$$\hat{O}_{C_2}(x, y) = (-x, -y) = (x, y) \cdot \underline{\mathbf{D}}^{(5)}(C_2) \Rightarrow \underline{\mathbf{D}}^{(5)}(C_2) = \begin{bmatrix} -1 & 0 \\ 0 & -1 \end{bmatrix}$$

and similarly for the remaining matrices.

Theorem 15 – Vector orthogonality *For the partner functions $\phi_k^{(\alpha)}(\mathbf{r})$ belonging to the αth IIR and $\psi_l^{(\beta)}(\mathbf{r})$ belonging to the βth IIR, the following relation holds:*

$$\langle \phi_k^{(\alpha)} | \psi_l^{(\beta)} \rangle = c_\alpha \delta_{kl} \delta_{\alpha\beta} \tag{3.49}$$

Proof From the definition of the partner functions:

$$\langle \hat{O}_U \phi_k^{(\alpha)} | = \sum_i \langle \phi_i^{(\alpha)} | D_{ik}^{(\alpha)*}(U), \quad |\hat{O}_U \psi_l^{(\beta)} \rangle = \sum_j |\psi_j^{(\beta)} \rangle D_{jl}^{(\beta)}(U) \Rightarrow$$

$$\langle \phi_k^{(\alpha)} | \psi_l^{(\beta)} \rangle = \frac{1}{\mathcal{N}} \sum_{n=1}^{\mathcal{N}} \langle \hat{O}_{U_n} \phi_k^{(\alpha)} | \hat{O}_{U_n} \psi_l^{(\beta)} \rangle = \frac{1}{\mathcal{N}} \sum_{ij} \langle \phi_i^{(\alpha)} | \psi_j^{(\beta)} \rangle \sum_{n=1}^{\mathcal{N}} D_{ik}^{*(\alpha)}(U_n) D_{jl}^{(\beta)}(U_n)$$

and using the grand orthogonality theorem (Theorem 5), we obtain:

$$\langle \phi_k^{(\alpha)} | \psi_l^{(\beta)} \rangle = \sum_{ij} \frac{1}{\eta_\alpha} \langle \phi_i^{(\alpha)} | \psi_j^{(\beta)} \rangle \delta_{\alpha\beta} \delta_{ij} \delta_{kl} = \frac{1}{\eta_\alpha} \sum_i \langle \phi_i^{(\alpha)} | \psi_i^{(\beta)} \rangle \delta_{\alpha\beta} \delta_{kl}$$

But we also have:

$$\frac{1}{\eta_\alpha} \sum_k \langle \phi_k^{(\alpha)} | \psi_k^{(\alpha)} \rangle = c_\alpha$$

for two sets of partner functions that belong to the *same* representation, and this relation, when substituted in the previous expression, proves the theorem. ∎

Theorem 16 – Vector orthogonality of hamiltonian matrix *For the partner functions $\phi_k^{(\alpha)}(\mathbf{r})$ and $\psi_l^{(\beta)}(\mathbf{r})$ of the αth and βth IIRs, the matrix elements of the hamiltonian obey:*

$$\langle \phi_k^{(\alpha)} | \hat{\mathcal{H}}^{\mathrm{sp}} | \psi_l^{(\beta)} \rangle = d_\alpha \delta_{kl} \delta_{\alpha\beta} \tag{3.50}$$

Proof The proof proceeds along the same steps as in the previous theorem, by substituting $|\psi_l^{(\beta)}\rangle \rightarrow \hat{\mathcal{H}}^{\mathrm{sp}}|\psi_l^{(\beta)}\rangle$, since the hamiltonian is invariant under any of the group operations. ∎

For any operator $\hat{\mathcal{F}}(\mathbf{r})$ corresponding to an observable, we can write its decomposition according to irreducible representations and partner functions:

$$\hat{\mathcal{F}}(\mathbf{r}) = \sum_{m,\gamma} \hat{\mathcal{F}}_m^{(\gamma)}(\mathbf{r})$$

With this decomposition, any value of a physical process will be expressed in terms of matrix elements of $\hat{\mathcal{F}}_m^{(\gamma)}(\mathbf{r})$ between partner functions of other irreducible representations of the form:

$$\langle \phi_k^{(\alpha)} | \hat{\mathcal{F}}_m^{(\gamma)} | \psi_l^{(\beta)} \rangle$$

This motivates the notion of the "direct product" of representations.

Definition 11 – Direct product of two representations *The direct product $\underline{\mathbf{A}} \otimes \underline{\mathbf{B}}$ of two representations $\underline{\mathbf{A}}$ and $\underline{\mathbf{B}}$ is given by the following matrix:*

$$\underline{\mathbf{A}} \otimes \underline{\mathbf{B}} = \begin{bmatrix} a_{11}\underline{\mathbf{B}} & a_{12}\underline{\mathbf{B}} & \cdots & a_{1n_A}\underline{\mathbf{B}} \\ a_{21}\underline{\mathbf{B}} & a_{22}\underline{\mathbf{B}} & \cdots & a_{2n_A}\underline{\mathbf{B}} \\ & & \cdot & \\ & & \cdot & \\ & & \cdot & \\ a_{n_A 1}\underline{\mathbf{B}} & a_{n_A 2}\underline{\mathbf{B}} & \cdots & a_{n_A n_A}\underline{\mathbf{B}} \end{bmatrix}$$

where $a_{ij}, i,j = 1,\ldots,n_A$, are the matrix elements of $\underline{\mathbf{A}}$, which has size $N_A \times N_A$; these multiply the entire matrix $\underline{\mathbf{B}}$, which has size $N_B \times N_B$, so that the resulting direct product matrix has size $(N_A N_B) \times (N_A N_B)$.

Theorem 17 – Product representation *Given two IIRs of a group, denoted by $\underline{\mathbf{D}}^{(\alpha)}, \underline{\mathbf{D}}^{(\beta)}$, the set of matrices*

$$\underline{\mathbf{D}}^{(\alpha \otimes \beta)}(U) \equiv \underline{\mathbf{D}}^{(\alpha)}(U) \otimes \underline{\mathbf{D}}^{(\beta)}(U)$$

forms a representation of the group, called the "product representation"; this, in general, is not an irreducible representation of the group.

Corollary *For the characters of the product representation, $\chi^{(\alpha\otimes\beta)}(U)$, the following relation holds in terms of the characters of the αth and βth representations, $\chi^{(\alpha)}(U)$ and $\chi^{(\beta)}(U)$:*

$$\chi^{(\alpha\otimes\beta)}(U) = \chi^{(\alpha)}(U) \cdot \chi^{(\beta)}(U)$$

as is evident from the definition of the matrices corresponding to the product representation.

Corollary *A set of basis functions can be constructed for the product representation from basis functions of the two representations, $\phi_k^{(\alpha)}(\mathbf{r})$ and $\psi_l^{(\beta)}(\mathbf{r})$:*

$$\Psi_{kl}^{(\alpha,\beta)}(\mathbf{r}) = \phi_k^{(\alpha)}(\mathbf{r}) \cdot \psi_l^{(\beta)}(\mathbf{r})$$

Theorem 18 – Vanishing matrix elements of operators *For the partner functions $\phi_k^{(\alpha)}(\mathbf{r})$ and $\psi_l^{(\beta)}(\mathbf{r})$ of the αth and βth IIRs, and the $\hat{\mathcal{F}}_m^{(\gamma)}(\mathbf{r})$ component of an operator belonging to the γth IIR, we will have:*

$$\langle\phi_k^{(\alpha)}|\hat{\mathcal{F}}_m^{(\gamma)}|\psi_l^{(\beta)}\rangle = 0 \quad \text{if} \quad \sum_{n=1}^{\mathcal{N}} \chi^{*(\alpha)}(U_n)\chi^{(\gamma)}(U_n)\chi^{(\beta)}(U_n) = 0$$

Proof The product $\hat{\mathcal{F}}_m^{(\gamma)}|\psi_l^{(\beta)}\rangle$ forms a representation which is reducible and can be decomposed into IIRs as:

$$\underline{\mathbf{D}}^{(\gamma\otimes\beta)}(U) = \sum_{\alpha} \mu_\alpha^{(\gamma\otimes\beta)}\underline{\mathbf{D}}^{(\alpha)}(U)$$

whose characters, from the corollary mentioned above, give the number of occurrences:

$$\mu_\alpha^{(\gamma\otimes\beta)} = \frac{1}{\mathcal{N}}\sum_{n=1}^{\mathcal{N}} \chi^{*(\alpha)}(U_n)\chi^{(\gamma\otimes\beta)}(U_n) = \frac{1}{\mathcal{N}}\sum_{n=1}^{\mathcal{N}} \chi^{*(\alpha)}(U_n)\chi^{(\gamma)}(U_n)\chi^{(\beta)}(U_n) \quad (3.51)$$

If the last sum of products of three characters is zero for a specific representation labeled α, then the number of occurrences $\mu_\alpha^{(\gamma\otimes\beta)}$ is zero, which proves the theorem. ∎

Definition 12 – Compatibility relations *Consider a group $\mathcal{G}$ and its subgroup $\mathcal{B}$, and an irreducible representation of $\mathcal{G}$ defined by the matrices $\underline{\mathbf{D}}^{(\alpha)}(U)$. The set of matrices $\underline{\mathbf{D}}^{(\alpha)}(W)$, where W is a member of $\mathcal{B}$, constitutes a representation of $\mathcal{B}$. In general, if the representation α is irreducible for $\mathcal{G}$, it is reducible for $\mathcal{B}$:*

$$\underline{\mathbf{D}}^{(\alpha)}(W) = \sum_{\beta} \mu_\beta^{(\alpha)}\underline{\mathbf{D}}^{(\beta)}(W)$$

where $\underline{\mathbf{D}}^{(\beta)}(W)$ are irreducible representations of $\mathcal{B}$ and μ_β their number of occurrences in the reducible representation $\underline{\mathbf{D}}^{(\alpha)}$, which is given by:

$$\mu_\beta^{(\alpha)} = \frac{1}{\mathcal{N}_\mathcal{B}}\sum_{n=1}^{\mathcal{N}_\mathcal{B}} \chi^{*(\beta)}(W_n)\chi^{(\alpha)}(W_n) \quad (3.52)$$

with $\mathcal{N}_{\mathcal{B}}$ the number of elements in the subgroup $\mathcal{B}$. If $\mu_\beta^{(\alpha)} \neq 0$ for a given IIR of $\mathcal{B}$ labeled β, then the representations $\underline{\mathbf{D}}^{(\beta)}$ of the subgroup and $\underline{\mathbf{D}}^{(\alpha)}$ of the full group are called "compatible."

Example 3.6   Compatible representations in bands of 2D square lattice

To illustrate how these concepts are used, we revisit the 2D square model with one atom per unit cell, with s, p_x, p_y orbitals, as discussed in detail in Chapter 2. In the present example, we will omit the p_z orbital for simplicity, and we assume parameters that lead to a clear separation of the s-like and p-like bands, as shown in Fig. 3.7. The high-symmetry points in the BZ are labeled $\Gamma, \Delta, X, Z, M, \Sigma$. Of these, Γ and M have the full symmetry of the C_{4v} group, and the rest have the smaller symmetry groups (subgroups of C_{4v}), composed of the operations: for $X \rightarrow \{E, C_2, \sigma_x, \sigma_y\}$, for $Z \rightarrow \{E, \sigma_y\}$, for $\Delta \rightarrow \{E, \sigma_x\}$, and for $\Sigma \rightarrow \{E, \sigma_1\}$. The character tables for Γ and M are those of the C_{4v} group derived in the previous example. The group of X has size $\mathcal{N} = 4$ and four classes (each group element is its own class), and therefore four IIRs each of dimension 1, while the groups for Δ, Σ, and Z have size $\mathcal{N} = 2$ and two classes (each element is its own class) and therefore two IIRs each of dimension 1. The character tables for the group of X and for the group of Δ, derived from the usual rules, are shown in Table 3.7.

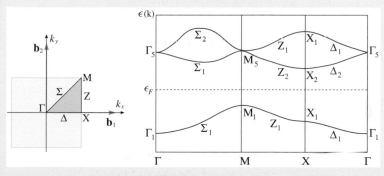

Fig. 3.7

Left: The Brillouin zone and its irreducible part (the shaded triangle) of the 2D square lattice with the high-symmetry points labeled $\Gamma, \Delta, X, Z, M, \Sigma$. **Right**: The energy bands of the square lattice with s, p_x, p_y orbitals on each site, along the high-symmetry directions, labeled according to the irreducible representations of the C_{4v} group, which give the compatibilities between bands (red band: s-like states, blue bands: p-like states). The horizontal dashed line indicates the Fermi level, ϵ_F, presumed for this example.

Table 3.8 Character tables for the irreducible representations corresponding to the X and Δ high-symmetry points of the 2D square lattice.

$\mathcal{C}_i$	$\mathcal{C}_1$	$\mathcal{C}_2$	$\mathcal{C}_3$	$\mathcal{C}_4$		$\mathcal{C}_i$	$\mathcal{C}_1$	$\mathcal{C}_2$
N_i	1	1	1	1		N_i	1	1
X_1	1	1	1	1		Δ_1	1	1
X_2	1	1	-1	-1		Δ_2	1	-1
X_3	1	-1	1	-1				
X_4	1	-1	-1	1				

The character tables for the groups of Σ and Z are identical to that of Δ. Since all these representations are 1D, the characters are also the matrices of each representation. We can now figure out the compatibility relations using Eq. (3.52). We begin with the compatibility between Γ_1 and Δ_1, Δ_2:

$$\mu_{\Delta_1}^{(\Gamma_1)} = \frac{1}{2}\left[\chi^{*(\Delta_1)}(E)\chi^{(\Gamma_1)}(E) + \chi^{*(\Delta_1)}(\sigma_x)\chi^{(\Gamma_1)}(\sigma_x)\right] = 1$$

$$\mu_{\Delta_2}^{(\Gamma_1)} = \frac{1}{2}\left[\chi^{*(\Delta_2)}(E)\chi^{(\Gamma_1)}(E) + \chi^{*(\Delta_2)}(\sigma_x)\chi^{(\Gamma_1)}(\sigma_x)\right] = 0$$

as we would expect, that is, the identical representation Δ_1 is compatible with the identical representation Γ_1, but Δ_2 is not. Similarly, we find for the compatibility between Γ_5 and the two representations Δ_1, Δ_2:

$$\mu_{\Delta_1}^{(\Gamma_5)} = \frac{1}{2}\left[\chi^{*(\Delta_1)}(E)\chi^{(\Gamma_5)}(E) + \chi^{*(\Delta_1)}(\sigma_x)\chi^{(\Gamma_5)}(\sigma_x)\right] = 1$$

$$\mu_{\Delta_2}^{(\Gamma_5)} = \frac{1}{2}\left[\chi^{*(\Delta_2)}(E)\chi^{(\Gamma_5)}(E) + \chi^{*(\Delta_2)}(\sigma_x)\chi^{(\Gamma_5)}(\sigma_x)\right] = 1$$

that is, the 2D representation Γ_5 is compatible with both 1D representations of Δ. By similar considerations we find that the compatibilities of the various representations at the high-symmetry points are:

$$\Gamma_1 \text{ or } M_1 \rightarrow \Delta_1, \Sigma_1, Z_1, \quad \Gamma_2 \text{ or } M_2 \rightarrow \Delta_2, \Sigma_2, Z_2, \quad \Gamma_3 \text{ or } M_3 \rightarrow \Delta_1, \Sigma_2, Z_1,$$

$$\Gamma_4 \text{ or } M_4 \rightarrow \Delta_2, \Sigma_1, Z_2, \quad \text{and} \quad \Gamma_5 \text{ or } M_5 \rightarrow (\Delta_1 + \Delta_2), (\Sigma_1 + \Sigma_2), (Z_1 + Z_2)$$

These relations allow us to label the bands along the high-symmetry directions according to the IIRs of each group, as shown in Fig. 3.7.

Finally, we can use the above results to figure out the allowed optical transitions between the s-like occupied band and the p-like unoccupied bands, as indicated in Fig. 3.7. To do this, we will need to calculate the matrix elements of the optical transition operator, which is proportional to the momentum operator or, equivalently, from commutation relations of the hamiltonian, to the position operator $\langle\psi_{\mathbf{k}}^{(n)}|\mathbf{r}|\psi_{\mathbf{k}}^{(n')}\rangle$, where $|\psi_{\mathbf{k}}^{(n)}\rangle, |\psi_{\mathbf{k}}^{(n')}\rangle$ are the two single-particle initial and final states involved in the transition (for details, see the discussion in Appendix C). We need first to decompose the operator $\mathbf{r}$ to the IIRs of the group of $\mathbf{k}$. We will work out two cases, the high-symmetry points Γ and Δ.

At Γ, the group of $\mathbf{k}$ is the full group, C_{4v}, and the operator $\hat{r}$ is the three-component vector $\mathbf{r} = x\hat{x} + y\hat{y} + z\hat{z}$, which contains Γ_1 (the z component transforms like this representation) and Γ_5 (the (x, y) vector transforms like this representation, as we showed explicitly in the previous example). We then need to calculate matrix elements of the general type:

$$\langle\psi^{(\Gamma_k)}|\hat{r}^{(\Gamma_m)}|\psi^{(\Gamma_l)}\rangle, \quad k, l, m = 1, \ldots, 5$$

and in this particular case, $k = 1$ for the occupied state, $l = 5$ for the unoccupied state, and $m = 1$ or 5 for the components of the operator $\hat{r}$. From Theorem 18, Eq. (3.51), we obtain the following decompositions of the direct product representations:

$$\Gamma_1 \otimes \Gamma_l = \Gamma_l \Rightarrow \langle \psi^{(\Gamma_k)} | \hat{r}^{(\Gamma_1)} | \psi^{(\Gamma_l)} \rangle \propto \delta_{kl}$$

so from the $\hat{r}^{(\Gamma_1)}$ component the transition between the $\langle \psi^{(\Gamma_1)} |$ (occupied, s-like) and $| \psi^{(\Gamma_5)} \rangle$ (unoccupied, p-like) states is *not* allowed, since:

$$\langle \psi^{(\Gamma_1)} | \hat{r}^{(\Gamma_1)} | \psi^{(\Gamma_5)} \rangle \propto \langle \psi^{(\Gamma_1)} | \psi^{(\Gamma_5)} \rangle = 0$$

while for the $\hat{r}^{(\Gamma_1)}$ component we have:

$$\Gamma_5 \otimes \Gamma_l = \Gamma_5, \quad l = 1, 2, 3, 4, \quad \text{and} \quad \Gamma_5 \otimes \Gamma_5 = \Gamma_1 + \Gamma_2 + \Gamma_3 + \Gamma_4$$
$$\Rightarrow \langle \psi^{(\Gamma_1)} | \hat{r}^{(\Gamma_5)} | \psi^{(\Gamma_5)} \rangle \neq 0$$

so the transition between $\langle \psi^{(\Gamma_1)} |$ and $| \psi^{(\Gamma_5)} \rangle$ is allowed through the $\hat{r}^{(\Gamma_5)}$ component.

In the case of the Δ point, σ_x leaves x and z invariant and it takes y to $-y$, which means that x and z transform like the Δ_1 representation and y transforms like the Δ_2 representation. As before, we then have:

$$\Delta_1 \otimes \Delta_l = \Delta_l \Rightarrow \langle \psi^{(\Delta_k)} | \hat{r}^{(\Delta_1)} | \psi^{(\Delta_l)} \rangle \propto \delta_{kl}$$

which means that only transition between the occupied Δ_1 s-like state and the unoccupied Δ_1 p-like state is allowed through the $\hat{r}^{(\Delta_1)}$ component of the operator. In a similar vein:

$$\Delta_2 \otimes \Delta_1 = \Delta_2, \quad \Delta_2 \otimes \Delta_2 = \Delta_1 \Rightarrow \langle \psi^{(\Delta_1)} | \hat{r}^{(\Delta_2)} | \psi^{(\Delta_1)} \rangle = 0,$$
$$\langle \psi^{(\Delta_1)} | \hat{r}^{(\Delta_2)} | \psi^{(\Delta_2)} \rangle \neq 0$$

that is, only transitions between the occupied Δ_1 s-like state and the unoccupied Δ_2 p-like state is allowed through the $\hat{r}^{(\Delta_2)}$ component of the operator. The usefulness of this analysis becomes more evident when dealing with light of a certain polarization; for instance, for light polarized in the z direction, only the Δ_1 component applies, which will affect the allowed transitions.

3.7 Application: The N-V-Center in Diamond

We demonstrate the use of the concepts developed in this chapter by applying them to an important material system, the nitrogen-vacancy defect in diamond. This defect consists of a missing C atom, referred to as a "vacancy" or V, and one of the C atoms adjacent to the V site being replaced by an N atom. The structure of this defect, which is known as the N-V-center, is shown in Fig. 3.8. This defect has become the focus of intense theoretical and experimental investigation, because it may be a good candidate for a quantum bit, the basic element of a quantum computer. Some of its most important properties are that it can absorb and emit photons of a specific frequency and polarization, which makes it possible

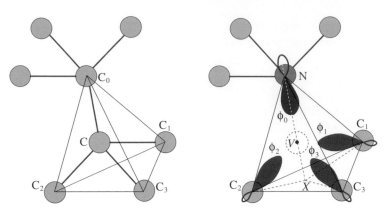

Fig. 3.8 Structure and symmetries of the N-V-center in diamond. **Left**: The ideal diamond structure, with the carbon atoms at the corners of a tetrahedron labeled C_0, C_1, C_2, C_3, and the one at the center labeled C. **Right**: The atom C at the center is removed and is now a vacancy, labeled V (dashed circle), and the atom C_0 at the apex has been substituted by an N atom (shown in blue). For each of the neighbors of the vacancy, the corresponding hybrids (the sp^3 dangling bonds for C_1, C_2, and C_3 and the p orbital for N) are also shown as colored lobes. X is the center of the equilateral triangle defined by the positions of C_1, C_2, C_3. The three planes defined by the triangles NXC_i, $i = 1, 2, 3$, are the mirror planes of the C_{3v} group.

to manipulate these effects for applications of quantum entanglement in cryptography, communications, and imaging. The absorption and emission of photons with these special features is made possible by electronic transitions between states that are created by the presence of this defect, and do not exist in the pure crystal. The possible transitions between states are determined by their symmetry, and the selection rules discussed above. We give here a simplified version of the N-V-center properties, based on the group-theoretical tools we have developed. The simplification in our description is that we will not take full account of the spin degrees of freedom, which considerably complicates the problem. For a full account, we refer the reader to the relevant literature.[2]

Our goal here is to construct the wavefunctions with the proper symmetry, which can then be used to calculate the allowed optical transitions. We begin with the basis of orbitals appropriate for the N-V-center. The three C atoms and the N atom that surround the vacancy all have three bonds to the other C atoms of the diamond lattice. These consist of the three "dangling bonds" of the C atoms next to the vacant site, that is, the hybrid orbitals that would normally form covalent bonds to the C atom that is missing, and one orbital of the N atom which is a "lone-pair" orbital. The three C dangling bonds have sp^3 character, whereas the N lone-pair orbital is closer to a p orbital. The system has C_{3v} symmetry around an axis that passes through the N position and the V position, as illustrated in Fig. 3.8. Thus, we have a basis of four orbitals labeled ϕ_1, ϕ_2, ϕ_3 for the three C dangling bonds, which are equivalent and related by threefold rotations around the C_{3v} axis, and the N lone-pair orbital labeled ϕ_0. There are three mirror planes, passing through a C atom, the N atom,

[2] See, for example, A. Gali, M. Fyta, and E. Kaxiras, *Phys. Rev. B* **77**, 155206 (2008); J. R. Maze, A. Gali, E. Togan, Y. Chu, A. Trifonov, E. Kaxiras, and M. D. Lukin, *New J. Phys.*, **13**, 025025 (2011).

and the V site. We can then construct the matrices that describe how each operation of the C_{3v} group transforms the four-component vector $(\phi_0, \phi_1, \phi_2, \phi_3)$. These matrices are:

$$\underline{\mathbf{D}}'(E) = \begin{bmatrix} 1 & 0 & 0 & 0 \\ 0 & 1 & 0 & 0 \\ 0 & 0 & 1 & 0 \\ 0 & 0 & 0 & 1 \end{bmatrix}, \quad \underline{\mathbf{D}}'(C_3) = \begin{bmatrix} 1 & 0 & 0 & 0 \\ 0 & 0 & 0 & 1 \\ 0 & 1 & 0 & 0 \\ 0 & 0 & 1 & 0 \end{bmatrix}, \quad \underline{\mathbf{D}}'(C_3^2) = \begin{bmatrix} 1 & 0 & 0 & 0 \\ 0 & 0 & 1 & 0 \\ 0 & 0 & 0 & 1 \\ 0 & 1 & 0 & 0 \end{bmatrix}$$

$$\underline{\mathbf{D}}'(\sigma_1) = \begin{bmatrix} 1 & 0 & 0 & 0 \\ 0 & 1 & 0 & 0 \\ 0 & 0 & 0 & 1 \\ 0 & 0 & 1 & 0 \end{bmatrix}, \quad \underline{\mathbf{D}}'(\sigma_2) = \begin{bmatrix} 1 & 0 & 0 & 0 \\ 0 & 0 & 0 & 1 \\ 0 & 0 & 1 & 0 \\ 0 & 1 & 0 & 0 \end{bmatrix}, \quad \underline{\mathbf{D}}'(\sigma_3) = \begin{bmatrix} 1 & 0 & 0 & 0 \\ 0 & 0 & 1 & 0 \\ 0 & 1 & 0 & 0 \\ 0 & 0 & 0 & 1 \end{bmatrix}$$

The first task is to break these 4×4 matrices down to a diagonal form that contains the IIRs of the C_{3v} group. We notice that the matrices are already in a form that helps this task, since they contain a 1×1 submatrix on the upper-left corner and 3×3 sub-matrices which are not diagonal. This is because the symmetry operations do not mix the lone-pair orbital of N with any of the other three basis functions, as it is physically different; this orbital forms by itself the 1×1 submatrix, which is also the identical representation, Γ_1. The set of 3×3 submatrices are exactly the $\underline{\mathbf{D}}'(U)$ matrices we studied in Example 3.3. We can then take the results of that example and apply them to break down the 3×3 submatrices into 1×1 and 2×2 submatrices, with use of the similarity transformation defined by the matrices

$$\underline{\mathbf{S}} = \frac{1}{\sqrt{6}} \begin{bmatrix} \sqrt{2} & \sqrt{2} & \sqrt{2} \\ 0 & \sqrt{3} & -\sqrt{3} \\ -2 & 1 & 1 \end{bmatrix}, \quad \underline{\mathbf{S}}^{-1} = \frac{1}{\sqrt{6}} \begin{bmatrix} \sqrt{2} & 0 & -2 \\ \sqrt{2} & \sqrt{3} & 1 \\ \sqrt{2} & -\sqrt{3} & 1 \end{bmatrix}$$

which gave the set of 3×3 matrices $\underline{\mathbf{D}}(U)$ in Example 3.3. The similarity transformation is also useful in finding the appropriate combinations of the basis orbitals that correspond to the block-diagonal group representation. Using Eq. (3.47), these are the following:

$$\mathbf{\Phi} = \mathbf{\Psi} \cdot \underline{\mathbf{S}} \Rightarrow (\psi_1, \psi_2, \psi_3) = (\phi_1, \phi_2, \phi_3)\, \underline{\mathbf{S}}^{-1}$$

giving the new symmetrized functions:

$$\psi_1' = \frac{1}{\sqrt{3}}(\phi_1 + \phi_2 + \phi_3), \quad \psi_2 = \frac{1}{\sqrt{2}}(\phi_2 - \phi_3), \quad \psi_3 = \frac{1}{\sqrt{6}}(2\phi_1 - \phi_2 - \phi_3) \quad (3.53)$$

To complete the picture, we must also include as one of the symmetrized functions the state corresponding to the nitrogen lone-pair orbital, which we call $\psi_0' = \phi_0$; this transforms like Γ_1, the identical IIR. Our choices for the states labeled "0" and "1" are not the final ones (hence the primes), for the reason we explain next. We know from the form of the 3×3 matrices $\underline{\mathbf{D}}(U)$ that this block-diagonal representation contains Γ_1, the 1D identical IIR, and the 2D Γ_3 representation, as we showed explicitly in Example 3.4. Thus, both ψ_0' and ψ_1' transform like Γ_1, therefore for our final choice of symmetrized basis functions we can choose any linear combination of those two. We construct a pair of functions that are linear combinations of ψ_0', ψ_1' and are orthogonal, which leads to the more useful, final choice of basis functions ψ_0, ψ_1:

$$\psi_0 = \alpha\phi_0 + \frac{\sqrt{1-\alpha^2}}{\sqrt{3}}(\phi_1 + \phi_2 + \phi_3), \quad \psi_1 = \sqrt{1-\alpha^2}\phi_0 - \frac{\alpha}{\sqrt{3}}(\phi_1 + \phi_2 + \phi_3) \quad (3.54)$$

where α is a parameter that we need to determine from considerations beyond symmetry.

With the help of electronic structure calculations, it can be established that ψ_0 is the state with the lowest energy, ψ_1 is the next one, and ψ_2, ψ_3 are a pair of degenerate states. This much we expected, since the symmetric ("bonding") combination ψ_0 of the two totally symmetric states ψ_0' and ψ_1', both of which transform like the identity representation, is typically the one with the lowest energy, and their antisymmetric ("anti-bonding") combination ψ_1 is the next one in energy, while the two states ψ_2, ψ_3 are partner functions of the Γ_3 2D IIR of C_{3v}, and their lower symmetry typically implies higher energy. To conform with literature conventions, we label the ψ_0 state as $a_1(1)$, the ψ_1 state as $a_1(2)$, and the ψ_2, ψ_3 states as e_x, e_y, because the latter transform like the (x, y) pair under the symmetry operations of C_{3v}.

What is left to do, is to include the occupancy of these single-particle states and the spin degrees of freedom in order to construct many-body wavefunctions relevant to the problem. Each C dangling bond contains one electron (half of the covalent bond content) and the N lone-pair orbital contains two electrons. The N-V-center is typically in a negatively charged state due to the presence of a nearby defect that gives up one electron, which is denoted as the N-V^--center. Thus, we have a total of six electrons to distribute to the available states. It turns out that the $a_1(1)$ state is well below the valence band maximum (VBM) of diamond, so it is fully occupied by two electrons of opposite spin and is not relevant to the optical properties of the N-V^--center. The ground state of the system then has the $a_1(2)$ state fully occupied by two electrons of opposite spin, and one electron of spin up in each of the degenerate e_x, e_y states, as shown in Fig. 3.9. The lowest-energy excitations from this state consist of promoting one or two electrons from the $a_1(2)$ state to one of the partially

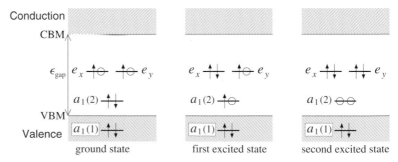

Fig. 3.9   Illustration of the occupation of states in the N-V^--center. The $a_1(1)$ single-particle state is well below the VBM and is fully occupied. In the ground state there are two electrons in the $a_1(2)$ state and one electron in each of the e_x, e_y states. In the first excited state there is one electron and one hole in the $a_1(2)$ state, and one of the e_x, e_y states holds two electrons while the other one has an electron and a hole. In the next excited state, $a_1(2)$ is empty (it has two holes) and the e_x, e_y states are fully occupied.

occupied e_x, e_y states. These excited states are shown schematically in Fig. 3.9. This picture is rather simplified, in two important ways. First, spin–orbit interactions split the spin-up and spin-down levels so that they are no longer degenerate, and the spin-up, spin-down channels must be treated separately. Second, a full treatment involves the construction of many-body wavefunctions for all the electrons participating in the initial and final states of the transition upon photon absorption; this is beyond what we have covered so far, and is one of the topics addressed in Chapter 4. The many-body wavefunctions will involve linear combinations of the single-particle states introduced above. Consequently, the transition probabilities will depend on matrix elements of the operator $\hat{\mathbf{p}}$ or equivalently $\hat{\mathbf{r}}$, as discussed in the previous example. Figuring out which of these matrix elements vanish or not will depend on the representations of the single-particle states and the transition operator. We leave this as an exercise for the reader (see Problem 6).

Further Reading

1. *Group Theory in Quantum Mechanics*, V. Heine (Pergamon Press, New York, 1960).
2. *Group Theory and Quantum Mechanics*, M. Tinkham (McGraw-Hill, New York, 1964).
3. *The Structure of Materials*, S. M. Allen and E. L. Thomas (Wiley, New York, 1998).

Problems

1. Show that the only rotations compatible with 3D periodicity are multiples of $2\pi/4$ and $2\pi/6$.

2. Find the symmetries and construct the group multiplication table for a 2D square lattice model, with two identical atoms per unit cell, at positions $\mathbf{t}_1 = 0, \mathbf{t}_2 = 0.5a\hat{\mathbf{x}} + 0.5a\hat{\mathbf{y}}$, where a is the lattice constant. Is this group symmorphic or non-symmorphic?

3. Consider the 2D honeycomb lattice with two identical atoms per unit cell, defined by the lattice vectors:

$$\mathbf{a}_1 = \frac{\sqrt{3}}{2}a\hat{\mathbf{x}} - \frac{1}{2}a\hat{\mathbf{y}}, \quad \mathbf{a}_2 = \frac{\sqrt{3}}{2}a\hat{\mathbf{x}} + \frac{1}{2}a\hat{\mathbf{y}} \tag{3.55}$$

(these form the 2D hexagonal lattice), and atomic positions $\mathbf{t}_1 = 0, \mathbf{t}_2 = 1/\sqrt{3}(a\hat{\mathbf{x}})$. Determine all the symmetries of this lattice, and construct its group multiplication table. Draw the irreducible Brillouin zone and indicate the high-symmetry points and the symmetry operations for each.

4. Find a special $\mathbf{k}$-point for the 2D square lattice and the 2D hexagonal lattice. How many shells can you make vanish with a single special $\mathbf{k}$-point in each case?

5. Show that the hamiltonian $\hat{\mathcal{H}}_{U\mathbf{k}}^{\text{sp}}$ defined in the last step of Eq. (3.17) leads to the relations

$$\epsilon_{U\mathbf{k}}^{(n)} = \epsilon_{\mathbf{k}}^{(n)}, \quad \psi_{U\mathbf{k}}^{(n)}(\mathbf{r}) = \hat{\mathcal{O}}_{\{U|\mathbf{t}\}} \psi_{\mathbf{k}}^{(n)}(\mathbf{r})$$

6. Determine the allowed optical transitions between the single-particle states involved in the description of the N-V^--center in diamond as discussed in Section 3.7.

4 From Many Particles to the Single-Particle Picture

Up to this point we have treated electrons in solids as essentially independent particles and solved the appropriate single-particle Schrödinger equations, exploiting only the symmetries imposed by the presence of the crystal lattice of ions. Given the strong and long-range interaction between electrons, that is, the Coulomb repulsion, and the fact that they are indistinguishable particles, we would expect a more complicated behavior, including some degree of correlation in the motion of these interacting particles. Thus, a better justification for the single-particle picture is required. We provide a comprehensive justification in this chapter.

We will do this in several stages. The first stage is the definition of the exact hamiltonian for the solid and a thorough discussion of the important terms. Next, we will introduce the simplest possible ansatz for the many-body wavefunction, namely as a product of single-particle states, called the Hartree approximation; this will be followed by a more sophisticated ansatz for the many-body wavefunction that includes the Pauli exclusion principle for electrons, called the Hartree–Fock approximation. Both of these approximations, through a variational argument that minimizes the total energy, lead to a set of single-particle equations that each of these states must satisfy. In the single-particle equations, each electron experiences the presence of all other electrons through an effective potential which encapsulates the many-body nature of the true system in an approximate way. Finally, we will discuss a formulation of the problem which is again based on a variational principle but using the electron *density* as the main physical quantity to describe the system; this approach is known as the density functional theory (DFT). This is an exact approach, but some of the expressions needed to carry out actual calculations are not known. Introducing physically motivated approximations for the unknown quantities, and mapping the problem onto a system of *non-interacting* fictitious particles that obey certain conditions, provides a path toward practical implementations of the theory. The advantages of this approach are twofold. Firstly, it gives a very accurate description of the total energy and the electron density of the system, based on which a wide range of properties of materials can be obtained with remarkable accuracy. Secondly, it involves no adjustable parameters, other than the approximation of an important term in the energy functional, which, however, can be derived systematically from exact calculations of idealized model systems, like the uniform electron gas. In this sense, it is a truly first-principles approach for the description of properties of matter.

There are important advantages to using the single-particle picture to describe the properties of materials. One is conceptual: it is much simpler to think of individual electrons moving in a field created by the presence of all charged particles present (ions and other electrons), even though this is not exact because electrons are identical particles and singling out one electron is a gross oversimplification of the real system. In fact, it is quite challenging to conceive of the entire system of identical and interacting electrons at once. Another reason for adopting the single-particle picture is practical: it represents a convenient way of doing calculations for real materials and for a very broad range of properties; had we been forced to deal with the true many-body character of the electrons, we would have been restricted to a much more narrow range of problems that could be realistically investigated. We have already experienced the power of this approach, by applying it to study the behavior of single-particle states in a crystal, in simple qualitative models, in Chapter 2. With the ideas and approximations developed in the present chapter, we will then be able to provide a quantitative description of the behavior of electrons in solids and its implications for the physical properties of solids.

4.1 The Hamiltonian of the Solid

An exact non-relativistic theory for a system of ions and interacting electrons is inherently quantum mechanical, and is based on solving a many-body Schrödinger equation of the form

$$\hat{\mathcal{H}}^{\text{tot}}(\{\mathbf{R};\mathbf{r}\})\Psi(\{\mathbf{R};\mathbf{r}\}) = E^{\text{tot}}\Psi(\{\mathbf{R};\mathbf{r}\}) \tag{4.1}$$

where $\hat{\mathcal{H}}^{\text{tot}}(\{\mathbf{R};\mathbf{r}\})$ is the hamiltonian of the *total* system, which contains as variables both the electronic degrees of freedom, denoted as $\{\mathbf{r}\}$, and the ionic degrees of freedom, denoted as $\{\mathbf{R}\}$, E^{tot} is the energy of the entire system and $\Psi(\{\mathbf{R};\mathbf{r}\})$ is the many-body wavefunction that describes the state of the entire system. In order to write down the full hamiltonian, we consider all the possible contributions to the energy. We will use the following symbols in the various expressions: $\hbar$ is Planck's constant divided by 2π, M_I is the mass of ion I and m_e is the mass of the electron, e is the magnitude of the charge of an electron (a positive quantity), Z_I is the valence charge (nucleus plus core electrons) of ion I; the indices I, J and i, j run over the ionic and electronic degrees of freedom, respectively.

The first type of contribution is kinetic energy from the electrons and the ions. For each particle, the quantum-mechanical operator acting on the wavefunction to produce the kinetic energy is expressed in terms of the momentum operator, $\hat{\mathbf{P}}_I = -i\hbar\nabla_{\mathbf{R}_I}$ (for ions) and $\hat{\mathbf{p}}_i = -i\hbar\nabla_{\mathbf{r}_i}$ (for electrons), and the mass of the particle, M_I for ions, m_e for electrons:

$$\text{kinetic energy}: \quad \frac{\hat{\mathbf{P}}_I^2}{2M_I} = -\frac{\hbar^2\nabla_{\mathbf{R}_I}^2}{2M_I}, \quad \frac{\hat{\mathbf{p}}_i^2}{2m_e} = -\frac{\hbar^2\nabla_{\mathbf{r}_i}^2}{2m_e}$$

There are three types of contributions to the potential energy. The first is the potential energy of the Coulomb interaction between electrons and ions due to their electric charges: an electron of charge $-e$ at position $\mathbf{r}_i$ is attracted to each positively charged ion of charge $Z_I e$ at position $\mathbf{R}_I$, producing a potential energy term

$$\text{electron–ion attraction energy}: \quad -\frac{Z_I e^2}{|\mathbf{R}_I - \mathbf{r}_i|}$$

so that the total potential energy of an electron due to the presence of all the ions at positions $\{\mathbf{R}\}$ in the system is:

$$V_{\{\mathbf{R}\}}(\mathbf{r}_i) = -\sum_I \frac{Z_I e^2}{|\mathbf{R}_I - \mathbf{r}_i|} \tag{4.2}$$

The second is the potential energy due to the Coulomb interaction between each pair of electrons: two electrons at positions $\mathbf{r}_i, \mathbf{r}_j$ repel one another, which produces a potential energy

$$\text{electron–electron repulsion energy}: \quad \frac{e^2}{|\mathbf{r}_i - \mathbf{r}_j|}$$

The third is the potential energy due to the Coulomb interaction between each pair of ions: two ions at positions $\mathbf{R}_I, \mathbf{R}_J$ with charges $Z_I e, Z_J e$ repel one another, giving rise to a potential energy term

$$\text{ion–ion repulsion energy}: \quad \frac{Z_I Z_J e^2}{|\mathbf{R}_I - \mathbf{R}_J|}$$

To obtain the final expression for the total hamiltonian, we collect all terms and sum over the electronic and ionic degrees of freedom. We will split the total hamiltonian into two parts: the first part contains as variables the electronic degrees of freedom, which we will call $\hat{\mathcal{H}}_{\{\mathbf{R}\}}^{\text{ele}}(\{\mathbf{r}\})$; the second part contains the remaining terms, in which the only variables left are the ionic degrees of freedom, which we will call $\hat{\mathcal{H}}^{\text{ion}}(\{\mathbf{R}\})$. This gives:

$$\hat{\mathcal{H}}^{\text{tot}}(\{\mathbf{R}; \mathbf{r}\}) = \hat{\mathcal{H}}_{\{\mathbf{R}\}}^{\text{ele}}(\{\mathbf{r}\}) + \hat{\mathcal{H}}^{\text{ion}}(\{\mathbf{R}\})$$

$$\hat{\mathcal{H}}_{\{\mathbf{R}\}}^{\text{ele}}(\{\mathbf{r}\}) = -\sum_i \frac{\hbar^2}{2m_e} \nabla_{\mathbf{r}_i}^2 + \sum_i V_{\{\mathbf{R}\}}(\mathbf{r}_i) + \frac{e^2}{2} \sum_{i \neq j} \frac{1}{|\mathbf{r}_i - \mathbf{r}_j|} \tag{4.3}$$

$$\hat{\mathcal{H}}^{\text{ion}}(\{\mathbf{R}\}) = -\sum_I \frac{\hbar^2}{2M_I} \nabla_{\mathbf{R}_I}^2 + \frac{e^2}{2} \sum_{I \neq J} \frac{Z_I Z_J}{|\mathbf{R}_I - \mathbf{R}_J|} \tag{4.4}$$

This separation will prove very useful in the following discussion. Notice that a factor of 1/2 must be included in the sums of the electron–electron and ion–ion pairs to take into account the double counting.

4.1.1 Born–Oppenheimer Approximation

In most situations, we can think of the ions as moving slowly in space and the electrons responding instantaneously to any ionic motion, so that $\Psi(\{\mathbf{R}; \mathbf{r}\})$ has an explicit dependence on the instantaneous electronic degrees of freedom alone: this is known as the Born–Oppenheimer approximation (BOA). Its validity is based on the huge difference in mass between ions and electrons (three to four orders of magnitude), making the former behave like classical particles. Invoking the BOA is extremely useful from a practical point of view, because it helps break down the problem into two parts that can be solved more easily as separate problems. It is also of great conceptual value, as it allows a better understanding of how electrons and ions behave in the solid, a picture which is quite realistic and very successful. Some exceptions are systems that involve the lightest element, hydrogen, or situations where the electronic and ionic degrees of freedom are interwoven, in which case the motion of the electrons and ions has to be solved as a single coupled system; we will encounter some examples in later chapters.

We next take a closer look at how the BOA can be implemented, and the consequences of this implementation. The basic ansatz is that the motion of ions and electrons can be separated out; that is, the total wavefunction $\Psi(\{\mathbf{R};\mathbf{r}\})$ can be written as a product of two wavefunctions, one that contains the ionic degrees of freedom explicitly, which we denote as $\Phi(\{\mathbf{R}\})$, and one that contains the dependence on all the electronic degrees of freedom for a given ionic configuration, which we denote as $\Psi_{\{\mathbf{R}\}}(\{\mathbf{r}\})$:

$$\Psi(\{\mathbf{R};\mathbf{r}\}) = \Phi(\{\mathbf{R}\})\Psi_{\{\mathbf{R}\}}(\{\mathbf{r}\}) \tag{4.5}$$

The essence of the BOA is to solve the electronic problem first for a given ionic configuration, considering the ions fixed in their positions $\{\mathbf{R}\}$, and then deal with the motion of ions separately. Our first task then is to solve for the electronic wavefunction $\Psi_{\{\mathbf{R}\}}(\{\mathbf{r}\})$ and corresponding energy $E_{\{\mathbf{R}\}}^{\text{ele}}$, using the hamiltonian $\hat{\mathcal{H}}_{\{\mathbf{R}\}}^{\text{ele}}$:

$$\hat{\mathcal{H}}_{\{\mathbf{R}\}}^{\text{ele}}(\{\mathbf{r}\})\Psi_{\{\mathbf{R}\}}(\{\mathbf{r}\}) = E_{\{\mathbf{R}\}}^{\text{ele}}\Psi_{\{\mathbf{R}\}}(\{\mathbf{r}\}) \tag{4.6}$$

and then treat the motion of ions separately. To investigate the consequences of this separation of the motion of electrons and ions from each other, we substitute in Eq. (4.1) the expression for $\Psi(\{\mathbf{R};\mathbf{r}\})$ from Eq. (4.5), and we multiply from the left both sides of the equation with $\Psi_{\{\mathbf{R}\}}^*(\{\mathbf{r}\})$, denoted as $\langle\Psi_{\{\mathbf{R}\}}|$, and integrate over all electronic coordinates, assuming that the electronic wavefunction is properly normalized, $\langle\Psi_{\{\mathbf{R}\}}|\Psi_{\{\mathbf{R}\}}\rangle = 1$. This leads to:

$$\langle\Psi_{\{\mathbf{R}\}}|\hat{\mathcal{H}}^{\text{tot}}|\Psi_{\{\mathbf{R}\}}\rangle\Phi(\{\mathbf{R}\}) = \left[\langle\Psi_{\{\mathbf{R}\}}|\hat{\mathcal{H}}^{\text{ion}}|\Psi_{\{\mathbf{R}\}}\rangle + E_{\{\mathbf{R}\}}^{\text{ele}}\right]\Phi(\{\mathbf{R}\}) = E^{\text{tot}}\Phi(\{\mathbf{R}\})$$

The term that contains the potential energy of the ions in the above expression is given by:

$$\langle\Psi_{\{\mathbf{R}\}}|\frac{e^2}{2}\sum_{I\neq J}\frac{Z_I Z_J}{|\mathbf{R}_I - \mathbf{R}_J|}|\Psi_{\{\mathbf{R}\}}\rangle = \frac{e^2}{2}\sum_{I\neq J}\frac{Z_I Z_J}{|\mathbf{R}_I - \mathbf{R}_J|}$$

since the wavefunction $|\Psi_{\{\mathbf{R}\}}\rangle$ is properly normalized and does not depend explicitly on the ionic coordinates. The only other term whose expectation value in the wavefunction $|\Psi_{\{\mathbf{R}\}}\rangle$ we need to obtain is the kinetic energy of the ions. To proceed, we examine the action of the momentum operator corresponding to ion I at position $\mathbf{R}_I$, $\hat{\mathbf{P}}_I = -i\hbar\nabla_{\mathbf{R}_I}$, on the total wavefunction, as expressed above:

$$-i\hbar\nabla_{\mathbf{R}_I}\Psi(\{\mathbf{R};\mathbf{r}\}) = -i\hbar\left(\nabla_{\mathbf{R}_I}\Phi(\{\mathbf{R}\})\right)\Psi_{\{\mathbf{R}\}}(\{\mathbf{r}\}) - i\hbar\Phi(\{\mathbf{R}\})\left(\nabla_{\mathbf{R}_I}\Psi_{\{\mathbf{R}\}}(\{\mathbf{r}\})\right) \Rightarrow$$

$$\langle\Psi_{\{\mathbf{R}\}}|(-i\hbar\nabla_{\mathbf{R}_I})\Psi(\{\mathbf{R};\mathbf{r}\})\rangle = -i\hbar\left(\nabla_{\mathbf{R}_I} + \langle\Psi_{\{\mathbf{R}\}}|\nabla_{\mathbf{R}_I}\Psi_{\{\mathbf{R}\}}\rangle\right)\Phi(\{\mathbf{R}\})$$

With the help of this last expression, we define the kinetic energy operator for the motion of ions, as it should appear in the expression $\langle\Psi_{\{\mathbf{R}\}}|\hat{\mathcal{H}}^{\text{ion}}|\Psi_{\{\mathbf{R}\}}\rangle\Phi(\{\mathbf{R}\})$, to be given by:

$$\hat{\mathcal{K}}^{\text{ion}} \equiv \sum_I \frac{1}{2M_I}\left(-i\hbar\nabla_{\mathbf{R}_I} - i\hbar\langle\Psi_{\{\mathbf{R}\}}|\nabla_{\mathbf{R}_I}\Psi_{\{\mathbf{R}\}}\rangle\right)^2 \tag{4.7}$$

which leads to the following equation for the motion of ions:

$$\left[\hat{\mathcal{K}}^{\text{ion}} + \frac{e^2}{2} \sum_{I \neq J} \frac{Z_I Z_J}{|\mathbf{R}_I - \mathbf{R}_J|} + E^{\text{ele}}_{\{\mathbf{R}\}} \right] \Phi(\{\mathbf{R}\}) = E^{\text{tot}} \Phi(\{\mathbf{R}\}) \qquad (4.8)$$

At this point, two additional approximations are commonly introduced. The first approximation is to take the wavefunction of the ions to be a product of δ-functions, that is:

$$\Phi(\{\mathbf{R}\}) = \prod_I \delta(\mathbf{R} - \mathbf{R}_I) \qquad (4.9)$$

This is equivalent to considering the ions as independent classical point particles. The other approximation is to neglect the second term in the expression for the kinetic energy operator of the ions, Eq (4.7); that is, we take

$$\langle \Psi_{\{\mathbf{R}\}} | \nabla_{\mathbf{R}_I} \Psi_{\{\mathbf{R}\}} \rangle = 0$$

which has no *a priori* justification. It turns out that this term is not important in most situations, but could lead to interesting effects, especially in case there is an external magnetic field. This term is of geometric nature, as it involves changes induced by the positions of ions, and is related to the Berry connection and phase. Briefly, the parametric dependence of the hamiltonian for the electrons $\hat{\mathcal{H}}^{\text{ele}}_{\{\mathbf{R}\}}(\{\mathbf{r}\})$ on the ionic positions produces an overall phase in the wavefunction (the Berry phase), when the system is transported through a closed loop in the configurational space of these parameters, which can lead to observable effects like the Aharonov–Bohm effect. We shall revisit these topics in more detail in Chapter 9.

We can now take advantage of the approximations introduced to calculate the total energy of the system. We start with the kinetic energy of the ions, obtained as the expectation value of the kinetic energy operator, $\hat{\mathcal{K}}^{\text{ion}}$, in the wavefunction of Eq. (4.9):

$$K^{\text{ion}} = \langle \Phi | \hat{\mathcal{K}}^{\text{ion}} | \Phi \rangle \approx \langle \Phi | \sum_I \frac{1}{2M_I} \left(-i\hbar \nabla_{\mathbf{R}_I} \right)^2 | \Phi \rangle = \langle \Phi | \sum_I \frac{\hat{\mathbf{P}}_I^2}{2M_I} | \Phi \rangle = \sum_I \frac{\mathbf{P}_I^2}{2M_I}$$

where $\mathbf{P}_I = \langle \Phi | \hat{\mathbf{P}}_I | \Phi \rangle$ is simply the momentum of a point particle. It is a trivial step to obtain the expectation values of the other two terms on the left-hand side of Eq. (4.8) in the ionic wavefunction of Eq. (4.9). These contributions, combined with the kinetic energy K^{ion}, give for the ionic motion:

$$\sum_I \frac{\mathbf{P}_I^2}{2M_I} + \frac{e^2}{2} \sum_{I \neq J} \frac{Z_I Z_J}{|\mathbf{R}_I - \mathbf{R}_J|} + E^{\text{ele}}_{\{\mathbf{R}\}} = E^{\text{tot}} \qquad (4.10)$$

This is the familiar expression for the total energy of a set of classical charged particles, with the conventional kinetic energy contribution (first term) and a potential energy due to the Coulomb repulsion between them (second term), as well as an additional contribution, $E^{\text{ele}}_{\{\mathbf{R}\}}$, coming from integrating out the motion of electrons. We shall return to a more detailed discussion of the total energy in Section 4.10.

In the following sections we will examine in detail how to solve the problem of the electronic motion as defined in Eq. (4.6), with the hamiltonian of Eq. (4.3). For a cleaner notation, in the following we omit the subscript $\{\mathbf{R}\}$ in the wavefunction of the electronic degrees of freedom, the energy, and the potential that the electrons experience due to the presence of the ions, with the understanding that the ions are at fixed positions while we deal with the electrons. Even with this simplification, however, solving for $\Psi(\{\mathbf{r}\})$ is an extremely difficult task, because of the nature of the electrons. If two electrons of the same spin interchange positions, $\Psi(\{\mathbf{r}\})$ must change sign; this is known as the "exchange" property, and is a manifestation of the Pauli exclusion principle. Moreover, each electron is affected by the motion of every other electron in the system; this is known as the "correlation" property. It is possible to produce a simpler, approximate picture, in which we describe the system as a collection of classical ions and essentially single quantum-mechanical particles that reproduce the behavior of the electrons; this is the single-particle or one-electron picture. It is an appropriate description when the effects of exchange and correlation are not crucial for describing the phenomena we are interested in. Such phenomena include, for example, optical excitations in solids, the electrical conductivity, and all properties of solids that are related to atomic cohesion (such as mechanical properties). Phenomena which are outside the scope of the single-particle picture include all the situations where electron exchange and correlation effects are crucial, such as superconductivity and transport in high magnetic fields (the quantum Hall effects). Even in those cases, though, the single-particle picture plays a very important role: it is used as the basis for constructing many-body wavefunctions that contain the correct physics by including correlations in the motion of electrons.

In developing the single-particle picture of solids, we will not neglect the exchange and correlation effects between electrons, we will simply take them into account in an average way; this is often referred to as a mean-field approximation for the electron–electron interactions. To do this, we have to pass from the many-body picture to an effective one-electron picture. We will first derive equations that look like single-particle equations, and then try to explore their meaning. Before we treat the general problem of electrons in solids, in order to demonstrate the difficulty of including explicitly all the interaction effects in a system with more than one electron, we discuss a model of the hydrogen molecule, the simplest realistic two-electron system.

4.2 The Hydrogen Molecule

This molecule consists of two protons and two electrons, so it is the simplest possible system for studying electron–electron interactions in a realistic manner. The hamiltonian of a single hydrogen atom is:

$$h_1(\mathbf{r}) = -\frac{\hbar^2 \nabla_{\mathbf{r}}^2}{2m_e} - \frac{e^2}{|\mathbf{R}_1 - \mathbf{r}|} \qquad (4.11)$$

where $\mathbf{R}_1$ is the position of the first proton. We denote the ground-state wavefunction for this hamiltonian as $s(\mathbf{r} - \mathbf{R}_1) = \phi_1(\mathbf{r})$, and its ground-state energy as ϵ_0. Similarly, an atom at position $\mathbf{R}_2$ (far from $\mathbf{R}_1$) will have the hamiltonian

$$h_2(\mathbf{r}) = -\frac{\hbar^2 \nabla_{\mathbf{r}}^2}{2m_e} - \frac{e^2}{|\mathbf{R}_2 - \mathbf{r}|} \tag{4.12}$$

and the wavefunction $s(\mathbf{r} - \mathbf{R}_2) = \phi_2(\mathbf{r})$. When the two protons are very far away, the two electrons do not interact and the two electronic wavefunctions are the same, only centered at different points in space. When the atoms are brought together, the new hamiltonian becomes:

$$\hat{\mathcal{H}}(\mathbf{r}_1, \mathbf{r}_2) = h_1(\mathbf{r}_1) + h_2(\mathbf{r}_2) - \frac{e^2}{|\mathbf{R}_1 - \mathbf{r}_2|} - \frac{e^2}{|\mathbf{R}_2 - \mathbf{r}_1|} + \frac{e^2}{|\mathbf{r}_1 - \mathbf{r}_2|} + \frac{e^2}{|\mathbf{R}_1 - \mathbf{R}_2|} \tag{4.13}$$

where the last four terms represent electron–proton attraction between the electron in one atom and the proton in the other (the cross terms), and electron–electron, and proton–proton repulsion. As discussed earlier, we will ignore the proton–proton repulsion [last term in Eq. (4.13)], since it is only a constant term as far as the electrons are concerned, and it does not change the character of the electronic wavefunction. This is equivalent to applying the BOA to the problem and neglecting the quantum nature of the protons, even though we mentioned in Chapter 1 that this may not be entirely appropriate for hydrogen. The justification for using this approximation here is that we are concentrating our attention on the electron–electron interactions in the simplest possible model in order to develop insight into the physics of many-electron systems, rather than attempting to give an exact treatment of the real hydrogen molecule. Solving for the wavefunction $\Psi(\mathbf{r}_1, \mathbf{r}_2)$ of this new hamiltonian analytically is already an impossible task. We will attempt to do this approximately, using the orbitals $\phi_1(\mathbf{r})$ and $\phi_2(\mathbf{r})$ as a convenient basis, as we did in the tight-binding approximation in Chapter 2. We will address the problem in two stages, always working within the basis of the two orbitals $\phi_1(\mathbf{r}), \phi_2(\mathbf{r})$. In the first stage, we solve the problem of a single electron in the hamiltonian produced by the presence of the two nuclei, and attempt to discuss the two-electron system based on this solution; this will prove instructive but insufficient. In the second stage, we try to construct an appropriate solution for the two-electron problem by defining a proper many-body wavefunction and explore the nature of the solution; this, which is actually a reasonable approximation, will give us additional insight.

A single electron would experience the following hamiltonian in the presence of the two protons:

$$\hat{\mathcal{H}}^{\mathrm{sp}}(\mathbf{r}) = -\frac{\hbar^2 \nabla_{\mathbf{r}}^2}{2m_e} - \frac{e^2}{|\mathbf{R}_1 - \mathbf{r}|} - \frac{e^2}{|\mathbf{R}_2 - \mathbf{r}|} \tag{4.14}$$

Since we are dealing with only one electron, and the hamiltonian contains no terms that affect its spin, we can ignore for the moment the spin degree of freedom; when we start treating the two-electron system, it will be necessary to treat the spin degrees of freedom explicitly. We must construct the single-particle states using as a basis $\phi_1(\mathbf{r})$ and $\phi_2(\mathbf{r})$, so that they reflect the symmetry of the problem. For the case of a single electron, the only relevant symmetry is the spatial symmetry of the hamiltonian, that is, inversion relative to

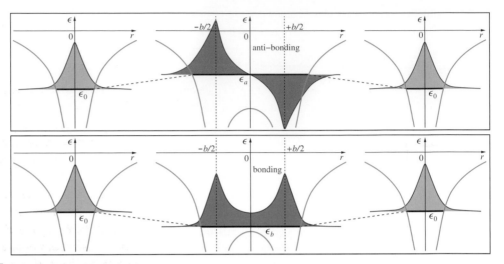

Fig. 4.1 Schematic representation of the hydrogen wavefunctions for isolated atoms (red-shaded curves, with energy ϵ_0) and the linear combinations that preserve the inversion symmetry with respect to the center of the molecule: a symmetric combination (dark-red-shaded curve, with energy $\epsilon_b < \epsilon_0$) and an antisymmetric combination (purple-shaded curve, with energy $\epsilon_a > \epsilon_0$); the latter two are the states defined in Eq. (4.15). The vertical dashed lines at $\pm b/2$ indicate the positions of the protons and the green lines indicate the Coulomb potential of the nuclei in the isolated atoms and in the molecule.

the midpoint of the distance between the two protons (the center of the molecule). There are two possibilities for single-particle states with the proper symmetry under inversion:

$$\psi_0(\mathbf{r}) = \frac{1}{\sqrt{2}} \left[\phi_1(\mathbf{r}) + \phi_2(\mathbf{r}) \right]$$

$$\psi_1(\mathbf{r}) = \frac{1}{\sqrt{2}} \left[\phi_1(\mathbf{r}) - \phi_2(\mathbf{r}) \right] \tag{4.15}$$

the first being a symmetric and the second an antisymmetric wavefunction, upon inversion with respect to the center of the molecule; these are illustrated in Fig. 4.1.

The energy that corresponds to these two states is obtained by calculating the expectation value of the hamiltonian $\hat{\mathcal{H}}^{\text{sp}}$ in the wavefunctions $\psi_0(\mathbf{r})$ and $\psi_1(\mathbf{r})$. Obviously, these expectation values will be expressed in terms of matrix elements of $\hat{\mathcal{H}}^{\text{sp}}$ in the basis orbitals, that is, $\langle \phi_i | \hat{\mathcal{H}}^{\text{sp}} | \phi_j \rangle$ $(i, j = 1, 2)$ (the bra $\langle \phi_i |$ and ket $| \phi_i \rangle$ notation for single-particle states and its extension to many-particle states constructed as products of single-particle states are discussed in Appendix C). In order to express the results in more compact and transparent notation, we define the matrix element of this hamiltonian in the orbital $\phi_1(\mathbf{r})$ or $\phi_2(\mathbf{r})$ to be:

$$\epsilon \equiv \langle \phi_1 | \hat{\mathcal{H}}^{\text{sp}} | \phi_1 \rangle = \langle \phi_2 | \hat{\mathcal{H}}^{\text{sp}} | \phi_2 \rangle \tag{4.16}$$

Notice that $\phi_1(\mathbf{r})$ or $\phi_2(\mathbf{r})$ are *not* eigenstates of $\hat{\mathcal{H}}^{\text{sp}}(\mathbf{r})$, and $\epsilon \neq \epsilon_0$, because $\phi_i(\mathbf{r})$ is an eigenstate of the hamiltonian $h_i(\mathbf{r})$ $(i = 1, 2)$, which contains only one proton (at the position $\mathbf{R}_i$) attracting the electron. However, the wavefunctions $\phi_i(\mathbf{r})$ are fairly localized and die exponentially as a function of the distance from the position of the proton where they are centered, therefore the extra terms that appear in the evaluation of ϵ due to the presence of the other proton in $\hat{\mathcal{H}}^{\text{sp}}$, that is, the terms

$$\langle\phi_1|\frac{-e^2}{|\mathbf{R}_2 - \mathbf{r}|}|\phi_1\rangle \quad \text{and} \quad \langle\phi_2|\frac{-e^2}{|\mathbf{R}_1 - \mathbf{r}|}|\phi_2\rangle$$

are very small, so to a good approximation we may take $\epsilon \approx \epsilon_0$ (to be more precise, since the contributions of the above two terms are strictly negative, $\epsilon < \epsilon_0$). We also define the so-called "hopping" matrix elements of the single-particle hamiltonian in the atomic orbital basis:

$$t \equiv -\langle\phi_1|\hat{\mathcal{H}}^{\text{sp}}|\phi_2\rangle = -\langle\phi_2|\hat{\mathcal{H}}^{\text{sp}}|\phi_1\rangle \tag{4.17}$$

where we can choose the phases in the wavefunctions $\phi_1(\mathbf{r})$, $\phi_2(\mathbf{r})$ to make sure that t is a real positive number. These matrix elements describe the probability of one electron "hopping" from state $\phi_1(\mathbf{r})$ to $\phi_2(\mathbf{r})$ (or vice versa), within the single-particle hamiltonian $\hat{\mathcal{H}}^{\text{sp}}(\mathbf{r})$; this hopping enhances the kinetic energy of the electrons. The expectation values of $\hat{\mathcal{H}}^{\text{sp}}$ in terms of the $\psi_i(\mathbf{r})$, which we define as ϵ_b and ϵ_a for $i = 0$ and $i = 1$, respectively, are:

$$\epsilon_b = \langle\psi_0|\hat{\mathcal{H}}^{\text{sp}}(\mathbf{r})|\psi_0\rangle = \epsilon - t$$
$$\epsilon_a = \langle\psi_1|\hat{\mathcal{H}}^{\text{sp}}(\mathbf{r})|\psi_1\rangle = \epsilon + t \tag{4.18}$$

We emphasize that the wavefunctions $\psi_0(\mathbf{r})$, $\psi_1(\mathbf{r})$ are *not* eigenfunctions of $\hat{\mathcal{H}}^{\text{sp}}$; they are simply single-particle states with the proper spatial symmetry, and as such more appropriate for describing the state of electrons in the single-particle hamiltonian than the original atomic orbitals $\phi_i(\mathbf{r})$ ($i = 1, 2$).

The results we have derived so far are quite instructive. As seen from Fig. 4.1, the state $\psi_0(\mathbf{r})$, with energy expectation value $\epsilon_b < \epsilon_0$, has more weight in the region near the center of the molecule, which contributes to its stability by enhancing the attraction between negative (electrons) and positive (protons) charges; this was designated as the "bonding" state in the general discussion on bonding in Chapter 1. In contrast to this, the other state, $\psi_1(\mathbf{r})$, with energy expectation value $\epsilon_a > \epsilon_0$, has a node at the midpoint between the two protons, and as a consequence has more electronic charge on the regions farther away from the proton positions; this is the state that was designated as "anti-bonding" in Chapter 1, and does not contribute to the cohesion of the molecule.

To complete the picture, we might consider putting two electrons in each of the single-particle states, taking advantage of their spin degree of freedom by assuming that the spins of the two electrons are opposite when they are in the same spatial wavefunction. More precisely, we take the total spin state to be antisymmetric (a "singlet," we return to this in more detail shortly) so that the only possibility for the spatial part of the two-body wavefunction is to be totally symmetric with respect to interchange of the electron coordinates, $\mathbf{r}_1, \mathbf{r}_2$. These assumptions lead to the following two-electron wavefunctions:

$$\Psi_0^{\text{H}}(\mathbf{r}_1, \mathbf{r}_2) = \psi_0(\mathbf{r}_1)\psi_0(\mathbf{r}_2) \tag{4.19}$$
$$\Psi_1^{\text{H}}(\mathbf{r}_1, \mathbf{r}_2) = \psi_1(\mathbf{r}_1)\psi_1(\mathbf{r}_2) \tag{4.20}$$

We use the superscript H to denote these two-body wavefunctions, in anticipation of the discussion in the next section; a generalization of this approach is the so-called "Hartree" approximation. Since we are now dealing with two electrons, we can write the total hamiltonian as:

$$\hat{\mathcal{H}}(\mathbf{r}_1, \mathbf{r}_2) = \hat{\mathcal{H}}^{\mathrm{sp}}(\mathbf{r}_1) + \hat{\mathcal{H}}^{\mathrm{sp}}(\mathbf{r}_2) + \frac{e^2}{|\mathbf{r}_1 - \mathbf{r}_2|} \tag{4.21}$$

We call the very last term in this expression, the electron–electron repulsion, the "interaction" term. The expectation values of the total hamiltonian in terms of the $\Psi_i^{\mathrm{H}}(\mathbf{r}_1, \mathbf{r}_2)$ two-electron wavefunctions, which we define as E_0^{H} and E_1^{H} for $i = 0, 1$, respectively, are:

$$E_0^{\mathrm{H}} = \langle \Psi_0^{\mathrm{H}} | \hat{\mathcal{H}}(\mathbf{r}_1, \mathbf{r}_2) | \Psi_0^{\mathrm{H}} \rangle = 2(\epsilon - t) + \frac{U}{2} \tag{4.22}$$

$$E_1^{\mathrm{H}} = \langle \Psi_1^{\mathrm{H}} | \hat{\mathcal{H}}(\mathbf{r}_1, \mathbf{r}_2) | \Psi_1^{\mathrm{H}} \rangle = 2(\epsilon + t) + \frac{U}{2} \tag{4.23}$$

where we have defined the "on-site" repulsive interaction between two electrons, which arises from the interaction term when the two electrons are placed at the same orbital, as:

$$U \equiv \langle \phi_i(\mathbf{r}_1)\phi_i(\mathbf{r}_2) | \frac{e^2}{|\mathbf{r}_1 - \mathbf{r}_2|} | \phi_i(\mathbf{r}_1)\phi_i(\mathbf{r}_2) \rangle, \quad i = 1, 2 \tag{4.24}$$

From its definition, U is a real, positive number. Other terms arising from the interaction term and one-electron states at *different* positions will be considered much smaller in magnitude than U and will be neglected. A model based on these physical quantities, the hopping matrix element t, and the on-site Coulomb repulsion energy U, was introduced originally by J. Hubbard.[1] The model contains the bare essentials for describing electron–electron interactions in solids, and has found many applications, especially in highly correlated electron systems. Despite its apparent simplicity, the Hubbard model has not been solved analytically and the effort to determine the physics of this simple but powerful model remains an active area of research.

The picture we have constructed so far is quite useful but not entirely satisfactory. The main problem is the very casual treatment of the spin degrees of freedom, and consequently of the symmetry of the two-electron wavefunction in terms of electron exchange properties. Specifically, we know that the two-electron wavefunction must be antisymmetric with respect to exchange of the electron degrees of freedom, including position and spin. This symmetry should not be confused with the symmetry that applies to spatial degrees of freedom, which is imposed by the nature of the hamiltonian, in the present case the inversion symmetry with respect to the center of the molecule. In fact, both types of symmetry should be obeyed by the two-electron wavefunction. In order to achieve this, we will start over, attempting to build these symmetries into the two-electron trial wavefunctions from the beginning, in all the ways that this is possible with our chosen basis of the two single-particle orbitals $\phi_1(\mathbf{r}), \phi_2(\mathbf{r})$. Having done this, we will then try to figure out the optimal linear combination of such trial wavefunctions which gives the lowest total energy for the two-electron system. The generalization of this type of approach is known as the "Hartree–Fock" approximation.

[1] J. Hubbard, *Proc. Roy. Soc. A* **276**, 238 (1963); *Proc. Roy. Soc. A* **277**, 237 (1964); *Proc. Roy. Soc. A* **281**, 401 (1964).

We begin with a discussion of the spin part of the two-electron wavefunction. As discussed in more detail in Appendix C, the spins of two spin-1/2 fermions can be combined in states with total spin S and z-projection S_z. Of these states, the one with total spin $S = 0$, called the spin "singlet," is antisymmetric with respect to exchange of the fermions, whereas the three states with total spin $S = 1$ and $S_z = 0, \pm 1$, called the spin "triplet," are symmetric with respect to exchange of the electrons. For example, the $S_z = 0$ states for the spin singlet and the spin triplet are:

$$\text{singlet} : \frac{1}{\sqrt{2}} [\uparrow\downarrow - \downarrow\uparrow], \quad \text{triplet} : \frac{1}{\sqrt{2}} [\uparrow\downarrow + \downarrow\uparrow]$$

with the convention that the first up or down arrow in each pair corresponds to the particle with label 1 and the second to the particle with label 2. Having built the spin part of the two-fermion wavefunction with a specific symmetry upon exchange, we must next build the spatial part so that the total wavefunction, which is a product of the spatial and spin parts, will be totally antisymmetric upon exchange. Consequently, the part accompanying the spin-singlet state must be symmetric in the spatial coordinates and the part accompanying the spin-triplet state must be antisymmetric in the spatial coordinates. Taking into account the symmetry imposed by the hamiltonian, that is, inversion symmetry upon exchange of the electron positions due to the symmetry of the molecule, we find that the following two-electron trial wavefunctions are allowed in terms of the single-particle orbitals, $\phi_1(\mathbf{r}), \phi_2(\mathbf{r})$:

$$\Phi_0^{(s)}(\mathbf{r}_1, \mathbf{r}_2) = \frac{1}{\sqrt{2}} [\phi_1(\mathbf{r}_1)\phi_2(\mathbf{r}_2) + \phi_1(\mathbf{r}_2)\phi_2(\mathbf{r}_1)] \tag{4.25}$$

$$\Phi_1^{(s)}(\mathbf{r}_1, \mathbf{r}_2) = \phi_1(\mathbf{r}_1)\phi_1(\mathbf{r}_2) \tag{4.26}$$

$$\Phi_2^{(s)}(\mathbf{r}_1, \mathbf{r}_2) = \phi_2(\mathbf{r}_1)\phi_2(\mathbf{r}_2) \tag{4.27}$$

$$\Phi^{(t)}(\mathbf{r}_1, \mathbf{r}_2) = \frac{1}{\sqrt{2}} [\phi_1(\mathbf{r}_1)\phi_2(\mathbf{r}_2) - \phi_1(\mathbf{r}_2)\phi_2(\mathbf{r}_1)] \tag{4.28}$$

The wavefunction $\Phi_0^{(s)}(\mathbf{r}_1, \mathbf{r}_2)$ is also known as the Heitler–London approximation. Since the three singlet states have exactly the same symmetries, both spatial and electron-exchange, any linear combination of them will also have the same symmetries. Therefore, in order to find the optimal combination which will give the lowest energy, we need to construct matrix elements of the two-particle hamiltonian $\hat{\mathcal{H}}_{ij}^{(s)} = \langle \Phi_i^{(s)} | \hat{\mathcal{H}} | \Phi_j^{(s)} \rangle$ for all possible values of the indices i, j, which produces a 3×3 matrix. We then diagonalize this matrix to find its eigenvalues and eigenstates. This exercise shows that the three eigenvalues are given by:

$$E_0^{(s)} = 2\epsilon + 2t \left[\lambda - \sqrt{\lambda^2 + 1} \right] = 2 \left(\epsilon - t\sqrt{\lambda^2 + 1} \right) + \frac{U}{2} \tag{4.29}$$

$$E_1^{(s)} = 2\epsilon + U \tag{4.30}$$

$$E_2^{(s)} = 2\epsilon + 2t \left[\lambda + \sqrt{\lambda^2 + 1} \right] = 2 \left(\epsilon + t\sqrt{\lambda^2 + 1} \right) + \frac{U}{2} \tag{4.31}$$

and the corresponding two-electron wavefunctions are:

$$\Psi_0^{(s)}(\mathbf{r}_1, \mathbf{r}_2) = \frac{1}{\mathcal{N}_0} \left[\sqrt{2}\Phi_0^{(s)}(\mathbf{r}_1, \mathbf{r}_2) + \frac{\Phi_1^{(s)}(\mathbf{r}_1, \mathbf{r}_2) + \Phi_2^{(s)}(\mathbf{r}_1, \mathbf{r}_2)}{\lambda + \sqrt{1 + \lambda^2}} \right] \tag{4.32}$$

$$\Psi_1^{(s)}(\mathbf{r}_1, \mathbf{r}_2) = \frac{1}{\sqrt{2}} \left[\Phi_1^{(s)}(\mathbf{r}_1, \mathbf{r}_2) - \Phi_2^{(s)}(\mathbf{r}_1, \mathbf{r}_2) \right] \tag{4.33}$$

$$\Psi_2^{(s)}(\mathbf{r}_1, \mathbf{r}_2) = \frac{1}{\mathcal{N}_2} \left[\sqrt{2}\Phi_0^{(s)}(\mathbf{r}_1, \mathbf{r}_2) + \frac{\Phi_1^{(s)}(\mathbf{r}_1, \mathbf{r}_2) + \Phi_2^{(s)}(\mathbf{r}_1, \mathbf{r}_2)}{\lambda - \sqrt{1 + \lambda^2}} \right] \tag{4.34}$$

where $\lambda = U/4t$ and $\mathcal{N}_0, \mathcal{N}_2$ are normalization constants. As far as the triplet state is concerned, since there is only one possibility for the spatial part, namely $\Phi^{(t)}(\mathbf{r}_1, \mathbf{r}_2)$, defined in Eq. (4.28), the corresponding energy is:

$$E^{(t)} = \langle \Phi^{(t)} | \hat{\mathcal{H}} | \Phi^{(t)} \rangle = 2\epsilon \tag{4.35}$$

The order of the total-energy values for the two-electron states, within what we called heuristically the Hartree–Fock approximation, is:

$$E_0^{(s)} < E^{(t)} < E_1^{(s)} < E_2^{(s)}$$

The true Hartree–Fock approximation, discussed in detail in the following section, involves a variational optimization of the single-particle orbitals, which can differ from the original atomic states $\phi_i(\mathbf{r})$ ($i = 1, 2$), used so far as a convenient basis. We note that the lowest of these total-energy values, $E_0^{(s)}$, is *lower* than the lowest energy value we had found in what we called the Hartree approximation, E_0^H, Eq. (4.22). The corresponding lowest-energy two-electron state, with wavefunction $\Psi_0^{(s)}(\mathbf{r}_1, \mathbf{r}_2)$, involves several two-particle trial wavefunctions: $\Phi_0^{(s)}(\mathbf{r}_1, \mathbf{r}_2)$, which places one electron in orbital $\phi_1(\mathbf{r})$ and the other electron in orbital $\phi_2(\mathbf{r})$, as well as $\Phi_i^{(s)}(\mathbf{r}_1, \mathbf{r}_2)$ ($i = 1, 2$), each of which places *both* electrons in the same single-particle orbital, $\phi_i(\mathbf{r})$ ($i = 1, 2$). This is a rather complex situation involving a delicate balance between placing the two electrons simultaneously around the same proton or placing them around a different proton each; the net effect of this elaborate two-electron "dance" around the proton positions is to produce the lowest total energy from all the other situations we have considered so far. This shows that taking into account the correlated electron motion, which tries to balance the cost of the Coulomb repulsion between them (the U term) and their propensity to move around and optimize their kinetic energy (the t term), gives the best possible solution for the total energy. A study of the two-electron ground-state energy and wavefunction as a function of the parameter $\lambda = (U/4t)$ gives further insight into the effects of correlation between the electrons in this simple model (see Problem 1).

As a final word on this model system we wish to emphasize that, despite all the effort expended so far, we have not found an actual *solution* to the problem; none of the two-electron states we have described is an eigenfunction of the total hamiltonian of the system – they are just reasonable approximations in the convenient single-particle basis $\phi_i(\mathbf{r})$ ($i = 1, 2$). A true solution to the problem is even more complicated and involves electronic states that differ from simple products of the type $\phi_i(\mathbf{r}_1)\phi_j(\mathbf{r}_2)$ ($i, j = 1, 2$). Such a solution can only be found computationally. For instance, a better description in terms of single-particle orbitals should include the excited states of electrons in each atom, but this increases significantly the size of the matrices involved. Extending this picture to more complex systems produces an essentially exponential increase of the computational cost as a function of system size.

4.3 The Hartree and Hartree–Fock Approximations

We will next generalize the concepts introduced in relation to the hydrogen molecule to a system consisting of many electrons and ions. Our goal is to construct reasonable starting many-body wavefunctions for the system using single-particle states as the basis, calculate the expectation value of the hamiltonian, and then improve on the original guess through a variational argument, that is, a minimization of the total energy with respect to variations in the single-particle states.

4.3.1 The Hartree Approximation

The simplest approach is to assume a specific form for the variational many-body wavefunction that explicitly ignores the Pauli exclusion principle and would be appropriate if the electrons were independent particles, namely

$$\Psi^H(\{\mathbf{r}\}) = \psi_1(\mathbf{r}_1)\psi_2(\mathbf{r}_2)\cdots\psi_N(\mathbf{r}_N) \tag{4.36}$$

with the index i running over all electrons and the $\psi_i(\mathbf{r})$ being orthonormal single-particle wavefunctions to be determined variationally. What about the Pauli exclusion principle? This will be added in by hand, by restricting the occupancy of each single-particle state to a pair of electrons with opposite spin, as was done for the case of the hydrogen molecule in our first attempt to build a many-body wavefunction. This is known as the Hartree approximation (hence the superscript H). With this approximation, the total energy of the system, given by $E^H = \langle \Psi^H | \hat{\mathcal{H}} | \Psi^H \rangle$, becomes:

$$E^H = \sum_i \langle \psi_i | \frac{-\hbar^2 \nabla_{\mathbf{r}}^2}{2m_e} + V(\mathbf{r}) | \psi_i \rangle + \frac{e^2}{2} \sum_{i \neq j} \langle \psi_i \psi_j | \frac{1}{|\mathbf{r} - \mathbf{r}'|} | \psi_i \psi_j \rangle$$

We will use a variational argument to obtain from this the single-particle Hartree equations. If we had reached the lowest-energy state within the form assumed for the many-body wavefunction of Eq. (4.36), this would be a stationary state of the system. Accordingly, any variation in the wavefunction will give a zero variation in the energy (equivalent to the statement that the derivative of a function at an extremum is zero). We can take the variation in the wavefunction to be of the form $\langle \delta \psi_i |$, subject to the constraint that $\langle \psi_i | \psi_i \rangle = 1$, which can be taken into account by introducing a Lagrange multiplier ϵ_i:

$$\delta \left[E^H - \sum_i \epsilon_i \left(\langle \psi_i | \psi_i \rangle - 1 \right) \right] = 0$$

Notice that the variations of the bra and the ket of ψ_i are considered to be independent of each other; this is allowed because the wavefunctions are complex quantities, so varying the bra and the ket independently is equivalent to varying the real and imaginary parts of a complex variable independently, which is legitimate since they represent independent components.[2] The above variation then produces:

$$\langle \delta\psi_i| - \frac{\hbar^2 \nabla_{\mathbf{r}}^2}{2m_e} + \mathcal{V}(\mathbf{r})|\psi_i\rangle + e^2 \sum_{j\neq i}\langle \delta\psi_i\psi_j|\frac{1}{|\mathbf{r}-\mathbf{r}'|}|\psi_i\psi_j\rangle - \epsilon_i\langle\delta\psi_i|\psi_i\rangle =$$

$$\langle\delta\psi_i|\left[-\frac{\hbar^2\nabla_{\mathbf{r}}^2}{2m_e} + \mathcal{V}(\mathbf{r}) + e^2\sum_{j\neq i}\langle\psi_j|\frac{1}{|\mathbf{r}-\mathbf{r}'|}|\psi_j\rangle - \epsilon_i \right]|\psi_i\rangle = 0$$

Since this has to be true for any variation $\langle\delta\psi_i|$, we conclude that:

$$\left[\frac{-\hbar^2\nabla_{\mathbf{r}}^2}{2m_e} + \mathcal{V}(\mathbf{r}) + e^2\sum_{j\neq i}\langle\psi_j|\frac{1}{|\mathbf{r}-\mathbf{r}'|}|\psi_j\rangle \right]\psi_i(\mathbf{r}) = \epsilon_i\psi_i(\mathbf{r}) \tag{4.37}$$

which is the Hartree single-particle equation. Each orbital $\psi_i(\mathbf{r}_i)$ can then be determined by solving the corresponding single-particle Schrödinger equation, if all the other orbitals $\psi_j(\mathbf{r}_j), j \neq i$ were known. We address the problem of self-consistency, that is, the fact that the equation for one ψ_i depends on all the other ψ_js, below. Before we do this, we will take a closer look at the physical meaning of the effective potential that each single-particle experiences.

Each $\psi_i(\mathbf{r})$ experiences the ionic potential $\mathcal{V}(\mathbf{r})$ as well as a potential due to the presence of all other electrons, $\mathcal{V}_i^{\mathrm{H}}(\mathbf{r})$, defined as:

$$\mathcal{V}_i^{\mathrm{H}}(\mathbf{r}) = +e^2\sum_{j\neq i}\langle\psi_j|\frac{1}{|\mathbf{r}-\mathbf{r}'|}|\psi_j\rangle \tag{4.38}$$

This potential includes only the Coulomb repulsion between electrons and is different for each particle. We next define the single-particle number density $n_i(\mathbf{r})$ as:

$$n_i(\mathbf{r}) = |\psi_i(\mathbf{r})|^2 \tag{4.39}$$

in terms of which the total number density $n(\mathbf{r})$ is:

$$n(\mathbf{r}) = \sum_{i=1}^{N} n_i(\mathbf{r}) = \sum_{i=1}^{N}|\psi_i(\mathbf{r})|^2 \tag{4.40}$$

where N is the total number of particles in the system. We also define the total potential produced by electron repulsion in the system as:

$$\mathcal{V}^{\mathrm{H}}(\mathbf{r}) = e^2 \int \frac{n(\mathbf{r}')}{|\mathbf{r}-\mathbf{r}'|}\mathrm{d}\mathbf{r}' \tag{4.41}$$

which is known as the Hartree potential. With these definitions, the Coulomb repulsion part of the effective potential in the single-particle Hartree equations takes the form:

---

[2] H. A. Bethe and R. W. Jackiw, *Intermediate Quantum Mechanics* (Benjamin/Cummings, Reading, MA, 1968).

$$\mathcal{V}_i^{\mathrm{H}}(\mathbf{r}) = e^2 \sum_{j \neq i} \int \frac{n_j(\mathbf{r}')}{|\mathbf{r} - \mathbf{r}'|} \mathrm{d}\mathbf{r}' = e^2 \int \frac{n(\mathbf{r}') - n_i(\mathbf{r}')}{|\mathbf{r} - \mathbf{r}'|} \mathrm{d}\mathbf{r}' = \mathcal{V}^{\mathrm{H}}(\mathbf{r}) - e^2 \int \frac{n_i(\mathbf{r}')}{|\mathbf{r} - \mathbf{r}'|} \mathrm{d}\mathbf{r}'$$

$$(4.42)$$

which is known as the "direct" potential, in distinction from the "exchange" potential that appears in the Hartree–Fock approximation (discussed in the next subsection). Thus, this potential describes the repulsive electrostatic interaction of one electron with all the other electrons in the system (the $n(\mathbf{r})$ term), excluding the interaction with itself (the $n_i(\mathbf{r})$ term). In other words, the Hartree approximation is a mean-field approximation to the electron–electron interaction, taking into account the electronic charge only and excluding self-interaction.

We now return to the self-consistency issue. We note first that we can assume that the single-particle states ψ are orthogonal, a condition that can easily be imposed on them by standard numerical schemes. From the preceding discussion, the potential in each single-particle equation is different and it depends on the total number density $n(\mathbf{r})$ as well as the density of the single-particle $n_i(\mathbf{r})$; this effective single-particle potential is defined as:

$$\mathcal{V}_i^{\mathrm{eff}}(\mathbf{r}, n(\mathbf{r})) = \mathcal{V}(\mathbf{r}) + e^2 \sum_{j \neq i} \langle \psi_j | \frac{1}{|\mathbf{r} - \mathbf{r}'|} | \psi_j \rangle = \mathcal{V}(\mathbf{r}) + e^2 \int \frac{n(\mathbf{r}') - n_i(\mathbf{r}')}{|\mathbf{r} - \mathbf{r}'|} \mathrm{d}\mathbf{r}' \quad (4.43)$$

We assume a set of ψ_is, use these to construct the number density $n(\mathbf{r})$, and with this the potential in the single-particle equations $\mathcal{V}_i^{\mathrm{eff}}(\mathbf{r}, n(\mathbf{r}))$, which allows us to solve the equations for each new ψ_i; we then compare the resulting ψ_is with the original ones, and modify the original ψ_is so that they resemble more the new ψ_is. This cycle is continued until input and output ψ_is are the same up to a tolerance δ_{tol}, as illustrated in the flowchart below. In this example, the comparison of input and output wavefunctions is made through the densities $n^{(\mathrm{in})}(\mathbf{r})$ and $n^{(\mathrm{out})}(\mathbf{r})$. This kind of iterative procedure is easily implemented by computer.

Iterative solution of coupled single-particle Hartree equations

1. CHOOSE $\psi_i^{(\mathrm{in})}(\mathbf{r})$

2. CONSTRUCT $n^{(\mathrm{in})}(\mathbf{r}) = \sum_i |\psi_i^{(\mathrm{in})}(\mathbf{r})|^2 \rightarrow \mathcal{V}_i^{\mathrm{eff}}(\mathbf{r}, n^{(\mathrm{in})}(\mathbf{r}))$

3. SOLVE $\left[-\frac{\hbar^2}{2m_e} \nabla_{\mathbf{r}}^2 + \mathcal{V}_i^{\mathrm{eff}}(\mathbf{r}, n^{(\mathrm{in})}(\mathbf{r})) \right] \psi_i^{(\mathrm{out})}(\mathbf{r}) = \epsilon_i^{(\mathrm{out})} \psi_i^{(\mathrm{out})}(\mathbf{r})$

4. CONSTRUCT $n^{(\mathrm{out})}(\mathbf{r}) = \sum_i |\psi_i^{(\mathrm{out})}(\mathbf{r})|^2$

5. COMPARE $n^{(\mathrm{out})}(\mathbf{r})$ to $n^{(\mathrm{in})}(\mathbf{r})$

    IF : $|n^{(\mathrm{in})}(\mathbf{r}) - n^{(\mathrm{out})}(\mathbf{r})| < \delta_{\mathrm{tol}}$ STOP

    ELSE : $\psi_i^{(\mathrm{in})}(\mathbf{r}) = \psi_i^{(\mathrm{out})}(\mathbf{r})$,   GOTO 2.

4.3.2 The Hartree–Fock Approximation

The next level of sophistication is to try to incorporate the fermionic nature of electrons in the many-body wavefunction $\Psi(\{\mathbf{r}\})$. To this end, we can choose a wavefunction which is a properly antisymmetrized version of the Hartree wavefunction, that is, it changes sign when the coordinates of two electrons are interchanged. This is known as the Hartree–Fock approximation. At the Hartree–Fock level it is possible to include explicitly the spin degrees of freedom, by considering all the single-particle states described by a spatial part and an up or down spin. This is the generalization of what we did in the case of the hydrogen molecule, where the spin degrees of freedom were handled separately and explicitly, in order to ensure the antisymmetric character of the many-body wavefunction, leaving a spatial part with a definite symmetry, namely symmetric for the singlet spin state and antisymmetric for the triplet spin case.

For simplicity, we will assume that the spin degrees of freedom are handled either explicitly or separately, and all we need to worry about is the symmetry of the spatial part of the many-body wavefunction with respect to electron exchange. Combining Hartree-type wavefunctions to form a properly antisymmetrized wavefunction for the system, we obtain the following determinant, first introduced by J. C. Slater[3] and known as the "Slater determinant," symbolized in the following by $S[\{\psi_j(\mathbf{r}_i)\}]$:

$$\Psi^{\mathrm{HF}}(\{\mathbf{r}\}) = S[\{\psi_j(\mathbf{r}_i)\}] = \frac{1}{\sqrt{N!}} \begin{vmatrix} \psi_1(\mathbf{r}_1) & \psi_1(\mathbf{r}_2) & \cdots & \psi_1(\mathbf{r}_N) \\ \psi_2(\mathbf{r}_1) & \psi_2(\mathbf{r}_2) & \cdots & \psi_2(\mathbf{r}_N) \\ \cdot & \cdot & & \cdot \\ \cdot & \cdot & & \cdot \\ \cdot & \cdot & & \cdot \\ \psi_N(\mathbf{r}_1) & \psi_N(\mathbf{r}_2) & \cdots & \psi_N(\mathbf{r}_N) \end{vmatrix} \tag{4.44}$$

where N is the total number of electrons. This has the desired antisymmetric electron-exchange property, since interchanging the position of two electrons is equivalent to interchanging the corresponding columns in the determinant, which changes its sign. It is important to realize that the exact many-body wavefunction would be a sum over all possible determinants with N occupied single-particle states; the Hartree–Fock approximation consists of taking only the lowest occupied single-particle states.

The total energy with the Hartree–Fock wavefunction, given by $E^{\mathrm{HF}} = \langle \Psi^{\mathrm{HF}} | \hat{\mathcal{H}} | \Psi^{\mathrm{HF}} \rangle$, takes the form

$$E^{\mathrm{HF}} = \sum_i \langle \psi_i | \frac{-\hbar^2 \nabla_{\mathbf{r}}^2}{2m_e} + \mathcal{V}(\mathbf{r}) | \psi_i \rangle$$

$$+ \frac{e^2}{2} \sum_{i \neq j} \langle \psi_i \psi_j | \frac{1}{|\mathbf{r} - \mathbf{r}'|} | \psi_i \psi_j \rangle - \frac{e^2}{2} \sum_{i \neq j} \langle \psi_i \psi_j | \frac{1}{|\mathbf{r} - \mathbf{r}'|} | \psi_j \psi_i \rangle \tag{4.45}$$

---

[3] J. C. Slater, *Phys. Rev.* **34**, 1293 (1929).

and the single-particle Hartree–Fock equations, obtained by a variational calculation, similar to that discussed for the Hartree approximation, are:

$$\left[\frac{-\hbar^2\nabla_{\mathbf{r}}^2}{2m_e} + V(\mathbf{r}) + V_i^{\mathrm{H}}(\mathbf{r})\right]\psi_i(\mathbf{r}) - e^2\sum_{j\neq i}\langle\psi_j(\mathbf{r}')|\frac{1}{|\mathbf{r}-\mathbf{r}'|}|\psi_i(\mathbf{r}')\rangle\psi_j(\mathbf{r}) = \epsilon_i\psi_i(\mathbf{r}) \quad (4.46)$$

This equation has one extra term compared to the Hartree equation, the last one, which is called the "exchange" term; this term describes the effects of exchange between electrons, which were incorporated into the Hartree–Fock many-particle wavefunction by construction.

It is instructive to consider the exchange term in some more detail, and compare it to the direct term of the Hartree single-particle equations, Eq. (4.42). This term involves a summation over all values of the index j, except for the value $j = i$ corresponding to the particular single-particle state $\psi_i(\mathbf{r})$ which we are trying to determine by solving the corresponding Schrödinger equation. We can remove this restriction by adding and subtracting the $j = i$ term. We also introduce the one-particle density matrix $\gamma(\mathbf{r}',\mathbf{r})$:

$$\gamma(\mathbf{r}',\mathbf{r}) = N\int\Psi^*(\mathbf{r}',\mathbf{r}_2,\ldots,\mathbf{r}_N)\Psi(\mathbf{r},\mathbf{r}_2,\ldots,\mathbf{r}_N)\,\mathrm{d}\mathbf{r}_2\cdots\mathrm{d}\mathbf{r}_N \quad (4.47)$$

discussed in detail in Appendix C, Eq. (C.14), which for the present choice of the many-body wavefunction, defined in Eq. (4.44), takes the form

$$\gamma(\mathbf{r}',\mathbf{r}) = \sum_j\psi_j^*(\mathbf{r}')\psi_j(\mathbf{r}) \quad (4.48)$$

With this definition, we arrive at the following expression for the exchange term:

$$\int V_i^{\mathrm{X}}(\mathbf{r},\mathbf{r}')\psi_i(\mathbf{r}')\,\mathrm{d}\mathbf{r}' = -e^2\int\frac{\gamma(\mathbf{r}',\mathbf{r})}{|\mathbf{r}-\mathbf{r}'|}\psi_i(\mathbf{r}')\,\mathrm{d}\mathbf{r}' + e^2\int\frac{n_i(\mathbf{r}')}{|\mathbf{r}-\mathbf{r}'|}\mathrm{d}\mathbf{r}'\,\psi_i(\mathbf{r}) \quad (4.49)$$

where we have also used the definition of the single-particle density $n_i(\mathbf{r})$, Eq. (4.39). This expression for the exchange term has the same general structure as the expression for the direct term, $V_i^{\mathrm{H}}(\mathbf{r})$, Eq. (4.42). Specifically, there is a term that describes exchange effects between all the particles in the system, which is the first term, containing $\gamma(\mathbf{r}',\mathbf{r})$. From this global term, the self-interaction term must be subtracted, which is the second term, containing $n_i(\mathbf{r}')$. Note that both the direct term, $V_i^{\mathrm{H}}(\mathbf{r})$, and the exchange term, $V_i^{\mathrm{X}}(\mathbf{r},\mathbf{r}')$, are state-dependent (they carry the index i for the state $\psi_i(\mathbf{r})$ to which they apply) because of the self-interaction terms that they contain. However, in the Hartree–Fock single-particle equations, the two self-interaction terms exactly cancel each other. We can then define the state-independent exchange potential $V^{\mathrm{X}}(\mathbf{r},\mathbf{r}')$, as we had done for the Hartree potential $V^{\mathrm{H}}(\mathbf{r})$, Eq. (4.41), through the following expression:

$$V^{\mathrm{X}}(\mathbf{r},\mathbf{r}') = -e^2\frac{\gamma(\mathbf{r}',\mathbf{r})}{|\mathbf{r}-\mathbf{r}'|} \quad (4.50)$$

With this expression, the single-particle Hartree–Fock equations take the form

$$\left[-\frac{\hbar^2\nabla_{\mathbf{r}}^2}{2m_e} + V(\mathbf{r}) + V^{\mathrm{H}}(\mathbf{r})\right]\psi_i(\mathbf{r}) + \int V^{\mathrm{X}}(\mathbf{r},\mathbf{r}')\psi_i(\mathbf{r}')\,\mathrm{d}\mathbf{r}' = \epsilon_i\psi_i(\mathbf{r}) \quad (4.51)$$

which contains only the global terms for electron–electron Coulomb repulsion, the Hartree potential $\mathcal{V}^H(\mathbf{r})$, and for electron exchange effects, the exchange potential $\mathcal{V}^X(\mathbf{r}, \mathbf{r}')$. Writing both terms as expressions of the single-particle density matrix $\gamma(\mathbf{r}, \mathbf{r}')$, we obtain:

$$\left[-\frac{\hbar^2 \nabla_\mathbf{r}^2}{2m_e} + V(\mathbf{r}) \right] \psi_i(\mathbf{r}) + e^2 \int \left[\frac{\gamma(\mathbf{r}', \mathbf{r}') \, \psi_i(\mathbf{r}) - \gamma(\mathbf{r}', \mathbf{r}) \, \psi_i(\mathbf{r}')}{|\mathbf{r} - \mathbf{r}'|} \right] \, d\mathbf{r}' = \epsilon_i \psi_i(\mathbf{r}) \quad (4.52)$$

The exchange potential is fundamentally different from the direct potential, because it cannot be cast into a simple expression which is a local function of the variable $\mathbf{r}$ and multiplies the single-particle state $\psi_i(\mathbf{r})$. Thus, the exchange term is a *non-local* integral operator that depends on two spatial variables $\mathbf{r}$ and $\mathbf{r}'$ simultaneously, and involves an integration over the variable $\mathbf{r}'$ of the state $\psi_i(\mathbf{r}')$. These additional features make the exchange term more difficult to handle analytically and more computationally expensive. These complications are not unexpected, because with this term we are trying to capture many-body effects, that is, the exchange property of the many-body wavefunction, through effective interactions at the level of the single-particle equations that determine the states $\psi_i(\mathbf{r})$. The following sections aim to provide insight into this problem and discuss ways of handling it.

4.4 Hartree–Fock Theory of Free Electrons

To elucidate the physical meaning of the approximations introduced above, we will consider the simplest possible case, that is, the free-electron model discussed in Chapter 1. With the electrons represented by plane waves, the electronic density must be uniform and equal to the ionic density. The Coulomb interaction of electrons with the uniform positive ionic charge (an attractive term) and the Coulomb interaction of electrons among themselves (a repulsive term), expressed through the uniform electronic charge of equal density to the ionic one, exactly cancel each other. The only terms remaining in the single-particle equation are the kinetic energy and the exchange term, defined in Eq. (4.50):

$$-\frac{\hbar^2 \nabla_\mathbf{r}^2}{2m_e} \psi_\mathbf{k}(\mathbf{r}) + \int \mathcal{V}^X(\mathbf{r}, \mathbf{r}') \psi_\mathbf{k}(\mathbf{r}') \, d\mathbf{r}' = \epsilon_\mathbf{k} \psi_\mathbf{k}(\mathbf{r}) \quad (4.53)$$

We have asserted above that the behavior of electrons in this system is described by plane waves; we prove this statement next. Plane waves are of course eigenfunctions of the kinetic energy operator:

$$-\frac{\hbar^2 \nabla^2}{2m_e} \frac{1}{\sqrt{V}} e^{i\mathbf{k} \cdot \mathbf{r}} = \left(\frac{\hbar^2 \mathbf{k}^2}{2m_e} \right) \frac{1}{\sqrt{V}} e^{i\mathbf{k} \cdot \mathbf{r}} \quad (4.54)$$

so all we need to show is that they are also eigenfunctions of the second term in the hamiltonian of Eq. (4.53). Using Eq. (4.50) and identifying $\mathbf{k}$ with the index i and $\mathbf{k}'$ with the index j, we obtain:

$$\int \mathcal{V}^X(\mathbf{r}, \mathbf{r}')\psi_{\mathbf{k}}(\mathbf{r}')\,d\mathbf{r}' = -e^2 \left[\sum_{\mathbf{k}'(k'<k_F)} \int \frac{e^{-i(\mathbf{k}-\mathbf{k}')\cdot(\mathbf{r}-\mathbf{r}')}}{V} \frac{1}{|\mathbf{r}-\mathbf{r}'|} d\mathbf{r}' \right] \psi_{\mathbf{k}}(\mathbf{r})$$

$$= -e^2 \left[\int_{k'<k_F} \left(\int \frac{e^{-i(\mathbf{k}-\mathbf{k}')\cdot(\mathbf{r}-\mathbf{r}')}}{|\mathbf{r}-\mathbf{r}'|} d\mathbf{r}' \right) \frac{d\mathbf{k}'}{(2\pi)^3} \right] \psi_{\mathbf{k}}(\mathbf{r}) \qquad (4.55)$$

where in the last step we have used the familiar expression to turn the summation over $\mathbf{k}'$ into an integral, Eq. (1.4), and made explicit the restriction $k' = |\mathbf{k}'| < k_F$, requiring wavevectors to be smaller in magnitude than the Fermi momentum. We can already see from this expression that this term will give a result which is a function of $\mathbf{k}$ at most, once the integration over $d\mathbf{r}'$ is carried out. The result, which we refer to as the "Coulomb integral," takes a few steps to derive (given in detail in Appendix A, Section A.6) and leads to:

$$\int \mathcal{V}^X(\mathbf{r}, \mathbf{r}')\psi_{\mathbf{k}}(\mathbf{r})\,d\mathbf{r}' = -e^2 \frac{k_F}{\pi} F(k/k_F)\,\psi_{\mathbf{k}}(\mathbf{r}) \qquad (4.56)$$

with $k = |\mathbf{k}|$, where the function $F(x)$ is defined as:

$$F(x) = 1 + \frac{1-x^2}{2x} \ln \left| \frac{1+x}{1-x} \right| \qquad (4.57)$$

This completes the proof that plane waves are eigenfunctions of the single-particle hamiltonian in Eq. (4.53).

With this result, the energy of the single-particle state $\psi_{\mathbf{k}}(\mathbf{r})$ is given by:

$$\epsilon_{\mathbf{k}} = \frac{\hbar^2 k^2}{2m_e} - \frac{e^2}{\pi} k_F\, F(k/k_F) \qquad (4.58)$$

which, using the variable r_s introduced in Eq. (1.9) and the definition of the energy unit Ry given in Eq. (1.11), can be rewritten in the following form:

$$\epsilon_{\mathbf{k}} = \left[\left(\frac{(9\pi/4)^{1/3}}{r_s/a_0} \right)^2 (k/k_F)^2 - \frac{2}{\pi} \left(\frac{(9\pi/4)^{1/3}}{r_s/a_0} \right) F(k/k_F) \right] \text{Ry} \qquad (4.59)$$

The behavior of the energy $\epsilon_{\mathbf{k}}$ as a function of the momentum (in units of k_F) is illustrated in Fig. 4.2 for two different values of r_s.

This is an intriguing result: it shows that, even though plane waves are eigenstates of this hypothetical system, due to the exchange interaction the energy of state $\psi_{\mathbf{k}}$ is not simply $\hbar^2 k^2/2m_e$, as might be expected for non-interacting particles; it also contains the term proportional to $F(k/k_F)$ in Eq. (4.58). This term has interesting behavior at $|\mathbf{k}| = k_F$, as is evident in Fig. 4.2. It also gives a lower energy and higher band width than the non-interacting electron case for all values of $\mathbf{k}$, an effect which is more pronounced for small values of r_s (see Fig. 4.2). Thus, the electron–electron interaction included at the

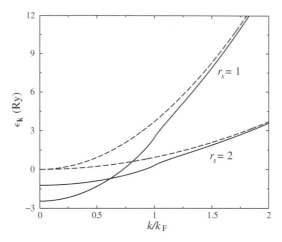

Fig. 4.2 Energy $\epsilon_\mathbf{k}$ (Ry) of single-particle states as a function of momentum k/k_F (with k_F the Fermi momentum), as given by Eq. (4.59), for two different values of $r_s = 1, 2$ (a_0). The dashed curves give the kinetic energy contribution [first term on the right-hand side of Eq. (4.59)].

Hartree–Fock level lowers the energy of the system significantly. We can calculate the total energy of this system by summing the single-particle energies over $\mathbf{k}$ up to momentum k_F:

$$E^{\mathrm{HF}} = 2 \sum_{\mathbf{k}(k \leq k_F)} \frac{\hbar^2 \mathbf{k}^2}{2m_e} - \frac{e^2 k_F}{\pi} \sum_{\mathbf{k}(k \leq k_F)} \left[1 + \frac{k_F^2 - k^2}{2kk_F} \ln \left| \frac{k_F + k}{k_F - k} \right| \right] \qquad (4.60)$$

Notice that we must include a factor of 2 for the spin of the electrons in both summations. This was done explicitly for the kinetic energy part, Eq. (1.8), to gave:

$$E^{\mathrm{kin}} = 2 \sum_{\mathbf{k}(k \leq k_F)} \frac{\hbar^2 \mathbf{k}^2}{2m_e} = \frac{3}{5} N \epsilon_F$$

For the second term, which represents the effective electron–electron interaction due to exchange, this factor of 2 is canceled by a factor of 1/2 needed to compensate double counting of the effective interaction in the sum of $\epsilon_\mathbf{k}$s; remember that this effective interaction is contained in the Hartree–Fock single-particle equations, Eq. (4.51), as the sum over all states, so if we simply sum all these contributions contained in the $\epsilon_\mathbf{k}$s we will be counting each contribution twice. Turning the second term in the above equation into an integral through the usual procedure, we can evaluate the sum to find:

$$\frac{E^{\mathrm{HF}}}{N} = \frac{3}{5} \epsilon_F - \frac{3}{4} \frac{e^2 k_F}{\pi} \qquad (4.61)$$

which quantifies by how much the effective electron–electron interaction due to exchange lowers the energy of the system relative to the kinetic energy alone. Using the expression of k_F in terms of r_s, Eq. (1.10), and expressing everything in units of Ry with the help of Eq. (1.11), we obtain:

$$\frac{E^{\mathrm{HF}}}{N} = \left[\frac{2.21}{(r_s/a_0)^2} - \frac{0.916}{(r_s/a_0)}\right]\mathrm{Ry} \tag{4.62}$$

This result should be compared to the expansion for the exact energy of the electron gas in the high-density limit (low r_s/a_0 values), first obtained by Gell–Mann and Brueckner[4]:

$$\frac{E}{N} = \left[\frac{2.21}{(r_s/a_0)^2} - \frac{0.916}{(r_s/a_0)} + 0.0622\ln(r_s/a_0) - 0.096 + \mathcal{O}(r_s/a_0)\right]\mathrm{Ry} \tag{4.63}$$

It is quite remarkable that the Hartree–Fock approximation, based on an ad-hoc expression for the many-body wavefunction, captures the first two terms in the exact expansion of the total energy. Of course, in real situations this may not be very helpful, since in typical metals (r_s/a_0) varies between 2 and 6.

Another interesting point is that we can express the potential due to exchange in a way that involves the density. This potential will give rise to the second term on the right-hand side of Eq. (4.62), namely:

$$\frac{E^{\mathrm{X}}}{N} = -\frac{0.916}{(r_s/a_0)}\mathrm{Ry} \tag{4.64}$$

which, using the expressions for r_s discussed earlier, can be written as:

$$\frac{E^{\mathrm{X}}}{N} = -\frac{3e^2}{4}\left(\frac{3}{\pi}\right)^{1/3}n^{1/3} = -1.477[a_0^3 n]^{1/3}\mathrm{Ry} \tag{4.65}$$

One of the most insightful proposals in the early calculations of the properties of solids, due to J. C. Slater,[5] was to generalize this term for situations where the density is not constant, that is, a system with non-homogeneous distribution of electrons. In this case the exchange energy would arise from a potential energy term in the single-particle hamiltonian, which will have the form:

$$\mathcal{V}^{\mathrm{X}}(\mathbf{r}) = -\frac{3e^2}{2}\left(\frac{3}{\pi}\right)^{1/3}[n(\mathbf{r})]^{1/3} = -\frac{3e^2}{2\pi}k_{\mathrm{F}}(\mathbf{r}) = -2.954[a_0^3 n(\mathbf{r})]^{1/3}\mathrm{Ry} \tag{4.66}$$

where an extra factor of 2 is introduced to account for the fact that a variational derivation gives rise to a potential term in the single-particle equations which is twice as large as the corresponding energy term; conversely, when one calculates the total energy by summing terms in the single-particle equations, a factor of 1/2 must be introduced to account for double counting of interactions. In the last two equations, the density, and hence the Fermi momentum, have become functions of $\mathbf{r}$, that is, they can be non-homogeneous. There is actually good justification to use such a term in single-particle equations in order to describe the exchange contribution, although the values of the constants involved are different from Slater's. This is discussed in more detail in the next section.

---

[4] For details, see A. L. Fetter and J. D. Walecka, *Quantum Theory of Many-Particle Systems* (McGraw-Hill, New York, 1971).
[5] J. C. Slater, *Phys. Rev.* **87**, 385 (1951).

4.5 Density Functional Theory

In two seminal papers, Hohenberg, Kohn, and Sham developed yet a different way of looking at the problem which has been called density functional theory (DFT). The original formulation contained in the papers of Hohenberg, Kohn, and Sham[6] is referred to as the Hohenberg–Kohn–Sham theorem. This theory has had a tremendous impact on realistic calculations of the properties of molecules and solids, and its applications to different problems continue to expand. A measure of its importance and success is that W. Kohn shared the 1998 Nobel Prize for Chemistry with J. A. Pople (who used DFT extensively in calculations for molecules). We will review here the essential ideas behind this theory. Before we do this, we shall take a quick detour to discuss the Thomas–Fermi theory, which embodied many of the ideas that were later put on a solid theoretical foundation by DFT.

4.5.1 Thomas–Fermi–Dirac Theory

The discussion of Hartree–Fock theory of free electrons showed that we can express the energy per particle as:

$$\frac{E^{\mathrm{HF}}}{N} = \frac{3\hbar^2}{10m_e}(3\pi^2)^{2/3}n^{2/3} - \frac{3e^2}{2}\left(\frac{3}{\pi}\right)^{1/3}n^{1/3} \tag{4.67}$$

with the first term representing the kinetic energy and the second term the exchange energy. This result was obtained for the uniform electron gas for which the electrostatic repulsion between electrons exactly canceled the electron–ion attraction for a uniform positive ionic charge that is equal to the electronic charge. In the more general case, when the ions are not represented by a uniform positive background, we can add these two terms back, to obtain:

$$\frac{E^{\mathrm{HF}}}{N} = \frac{3\hbar^2}{10m_e}(3\pi^2)^{2/3}n^{2/3} - \frac{3e^2}{2}\left(\frac{3}{\pi}\right)^{1/3}n^{1/3} + \langle \mathcal{V}(\mathbf{r})\rangle + \langle\frac{e^2}{2}\int\frac{n(\mathbf{r}')}{|\mathbf{r}-\mathbf{r}'|}d\mathbf{r}'\rangle \tag{4.68}$$

where $\mathcal{V}(\mathbf{r})$ is the external potential that each particle experiences due to the presence of the ions and the last term is the Coulomb repulsion between an electron at position $\mathbf{r}$ with all other electrons described by the density $n(\mathbf{r}')$, with a factor of $1/2$ to compensate for double counting of this interaction. The last two terms are appropriately averaged over all space to give the energy per particle (this is the meaning of the $\langle\ \rangle$ symbols). We are also assuming that the external potential is weak, in the sense that it does not significantly change the Hartree–Fock results for kinetic and exchange energy contributions that we derived in the context of the free-electron model. We can now rewrite the total number of particles as:

$$N = \int n(\mathbf{r})d\mathbf{r}$$

---

[6] P. Hohenberg and W. Kohn, *Phys. Rev.* **136**, B864 (1964); W. Kohn and L. Sham, *Phys. Rev.* **140**, A1133 (1965).

where we have allowed the density $n(\mathbf{r})$ to be a function of position $\mathbf{r}$ rather than the constant value it assumed in the simplified free-electron model, and with this we can replace the averages as summations over all the values of the relevant quantities, that is:

$$N\langle \mathcal{V}(\mathbf{r}) \rangle \rightarrow \int \mathcal{V}(\mathbf{r}) n(\mathbf{r}) d\mathbf{r}$$

and similarly for the other terms. We can then express the total energy of the system as:

$$E^{\text{TFD}} = C_1 \int [n(\mathbf{r})]^{5/3} d\mathbf{r} + C_2 \int [n(\mathbf{r})]^{4/3} d\mathbf{r} + \int \mathcal{V}(\mathbf{r}) n(\mathbf{r}) d\mathbf{r} + \frac{e^2}{2} \int \frac{n(\mathbf{r}) n(\mathbf{r}')}{|\mathbf{r} - \mathbf{r}'|} d\mathbf{r} d\mathbf{r}'$$

$$(4.69)$$

This is an expression derived by L. Thomas, E. Fermi, and P. A. M. Dirac, to describe the properties of many-electron systems in an external potential $\mathcal{V}(\mathbf{r})$. In the above exrpession, with

$$C_1 = \frac{3\hbar^2}{10 m_e} (3\pi^2)^{2/3}, \quad C_2 = -\frac{3e^2}{2} \left(\frac{3}{\pi} \right)^{1/3}$$

we recognize the first term as representing the kinetic energy, the second term the exchange energy, the third the external potential, and the fourth the Coulomb repulsion between electrons. We can now think of the constants C_1, C_2 as being adjustable parameters that could be tuned to represent the energy of the electrons in the general case, as opposed to taking the values corresponding to the free-electron case. This is quite remarkable, because it was done before the development of the Hartree–Fock theory. It highlighted the fact that the density is a crucial physical quantity involved in determining the physics of the many-electron system. However, attempts to apply this theory to realistic systems were met with failure. The main problem is the inadequacy of the first term in the expression of Eq. (4.69) to represent the kinetic energy. Another important shortcoming is the fact that, since we only kept the density as the important quantity, there is no more information about single-particle states (referred to as "orbitals"); for this reason, this approach cannot reproduce the shell structure of atoms. Finally, the density obtained from minimizing the energy with respect to density variations does not decay exponentially away from the ions, as it does in realistic systems (see, for example, the simple model of the hydrogen molecule discussed earlier). Nevertheless, versions of this theory (with improved terms for the exchange contribution and inclusion of correlation effects) are still in use in the literature and can be quite useful in capturing the physics of systems that are not far from the free-electron model; these theoretical methods are referred to as "orbital-free density functional theory."

4.5.2 General Formulation of DFT

The basic concept is that instead of dealing with the many-body Schrödinger equation, Eq. (4.1), which involves the many-body wavefunction $\Psi(\{\mathbf{r}\})$, one deals with an exact formulation of the problem that involves the total number density of electrons $n(\mathbf{r})$. This is a huge simplification, since the many-body wavefunction need never be explicitly specified, as was done in the Hartree and Hartree–Fock approximations. Thus, instead of starting with a drastic approximation for the behavior of the system (which is what the Hartree and Hartree–Fock wavefunctions represent), one can develop the appropriate single-particle equations in an exact manner, and then introduce approximations as needed.

In the following discussion the main physical quantity of interest is the electronic number density $n(\mathbf{r})$. This density can be expressed in terms of the many-body wavefunction as:

$$n(\mathbf{r}) = N \int \Psi^*(\mathbf{r}, \ldots, \mathbf{r}_N) \Psi(\mathbf{r}, \ldots, \mathbf{r}_N) d\mathbf{r}_2 \cdots d\mathbf{r}_N$$

[see Appendix C, Eq. (C.13)]. First, we will show that the density $n(\mathbf{r})$ is uniquely defined given an external potential $\mathcal{V}(\mathbf{r})$ for the electrons (this, of course, is identified with the ionic potential). To prove this, suppose that two different external potentials, $\mathcal{V}(\mathbf{r})$ and $\mathcal{V}'(\mathbf{r})$, give rise to the same density $n(\mathbf{r})$. We will show that this is impossible. We assume that $\mathcal{V}(\mathbf{r})$ and $\mathcal{V}'(\mathbf{r})$ are different in a non-trivial way, that is, they do not differ merely by a constant. Let E and Ψ be the total energy and wavefunction of the ground state of hamiltonian $\hat{\mathcal{H}}$ and E' and Ψ' be the total energy and wavefunction of the ground state of the hamiltonian $\hat{\mathcal{H}}'$, where the first hamiltonian contains $\mathcal{V}(\mathbf{r})$ and the second $\mathcal{V}'(\mathbf{r})$ as external potential:

$$\hat{\mathcal{H}} = \hat{\mathcal{F}} + \mathcal{V}, \quad \hat{\mathcal{H}}' = \hat{\mathcal{F}} + \mathcal{V}', \quad E = \langle \Psi | \hat{\mathcal{H}} | \Psi \rangle, \quad E' = \langle \Psi' | \hat{\mathcal{H}}' | \Psi' \rangle$$

We will assume that the ground states of the two hamiltonians are different, since the potentials are different, which implies $E \neq E$. In the above expressions the operator $\hat{\mathcal{F}}$ includes all the terms in the hamiltonian other than the external potential, and is therefore common to the two hamiltonians. This term is in fact universal, because it contains the kinetic energy and electron–electron interaction terms, the forms of which are the same for all systems. According to the variational principle, the expectation value of the hamiltonian $\hat{\mathcal{H}}$ in a wavefunction $|\Psi'\rangle$, that is different from its ground state $|\Psi\rangle$, is higher than the ground-state energy E of $\hat{\mathcal{H}}$:

$$
\begin{aligned}
E < \langle \Psi' | \hat{\mathcal{H}} | \Psi' \rangle &= \langle \Psi' | \hat{\mathcal{F}} + \mathcal{V} + \mathcal{V}' - \mathcal{V}' | \Psi' \rangle = \langle \Psi' | \hat{\mathcal{H}}' + \mathcal{V} - \mathcal{V}' | \Psi' \rangle \\
&= \langle \Psi' | \hat{\mathcal{H}}' | \Psi' \rangle + \langle \Psi' | (\mathcal{V} - \mathcal{V}') | \Psi' \rangle = E' + \langle \Psi' | (\mathcal{V} - \mathcal{V}') | \Psi' \rangle \quad (4.70)
\end{aligned}
$$

where the strict inequality of the first step is a consequence of the fact that the two potentials are different in a non-trivial way and their corresponding ground-state energies are different. We also note that, for the same reason, the wavefunctions $|\Psi\rangle$ and $|\Psi'\rangle$ must be different. If they were the same, we could write the corresponding Schrödinger equations as:

$$\hat{\mathcal{H}} | \Psi \rangle = E | \Psi \rangle, \quad \hat{\mathcal{H}}' | \Psi \rangle = E' | \Psi \rangle$$

and subtracting the two equations we would obtain:

$$(\mathcal{V} - \mathcal{V}') | \Psi \rangle = (E - E') | \Psi \rangle$$

which implies that $\mathcal{V}$ and $\mathcal{V}'$ differ by a constant, contrary to our initial assumption. By following exactly the same steps that led us to Eq. (4.70), we can prove:

$$E' < E - \langle \Psi | (\mathcal{V} - \mathcal{V}') | \Psi \rangle \quad (4.71)$$

Adding Eqs (4.70) and (4.71), we obtain:

$$(E + E') < (E + E') + \langle \Psi'|(\mathcal{V} - \mathcal{V}')|\Psi' \rangle - \langle \Psi|(\mathcal{V} - \mathcal{V}')|\Psi \rangle \qquad (4.72)$$

But the last two terms on the right-hand side of Eq. (4.72) give:

$$\int n'(\mathbf{r})[\mathcal{V}(\mathbf{r}) - \mathcal{V}'(\mathbf{r})]d\mathbf{r} - \int n(\mathbf{r})[\mathcal{V}(\mathbf{r}) - \mathcal{V}'(\mathbf{r})]d\mathbf{r} = 0$$

because by assumption the densities $n(\mathbf{r})$ and $n'(\mathbf{r})$ corresponding to the two potentials are the same. This leads to the relation $E + E' < E + E'$, which is obviously wrong, therefore we conclude that our assumption about the densities being the same cannot be correct. This proves that there is a one-to-one correspondence between an external potential $\mathcal{V}(\mathbf{r})$ and the density $n(\mathbf{r})$. But the external potential determines the wavefunction, so the wavefunction must be a unique functional of the density. We then conclude that the expression

$$\langle \Psi|(\hat{\mathcal{H}} - \mathcal{V})|\Psi \rangle = \langle \Psi|\hat{\mathcal{F}}|\Psi \rangle = \mathcal{F}[n(\mathbf{r})] \qquad (4.73)$$

must be a universal functional of the density, because, as we argued above, it contains only terms that are common to all systems, and therefore this functional, through its dependence on the wavefunction, depends only on the electron density. This argument is summarized below.

Fundamental argument of density functional theory

(i) $\mathcal{V}(\mathbf{r}) \leftrightarrow n(\mathbf{r})$ one-to-one

         correspondence $\Big\} \Rightarrow |\Psi\rangle$ unique functional of $n(\mathbf{r}) \rightarrow \mathcal{F}[n(\mathbf{r})]$: universal

(ii) $\mathcal{V}(\mathbf{r})$ determines $|\Psi\rangle$

From these considerations we then conclude that the total energy of the system is a functional of the density, and is given by:

$$\mathcal{E}[n(\mathbf{r})] = \langle \Psi|\hat{\mathcal{H}}|\Psi \rangle = \mathcal{F}[n(\mathbf{r})] + \int \mathcal{V}(\mathbf{r})n(\mathbf{r})d\mathbf{r} \qquad (4.74)$$

[we use the symbol $\mathcal{E}[n(\mathbf{r})]$ for the general functional of the energy and the symbol E for a specific value of this functional for a given density $n(\mathbf{r})$]. From the variational principle, we can deduce that this functional attains its minimum for the correct density $n(\mathbf{r})$ corresponding to $\mathcal{V}(\mathbf{r})$, since for a given $\mathcal{V}(\mathbf{r})$ and any other density $n'(\mathbf{r})$ we would have:

$$\mathcal{E}[n'(\mathbf{r})] = \mathcal{F}[n'(\mathbf{r})] + \int \mathcal{V}(\mathbf{r})n'(\mathbf{r})d\mathbf{r} = \langle \Psi'|\hat{\mathcal{H}}|\Psi' \rangle > \langle \Psi|\hat{\mathcal{H}}|\Psi \rangle = \mathcal{E}[n(\mathbf{r})] = E \quad (4.75)$$

Having established that the external potential uniquely defines the density and that there is a functional of the density which attains its minimum value for the correct density, we can proceed to derive single-particle equations using variational methods, as in the case of the Hartree and Hartree–Fock approximations. The difference now is that the single-particle states, whatever equations they obey, must also produce the correct density that gives the minimum value of the energy functional, for the ground state of the system.

4.5.3 Single-Particle Equations in DFT

In order to proceed, we will find it helpful to introduce terms that can conveniently be expressed in the single-particle states. We have already encountered two such expressions, for the density $n(\mathbf{r})$, which was expressed in terms of single-particle states in Eq. (4.40), and for the one-particle density matrix $\gamma(\mathbf{r}, \mathbf{r}')$, also expressed in terms of single-particle states in Eq. (4.48). We introduce next the two-particle density matrix

$$\Gamma(\mathbf{r}_1, \mathbf{r}'_1 | \mathbf{r}_2, \mathbf{r}'_2) = \frac{N(N-1)}{2} \int \Psi^*(\mathbf{r}_1, \mathbf{r}'_1, \mathbf{r}_3, \ldots, \mathbf{r}_N) \Psi(\mathbf{r}_2, \mathbf{r}'_2, \mathbf{r}_3, \ldots, \mathbf{r}_N) \, d\mathbf{r}_3 \cdots d\mathbf{r}_N \tag{4.76}$$

which is a generalization of the one-particle density matrix $\gamma(\mathbf{r}, \mathbf{r}')$, defined earlier, in the context of the Hartree–Fock approximation, Eq. (4.47). These quantities and their physical meaning are discussed in detail in Appendix C, Eqs (C.14) and (C.15). Using these, we can obtain explicit expressions for $\mathcal{E}[n]$ and $\mathcal{F}[n]$:

$$\mathcal{E}[n(\mathbf{r})] = \langle \Psi | \hat{\mathcal{H}} | \Psi \rangle = -\frac{\hbar^2}{2m_e} \int \nabla^2_{\mathbf{r}'} \gamma(\mathbf{r}, \mathbf{r}')|_{\mathbf{r}'=\mathbf{r}} d\mathbf{r}$$
$$+ \int \int \frac{e^2}{|\mathbf{r} - \mathbf{r}'|} \Gamma(\mathbf{r}, \mathbf{r}' | \mathbf{r}, \mathbf{r}') d\mathbf{r} d\mathbf{r}' + \int \mathcal{V}(\mathbf{r}) n(\mathbf{r}) d\mathbf{r} \tag{4.77}$$

Now we can attempt to reduce these expressions to a set of single-particle equations, as we did in the case of the Hartree and Hartree–Fock approximations, when we were working explicitly with sets of single-particle states $\psi_i(\mathbf{r})$ which were independent. The important difference in the present case is that we do not have to interpret these single-particle states as corresponding to electrons. They represent fictitious fermionic particles with the only requirement that their density is identical to the density of the real electrons. These particles can be considered to be *non-interacting*.

We digress from the derivation of the single-particle equations briefly, in order to discuss an important fact: even though there is no formal mathematical justification to relate these fictitious independent states to electron states, there is mounting circumstantial evidence based on a wide range of results that the fictitious states are actually quite reasonable descriptions of electron states. Examples of such evidence are:

- The density of states that is obtained from the fictitious single-particle states corresponds very closely to experimental measurements.
- Optical spectra of solids and molecules calculated using as a basis the fictitious single-particle states can be very close to experimental measurements, especially when hybrid exchange–correlation functionals are used, or the time-dependent version of DFT is employed.
- Experimental measurements of bulk and surface electronic charge densities, especially results obtained with scanning tunneling microscopy (STM) on a wide range of solid surfaces, are very close to those calculated from the single-particle fictitious states.
- Band-structure features obtained, for example, through angle resolved photo-emission spectroscopy (ARPS) are very close to calculations based on the fictitious single-particle picture.

It therefore seems reasonable to take the fictitious single-particle states more seriously than as mere abstract tools for recasting the many-body problem into the single-particle picture, and to assign to them physical meaning, namely that they represent (approximately) true electron states, even though all the evidence mentioned above cannot provide firm justification for this view. It is also interesting that formulations of the problem that go beyond the single-particle picture often rely on the fictitious single-particle states obtained by the procedure discussed below, in order to provide a more accurate description of the physics. We return to these issues throughout the rest of this book, and discuss them on several occasions as specific examples of physical properties permit.

Our next goal is to express the various terms in the total energy using the single-particle states, denoted in the following as $\phi_i(\mathbf{r})$. Since these are also independent particles, we have:

$$n(\mathbf{r}) = \sum_i |\phi_i(\mathbf{r})|^2 \tag{4.78}$$

$$\gamma(\mathbf{r}, \mathbf{r}') = \sum_i \phi_i^*(\mathbf{r})\phi_i(\mathbf{r}') \tag{4.79}$$

$$\Gamma(\mathbf{r}, \mathbf{r}'|\mathbf{r}, \mathbf{r}') = \frac{1}{2}\left[n(\mathbf{r})n(\mathbf{r}') - |\gamma(\mathbf{r}, \mathbf{r}')|^2\right] \tag{4.80}$$

as would be the case for a Slater determinant composed of such states (see Appendix C). The universal functional $\mathcal{F}[n(\mathbf{r})]$ takes the form

$$\mathcal{F}[n(\mathbf{r})] = \frac{e^2}{2} \int \int \frac{n(\mathbf{r})n(\mathbf{r}')}{|\mathbf{r} - \mathbf{r}'|} d\mathbf{r} d\mathbf{r}' + \mathcal{K}[\gamma(\mathbf{r}, \mathbf{r}')] \tag{4.81}$$

where we have separated out the term from $\Gamma(\mathbf{r}, \mathbf{r}'|\mathbf{r}, \mathbf{r}')$ that explicitly involves the density $n(\mathbf{r})$ and combined all other terms which involve $\gamma(\mathbf{r}, \mathbf{r}')$ into the many-body term we called $\mathcal{K}$. The first term represents the electrostatic Coulomb repulsion between electrons. We further split the second term $\mathcal{K}$ into two parts:

$$\mathcal{K}[\gamma(\mathbf{r}, \mathbf{r}')] = \mathcal{K}^S[n(\mathbf{r})] + \mathcal{E}^{XC}[n(\mathbf{r})] \tag{4.82}$$

In this expression, the first term represents the kinetic energy of single-particle states which can be written exactly as:

$$\mathcal{K}^S[n(\mathbf{r})] = \sum_i \langle\phi_i| - \frac{\hbar^2}{2m_e}\nabla_\mathbf{r}^2 |\phi_i\rangle \tag{4.83}$$

This is formally the same expression as would be obtained from a Slater determinant (hence the superscript S). The second term on the right-hand side of Eq. (4.82) represents all other contributions to the many-body energy of electrons; this is referred to as the "exchange–correlation" functional. Note that we wrote the two terms $\mathcal{K}^S$ and $\mathcal{E}^{XC}$ as explicit functionals of the density, since we know from our earlier discussion that their sum must be such a functional. The term $\mathcal{K}^S$ is not the exact kinetic energy of the many-body system of electrons; it is a convenient approximation accounting only for the kinetic energy contribution of the independent fictitious particles. The term $\mathcal{E}^{XC}$ is the correction to this approximation, which is needed to reproduce the correct functional $\mathcal{K}$. This latter is added to the Coulomb repulsion term to produce the universal functional $\mathcal{F}$. Thus, $\mathcal{K}^S$ and $\mathcal{E}^{XC}$ are intimately connected. What this approach has yielded so far is a systematic way of expressing the various contributions to the total energy of the system in terms of the density $n(\mathbf{r})$ and the single-particle states $\phi_i(\mathbf{r})$.

Putting the above terms together, we now have the following total-energy functional, referred to as the "Kohn–Sham" functional:

$$\mathcal{E}^{KS}[n(\mathbf{r})] = \sum_i \langle \phi_i | - \frac{\hbar^2 \nabla_{\mathbf{r}}^2}{2m_e} | \phi_i \rangle + \int \int \frac{e^2 n(\mathbf{r}) n(\mathbf{r}')}{2|\mathbf{r} - \mathbf{r}'|} d\mathbf{r} d\mathbf{r}' + \mathcal{E}^{XC}[n(\mathbf{r})] + \int \mathcal{V}(\mathbf{r}) n(\mathbf{r}) d\mathbf{r} \tag{4.84}$$

which we need to minimize with respect to variations in the density in order to find the ground-state energy. We consider a variation in the density, which we choose to be

$$\delta n(\mathbf{r}) = \delta \phi_i^*(\mathbf{r}) \phi_i(\mathbf{r}) \tag{4.85}$$

with the restriction that

$$\int \delta n(\mathbf{r}) d\mathbf{r} = \int \delta \phi_i^*(\mathbf{r}) \phi_i(\mathbf{r}) d\mathbf{r} = 0 \tag{4.86}$$

so that the total number of particles does not change; note that $\phi_i(\mathbf{r})$ and $\phi_i^*(\mathbf{r})$ are treated as independent, as far as their variation is concerned. With this choice, and taking the restriction into account through a Lagrange multiplier ζ_i, we arrive at the following single-particle equations, through a variational argument:

$$\left[- \frac{\hbar^2}{2m_e} \nabla_{\mathbf{r}}^2 + \mathcal{V}^{\text{eff}}(\mathbf{r}, n(\mathbf{r})) \right] \phi_i(\mathbf{r}) = \zeta_i \phi_i(\mathbf{r}) \tag{4.87}$$

where the effective potential is given by:

$$\mathcal{V}^{\text{eff}}(\mathbf{r}, n(\mathbf{r})) = \mathcal{V}(\mathbf{r}) + e^2 \int \frac{n(\mathbf{r}')}{|\mathbf{r} - \mathbf{r}'|} d\mathbf{r}' + \frac{\delta \mathcal{E}^{XC}[n(\mathbf{r})]}{\delta n(\mathbf{r})} \tag{4.88}$$

with the first term, $\mathcal{V}(\mathbf{r})$, being the external potential due to the ions and the second term being the familiar Hartree potential, defined in Eq. (4.41); the last term is the variational functional derivative[7] of the as yet unspecified functional $\mathcal{E}^{XC}[n(\mathbf{r})]$, which we define as the exchange–correlation potential:

$$\mathcal{V}^{XC}[n(\mathbf{r})] = \frac{\delta \mathcal{E}^{XC}[n(\mathbf{r})]}{\delta n(\mathbf{r})} \tag{4.89}$$

The single-particle equations of Eq. (4.87) are referred to as Kohn–Sham equations and the single-particle orbitals $\phi_i(\mathbf{r})$ that are their solutions are called Kohn–Sham orbitals. We can

---

[7] For the definition of this term in the context of the present theory, see Appendix A.

also define the single-particle hamiltonian $\hat{\mathcal{H}}^{\mathrm{sp}}$ and the universal functional that contains all terms except for the external potential $\hat{\mathcal{F}}^{\mathrm{sp}}$ for the Kohn–Sham equations as:

$$\hat{\mathcal{H}}^{\mathrm{sp}}[\mathbf{r}, n(\mathbf{r})] \equiv -\frac{\hbar^2}{2m_e}\nabla_{\mathbf{r}}^2 + \mathcal{V}(\mathbf{r}) + e^2 \int \frac{n(\mathbf{r}')}{|\mathbf{r}-\mathbf{r}'|}\mathrm{d}\mathbf{r}' + \mathcal{V}^{\mathrm{XC}}[n(\mathbf{r})] \qquad (4.90)$$

$$\hat{\mathcal{F}}^{\mathrm{sp}}[\mathbf{r}, n(\mathbf{r})] \equiv -\frac{\hbar^2}{2m_e}\nabla_{\mathbf{r}}^2 + e^2 \int \frac{n(\mathbf{r}')}{|\mathbf{r}-\mathbf{r}'|}\mathrm{d}\mathbf{r}' + \mathcal{V}^{\mathrm{XC}}[n(\mathbf{r})] \qquad (4.91)$$

We compare below the salient features of Hartree–Fock theory and density functional theory.

Comparison of Hartree-Fock theory and density functional theory

	Hartree–Fock theory	Density functional theory
fundamental ansatz	$\Psi^{\mathrm{HF}}(\{\mathbf{r}_i\}) = \mathrm{S}[\{\psi_j(\mathbf{r}_i)\}]$	$E = \mathcal{E}[n(\mathbf{r})], n(\mathbf{r}) = \sum_j |\phi_j(\mathbf{r})|^2$
main limitation	$\Psi^{\mathrm{HF}}(\{\mathbf{r}_i\})$: no correlations	$\mathcal{E}^{\mathrm{XC}}[n(\mathbf{r})]$: unknown
single-particle	$\psi_j(\mathbf{r})\ (j = 1,\ldots,N)$:	$\phi_j(\mathbf{r})\ (j = 1,\ldots,N)$:
states	real, interacting	fictitious, non-interacting
Hartree term	$\mathcal{V}^{\mathrm{H}}(\mathbf{r}) = e^2 \int \frac{n(\mathbf{r}')}{|\mathbf{r}-\mathbf{r}'|}\,\mathrm{d}\mathbf{r}'$	$\mathcal{V}^{\mathrm{H}}(\mathbf{r}) = e^2 \int \frac{n(\mathbf{r}')}{|\mathbf{r}-\mathbf{r}'|}\,\mathrm{d}\mathbf{r}'$
exchange term	$\mathcal{V}^{\mathrm{X}}(\mathbf{r}, \mathbf{r}') = -e^2 \frac{\gamma(\mathbf{r}',\mathbf{r})}{|\mathbf{r}-\mathbf{r}'|}$	$\mathcal{V}^{\mathrm{XC}}(\mathbf{r}) = \frac{\delta \mathcal{E}^{\mathrm{XC}}[n(\mathbf{r})]}{\delta n(\mathbf{r})}$
exchange	exact	exact but unknown
correlation	0	exact but unknown

Since the effective potential is a function of the density, which is obtained from Eq. (4.78) and hence depends on all the single-particle states, we will need to solve these equations by iteration until we reach self-consistency. As mentioned earlier in connection to the Hartree and Hartree–Fock equations, this is not a significant problem and can be solved numerically. A more pressing issue is the exact form of $\mathcal{E}^{\mathrm{XC}}[n(\mathbf{r})]$, which is unknown, and without explicit knowledge of this term the single-particle equations cannot be solved. Much effort has been devoted to obtaining reasonable expressions for this term, all of them approximate, since there is no analytic solution, even in the simplest case of a uniform electron gas.

4.5.4 The Exchange–Correlation Term in DFT

We can consider the simplest situation, in which the true electronic system is endowed with only one aspect of electron interactions (beyond Coulomb repulsion), that is, the exchange property. As we saw in the case of the Hartree–Fock approximation, which takes into account exchange explicitly, in a uniform system the contribution of exchange to the total energy is:

$$E^{\mathrm{X}} = -\frac{3}{4}\frac{e^2}{\pi}k_{\mathrm{F}}N \tag{4.92}$$

Since the total number of electrons in the system can be written as $N = \int n\mathbf{dr}$, we can write:

$$E^{\mathrm{X}}[n] = -\frac{3}{4}\frac{e^2}{\pi}\int k_{\mathrm{F}}n\mathbf{dr} = -\frac{3}{4}e^2\left(\frac{3}{\pi}\right)^{1/3}\int [n]^{1/3}\,n\mathbf{dr} \tag{4.93}$$

Let us now try to generalize this to situations where the density is not uniform. If $n(\mathbf{r})$ is slowly varying, it may not be too bad an approximation to assume that:

$$\mathcal{E}^{\mathrm{X}}[n(\mathbf{r})] = \int \epsilon^{\mathrm{X}}[n(\mathbf{r})]n(\mathbf{r})\mathbf{dr} \tag{4.94}$$

$$\text{with} \quad \epsilon^{\mathrm{X}}[n(\mathbf{r})] = -\frac{3}{4}e^2\left(\frac{3}{\pi}\right)^{1/3}[n(\mathbf{r})]^{1/3} \tag{4.95}$$

This allows us to calculate the expression for $\delta\mathcal{E}^{\mathrm{XC}}[n]/\delta n$ in the case where we are considering only the exchange aspect of the many-body character. We obtain:

$$\frac{\delta\mathcal{E}^{\mathrm{X}}[n(\mathbf{r})]}{\delta n(\mathbf{r})} = \frac{\partial}{\partial n(\mathbf{r})}\left[\epsilon^{\mathrm{X}}[n(\mathbf{r})]n(\mathbf{r})\right] = \frac{4}{3}\epsilon^{\mathrm{X}}[n(\mathbf{r})] = -e^2\left(\frac{3}{\pi}\right)^{1/3}[n(\mathbf{r})]^{1/3} \tag{4.96}$$

This is remarkably similar to Slater's exchange potential, Eq. (4.66), which was based on an ad-hoc assumption, and differs from it only by a factor of $2/3$.

The discussion so far provides some guidance on how to express the exchange–correlation functional in terms of the density, assuming that the density is slowly varying in space. In particular, we attempted to take into account electron exchange effects drawing on our experience from the Hartree–Fock approximation, which treats exchange exactly, by incorporating its effects explicitly into the many-body wavefunction. What has been left out is correlation effects (recall the discussion of the hydrogen molecule example). In an early attempt to include correlation effects, Slater introduced a "fudge factor" in his expression for the exchange potential, denoted by α (hence the expression $X - \alpha$ potential). This factor is usually taken to be close to, but somewhat smaller than, unity, $\alpha = 0.75$ being a typical choice. If $\alpha \neq 2/3$, the value required for the potential arising from pure exchange by analogy to Hartree–Fock theory for free electrons, it is thought that the $X - \alpha$ expression includes in some crude way the effects of both exchange and correlation. It is easy to extract the part that corresponds to correlation in the $X - \alpha$ expression, by comparing Slater's exchange potential, multiplied by the fudge factor α, to the potential involved in the single-particle equations derived from DFT (see Table 4.1).

What should $\mathcal{E}^{\mathrm{XC}}[n(\mathbf{r})]$ actually be in order to capture accurately the many-body exchange and correlation effects? So far, no completely satisfactory answer has emerged from the many attempts to provide a good enough approximation. There are many

Table 4.1 Correlation energy functionals $\epsilon^{\text{cor}}[n(\mathbf{r})]$ and exchange–correlation potentials $\mathcal{V}^{\text{XC}}[n(\mathbf{r})]$ in various models (H–L = Hedin–Lundqvist, P–Z = Perdew–Zunger). ϵ^{X} is the pure exchange energy from Eq. (4.95). r_s is measured in units of a_0 and the energy is in rydbergs. The numerical constants have units which depend on the factor of r_s involved with each, and values Wigner: $A = 0.884, B = 7.8$; H–L: $A = 21, B = 0.0368$; P–Z: $A_1 = 0.096, A_2 = 0.0232, A_3 = 0.0622, A_4 = 0.004,$ $B_1 = 0.2846, B_2 = 1.0529, B_3 = 0.3334.$

Model	$\epsilon^{\text{cor}}[n(\mathbf{r})]$	$\mathcal{V}^{\text{XC}}[n(\mathbf{r})]$
Exchange	0	$\frac{4}{3}\epsilon^{\text{X}}$
Slater	$(\frac{3}{2}\alpha - 1)\epsilon^{\text{X}}$	$2\alpha\epsilon^{\text{X}}$
Wigner	$-A(B + r_s)^{-1}$	
H–L		$\frac{4}{3}\epsilon^{\text{X}}\left[1 + Br_s \ln\left(1 + A/r_s\right)\right]$
P–Z:		
$r_s < 1$	$-A_1 - A_2 r_s + \left[A_3 + A_4 r_s\right]\ln(r_s)$	
$r_s \geq 1$	$-B_1\left[1 + B_2\sqrt{r_s} + B_3 r_s\right]^{-1}$	

interesting models, of which we mention a few so that the reader can get a feeling of what is typically involved, but the problem remains an area of active research. In fact, it is not likely that any expression which depends on $n(\mathbf{r})$ in a *local* fashion will suffice, since the exchange and correlation effects are inherently non-local in an interacting electron system. A collection of proposed expressions for the correlation energies and exchange–correlation potentials is given in Table 4.1. In these expressions, the exchange–correlation functional is written as:

$$\mathcal{E}^{\text{XC}}[n(\mathbf{r})] = \int \left(\epsilon^{\text{X}}[n(\mathbf{r})] + \epsilon^{\text{cor}}[n(\mathbf{r})]\right) n(\mathbf{r})d\mathbf{r} \tag{4.97}$$

from which the exchange–correlation potential $\mathcal{V}^{\text{XC}}[n(\mathbf{r})]$ that appears in the single-particle equations, defined in Eq. (4.89), can be obtained directly by functional differentiation; the pure exchange energy $\epsilon^{\text{X}}[n]$ is the expression given in Eq. (4.95). These expressions are usually given in terms of r_s, which is related to the density through Eq. (1.9). The expression proposed by Winer extrapolates between known limits in r_s, obtained by series expansions (see Problem 8). The parameters that appear in the expression proposed by Hedin and Lundqvist[8] are determined by fitting to the energy of the uniform electron gas, obtained by numerical methods at different densities. A similar type of expression was proposed by Perdew and Zunger,[9] which captures the more sophisticated numerical

[8] L. Hedin and B. I. Lundqvist, *J. Phys. C* **4**, 2064 (1971).
[9] J. P. Perdew and A. Zunger, *Phys. Rev. B* **23**, 5048 (1981).

calculations for the uniform electron gas at different densities performed by Ceperley and Alder.[10]

The common feature in all these approaches is that $\mathcal{E}^{XC}$ depends on $n(\mathbf{r})$ in a *local* fashion, that is, n needs to be evaluated at one point in space at a time. For this reason they are referred to as the local density approximation (LDA) to DFT. This is actually a severe restriction, because even at the exchange level, the functional should be *non-local*, that is, it should depend on $\mathbf{r}$ and $\mathbf{r}'$ simultaneously [recall, for example, the expressions for the exchange potential, $\mathcal{V}^X(\mathbf{r}, \mathbf{r}')$, Eq. (4.50)]. It is a much more difficult task to develop non-local exchange–correlation functionals. More recently, a concentrated effort has been directed toward producing expressions for $\mathcal{E}^{XC}$ that depend not only on the density $n(\mathbf{r})$, but also on its gradients, referred to as the generalized gradient approximation.[11] These expansions tend to work better for finite systems (molecules, surfaces, etc.), but still represent a local approximation to the exchange–correlation functional. Including correlation effects in a realistic manner is exceedingly difficult, as we have already demonstrated for the hydrogen molecule.

Despite the seemingly severe approximations required to obtain a workable expression for the exchange–correlation functional, applications of DFT to a wide range of physical and chemical properties of materials have proven very successful. In fact, in the last few years DFT-based computational methods have become the preferred approach to describe properties such as mechanical, electrical, optical, magnetic, and thermal, for very diverse materials ranging from the semiconductor components of electronic devices, to photovoltaics, batteries, structural elements in mechanical systems, superconductors, many proteins, DNA, and so on. This is a strong indication that the single-particle picture of the behavior of electrons in matter captures much of the physical reality, and the approximations required to solve the corresponding single-particle equations, like the LDA to the exchange–correlation term in DFT, are quite reasonable. Several examples are discussed in other parts of the book; in anticipation of some of these discussions, we show in Fig. 4.3 the density of valence electrons in the silicon crystal, a prototypical covalently bonded material and representative semiconductor, widely used in electronic devices (for example, computer chips). The traditional representation of this crystal in "ball-and-stick" models is superimposed with the electronic density representation. From this figure it is evident that the "sticks" representing the covalent bonds between atoms correspond to a very high density of valence electrons (the blue-colored isosurface) of approximately cylindrical shape, concentrated precisely along the direction between neighboring ions. In the rest of the material the electronic density is much lower to vanishingly small. This provides justification for the simpler ball-and-stick representation of the crystal, but contains in addition much more interesting and important information about the behavior of the real material, as will be explored in later chapters.

---

[10] D. M. Ceperley and B. J. Alder, *Phys. Rev. Lett.* **45**, 566 (1980).
[11] J. P. Perdew and Y. Wang, *Phys. Rev. B* **33**, 8800 (1986).

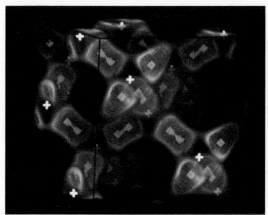 

Fig. 4.3 **Left**: The charge density of valence electrons in the silicon crystal within the conventional cubic cell. The positions of the ions are indicated by white crosses. The valence electron density is indicated by colored isosurfaces with the highest value corresponding to the green areas [adapted from R. H. Wolfe, M. Needles, T. Arias, and J. D. Joannopoulos, "Visual revelations from silicon ab initio calculations," *IEEE Computer Graphics and Applications* **12**, 45 (1992)]. **Right**: A conventional ball-and-stick representation of the diamond structure of bulk Si, in a cube containing the same number of atoms and from a similar perspective.

4.5.5 Time-Dependent DFT

The original DFT approach was developed for a time-independent external potential $\mathcal{V}(\mathbf{r})$. The theory was extended to time-dependent potentials by Runge and Gross,[12] for potentials of the form $\mathcal{V}(\mathbf{r}, t)$ which can be Taylor-expanded around some time $t = t_0$. The proof that a given time-dependent potential produces a unique time-dependent density $n(\mathbf{r}, t)$ follows the same logic as in the time-independent case, that is, by proving that potentials differing by more than a mere time-dependent function give rise to densities that are also different:

$$\mathcal{V}(\mathbf{r}, t) - \mathcal{V}'(\mathbf{r}, t) \neq c(t) \Rightarrow n(\mathbf{r}, t) \neq n'(\mathbf{r}, t) \tag{4.98}$$

The proof proceeds by calculating the quantum-mechanical current associated with the wavefunctions corresponding to the two potentials, and relating the time derivative of the density to the divergence of the current, through the continuity equation. Then, the assumption that the densities are the same leads to a contradiction, which shows that the densities arising from the two different potentials cannot be the same. Note that if the potentials are different by purely a function of time, the result would be a phase factor difference in the many-body wavefunctions, which would cancel when taking expectation values of operators that describe physical observables.

The density in this case is given by the expression

$$n(\mathbf{r}, t) = \sum_{i=1}^{N} |\phi_i(\mathbf{r}, t)|^2 \tag{4.99}$$

---

[12] E. Runge and E. K. U. Gross, *Phys. Rev. Lett.* **52**, 997 (1984).

where the single-particle states $\phi(\mathbf{r}, t)$ obey the Schrödinger equation

$$i\hbar \frac{\partial}{\partial t}\phi_i(\mathbf{r}, t) = \left[-\frac{\hbar^2 \nabla_{\mathbf{r}}^2}{2m_e} + \mathcal{V}^{\text{eff}}(\mathbf{r}, t) \right] \phi_i(\mathbf{r}, t) \tag{4.100}$$

with the time-dependent effective potential given by:

$$\mathcal{V}^{\text{eff}}(\mathbf{r}, t) = \mathcal{V}(\mathbf{r}, t) + \mathcal{V}^{\text{H}}(\mathbf{r}, t) + \mathcal{V}^{\text{XC}}[n(\mathbf{r}, t), \Psi_0(\{\mathbf{r}\})] \tag{4.101}$$

where the different terms are the familiar external potential, Coulomb repulsion, and exchange–correlation contribution. Interestingly, the last term depends not only on the instantaneous density $n(\mathbf{r}, t)$, and is non-local in space (as in the time-independent case), but also on previous times $t' < t$, through the initial condition of the many-body wavefunction at $t = 0$, $\Psi_0(\{\mathbf{r}\})$, and through the value of the density $n(\mathbf{r}, t')$ at previous times. This temporal non-locality, in addition to the spatial non-locality we encountered earlier, makes the calculations more complicated. In practice, it is not taken into account explicitly, on the assumption of an adiabatic process, that is, a slow introduction of the time dependence of the potential. Indeed, a common approximation for the exchange–correlation potential is the usual one from the homogeneous electron gas, evaluated for the local value of the density at time t, $n(\mathbf{r}, t)$, known as the adiabatic local density approximation (ALDA):

$$\mathcal{V}^{\text{XC−ALDA}}(\mathbf{r}, t) = \mathcal{V}^{\text{XC−LDA}}[n(\mathbf{r}, t)] \tag{4.102}$$

where the exchange–correlation potential on the right-hand side is one of the approximations discussed earlier as a functional of the static density $n(\mathbf{r})$.

The significance of this proof is that DFT can be applied to calculate the time evolution of a system (known as time-dependent density functional theory, or TDDFT), which opens new horizons for studying interesting properties of materials. A particularly successful type of application employing TDDFT is that of optical properties of materials. To this end, a common useful approach is to evolve the single-particle wavefunctions in time, after calculating the ground state of the system and inducing an excitation by promoting an electron from an occupied to an unoccupied state. The time evolution of the single-particle states takes the form

$$\phi_i(\mathbf{r}, t) = \hat{U}(t, t_0)\phi(\mathbf{r}, t_0), \quad \hat{U}(t, t_0) = \hat{T} \exp\left[-\frac{i}{\hbar} \int_{t_0}^{t} \hat{\mathcal{H}}_{\text{KS}}(t') dt' \right] \tag{4.103}$$

where $\hat{U}$ is the time-evolution operator, $\hat{T}$ the time-ordering operator ensuring that $t > t_0$, and $\hat{\mathcal{H}}_{\text{KS}}$ is the Kohn–Sham single-particle hamiltonian. From the time propagation of the single-particle states, we can extract the response of the system to external perturbations; for example, by considering the external perturbation to be a time-dependent electric field, the time-dependent induced dipole moment is calculated from the time evolution of the system, and from this the optical absorption is obtained that can be directly compared to experiment.

4.6 Quasiparticles and Collective Excitations

So far we have examined how one can justify the reduction of the many-body equation, Eq. (4.1), to a set of single-particle equations. This was done by introducing certain approximations. In the case of the Hartree and Hartree–Fock approximations, we started with a guess for the many-body wavefunction, expressed in terms of single-particle states. The resulting single-particle equations describe the behavior of electrons as independent particles in an external potential defined by the ions, as well as an external field produced by the presence of all other electrons. In the case of DFT, we derived exactly the single-particle equations for *non-interacting, fictitious* particles, whose density is the same as the density of real electrons. However, these equations cannot be solved exactly, because the exchange–correlation functional which appears in them is not known explicitly. We then constructed approximations to this functional, by comparison to the results of numerical calculations for the electron gas. With this, the single-particle equations can be solved, and the wavefunctions of the fictitious particles can be determined.

Why is the single-particle approximation successful? We can cite the following general arguments:

- **The variational principle**. Even when the wavefunctions are not accurate, as in the case of the Hartree–Fock approximation, the total energy is not all that bad. Energy differences between states of the system (corresponding to different atomic configurations, for which the single-particle equations are solved self-consistently each time) turn out to be remarkably good. This is because the optimal set of single-particle states contains most of the physics related to the motion of ions. The case of coherent many-body states for the electrons, where the single-particle approximation fails, concerns much more delicate phenomena in solids. For example, the phenomenon of superconductivity involves energy scales of a few kelvin (or at most ~ 100 K), whereas the motion of atoms in solids (involving changes in atomic positions) involves energies of the order of 1 eV = 11,604 K.

- **Screening**. The Coulomb interaction between real electrons is "screened" by the correlated motion of all other electrons: each electron is surrounded by a region where the density of electronic charge is depleted; this is referred to as the "exchange–correlation hole." This forms an effective positive charge cloud which screens an electron from all its neighbors. There is no real hole in the system; the term is just a figurative way of describing the many-body effects in the single-particle picture. The net effect is a weakened interaction between electrons; the weaker the interaction, the closer the system is to the picture of non-interacting single particles. These particles, however, can no longer be identified with individual electrons, since they carry with them the effects of interaction with all other electrons (the exchange–correlation hole). Thus, not only are the total energies in the single-particle picture quite good, but the description of electronic behavior is also reasonable.

The screened electrons behave as "quasiparticles," a notion introduced by L. D. Landau: in a complicated system of *strongly* interacting particles, it may still be possible to describe the properties of the system in terms of *weakly* interacting particles, experiencing a weaker effective interaction than the original one due to many-body effects. Dealing with these weakly interacting particles is much simpler, so describing the system in this language can be very advantageous. In particular, perturbation theory can be used when one deals with a weakly interacting system of particles, beginning with the non-interacting particles as the unperturbed state. Quasiparticles usually maintain several of the key aspects of the original, physical particles they describe in an effective manner; in the case of screened electrons, these have the same charge as real electrons but their interactions are weaker interactions than the actual Coulomb repulsion.

There is another notion which is also very important for the description of the properties of solids, that of "collective excitations." In contrast to quasiparticles, these are bosons, they bear no resemblance to constituent particles of a real system, and they involve collective (that is, coherent) motion of many physical particles.

We summarize here the most common quasiparticles and collective excitations encountered in solids:

(a) **Electron**. As discussed already, this is a quasiparticle consisting of a real electron and the exchange–correlation hole, a cloud of effective charge of opposite sign due to exchange and correlation effects arising from interaction with all other electrons in the system. The electron is a fermion with spin 1/2. The Fermi energy (highest occupied state) is of order 5 eV, and the Fermi velocity ($v_F = \hbar k_F / m_e$) is $\sim 10^8$ cm/sec, that is, it can be treated as a non-relativistic particle. Notice that the mass of this quasiparticle can be different from that of the free electron.

(b) **Hole**. This is a quasiparticle like the electron, but of opposite charge; it corresponds to the absence of an electron from a single-particle state which lies below the Fermi level. The notion of a hole is particularly convenient when the ground state consists of quasiparticle states that are fully occupied and are separated by an energy gap from the unoccupied states. Perturbations with respect to this ground state, such as missing electrons, can be conveniently discussed in terms of holes. This is, for example, the situation in p-doped semiconductor crystals.

(c) **Polaron**. This is a quasiparticle like the electron, tied to a distortion of the lattice of ions. Polarons are invoked to describe polar crystals, where the motion of a negatively charged electron distorts the lattice of positive and negative ions around it. Because its motion is coupled to the motion of ions, the polaron has a different mass from the electron.

(d) **Exciton**. This is a collective excitation, corresponding to a bound state of an electron and a hole. The binding energy is of order $e^2/(\varepsilon a)$, where ε is the dielectric constant of the material (typically of order 10) and a is the distance between the two quasiparticles (typically a few lattice constants, of order 10 Å), which give for the binding energy ~ 0.1 eV.

(e) **Phonon**. This is a collective excitation, corresponding to coherent motion of all the atoms in the solid. It is a quantized lattice vibration, with a typical energy scale of $\hbar\omega \sim 0.1$ eV.

(f) **Plasmon**. This is a collective excitation of the entire electron gas relative to the lattice of ions; its existence is a manifestation of the long-range nature of the Coulomb interaction. The energy scale of plasmons is $\hbar\omega \sim \hbar\sqrt{4\pi ne^2/m_e}$, where n is the density; for typical densities, this gives an energy of order 5–20 eV.

(g) **Magnon**. This is a collective excitation of the spin degrees of freedom on the crystalline lattice. It corresponds to a spin wave, with an energy scale of $\hbar\omega \sim 0.001$– 0.1 eV.

In the following chapters we will explore in detail the properties of some of these quasiparticles and collective excitations.

4.7 Screening: The Thomas–Fermi Model

To put a more quantitative expression to the idea of screened electrons we will calcuate, using a very simple model, the dielectric function which describes the response of a system of charged particles to an external potential Φ^{ext}. The usual definition of the dielectric constant ε is:

$$\Phi^{\text{tot}} = \frac{\Phi^{\text{ext}}}{\varepsilon}$$

that is, it is a number by which we divide the value of the external potential to obtain the total potential Φ^{tot}. In the general case, both the external and total potentials depend on the position $\mathbf{r}$, so the dielectric constant must be a function of space as well. In fact, we must take the latter to be a non-local function of space, $\varepsilon(\mathbf{r}, \mathbf{r}')$, because it describes how the external potential is affected throughout space to produce the total potential $\Phi^{\text{tot}}(\mathbf{r})$; the non-locality comes from the fact that what happens at position $\mathbf{r}$ is affected by how electric charges and potentials are distributed at every other position in space $\mathbf{r}'$. For an infinite solid, the dielectric function must be translationally invariant; that is, if both points $\mathbf{r}$ and $\mathbf{r}'$ are shifted by a constant vector, this function should remain invariant. This implies that it is a function of the difference $(\mathbf{r} - \mathbf{r}')$. The definition of ε in real space is then given through the relation

$$\Phi^{\text{tot}}(\mathbf{r}) = \int \varepsilon^{-1}(\mathbf{r} - \mathbf{r}')\Phi^{\text{ext}}(\mathbf{r}') \, d\mathbf{r}' \tag{4.104}$$

It is more convenient to work in Fourier space, which makes the expressions simpler to handle. Note that here we are omitting the time dependence of the external potential, since the system we will consider is time independent. More generally, the external potential can have a time dependence, which would introduce a time dependence to the dielectric function, or equivalently a frequency (energy) dependence in its Fourier components. We introduce the Fourier transforms of the dielectric function and the potentials and use the convolution theorem [see Appendix A, Eq. (A.51)] to arrive at:

$$\Phi^{\text{tot}}(\mathbf{k}) = \varepsilon^{-1}(\mathbf{k})\Phi^{\text{ext}}(\mathbf{k}) \Rightarrow \varepsilon(\mathbf{k}) = \frac{\Phi^{\text{ext}}(\mathbf{k})}{\Phi^{\text{tot}}(\mathbf{k})} \tag{4.105}$$

We next define the induced potential as the difference between the total and external potentials. Write the total potential as a sum of two contributions, the external potential

Φ^{ext} and the induced potential Φ^{ind}:

$$\Phi^{\text{ind}}(\mathbf{r}) \equiv \Phi^{\text{tot}}(\mathbf{r}) - \Phi^{\text{ext}}(\mathbf{r})$$

and the corresponding induced charge density $\rho^{\text{ind}}(\mathbf{r})$ through the Poisson equation:

$$\nabla^2 \Phi^{\text{ind}}(\mathbf{r}) = -4\pi \rho^{\text{ind}}(\mathbf{r}) \tag{4.106}$$

Using the Fourier transforms of $\Phi^{\text{ind}}(\mathbf{r})$ and $\rho^{\text{ind}}(\mathbf{r})$, we obtain:

$$\nabla^2 \sum_{\mathbf{k}} \Phi^{\text{ind}}(\mathbf{k}) e^{-i\mathbf{k}\cdot\mathbf{r}} = -4\pi \sum_{\mathbf{k}} \rho^{\text{ind}}(\mathbf{k}) e^{-i\mathbf{k}\cdot\mathbf{r}} \implies \Phi^{\text{ind}}(\mathbf{k}) = \frac{4\pi}{|\mathbf{k}|^2} \rho^{\text{ind}}(\mathbf{k}) \tag{4.107}$$

which, when inserted in Eq. (4.105), gives for the dielectric function in Fourier space:

$$\varepsilon(\mathbf{k}) = 1 - \frac{4\pi}{k^2} \frac{\rho^{\text{ind}}(\mathbf{k})}{\Phi^{\text{tot}}(\mathbf{k})} \tag{4.108}$$

For sufficiently weak potentials, we take the response of the system, as described by the induced charge, to be linear in the total potential:

$$\rho^{\text{ind}}(\mathbf{k}) = \chi(\mathbf{k}) \Phi^{\text{tot}}(\mathbf{k}) \tag{4.109}$$

where the function $\chi(\mathbf{k})$ is the susceptibility or response function. This gives, for the dielectric function:

$$\varepsilon(\mathbf{k}) = 1 - \frac{4\pi}{k^2} \chi(\mathbf{k}) \tag{4.110}$$

Using perturbation theory (see Problem 9), we find that for a free-electron system with energy $\epsilon_{\mathbf{k}} = \hbar^2 k^2 / 2m_e$ and Fermi occupation numbers (see Appendix D):

$$f(\mathbf{k}) = \frac{1}{e^{(\epsilon_{\mathbf{k}} - \mu)/k_B T} + 1} \tag{4.111}$$

where μ is the chemical potential, in this case equal to the Fermi energy (see Appendix D), and T the temperature. The response function $\chi(\mathbf{k})$ takes the form

$$\chi(\mathbf{k}) = -e^2 \int \left[\frac{f(\mathbf{k}' - \mathbf{k}/2) - f(\mathbf{k}' + \mathbf{k}/2)}{\hbar^2 (\mathbf{k} \cdot \mathbf{k}')/2m_e} \right] \frac{d\mathbf{k}'}{(2\pi)^3} \tag{4.112}$$

This is called the Lindhard dielectric response function. Notice that, at $T = 0$, in order to have a non-vanishing integrand, one of the occupation numbers must correspond to a state below the Fermi level, and the other to a state above the Fermi level; that is, we must have $|\mathbf{k}' - \mathbf{k}/2| < k_F$ and $|\mathbf{k}' + \mathbf{k}/2| > k_F$ (or vice versa), since otherwise the occupation numbers are either both 0 or both 1. These considerations indicate that the contributions to the integral will come from electron–hole excitations with total momentum $\mathbf{k}$. At $T = 0$, the integral over $\mathbf{k}'$ in the Lindhard dielectric function can be evaluated to yield:

$$\chi_0(\mathbf{k}) = -e^2 \frac{m_e k_F}{2\hbar^2 \pi^2} F(k/2k_F) \tag{4.113}$$

where $F(x)$ is the same function as the one encountered earlier in the discussion of Hartree–Fock single-particle energies, defined in Eq. (4.57). The study of this function provides insight into the behavior of the system of single particles (see Problem 9).

A special case is known as the Thomas–Fermi screening. This corresponds to the limit of the Lindhard function for $|\mathbf{k}| \ll k_F$; that is, electron–hole excitations with small total momentum relative to the Fermi momentum. In this case, expanding the occupation numbers $f(\mathbf{k}' \pm \mathbf{k}/2)$ through their definition in Eq. (4.111) about $\mathbf{k} = 0$ we obtain, to first order in $\mathbf{k}$:

$$f(\mathbf{k}' \pm \mathbf{k}/2) = f(\mathbf{k}') \mp \frac{\hbar^2}{2m_e}(\mathbf{k}' \cdot \mathbf{k})\frac{\partial f(\mathbf{k}')}{\partial \mu} \tag{4.114}$$

This expression, when substituted in the equation for the dielectric constant, gives:

$$\varepsilon(\mathbf{k}) = 1 + \left(\frac{k_s}{k}\right)^2, \quad k_s(T) = \sqrt{4\pi e^2 \frac{\partial f_T}{\partial \mu}} \tag{4.115}$$

where k_s is the so-called Thomas–Fermi screening inverse length, which depends on the temperature T through the occupation f_T, defined as:

$$f_T = 2 \int \frac{d\mathbf{k}}{(2\pi)^3} \frac{1}{e^{(\epsilon_\mathbf{k} - \mu)/k_\mathrm{B}T} + 1} \tag{4.116}$$

This quantity represents the total occupation of states in the ground state of the single-particle system at temperature T, and is evidently a function of μ.

We can rewrite the expression for the dielectric function in the Thomas–Fermi model as:

$$\varepsilon(\mathbf{k}) = 1 + \left(\frac{k_s}{k}\right)^2 = \frac{(-e)4\pi/k^2}{(-e)4\pi/(k^2 + k_s^2)} = \frac{\Phi^{\mathrm{ext}}(\mathbf{k})}{\Phi^{\mathrm{tot}}(\mathbf{k})} \tag{4.117}$$

where we have also invoked Eq. (4.105). We can now use the fact that the Fourier transform of the Coulomb potential is:

$$\Phi^{\mathrm{ext}}(\mathbf{r}) = \frac{-e}{r} \Rightarrow \Phi^{\mathrm{ext}}(\mathbf{k}) = \frac{(-e)4\pi}{k^2}$$

and the above equation, to obtain by inverse Fourier transform (see Appendix A):

$$\Phi^{\mathrm{tot}}(\mathbf{k}) = \frac{(-e)4\pi}{k^2 + k_s^2} \Rightarrow \Phi^{\mathrm{tot}}(\mathbf{r}) = \frac{(-e)}{r}e^{-k_s r} \tag{4.118}$$

Thus, the true Coulomb potential of an electron $(-e)/r$ is quickly suppressed (screened) as we move away from the electron's position (taken to be the origin of the coordinate system) by the exponential factor $e^{-k_s r}$, where k_s is the Thomas–Fermi screening inverse length. The effective potential is negligible at distances larger than a few times k_s^{-1}, as shown in Fig. 4.4.

4.8 Quasiparticle Energies: GW Approximation

The power of the quasiparticle approach becomes evident when we try to compare the energy of electrons in a system to values obtained experimentally. We return to the issue of single-particle equations for electrons in solids, that experience the many-body hamiltonian

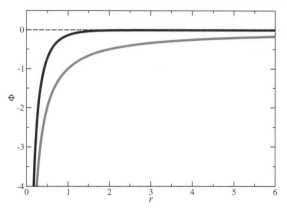

Fig. 4.4 The original Coulomb potential $\Phi^{\mathrm{ext}}(\mathbf{r}) = (-e)/r$, shown as a green line, and the total (screened) potential $\Phi^{\mathrm{tot}}(\mathbf{r}) = (-e)\exp(-k_s r)/r$, shown as a red line, for $k_s = 2a_0^{-1}$. The two potentials are in units of e/a_0 and the distance r in units of a_0. The screened potential is negligible beyond $r \sim 2a_0 = 4k_s^{-1}$.

given in Eq. (4.3). Adopting a quasiparticle picture, the single particles in such a physical system will obey the equation

$$\left[-\frac{\hbar^2}{2m_e}\nabla_{\mathbf{r}}^2 + V(\mathbf{r}) + V^{\mathrm{H}}(\mathbf{r}) \right]\psi_i(\mathbf{r}) + \int \Sigma(\mathbf{r}, \mathbf{r}'; \epsilon_i)\psi_i(\mathbf{r}')d\mathbf{r}' = \epsilon_i\psi_i(\mathbf{r}) \qquad (4.119)$$

where $V(\mathbf{r})$ is the external potential due to the presence of ions, $V^{\mathrm{H}}(\mathbf{r})$ is the Hartree potential, Eq. (4.41), and we have introduced a new term, the self-energy operator $\Sigma(\mathbf{r}, \mathbf{r}'; \epsilon)$, to account for all the effects that the other particles have on particle i, beyond Coulomb repulsion (which is taken into account by the Hartree term). Using many-body perturbation theory, Hedin[13] showed that the self-energy can be expressed as:

$$\Sigma(\mathbf{r}, \mathbf{r}'; \epsilon) = iG(\mathbf{r}, \mathbf{r}'; \epsilon)W(\mathbf{r}, \mathbf{r}'; \epsilon) \qquad (4.120)$$

where $G(\mathbf{r}, \mathbf{r}'; \epsilon)$ is the Green's function (see Appendix A) and $W(\mathbf{r}, \mathbf{r}'; \epsilon)$ is the so-called "screened" Coulomb potential, given by:

$$W(\mathbf{r}, \mathbf{r}'; \epsilon) = \int \varepsilon^{-1}(\mathbf{r}, \mathbf{r}''; \epsilon)\frac{e^2}{|\mathbf{r}'' - \mathbf{r}'|}d\mathbf{r}'' \qquad (4.121)$$

with $\varepsilon(\mathbf{r}, \mathbf{r}'; \epsilon)$ the dielectric function. This is an approximation, in which higher-order terms in the Coulomb interaction of many-particle systems are neglected, and is known as the "GW" approximation. The dielectric function can be expressed in terms of the polarizability $P(\mathbf{r}, \mathbf{r}'; \epsilon)$:

$$\varepsilon(\mathbf{r}, \mathbf{r}'; \epsilon) = \delta(\mathbf{r} - \mathbf{r}') - \int \frac{e^2}{|\mathbf{r} - \mathbf{r}''|}P(\mathbf{r}'', \mathbf{r}'; \epsilon)d\mathbf{r}'', \quad P(\mathbf{r}, \mathbf{r}'; \epsilon) = \frac{\delta n(\mathbf{r})}{\delta V(\mathbf{r}'; \epsilon)} \qquad (4.122)$$

the latter describing the effect of external perturbations, that is, a small change in the density $\delta n(\mathbf{r})$ when the system is perturbed by a small potential $\delta V(\mathbf{r}; \epsilon)$.

[13] L. Hedin, *Phys. Rev.* **139**, A796 (1965).

We now have a complete recipe for solving the single-particle equations that describe the quasiparticles: we start with the calculation of the polarizability $P(\mathbf{r}, \mathbf{r}'; \epsilon)$ by using a small perturbing potential $\delta V(\mathbf{r}'; \epsilon)$ to obtain the corresponding small change in the density $\delta n(\mathbf{r})$ (the energy dependence of the perturbing potential can be included as $\exp[-i\epsilon t/\hbar]$). From this we can obtain the dielectric function $\varepsilon(\mathbf{r}, \mathbf{r}'; \epsilon)$ and find its inverse, which then give us the screened Coulomb potential $W(\mathbf{r}, \mathbf{r}'; \epsilon)$. With this, and the Green's function obtained from Eq. (A.95), we find the self-energy from Eq. (4.120), and use it in the quasiparticle equations, Eq. (4.119), to calculate their eigenfunctions $\psi_i(\mathbf{r})$ and eigenvalues ϵ_i.

This process may sound straightforward, but can be mired in a swamp of computational difficulties. First, it is far from clear how to calculate the polarizability from its general expression by considering all possible (or at least all relevant) perturbations $\delta V(\mathbf{r}'; \epsilon)$. Second, we don't know the wavefunctions $\psi_i(\mathbf{r})$ and eigenvalues ϵ_i needed to obtain the Green's function from Eq. (A.95), and in fact these are the very quantities we are trying to find! Finally, the self-energy in the quasiparticle equations must be evaluated at $\epsilon = \epsilon_i$, which are the poles of the Green's function. A number of tricks and approximations need to be employed to obtain a solution to the quasiparticle equations, the discussion of which is beyond the scope of the present treatment (the interested reader is directed to the several reviews of the subject given at the end of this chapter).

To give a flavor of what is involved in these approximations, it is common practice to use the eigenvalues and eigenfunctions of single-particle equations that can be solved more easily, such as those of the Kohn–Sham equations in the LDA, $\phi_i(\mathbf{r})$ and ζ_i, respectively, to construct both the polarizability and the Green's function. Specifically, the polarizability is given by:

$$P_0(\mathbf{r}, \mathbf{r}'; \epsilon) = \sum_{i,j} \phi_i^*(\mathbf{r})\phi_j(\mathbf{r})\phi_j^*(\mathbf{r}')\phi_i(\mathbf{r}') \left[\frac{f_i((1-f_j)}{\zeta_i - \zeta_j + \epsilon + i\eta} + \frac{f_j((1-f_i)}{\zeta_j - \zeta_i - \epsilon + i\eta} \right] \quad (4.123)$$

with η a small positive constant and f_i the Fermi occupation of state i ($f_i = 1$ for $\zeta_i \leq \epsilon_F$, $f_i = 0$ for $\zeta_i > \epsilon_F$ at $T = 0$). This has the familiar form of summing over all possible particle–hole excitations (state i must be occupied and state j empty, or vice versa, to get non-vanishing contributions to the sum), with energy denominators that go to zero when the particle–hole energy difference $(\zeta_i - \zeta_j)$ is equal to $\pm\epsilon$. In physical terms, the polarizability is given by all possible virtual excitations of a particle–hole pair that has an energy difference in resonance with the energy of the perturbing potential. This is very similar to what we found for the response function $\chi(\mathbf{k})$ of the uniform electron gas in the previous section, see Eq. (4.112). With $P_0(\mathbf{r}, \mathbf{r}'; \epsilon)$ we can then obtain the inverse of the dielectric function $\varepsilon^{-1}(\mathbf{r}, \mathbf{r}'; \epsilon)$ and the screened potential $W_0(\mathbf{r}, \mathbf{r}'; \epsilon)$. In the same spirit, the Green's function is expressed as:

$$G_0(\mathbf{r}, \mathbf{r}'; \epsilon) = \sum_i \frac{\phi_i(\mathbf{r})\phi_i^*(\mathbf{r}')}{\epsilon - \zeta_i} \quad (4.124)$$

Finally, the self-energy operator is evaluated from Eq. (4.120) with the approximate expressions for $G_0(\mathbf{r}, \mathbf{r}'; \epsilon)$, $W_0(\mathbf{r}, \mathbf{r}'; \epsilon)$; this quantity is then used in the quasiparticle self-energy equations, Eq. (4.119), which must be solved self-consistently, since the Hartree

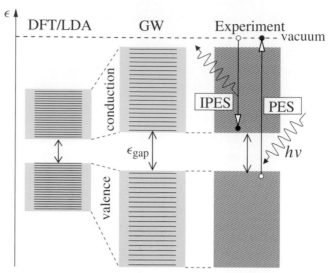

Fig. 4.5 Schematic illustration of typical calculated energy eigenvalues for fictitious particles in the DFT/LDA approach and in the GW approach, and comparison with typical experimenal measurements from photoemission spectroscopy (PES) and inverse photoemission spectroscopy (IPES): in PES, an electron from an occupied state absorbs a photon of energy $h\nu$ and is ejected to the vacuum (or higher-energy state); in IPES, an electron from the vacuum (or higher-energy) state is captured into one of the unoccupied states of the solid, emitting a photon in the process.

potential $V^H(\mathbf{r})$ depends on the density $n(\mathbf{r})$, which is given in terms of the single-particle wavefunctions: $n(\mathbf{r}) = \sum_i |\psi_i(\mathbf{r})|^2$.

The payoff of these approximations and of the overall much more elaborate way of calculating the quasiparticle spectrum of energies is that comparison with experiment is vastly improved over the simpler DFT results.[14] More specifically, both the gap between occupied (valence) and unoccupied (conduction) states of semiconductors, as well as the range of energy of those states, are significantly increased (often by a factor of 2) in GW calculations compared to the values for the same quantities in DFT/LDA calculations, bringing the results to much closer agreement with experimentally measured quantities by photo-emission and inverse photo-emission, as indicated schematically in Fig. 4.5.

4.9 The Pseudopotential

In the solid, we are mostly concerned with the valence electrons of atoms, because the core electrons are largely unaffected by the environment. The separation of the atomic electron density into its core and valence parts is shown in Fig. 4.6 for four elements of column IV of the Periodic Table, namely C, Si, Ge, and Pb. It is clear from these figures that the contribution of the valence states to the total electron density is negligible within

---

[14] M. S. Hybertsen and S. G. Louie, *Phys. Rev. Lett.* **55**, 1418 (1985).

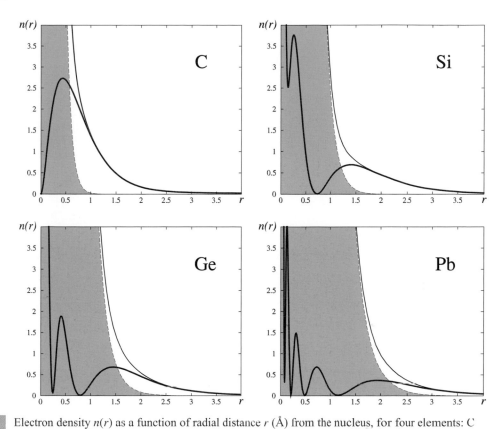

Fig. 4.6 Electron density $n(r)$ as a function of radial distance r (Å) from the nucleus, for four elements: C ($Z = 6$, [He] $2s^2 2p^2$), Si ($Z = 14$, [Ne] $3s^2 3p^2$), Ge ($Z = 32$, [Ar $+ 3d^{10}$] $4s^2 4p^2$), and Pb ($Z = 82$, [Xe $+ 4f^{14} 5d^{10}$] $6s^2 6p^2$); the core states are given inside square brackets. In each case, the dashed black line with the pink-shaded area underneath it represents the density of core electrons, while the solid black line represents the total density of electrons (core plus valence); the blue line represents the density of valence electrons. The core electron density for C is confined approximately below 1.0 Å, for Si below 1.5 Å, for Ge below 2.0 Å, and for Pb below 2.5 Å. In all cases the valence electron density extends well beyond the range of the core electron density and is relatively small, compared to the total, within the core. The wiggles that develop in the valence electron densities for Si, Ge, and Pb are due to the nodes of the corresponding wavefunctions, which acquire oscillations in order to become orthogonal to the core states. These wiggles are eliminated by the construction of the pseudo-wavefunction.

the core region and dominant beyond it. Because of this difference between valence and core electrons, a highly effective approach has been developed to separate the two sets of states. This approach, known as the *pseudopotential method*, allows us to take the core electrons out of the picture, and at the same time to create a smoother potential for the valence electrons; the work of Phillips and Kleinman first established the theoretical basis of the pseudopotential method.[15]

In order to develop the pseudopotential for a specific atom, we consider it as isolated, and denote by $|\psi^{(n)}\rangle$ the single-particle states which are the solutions of the single-particle hamiltonian. In principle, we need to calculate these states for all the electrons of the

[15] J. C. Phillips and L. Kleinman, *Phys. Rev.* **116**, 287 (1959).

atom, using as an external potential that of its nucleus. Let us separate explicitly the single-particle states into valence and core sets, identified as $|\psi^{(v)}\rangle$ and $|\psi^{(c)}\rangle$, respectively. These satisfy the Schrödinger equations

$$\hat{\mathcal{H}}^{\text{sp}}|\psi^{(v)}\rangle = \epsilon^{(v)}|\psi^{(v)}\rangle \tag{4.125}$$

$$\hat{\mathcal{H}}^{\text{sp}}|\psi^{(c)}\rangle = \epsilon^{(c)}|\psi^{(c)}\rangle \tag{4.126}$$

where $\hat{\mathcal{H}}^{\text{sp}}$ is the appropriate single-particle hamiltonian for the atom. It contains a potential $\mathcal{V}^{\text{sp}}$ which includes the external potential due to the nucleus, as well as all the other terms arising from electron–electron interactions. Now let us define a new set of single-particle valence states $|\tilde{\phi}^{(v)}\rangle$ through the following relation:

$$|\psi^{(v)}\rangle = |\tilde{\phi}^{(v)}\rangle - \sum_c \langle\psi^{(c)}|\tilde{\phi}^{(v)}\rangle|\psi^{(c)}\rangle \tag{4.127}$$

which is the same relation as in Eq. (2.102). Applying the single-particle hamiltonian $\hat{\mathcal{H}}^{\text{sp}}$ to this equation, we obtain:

$$\hat{\mathcal{H}}^{\text{sp}}|\tilde{\phi}^{(v)}\rangle - \sum_c \langle\psi^{(c)}|\tilde{\phi}^{(v)}\rangle\hat{\mathcal{H}}^{\text{sp}}|\psi^{(c)}\rangle = \epsilon^{(v)}\left[|\tilde{\phi}^{(v)}\rangle - \sum_c \langle\psi^{(c)}|\tilde{\phi}^{(v)}\rangle|\psi^{(c)}\rangle\right] \tag{4.128}$$

which, taking into account that $\hat{\mathcal{H}}^{\text{sp}}|\psi^{(c)}\rangle = \epsilon^{(c)}|\psi^{(c)}\rangle$, gives:

$$\left[\hat{\mathcal{H}}^{\text{sp}} - \sum_c \epsilon^{(c)}|\psi^{(c)}\rangle\langle\psi^{(c)}|\right]|\tilde{\phi}^{(v)}\rangle = \epsilon^{(v)}\left[1 - \sum_c |\psi^{(c)}\rangle\langle\psi^{(c)}|\right]|\tilde{\phi}^{(v)}\rangle \Rightarrow$$

$$\left[\hat{\mathcal{H}}^{\text{sp}} + \sum_c (\epsilon^{(v)} - \epsilon^{(c)})|\psi^{(c)}\rangle\langle\psi^{(c)}|\right]|\tilde{\phi}^{(v)}\rangle = \epsilon^{(v)}|\tilde{\phi}^{(v)}\rangle \tag{4.129}$$

Therefore, the new states $|\tilde{\phi}^{(v)}\rangle$ obey a single-particle equation with a modified potential, but have the same eigenvalues $\epsilon^{(v)}$ as the original valence states $|\psi^{(v)}\rangle$. The modified potential for these states is called the "pseudopotential," given by:

$$\tilde{\mathcal{V}}^{\text{ion}} = \mathcal{V}^{\text{sp}} + \sum_c (\epsilon^{(v)} - \epsilon^{(c)})|\psi^{(c)}\rangle\langle\psi^{(c)}| \tag{4.130}$$

and correspondingly, the $|\tilde{\phi}^{(v)}\rangle$s are called "pseudo-wavefunctions."

Why is this a useful approach? First, consider the definition of the pseudo-wavefunctions through Eq. (4.127): what this definition amounts to is projecting out of the valence wavefunctions any overlap they have with the core wavefunctions. In fact, the quantity

$$\sum_c |\psi^{(c)}\rangle\langle\psi^{(c)}| \tag{4.131}$$

is a projection operator that achieves exactly this result. So, the new valence states defined through Eq. (4.127) have zero overlap with the core states, but they have the same eigenvalues as the original valence states. Moreover, the potential that these states experience includes, in addition to the regular potential $\mathcal{V}^{\text{sp}}$, the term

$$\sum_c (\epsilon^{(v)} - \epsilon^{(c)})|\psi^{(c)}\rangle\langle\psi^{(c)}| \tag{4.132}$$

which is strictly positive, because $\epsilon^{(v)} > \epsilon^{(c)}$ (valence states have, by definition, higher energy than core states). Thus, this term is repulsive and tends to push the corresponding states $|\tilde{\phi}^{(v)}\rangle$ outside the core. In this sense, the pseudopotential represents the effective potential that valence electrons feel, if the only effect of core electrons were to repel them from the core region. Therefore, the pseudo-wavefunctions experience an attractive Coulomb potential which is shielded near the position of the nucleus by the core electrons, so it should be a much smoother potential without the $1/r$ singularity due to the nucleus with Z protons at the origin. Farther away from the core region, where the core states die exponentially, the potential that the pseudo-wavefunctions experience is the same as the Coulomb potential of an ion, consisting of the nucleus plus the core electrons. In other words, through the pseudopotential formulation we have created a new set of valence states, which experience a weaker potential near the atomic nucleus, but the proper ionic potential away from the core region. Since it is this region in which the valence electrons interact to form bonds that hold the solid together, the pseudo-wavefunctions preserve all the important physics relevant to the behavior of the solid. The fact that they also have exactly the same eigenvalues as the original valence states indicates that they faithfully reproduce the behavior of true valence states.

There are some aspects of the pseudopotential, at least as formulated above, that make it somewhat suspicious. First, it is a non-local potential: applying it to the state $|\tilde{\phi}^{(v)}\rangle$ gives

$$\sum_c (\epsilon^{(v)} - \epsilon^{(c)})|\psi^{(c)}\rangle\langle\psi^{(c)}|\tilde{\phi}^{(v)}\rangle = \int \tilde{V}(\mathbf{r},\mathbf{r}')\tilde{\phi}^{(v)}(\mathbf{r}')d\mathbf{r}'$$

$$\Rightarrow \tilde{V}(\mathbf{r},\mathbf{r}') = \sum_c (\epsilon^{(v)} - \epsilon^{(c)})\psi^{(c)*}(\mathbf{r}')\psi^{(c)}(\mathbf{r})$$

This non-locality certainly complicates things. The pseudopotential also depends on the energy $\epsilon^{(v)}$, as the above relationship demonstrates, which is an unknown quantity if we view Eq. (4.129) as the Scrhödinger equation that determines the pseudo-wavefunctions $|\tilde{\phi}^{(v)}\rangle$ and their eigenvalues. Finally, the pseudopotential is not unique. This can be demonstrated by adding any linear combination of $|\psi^{(c)}\rangle$ states to $|\tilde{\phi}^{(v)}\rangle$ to obtain a new state $|\hat{\phi}^{(v)}\rangle$:

$$|\hat{\phi}^{(v)}\rangle = |\tilde{\phi}^{(v)}\rangle + \sum_{c'} \gamma_{c'}|\psi^{(c')}\rangle \Rightarrow |\tilde{\phi}^{(v)}\rangle = |\hat{\phi}^v\rangle - \sum_{c'} \gamma_{c'}|\psi^{(c')}\rangle \qquad (4.133)$$

where $\gamma_{c'}$ are constants. Using the new expression for $|\tilde{\phi}^{(v)}\rangle$ in Eq. (4.129), we obtain:

$$\left[\hat{\mathcal{H}}^{\mathrm{sp}} + \sum_c (\epsilon^{(v)} - \epsilon^{(c)})|\psi^{(c)}\rangle\langle\psi^{(c)}|\right]\left[|\hat{\phi}^{(v)}\rangle - \sum_{c'} \gamma_{c'}|\psi^{(c')}\rangle\right]$$

$$= \epsilon^{(v)}\left[|\hat{\phi}^{(v)}\rangle - \sum_{c'} \gamma_{c'}|\psi^{(c')}\rangle\right]$$

We can now use $\langle\psi^{(c)}|\psi^{(c')}\rangle = \delta_{cc'}$ to reduce the double sum $\sum_c \sum_{c'}$ on the left-hand side of this equation to a single sum, and eliminate common terms from both sides to arrive at:

$$\left[\hat{\mathcal{H}}^{\mathrm{sp}} + \sum_c (\epsilon^{(v)} - \epsilon^{(c)})|\psi^{(c)}\rangle\langle\psi^{(c)}|\right]|\hat{\phi}^{(v)}\rangle = \epsilon^{(v)}|\hat{\phi}^{(v)}\rangle \qquad (4.134)$$

This shows that the state $|\hat{\phi}^{(v)}\rangle$ obeys exactly the same single-particle equation as the state $|\tilde{\phi}^{(v)}\rangle$, which means it is not uniquely defined, and therefore the pseudopotential is not uniquely defined. Practice, however, has shown that these features can actually be exploited to define pseudopotentials that work very well in reproducing the behavior of the valence wavefunctions in the regions outside the core, which are precisely the regions of interest for the physics of solids.

As an example, we discuss next how typical pseudopotentials are constructed for modern calculations of the properties of solids.[16] We begin with a self-consistent solution of the single-particle equations for all the electrons in an atom (core and valence). For each valence state of interest, we take the calculated radial wavefunction and keep the tail starting at some point slightly before the last extremum. When atoms are placed at usual interatomic distances in a solid, these valence tails overlap significantly, and the resulting interaction between the corresponding electrons produces binding between the atoms. We want therefore to keep this part of the valence wavefunction as realistic as possible, and we identify it with the tail of the calculated atomic wavefunction. We call the radial distance beyond which this tail extends the "cutoff radius" r_c, so that the region $r < r_c$ corresponds to the core. Inside the core, the behavior of the wavefunction is not as important for the properties of the solid. Therefore, we can construct the pseudo-wavefunction to be a smooth function, which has no nodes and goes to zero at the origin, as shown in Fig. 4.7. We can achieve this by taking some combination of smooth functions which we can fit to match the true wavefunction and its first derivative at r_c, and approach smoothly to zero at the origin. This hypothetical wavefunction must be normalized properly. Having defined the pseudo-wavefunction, we can invert the Schrödinger equation to obtain the potential which would produce such a wavefunction. This is by definition the desired pseudopotential: it is guaranteed by construction to produce a wavefunction which matches exactly the real atomic wavefunction beyond the core region ($r > r_c$), and is smooth and

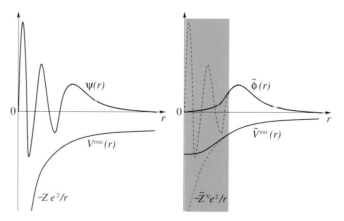

Fig. 4.7 Schematic representation of the construction of the atomic pseudo-wavefunction $\tilde{\phi}(r)$ and the atomic pseudopotential $\tilde{\mathcal{V}}^{\text{ion}}(r)$ (right panel), beginning with the valence wavefunction $\psi(r)$ and the Coulomb potential of the nucleus $\mathcal{V}^{\text{nuc}}(r)$ (left panel); the range of the core, beyond which $\psi(r)$ matches $\tilde{\phi}(r)$ and $\tilde{\mathcal{V}}^{\text{ion}}(r)$ matches the Coulomb potential of the ion, is shown shaded in pink.

[16] D. R. Hamann, M. Schlüter, and C. Chang, *Phys. Rev. Lett.* **43**, 1494 (1979).

nodeless inside the core region, giving rise to a smooth pseudopotential. We can then use this pseudopotential as the appropriate potential for the valence electrons in the solid. This is illustrated schematically below.

Basic steps in constructing a pseudopotential

1. Solve $\hat{\mathcal{H}}^{\text{sp}}\psi^{(v)}(r) = [\hat{\mathcal{F}}^{\text{sp}} + \mathcal{V}^{\text{nuc}}(r)]\psi^{(v)}(r) = \epsilon^{(v)}\psi^{(v)}(r)$

2. Fix $\phi^{(v)}(r) = \psi^{(v)}(r)$ for $r \geq r_c$

3. Construct $\phi^{(v)}(r)$ for $0 \leq r < r_c$, under the following conditions:

   (i) $\phi^{(v)}(r)$: smooth, nodeless; (ii) $\dfrac{d\phi^{(v)}}{dr}, \dfrac{d^2\phi^{(v)}}{dr^2}$: continuous at r_c

4. Normalize $\phi^{(v)}(r)$ for $0 \leq r < \infty$

5. Invert $[\hat{\mathcal{F}}^{\text{sp}} + \tilde{\mathcal{V}}^{\text{ion}}(r)]\phi^{(v)}(r) = \epsilon^{(v)}\phi^{(v)}(r) \Rightarrow \tilde{\mathcal{V}}^{\text{ion}}(r) = \epsilon^{(v)} - \dfrac{[\hat{\mathcal{F}}^{\text{sp}}\phi^{(v)}(r)]}{\phi^{(v)}(r)}$

Recall that $\hat{\mathcal{F}}^{\text{sp}}$ is the operator in the single-particle hamiltonian $\hat{\mathcal{H}}^{\text{sp}}$ that contains all other terms except the external potential: it consists of the kinetic energy operator, the Hartree potential term, and the exchange–correlation term. $\mathcal{V}^{\text{nuc}}(r)$ and $\tilde{\mathcal{V}}^{\text{ion}}(r)$ are the Coulomb potential of the nucleus with Z protons and the pseudopotential of the ion with a valence charge $\tilde{Z}^V$, respectively.

We note here two important points:

(1) The pseudo-wavefunctions can be chosen to be nodeless inside the core, due to the non-uniqueness in the definition of the pseudopotential and the fact that their behavior inside the core is not relevant to the physics of the solid. The true valence wavefunctions have many nodes in order to be orthogonal to core states.

(2) The nodeless and smooth character of the pseudo-wavefunctions guarantees that the pseudopotentials produced by inversion of the Schrödinger equation are finite and smooth near the origin, instead of having a $1/r$ singularity like the Coulomb potential of the nucleus with Z protons.

Of course, each valence state will give rise to a different pseudopotential, but this is not a serious complication as far as actual calculations are concerned. All the pseudopotentials corresponding to an atom will have tails that behave like $-\tilde{Z}^V e^2/r$, where $\tilde{Z}^V$ is the valence charge of the atom, that is, the ionic charge for an ion consisting of the nucleus and the core electrons. The huge advantage of the pseudopotential is that now we have to deal with the valence electrons only in the solid (the core electrons are essentially frozen in their atomic wavefunctions), and the pseudopotentials are smooth so that standard numerical methods can be applied (like Fourier expansions) to solve the single-particle equations. There are several details of the construction of the pseudopotential that require special attention in order to obtain potentials that work and actually simplify the calculations of the properties of solids, but we will not go into these details here. Suffice to say that pseudopotential construction is one of the arts of performing reliable and accurate calculations for solids, but through the careful work of many researchers in this field over the last couple of decades there now exist very good pseudopotentials for essentially all elements of interest in the Periodic Table.[17] The modern practice of pseudopotentials has strived to produce, in a systematic way, potentials that are simultaneously smoother, more accurate, and more transferable, for a wide range of elements.[18]

4.10 Energetics and Ion Dynamics

The discussion up to this point has focused on the behavior of electrons for a given configuration of the ions. Allowing the ions to move brings into the picture a host of new issues that concern the thermodynamic and mechanical behavior of materials. Modeling of these phenomena from the microscopic structure of materials requires an accurate calculation of the total energy of the solid and the forces acting on the ions, for a given atomic configuration. For example:

- phase transitions as a function of pressure can be predicted if the total energy of different phases is known as a function of volume;
- alloy phase diagrams as a function of temperature and composition can be constructed by calculating the total energy of various structures with different types of elements;
- the relative stability of competing surface structures can be determined through their total energies, and from those one can predict the shape of solids;
- the dynamics of ions in the interior and on the surface of a solid can be described by calculating the total energy of relevant configurations, which can elucidate complex phenomena like bulk diffusion and surface growth;
- the energetics of extended deformations, like shear or cleavage of a solid, are crucial in understanding its mechanical response, such as brittle or ductile behavior;
- the properties of defects of dimensionality zero, one, and two can be elucidated through total-energy calculations of model structures, which in turn provides insight into complex phenomena like fracture, catalysis, corrosion, adhesion, etc.

Calculations of this type have proliferated since the early 1980s, providing a wealth of useful information on the behavior of real materials. It is impossible to provide a comprehensive review of such applications here, which are being expanded and refined at a very rapid pace by many practitioners worldwide.

4.10.1 The Total Energy

We will describe the calculation of the total energy and its relation to the band structure in the framework of DFT. The reason is that this formulation has proven the most successful

---

[17] G. Bachelet, D. R. Hamann, and M. Schlüter, *Phys. Rev. B* **26**, 4199 (1982).
[18] A. M. Rappe, K. M. Rabe, E. Kaxiras, and J. D. Joannopoulos, *Phys. Rev. B* **41**, 1227 (1990); D. Vanderbilt, *Phys. Rev. B* **41**, 7892 (1990); N. Troullier and J. L. Martins, *Phys. Rev. B* **43**, 8861 (1991).

compromise between accuracy and efficiency for total-energy calculations in a very wide range of solids. In the following we will also adopt the pseudopotential method for describing the ionic cores, which allows for an efficient treatment of the valence electrons only. Furthermore, we will assume that the ionic pseudopotentials are the same for all electronic states in the atom, an approximation known as the "local pseudopotential"; this will help keep the discussion simple. In realistic calculations the pseudopotential typically depends on the angular momentum of the atomic state it represents, which is known as a "non-local pseudopotential."

We begin with the expression for the total energy of a solid for a particular configuration of the ions, $\{\mathbf{R}\}$, that we derived earlier, Eq. (4.10):

$$E^{\text{tot}} = K^{\text{ion}} + U^{\text{ion}} + E^{\text{ele}} = K^{\text{ion}} + U^{\text{ion}} + \left[K^{\text{ele}} + U^{\text{ion}-\text{ele}} + U^{\text{ele}-\text{ele}} \right] \quad (4.135)$$

where K^{ion} is the familiar kinetic energy of classical point particles with mass M_I and U^{ion} is the energy due to the ion–ion repulsive interaction; this term is referred to as the "Madelung energy," and can be calculated by the Ewald summation method. We will deal with this term first, and then turn our attention to all the terms that involve electronic degrees of freedom.

Ionic Contributions: Madelung Energy

The Madelung energy is given explicitly by the expression

$$U^{\text{ion}} = \frac{1}{2} \sum_{I \neq J} \frac{Z_I Z_J e^2}{|\mathbf{R}_I - \mathbf{R}_J|} \quad (4.136)$$

where Z_I, Z_J are point charges, corresponding to the valence charge of ions labeled I and J, located at positions $\mathbf{R}_I, \mathbf{R}_J$, which are not necessarily lattice vectors. This sum converges slowly and cannot easily be calculated by truncating the summation. The periodic nature of crystals allows us to address this problem in an efficient manner, through a clever manipulation of the sum. This consists essentially of adding and subtracting artificial charges to the real ionic charges, which produces two series that converge faster than the original one.

The method we will discuss here for obtaining accurate values of the Madelung energy is due to Ewald; it relies on exploiting Poisson's equation

$$\nabla_{\mathbf{r}}^2 \Phi(\mathbf{r}) = -4\pi \rho(\mathbf{r})$$

with $\Phi(\mathbf{r})$ the electrostatic potential due to the charge distribution $\rho(\mathbf{r})$, and takes advantage of the reciprocal space vectors $\mathbf{G}$ to express the electrostatic potential of a set of point charges. The Madelung energy can be expressed as:

$$U^{\text{ion}} = \frac{1}{2} \sum_J Z_J e \Phi_J(\mathbf{R}_J) \quad (4.137)$$

where $\Phi_J(\mathbf{r})$ is the electrostatic potential due to the ions:

$$\Phi_J(\mathbf{r}) = \sum_{I \neq J} \frac{Z_I e}{|\mathbf{R}_I - \mathbf{r}|} \tag{4.138}$$

with the contribution of ion J excluded because it will lead to infinity when evaluated at $\mathbf{r} = \mathbf{R}_J$. For crystals bonded through metallic or ionic bonds, the Madelung energy is positive (repulsive) and the point charges represent the ions stripped of their valence electrons, which have been distributed to bonds. In the opposite extreme of a purely ionic crystal, this energy is negative (attractive), and is the reason for the stability of the solid. In either case, the crystal is considered to be overall neutral when the charges of the ions and the valence electrons are taken into account, so that the average charge and the average value of the electrostatic potential are both zero. In the Fourier transform of the potential, the term that corresponds to $\mathbf{G} = 0$ represents the average potential, so this term can be taken to be zero.

The Fourier transform of the charge distribution $\rho(\mathbf{r})$ for a crystal with reciprocal space vectors $\mathbf{G}$ is given by:

$$\rho(\mathbf{r}) = \sum_{\mathbf{G}} \rho(\mathbf{G}) e^{i\mathbf{G} \cdot \mathbf{r}} \rightarrow \rho(\mathbf{G}) = \frac{1}{V} \int \rho(\mathbf{r}) e^{-i\mathbf{G} \cdot \mathbf{r}} \, d\mathbf{r} \tag{4.139}$$

with V the volume of the crystal. When this is substituted in the Poisson equation, it leads to:

$$\Phi(\mathbf{r}) = 4\pi \sum_{\mathbf{G}} \rho(\mathbf{G}) e^{i\mathbf{G} \cdot \mathbf{r}} \frac{1}{|\mathbf{G}|^2} \tag{4.140}$$

For a set of point charges $Z_I e$ located at $\mathbf{R}_I$, these expressions give, for the charge density:

$$\rho(\mathbf{r}) = \sum_I Z_I e \delta(\mathbf{r} - \mathbf{R}_I) \rightarrow \rho(\mathbf{G}) = \frac{1}{V} \sum_I Z_I e e^{-i\mathbf{G} \cdot \mathbf{R}_I} \tag{4.141}$$

while the total potential takes the form

$$\Phi(\mathbf{r}) = \frac{4\pi}{V} \sum_I \sum_{\mathbf{G} \neq 0} Z_I e \frac{e^{i\mathbf{G} \cdot (\mathbf{r} - \mathbf{R}_I)}}{|\mathbf{G}|^2} \tag{4.142}$$

where we have excluded the $\mathbf{G} = 0$ term from the summation, consistent with our earlier analysis of the average charge and potential, which both vanish. If we wanted to calculate the electrostatic energy using this expression for the potential, we would obtain

$$U^{\text{ion}} = \frac{1}{2} \sum_J Z_J e \Phi(\mathbf{R}_J) = \frac{2\pi}{V} \sum_{I,J} \sum_{\mathbf{G} \neq 0} Z_I Z_J e^2 \frac{e^{i\mathbf{G} \cdot (\mathbf{R}_J - \mathbf{R}_I)}}{|\mathbf{G}|^2} \tag{4.143}$$

where now, having expressed the potential by its Fourier transform, we need not explicitly exclude the $I = J$ term since it does not lead to infinities; this, however, introduces an additional term into the energy, which we will have to eliminate later. This expression does not converge any faster than the one in Eq. (4.136), because the summation over $\mathbf{G}$ can be thought of as equivalent to an integral

$$\sum_{\mathbf{G}} \rightarrow \int \int |\mathbf{G}|^2 dG d\hat{\Omega}_{\mathbf{G}}$$

with $G = |\mathbf{G}|$ the magnitude and $\hat{\Omega}_{\mathbf{G}}$ the angular term of the vector $\mathbf{G}$, so that the $1/|\mathbf{G}|^2$ term in Eq. (4.143) cancels out and the remaining summation over I, J converges again slowly (the factor $\exp[i\mathbf{G} \cdot (\mathbf{R}_J - \mathbf{R}_I)]$ has magnitude 1).

Ewald's trick was to replace the ionic point charges with gaussians normalized to unity (see Appendix A):

$$Z_I e \delta(\mathbf{r} - \mathbf{R}_I) \rightarrow Z_I e \left(\frac{\alpha}{\sqrt{\pi}} \right)^3 e^{-\alpha^2 |\mathbf{r} - \mathbf{R}_I|^2} \tag{4.144}$$

where α is a parameter; with $\alpha \rightarrow \infty$ the normalized gaussian approaches a δ-function. With this change, the Fourier transform of the charge density and the corresponding potential become:

$$\rho_\alpha(\mathbf{G}) = \frac{1}{V} \sum_I Z_I e \, e^{-|\mathbf{G}|^2/4\alpha^2} e^{-i\mathbf{G} \cdot \mathbf{R}}$$

$$\Psi_\alpha(\mathbf{r}) = \frac{4\pi}{V} \sum_I \sum_{\mathbf{G} \neq 0} Z_I e \frac{e^{-|\mathbf{G}|^2/4\alpha^2}}{|\mathbf{G}|^2} e^{i\mathbf{G} \cdot (\mathbf{r} - \mathbf{R}_I)} \tag{4.145}$$

which both depend on the parameter α. In the limit $\alpha \rightarrow \infty$, these expressions reduce exactly to the ones derived above, Eqs (4.141) and (4.142). If α is finite, we need to add the proper terms to restore the physical situation of point charges. This is accomplished by adding to the potential, for each ion $Z_I e$ at $\mathbf{R}_I$, a term equal to the contribution of a point charge $Z_I e/|\mathbf{r} - \mathbf{R}_I|$ plus the contribution of a *negative* gaussian, that is, the opposite of the term in Eq. (4.144), to cancel the contribution of that artificial term. The potential generated by a gaussian charge distribution is described by the error function, as shown in Appendix A [cf. Eq. (A.87)]. Consequently, the potential of the point charge $Z_I e$ and the accompanying negative gaussian, both centered at $\mathbf{R}_I$, takes the form:

$$Z_I e \frac{1 - \text{erf}(\alpha |\mathbf{r} - \mathbf{R}_I|)}{|\mathbf{r} - \mathbf{R}_I|} \tag{4.146}$$

This gives for the total potential:

$$\Phi(\mathbf{r}) = \frac{4\pi}{V} \sum_I \sum_{\mathbf{G} \neq 0} Z_I e \frac{e^{-|\mathbf{G}|^2/4\alpha^2} e^{i\mathbf{G} \cdot (\mathbf{r} - \mathbf{R}_I)}}{|\mathbf{G}|^2} + \sum_I Z_I e \frac{1 - \text{erf}(\alpha |\mathbf{r} - \mathbf{R}_I|)}{|\mathbf{r} - \mathbf{R}_I|} \tag{4.147}$$

which is a generally useful expression because it is valid for every point $\mathbf{r}$ in the crystal, except at the position of the ions, because of the infinities introduced by the term $1/|\mathbf{r} - \mathbf{R}_I|$ in the second sum. It is also a convenient form for actual calculations of the Madelung energy. The reason is that the potential in Eq. (4.147) involves two sums, one over $\mathbf{G}$ and I

and one over only the ionic positions $\mathbf{R}_I$, and both summands in these sums converge very quickly, because the terms

$$\frac{e^{-|\mathbf{G}|^2/4\alpha^2}}{|\mathbf{G}|^2}, \quad \frac{1 - \mathrm{erf}(\alpha r)}{r}$$

fall off very fast with increasing $|\mathbf{G}|$ and r, for finite α. The advantage of this approach is that the Madelung energy can now be obtained accurately by including a relatively small number of terms in the summations over $\mathbf{G}$ and I. The number of terms to be included is determined by the shells in reciprocal space (sets of reciprocal space vectors of equal magnitude) and the shells in real space (sets of Bravais lattice vectors of equal magnitude) required to achieve the desired level of convergence. Typical values of α for such calculations are between 1 and 10.

In the final expression for the Madelung energy we must take care to exclude the contribution to the energy of the true Coulomb potential due to ion I evaluated at $\mathbf{R}_I$ to avoid infinities. This term arises from the $1/|\mathbf{r} - \mathbf{R}_I|$ part of the second sum in Eq. (4.147). We must also subtract separately the term which has $\mathbf{r} = \mathbf{R}_I$ in the argument of the error function in the second sum. This is the term needed to cancel the contribution of the $I = J$ term in the first sum of Eq. (4.147), which had allowed us to extend that sum over all values of I, but was not present in the original summation for the electrostatic energy, Eq. (4.136). From the definition of the error function, we have:

$$\mathrm{erf}(r) = \frac{2}{\sqrt{\pi}} \int_0^r e^{-t^2} \mathrm{d}t = \frac{2}{\sqrt{\pi}} \int_0^r (1 - t^2 + \cdots) \mathrm{d}t = \frac{2}{\sqrt{\pi}} \left[r - \frac{1}{3} r^3 + \cdots \right]$$

which is valid in the limit of $r \to 0$; this shows that the term in question becomes:

$$\lim_{\mathbf{r} \to \mathbf{R}_J} \left[\frac{\mathrm{erf}(\alpha |\mathbf{r} - \mathbf{R}_J|)}{|\mathbf{r} - \mathbf{R}_J|} \right] \to \frac{2}{\sqrt{\pi}} \alpha$$

Putting this result together with the previous expressions, we obtain for the Madelung energy:

$$U^{\mathrm{ion}} = \frac{2\pi}{V} \sum_{I,J} \sum_{\mathbf{G} \neq 0} Z_I Z_J e^2 \frac{e^{-|\mathbf{G}|^2/4\alpha^2} e^{i\mathbf{G} \cdot (\mathbf{R}_J - \mathbf{R}_I)}}{|\mathbf{G}|^2} - \sum_J Z_J^2 e^2 \frac{\alpha}{\sqrt{\pi}}$$

$$+ \frac{1}{2} \sum_{I \neq J} Z_I Z_J e^2 \frac{1 - \mathrm{erf}(\alpha |\mathbf{R}_J - \mathbf{R}_I|)}{|\mathbf{R}_J - \mathbf{R}_I|} \tag{4.148}$$

As a final step, we replace the summations over the positions of ions in the entire crystal by summations over Bravais lattice vectors $\mathbf{R}, \mathbf{R}'$ and positions of ions $\mathbf{t}_I, \mathbf{t}_J$ *within* the primitive unit cell:

$$\mathbf{R}_I \to \mathbf{R} + \mathbf{t}_I, \quad \mathbf{R}_J \to \mathbf{R}' + \mathbf{t}_J, \quad \mathbf{t}_I, \mathbf{t}_J \in \mathrm{PUC}$$

Since for all Bravais lattice vectors $\mathbf{R}$ and all reciprocal lattice vectors $\mathbf{G}$: $\exp[\pm i\mathbf{G}\cdot\mathbf{R}] = 1$, the expression for the Madelung energy takes the form:

$$U^{\text{ion}} = \frac{2\pi}{V} \sum_{\mathbf{R},\mathbf{R}'} \sum_{I,J\in\text{PUC}} \sum_{\mathbf{G}\neq 0} Z_I Z_J e^2 \frac{e^{-|\mathbf{G}|^2/4\alpha^2} e^{i\mathbf{G}\cdot(\mathbf{t}_J - \mathbf{t}_I)}}{|\mathbf{G}|^2} - \sum_{J} Z_J^2 e^2 \frac{\alpha}{\sqrt{\pi}}$$

$$+ \frac{1}{2} \sum_{\mathbf{R},\mathbf{R}'} \sum_{I,J\in\text{PUC}}^{\prime} Z_I Z_J e^2 \frac{1 - \text{erf}(\alpha|(\mathbf{R}' + \mathbf{t}_J) - (\mathbf{R} + \mathbf{t}_I)|)}{|(\mathbf{R}' + \mathbf{t}_J) - (\mathbf{R} + \mathbf{t}_I)|}$$

where the symbol $\sum^{\prime}$ implies summation over values of I, J within the PUC, such that $I \neq J$ when $\mathbf{R} = \mathbf{R}'$ (the pair $I = J$ is allowed for $\mathbf{R} \neq \mathbf{R}'$). Since the vector $\mathbf{R} - \mathbf{R}'$ is simply another Bravais lattice, we can express the Madelung energy per primitive unit cell as:

$$\frac{U^{\text{ion}}}{N} = \frac{2\pi}{V_{\text{PUC}}} \sum_{I,J\in\text{PUC}} \sum_{\mathbf{G}\neq 0} Z_I Z_J e^2 \frac{e^{-|\mathbf{G}|^2/4\alpha^2} e^{i\mathbf{G}\cdot(\mathbf{t}_J - \mathbf{t}_I)}}{|\mathbf{G}|^2} - \sum_{J\in\text{PUC}} Z_J^2 e^2 \frac{\alpha}{\sqrt{\pi}}$$

$$+ \frac{1}{2} \sum_{\mathbf{R}} \sum_{I,J\in\text{PUC}}^{\prime} Z_I Z_J e^2 \frac{1 - \text{erf}(\alpha|(\mathbf{t}_J - \mathbf{t}_I) + \mathbf{R}|)}{|(\mathbf{t}_J - \mathbf{t}_I) + \mathbf{R}|}$$

with $\sum^{\prime}$ implying summation over values of $I \neq J$ for $\mathbf{R} = 0$; in the above expressions N is the total number of primitive unit cells in the crystal and $V_{\text{PUC}} = V/N$ the volume of the primitive unit cell.

Electronic Contributions

We turn next to the contributions to the total energy that involve electronic degrees of freedom. We have split the energy arising from electronic motion into three parts: K^{ele} is the kinetic energy of electrons, $U^{\text{ion}-\text{ele}}$ is the energy due to the ion–electron attractive interaction, and $U^{\text{ele}-\text{ele}}$ is the energy due to the electron–electron interaction including the Coulomb repulsion and exchange and correlation effects. Within DFT, these terms take the form

$$K^{\text{ele}} = \sum_{\kappa} \langle \psi_{\kappa} | - \frac{\hbar^2 \nabla_{\mathbf{r}}^2}{2m_e} | \psi_{\kappa} \rangle \tag{4.149}$$

$$U^{\text{ion}-\text{ele}} = \sum_{\kappa} \langle \psi_{\kappa} | \mathcal{V}^{\text{ps}}(\mathbf{r}) | \psi_{\kappa} \rangle \tag{4.150}$$

$$U^{\text{ele}-\text{ele}} = \frac{1}{2} \sum_{\kappa\kappa'} \langle \psi_{\kappa} \psi_{\kappa'} | \frac{e^2}{|\mathbf{r} - \mathbf{r}'|} | \psi_{\kappa} \psi_{\kappa'} \rangle + \mathcal{E}^{\text{XC}}[n(\mathbf{r})] \tag{4.151}$$

where $|\psi_{\kappa}\rangle$ are the single-particle states obtained from a self-consistent solution of the set of single-particle Schrödinger equations and $n(\mathbf{r})$ is the electron number density. For simplicity, we used a single index κ to identify the single-particle wavefunctions, with the understanding that it encompasses both the wave-vector and the band index. In terms of these states, the density is given by:

$$n(\mathbf{r}) = \sum_{\kappa,\epsilon_\kappa < \epsilon_F} |\psi_\kappa(\mathbf{r})|^2$$

with the summation running over all occupied states with energy ϵ_κ below the Fermi level ϵ_F. $\mathcal{V}^{\mathrm{ps}}(\mathbf{r})$ is the external potential that each valence electron in the solid experiences due to the presence of ions described by pseudopotentials, and $\mathcal{E}^{\mathrm{XC}}[n(\mathbf{r})]$ is the exchange and correlation contribution to the total energy, which, in the framework of DFT, depends on the electron density only. We will adopt the LDA for the exchange–correlation functional in terms of the electron density, with $\epsilon^{\mathrm{XC}}[n]$ the local function of the density that accounts for exchange and correlation effects, which gives for the exchange–correlation potential $\mathcal{V}^{\mathrm{XC}}(\mathbf{r})$:

$$\mathcal{E}^{\mathrm{XC}}[n(\mathbf{r})] = \int \epsilon^{\mathrm{XC}}[n(\mathbf{r})]n(\mathbf{r})\mathrm{d}\mathbf{r} \Rightarrow \mathcal{V}^{\mathrm{XC}}(\mathbf{r}) = \frac{\partial \mathcal{E}^{\mathrm{XC}}[n(\mathbf{r})]}{\partial n(\mathbf{r})} = \epsilon^{\mathrm{XC}}[n(\mathbf{r})] + \frac{\partial \epsilon^{\mathrm{XC}}[n(\mathbf{r})]}{\partial n(\mathbf{r})}n(\mathbf{r})$$

and the single-particle equations take the form

$$\left[-\frac{\hbar^2 \nabla_{\mathbf{r}}^2}{2m_e} + \mathcal{V}^{\mathrm{ps}}(\mathbf{r}) + \int \frac{e^2 n(\mathbf{r}')}{|\mathbf{r} - \mathbf{r}'|}\mathrm{d}\mathbf{r}' + \mathcal{V}^{\mathrm{XC}}(\mathbf{r}) \right] |\psi_\kappa\rangle = \epsilon_\kappa |\psi_\kappa\rangle \qquad (4.152)$$

Multiplying this equation by $\langle \psi_\kappa |$ from the left and summing over all occupied states, we obtain:

$$\sum_{\kappa,\epsilon_\kappa < \epsilon_F} \langle \psi_\kappa | \left[-\frac{\hbar^2 \nabla_{\mathbf{r}}^2}{2m_e} + \mathcal{V}^{\mathrm{ps}}(\mathbf{r}) \right] |\psi_\kappa\rangle + \int\int \frac{e^2 n(\mathbf{r})n(\mathbf{r}')}{|\mathbf{r} - \mathbf{r}'|}\mathrm{d}\mathbf{r}\mathrm{d}\mathbf{r}' + \int \mathcal{V}^{\mathrm{XC}}(\mathbf{r})n(\mathbf{r})\mathrm{d}\mathbf{r} = \sum_{\kappa,\epsilon_\kappa < \epsilon_F} \epsilon_\kappa$$

Comparing with our earlier expression for the total energy, Eqs (4.135) and (4.151), we find:

$$E^{\mathrm{tot}} = K^{\mathrm{ion}} + U^{\mathrm{ion}} + \sum_{\kappa,\epsilon_\kappa < \epsilon_F} \epsilon_\kappa - \frac{1}{2}\int \mathcal{V}^{\mathrm{H}}(\mathbf{r})n(\mathbf{r})\mathrm{d}\mathbf{r} - \int \Delta\mathcal{V}^{\mathrm{XC}}(\mathbf{r})n(\mathbf{r})\mathrm{d}\mathbf{r} \qquad (4.153)$$

with $\mathcal{V}^{\mathrm{H}}(\mathbf{r})$ the Hartree potential, defined in Eq. (4.42), representing the electron–electron Coulomb repulsion and the difference in exchange and correlation potential as:

$$\Delta\mathcal{V}^{\mathrm{XC}}(\mathbf{r}) = \frac{\partial \epsilon^{\mathrm{XC}}[n(\mathbf{r})]}{\partial n(\mathbf{r})} = \mathcal{V}^{\mathrm{XC}}(\mathbf{r}) - \epsilon^{\mathrm{XC}}[n(\mathbf{r})] \qquad (4.154)$$

For a periodic solid, the Fourier transforms of the density and the potentials provide a particularly convenient platform for evaluating the various terms of the total energy. We follow here the original formulation of the total energy in terms of the plane waves $\exp[\mathrm{i}\mathbf{G} \cdot \mathbf{r}]$, defined by the reciprocal-space lattice vectors $\mathbf{G}$, as derived by Ihm, Zunger, and Cohen.[19] The Fourier transforms of the density and the potentials are given by:

$$n(\mathbf{r}) = \sum_{\mathbf{G}} e^{\mathrm{i}\mathbf{G}\cdot\mathbf{r}}n(\mathbf{G}), \quad \mathcal{V}^{\mathrm{H}}(\mathbf{r}) = \sum_{\mathbf{G}} e^{\mathrm{i}\mathbf{G}\cdot\mathbf{r}}\mathcal{V}^{\mathrm{H}}(\mathbf{G}), \quad \Delta\mathcal{V}^{\mathrm{XC}}(\mathbf{r}) = \sum_{\mathbf{G}} e^{\mathrm{i}\mathbf{G}\cdot\mathbf{r}}\Delta\mathcal{V}^{\mathrm{XC}}(\mathbf{G})$$

[19] J. Ihm, A. Zunger, and M. L. Cohen, *J. Phys. C: Sol. St. Phys.* **12**, 4409 (1979).

in terms of which we obtain, for the exchange and correlation term in Eq. (4.153):

$$\int \Delta V^{XC}(\mathbf{r})n(\mathbf{r})d\mathbf{r} = V\sum_{\mathbf{G}} \Delta V^{XC}(\mathbf{G})n(\mathbf{G})$$

with V the total volume of the solid. For the Coulomb term, the calculation is a bit more tricky. First, we can use the identity (discussed in Appendix A, Section A.6):

$$\int \frac{e^{i\mathbf{q}\cdot\mathbf{r}}}{r}d\mathbf{r} = \frac{4\pi}{|\mathbf{q}|^2}$$

and the relation between electrostatic potential and electrical charge (Poisson equation):

$$\nabla_{\mathbf{r}}^2 V^H(\mathbf{r}) = -4\pi e^2 n(\mathbf{r}) \Rightarrow V^H(\mathbf{G}) = \frac{4\pi e^2}{|\mathbf{G}|^2}n(\mathbf{G}) \tag{4.155}$$

to rewrite the Coulomb term in Eq. (4.153) as:

$$\int V^H(\mathbf{r})n(\mathbf{r})d\mathbf{r} = V\sum_{\mathbf{G}} V^H(\mathbf{G})n(\mathbf{G}) = 4\pi e^2 V\sum_{\mathbf{G}} \frac{[n(\mathbf{G})]^2}{|\mathbf{G}|^2} \tag{4.156}$$

In a solid which is overall neutral, the total positive charge of the ions is canceled by the total negative charge of the electrons. This means that the average potential is zero. The $\mathbf{G} = 0$ term in the Fourier transform corresponds to the average over all space, which implies that due to charge neutrality we can omit the $\mathbf{G} = 0$ term from all the Coulomb contributions. However, we have introduced an alteration of the physical system by representing the ions with pseudopotentials in the solid: the pseudopotential matches exactly the Coulomb potential beyond the cutoff radius, but deviates from it inside the core. When accounting for the infinite terms, we have to compensate for the alteration introduced by the pseudopotential. This is done by adding to the total energy the following term:

$$\Delta U^{ps} = \sum_I Z_I \int \left[V_I^{ps}(\mathbf{r}) + \frac{Z_I e^2}{r} \right] d\mathbf{r} \tag{4.157}$$

where the summation is over all the ions and $V_I^{ps}(\mathbf{r})$ is the pseudopotential corresponding to ion I; the integrand in this term does not vanish inside the core region, where the pseudopotential is different from the Coulomb potential. With these contributions, the total energy of the solid takes the form

$$E^{tot} = K^{ion} + U^{ion} + \sum_{\kappa(\epsilon_\kappa < \epsilon_F)} \epsilon_\kappa - 2\pi e^2 V\sum_{\mathbf{G}\neq 0} \frac{[n(\mathbf{G})]^2}{|\mathbf{G}|^2} - V\sum_{\mathbf{G}} \Delta V^{XC}(\mathbf{G})n(\mathbf{G}) + \Delta U^{ps} \tag{4.158}$$

This final expression has an appealing form. The first two terms are the classical contributions of the ions as point charges, that is, their kinetic energy and Coulomb repulsion. The next four terms constitute the contribution of the electronic motion to the total energy, expressed as the sum of the energies of all occupied single-particle states (third term), corrected for double counting the Coulomb and exchange–correlation contributions (fourth and fifth terms), and adjusted for the infinite terms due to Coulomb interactions between the various charges (sixth term).

This expression for the total energy has inspired the derivation of semi-empirical schemes, which involve much smaller computational cost than fully self-consistent methods like DFT. A standard approach is to use a tight-binding hamiltonian which, when properly parametrized, can provide an accurate estimate of the first term in the total energy, the sum of single-particle energies. This is actually the dominant term in the total energy. The three other terms are then viewed as corrections, which can be approximated by an empirical term in the form of a classical interatomic potential that depends only on the distance between atoms. This "correction potential" is fitted to reproduce the exact energies of a few possible structures, and can then be used to calculate the energy of a variety of other structures. Although not as accurate as a fully self-consistent approach, this method is quite promising because it makes feasible the application of total-energy calculations to systems involving large numbers of atoms. In particular, this method captures the essential aspects of bonding in solids because it preserves the quantum-mechanical nature of bond formation and destruction through the electronic states with energies ϵ_κ. There is certainly a loss in accuracy due to the very approximate treatment of the terms other than the sum of the electronic eigenvalues; this can be tolerated for certain classes of problems, where the exact changes in the energy are not crucial, but a reasonable estimate of the energy associated with various processes is important.

The cohesive energy of the solid is given by the difference between the total energy per atom at the minimum of the curve and the total energy of a free atom of the same element (for elemental solids). While this is a simple definition, it actually introduces significant errors because of certain subtleties: the calculation of the total energy of the free atom cannot be carried out within the framework developed earlier for solids, because in the case of the free atom it is not possible to define a real-space lattice and the corresponding reciprocal space of $\mathbf{G}$ vectors for expanding the wavefunctions and potentials, as was done for the solid. The calculation of the total energy of the free atom has to be carried out in a real-space approach, without involving reciprocal-space vectors and summations. The difference in computational methods for the solid and the free-atom total energies introduces numerical errors which are not easy to eliminate. Even if both total-energy calculations were carried out with the same computational approach (for example, by placing an atom at the center of a large box and repeating this box periodically in 3D space, thus artificially creating a periodic solid that approximates the isolated atom), the underlying formalism does not give results of equal accuracy for the two cases. This is because the DFT/LDA formalism is well suited for solids, where the variations in the electron density are not too severe. In the case of a free atom, the electron density goes to zero at a relatively short distance from the nucleus (the electronic wavefunctions decay exponentially), which presents a greater challenge to the formalism. Modifications of the formalism that include gradient corrections to the density functionals can improve the situation considerably.

To illustrate the use of total-energy calculations in elucidating the properties of solids, we discuss the behavior of the energy as a function of volume for different phases of the

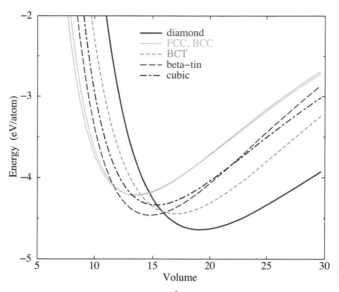

Fig. 4.8 Total energy (eV/atom) as a function of volume (Å^3) for various crystal phases of silicon: diamond (a semiconducting phase with four nearest neighbors), BCC, FCC, BCT, beta-tin, and cubic (all metallic phases with higher coordination). The zero of the energy scale corresponds to isolated silicon atoms.

same element. This type of analysis using DFT calculations was pioneered by Yin and Cohen.[20] The classic case is silicon which, as we have mentioned before, has as its ground state the diamond lattice and is a semiconductor (see Chapter 1).

The question we want to address is: what is the energy of other possible crystalline phases of silicon relative to its ground state? The answer to this question is shown in Fig. 4.8, where we present the energy per atom of the simple cubic lattice with six nearest neighbors for each atom, the FCC and BCC high-coordination lattices with 12 and eight nearest neighbors, respectively, a body centered tetragonal (BCT) structure with five nearest neighbors,[21] the so-called beta-tin structure which has six nearest neighbors, and the ground-state diamond lattice which has four nearest neighbors. We must compare the energy of the various structures as a function of the volume per atom, since we do not know *a priori* at which volume the different crystalline phases will have their minimum value. The minimum in each curve corresponds to a different volume, as the variations of coordination in the different phases suggest. At their minimum, all other phases have higher energy than the diamond lattice. This comparison shows that it would not be possible to form any of these phases under equilibrium conditions. However, if the volume per atom of the solid were somehow reduced, for example by applying pressure, the results of Fig. 4.8 indicate that the lowest-energy structure would become the beta-tin phase. It turns out that when pressure is applied to silicon, the structure indeed becomes beta-tin. A Maxwell (common tangent) construction between the energy-vs.-volume curves of the two lowest-energy phases gives the critical pressure at which the diamond phase begins to transform

[20] M. T. Yin and M. L. Cohen, *Phys. Rev. B* **26**, 5668 (1982).
[21] L. L. Boyer, E. Kaxiras, J. L. Feldman, J. Q. Broughton, and M. J. Mehl, *Phys. Rev. Lett.* **67**, 715 (1991).

to the beta-tin; the value of this critical pressure predicted from the curves in Fig. 4.8 is in excellent agreement with experimental measurements.[22]

4.10.2 Forces and Ion Dynamics

Having calculated the total energy of a solid, it should be feasible to calculate forces on the ions, by simply taking the derivative of the total energy with respect to individual ionic positions. This type of calculation is indeed possible, and opens up a very broad field, that is, the study of the dynamics of ions in the solid state. Here, by dynamics we refer to the behavior of ions as classical particles, obeying Newton's equations of motion, with the calculated forces at each instant in time. We discuss next the details of the calculation of forces in the context of DFT. First, we note that there are two contributions, one from the ion–ion interaction energy, and one from the ion–electron interaction energy. For the contribution from the ion–ion interaction energy, the derivative of U^{ion}, given in Eq. (4.136), with respect to the position of a particular ion, $\mathbf{R}_I$, gives for the force on this ion:

$$\mathbf{F}_I^{\mathrm{ion}} = -\frac{\partial}{\partial \mathbf{R}_I} U^{\mathrm{ion}} = \sum_{J \neq I} \frac{Z_I Z_J e^2}{|\mathbf{R}_I - \mathbf{R}_J|^3} (\mathbf{R}_I - \mathbf{R}_J) \tag{4.159}$$

which can be evaluated by methods analogous to those used for the Madelung energy.

In order to calculate the contribution from the ion–electron interaction, we first prove a very useful general theorem, known as the Hellmann–Feynman theorem. This theorem states that the force on an atom I at position $\mathbf{R}_I$ is given by:

$$\mathbf{F}_I^{\mathrm{ion-ele}} = -\langle \Psi_0 | \frac{\partial \mathcal{H}}{\partial \mathbf{R}_I} | \Psi_0 \rangle \tag{4.160}$$

where $\mathcal{H}$ is the hamiltonian of the system and $|\Psi_0\rangle$ the normalized ground-state wavefunction (we assume that the hamiltonian does not include ion–ion interactions, which have been taken into account explicitly with the $\mathbf{F}^{\mathrm{ion}}$ term). To prove this theorem, we begin with the standard definition of the force on ion I, as the derivative of the total energy $\langle \Psi_0 | \mathcal{H} | \Psi_0 \rangle$ with respect to $\mathbf{R}_I$:

$$\mathbf{F}_I^{\mathrm{ion-ele}} = -\frac{\partial \langle \Psi_0 | \mathcal{H} | \Psi_0 \rangle}{\partial \mathbf{R}_I}$$

$$= -\langle \frac{\partial \Psi_0}{\partial \mathbf{R}_I} | \mathcal{H} | \Psi_0 \rangle - \langle \Psi_0 | \frac{\partial \mathcal{H}}{\partial \mathbf{R}_I} | \Psi_0 \rangle - \langle \Psi_0 | \mathcal{H} | \frac{\partial \Psi_0}{\partial \mathbf{R}_I} \rangle$$

Now, using $\mathcal{H} | \Psi_0 \rangle = E_0 | \Psi_0 \rangle$ and $\langle \Psi_0 | \mathcal{H} = E_0 \langle \Psi_0 |$, where E_0 is the energy corresponding to the eigenstate $|\Psi_0\rangle$, we can rewrite the second and third terms in the expression for the ion–electron contribution to the force as:

$$\langle \frac{\partial \Psi_0}{\partial \mathbf{R}_I} | \Psi_0 \rangle E_0 + E_0 \langle \Psi_0 | \frac{\partial \Psi_0}{\partial \mathbf{R}_I} \rangle = E_0 \frac{\partial \langle \Psi_0 | \Psi_0 \rangle}{\partial \mathbf{R}_I} = 0 \tag{4.161}$$

where the last result is due to the normalization of the wavefunction, $\langle \Psi_0 | \Psi_0 \rangle = 1$. This then proves the Hellmann–Feynman theorem, as stated in Eq. (4.160).

---

[22] M. T. Yin and M. L. Cohen, *Phys. Rev. B* **26**, 3259 (1982).

The importance of the Hellmann–Feynman theorem is the following: the only terms needed in order to calculate the contribution of ion–electron interactions to the force are those terms in the hamiltonian that depend *explicitly* on the atomic positions **R**. From our analysis of the various terms that contribute to the total energy, it is clear that there is only one such term in the hamiltonian, namely $U^{\text{ion–ele}}$, defined in Eq. (4.151). The expectation value of this term in the ground-state wavefunction, expressed in terms of the single-particle wavefunctions $|\psi_\kappa\rangle$, is given by:

$$\sum_{\kappa,I} \langle \psi_\kappa | V_I^{\text{ps}}(\mathbf{r} - \mathbf{R}_I) | \psi_\kappa \rangle \tag{4.162}$$

Taking the derivative of this term with respect to $\mathbf{R}_I$, we obtain the force contribution $\mathbf{F}_I^{\text{ion–ele}}$:

$$\mathbf{F}_I^{\text{ion–ele}} = -\frac{\partial}{\partial \mathbf{R}_I} \sum_{\kappa,I} \langle \psi_\kappa | V_I^{\text{ps}}(\mathbf{r} - \mathbf{R}_I) | \psi_\kappa \rangle \tag{4.163}$$

We will derive the ion–electron contribution to the force explicitly for the case where there is only one atom per unit cell, and the origin of the coordinate system is chosen so that the atomic positions coincide with the lattice vectors. Following the derivation of the expression for the total energy, we use again the Fourier transforms of the potentials and single-particle wavefunctions, with Fourier components $\alpha_\kappa(\mathbf{G})$ and $\mathcal{V}^{\text{ps}}(\mathbf{G})$, respectively. We will also drop the index I from the pseudopotential for simplicity in notation, assuming there is only one type of ion in our example. With these definitions, we obtain for the ion–electron contribution to the force:

$$
\begin{aligned}
\mathbf{F}_I^{\text{ion–ele}} &= -\frac{\partial}{\partial \mathbf{R}_I} V \sum_{\kappa,\mathbf{G},\mathbf{G}'} \alpha_\kappa^*(\mathbf{G})\alpha_\kappa(\mathbf{G}')\mathcal{V}^{\text{ps}}(\mathbf{G}' - \mathbf{G})S(\mathbf{G}' - \mathbf{G}) \\
&= -\mathrm{i}V_{\text{at}} \sum_{\mathbf{G}} \mathbf{G}\,\mathrm{e}^{\mathrm{i}\mathbf{G}\cdot\mathbf{R}_I} \mathcal{V}^{\text{ps}}(\mathbf{G})\gamma(\mathbf{G})
\end{aligned}
\tag{4.164}
$$

where $S(\mathbf{G}) = (1/N)\sum_I \mathrm{e}^{\mathrm{i}\mathbf{G}\cdot\mathbf{R}_I}$ is the structure factor, V_{at} is the atomic volume $V_{\text{at}} = V/N_{\text{at}}$ (N_{at} being the total number of atoms in the solid), and $\gamma(\mathbf{G})$ is defined as:

$$\gamma(\mathbf{G}) = \sum_{\kappa,\mathbf{G}'} \alpha_\kappa^*(\mathbf{G}')\alpha_\kappa(\mathbf{G}' - \mathbf{G}) \tag{4.165}$$

We see from Eq. (4.164) that when the single-particle wavefunctions have been calculated [i.e. the coefficients $\alpha_\kappa(\mathbf{G})$ have been determined, when working with the basis of plane waves $\mathbf{G}$], we have all the necessary information to calculate the forces on the ions.

This last statement can be exploited to devise a scheme for simulating the dynamics of ions in a computationally efficient manner. This scheme was first proposed by Car and Parrinello,[23] and consists of evolving simultaneously the electronic and ionic degrees of freedom. The big advantage of this method is that the single-particle electronic wavefunctions, from which the forces on the ions can be calculated, do not need to be obtained from scratch through a self-consistent calculation when the ions have been moved.

[23] R. Car and M. Parrinello, *Phys. Rev. Lett.* **55**, 2471 (1985).

This is achieved by coupling the electron and ion dynamics through an effective lagrangian, defined as:

$$\mathcal{L} = \sum_{\kappa} \frac{\mu}{2} \langle \frac{\mathrm{d}\psi_{\kappa}}{\mathrm{d}t} | \frac{\mathrm{d}\psi_{\kappa}}{\mathrm{d}t} \rangle + \sum_{I} \frac{M_I}{2} \left(\frac{\mathrm{d}\mathbf{R}_I}{\mathrm{d}t} \right)^2 - E^{\mathrm{tot}}[\{\psi_{\kappa}\}, \{\mathbf{R}_I\}] + \sum_{\kappa\kappa'} \lambda_{\kappa\kappa'} (\langle \psi_{\kappa} | \psi_{\kappa'} \rangle - \delta_{\kappa\kappa'})$$

$$(4.166)$$

Here the first two terms represent the kinetic energy of electrons and ions (M_I and $\mathbf{R}_I$ are the mass and position of ion I), the third term is the total energy of the electron–ion system, which plays the role of the potential energy in the lagrangian, and the last term contains the Laplace multipliers $\lambda_{\kappa\kappa'}$, which insure the orthogonality of the wavefunctions $|\psi_{\kappa}\rangle$.

It is important to note that the kinetic energy of the electron wavefunctions in the Car–Parrinello lagrangian is a fictitious term, introduced for the sole purpose of coupling the dynamics of ionic and electronic degrees of freedom; it has nothing to do with the true quantum-mechanical kinetic energy of electrons, which is of course part of the total energy $E^{\mathrm{tot}}[\{\psi_{\kappa}\}, \{\mathbf{R}_I\}]$. Accordingly, the "mass" μ associated with the kinetic energy of the electron wavefunctions is a fictitious mass, and is essentially a free parameter that determines the coupling between electron and ion dynamics. When the velocities associated with both the electronic and ionic degrees of freedom are reduced to zero, the system is in its equilibrium state, with $E[\{\psi_{\kappa}\}, \{\mathbf{R}_I\}]$ a minimum at zero temperature. This method can be applied to the study of solids in two ways:

(a) It can be applied under conditions where the velocities of electrons and ions are steadily reduced so that the true ground state of the system is reached, in the sense that both ionic and electronic degrees of freedom have been optimized; in this case the temperature is reduced to zero.

(b) It can be applied at finite temperature to study the time evolution of the system, which is allowed to explore the phase space consistent with the externally imposed temperature and pressure conditions.

The value of the electron effective mass μ, together with the value of the time step $\mathrm{d}t$ used to update the electronic wavefunctions and the ionic positions, are adjusted to make the evolution of the system stable as well as computationally efficient. Specifically, the time step $\mathrm{d}t$ should be as large as possible in order to either reach the true ground state of the system in the fewest possible steps, or simulate the evolution of the system for the largest possible time interval at finite temperature with the total number of steps in the simulation determined by the available computational resources. However, a large time step might throw the system out of balance, in the sense that the time-evolved wavefunctions may not correspond to the ground state of the ionic system. For small enough values of $\mathrm{d}t$, the balance between the time-evolved wavefunctions and positions of the ions is maintained – this is referred to as the system remaining on the Born–Oppenheimer surface. In this way, a fully self-consistent solution of the electronic problem for a given ionic configuration, which is the most computationally demanding part of the calculation, is avoided except at the initial step. The ensuing dynamics of the ions are always governed by accurate quantum-mechanical forces, obtained by the general method outlined above. This produces very realistic simulations of the dynamic evolution of complicated physical systems.

The total-energy and force calculations which are able to determine the lowest-energy configuration of a system, taken together with the electronic structure information inherent in these calculations, provide a powerful method for a deeper understanding of the properties of solids. We have already mentioned above some examples where such calculations can prove very useful. The computational approach we have discussed, based on DFT/LDA and its extensions using gradient corrections to the exchange–correlation density functional, in conjunction with atomic pseudopotentials to eliminate core electrons, has certain important advantages. Probably its biggest advantage is that it does not rely on any empirical parameters: the only input to the calculations is the atomic number of constituent atoms and the number of valence electrons, that is, the atomic number of each element. These calculations are not perfect: for example, calculated lattice constants differ from experimental ones by a few percent (typically 1–2%), while calculated phonon frequencies, bulk moduli, and elastic constants may differ from experimental values by 5–10%. Moreover, in the simplest version of the computations, the calculated band gap of semiconductors and insulators can be off from the experimental value by as much as 50%, but this can be corrected, either within the same theoretical framework or by more elaborate calculations, as already mentioned earlier. Nevertheless, the ability of these calculations to address a wide range of properties for systems composed of almost any element in the Periodic Table, without involving any empirical parameters, makes them a truly powerful tool for the study of the properties of solids, as illustrated by several examples in later chapters.

Further Reading

1. *Density Functional Theory of Atoms and Molecules*, R. G. Parr and W. Yang (Oxford University Press, Oxford, 1989).

2. *Electron Correlations in Molecules and Solids*, P. Fulde (Springer-Verlag, Berlin, 1991).

3. *Density Functional Theory*, E. K. U. Gross and R. M. Dreizler (Eds) (Plenum Press, New York, 1995).

4. *Electronic Structure – Basic Theory and Practical Methods*, R. M. Martin (Cambridge University Press, Cambridge, 2004).

5. *Planewaves, Pseudopotentials and the LAPW Method*, D. Singh (Kluwer Academic, Boston, MA, 1994).

6. *Ab Initio Molecular Dynamics*, D. Marx and J. Hutter (Cambridge University Press, Cambridge, 2009).

7. *Fundamentals of Condensed Matter Physics*, M. L. Cohen and S. G. Louie (Cambridge University Press, Cambridge, 2016).

8. *Interacting Electrons – Theory and Computational Approaches*, R. M. Martin, L. Reining, and D. M. Ceperley (Cambridge University Press, Cambridge, 2016).

Problems

1. We will investigate the model of the hydrogen molecule discussed in the text.

    (a) Consider first the single-particle hamiltonian given in Eq. (4.14); show that its expectation values in terms of the single-particle wavefunctions ϕ_i ($i = 0, 1$) defined in Eq. (4.15) are those given in Eq. (4.18).

    (b) Consider next the two-particle hamiltonian $\hat{\mathcal{H}}(\mathbf{r}_1, \mathbf{r}_2)$, given in Eq. (4.21), which contains the interaction term; show that its expectation values in terms of the Hartree wavefunctions Ψ_i^{H} ($i = 0, 1$) defined in Eqs (4.19) and (4.20) are those given in Eqs (4.22) and (4.23), respectively. To derive these results, certain matrix elements of the interaction term need to be neglected; under what assumptions is this a reasonable approximation?

    (c) Using the Hartree–Fock-type wavefunctions defined in Eqs (4.25)–(4.27), construct the matrix elements of the hamiltonian $\hat{\mathcal{H}}_{ij}^{(s)} = \langle \Phi_i^{(s)} | \hat{\mathcal{H}} | \Phi_j^{(s)} \rangle$ and diagonalize the resulting 3×3 matrix to find the eigenvalues and eigenstates; verify that the ground-state energy and wavefunction are those given in Eqs (4.29) and (4.32), respectively. Here we will assume that the same approximations as those involved in part (b) are applicable.

    (d) Find the probability that the two electrons in the ground state, defined by Eq. (4.32), are on the same proton. Give a plot of this result as a function of $\lambda = U/4t$ and explain the physical meaning of the answer for the behavior at the small and large limits of this parameter.

    (e) Find the order of the expectation values of the energy for the triplet and singlet states in the two limits $\lambda \ll 1$ and $\lambda \gg 1$ and interpret the results in terms of the behavior of the two-electron system. In which of the two cases does one recover the results of the Hartree approximation, Eqs (4.22) and (4.23)?

2. We revisit the N-V^--center in diamond, discussed in Section 3.7, in order to provide a more complete description of states involved in the optical transitions. It is more convenient to work in the basis of the two holes, that is, the absence of two electrons from a completely filled system, rather than that of the four electrons occupying the states $a_1(2)$ and e_x, e_y. The reason is that, in a system with all available defect states completely filled by electrons, no transitions into any of those states would be allowed and the system would not absorb any photons; thus, a description of the system in terms of the electrons missing from a completely filled state (the holes) is equivalent to a description in terms of the actual number of electrons that occupy the states of interest.

    (a) In the basis of two holes, construct the two-body wavefunctions of the form

    $$\Psi_{S,S_z}^{(\Gamma)}(\mathbf{r}_1, \mathbf{r}_2; s_1, s_2) = \Psi^{(\Gamma)}(\mathbf{r}_1, \mathbf{r}_2) \times \Psi_{S_z}^{(S)}(s_1, s_2)$$

    where $\mathbf{r}_1, \mathbf{r}_2$ are the hole spatial coordinates and s_1, s_2 their spins, and Γ is the symmetry of the total spatial wavefunction and S, S_z is the total spin and its z

component. To this end, start with the spin part of the wavefunction $\Psi_{S_z}^{(S)}(s_1, s_2)$ and write the singlet ($S = 0$) or triplet ($S = 1$) state in terms of $s_1, s_2 = \uparrow$, $\downarrow$, then determine all possible combinations of the real-space part $\Psi^{(\Gamma)}(\mathbf{r}_1, \mathbf{r}_2)$ that produce totally antisymmetric two-body wavefunctions. In the absence of any spin terms in the hamiltonian, the singlet and triplet states are degenerate in energy. When spin interactions are included, there is a split in the energy of the singlet and triplet states, with the more symmetric triplet state being lower in energy (this is the analog of Hund's rule in atoms, see Chapter 1).

(b) Once these wavefunctions have been constructed, we can then calculate the matrix elements of the dipole operator between initial and final many-body states. At this stage, we need to take into account selection rules for allowed or forbidden transitions between the single-particle states, which will determine the dominant processes for excitation and the related photon frequencies and polarizations that can be observed (see Problem 6 of Chapter 3). Using these results, find the dominant optical transitions.

3. Starting with the Hartree–Fock many-body wavefunction defined in Eq. (4.44), derive the expression for the total energy given by Eq. (4.45). Then, use a variational calculation to obtain the Hartree–Fock single-particle equations, Eq. (4.46), by minimizing the total energy.

4. Show that the quantities ϵ_i appearing in the Hartree–Fock equations, which were introduced as the Lagrange multipliers to preserve the normalization of state $\psi_i(\mathbf{r}_i)$, have the physical meaning of the energy required to remove this state from the system. To do this, find the energy difference between two systems, one with and one without the state $\psi_i(\mathbf{r}_i)$, which have different numbers of electrons, N and $N - 1$, respectively; you may assume that N is very large, so that removing the electron in state $\psi_i(\mathbf{r}_i)$ does not affect the other states $\psi_j(\mathbf{r}_j)$.

5. The bulk modulus B of a solid is defined as:

$$B = -V \frac{\partial P}{\partial V} = V \frac{\partial^2 E}{\partial V^2} \tag{4.167}$$

where V is the volume, P the pressure, and E the total energy; this quantity describes how the solids respond to external pressure by changes in volume. Show that for the uniform electron gas with kinetic energy and exchange energy terms only, Eqs (1.8) and (4.64), respectively, the bulk modulus is given by:

$$B = \left[\frac{5}{6\pi} \frac{2.21}{(r_s/a_0)^5} - \frac{2}{6\pi} \frac{0.916}{(r_s/a_0)^4} \right] (\mathrm{Ry}/a_0^3) \tag{4.168}$$

or equivalently, in terms of the kinetic and exchange energies:

$$B = \frac{1}{6\pi(r_s/a_0)^3} \left[5 \frac{E^{\mathrm{kin}}}{N} + 2 \frac{E^{\mathrm{X}}}{N} \right] (1/a_0^3) \tag{4.169}$$

Discuss the physical implications of this result for a hypothetical solid that might reasonably be described in terms of the uniform electron gas, and in which the value of (r_s/a_0) is relatively small $[(r_s/a_0) < 1]$.

6.   (a) For a system of $N = 3$ particles, prove the relationship between the one-particle and two-particle density matrices, as given in Eq. (4.80), for the case where the four variables in the two-particle density matrix are related by $\mathbf{r}_1 = \mathbf{r}_1' = \mathbf{r}, \mathbf{r}_2 = \mathbf{r}_2' = \mathbf{r}'$. Write this relationship as a determinant in terms of the density matrices $\gamma(\mathbf{r}, \mathbf{r})$ and $\gamma(\mathbf{r}, \mathbf{r}')$, and generalize it to the case where the four variables in the two-particle density matrix are independent.
     (b) Write the expression for the n-particle density matrix in terms of the many-body wavefunction $\Psi(\mathbf{r}_1, \mathbf{r}_2, ..., \mathbf{r}_N)$, and express it as a determinant in terms of the density matrix $\gamma(\mathbf{r}, \mathbf{r}')$; this relation shows the isomorphism between the many-body wavefunction and the density matrix representations.[24]

7.   We want to determine the physical meaning of the quantities ϵ_i in the density functional theory single-particle equations, Eq. (4.87). To do this, we express the density as:

$$n(\mathbf{r}) = \sum_i f_i |\phi_i(\mathbf{r})|^2$$

where the f_i are real numbers between 0 and 1, called the "filling factors." We take a partial derivative of the total energy with respect to f_i and relate it to ζ_i. Then we integrate this relation with respect to f_i. What is the physical meaning of the resulting equation?

8.   In the extremely low density limit, a system of electrons will form a regular lattice, with each electron occupying a unit cell; this is known as the Wigner crystal. The energy of this crystal has been calculated to be:

$$E^{W} = \left[-\frac{3}{(r_s/a_0)} + \frac{3}{(r_s/a_0)^{3/2}} \right] \text{Ry}$$

This can be compared to the energy of the electron gas in the Hartree–Fock approximation, Eq. (4.62), to which we must add the electrostatic energy (this term is canceled by the uniform positive background of the ions, but here we are considering the electron gas by itself). The electrostatic energy turns out to be:

$$E^{es} = -\frac{6}{5} \frac{1}{(r_s/a_0)} \text{Ry} \qquad (4.170)$$

Taking the difference between the two energies, E^{W} and $E^{HF} + E^{es}$, we obtain the correlation energy, which is by definition the interaction energy after we have taken into account all the other contributions, kinetic, electrostatic, and exchange. Show that the result is compatible with the Wigner correlation energy given in Table 4.1, in the low-density [high-(r_s/a_0)] limit.

---

[24] See, for example, P. O. Lowdin, *Phys. Rev.* **97**, 1490 (1955).

9. We want to derive the Lindhard dielectric response function for the free-electron gas, using perturbation theory. The charge density is defined in terms of the single-particle wavefunctions as:

$$n(\mathbf{r}) = (-e)2 \sum_{\mathbf{k}} f(\mathbf{k})|\psi_{\mathbf{k}}(\mathbf{r})|^2$$

with $f(\mathbf{k})$ the Fermi occupation numbers. From first-order perturbation theory (see Appendix C), the change in wavefunction of state $\mathbf{k}$ due to a perturbation represented by the potential $\mathcal{V}^{\text{int}}(\mathbf{r})$ is given by:

$$|\delta\psi_{\mathbf{k}}\rangle = \sum_{\mathbf{k}'} \frac{\langle \psi_{\mathbf{k}'}^{(0)}|\mathcal{V}^{\text{int}}|\psi_{\mathbf{k}}^{(0)}\rangle}{\epsilon_{\mathbf{k}}^{(0)} - \epsilon_{\mathbf{k}'}^{(0)}}|\psi_{\mathbf{k}'}^{(0)}\rangle$$

with $|\psi_{\mathbf{k}}^{(0)}\rangle$ the unperturbed wavefunctions and $\epsilon_{\mathbf{k}}^{(0)}$ the corresponding energies. These changes in the wavefunctions give rise to the induced charge density $\rho^{\text{ind}}(\mathbf{r})$ to first order in $\mathcal{V}^{\text{int}}$.

(a) Derive the expression for the Lindhard dielectric response function, given in Eq. (4.112), for free electrons with energy $\epsilon_{\mathbf{k}}^{(0)} = \hbar^2|\mathbf{k}|^2/2m_e$ and with chemical potential μ, by keeping only first-order terms in $\mathcal{V}^{\text{int}}$ in the perturbation expansion.

(b) Evaluate the zero-temperature Lindhard response function, Eq. (4.113), at $k = 2k_F$, and the corresponding dielectric constant $\varepsilon = 1 - 4\pi\chi/k^2$; interpret their behavior at $k = 2k_F$ in terms of the single-particle picture.

10. Show that at zero temperature the Thomas–Fermi inverse screening length k_s, defined in Eq. (4.115), with total occupation f_T given by Eq. (4.116), takes the form

$$k_s(T = 0) = \frac{2}{\sqrt{\pi}}\frac{1}{a_0}\sqrt{k_F a_0}$$

with k_F the Fermi momentum and a_0 the Bohr radius.

11. Consider a fictitious atom which has a harmonic potential for the radial equation

$$\left[-\frac{\hbar^2}{2m_e}\frac{d^2}{dr^2} + \frac{1}{2}m_e\omega^2 r^2\right]\phi_i(r) = \epsilon_i\phi_i(r) \qquad (4.171)$$

and has nine electrons. The harmonic oscillator potential is discussed in detail in Appendix C. The first four states, $\phi_0(r), \phi_1(r), \phi_2(r), \phi_3(r)$, are fully occupied core states and the last state, $\phi_4(r)$, is a valence state with one electron. We want to construct a pseudopotential which gives a state $\psi_4(r)$ that is smooth and nodeless in the core region. Choose as the cutoff radius r_c the position of the last extremum of $\phi_4(r)$, and use the simple expression

$$\psi_4(r) = Az^2 e^{-Bz^2} \qquad r \le r_c$$
$$= \phi_4(r) \qquad r > r_c$$

for the pseudo-wavefunction, where $z = r\sqrt{m_e \omega/\hbar}$. Determine the parameters A, B so that the pseudo-wavefunction $\psi_4(r)$ and its derivative are continuous at r_c. Then invert the radial Schrödinger equation to obtain the pseudopotential which has $\psi_4(r)$ as its solution. Plot the pseudopotential you obtained as a function of r. Does this procedure produce a physically acceptable pseudopotential?

12. Show that the potential calculated with the Ewald method for a set of ions $Z_I e$ at positions $\mathbf{R}_I$, Eq. (4.147), is indeed independent of the parameter α, when the summations over reciprocal lattice vectors and ion shells are truncated. [*Hint:* calculate $d\Phi/d\alpha$ for the expression in Eq. (4.147) and show that it vanishes.]

13. Show that an equivalent expression to Eq. (4.148) for the calculation of the Madelung energy is:

$$U^{\text{ion}} = \sum_{I \neq J} \sum_{\mathbf{G} \neq 0} \frac{2\pi Z_I Z_J e^2}{V} \frac{e^{i\mathbf{G} \cdot (\mathbf{R}_J - \mathbf{R}_I) - |\mathbf{G}|^2/4\alpha^2}}{|\mathbf{G}|^2} + \sum_{I \neq J} \frac{Z_I Z_J e^2}{2} \frac{1 - \text{erf}(\alpha |\mathbf{R}_J - \mathbf{R}_I|)}{|\mathbf{R}_J - \mathbf{R}_I|}$$

The reason why Eq. (4.148) is often preferred is that it is based on the total potential $\Phi(\mathbf{r})$ obtained from Eq. (4.147), which is valid everywhere in the crystal except at the positions of the ions.

14. Use the Ewald summation method to calculate the Madelung energy per unit cell of the NaCl crystal. How many shells do you need to include for convergence of the Madelung energy to 1 part in 10^3? Verify, by using a few different values of α, that the final answer does not depend on this parameter (but the number of terms required to reach the desired convergence may depend on it).

15. Derive the equations of motion for the electronic and ionic degrees of freedom from the Car–Parrinello lagrangian, Eq. (4.166).

5

Electronic Properties of Crystals

Our goal in this chapter is to examine the electronic properties of several representative solids. To obtain the electronic band structure of these solids, we employ the density functional theory approach discussed in Chapter 4, which we solve numerically using the methods discussed in Chapters 2, 3, and 4. We will also rely heavily on ideas from the tight-binding approximation, which will help us understand the physics in a more transparent way.

Typical solids exist as 3D crystals; the last decade has witnessed a surge of interest in solids that exist in 2D form, something that was unexpected, until the discovery of graphene. We will look at some representative 2D solids and we will devote a substantial part of this chapter to the discussion of 3D solids. To set the stage for these discussions, we examine first two idealized systems, a small molecule, benzene, which we can think of as a *finite* "1D solid" and a macromolecule, polyacetylene, which we can think of as an *infinite* "1D solid." The reason for studying these two systems is that they bring to the fore many of the concepts we laid out in previous chapters, without all the complexity that 2D or 3D crystals possess. Specifically, these idealized structures have periodicity that is easy to see, and its consequences on the electronic structure are manifest in an intuitive manner. Moreover, for these idealized solids, we can make comparisons between a simple treatment based on the tight-binding approximation, and full-blown, realistic electronic structure calculations, which will show that the results based on basic models and on more accurate calculations exhibit a very close correspondence.

5.1 Band Structure of Idealized 1D Solids

5.1.1 A Finite "1D Solid": Benzene

Benzene is a small molecule which encompasses some features of a periodic solid. It consists of six carbon and six hydrogen atoms, with chemical formula C_6H_6, with the C atoms forming a perfect hexagon, as shown in Fig. 5.1. As discussed in Chapter 1, we can form sp^2 hybrids out of the s, p_x, p_z orbitals of the C atoms, which combine with neighboring sp^2 hybrids, as well as with the s states of the H atoms, to form three covalent bonds of σ character for each C atom on the plane of the molecule. These are clearly visible in the charge density plots of Fig. 5.1, with the covalent bonds identified by the high density of electrons between pairs of atoms. The bonds correspond to the symmetric (bonding)

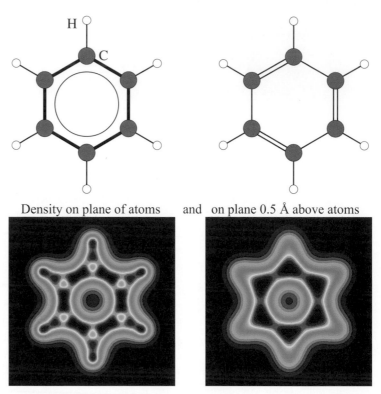

Density on plane of atoms   and   on plane 0.5 Å above atoms

Fig. 5.1   **Top row**: The structure of benzene, C_6H_6, with the C and H atoms shown as gray and white circles. Left: All bonds between C atoms are shown as equivalent due to resonance, as suggested by the circle in the middle of the hexagonal ring. Right: The common representation of benzene in terms of alternating single and double bonds between pairs of C atoms, distributed in one of the two equivalent ways. **Bottom row**: The electronic density on a plane through the C and H atoms (left), and on a plane situated 0.5 Å above the atoms (right); the highest density corresponds to red and the zero value to blue (from DFT calculations performed by D. T. Larson).

combinations of sp^2 hybrids in nearest-neighbor C–C and C–H atoms, these are the lowest-energy states. The antisymmetric (anti-bonding) combinations of the same hybrid orbitals have much higher energy. The remaining p_z orbitals on the C atoms form the π (bonding) and π^* (anti-bonding) states. For each C atom, there are two σ bonds to its two neighboring C atoms and one σ bond to its nearest H atom, for a total of six C–C σ bonds and six C–H σ bonds. By counting the number of available valence electrons (four per C atom, one per H atom), we find that the σ bonding states are completely full, accommodating 24 of the available 30 valence electrons; the remaining six electrons occupy the π bonding states, while the higher-energy σ^* and π^* anti-bonding states are empty.

We focus next on the π and π^* states that are close to the Fermi level and dominate the physics. The π bonds can be thought of as additional bonds between specific pairs of C atoms. Three such bonds would be needed to accommodate the six electrons that are not contributing to the σ bonds; the resulting three "double" bonds, due to the additional bonding of the π bonds, can be distributed in two equivalent ways, as illustrated in Fig. 5.1. The actual structure of benzene, as confirmed by experiment, is a *resonance* between these

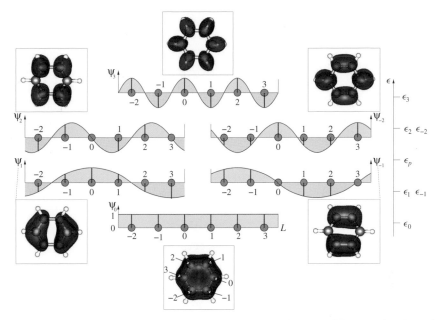

Fig. 5.2 Interpretation of orbitals in benzene, C_6H_6, as "Bloch" states. The wave diagrams give the real part of the phases for each orbital in the wavefunctions $\psi_m(x)$, Eq. (5.2); an overall phase shift is required for each state to make this correspondence. The figures inside boxes show the actual wavefunctions for each state, obtained from first-principles electronic structure calculations, based on DFT (performed by J. Ren and S. Meng); positive values are shown in blue and negative in red. The scale on the right indicates the relative energy of each state, Eq. (5.3).

two possibilities, with all the bonds between nearest-neighbor C atoms, including the σ and π contributions, being exactly equivalent. The concept of resonance is very important in chemistry.[1]

For a deeper understanding of these ideas, we will treat benzene as a linear chain of $N = 6$ atoms with periodic boundary conditions, in the general framework of the Bloch theorem and corresponding wave-like states that extend throughout the system. We will construct a picture of π bonds in benzene based on the ideas we have developed for infinite solids. In the case of the six-atom ring of benzene, this picture takes the following form:

$$N = 6, \quad n = 0, \pm 1, \pm 2, 3, \Rightarrow k = \frac{2\pi}{b}\frac{n}{N} = 0, \pm\frac{\pi}{3b}, \pm\frac{2\pi}{3b}, \frac{\pi}{b} \qquad (5.1)$$

with b the C–C bond length and the atomic positions along the chain given by $x_m = mb$, with $m = -2, \ldots, 3$ as shown in Fig. 5.2. There are six wave-like states associated with this model, given by:

$$\psi_n(x) = \sum_{m=-2}^{3} e^{i(2\pi n/6)m} \phi_{p_z}(x - mb), \quad n = 0, \pm 1, \pm 2, 3 \qquad (5.2)$$

---

[1] For a classic treatment of the subject, see L. C. Pauling, *The Nature of the Chemical Bond and the Structure of Molecules and Crystals* (Cornell University Press, Ithaca, NY, 1960).

The weights associated with each orbital in the wavefunctions can be visualized as corresponding to the value of a wave that fits in the box of length $L = 6b$: the wave-vector k can take the values $k = 0$ or $k = 2\pi/\lambda$ with wavelength $\lambda = L, L/2, L/3$, which are compatible with the periodicity of the structure. There is only one way to fit the $k = 0$ and the $\lambda = L/3$ waves in the box, but two different ways to fit the waves with $\lambda = L$ and $\lambda = L/2$, relative to the positions of the atoms. Specifically, for $\lambda = L$ and $\lambda = L/2$, one choice corresponds to the atom labeled "0" having maximum weight (equal to unity), and the other to zero weight; the weights of the remaining atoms are then determined by the value of the wave at their positions. By contrast, for the waves corresponding to $k = 0$ or $\lambda = L/3$, only one choice of the weight of the atom labeled "0" is valid, namely unity, because the choice of zero leads to wavefunctions that vanish identically. This assignment of weights, which provides an intuitive explanation of the allowed wave-like states, requires the introduction of an overall phase for each wavefunction relative to its definition in Eq. (5.2), which does not change the physics.

We now shift language to a TBA picture of the system. In this picture, using nearest-neighbor hopping matrix element t_0 between the p_z orbitals and an on-site energy ϵ_p, we obtain for the eigenvalues:

$$\epsilon_n = \epsilon_p + 2t_0 \cos(2\pi n/6), \quad n = 0, \pm 1, \pm 2, 3 \tag{5.3}$$

which leads to the set of values:

$$\epsilon_0 = \epsilon_p + 2t_0, \quad \epsilon_{\pm 1} = \epsilon_p + t_0, \quad \epsilon_{\pm 2} = \epsilon_p - t_0, \quad \epsilon_3 = \epsilon_p - 2t_0$$

We can make a formal correspondence with band theory by considering a "band" of energies, with the wave-vector k in the range $-\pi/b \le k \le \pi/b$, as in a 1D system with period b, but allowed to take only the discrete values given by Eq. (5.1), and the energy ϵ_k allowed to take only the discrete values given by Eq. (5.3); this is shown in Fig. 5.3. The states at $k = \pi/b$ and $k = -\pi/b$ are equivalent, being at the edge of the "Brillouin zone," and thus related by a "reciprocal lattice vector" or magnitude $2\pi/b$. In this picture, the Fermi level is at $\epsilon_F = \epsilon_p$, as we discuss below. The correspondence between the Bloch

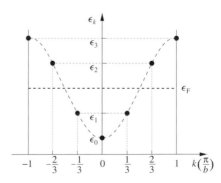

Fig. 5.3 Energy levels of benzene represented as a band structure, with the wave-vector k taking discrete values, given by Eq. (5.1), and the energy ϵ_k taking the corresponding discrete values, given by Eq. (5.3). The Fermi level is at $\epsilon_F = \epsilon_p$.

waves and the TBA expression nicely explains the degeneracy of the states in the TBA model.

From the TBA picture, we can also make better sense of the bonding arrangement in benzene: of the six possible π states, only three are occupied by the available valence electrons, namely $\psi_0(x)$ and $\psi_{\pm 1}(x)$, while the other three are unoccupied. The three occupied ones have most of the weight *between* pairs of C–C atoms, thus contributing to bonding. In fact, if we construct the total charge density from the three occupied states:

$$\rho(x) = |\psi_0(x)|^2 + |\psi_{-1}(x)|^2 + |\psi_1(x)|^2$$

we find that the amount of charge between any pair of carbon atoms is the same, leading to exactly six equal C–C bonds. Moreover, the unoccupied states, $\psi_{\pm 2}(x)$ and $\psi_3(x)$, have nodes between pairs of carbon atoms, and are therefore anti-bonding in character. There is a gap between the highest occupied molecular orbital (HOMO) and the lowest unoccupied molecular orbital (LUMO), which according to our simple model is equal to:

$$\epsilon_{\text{gap}} = \epsilon_{\pm 2} - \epsilon_{\pm 1} = -2t_0$$

Calculations based on DFT give a value for this gap of 6.95 eV, corresponding to the optical excitation from the HOMO to the LUMO, a value that is in good agreement with experiment (see the detailed discussion in Chapter 6); this implies a value for $t_0 \approx -3.5$ eV.

5.1.2 An Infinite "1D Solid": Polyacetylene

An extension of the features found in benzene is the macromolecule polyacetylene, $[-C_2H_2-]_n$. In this case the actual repeat unit consists of two C atoms and two H atoms, in the structure shown in Fig. 5.4, called trans-polyacetylene, with the distance between equivalent atoms along the chain defined as a; this will be the lattice constant in a 1D model of this structure. There is another possible structure, that preserves all the essential features of benzene in an infinite linear chain with four C and four H atoms per unit cell, called cis-polyacetylene (see Problem 1). It is tempting to consider all the C–C bonds in trans-polyacetylene as being equivalent, as was the case in benzene. This structure can then be projected onto a linear chain. We focus again on the p_z orbitals of C, which will give a period of $a/2$ between equivalent sites projected on the linear chain. This can be rationalized as the electrons in these orbitals not caring about the actual geometry of the structure and simply seeing nearest-neighbor orbitals at distance $a/2$ along the projected chain. The resulting band structure is given by the single energy band

$$\epsilon_k = \epsilon_p + 2t_0 \cos\left(\frac{ka}{2}\right), \quad -\frac{2\pi}{a} \le k \le \frac{2\pi}{a}$$

with t_0 the nearest-neighbor hopping matrix element. If we insist on preserving the periodicity imposed by the actual structure, with a two C atom unit cell of period a, this would simply produce two energy bands:

$$\epsilon_k^{(\pm)} = \epsilon_p \pm 2t_0 \cos\left(\frac{ka}{2}\right), \quad -\frac{\pi}{a} \le k \le \frac{\pi}{a}$$

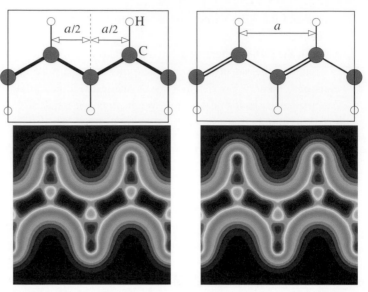

Fig. 5.4 **Top row**: Illustration of the structure of trans-polyacetylene, $[-C_2H_2-]_n$, with the C and H atoms shown as gray and white circles, in the symmetric (left) and asymmetric (right) configurations. **Bottom row**: The total charge density obtained from DFT calculations for the two models; red corresponds to highest electron density and blue to zero density (calculations performed by D. T. Larson).

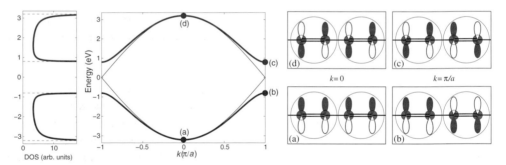

Fig. 5.5 **Left**: Energy bands of polyacetylene, $[-C_2H_2-]_n$, and density of states (far left), with the band extrema denoted by horizontal dashed lines. The thin black band corresponds to all bonds being the same [$\delta = 0$ in Eq. (5.4)], the red and blue bands correspond to energies $\epsilon_k^{(\perp)}$ for the alternating double/single bonds along the chain ($\delta \neq 0$). Parameter values for this example: $\epsilon_p = 0$ eV, $t_0 = -1.6$ eV, $\Delta t = 0.4$ eV. **Right**: Wavefunctions of polyacetylene shown as linear combinations of p_z orbitals, indicated by filled (positive) and open (negative) lobes, at $k = 0$ and $k = \pi/a$, corresponding to the states labeled (a), (b) for the valence band (in red boxes) and (c), (d) for the conduction band (in blue boxes); pairs in a single unit cell are shown inside circles.

which is equivalent to folding the reciprocal space by a factor of 2 and having two bands in the first BZ, as shown in Fig. 5.5.

The alternative picture is to have a sequence of alternating double and single bonds along the chain, which would break the symmetry and require the model to have a two C atom unit cell of period a, even along the projected linear chain. Note that due to symmetry, in the infinite chain model the choice of placement of the double bond is not important, as the two possible choices give physical models that are exactly equivalent by symmetry, just like in the case of benzene. To be specific, if we take the C atom with the H bond point upward in Fig. 5.4, the double bond could be placed to its right or to its left, the two possibilities being exactly equivalent and related by mirror symmetry; these two arrangements are referred to as A pattern and B pattern. For simplicity, we assume that the hopping matrix elements for the single and double bonds are given by t_1, t_2, differing by the same amount, $\pm \Delta t$, from the hopping matrix element t_0 of the one C atom unit cell:

$$t_1 = t_0 + \Delta t, \quad t_2 = t_0 - \Delta t, \quad \delta = \frac{\Delta t}{t_0}$$

where we have also defined δ to be the ratio of the difference Δt to the original value of t_0. The resulting band structure in this case is:

$$\epsilon_k^{(\pm)} = \epsilon_p \pm 2t_0 \left[1 - (1 - \delta^2) \sin^2\left(\frac{ka}{2}\right) \right]^{1/2}, \quad -\frac{\pi}{a} \le k \le \frac{\pi}{a} \tag{5.4}$$

shown in Fig. 5.5. This band structure is qualitatively different from the one C atom model, because it has a band gap of

$$\epsilon_{\text{gap}} = \epsilon_{\pi/a}^{(-)} - \epsilon_{\pi/a}^{(+)} = -4\Delta t$$

at $k = \pm \pi/a$. Indeed, experimental measurements indicate that polyacetylene has a band gap of about 1.7 eV, which implies a value $|\Delta t| \approx 0.425$ eV. Additional evidence comes from the fact that DFT calculations show the alternating double/single-bond pattern to be *lower* in energy, with the double bond having clearly higher electron density than the single bond, as shown in Fig. 5.4. The doubling of the unit cell and the creation of a gap between occupied and unoccupied states, which is accompanied by a reduction in the energy of the system toward a more stable configuration, is referred to as the "Peierls instability." The measured bond lengths of single and double C–C bonds are $b_1 = 1.54$ Å and $b_2 = 1.34$ Å, respectively; for comparison, the C–C bond length in benzene is $b_0 = 1.40$ Å, while in graphene, the corresponding infinite 2D structure (discussed in detail in Section 5.2), it is 1.42 Å, both values being close to the average bond length of the single and double bonds, 1.44 Å, as we would expect. For comparison, we show in Fig. 5.6 the energy bands for polyacetylene, in the symmetric (all C–C bonds equal) and asymmetric (unequal single and double bonds) models, as obtained from first-principles calculations based on DFT. The π and π^* bands are clearly distinguishable and dominate around the Fermi level; the symmetric mode has no gap, whereas the asymmetric model has a large gap, as expected from the simple models we studied. Additional bands, not included in the simple models that used only p_z orbitals as the basis, are also present, arising from states of σ, σ^* character. These states are far from the Fermi level and do affect the properties of the solid. Note that, because of their different symmetry, the π and σ bands do not mix, and therefore cross each other at various points. However, the π^* band hybridizes with some of the other unoccupied bands of similar character, and therefore does not have the simple shape expected from the tight-binding model based on p_z orbitals only. We also note that near the center of the BZ, the occupied σ bands change significantly when the symmetry

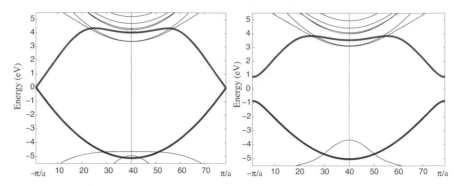

Fig. 5.6 Band structure of symmetric (left) and asymmetric (right) models of polyacetylene, obtained from DFT calculations (performed by D. T. Larson). In the asymmetric model, the C–C single and double bonds are held fixed to the measured values of $b_1 = 1.54$ Å and $b_2 = 1.34$ Å, respectively. The red and blue highlighted bands correspond to the π and π^* bands; the remaining bands (black) correspond to σ and σ^* bands.

of the bonds is broken and, because they have the same symmetry, they hybridize instead of crossing, which is referred to as "band repulsion."

It is also instructive to consider the behavior of the wavefunctions corresponding to the π and π^* states, which helps us understand the energy of these states. In Fig. 5.5 we show the wavefunctions as linear combinations of the p_z orbitals that correspond to states at $k = 0, \pi/a$ for the two bands. At $k = 0$, the wavefunctions are in-phase for pairs that reside in neighboring cells. For the lower-energy band, $\epsilon_k^{(+)}$, a pair in the first unit cell has a bond between the atoms, corresponding to the lowest-energy state, labeled (a), while for the higher-energy band, $\epsilon_k^{(-)}$, a pair in the first unit cell has the two p_z orbitals with opposite sign. This corresponds to an anti-bonding configuration both *within* the unit cell as well as *between* neighboring cells, and produces the state with the highest energy, labeled (d). At $k = \pi/a$, the wavefunctions have a phase difference of π for pairs that reside in neighboring unit cells. For the symmetric combination, $\epsilon_k^{(+)}$, this results in a node between the orbitals that belong to *adjacent* cells, labeled (c), and therefore has higher energy than at $k = 0$. For the antisymmetric combination, $\epsilon_k^{(-)}$, labeled (c), it leads to orbitals in adjacent unit cells that have the same orientation, which restores bonding between them, producing a state of lower energy than at $k = 0$.

The picture we have described so far predicts an insulating behavior for polyacetylene in its neutral form, due to the presence of the band gap. It turns out that it is easy to produce metallic behavior by doping the macromolecule, that is, by adding electrons to the conduction band or removing electrons from the valence band, through the introduction of foreign atoms in the chain. We provide a more thorough discussion of doping of semiconductors and insulators in Section 5.5. A more surprising phenomenon is the conducting (metallic) behavior of *undoped* polyacetylene. This was explained by the existence of "solitons," that is, solitary waves that can easily move through the structure without dispersion; solitons are low-energy excitations due to the presence of a "domain wall" between two domains, as shown schematically in Fig. 5.7. In this example, in the ideal structure each even atom has a single bond to the left and a double bond to the right,

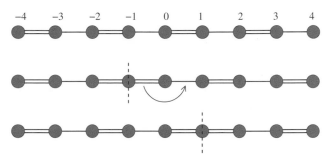

Fig. 5.7 Schematic illustration of the soliton motion in polyacetylene. The structure is represented as a linear chain of atoms bonded by an alternating pattern of double and single bonds. **Top**: The ideal chain. **Middle**: A chain with a domain wall at position -1, with this atom participating in two double bonds. **Bottom**: The domain wall has moved to position 1, by a small displacement of atom 0 to the right, which changes its left bond to single and its right bond to double.

while an odd atom has a single bond to its right and a double bond to its left. At the domain wall, one atom participates in two bonds of the same type (in our example, atom -1 has two double bonds). It takes only a small displacement of atom 0 to its right, to switch its bonds, so that the single bond is on its left (slightly longer) while the double bond is on its right (slightly shorter). But now, due to this small displacement of atom 0, the domain wall has moved to position 1, since this atom has two double bonds in the new configuration. This slight motion of an atom at the domain wall can propagate through the chain with relative ease, creating the motion of the soliton through the structure at very low energy cost. This phenomenon, first explained by Su, Schrieffer, and Heeger,[2] is of great theoretical interest; it is also of practical importance, since many electronic and optical devices are based on organic materials like polyacetylene and related carbon-based polymers.

5.2 2D Solids: Graphene and Beyond

In the last few years, there has been much attention paid to 2D solids. This has been motivated both by the possibility of observing novel physical phenomena in these materials, as well as by interest in their potential applications to electronic and optoelectronic devices. As we have seen in Chapter 1, graphite consists of a stack of sheets of threefold-coordinated carbon atoms. On a single sheet of graphite the C atoms form a honeycomb lattice, that is, a hexagonal Bravais lattice with a two-atom basis. The interaction between planes is rather weak, of van der Waals type, and the overlap of wavefunctions on different planes is essentially non-existent. We present in Fig. 5.8 the band structure for a single, periodic, infinite graphitic sheet, referred to as "graphene." In this and following figures, to facilitate the discussion of the electronic properties that emerge from the band structure calculations, we adopt a color scheme to identify the character of the different states: red

[2] W. P. Su, J. R. Schrieffer, and A. J. Heeger, *Phys. Rev. Lett.* **42**, 1698 (1979).

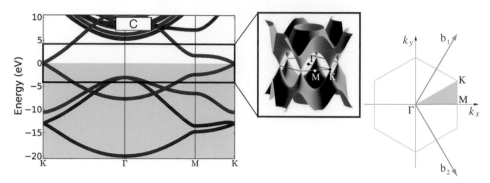

Fig. 5.8   Band structure of graphene, as obtained from DFT calculations. The pink-shaded area represents the filled bands, with the position of the Fermi level set at zero. The color scheme for the bands is: red, s-like character; blue, p-like character. The Brillouin zone with high-symmetry points Γ, K, M is shown on the right, along with a 3D representation of the π bands, illustrating the Dirac cones at the six equivalent K points (calculations performed by G. A. Tritsaris).

represents s-like states, blue p-like states, and green d-like states. This character is revealed by projecting the calculated wavefunctions of different single-particle states onto s, p, and d atomic orbitals.

In this plot, we easily recognize the lowest band as arising from a bonding state of s character; this corresponds to σ bonds between C atoms. The next three bands intersect each other at several points in the BZ. The two bands that are degenerate at Γ represent a p-like bonding state. There are two p states participating in this type of bonding, the two p orbitals that combine to form the sp^2 hybrids involved in the σ bonds in the plane. The single band intersecting the other two is a state with p character, arising from the p_z orbitals that contribute to the π bonding; it is the symmetric (bonding) combination of these two p_z orbitals. The antisymmetric (anti-bonding) combination has the reverse dispersion and lies higher in energy; it is almost the mirror image of the π-bonding state with respect to the Fermi level.

Since, in this crystal, there are two C atoms per unit cell with four valence electrons each, that is, a total of eight valence electrons per unit cell, we need four completely filled bands in the BZ to accommodate them. Indeed, the Fermi level must be at a position which makes the three σ-bonding states (one s-like and two p-like) and the π-bonding state completely full, while the π-anti-bonding state is completely empty. Similarly, anti-bonding states arising from antisymmetric combinations of s and p_x, p_y orbitals lie even higher in energy and are completely empty.

The bonding and anti-bonding combinations of the p_z states are degenerate at the K point of the BZ. The dispersion of the energy near this point is *linear*, which is rather unusual and has to do with the symmetry of the crystal lattice and the 2D nature of this structure. The linear dispersion of the energy bands near K, referred to as the "Dirac points" and the "Dirac cone" (see Fig. 5.8), has produced significant interest in this structure and the possibilities of using the special properties of linear bands for electronic applications.

At zero temperature, electrons obey a Fermi distribution, with an abrupt step cutoff at the Fermi level ϵ_F. At non-zero temperature T, the distribution will be smoother around the Fermi level, with a width at the step of order $k_B T$. This means that some states below the Fermi level will be unoccupied and some above will be occupied. This is the hallmark of metallic behavior, that is, the availability of states immediately below and immediately above the Fermi level, which makes it possible to thermally excite electrons. Placing electrons in unoccupied states at the bottom of empty bands allows them to move freely in these bands, as we discussed for the free-electron model. In the case of graphene, the number of states immediately below and above the Fermi level is actually very small: the π-bonding and anti-bonding bands do not overlap, but simply touch each other at the K point of the BZ. Accordingly, graphene is considered a semimetal, barely exhibiting the characteristics of metallic behavior even though, strictly speaking according to the definition given above, it cannot be described as anything else.

Another interesting aspect is the distribution of electrons in real space, for graphene, shown in Fig. 5.9. In this plot, we can identify the high concentration of electrons between the ions as the covalent bonds. The nature of the σ bonds is clear in both the plot on the plane of the atoms as well as on the plane perpendicular to it that goes through pairs of C atoms. The nature of the π bonds is also clear in the plane perpendicular to the atomic plane, as a high concentration of electrons in regions above and below the atoms, between the nearest-neighbor C atoms. The regions at the center of the hexagonal rings formed by the atoms, and above and below the plane of atoms, are devoid of any electron density.

The hexagonal boron nitride (h-BN) layered crystal has the same crystal structure as graphene, and is an insulator. A structure similar to graphene but composed of Si atoms, named "silicene," has been formed on metallic substrates but not in isolation. Its electronic structure and properties are very similar to those of graphene. Finally, a broad class of

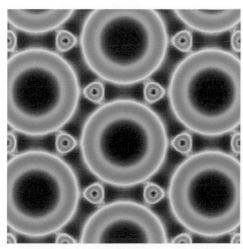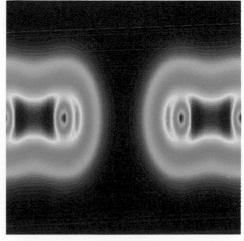

Fig. 5.9   Valence electron charge density plots for graphene, on the plane of the atoms (left) and on a plane perpendicular to it. Red corresponds to the highest value of the charge density and blue to zero.

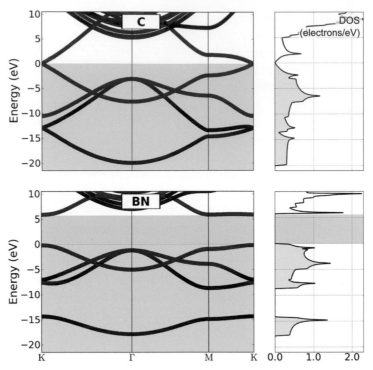

Fig. 5.10 Density of states of representative 2D crystals: graphene (top), a semimetal and BN (bottom), an insulator. The color scheme for the bands is: red, *s*-like character; blue, *p*-like character (calculations performed by G. A. Tritsaris).

2D crystals are formed by combinations of transition metal (M) atoms (M = Mo, W, Ti) and chalcogen (X) atoms (X = S, Se, Te), with chemical formula MX_2. The bonding of atoms in these structures is both covalent and ionic, with the chalcogen atoms attracting some excess electronic charge as the more electronegative ones. The crystals are typically semiconductors with large gaps (~ 2 eV) and strong excitonic effects.

In Fig. 5.10 we show the DOS of the 2D semimetal graphene and the 2D insulator BN, and in Fig. 5.11 the DOS of the 2D semimetal silicene and the 2D layer of MoS_2, a representative of the metal–dichalcogenide family of layered structures. Graphene has no band gap, and the DOS near the minimum of the valence bands has the characteristic behavior expected of a 2D crystal. Near the Fermi level, the behavior is different than analyzed above for the free-electron case, because the band dispersion is linear rather than quadratic. Very similar features are found in the band structure and DOS of silicene. Finally, the DOS of BN shows all the characteristic features of a 2D insulator, including a large band gap of more than 5 eV, and the dependence of the DOS on the energy for the free-electron case near the band extrema. Similarly, the DOS of MoS_2 shows the characteristic features of a 2D semiconductor, with a band gap of about 1.8 eV and free-electron-like behavior near the band extrema.

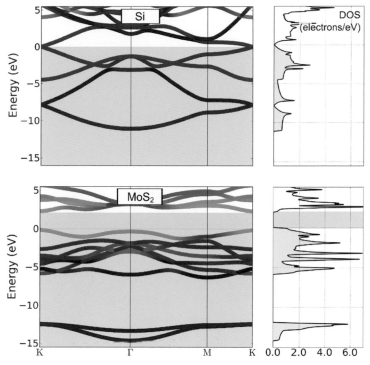

Fig. 5.11 Density of states of representative 2D crystals: silicene (top), a structure similar to graphene but consisting of Si atoms, a semimetal and MoS_2 (bottom), a semiconductor, representative of the metal–dichalcogenide family of layered solids. The color scheme for the bands is: red, s-like character; blue, p-like character; green, d-like character (calculations performed by G. A. Tritsaris).

5.2.1 Carbon Nanotubes

An interesting variation on the theme of graphene are sections of the 2D honeycomb lattice rolled up into hollow cylinders, which are called "carbon nanotubes" (CNTs), first observed by Iijima.[3] Another interesting variation is closed-shell structures composed of hexagons and pentagons of C atoms, the latter units needed to produce a curved surface that can be closed up; these are referred to as "carbon fullerenes" due to their resemblance to the geometric domes designed by Buckminster Fuller. The most stable fullerene is C_{60}, resembling a soccer ball, which was discovered by R. F. Curl, H. W. Kroto, and R. E. Smalley.

We discuss here the physics of CNTs, which are highly stable, versatile, and exhibit intriguing 1D physics phenomena, possibly even high-temperature superconductivity. In experimental observations, the nanotubes are often nested within each other as coaxial cylinders and they have closed ends. Their ends are typically half cages formed from a half fullerene whose diameter and structure are compatible with the diameter of the tube. Given that the inter-tube interactions are weak, similar to the interaction between graphene

---

[3] S. Iijima, *Nature* **354**, 56 (1991).

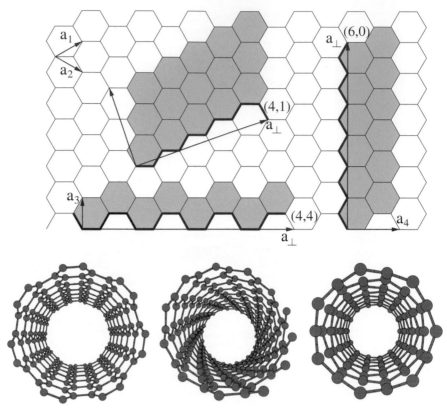

Fig. 5.12 **Top**: A graphene sheet with the ideal lattice vectors denoted as $\mathbf{a}_1, \mathbf{a}_2$. The thicker blue lines show the edge profile of nanotube examples: a $(4, 4)$ (armchair), a $(4, 1)$ (chiral), and a $(6, 0)$ (zigzag) tube. Note that the pair of vectors $\mathbf{a}_3, \mathbf{a}_4$, defined in Eq. (9.60), makes an equally convenient set for labeling the tubes because they are mutually orthogonal (but not of equal length). The tubes are formed by matching the end points of these profiles, as indicated by the blue arrows. The hexagons that form the basic repeat unit of each tube are shown shaded, with the red arrows indicating the repeat vectors along the axis of the tube. **Bottom**: Perspective views of the $(7, 7)$ armchair tube, the $(8, 4)$ chiral tube, and the $(7, 0)$ zigzag tube, along their axes.

sheets, and that their length is of order $\sim 1\,\mu\text{m}$, it is reasonable to consider them as single infinite tubes in order to gain a basic understanding of their properties. This is the picture we adopt below.

The simplest way to visualize the structure of CNTs is by considering a graphene sheet and considering how it can be rolled into a cylindrical shape. This is illustrated in Fig. 5.12. The types of cylindrical shapes that can be formed in this way are described in terms of the multiples n and m of the in-plane primitive lattice vectors $\mathbf{a}_1, \mathbf{a}_2$ that form the vector connecting two atoms which become identical under the operation of rolling the graphene into a cylinder. We define this vector, which is perpendicular to the tube axis, as $\mathbf{a}_\perp^{(n,m)}$. The vector perpendicular to it, which is along the tube axis, is defined as $\mathbf{a}_\parallel^{(n,m)}$. In terms of cartesian components on the graphene plane, in the orientation shown in Fig. 5.12, these two vectors are expressed as:

$$\mathbf{a}_\perp^{(n,m)} = n\mathbf{a}_1 + m\mathbf{a}_2 = na\left[\frac{(1+\kappa)\sqrt{3}}{2}\hat{\mathbf{x}} + \frac{(1-\kappa)}{2}\hat{\mathbf{y}}\right] \tag{5.5}$$

$$\mathbf{a}_\parallel^{(n,m)} = a\left[\frac{(\kappa-1)}{2}\hat{\mathbf{x}} + \frac{(\kappa+1)\sqrt{3}}{2}\hat{\mathbf{y}}\right]\frac{\sqrt{3}\lambda}{(1+\kappa+\kappa^2)}$$

where we have defined two variables, $\kappa = n/m$ and λ, to produce a more compact notation. The second variable is defined as the smallest rational number that produces a vector $\mathbf{a}_\parallel^{(n,m)}$ which is a graphene lattice vector. Such a rational number can always be found; for example, the choice $\lambda = (1+\kappa+\kappa^2)$, which is rational from the definition of κ, always produces a graphene lattice vector. The reason for introducing λ as an additional variable is to allow for the possibility that a smaller number than $(1+\kappa+\kappa^2)$ can be found which makes $\mathbf{a}_\parallel^{(n,m)}$ a graphene lattice vector, thus reducing the size of the basic repeat unit that produces the tube. The length of the $\mathbf{a}_\perp^{(n,m)}$ vector cannot be reduced and corresponds to the perimeter of the tube. The diameter of the tube can be inferred from this length divided by 2π. We can also define the corresponding vectors in reciprocal space:

$$\mathbf{b}_\perp^{(n,m)} = \frac{2\pi}{na}\left[\frac{(1+\kappa)\sqrt{3}}{2}\hat{\mathbf{x}} + \frac{(1-\kappa)}{2}\hat{\mathbf{y}}\right]\frac{1}{(1+\kappa+\kappa^2)} \tag{5.6}$$

$$\mathbf{b}_\parallel^{(n,m)} = \frac{2\pi}{a}\left[\frac{(\kappa-1)}{2}\hat{\mathbf{x}} + \frac{(\kappa+1)\sqrt{3}}{2}\hat{\mathbf{y}}\right]\frac{1}{\sqrt{3}\lambda}$$

With these definitions, we can then visualize both the atomic structure and the electronic structure of C nanotubes.

There are three types of tubular structures: the first corresponds to $m = 0$ or $(n,0)$, which are referred to as "zigzag" tubes; the second corresponds to $m = n$ or (n,n), which are referred to as "armchair" tubes; and the third corresponds to $m \neq n$ or (n,m), which are referred to as "chiral" tubes. Since there are several ways to define the same chiral tube with different sets of indices, we will adopt the convention that $m \leq n$, which produces a unique identification for every tube. Examples of the three types of tubes and the corresponding vectors along the tube axis and perpendicular to it are shown in Fig. 5.12. The first two types of tubes are quite simple and correspond to regular cylindrical shapes with small basic repeat units. The third type is more elaborate, because the hexagons on the surface of the cylinder form a helical structure. This is the reason why the basic repeat units are larger for these tubes.

The fact that the tubes can be described in terms of the two new vectors that are parallel and perpendicular to the tube axis and are both multiples of the primitive lattice vectors of graphene also helps determine the electronic structure of the tubes. To first approximation, this will be the same as the electronic structure of graphene folded into the Brillouin zone determined by the reciprocal lattice vectors $\mathbf{b}_\parallel^{(n,m)}$ and $\mathbf{b}_\perp^{(n,m)}$. Since these vectors are uniquely defined for a pair of indices (n,m) with $m \leq n$, it is in principle straightforward to take the band structure of graphene and fold it into the appropriate part of the original BZ to produce the desired band structure of the tube (n, m). This becomes somewhat complicated for the general case of chiral tubes, but it is quite simple for zigzag and armchair tubes. To facilitate the discussion, we start from the primitive lattice vectors $\mathbf{a}_1, \mathbf{a}_2$ and corresponding reciprocal-space vectors $\mathbf{b}_1, \mathbf{b}_2$, and introduce two new vectors which we call $\mathbf{a}_{ZZ} = \mathbf{a}_3$ for the zigzag tube (ZZ) and $\mathbf{a}_{AC} = \mathbf{a}_4$ for the armchair tube (AC):

$$\mathbf{a}_{ZZ} = \mathbf{a}_3 = \mathbf{a}_1 - \mathbf{a}_2, \quad \mathbf{a}_{AC} = \mathbf{a}_4 = \mathbf{a}_1 + \mathbf{a}_2$$

with corresponding reciprocal lattice vectors

$$\mathbf{b}_{ZZ} = \mathbf{b}_3 = \frac{1}{2}(\mathbf{b}_1 - \mathbf{b}_2), \quad \mathbf{b}_{AC} = \mathbf{b}_4 = \frac{1}{2}(\mathbf{b}_1 + \mathbf{b}_2)$$

In terms of these vectors, the zigzag and armchair tubes are described by the real and reciprocal-space vectors parallel and perpendicular to their axes, as follows:

$$ZZ: \quad \mathbf{a}_{\parallel}^{(n,0)} = \mathbf{a}_4, \ \mathbf{b}_{\parallel}^{(n,0)} = \mathbf{b}_4, \quad \mathbf{a}_{\perp}^{(n,0)} = n\mathbf{a}_3, \ \mathbf{b}_{\perp}^{(n,0)} = \frac{1}{n}\mathbf{b}_3$$

$$AC: \quad \mathbf{a}_{\parallel}^{(n,n)} = \mathbf{a}_3, \ \mathbf{b}_{\parallel}^{(n,n)} = \mathbf{b}_3, \quad \mathbf{a}_{\perp}^{(n,n)} = n\mathbf{a}_4, \ \mathbf{b}_{\perp}^{(n,n)} = \frac{1}{n}\mathbf{b}_4$$

The smallest possible folding, which *cannot* be realized physically, is $(1, 0)$ and $(1, 1)$, and produces a BZ which is half the size of the graphene BZ. The larger foldings, $(n, 0)$ and (n, n), with $n > 1$, further reduce the size of the tube BZ by creating stripes parallel to the $\mathbf{b}_3$ vector for the zigzag tubes or to the $\mathbf{b}_4$ vector for the armchair tubes.

From this analysis and the band structure of graphene discussed earlier, we can draw several conclusions about the electronic structure of nanotubes. Graphene is a semimetal, with the occupied and unoccupied bands of π character meeting at the K point of the BZ (see Fig. 5.8). From Fig. 5.13, it is evident that this point is mapped onto the point $(k_x, k_y) = \mathbf{b}_4/3$, which is always within the first BZ of the (n, n) armchair tubes. Therefore, all the armchair tubes are metallic, with two bands crossing the Fermi level at $k_y = 2\pi/3a$,

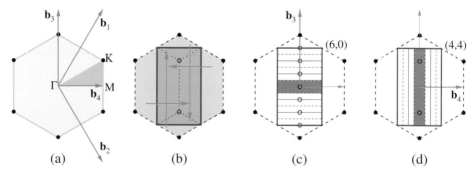

(a)   (b)   (c)   (d)

Fig. 5.13 (a) The graphene Brillouin zone, with reciprocal lattice vectors $\mathbf{b}_1, \mathbf{b}_2$ and tube-related vectors $\mathbf{b}_3, \mathbf{b}_4$. (b) The folding of the full zone into the reduced zone, determined by the vectors $\pm\mathbf{b}_3, \pm\mathbf{b}_4$. (c) The Brillouin zone for $(n, 0)$ tubes, and the example of the $(6, 0)$ tube: solid lines indicate sets of points equivalent to the $k_y = 0$ line, while dashed lines indicate zone boundaries. (d) The Brillouin zone for (n, n) tubes, and the example of the $(4, 4)$ tube: solid lines indicate sets of points equivalent to the $k_x = 0$ line, while dashed lines indicate zone boundaries. The black dots in (c) and (d) are the images of the K point of the graphene BZ under the folding introduced by the tube structure.

that is, two-thirds of the way from the center to the edge of their BZ in the direction parallel to the tube axis. It is also evident from Fig. 5.13 that in the $(n, 0)$ zigzag tubes, if n is a multiple of 3, the K point is mapped onto the center of the BZ, which makes these tubes metallic, whereas if n is not a multiple of 3 these tubes can have semiconducting character. Analogous considerations applied to the chiral tubes of small diameter (10–20 Å) lead to the conclusion that about one-third of them have metallic character while the other two-thirds are semiconducting. The chiral tubes of metallic character are those in which the indices n and m satisfy the relation $2n + m = 3l$, with l an integer. What we have described so far is a simplified but essentially correct picture of C nanotube electronic states. The true band structure is also affected by the curvature of the tube and variations in the bond lengths, which are not all equivalent. The effects are more pronounced in tubes of small diameter, but do not alter significantly the simple picture based on the electronic structure of graphene.

Examples of the band structures of various tubes are shown in Fig. 5.14. These band structures, plotted from the center to the edge of the BZ along the direction parallel to the tube axis, were obtained using a TBA,[4] which reproduces well the band structures of

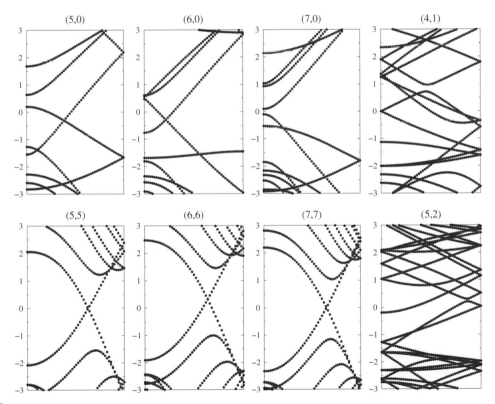

Fig. 5.14 Examples of CNT band structures obtained with the tight-binding approximation. The horizontal axis in each case runs from the center to the edge of the BZ in the direction parallel to the tube axis and the Fermi level is at 0 (energy scale, eV).

[4] M. J. Mehl and D. A. Papaconstantopoulos, *Phys. Rev. B* **54**, 4519 (1996).

graphene and diamond; for these calculations, the bond length was kept fixed at 1.42 Å (the bond length of graphene). The band structures contain both σ and π-state bonding and anti-bonding states. In the examples shown in Fig. 5.14, the (5,0) tube is metallic by accident (by which we mean that the metallic character is not dictated by symmetry), the (6,0) tube is metallic by necessity, as mentioned above, and the (7,0) tube is semiconducting, with a small band gap of about 0.2 eV. The three (n, n) tubes exhibit the characteristic crossing of two bands which occurs two-thirds of the way from the center to the edge of the BZ, a feature which renders them metallic. Finally, the $(4, 1)$ and $(5, 2)$ tubes, for which the rule $2n + m = 3l$ holds, are metallic in character with two bands meeting at the Fermi level at the center [for the $(4, 1)$ tube] or at the edge [for the $(5, 2)$ tube] of the BZ, as predicted by the simple analysis of zone folding in graphene (see also Problem 4).

5.3 3D Metallic Solids

We first discuss some representative 3D metallic solids. The examples we will consider correspond to simple crystals with one atom per unit cell in the close-packed (FCC or HCP) or almost close-packed (BCC) crystal structures. The nature of electronic states is close to the prototypical qualitative model we have studied several times, that is, the free-electron model, at least for the electrons with s and p character.

The first case is a simple metal, Al, in which only s and p orbitals are involved in the valence bands: the corresponding atomic states are $3s, 3p$. The band structure of this crystal, which we already encountered and compared to the free-electron model in Fig. 2.12, has all the characteristics expected of free electrons in the corresponding 3D FCC lattice. Indeed, Al is the prototypical solid with behavior close to that of free electrons. The dispersion near the bottom of the lowest band is nearly a perfect parabola, as would be expected for free electrons. Since there are only three valence electrons per atom and one atom per unit cell, we expect that on average 1.5 bands will be occupied throughout the BZ. As seen in Fig. 5.15, the Fermi level is at a position which makes the lowest band completely full throughout the BZ, and small portions of the second band full, especially along the X–W–L high-symmetry lines.

The next example is the so-called "noble" metals, Cu, Ag, Au, all in the FCC crystal structure. Their band structure is more complicated because it involves s and d electrons. In this case we have 11 valence electrons, and we expect 5.5 bands to be filled on average in the BZ. Indeed, we see five bands with little dispersion near the bottom of the energy range, all of which are filled states below the Fermi level. There is also one band with large dispersion, which intersects the Fermi level at several points, and is on average half filled. The five low-energy occupied bands are essentially bands arising from the $4d$ states. Their low dispersion is indicative of weak interactions among these orbitals. The next band can be identified with the s-like bonding band. This band interacts and hybridizes with the d bands, as the mixing of the spectrum near Γ suggests. In fact, if we were to neglect the five d bands, the rest of the band structure looks remarkably similar to that of Al. In both

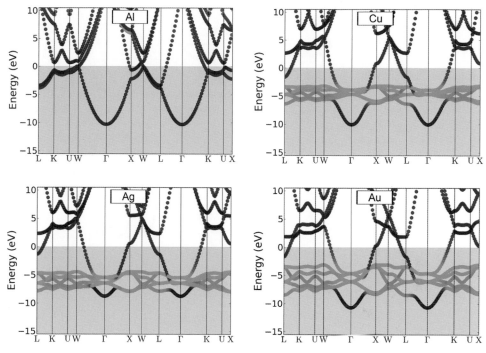

Fig. 5.15 Band structure of four metals, Al, Cu, Ag, Au, all in the FCC structure, along high-symmetry directions of the BZ shown in Fig. 2.12. The color scheme for the bands is: red, s-like character; blue, p-like character; green, d-like character (calculations performed by G. A. Tritsaris).

cases, the Fermi level intersects bands with high dispersion at several points, and thus there are plenty of states immediately below and above the Fermi level for thermal excitation of electrons. This indicates that these solids will act as good metals, being able to carry current when placed in an external electric field.

Very similar behavior is observed in three other typical metallic solids, V, Cr, and Fe, all of them in the BCC crystal structure, as shown in Fig. 5.16. The dispersion of the bands is different because of the different crystal structure and corresponding Brillouin zone, but the general features are similar to the noble metals: the s and p bands have large dispersion while the d bands have smaller dispersion and are clearly identified. In these three metals the Fermi level is somewhere in the range of d bands, and includes more of those bands as the atomic number increases (more valence electrons are available), from V to Cr to Fe.

The total valence charge density for Al and Ag, shown in Fig. 5.17, reveals some interesting features. In Al the valence electrons are mostly distributed around the ions, and show remarkable non-uniformity, which is surprising given the resemblance of the band structure of this solid to the free-electron model (for which the density should be uniform). As far as the representative noble metal Ag is concerned, notice how the total charge density is mostly concentrated around the atoms, and there seems to be little interaction between these atoms. This is consistent with the picture we had discussed of the noble metals, namely that they have an essentially full electronic shell (the $4d$ shell in Ag) and

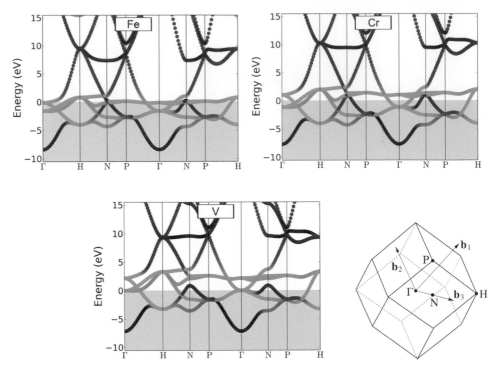

Fig. 5.16 Band structure of three metals, V, Cr, Fe, all in the BCC structure, along high-symmetry directions in the BZ. The color scheme for the bands is: red, s-like character; blue, p-like character; green, d-like character. The figure on the lower right shows the BZ of the BCC structure with high-symmetry points Γ, N, P, H and reciprocal lattice vectors $\mathbf{b}_1, \mathbf{b}_2, \mathbf{b}_3$ identified (calculations performed by G. A. Tritsaris).

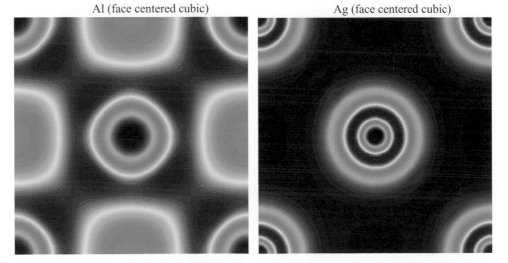

Fig. 5.17 Valence electron charge density plots on the (100) plane of the bulk (FCC) crystal structure for two metallic solids: Al (left) and Ag (right). Red corresponds to the highest value of the charge density and blue to zero.

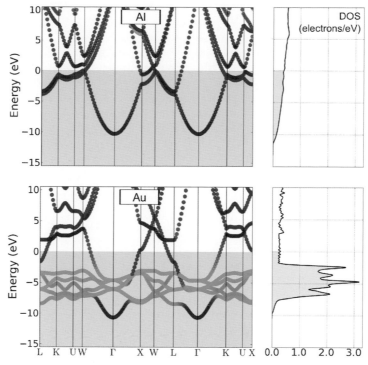

Fig. 5.18   Density of states of representative 3D metallic crystals: a simple, free-electron metal, Al (top) and a noble, *d*-electron metal, Au (bottom). The color scheme for the bands is: red, *s*-like character; blue, *p*-like character; green, *d*-like character (calculations performed by G. A. Tritsaris).

one additional *s* electron which is shared among all the atoms in the crystal. When the charge of the *d* orbitals, which are full, is added up, it produces the completely spherically symmetric distribution shown in the total density, as expected for closed-shell atoms. The lack of interaction among these states is reflected in the lack of dispersion of their energies in the band-structure plot, Fig. 5.15. Only the *s* state, which is shared among all atoms, shows significant dispersion and contributes to the metallic bonding in this solid.

In Fig. 5.18 we show the DOS of the 3D representative metals Al and Au. Al has an almost featureless DOS, corresponding to the behavior of free electrons with the characteristic $\epsilon^{1/2}$ dependence at the bottom of the energy range. Au has a large DOS with significant structure in the range of energies where the *d*-like states lie, but has very low and featureless DOS beyond that range, corresponding to the *s*-like state which has free-electron behavior.

Finally, in Table 5.1 we give values of the basic structural and electronic properties of common metals, grouped in two categories, those with *s/p* valence electrons and those with *d* valence electrons. An interesting feature is the density of states at the Fermi level, $g(\epsilon_F)$, in units of states/eV-atom; this is a measure of how many electrons are available to contribute to conduction. The conductivity and related properties of solids are discussed in more detail in Chapter 6.

Table 5.1 Structural and electronic properties of common metals, grouped according to crystal structure (BCC, HCP, FCC) and the type of valence electrons (s, p, or d). a is the lattice constant (Å) (including also the c/a ratio for HCP crystals) and $g(\epsilon_F)$ is the density of states (/eV-atom) at the Fermi energy ϵ_F (values marked $*$ are for the FCC structure). All values are from experiment. *Source:* C. Kittel, *Introduction to Solid State Physics* (7th edn, Wiley, New York, 1996) [except for $g(\epsilon_F)$, *source:* V. L. Moruzzi, J. F. Janak, and A. R. Williams, *Calculated Electronic Properties of Metals* (Pergamon Press, New York, 1978)].

BCC			HCP				FCC		
Solid	a (Å)	$g(\epsilon_F)$	Solid	a (Å)	(c/a)	$g(\epsilon_F)$	Solid	a (Å)	$g(\epsilon_F)$
s			s				s		
Li	3.491	0.48	Be	2.27	(1.5815)	0.05*	Ca	5.58	1.56
Na	4.225	0.45	Mg	3.21	(1.6231)	0.45*	Sr	6.08	0.31
K	5.225	0.73							
Rb	5.585	0.90	d				p		
			Sc	3.31	(1.5921)	1.73*	Al	4.05	0.41
			Y	3.65	(1.5699)	1.41*			
d							d		
V	3.03	1.64	Ti	2.95	(1.5864)	1.49*			
Nb	3.30	1.40	Zr	3.23	(1.5944)	1.28*	Rh	3.80	1.35
Cr	2.88	0.70	Ru	2.71	(1.5793)	1.13*	Ni	3.52	4.06
Mo	3.15	0.65	Co	2.51	(1.6215)	2.01*	Pd	3.89	2.31
Fe	2.87	3.06	Zn	2.66	(1.8609)	0.30*	Cu	3.61	0.29
			Cd	2.98	(1.8859)	0.36*	Ag	4.09	0.27

5.4 3D Ionic and Covalent Solids

We next discuss the band structure of ionic and covalent solids, which typically correspond to more complicated crystals, that is crystals with several basis atoms in each unit cell. These crystals usually behave as semiconductors or insulators. For simplicity and ease of comparison, we shall consider only some representative materials with characteristic structural themes. In the first theme, the structure is based on the diamond crystal and its variants. For this case, the crystal structure consists of two interpenetrating FCC lattices displaced relative to each other by 1/4 of the main diagonal of the cube (see Fig. 1.22), and a two-atom basis in the PUC. When the two atoms in the basis are the same, the structure is that of the diamond crystal; when the two atoms are different, the structure is that of the zincblende crystal. The second structural theme, relevant to ionic solids, is the rocksalt crystal structure, also consisting of two interpenetrating FCC lattices with a two-atom unit cell but in this case displaced by 1/2 of the main diagonal of the cube (see Fig. 1.25). The specific materials we will consider are the insulators LiF and NaCl (both in the rocksalt structure), C (diamond structure), and BN (zincblende structure), and the semiconductors Si (diamond), AlP (zincblende), SiC (zincblende), Ge (diamond), GaAs (zincblende), and ZnS (zincblende). In all of these examples there are eight valence electrons per PUC of the

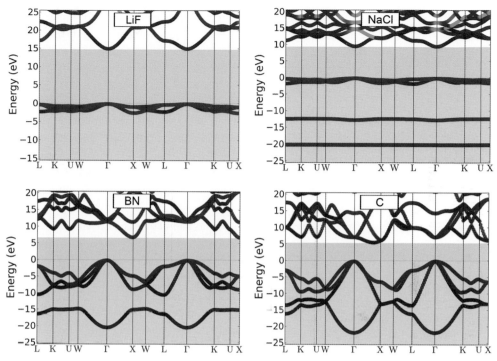

Fig. 5.19   Band structure of two ionic insulators, LiF and NaCl (both in rocksalt structure), a partially ionic/partially covalent insulator, BN (zincblende structure), and a covalently bonded insulator, C (diamond structure). The color scheme for the bands is: red, *s*-like character; blue, *p*-like character; green, *d*-like character (calculations performed by G. A. Tritsaris).

crystal. The elemental solids (C, Si, Ge) are characteristic examples of covalent bonding, the rest have partially ionic and partially covalent bonding character, except for LiF and NaCl, which are purely ionic solids (see also the discussion in Chapter 1).

We begin with four insulators whose band structures are shown in Fig. 5.19 in order of decreasing ionic character. The energy of the highest occupied band is taken to define the zero of the energy scale in each case. In all cases there is an important characteristic of the band structure, namely a range of energies where there are no electronic states across the entire BZ; this is the band gap, denoted by ϵ_{gap}. The states below the Fermi level are called "valence bands," while those above the Fermi level are called "conduction bands." The ramifications of this feature are very important to the behavior of the solid. We notice first that in every case there are four valence bands with energy lower than the band gap, which means that all four of these bands are fully occupied, since there are eight valence electrons in the PUC. Naively, we might expect that the Fermi level can be placed anywhere within the band gap, since for any such position all states below it remain occupied and all states above it remain unoccupied. A more detailed analysis reveals that for an ideal crystal, that is, one in which there are no intrinsic defects or impurities, the Fermi level is at the middle of the gap (see Problem 6). This means that there are no states immediately above or below the Fermi level, for an energy range of $\pm\epsilon_{\text{gap}}/2$. Thus, it will not be possible to thermally excite appreciable numbers of electrons from occupied to unoccupied bands, until the temperature reaches $\sim \epsilon_{gap}/2$. Since the band gap of these solids is well above 1 eV, and 1 eV is equivalent to 11,604 K, we conclude that for all practical purposes the states above the Fermi level remain unoccupied (these solids melt well below 5,800 K). For the insulators considered here, the band gap is quite large: > 5 eV for C-diamond, > 6 eV for BN, and of order 10 eV for the ionic crystals LiF and NaCl. This is the hallmark of insulating behavior, that is, the absence of any states above the Fermi level to which electrons can be thermally excited for temperatures well below the melting temperature of the solid. This makes it difficult for these solids to respond to external electric fields, since a filled band cannot carry current and a huge energy is required to transfer electrons from the filled bands to empty bands, which would create the conditions for current-carrying capability.

For semiconductors, where the band gap is of order 1 eV or smaller, there is a way out of the conundrum: when imperfections (defects) or impurities (dopants) are introduced in the crystal, the solid acquires the ability to respond to external electric fields. All semiconductors in use in electronic devices are of this type, that is, crystals with impurities. This is a topic of greater practical importance for electronic device applications, which we analyze in detail in the following section of this chapter.

It is also worth remarking that the larger the difference in electronegativity between the two elements in the crystal, the larger the band gap: in LiF this difference is 3.0 ($\epsilon_{gap} \approx 13$ eV), in Nacl it is 2.2 ($\epsilon_{gap} \approx 9$ eV), and in BN it is 1.0 ($\epsilon_{gap} \approx 7$ eV). In Fig. 5.19 we have included for comparison the band structure of C in the diamond crystal, which has a substantial gap of about 5.5 eV, even though it is not an ionic solid. In fact, C-diamond is one of the representative covalent solids, which we discuss next; even BN is not a purely ionic material but has partially ionic/partially covalent bonding. Another striking difference between the two purely ionic solids, LiF and NaCl, and the other two, BN and C-diamond, which involve significant (in the case of BN) or exclusive (in the case of C-diamond) covalent bonding, is that the bands of the ionic solids are very flat, showing almost no dispersion, while those of the solids with covalent bonding show very significant dispersion. We return to this issue below. Examples of semiconductor band structures are shown in Fig. 5.20

Several features of the semiconductor band structures are of interest. First we note that the highest occupied state (valence band maximum, VBM) for the examples shown here is at the Γ point (center of the BZ). The lowest unoccupied state (conduction band minimum, CBM) can be at different positions in the BZ. For C and Si it is somewhere between the Γ and X points, while for BN, AlP, SiC it is at the X point. Only for Ge, GaAs, and ZnS, as well as for the ionic crystals LiF and NaCl, is the CBM at Γ: this is referred to as a direct gap; the other cases discussed here, where the VBM and CBM occur at different points in the BZ, are indirect-gap semiconductors or insulators. The nature of the gap has important consequences for optical properties, as discussed in the next chapter.

Another interesting feature is the "band width," that is, the range of energies covered by the valence states. For example, in Si and GaAs it is about 12.5 eV, in SiC it is about 16 eV, and in C it is considerably larger, about 23 eV. There are two factors that influence the band width: the relative energy difference between the s and p atomic valence states, and

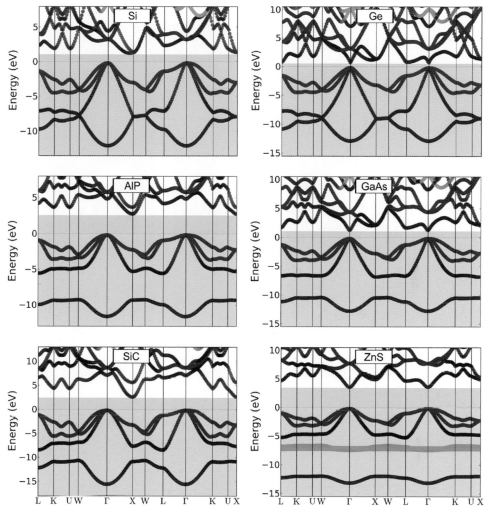

Fig. 5.20 Band structure of representative semiconductors: Si and Ge in the diamond structure, AlP, GaAs, SiC, and ZnS in the zincblende structure. The color scheme for the bands is: red, *s*-like character; blue, *p*-like character; green, *d*-like character (calculations performed by G. A. Tritsaris).

the interaction between the hybrid orbitals in the solid. For instance, in C, where we are dealing with $2s$ and $2p$ atomic states, both their energy difference and the interaction of the hybrid orbitals is large, giving a large band width, almost twice that of Si.

In all the examples considered here, it is easy to identify the lowest band at Γ as the *s*-like state of the bonding orbitals, which arise from the interaction of sp^3 hybrids in nearest-neighbor atoms. Since the sp^3 orbitals involve one *s* and three *p* states, the corresponding *p*-like states of the bonding orbitals are at the top of the valence manifold at Γ. This is illustrated in Fig. 1.21: in that example, we had shown the relative energy of atomic-like *s* and *p* orbitals for the two atoms in the unit cell as identical, since we were dealing with two atoms of the same type. When two different tetravalent elements (for example, Si and C) combine to form a solid in the zincblende lattice, the corresponding energy of atomic-like

Si (diamond)   C (diamond)

GaAs (zincblende)   SiC (zincblende)

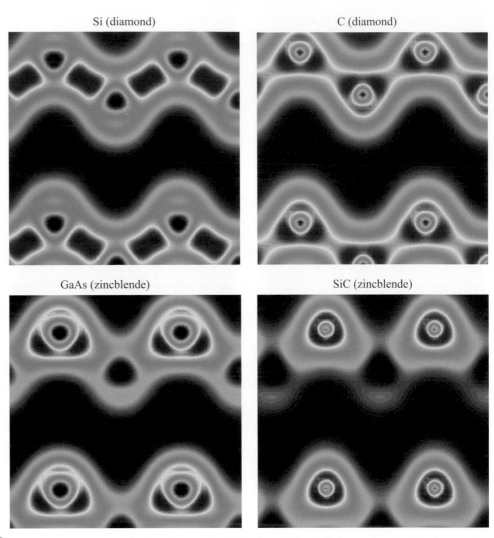

Fig. 5.21   Valence electron charge density plots on the (110) plane of the bulk diamond or zincblende crystal structure for representative solids. **Top row**: Two covalently bonded solids, Si (left) and C (right), both in the diamond structure. **Bottom row**: Two partially covalent/partially ionic solids, GaAs (left) and SiC (right), both in the zincblende structure. Red corresponds to the highest value of the charge density and blue to zero.

s and p orbitals for the two atoms will be different, with those of the more electronegative element being lower in energy. A similar diagram to that of Fig. 1.21, but with different occupation of the atomic orbitals and different energy spacings, would also apply to BN, AlP, or GaAs, which have three electrons in the group-III element (B, Al, Ga) orbitals and five electrons in the group-V element (N, P, As) orbitals, to ZnS (with two electrons in the Zn and six electrons in the S orbitals), and to LiF and NaCl (with one electron in the Li or Na and seven electrons in the F or Cl orbitals). This explains why the states near VBM have p bonding character and are associated with the more electronegative element (N, P, As, S, Na, Cl) in the solid, while those near the CBM have p anti-bonding character and are associated with the less electronegative element in the solid (B, Al, Ga, Zn, Li, Na). In other words, the character of the bands derives from the atomic states which are closest in energy to those bands. In the case of a homopolar solid, the two sets of atomic-like orbitals are the same and the character of bonding and anti-bonding states near the VBM and CBM is not differentiated among the two atoms in the unit cell. The VBM in all cases is threefold degenerate at Γ. These states disperse and their s and p character becomes less clear away from Γ. It is also interesting that the bottom s-like band is split off from the other three valence bands in the solids with two types of atoms, for example, compare the bands of BN and C in Fig. 5.19, or the bands of GaAs and Ge in Fig. 5.20. In these cases the s-like state bears more resemblance to the corresponding atomic state in the more electronegative element (N in the case of BN, As in the case of GaAs).

The implications of the band structure in what concerns the electron states are also borne out in electronic charge density plots. Representative cases are shown in Fig. 5.21. In these plots the positions of the atoms are identified by the round holes of very low charge density, due to the pseudopotential used to represent the ions. In the regions away from the atomic positions, the charge density can be measured quite accurately by experiments, and the agreement between the calculated and measured values is remarkably good.[5] In the case of Si, for instance, this distribution is the same relative to all atomic positions with the charge concentrated predominantly and symmetrically in the regions between atoms. In contrast to this, in GaAs the charge distribution is polarized closer to the more electronegative atoms (As). The high concentration of electrons in the regions between the nearest-neighbor atomic sites represents the covalent bonds between these atoms. Regions far from the bonds are completely devoid of charge. Moreover, there are just enough of these bonds to accommodate all the valence electrons. Specifically, there are four covalent bonds emanating from each atom (only two of which are contained on the plane shown in Fig. 5.21), and since each bond is shared by two atoms, there are on average two covalent bonds per atom. Since each bonding state can accommodate two electrons due to spin degeneracy, these covalent bonds take up all the valence electrons in the solid.

Certain features of the bonds are quite interesting. For instance, the charge density of the Si–Si covalent bonds is peaked exactly at the center of the bond, with the overall charge distribution almost cylindrical around the axis of the bond, pretty much like the bonds are depicted in simple ball-and-stick models of the solid. In the case of diamond, though, the charge density has a local *minimum* at the center of the C–C bond, which has to do with the special nature of C; it is this nature that leads to the highly flexible bonding properties of C with itself and with other elements, which is ultimately responsible for the plethora of molecular and solid structures it forms, including the wide variety of biomolecules.

In a similar vein, we note that the bonds can be highly polar not only for solids with two elements of different valence (heterovalent), like Ga and As, but even when the two elements comprising the solid are homovalent: a comparison of the GaAs and SiC charge densities makes this evident. In GaAs the two atoms are similar in size but the electronic density is polarized toward the As sites due to its higher electronegativity and higher valence. In SiC, the two elements are homovalent (both tetravalent), but the bonds are highly polarized with most of the charge closer to the C atoms. The degree of polarization

---

[5] See, for example, Z. W. Lu, A. Zunger, and M. Deusch, *Phys. Rev. B* **47**, 9385 (1993).

in SiC is even greater than in GaAs. The reason of course is the higher electronegativity of C as well as its smaller size (see Fig. 1.4), both contributing to the stronger attraction of valence electrons by the ion core of C. In fact, the SiC charge density resembles that of an ionic solid, with most of the charge around the more electronegative atom and a relatively small fraction in the interatomic region. In this sense, of the four examples shown in Fig. 5.21, SiC is the most ionic in character, even though the electronegativity difference between C and Si is only 0.65; for comparison, the electronegativity difference between As and Ga is 0.37. Judging from this, we conclude that in cases of larger electronegativity difference, like the highly ionic solids LiF and NaCl discussed earlier, the electron distribution will be highly polarized, with the valence electrons almost entirely bound to the halide atoms (F and Cl). This implies that there is not much interaction between the ions in these solids, other than the electrostatic attraction, and as a consequence electrons are not spread throughout the solid, which explains the almost complete lack of dispersion in the valence bands of these solids (see Fig. 5.19), which was noted earlier.

In Fig. 5.22 we show the DOS of the 3D semiconductors Si and GaAs. The valence bands in Si show a low-energy hump associated with the s-like states and a broader set of features associated with the p-like states, which have larger dispersion; the DOS also reflects the presence of the band gap, with valence and conduction bands clearly separated.

Finally, in Table 5.2 we list values for the basic structural and electronic properties of common semiconductors. These materials typically have crystal structures with a two-atom

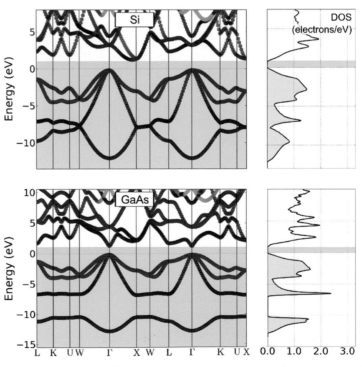

Fig. 5.22   Density of states of representative 3D semiconducting crystals: top, Si; bottom, GaAs. The color scheme for the bands is: red, s-like character; blue, p-like character; green, d-like character (calculations performed by G. A. Tritsaris).

Table 5.2 Structural and electronic properties of common semiconductors, grouped according to the type of elements that form the crystal. The crystal structures are denoted as DIA = diamond, ZBL = zincblende, WRZ = wurtzite; a is the lattice constant (Å) (and the c/a ratio for the wurtzite structures); ϵ_{gap} is the minimal band gap (eV), and k is the point where the CBM occurs for indirect gaps (the VBM is always at Γ, the center of the Brillouin zone, so when no k point is indicated, the gap is direct at Γ); X_1 is a point 0.76 of the distance $\Gamma - X$, X_2 is a point 0.85 of the distance $\Gamma - X$; $\tilde{X}$ means very close to the X point. $\bar{\varepsilon}_0$ is the room-temperature value of the static dielectric function, see Chapter 6 for details. *Sources*: D. R. Lide (Ed.), *CRC Handbook of Chemistry and Physics* (CRC Press, Boca Raton, FL, 1999–2000); *Semiconductors*, Landolt-Börnstein: Numerical Data and Functional Relationships in Science and Technology, New Series, Vol. 17, O. Madelung, M. Scholz, and H. Weiss (Eds) (Springer-Verlag, Berlin, 1982), p. 135.

Solid	a (Å) (c/a)	ϵ_{gap} (k)	$\bar{\varepsilon}_0$	Solid	a (Å) (c/a)	ϵ_{gap} (k)	$\bar{\varepsilon}_0$
IV–IV group				*III–V group*			
C (DIA)	3.57	5.48 (X_1)	5.9	BN (ZBL)	3.62	6.40 (X)	7.1
Si (DIA)	5.43	1.17 (X_2)	12.1	BP (ZBL)	4.54	2.40 (X)	
Ge (DIA)	5.66	0.74 (L)	15.8	AlN (WRZ)	3.11	6.28	9.1
SiC (ZBL)	4.34	2.39 (X)	9.7		(1.60)		
				AlP (ZBL)	5.45	2.50 (X)	
				AlAs (ZBL)	5.62	2.32 ($\tilde{X}$)	
II–VI group				AlSb (ZBL)	6.13	1.65 ($\tilde{X}$)	14.4
ZnS (ZBL)	5.40	3.84	8.3	GaN (ZBL)	4.69	3.52	
ZnSe (ZBL)	5.66	2.83	9.1	GaP (ZBL)	5.45	2.39 ($\tilde{X}$)	11.1
ZnTe (ZBL)	6.10	2.39	10.1	GaAs (ZBL)	5.66	1.52	13.0
CdS (ZBL)	5.82	2.58	9.7	GaSb (ZBL)	6.12	0.80	15.7
CdSe (ZBL)	6.08	1.73	9.7	InN (WRZ)	3.54	1.89	
CdSe (WRZ)	4.30	1.74			(1.61)		
	(1.63)			InP (ZBL)	5.87	1.42	12.6
CdTe (ZBL)	6.48	1.61	10.2	InAs (ZBL)	6.04	0.41	14.6
				InSb (ZBL)	6.48	0.23	17.9

basis in the PUC, since it takes at least two atoms per unit cell to produce a band gap, as discussed on general grounds in Chapter 1. In the examples we show, the two atoms are either both from column IV of the Periodic Table, or from columns III and V, or from columns II and VI. In all cases, there are eight valence electrons per unit cell, distributed among the covalent bonds between the nearest neighbors (for more details on the crystal structures, see Chapter 1). An important feature is the band gap, which can be direct or indirect, as discussed above; our examples in Table 5.2 include both direct and indirect semiconductors. We also list values for the static dielectric constant, a property of semiconductors and insulators that makes them useful for applications in electrical devices; this is a measure of the ability of the material to shield external electric fields due to polarization. We expand on these properties and their microscopic origin in Chapter 6.

5.5 Doping of Ideal Crystals

The presence of foreign atoms in a crystal in small concentration (typically one per billion) can have drastic effects on its electronic and optical properties. When the natural size of one of these "impurities" is compatible with the volume it occupies in the host crystal (that is, the distance between the impurity and its nearest neighbors in the crystal is not too different from its usual bond length), the substitution of the impurity for a regular crystal atom results in a stable structure with interesting physical properties. In certain cases, the impurity behaves essentially like an isolated atom in an external field possessing the point-group symmetries of the lattice. This external field usually induces splitting of levels that would normally be degenerate in a free atom. The impurity levels are in general different from the host crystal levels. Electronic transitions between impurity levels, through absorption or emission of photons, reveal a signature that is characteristic of, and therefore can uniquely identify, the impurity. Such transitions are often responsible for the color of crystals that otherwise would be transparent. This type of impurity is called a color center.

The doping of semiconductor crystals is a physical process of crucial importance to the operation of modern electronic devices, so we discuss its basic aspects next. We begin with a general discussion of the nature of impurity states in semiconductors. The energy of states introduced by the impurities, which are relevant to doping, lies within the band gap of the semiconductor. If the energy of the impurity-related states is near the middle of the band gap, these are called "deep" states; if it lies near the band extrema (VBM or CBM), they are called "shallow" states. It is the latter type of state that is extremely useful for practical applications, so we will assume from now on that the impurity-related states of interest have energies close to the band extrema. If the impurity atom has more valence electrons than the host atom, in which case it is called a donor, these extra electrons are in states near the CBM. In the opposite case, the impurity creates states near the VBM, which, when occupied, leave empty states in the VBM of the crystal called "holes"; this type of impurity is called an acceptor. These features are illustrated in Fig. 5.23.

5.5.1 Envelope Function Approximation

To describe in more detail the nature of states near the band extrema introduced by dopants, we will study a simple theoretical model that describes these situations in terms of the familiar Bloch states of the unperturbed crystal; this theory is called "envelope function" approximation, for reasons that will become obvious shortly.

One important aspect of the impurity-related states is the effective mass of charge carriers (extra electrons or holes): due to band-structure effects, these can have an effective mass which is smaller or larger than the electron mass, described as "light" or "heavy" electrons or holes. We recall from our earlier discussion of the behavior of electrons in a periodic potential (Chapter 2) that an electron in a crystal has an inverse effective mass which is a tensor, written as $[(\overline{m})^{-1}]_{\alpha\beta}$. The inverse effective mass depends on matrix

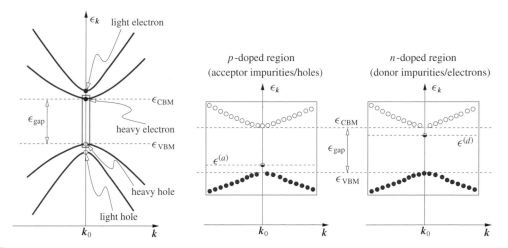

Fig. 5.23 **Left**: Schematic illustration of light and heavy electron and hole states; the bands corresponding to the light and heavy masses are split in energy for clarity. **Right**: Schematic illustration of shallow donor and acceptor impurity states in a semiconductor with direct gap.

elements of the momentum operator at the point of the BZ where it is calculated. At a given point $\mathbf{k}_0$ in the BZ we can always identify the principal axes which make the inverse effective mass a diagonal tensor, that is, $[(\overline{m})^{-1}]_{\alpha\beta} = \delta_{\alpha\beta}(1/\overline{m}_\alpha)$, as discussed in Chapter 2. The energy of electronic states around this point will be given by quadratic expressions in $(\mathbf{k} - \mathbf{k}_0)$, from $\mathbf{k} \cdot \mathbf{p}$ perturbation theory (see Chapter 2). Applying this to the VBM and the CBM of a semiconductor yields, for the energy of states near the highest valence and the lowest conduction states:

$$\epsilon_{\mathbf{k}}^{(n)} = \epsilon_{\mathbf{k}_0}^{(n)} + \frac{\hbar^2}{2}\left[\frac{(k_1 - k_{0,1})^2}{\overline{m}_1^{(n)}} + \frac{(k_2 - k_{0,2})^2}{\overline{m}_2^{(n)}} + \frac{(k_3 - k_{0,3})^2}{\overline{m}_3^{(n)}}\right] \tag{5.7}$$

where n stands for either the valence maximum ($n = v$) or the conduction minimum ($n = c$). For the effective masses of states at the VBM (hole states): $\overline{m}_i^{(v)} < 0, i = 1, 2, 3$, while for those of states at the CBM (electron states): $\overline{m}_i^{(c)} > 0, i = 1, 2, 3$.

We will assume the impurity has an effective charge $\overline{Z}$ which is the difference between the valence of the impurity and the valence of the crystal atom it replaces; with this definition, $\overline{Z}$ can be positive or negative. The potential that the presence of the impurity creates is then the Coulomb potential of a point charge $\overline{Z}e$, screened by the other electrons in the solid. For simplicity, if we take this screening to be just $\overline{\varepsilon}_0$, the static dielectric constant, then

$$\overline{\mathcal{V}}(\mathbf{r}) = -\frac{\overline{Z}e^2}{\overline{\varepsilon}_0 r} \tag{5.8}$$

We expect that in such a potential, the impurity-related states will be similar to those of an atom with effective nuclear charge $\overline{Z}e/\overline{\varepsilon}_0$. In order to examine the behavior of the extra electrons or holes associated with the presence of donor or acceptor dopants in the crystal, we will employ what is referred to as "effective-mass theory."[6] We will consider the potential introduced by the impurity to be a small perturbation to the crystal potential, and use the eigenfunctions of the perfect crystal as the basis for expanding the perturbed wavefunctions of the extra electrons or holes. In order to keep the notation simple, we will assume that we are dealing with a single band in the host crystal, the generalization to many bands being straightforward. The wavefunction of the impurity state will then be given by:

$$\overline{\psi}(\mathbf{r}) = \sum_{\mathbf{k}} \beta_{\mathbf{k}} \psi_{\mathbf{k}}^{(n)}(\mathbf{r}) = \sum_{\mathbf{k}} \beta_{\mathbf{k}} \frac{e^{i\mathbf{k}\cdot\mathbf{r}}}{\sqrt{V}} u_{\mathbf{k}}^{(n)}(\mathbf{r}) \tag{5.9}$$

where $\psi_{\mathbf{k}}^{(n)}(\mathbf{r})$ are the relevant crystal wavefunctions (conduction states for the extra electrons, valence states for the holes). We note that the relevant values of the wave-vector $\mathbf{k}$ are those close to $\mathbf{k}_0$, the position of the VBM (for holes) or the CBM (for electrons). We can take advantage of this fact to write:

$$\overline{\psi}(\mathbf{r}) = \sum_{\mathbf{k}} \beta_{\mathbf{k}} e^{i(\mathbf{k}-\mathbf{k}_0)\cdot\mathbf{r}} \frac{e^{i\mathbf{k}_0\cdot\mathbf{r}}}{\sqrt{V}} u_{\mathbf{k}}^{(n)}(\mathbf{r}) \approx \frac{e^{i\mathbf{k}_0\cdot\mathbf{r}}}{\sqrt{V}} u_{\mathbf{k}_0}^{(n)}(\mathbf{r}) \sum_{\mathbf{k}} \beta_{\mathbf{k}} e^{i(\mathbf{k}-\mathbf{k}_0)\cdot\mathbf{r}} = \psi_{\mathbf{k}_0}^{(n)} \sum_{\mathbf{k}} \beta_{\mathbf{k}} e^{i(\mathbf{k}-\mathbf{k}_0)\cdot\mathbf{r}}$$

To obtain the final expression above, we have also taken advantage of the fact that the functions $u_{\mathbf{k}}^{(n)}(\mathbf{r})$ are slowly varying with respect to the wave-vector $\mathbf{k}$, so since $\mathbf{k}$ is close to $\mathbf{k}_0$, we have replaced $u_{\mathbf{k}}^{(n)}(\mathbf{r})$ by $u_{\mathbf{k}_0}^{(n)}(\mathbf{r})$, which allowed us to take this factor outside the summation over $\mathbf{k}$. We next define the envelope function

$$f(\mathbf{r}) = \sum_{\mathbf{k}} \beta_{\mathbf{k}} \, e^{i(\mathbf{k}-\mathbf{k}_0)\cdot\mathbf{r}} \tag{5.10}$$

in terms of which the wavefunction $\overline{\psi}(\mathbf{r})$ becomes:

$$\overline{\psi}(\mathbf{r}) \approx \psi_{\mathbf{k}_0}^{(n)}(\mathbf{r}) f(\mathbf{r})$$

This expression is quite revealing: the envelope function $f(\mathbf{r})$ contains all the coefficients that we need to find in order to determine the impurity-state wavefunction $\overline{\psi}(\mathbf{r})$, with each $\beta_{\mathbf{k}}$ multiplied by a plane wave of very small wave-vector, since $\mathbf{k}$ is close to $\mathbf{k}_0$. This means that the envelope function is a very smooth function, because it involves only Fourier components of small magnitude $|\mathbf{k} - \mathbf{k}_0|$. We see from this analysis that the impurity state $\psi(\mathbf{r})$ can be approximated as the nearest state of the unperturbed crystal, $\psi_{\mathbf{k}_0}^{(n)}(\mathbf{r})$, multiplied by the envelope function $f(\mathbf{r})$. Our goal then is to derive the equation which determines $f(\mathbf{r})$.

The wavefunction $\overline{\psi}(\mathbf{r})$ will obey the single-particle equation

$$\left[\mathcal{H}_0 + \overline{\mathcal{V}}(\mathbf{r})\right] \overline{\psi}(\mathbf{r}) = \overline{\epsilon}\,\overline{\psi}(\mathbf{r}) \tag{5.11}$$

with $\mathcal{H}_0$ the single-particle hamiltonian for the ideal crystal, which of course when applied to the crystal wavefunctions $\psi_{\mathbf{k}}^{(n)}(\mathbf{r})$ gives:

$$\mathcal{H}_0 \psi_{\mathbf{k}}^{(n)}(\mathbf{r}) = \epsilon_{\mathbf{k}}^{(n)} \psi_{\mathbf{k}}^{(n)}(\mathbf{r})$$

---

[6] M. Lannoo, "Deep and shallow impurities in semiconductors," in P.T. Landsberg (Ed.), *Handbook on Semiconductors*, Vol. 1 (Amsterdam, North Holland, 1992), p. 113.

Substituting the expansion (5.9) in Eq. (5.11) and using the orthogonality of the crystal wavefunctions, we obtain:

$$\epsilon_{\mathbf{k}}^{(n)} \beta_{\mathbf{k}} + \sum_{\mathbf{k}'} \langle \psi_{\mathbf{k}}^{(n)} | \overline{\mathcal{V}} | \psi_{\mathbf{k}'}^{(n)} \rangle \beta_{\mathbf{k}'} = \overline{\epsilon} \beta_{\mathbf{k}} \tag{5.12}$$

For the matrix elements of the impurity potential that appear in Eq. (5.12), we obtain:

$$\langle \psi_{\mathbf{k}}^{(n)} | \overline{\mathcal{V}} | \psi_{\mathbf{k}'}^{(n)} \rangle = \frac{1}{V} \int e^{i(\mathbf{k}'-\mathbf{k}) \cdot \mathbf{r}} u_{\mathbf{k}}^{*(n)}(\mathbf{r}) u_{\mathbf{k}'}^{(n)}(\mathbf{r}) \overline{\mathcal{V}}(\mathbf{r}) d\mathbf{r}$$

Using again the fact that the functions $u_{\mathbf{k}}^{(n)}(\mathbf{r})$ vary slowly with respect to the wave-vector $\mathbf{k}$, and since both $\mathbf{k}$ and $\mathbf{k}'$ are close to $\mathbf{k}_0$, we can approximate the two functions that appear under the integral by $u_{\mathbf{k}_0}^{(n)}(\mathbf{r})$ and its complex conjugate. We also recall that these are smooth functions of $\mathbf{r}$, for example, in the limit of the free-electron or nearly free-electron approximation they are simply constants; it is then reasonable to replace them by their spatial average, denoted by $\langle \cdots \rangle$, which we take outside the integral to arrive at:

$$\langle \psi_{\mathbf{k}}^{(n)} | \overline{\mathcal{V}} | \psi_{\mathbf{k}'}^{(n)} \rangle \approx \left\langle u_{\mathbf{k}_0}^{*(n)}(\mathbf{r}) u_{\mathbf{k}_0}^{(n)}(\mathbf{r}) \right\rangle \overline{\mathcal{V}}(\mathbf{k}' - \mathbf{k})$$

where $\overline{\mathcal{V}}(\mathbf{q})$ is the Fourier transform of the impurity potential evaluated at $\mathbf{q} = \mathbf{k}' - \mathbf{k}$:

$$\overline{\mathcal{V}}(\mathbf{q}) = \frac{1}{V} \int \overline{\mathcal{V}}(\mathbf{r}) \, e^{i\mathbf{q} \cdot \mathbf{r}} \, d\mathbf{r} \Leftrightarrow \overline{\mathcal{V}}(\mathbf{r}) = \sum_{\mathbf{q} \in BZ} \overline{\mathcal{V}}(\mathbf{q}) \, e^{-i\mathbf{q} \cdot \mathbf{r}} \tag{5.13}$$

with the second relation being the familiar form of the inverse Fourier transform for $\mathbf{q}$ spanning the first BZ (see Appendix A). For properly normalized crystal wavefunctions, the spatial average introduced above becomes:

$$\left\langle u_{\mathbf{k}_0}^{*(n)}(\mathbf{r}) u_{\mathbf{k}_0}^{(n)}(\mathbf{r}) \right\rangle = \int \left| u_{\mathbf{k}_0}^{(n)}(\mathbf{r}) \right|^2 d\mathbf{r} = 1$$

Using these results in Eq. (5.12), we arrive at the following equation for the coefficients $\beta_{\mathbf{k}}$:

$$\epsilon_{\mathbf{k}}^{(n)} \beta_{\mathbf{k}} + \sum_{\mathbf{k}'} \overline{\mathcal{V}}(\mathbf{k}' - \mathbf{k}) \beta_{\mathbf{k}'} = \overline{\epsilon} \beta_{\mathbf{k}} \tag{5.14}$$

We next multiply both sides of Eq. (5.14) by $\exp[i(\mathbf{k} - \mathbf{k}_0) \cdot \mathbf{r}]$ and sum over $\mathbf{k}$ in order to create the expression for the envelope function $f(\mathbf{r})$ on the right-hand side, obtaining:

$$\sum_{\mathbf{k}} \epsilon_{\mathbf{k}}^{(n)} \beta_{\mathbf{k}} \, e^{i(\mathbf{k}-\mathbf{k}_0) \cdot \mathbf{r}} + \sum_{\mathbf{k},\mathbf{k}'} \overline{\mathcal{V}}(\mathbf{k}' - \mathbf{k}) \, e^{i(\mathbf{k}-\mathbf{k}_0) \cdot \mathbf{r}} \, \beta_{\mathbf{k}'} = \overline{\epsilon} f(\mathbf{r}) \tag{5.15}$$

We rewrite the second term on the left-hand side of this equation as:

$$\sum_{\mathbf{k}'} \left[\sum_{\mathbf{k}} \overline{\mathcal{V}}(\mathbf{k}' - \mathbf{k}) \, e^{-i(\mathbf{k}'-\mathbf{k}) \cdot \mathbf{r}} \right] e^{i(\mathbf{k}'-\mathbf{k}_0) \cdot \mathbf{r}} \, \beta_{\mathbf{k}'}$$

In the square bracket we recognize an expression similar to the inverse Fourier transform of $\overline{\mathcal{V}}(\mathbf{q})$, as in Eq. (5.13). In order to have a proper inverse Fourier transform, the sum over $\mathbf{k}$ must extend over all values in the BZ, whereas in our discussion we have restricted $\mathbf{k}$ to the neighborhood of $\mathbf{k}_0$. Since the screened Coulomb potential $\overline{\mathcal{V}}(\mathbf{r})$ due to the presence of the impurity is a smooth function, its Fourier transform $\overline{\mathcal{V}}(\mathbf{k})$ vanishes for large values of the argument $(\mathbf{k} - \mathbf{k}_0)$; based on this observation, we can extend the summation in the above equation to all values of $\mathbf{k}$ with little change in the result of the summation, which leads to:

$$\sum_{\mathbf{k}} \overline{\mathcal{V}}(\mathbf{k}' - \mathbf{k})\, e^{-i(\mathbf{k}'-\mathbf{k})\cdot\mathbf{r}} \approx \sum_{\mathbf{k}\in\mathrm{BZ}} \overline{\mathcal{V}}(\mathbf{k}' - \mathbf{k}) e^{-i(\mathbf{k}'-\mathbf{k})\cdot\mathbf{r}} = \overline{\mathcal{V}}(\mathbf{r})$$

With this result, Eq. (5.15) takes the form

$$\sum_{\mathbf{k}} \epsilon_{\mathbf{k}}^{(n)} \beta_{\mathbf{k}}\, e^{i(\mathbf{k}-\mathbf{k}_0)\cdot\mathbf{r}} + \overline{\mathcal{V}}(\mathbf{r})f(\mathbf{r}) = \overline{\epsilon} f(\mathbf{r}) \tag{5.16}$$

We next use the expansion of Eq. (5.7) for the energy of crystal states near the band extremum, with the help of which the equation for the coefficients $\beta_{\mathbf{k}}$ takes the form

$$\sum_{\mathbf{k}} \sum_{i=1}^{3} \frac{\hbar^2}{2\overline{m}_i^{(n)}} (k_i - k_{0,i})^2\, \beta_{\mathbf{k}}\, e^{i(\mathbf{k}-\mathbf{k}_0)\cdot\mathbf{r}} + \overline{\mathcal{V}}(\mathbf{r})f(\mathbf{r}) = \left[\overline{\epsilon} - \epsilon_{\mathbf{k}_0}^{(n)}\right]f(\mathbf{r})$$

Finally, using the identity

$$(k_i - k_{0,i})^2 e^{i(\mathbf{k}-\mathbf{k}_0)\cdot\mathbf{r}} = -\frac{\partial^2}{\partial x_i^2} e^{i(\mathbf{k}-\mathbf{k}_0)\cdot\mathbf{r}}, \quad i = 1, 2, 3$$

we arrive at the following equation for the function $f(\mathbf{r})$:

$$\left[\sum_{i=1}^{3} -\frac{\hbar^2}{2\overline{m}_i^{(n)}} \frac{\partial^2}{\partial x_i^2} + \overline{\mathcal{V}}(\mathbf{r})\right]f(\mathbf{r}) = \left[\overline{\epsilon} - \epsilon_{\mathbf{k}_0}^{(n)}\right]f(\mathbf{r})$$

This equation is equivalent to a Schrödinger equation for an atom with a nuclear potential $\overline{\mathcal{V}}(\mathbf{r})$, precisely the type of equation we were anticipating. In this case, not only is the effective charge of the nucleus modified by the dielectric constant of the crystal, but the effective mass of the particles $\overline{m}_i^{(n)}$ also bears the signature of the presence of the crystal.

With the impurity potential $\overline{\mathcal{V}}(\mathbf{r})$ given by Eq. (5.8), and assuming for simplicity an average effective mass $\overline{m}^{(n)}$ for all directions i (an isotropic crystal), the equation for the envelope function takes the form

$$\left[-\frac{\hbar^2 \nabla_{\mathbf{r}}^2}{2\overline{m}^{(n)}} - \frac{\overline{Z}e^2}{\overline{\epsilon}_0 r}\right]f(\mathbf{r}) = \left[\overline{\epsilon} - \epsilon_{\mathbf{k}_0}^{(n)}\right]f(\mathbf{r}) \tag{5.17}$$

This is identical to the equation for an electron in an atom with nuclear charge $\overline{Z}e/\overline{\epsilon}_0$. If the charge is negative, that is, the impurity state is a hole, the effective mass is also negative, so the equation is formally the same as that of an electron associated with a positive ion.

The solutions are hydrogen-like wavefunctions and energy eigenvalues, scaled appropriately by the constants that appear in (Eq. (5.17)) relative to those that appear in the free hydrogen atom (typically the impurity atoms have an effective charge of $\overline{Z} = \pm1$). Specifically, the eigenvalues of the extra electron or hole states are given by:

$$\overline{\epsilon}_\nu = \epsilon_{\mathbf{k}_0}^{(n)} - \frac{\overline{m}^{(n)}\overline{Z}^2 e^4}{2\hbar^2 \overline{\epsilon}_0^2} \frac{1}{\nu^2}, \quad \nu = 1, 2, \ldots \tag{5.18}$$

while the corresponding solutions for the envelope function are hydrogen-like wavefunctions with a length scale $\bar{a} = \hbar^2 \bar{\varepsilon}_0 / \overline{m}^{(n)} |\overline{Z}| e^2$; for instance, the wavefunction of the first state ($\nu = 1$) is:

$$f_1(\mathbf{r}) = A_1 e^{-r/\bar{a}}$$

with A_1 a normalization factor. Thus, the energy of impurity-related states is scaled by a factor

$$\frac{\overline{m}^{(n)} (\overline{Z} e^2 / \bar{\varepsilon}_0)^2}{m_e e^4} = \frac{\overline{Z}^2 (\overline{m}^{(n)} / m_e)}{\bar{\varepsilon}_0^2}$$

while the characteristic radius of the wavefunction is scaled by a factor

$$\frac{m_e e^2}{\overline{m}^{(n)} (|\overline{Z}| e^2 / \bar{\varepsilon}_0)} = \frac{\bar{\varepsilon}_0}{|\overline{Z}| (\overline{m}^{(n)} / m_e)}$$

As is evident from Eq. (5.18), the energy of donor electron states relative to the CBM is negative ($\overline{m}^{(n)}$ being positive in this case), while the energy of holes relative to the VBM is positive ($\overline{m}^{(n)}$ being negative in this case); this energy difference, $\left[\bar{\epsilon}_j - \epsilon_{\mathbf{k}_0}^{(n)} \right]$, is referred to as the "binding energy" of the electron or hole related to the presence of the impurity. Typical values of $\overline{m}^{(n)} / m_e$ in semiconductors are of order 10^{-1} (see Table 5.3), and typical values of $\bar{\varepsilon}_0$ are of order 10. Using these values, we find that the binding energy of electrons or holes is scaled by a factor of $\sim 10^{-3}$ (see examples in Table 5.3), while the radius of the wavefunction is scaled by a factor of $\sim 10^2$, relative to the values of the electron energy states and wavefunction radius in a free hydrogen atom. This indicates that in impurity-related states the electrons or holes are loosely bound, both in energy and in wavefunction spread around the impurity atom.

To illustrate how these notions play out in a realistic situation, we show in Fig. 5.24 the calculated wavefunction of the electron state in the case of a P dopant in bulk Si. This is a prototypical system, because P atoms are very close in size to Si atoms, so doping of Si by P causes almost no lattice distortion, and simply provides one extra electron per dopant atom. The wavefunction of this extra state is actually a superposition of conduction band states of bulk Si, and the corresponding binding energy (difference from the CBM) is $E_b = 0.45$ meV (see Table 5.3). This wavefunction is delocalized and spreads over many unit cells around the position of the P atom. For comparison, we also show in the same figure the valence s state of Na, an atom in the same row of the Periodic Table as Si, and since it is an alkali we expect this state to be rather weakly bound (see discussion in Chapter 1). The comparison shows that the dopant wavefunction extends to much larger distances away from the position of the P atom, even compared to the most weakly bound atomic state of an element of similar size as Si.

5.5.2 Effect of Doping in Semiconductors

To appreciate the importance of the results of the preceding discussion, we calculate first the number of conduction electrons or holes that exist in a semiconductor at non-zero

Table 5.3 Effective masses at the CBM and VBM of representative elemental (Si and Ge) and compound semiconductors (GaAs). The effective masses are given in units of the electron mass m_e, and are distinguished in longitudinal $\overline{m}_L$ and transverse $\overline{m}_T$ states. The **k**-points corresponding to the CBM and VBM are given in parentheses; for more details, see Table 5.2. E_b is the binding energy of donor and acceptor states from the band extrema for various impurity atoms (meV). In the case of compound semiconductors, we also indicate which atom is substituted by the impurity. *Source:* D. R. Lide (Ed.), *CRC Handbook of Chemistry and Physics* (CRC Press, Boca Raton, FL, 1999–2000).

		$\overline{m}_L/m_e$	$\overline{m}_T/m_e$	E_b (meV)	Impurity	$\overline{Z}$
				45	P	+1
Si:	CBM (X_2)	0.98	0.19	54	As	+1
				39	Sb	+1
				46	B	−1
	VBM (Γ)	0.52	0.16	67	Al	−1
				72	Ga	−1
				12	P	+1
Ge:	CBM (L)	1.60	0.08	13	As	+1
				10	Sb	+1
				10	B	−1
	VBM (Γ)	0.34	0.04	10	Al	−1
				11	Ga	−1
				6	S/As	+1
				6	Se/As	+1
GaAs:	CBM (Γ)	0.066	0.066	30	Te/As	+1
				6	Si/Ga	+1
				6	Sn/Ga	+1
				28	Be/Ga	−1
				28	Mg/Ga	−1
	VBM (Γ)	0.80	0.12	31	Zn/Ga	−1
				35	Cd/Ga	−1
				35	Si/As	−1

temperature T, in the absence of any dopant impurities. The number of conduction electrons, denoted by $n_c(T)$, will be given by the sum over occupied states with energy above the Fermi level ϵ_F. In a semiconductor, this will be equal to the integral over all conduction states of the density of conduction states $g_c(\epsilon)$ multiplied by the Fermi occupation number at temperature T:

$$n_c(T) = \int_{\epsilon_{CBM}}^{\infty} g_c(\epsilon) \frac{1}{e^{(\epsilon - \epsilon_F)/k_B T} + 1} d\epsilon \tag{5.19}$$

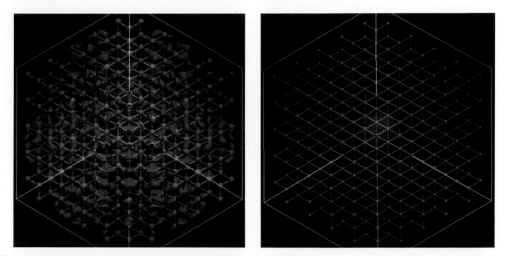

Phosphorus dopant state in bulk silicon. **Left**: The wavefunction of the gap state introduced by a P impurity in bulk Si, calculated in a cubic supercell (a $5 \times 5 \times 5$ multiple of the conventional cubic cell with an eight-atom basis), containing 1000 atoms shown as faint yellow spheres, with the atom at the center replaced by a P atom. **Right**: The wavefunction of the s state of a free Na atom, shown in the same volume as the Si crystal supercell and at the same charge contour level (calculations performed by G. A. Tritsaris).

Similarly, the number of holes, denoted by $p_v(T)$, will be equal to the integral over all valence states of the density of valence states $g_v(\epsilon)$ multiplied by one minus the Fermi occupation number:

$$p_v(T) = \int_{-\infty}^{\epsilon_{\mathrm{VBM}}} g_v(\epsilon) \left[1 - \frac{1}{e^{(\epsilon-\epsilon_F)/k_B T} + 1} \right] d\epsilon = \int_{-\infty}^{\epsilon_{\mathrm{VBM}}} g_v(\epsilon) \frac{1}{e^{(\epsilon_F-\epsilon)/k_B T} + 1} d\epsilon \quad (5.20)$$

In the above equations we have used the traditional symbols n_c for electrons and p_v for holes, which come from the fact that the former represent *negatively* charged carriers (with the subscript c to indicate they are related to conduction bands) and the latter represent *positively* charged carriers (with the subscript v to indicate they are related to valence bands). We have mentioned before (see Chapter 6) that in a perfect semiconductor the Fermi level lies in the middle of the band gap, a statement that can easily be proved from the arguments that follow. For the moment, we will use this fact to justify an approximation for the number of electrons or holes. Since the band gap of semiconductors is of order 1 eV and the temperatures of operation are of order 300 K, that is, $k_B T \sim 1/40$ eV, we can use the approximations

$$\frac{1}{e^{(\epsilon-\epsilon_F)/k_B T} + 1} \approx e^{-(\epsilon-\epsilon_F)/k_B T}, \quad \frac{1}{e^{(\epsilon_F-\epsilon)/k_B T} + 1} \approx e^{-(\epsilon_F-\epsilon)/k_B T}$$

These approximations are very good for the values $\epsilon = \epsilon_{\mathrm{CBM}}$ and $\epsilon = \epsilon_{\mathrm{VBM}}$ in the expressions for the number of electrons and holes, Eqs (5.19) and (5.20), respectively, and become even better for values higher than ϵ_{CBM} or lower than ϵ_{VBM}, so that we can use them for the entire range of values in each integral. This gives:

$$n_c(T) = \bar{n}_c(T)e^{-(\epsilon_{\mathrm{CBM}}-\epsilon_{\mathrm{F}})/k_{\mathrm{B}}T}, \quad \bar{n}_c(T) = \int_{\epsilon_{\mathrm{CBM}}}^{\infty} g_c(\epsilon)e^{-(\epsilon-\epsilon_{\mathrm{CBM}})/k_{\mathrm{B}}T}\,\mathrm{d}\epsilon$$

$$p_v(T) = \bar{p}_v(T)e^{-(\epsilon_{\mathrm{F}}-\epsilon_{\mathrm{VBM}})/k_{\mathrm{B}}T}, \quad \bar{p}_v(T) = \int_{-\infty}^{\epsilon_{\mathrm{VBM}}} g_v(\epsilon)e^{-(\epsilon_{\mathrm{VBM}}-\epsilon)/k_{\mathrm{B}}T}\,\mathrm{d}\epsilon$$

where the quantities $\bar{n}_c(T)$ and $\bar{p}_v(T)$ are obtained by extracting from the integrand a factor independent of the energy ϵ; we will refer to these quantities as the reduced number of electrons or holes. The exponential factors in the integrals for $\bar{n}_c(T)$ and $\bar{p}_v(T)$ fall very fast when the values of ϵ are far from ϵ_{CBM} and ϵ_{VBM}; accordingly, only values of ϵ very near these end points contribute significantly to each integral. We have seen above that near these values we can use $\mathbf{k} \cdot \mathbf{p}$ perturbation theory to express the energy near the CBM and VBM as quadratic expressions in $\mathbf{k}$. Moreover, from our analysis of the DOS near a minimum (such as the CBM) or a maximum (such as the VBM), Eqs (2.136) and (2.137), we can use for $g_c(\epsilon)$ and $g_v(\epsilon)$ expressions which are proportional to $(\epsilon - \epsilon_{\mathrm{CBM}})^{1/2}$ and $(\epsilon_{\mathrm{VBM}} - \epsilon)^{1/2}$, respectively. These considerations make it possible to evaluate explicitly the integrals in the above expressions for $\bar{n}_c(T)$ and $\bar{p}_v(T)$, giving:

$$\bar{n}_c(T) = \frac{1}{4}\left(\frac{2\bar{m}^{(c)}k_{\mathrm{B}}T}{\pi\hbar^2}\right)^{3/2}, \quad \bar{p}_v(T) = \frac{1}{4}\left(\frac{2\bar{m}^{(v)}k_{\mathrm{B}}T}{\pi\hbar^2}\right)^{3/2} \tag{5.21}$$

where $\bar{m}^{(c)}, \bar{m}^{(v)}$ are the appropriate effective masses at the CBM and VBM.

Now we can derive simple expressions for the number of electrons or holes under equilibrium conditions at temperature T in a crystalline semiconductor without dopants. We note that the product of electrons and holes at temperature T is:

$$n_c(T)p_v(T) = \frac{1}{16}\left(\frac{2k_{\mathrm{B}}T}{\pi\hbar^2}\right)^3 (\bar{m}^{(c)}\bar{m}^{(v)})^{3/2}e^{-(\epsilon_{\mathrm{CBM}}-\epsilon_{\mathrm{VBM}})/k_{\mathrm{B}}T}$$

and the two numbers must be equal because the number of electrons thermally excited to the conduction band is the same as the number of holes left behind in the valence band. Therefore, each number is equal to:

$$n_c(T) = p_v(T) = \frac{1}{4}\left(\frac{2k_{\mathrm{B}}T}{\pi\hbar^2}\right)^{3/2} (\bar{m}^{(c)}\bar{m}^{(v)})^{3/4}e^{-\epsilon_{\mathrm{gap}}/2k_{\mathrm{B}}T} \tag{5.22}$$

where $\epsilon_{\mathrm{gap}} = \epsilon_{\mathrm{CBM}} - \epsilon_{\mathrm{VBM}}$. Thus, the number of electrons or holes is proportional to the factor $\exp[-\epsilon_{\mathrm{gap}}/2k_{\mathrm{B}}T]$, which for temperatures of order room temperature is extremely small, the gap being of order 1 eV (11,604 K).

It is now easy to explain why the presence of defects is so crucial for the operation of real electronic devices. This operation depends on the presence of carriers that can be easily excited and made to flow in the direction of an external electric field (or opposite to it, depending on their sign). For electrons and holes which are due to the presence of impurities, we have seen that the binding energy is of order 10^{-3} of the binding energy of electrons in the hydrogen atom, that is, of order a few millielectron-volts (the binding energy of the $1s$ electron in hydrogen is 13.6 eV). The meaning of binding energy here is the amount by which the energy of the donor-related electron state is below the lowest unoccupied crystal state, the CBM, or the amount by which the energy of the acceptor-related electron state is above the highest occupied crystal state, the VBM. Excitation of electrons from the impurity state to a conduction state gives rise to a delocalized state that can carry current; this excitation is much easier than excitation of electrons from the VBM to the CBM across the entire band gap, in the undoped crystal. Thus, the presence of donor impurities in semiconductors makes it possible to have a reasonable number of thermally excited carriers at room temperature. Similarly, excitation of electrons from the top of the valence band into the acceptor-related state leaves behind a delocalized hole state that can carry current.

5.5.3 The p–n Junction

Finally, we will discuss in very broad terms the operation of electronic devices which are based on doped semiconductors (for details, see the books mentioned in Further Reading). The basic feature of such a device is the presence of two parts, one that is doped with donor impurities and has an excess of electrons, that is, it is negatively (n) doped and one that is doped with acceptor impurities and has an excess of holes, that is, it is positively (p) doped. The two parts are in contact, as shown schematically in Fig. 5.25. Because electrons and holes are mobile and can diffuse in the system, some electrons will move from the n-doped side to the p-doped side, leaving behind positively charged donor impurities. Similarly, some holes will diffuse from the p-doped side to the n-doped side, leaving behind negatively charged acceptor impurities. An alternative way of describing this effect is that the electrons which have moved from the n-doped side to the p-doped side are captured by the acceptor impurities, which then lose their holes and become negatively charged; the reverse applies to the motion of holes from the p-doped to the n-doped side. In either case, the carriers that move to the opposite side are no longer mobile. Once enough holes have passed to the n-doped side and enough electrons to the p-doped side, an electric field is set up due to the imbalance of charge, which prohibits further diffusion of electric charges. The potential $\Phi(x)$ corresponding to this electric field is also shown in Fig. 5.25. The region near the interface from which holes and electrons have left to go to the other side, and which is therefore depleted of carriers, is called the "depletion region." This arrangement is called a p–n junction.

The effect of the p–n junction is to rectify the current: when the junction is hooked to an external voltage bias with the plus pole connected to the p-doped side and the minus pole connected to the n-doped side, an arrangement called "forward bias," current flows because the holes are attracted to the negative pole and the electrons are attracted to the positive pole, freeing up the depletion region for additional charges to move into it. In this case, the external potential introduced by the bias voltage counteracts the potential due to the depletion region, as indicated in Fig. 5.25. If, on the other hand, the positive pole of the external voltage is connected to the n-doped region and the negative pole to the p-doped region, an arrangement called "reverse bias," then current cannot flow because the motion of charges would be against the potential barrier. In this case, the external potential introduced by the bias voltage enhances the potential due to the depletion region,

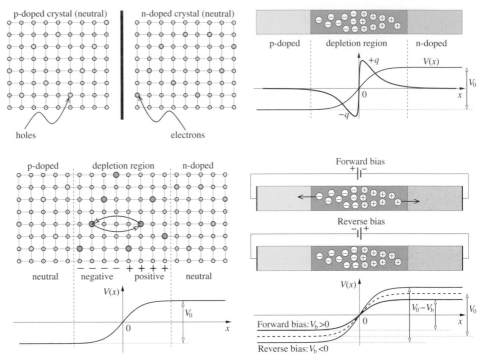

Fig. 5.25 Schematic representation of p–n junction elements. **Left**: The p-doped and n-doped regions before contact with the impurities that produce holes and electrons, and after contact with the charge transfer that eliminates many electrons and holes in the depletion region, giving rise to the intrinsic potential $\mathcal{V}(x)$ across the interface. **Right**: the charge distribution (blue curve labeled $+q$ and red curve labeled $-q$ for positive and negative charges, respectively), the corresponding contact potential $\mathcal{V}(x)$, and its variations in forward or reverse bias conditions, by the application of an external bias potential $\mathcal{V}_\text{b}$.

as indicated in Fig. 5.25. In reality, even in a reverse-biased p–n junction there is a small amount of current that can flow due to thermal generation of carriers in the doped regions; this is called the saturation current. For forward bias, by convention taken as positive applied voltage, the current flow increases with applied voltage, while for reverse bias, by convention taken as negative applied voltage, the current is essentially constant and very small (equal to the saturation current). Thus, the p–n junction preferentially allows current flow in one bias direction, leading to rectification.

The formation of the p–n junction has interesting consequences on the electronic structure. We consider first the situation when the two parts, p-doped and n-doped, are well separated. In this case, the band extrema (VBM and CBM) are at the same position for the two parts, but the Fermi levels are not: the Fermi level in the p-doped part, $\epsilon_\text{F}^{(p)}$, is near the VBM, while that of the n-doped part, $\epsilon_\text{F}^{(n)}$, is near the CBM (these assignments are explained below). When the two parts are brought into contact, the two Fermi levels must be aligned, since charge carriers move across the interface to establish a common Fermi level. When this happens, the bands on the two sides of the interface are distorted to accommodate the common Fermi level and maintain the same relation of the Fermi level to the band extrema on either side far from the interface, as shown in Fig. 5.26. This distortion

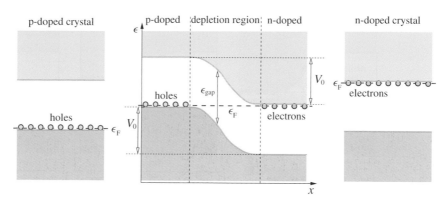

Fig. 5.26 Band bending associated with a p–n junction. The bands of the p-doped and n-doped parts when they are separated (far left and far right), with different Fermi levels $\epsilon_F^{(p)}$ and $\epsilon_F^{(n)}$. The bands when the two sides are brought together (middle), with a common Fermi level ϵ_F; the band bending in going from the p-doped side to the n-doped side is shown, with the energy change due to the contact potential eV_0.

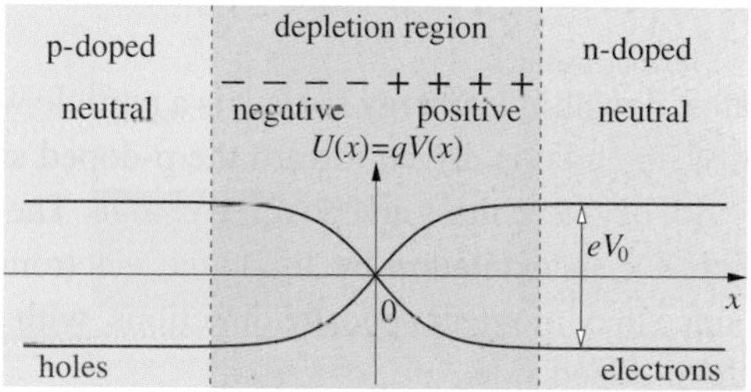

Fig. 5.27 Origin of the intrinsic potential in the p–n junction: the accumulation of negative and positive excess charge in the depletion region produces an electric potential $V(x)$ that results in a potential energy difference $U(x) = qV(x)$ for particles of charge q, across the interface. This potential energy difference is different for positively and negatively charged particles (holes and electrons, respectively).

of the bands is referred to as "band bending." The reason behind the band bending is the presence of the potential $\Phi(x)$ in the depletion region, as shown in Fig. 5.27. In fact, the amount by which the bands are bent upon forming the contact between the p-doped and n-doped regions is exactly equal to the potential difference far from the interface:

$$V_0 = \Phi(+\infty) - \Phi(-\infty)$$

which is called the "contact potential."[7]. The difference in the band extrema on the two sides far from the interface is then equal to eV_0.

The physics of the p–n junction discussed so far is already enough to describe the principle of operation of two important devices, the light-emitting diode (LED) and the photovoltaic device (PVD), as shown in Fig. 5.28. In the PV, a photon is absorbed by an electron in the depletion region, leaving a hole where the electron was. The electrons and holes experience different potential energies, which drive the electron toward the n-doped

---

[7] Sometimes this is also referred to as the "bias potential," but we avoid this term here in order to prevent any confusion with the externally applied bias potential.

since this is the amount by which the position of the Fermi levels in the p-doped and n-doped parts differs before contact.

To determine the lengths over which the depletion region extends into each side, l_p and l_n, respectively, we will consider uniform charge densities ρ_p and ρ_n in the p-doped and n-doped sides. In terms of the dopant concentrations, these charge densities will be given by:

$$\rho_n = eN^{(d)}, \quad \rho_p = -eN^{(a)}$$

assuming that within the depletion region all the dopants have been stripped of their carriers. The assumption of uniform charge densities is rather simplistic, but leads to correct results which are consistent with more realistic assumptions (see Problem 8). We define the direction perpendicular to the interface as the x-axis and take the origin to be at the interface, as indicated in Fig. 5.25. We also define the zero of the potential $\Phi(x)$ to be at the interface, $\Phi(0) = 0$. We then use Poisson's equation, which for this 1D problem gives, in the range $-l_p \leq x \leq l_n$:

$$\frac{d^2\Phi(x)}{dx^2} = -\frac{4\pi}{\bar{\varepsilon}_0}\rho_n, \quad x > 0$$

$$= -\frac{4\pi}{\bar{\varepsilon}_0}\rho_p, \quad x < 0$$

where $\bar{\varepsilon}_0$ is the static dielectric constant of the material. Integrating once and requiring that $d\Phi/dx$ vanishes at $x = +l_n$ and at $x = -l_p$, the edges of the depletion region where the potential has reached its asymptotic value and becomes constant, we find:

$$\frac{d\Phi(x)}{dx} = -\frac{4\pi}{\bar{\varepsilon}_0}\rho_n(x - l_n), \quad x > 0$$

$$= -\frac{4\pi}{\bar{\varepsilon}_0}\rho_p(x + l_p), \quad x < 0$$

The derivative of the potential, which is related to the electric field, must be continuous at the interface since there is no charge build up there and hence no discontinuity in the electric field (see Appendix B). This condition gives:

$$N^{(d)}l_n = N^{(a)}l_p \tag{5.23}$$

where we have also used the relation between the charge densities and the dopant concentrations mentioned above. Integrating the Poisson equation once again, and requiring the potential to vanish at the interface, leads to:

$$\Phi(x) = -\frac{2\pi}{\bar{\varepsilon}_0}\rho_n(x - l_n)^2 + \frac{2\pi}{\bar{\varepsilon}_0}\rho_n l_n^2, \quad x > 0$$

$$= -\frac{2\pi}{\bar{\varepsilon}_0}\rho_p(x + l_p)^2 + \frac{2\pi}{\bar{\varepsilon}_0}\rho_p l_p^2, \quad x < 0$$

for x in the range $-l_p \leq x \leq l_n$. From this expression we can calculate the contact potential as:

$$V_0 = \Phi(l_n) - \Phi(-l_p) = \frac{2\pi e}{\bar{\varepsilon}_0}\left[N^{(d)}l_n^2 + N^{(a)}l_p^2\right]$$

and using the relation of Eq. (5.23), we can solve for l_n and l_p:

$$l_n = \left[\left(\frac{\bar{\varepsilon}_0 V_0}{2\pi e}\right)\frac{N^{(a)}}{N^{(d)}}\frac{1}{N^{(a)} + N^{(d)}}\right]^{1/2}, \quad l_p = \left[\left(\frac{\bar{\varepsilon}_0 V_0}{2\pi e}\right)\frac{N^{(d)}}{N^{(a)}}\frac{1}{N^{(a)} + N^{(d)}}\right]^{1/2}$$

From these expressions we find that the total size of the depletion layer l_D is given by:

$$l_D = l_p + l_n = \left[\left(\frac{\bar{\varepsilon}_0 V_0}{2\pi e}\right)\frac{N^{(a)} + N^{(d)}}{N^{(a)}N^{(d)}}\right]^{1/2} \tag{5.24}$$

In the limiting cases when one of the two dopant concentrations dominates, we have:

$$N^{(a)} \gg N^{(d)} \Rightarrow l_D = \left(\frac{\bar{\varepsilon}_0 V_0}{2\pi e}\right)^{1/2}\sqrt{N^{(d)}}$$

$$N^{(d)} \gg N^{(a)} \Rightarrow l_D = \left(\frac{\bar{\varepsilon}_0 V_0}{2\pi e}\right)^{1/2}\sqrt{N^{(a)}}$$

that is, the size of the depletion region is determined by the *lowest* dopant concentration.

Up to this point we have been discussing electronic features of semiconductor junctions in which the two doped parts consist of the same material. It is also possible to create p–n junctions in which the two parts consist of different semiconducting materials; these are called "heterojunctions." In these situations the band gap and position of the band extrema are different on each side of the junction. For doped semiconductors, the Fermi levels on the two sides of a heterojunction will also be at different positions before contact. Two typical situations are shown in Fig. 5.29: in both cases the material on the left has a smaller band gap than the material on the right (this could represent, for example, a junction between GaAs on the left and $Al_xGa_{1-x}As$ on the right). When the Fermi levels of the two sides are aligned upon forming the contact, the bands are bent as usual to accommodate the common Fermi level. However, in these situations the contact potential is not the same for the electrons (conduction states) and the holes (valence states). As indicated in Fig. 5.29, in the case of a heterojunction with p-doping in the small-gap material and n-doping in the large-gap material, the contact potential for the conduction and valence states will be given by:

$$eV_0^{(c)} = \Delta\epsilon_F - \Delta\epsilon_{CBM}, \quad eV_0^{(v)} = \Delta\epsilon_F + \Delta\epsilon_{VBM}$$

whereas in the reverse case of n-doping in the small-gap material and p-doping in the large-gap material, the two contact potentials will be given by:

$$eV_0^{(c)} = \Delta\epsilon_F + \Delta\epsilon_{CBM}, \quad eV_0^{(v)} = \Delta\epsilon_F - \Delta\epsilon_{VBM}$$

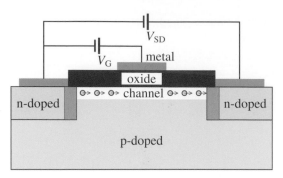

Fig. 5.30 The basic features of a MOSFET: the source and drain, both n-doped regions, buried in a larger
p-doped region and connected through two metal electrodes and an external voltage. The metal
electrodes are separated by the oxide layer. Two additional metal electrodes, the gate and body, are
attached to the oxide layer and to the bottom of the p-doped layer and are connected through the
bias voltage. The conducting channel is between the two n-doped regions.

where in both cases $\Delta\epsilon_F$ is the difference in Fermi level positions and $\Delta\epsilon_{CBM}$, $\Delta\epsilon_{VBM}$
are the differences in the positions of the band extrema before contact. It is evident from
Fig. 5.29 that in both situations there are discontinuities in the potential across the junction
due to the different band gaps on the two sides. Another interesting and very important
feature is the presence of energy wells due to these discontinuities; such a well for electron
states is created on the p-doped side in the first case and a similar well for hole states is
created on the n-doped side in the second case. There is a discrete set of quantized energy
levels associated with these wells in the x-direction, so if charge carriers are placed in the
wells they will be localized in these discrete states and form a 2D electron gas on a plane
parallel to the interface. Indeed, it is possible to populate these wells with carriers from
additional dopant atoms far from the interface, or by properly biasing the junction. The
2D gas of carriers can then be subjected to external magnetic fields, giving rise to very
interesting quantum behavior.

We discuss next three important applications of the p–n junction. The first is the
electronic transistor, the basic element of modern computer chips. In this and all other real
devices, the arrangement of n-doped and p-doped regions is more complicated. The basic
design for a transistor is the so-called metal–oxide–semiconductor field-effect transistor
(MOSFET). This allows the operation of a rectifying channel with very little loss of power.
A MOSFET is illustrated in Fig. 5.30: there are two n-doped regions buried in a larger p-
doped region. The two n-doped regions act as source (S) and drain (D) of electrons. An
external voltage is applied between the two n-doped regions with the two opposite poles
attached to the source and drain through two metal electrodes which are separated by an
insulating oxide layer. A different bias voltage is connected to an electrode placed at the
bottom of the p-doped layer, called the body (B), and to another electrode placed above the
insulating oxide layer, called the gate (G). When a sufficiently large bias voltage is applied
across the body and the gate electrodes, the holes in a region of the p-doped material
below the gate are repelled, leaving a channel through which the electrons can travel from
the source to the drain. The advantage of this arrangement is that no current flows between
the body and the gate, even though it is this pair of electrodes to which the bias voltage is

applied. Instead, the current flow is between the source and the drain, which takes much less power to maintain, with correspondingly lower generation of heat. In modern devices there are several layers of this and more complicated arrangements of p-doped and n-doped regions interconnected by complex patterns of metal wires.

The two other main applications of the p–n junction are the LED and the PVD, both illustrated in Fig. 5.28. In an LED, an electron and a hole are induced to move in opposite directions by the application of external forward bias to the p–n junction. When they collide somewhere in the depletion region they annihilate each other: the electron moves to the lower-energy state occupied by the hole, in a sense filling the hole, and the hole disappears, but this process of annihilation needs to conserve energy, which is accomplished by the emission of a photon (quantum of light). As long as the external field is being applied to the p–n junction in this forward-bias mode, electrons and holes can keep meeting and annihilating each other, producing light. However, light emission does not always take place in a p–n junction under external forward bias. The bias has to be strong enough to make it possible for holes and electrons to meet, so they can annihilate and emit light.

The exact opposite of this process is the basis for the operation of photovoltaic devices: when an electron in the depletion region is excited by absorbing a photon to a state of higher energy, it leaves behind a hole which leads to the creation of an electron–hole pair. However, due to the presence of the contact potential in the depletion region, the electron and the hole will move in opposite directions if the electron–hole annihilation (which would emit back the photon) is prevented. The electron, being a negative charge, wants to move toward higher values of the electric potential and thus will move toward the n-doped region. The hole, on the other hand, being a positively charged particle, wants to move toward lower values of the electric potential and thus will move toward the p-doped side. Therefore, the electron–hole pair will split up, with electrons moving in one direction and holes moving in the opposite direction. Both of these motions contribute to electrical current moving through the device, which can be used to do useful work. The crucial point is that the presence of the electric field which has built up in the depletion region automatically leads to electron–hole separation and therefore to electrical current, once the electron has been excited by the absorption of light. As a net result, the absorption of light leads to electrical current.

5.5.4 Metal–Semiconductor Junction

The connection between the metal electrodes and the semiconductor is of equal importance to the p–n junction for the operation of an electronic device. In particular, this connection affects the energy and occupation of electronic states on the semiconductor side, giving rise to effective barriers for electron transfer between the two sides. A particular realization of this behavior is shown schematically in Fig. 5.31: when the metal and semiconductor are separated, each has a well-defined Fermi level denoted by $\epsilon_F^{(m)}$ for the metal and $\epsilon_F^{(s)}$ for the semiconductor. For an n-doped semiconductor, the Fermi level lies close to the conduction band, as discussed above. The energy difference between the vacuum level, which is common to both systems, and the Fermi level is defined as the work function (the

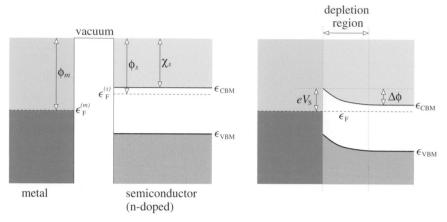

Fig. 5.31 Band alignment in metal–semiconductor junction: ϕ_m, ϕ_s are the work functions of the metal and the semiconductor; χ_s is the electron affinity; ϵ_{VBM} and ϵ_{CBM} are the semiconductor top of the valence band and bottom of the conduction band; $\epsilon_F^{(s)}, \epsilon_F^{(m)}$ represent the Fermi-level positions in the semiconductor and the metal; $\Delta\phi = \phi_m - \phi_s$ is the shift in work function; and $\mathcal{V}_S$ is the potential (Schottky) barrier.

energy cost to remove electrons from the system); it is denoted by ϕ_m for the metal and ϕ_s for the semiconductor. The energy difference between the conduction band minimum (ϵ_{CBM}) and the vacuum level, denoted by χ_s, is called the electron affinity.

When the metal and the semiconductor are brought into contact, the Fermi level on the two sides becomes aligned. This is achieved by transferring electrons from one side to the other, depending on the relative positions of the Fermi level. For the case illustrated in Fig. 5.31, when the two Fermi levels are aligned electrons have moved from the semiconductor (which originally had a higher Fermi level) to the metal. This creates a layer near the interface which has fewer electrons than usual on the semiconductor side, producing a "charge depletion" region on that side. Far from the interface, where the effect of the depletion region has died out, the relation of the semiconductor band edges (CBM and VBM) to the Fermi level must be the same as before the contact is formed. As a consequence of the charge transfer, the semiconductor bands are shifted down by an amount equal to

$$\Delta\phi = \phi_m - \phi_s$$

far from the interface. To achieve this, the electron energy bands of the semiconductor must bend, just like in the case of the p–n junction, since at the interface they must maintain their original relation to the metal bands. The presence of the depletion region makes it more difficult for electrons to flow across the interface. The depletion region and the corresponding bending of the semiconductor bands lead to a potential barrier $\mathcal{V}_S$, called the Schottky barrier.[8] In the case of a junction between a metal and an n-type semiconductor,

---

[8] W. Schottky, *Naturwissenschaften* **26**, 843 (1938).

the Schottky barrier is determined by the amount of extra energy that an electron needs in order to pass from the metal side to the semiconductor side: this barrier is given by

$$eV_S^{(n)} = \phi_m - \chi_s$$

as is evident from Fig. 5.31. In the case of a junction between a metal and a p-type semiconductor, the band bending is in the opposite sense and the corresponding Schottky barrier is given by

$$eV_S^{(p)} = \epsilon_{gap} - (\phi_m - \chi_s)$$

Combining the two expressions for the metal/n-type and metal/p-type semiconductor contacts, we obtain:

$$e\left[V_S^{(p)} + V_S^{(n)}\right] = \epsilon_{gap}$$

Two features of this picture of metal–semiconductor contact are worth emphasizing. First, it assumes there are no changes in the electronic structure of the metal or the semiconductor due to the presence of the interface between them, other than the band bending which comes from equilibrating the Fermi levels on both sides. Second, the Schottky barrier is proportional to the work function of the metal. Neither of these features is very realistic. The interface can induce dramatic changes in the electronic structure which alter the simple picture described above. Moreover, experiments indicate that measured Schottky barriers are indeed roughly proportional to the metal work function for large-gap semiconductors (ZnSe, ZnS), but they tend to be almost independent of the metal work function for small-gap semiconductors (Si, GaAs).[9]

The situation is further complicated by the presence of the insulating oxide layer between the metal and the semiconductor. Band bending occurs in this case as well, as shown for instance in Fig. 5.32 for a p-doped semiconductor. The interesting new feature is that the oxide layer can support an externally applied bias voltage, which we will refer to as the gate voltage V_G (see also Fig. 5.30). The gate voltage moves the electronic states on the metal side down in energy by eV_G relative to the common Fermi level. This produces additional band bending, which lowers both the valence and the conduction bands of the semiconductor in the immediate neighborhood of the interface. When the energy difference eV_G is sufficiently large it can produce an "inversion region," that is, a narrow layer near the interface between the semiconductor and the oxide where the bands have been bent to point where some conduction states of the semiconductor have moved below the Fermi level. When these states are occupied by electrons, current can flow from the source to the drain in the MOSFET. Under the right conditions, the confining potential created by the distorted bands can support only one occupied level below the Fermi level. In this situation, the electrons in the occupied semiconductor conduction bands form a 2D system of charge carriers confined in the inversion layer. In such systems, interesting phenomena which are particular to two dimensions, like the quantum Hall effect (integer and fractional) can be observed (see Chapter 9).

---

[9] S. G. Louie, J. Chelikowsky, and M. L. Cohen, *Phys. Rev. B* **15**, 2154 (1977).

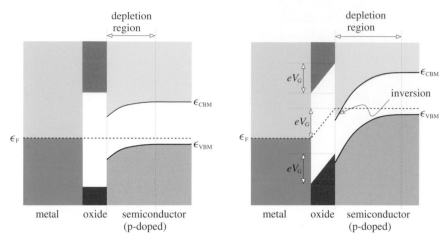

Fig. 5.32 Band alignment in metal–oxide–semiconductor junction, for a p-doped semiconductor. V_G is the gate (bias) voltage, which lowers the energy of electronic states in the metal by eV_G relative to the common Fermi level. Compare the band energy in the inversion region to the confining potential in Fig. 9.1.

Further Reading

1. *Bonds and Bands in Semiconductors*, J. C. Phillips (Academic Press, New York, 1973). This is a classic work, containing a wealth of physical insight into the origin and nature of energy bands in semiconductors.
2. *Electronic Structure and Optical Properties of Semiconductors*, M. L. Cohen and J. R. Chelikowsky (Springer-Verlag, Berlin, 1988). This monograph contains a wealth of examples of band structures and charge densities for many types of semiconductors.
3. *Physics of Semiconductor Devices*, S. M. Sze (Wiley, New York, 1981). This is a standard reference with extensive discussion of all aspects of semiconductor physics from the point of view of application in electronic devices.
4. *The Physics of Semiconductors*, K. F. Brennan (Cambridge University Press, Cambridge, 1999). This is a modern account of the physics of semiconductors, with extensive discussion of the basic methods for studying solids in general.
5. *Calculated Electronic Properties of Metals*, V. L. Moruzzi, J. F. Janak, and A. R. Williams (Pergamon Press, New York, 1978).
6. "Solitons in conducting polymers," A. J. Heeger, S. Kivelson, J. R. Schrieffer, and W.-P. Su, *Rev. Mod. Phys.* **60**, 781 (1988).

Problems

1.   The cis-polyacetylene structure (Fig. 5.33) is different from the trans-polyacetylene case discussed in Section 5.1. Explain why the two different arrangements of the

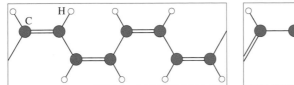 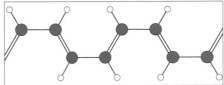

Fig. 5.33   Illustration of the structure of cis-polyacetylene, $[-C_4H_4-]_n$, in the two possible arrangements of the singe/double bond sequence along nearest-neighbor C atoms.

single/double bonds between nearest-neighbor C atoms are not equivalent in the cis-polyacetylene case, in contrast to the trans-polyacetylene case. Calculate the energy bands of this structure using a simple model with one p_z orbital per C site, with first-neighbor single-bond and double-bond hopping matrix elements (t_1, t_2) as in the case for trans-polyacetylene. The model can be made more realistic by the inclusion of second-neighbor hopping matrix elements, t_3, which can be chosen by scaling as d^{-4} relative to t_0, the average of t_1 and t_2, where d is the distance between C atoms. For simplicity, take all bond angles to be $2\pi/3$, and the bond lengths of single and double C–C bonds to be $b_1 = 1.54$ Å and $b_2 = 1.33$ Å, respectively. Describe the nature of the wavefunctions at the center ($k = 0$) and the boundary ($k = \pi/a$) of the BZ.

2. Consider a single layer of graphene, as described in Chapter 2.

    (a) Take a basis for each atom which consists of four orbitals, s, p_x, p_y, p_z. Determine the hamiltonian matrix for this system at each high-symmetry point of the IBZ, with nearest-neighbor interactions in the TBA, assuming an orthogonal overlap matrix. Use proper combinations of the atomic orbitals to take advantage of the symmetry of the problem (see also Chapter 1).

    (b) Choose the parameters that enter into the hamiltonian using arguments similar to those for the 2D square lattice, scaled appropriately to reflect the interactions between carbon orbitals. Compare with the schematic diagram of Fig. 1.18.

    (c) Show that the π bands between the p_z orbitals at nearest-neighbor sites can be described reasonably well by a model consisting of a single orbital per atom, with on-site energy ϵ_0 and nearest-neighbor hamiltonian matrix element t, which yields the expression of Eq. (2.87). Using this expression, calculate the crystal velocity of electrons at the Dirac points, given in terms of the parameter t. Compare with experimental results and comment on the value of this velocity.

    (d) Calculate the DOS for the 2D graphene lattice using the TBA. Identify the peaks that come from the σ bands and π bands.

3. Using the conventions of Problem 2 for the band structure of graphene, show that the π bands of the carbon nanotubes can be approximated by the following expressions.

(a) For the $(n, 0)$ zigzag tubes:

$$\epsilon_l^{(n,0)}(k) = \epsilon_0 \pm t \left[1 \pm 4\cos\left(\frac{\sqrt{3}a}{2}k\right)\cos\left(\frac{l\pi}{n}\right) + 4\cos^2\left(\frac{l\pi}{n}\right) \right]^{1/2}$$

with $-\pi/\sqrt{3}a < k < \pi/\sqrt{3}a$, where $l = 1, \ldots, n$.

(b) For the (n, n) armchair tubes:

$$\epsilon_l^{(n,n)}(k) = \epsilon_0 \pm t \left[1 \pm 4\cos\left(\frac{a}{2}k\right)\cos\left(\frac{l\pi}{n}\right) + 4\cos^2\left(\frac{a}{2}k\right) \right]^{1/2}$$

with $-\pi/a < k < \pi/a$, where $l = 1, \ldots, n$.

(c) For the (n, m) chiral tubes:

$$\epsilon_l^{(n,m)}(k) = \epsilon_0 \pm t \left[1 \pm 4\cos\left(\frac{l\pi}{n} - \frac{mka}{2n}\right)\cos\left(\frac{a}{2}k\right) + 4\cos^2\left(\frac{a}{2}k\right) \right]^{1/2}$$

with $-\pi/a < k < \pi/a$, where l is an integer determined by the condition

$$\sqrt{3}nk_x a + mk_y a = 2\pi l$$

and we have set $k = k_y$.

Compare these approximate bands to the band structure obtained from the tight-binding calculation for the tubes (6,0), (6,6).

4. Describe the basic repeat unit on the graphene plane of the chiral tubes with $(n, m) = (4, 1), (4, 2), (4, 3)$. Determine the BZ of the chiral $(n, m) = (4, 1), (4, 2), (4, 3)$ tubes using the zone-folding scheme described in relation to Fig. 5.13. Comment on why the tube for which the relation $2n + m = 3l$ (l an integer) holds must exhibit metallic character. Calculate the band structure of the (4,1) tube using the results of the previous problem and compare it to the tight-binding band structure shown in Fig. 5.14 (note that the σ bands will be missing from the approximate description that involves only the p_z orbitals).

5. Prove the expressions for the reduced number of electrons $\bar{n}_c(T)$ or holes $\bar{p}_v(T)$ given by Eq. (5.21).

6. Using the fact that the number of electrons in the conduction band is equal to the number of holes in the valence band for an intrinsic semiconductor (containing no dopants), show that in the zero-temperature limit the Fermi level lies exactly at the middle of the band gap. Show that this result also holds for finite temperature if the densities of states at the VBM and CBM are the same.

7. Calculate the number of available carriers at room temperature in undoped Si and compare it to the number of carriers when it is doped with P donor impurities at a concentration of 10^{16} cm^{-3} or with As donor impurities at a concentration of 10^{18} cm^{-3}.

8. We will analyze the potential at a p–n junction employing a more realistic set of charge distributions than the uniform distributions assumed in the text. Our starting

point is the following expressions for the charge distributions in the n-doped and p-doped regions:

$$\rho_n(x) = \tanh\left(\frac{x}{l_n}\right)\left[1 - \tanh^2\left(\frac{x}{l_n}\right)\right] eN^{(d)}, \quad x > 0$$

$$\rho_p(x) = \tanh\left(\frac{x}{l_p}\right)\left[1 - \tanh^2\left(\frac{x}{l_p}\right)\right] eN^{(a)}, \quad x < 0$$

(a) Plot these functions and show that they correspond to smooth distributions with no charge build-up at the interface, $x = 0$.

(b) Integrate Poisson's equation once to obtain the derivative of the potential $d\Phi/dx$ and determine the constants of integration by physical considerations. Show that from this result the relation of Eq. (5.23) follows.

(c) Integrate Poisson's equation again to obtain the potential $\Phi(x)$ and determine the constants of integration by physical considerations. Calculate the contact potential by setting a reasonable cutoff for the asymptotic values, for example, the point at which 99% of the charge distribution is included on either side. From this, derive expressions for the total size of the depletion region $l_D = l_p + l_n$, analogous to Eq. (5.24).

9. Describe in detail the nature of band bending at a metal–semiconductor interface for all possible situations: there are four possible cases, depending on whether the semiconductor is n-doped or p-doped and on whether $\epsilon_F^{(s)} > \epsilon_F^{(m)}$ or $\epsilon_F^{(s)} < \epsilon_F^{(m)}$; of these only one was discussed in the text, as shown in Fig. 5.31.

6

Electronic Excitations

Up to this point we have been dealing with the ground-state properties of electrons in solids. Even in the case of doped semiconductors, the presence of extra electrons or of holes relative to the undoped ideal crystal was due to the introduction of additional electrons (or removal of some electrons) as a result of the presence of impurities, with the additional charges still corresponding to the ground state of the solid. Some of the most important applications of materials result from exciting electrons out of the ground state. These include the optical properties of solids and the dielectric behavior (shielding of external electric fields). These phenomena are also some of the more interesting physical processes that can take place in solids when they interact with external electromagnetic fields. We turn our attention to these issues next.

The band structure of the solid can elucidate the way in which the electrons will respond to external perturbations, such as absorption or emission of light. This response is directly related to the optical and electrical properties of the solid. For example, using the band structure one can determine the possible optical excitations which in turn determine the color, reflectivity, and dielectric response of the solid. When electrons are excited from occupied to unoccupied states in a solid with a band gap, the excited electron and the empty state left behind (the hole) can be bound by their Coulomb attraction. This electron–hole bound state is called an "exciton" and, depending on the amount of binding, can have important consequences. We study these properties here, building on the formalism that we developed for the electronic structure of solids.

6.1 Optical Excitations

We consider first what happens when an electron in a crystalline solid absorbs a photon and jumps to an excited state. The transition rate for such an excitation is given by Fermi's golden rule (see Appendix C, Eq. (C.64)):

$$P_{i \to f}(\omega) = \frac{2\pi}{\hbar} \left| \langle \Psi_{\mathbf{K}'}^{(f)} | \hat{\mathcal{H}}^{\text{int}} | \Psi_{\mathbf{K}}^{(i)} \rangle \right|^2 \delta(E_{\mathbf{K}'}^{(f)} - E_{\mathbf{K}}^{(i)} - \hbar\omega)$$

where $\langle f | = \langle \Psi_{\mathbf{K}'}^{(f)} |, |i\rangle = |\Psi_{\mathbf{K}}^{(i)}\rangle$ are the final and initial states of the many-body system with the corresponding energy eigenvalues $E_{\mathbf{K}'}^{(f)}, E_{\mathbf{K}}^{(i)}$ and $\hat{\mathcal{H}}^{\text{int}}(\mathbf{r}, t)$ is the interaction hamiltonian that describes the interaction of the electron with the electromagnetic field. The change in the many-body system energy when an electron is excited from an occupied single-particle state to an unoccupied single-particle state involves expectation values of the appropriate interaction hamiltonian in these single-particle states. We will ignore for the moment the Coulomb interaction between the excited electron and the hole left behind, which can lead to interesting behavior that we explore in the next section. We can then express all the matrix elements and energy conservation rules in terms of the single-particle wavefunctions and eigenvalues involved in the excitation. In the following, we work with these quantities directly. Specifically, we will use the single-particle state $\langle \psi_{\mathbf{k}'}^{(n')}|$ instead of the many-body state $\langle \Psi_{\mathbf{K}'}^{(f)}|$ to represent the electron, and the single-particle state $|\psi_{\mathbf{k}}^{(n)}\rangle$ instead of the many-body state $|\Psi_{\mathbf{K}}^{(i)}\rangle$ to represent the hole for absorption of a photon (or the reverse for emission), and their corresponding eigenvalues $\epsilon_{\mathbf{k}'}^{(n')}, \epsilon_{\mathbf{k}}^{(n)}$ instead of $E_{\mathbf{K}'}^{(f)}, E_{\mathbf{K}}^{(i)}$ to ensure energy conservation. With these expressions the probability for absorption of a light quantum of frequency ω then becomes:

$$P_{i \to f}(\omega) = \frac{2\pi}{\hbar} \sum_{\mathbf{k},\mathbf{k}',n,n'} \left| \langle \psi_{\mathbf{k}'}^{(n')} | \hat{\mathcal{H}}^{\text{int}} | \psi_{\mathbf{k}}^{(n)} \rangle \right|^2 \delta(\epsilon_{\mathbf{k}'}^{(n')} - \epsilon_{\mathbf{k}}^{(n)} - \hbar\omega) \tag{6.1}$$

In order to have non-vanishing matrix elements in the above expression, the initial single-particle state $|\psi_{\mathbf{k}}^{(n)}\rangle$ must be occupied (a valence state), while the final state $\langle \psi_{\mathbf{k}'}^{(n')}|$ must be unoccupied (conduction state) and their energy difference $\epsilon_{\mathbf{k}'}^{(n')} - \epsilon_{\mathbf{k}}^{(n)}$ must be equal to $\hbar\omega$, where ω is the frequency of the radiation. Note that since the expression is symmetric in the indices $\mathbf{k}, \mathbf{k}'$ and n, n', there are two possibilities for assigning the valence and conduction states, that is, $\epsilon_{\mathbf{k}'}^{(n')} - \epsilon_{\mathbf{k}}^{(n)} > 0$ or $\epsilon_{\mathbf{k}'}^{(n')} - \epsilon_{\mathbf{k}}^{(n)} < 0$, which correspond to positive or negative values of the frequency ω. The physical meaning of this is that one case $\omega > 0$ describes the absorption of radiation and the other case $\omega < 0$ the emission of radiation of frequency ω. In the following we will develop expressions that describe the response of the solid to the absorption or emission of light of frequency ω in terms of quantities that can be measured experimentally, that is, the conductivity and the dielectric function. As expected from the above discussion, these quantities will involve the single-particle states that capture the electronic structure of the solid and energy-conserving δ-functions. We also expect that the response of the solid will be most pronounced for frequencies ω that resonate with energy differences between occupied and unoccupied states, $\epsilon_{\mathbf{k}'}^{(n')} - \epsilon_{\mathbf{k}}^{(n)} = \hbar\omega$.

The relevant interaction term $\hat{\mathcal{H}}^{\text{int}}(\mathbf{r}, t)$ is that which describes absorption or emission of photons, the carriers of the electromagnetic field:

$$\hat{\mathcal{H}}^{\text{int}}(\mathbf{r}, t) = \frac{e}{m_e c} \mathbf{A}(\mathbf{r}, t) \cdot \mathbf{p} \tag{6.2}$$

with $\mathbf{A}(\mathbf{r}, t)$ the vector potential of the electromagnetic radiation field and $\mathbf{p} = (\hbar/i)\nabla_{\mathbf{r}}$ the momentum operator, as discussed in more detail in Appendix C. To understand better what is involved in the photon absorption/emission processes, we will use the relations

$$[\hat{\mathcal{H}}_0, \mathbf{r}] = -\frac{i\hbar}{m_e} \mathbf{p}, \quad \hat{\mathcal{H}}_0 |\psi_{\mathbf{k}}^{(n)}\rangle = \epsilon_{\mathbf{k}}^{(n)} |\psi_{\mathbf{k}}^{(n)}\rangle \tag{6.3}$$

to rewrite the matrix elements of $\hat{\mathcal{H}}^{\text{int}}(\mathbf{r}, t)$ in Eq. (6.1) as:

$$\langle \psi_{\mathbf{k}'}^{(n')} | \hat{\mathcal{H}}^{\text{int}} | \psi_{\mathbf{k}}^{(n)} \rangle = i \frac{e}{\hbar c} \left(\epsilon_{\mathbf{k}'}^{(n')} - \epsilon_{\mathbf{k}}^{(n)} \right) \mathbf{A}(t) \cdot \langle \psi_{\mathbf{k}'}^{(n')} | \mathbf{r} | \psi_{\mathbf{k}}^{(n)} \rangle$$

which gives for the absorption probability of Eq. (6.1):

$$P_{i \to f}(\omega) = \frac{2\pi}{\hbar} \frac{\omega^2}{c^2} \sum_{\mathbf{k},\mathbf{k}',n,n'} \left| \mathbf{A}(t) \cdot \langle \psi_{\mathbf{k}'}^{(n')} | (e\mathbf{r}) | \psi_{\mathbf{k}}^{(n)} \rangle \right|^2 \delta(\epsilon_{\mathbf{k}'}^{(n')} - \epsilon_{\mathbf{k}}^{(n)} - \hbar\omega) \tag{6.4}$$

This shows that the relevant quantity at the microscopic level is the expectation value of the dipole moment, expressed as the matrix elements of the operator $\hat{\mathbf{d}} = e\mathbf{r}$ between intital and final states of the electons that are excited by the interaction with light.

What is actually measured experimentally is the response of a physical system, such as a molecule or a solid, to the external field, assuming that the field produces a relatively small change to the state of the system, so that the response is linear in the field. For an electromagnetic field with electric component $\mathbf{E}(\omega)$, the linear response is described by the polarizability $\underline{\alpha}(\omega)$, which relates the dipole moment $\mathbf{d}(\omega)$ to the applied field:

$$\mathbf{d}(\omega) = \underline{\alpha}(\omega) \cdot \mathbf{E}(\omega) \Rightarrow d_i(\omega) = \sum_j \alpha_{ij}(\omega) E_j(\omega)$$

In the last expression we have written out the polarizability as a tensor, with d_i, E_i the components of the dipole and the field vectors. By using a field with components only in one direction at a time, we can pick up the diagonal components $\alpha_{ii}, i = 1, 2, 3$ of the polarizability, and deal with the average polarizability given by:

$$\alpha(\omega) = \frac{1}{3} \text{Tr}\{\alpha_{ij}(\omega)\} = \frac{1}{3} \sum_i \alpha_{ii}(\omega)$$

which is independent of direction because the trace of the tensor is invariant under rotations. To measure the response of the system, as embodied in $\alpha(\omega)$, we can excite the system by applying fields of strength E_0 and fequency ω, and measure the probability of absorption of this radiation field, which experimentally is determined by the photoabsorption cross-section $\sigma(\omega)$:[1]

$$\sigma(\omega) \sim \frac{2m_e}{\pi e^2 \hbar} \omega \text{Im}[\alpha(\omega)]$$

This quantity will have sharp peaks when the frequency ω of the radiation field coincides with a resonance of the physical system.

Alternatively, we can excite the system with a field that is a step-function in time,[2] $E(t) = E_0 \theta(t)$, where $\theta(t)$ is the Heaviside function (see Appendix A). Because this function changes very abruptly at $t = 0$, its Fourier transform contains all frequencies:

$$E(\omega) = \int_{-\infty}^{\infty} E(t) e^{i\omega t} \, dt = E_0 \left(\pi \delta(\omega) - i\frac{1}{\omega} \right)$$

With this expression for the field, the imaginary part of the polarizability becomes:

$$\text{Im}[\alpha(\omega)] = \omega \frac{\text{Re}[d(\omega)]}{E_0}$$

---

[1] See, for example, H. A. Bethe and E. E. Salpeter, *Quantum Mechanics of One- and Two-Electron Atoms and Molecules* (Springer-Verlag, Berlin, 1957).
[2] A. Tsolakidis, D. Sanchez-Portal, and R. M. Martin, *Phys. Rev. B* **66**, 235416 (2002).

This last expression gives us a recipe for determining the photoabsorption cross-section, through the Fourer transform $\mathbf{d}(\omega)$ of the time-dependent dipole moment $\mathbf{d}(t)$ defined by:

$$\mathbf{d}(t) = \int \mathbf{r}\, n(\mathbf{r}, t)\mathrm{d}\mathbf{r} = \sum_{\mathbf{k}, n, \epsilon_{\mathbf{k}}^{(n)} < \epsilon_F} \langle \psi_{\mathbf{k}}^{(n)}(\mathbf{r}, t) | \mathbf{r} | \psi_{\mathbf{k}}^{(n)}(\mathbf{r}, t) \rangle$$

In this expression we have a similar physical quantity as in Eq. (6.4), that is, matrix elements of the dipole operator between microscopic states (we have omitted here the charge of the particles for simplicity). The main difference is that the matrix elements are now between occupied states ($\epsilon_{\mathbf{k}}^{(n)} < \epsilon_F$), rather than between pairs of occupied and unoccupied states, as they were in Eq. (6.4). The excitation in this case is reflected by the fact that the microscopic states are time dependent, and are solutions of the time-dependent Schrödinger equation resulting from the time dependence of the external electic field described above. This system of equations can be propagated in time as described in Chapter 4, using the time-dependent density functional theory (TDDFT) approach, Eq. (4.103).

As an example of how this approach performs in a real system, we show in Fig. 6.1 the intensity of photoabsorption of benzene. The energies where the main peaks of intensity occur are remarkably close to experimental values. Moreover, by calculating the matrix elements of the dipole operator between pairs of occupied–unoccupied states separately,

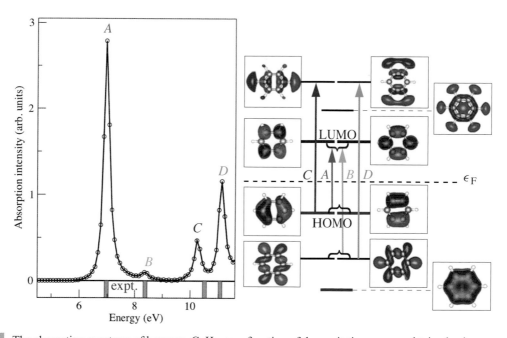

Fig. 6.1 The absorption spectrum of benzene, C_6H_6, as a function of the excitation energy, obtained using TDDFT. On the right we show the transitions between single-particle states responsible for the first four lowest energy peaks, labeled A, B, C, D; in each case, the transitions are between one of the doubly degenerate single-particle states and another doubly degenerate single-particle state. The wavefunctions of these states are represented by the blue (positive) and red (negative) isosurfaces of constant value. The red bars below the zero value of absorption indicate the positions of major peaks in the spectrum as measured by experiment (calculations performed by J. Ren and S. Meng).

we can identify the origin of each peak. The optical transitions between pairs of states responsible for the peaks at different frequencies $\hbar\omega$ (the exciation energies) are shown on the right of the absorption spectrum. The transition labeled A at $\hbar\omega = 6.95$ eV is between the first state below the Fermi level (referred to as "highest occupied molecular orbital," or HOMO) and the first state above it (referred to as "lowest unoccupied molecular orbital," or LUMO). The next, much weaker, peak labeled B corresponds to transitions between the state below the HOMO (referred to as HOMO−1) to the LUMO, the peak labeled C to transitions between the HOMO and the second state above the LUMO (referred to as LUMO+2), and the peak labeled D to transitions between HOMO−1 and LUMO+2. Note that the state immediately above the LUMO (LUMO+1) does not participate in any transitions, which can be explained by the fact that its dipole moment is zero due to symmetry. Similarly, there is another state in the same energy range around the Fermi level, HOMO−2, which also has zero dipole moment and hence does not participate in transitions.

Comparing to the wavefunctions of benzene states that we discussed in Chapter 5, we realize that the HOMO and LUMO states correspond to the p_z linear combinations of π and π^* character that are on either side of the Fermi level, while all other states have more complicated character, involving contributions from σ and σ^* states, except for the HOMO−2 state, which is the totally symmetric π state.

6.2 Conductivity and Dielectric Function

In a typical experimental setup to probe the optical properties of a solid, one measures the fraction of reflected radiation power, incident on the solid at normal angle, which is known as the "reflectance" R. This quantity is expressed in terms of the dielectric constant ε as:

$$R = \left| \frac{1 - \sqrt{\varepsilon}}{1 + \sqrt{\varepsilon}} \right|^2$$

as derived from the classical theory of electrodynamics [see Appendix B, Eq. (B.41)]. The real and imaginary parts of the dielectric function, $\varepsilon = \varepsilon_1 + i\varepsilon_2$, are related by:

$$\varepsilon_1(\omega) = 1 + \frac{1}{\pi} \mathcal{P} \int_{-\infty}^{\infty} \frac{\varepsilon_2(\omega')}{\omega' - \omega} \, d\omega', \quad \varepsilon_2(\omega) = -\frac{1}{\pi} \mathcal{P} \int_{-\infty}^{\infty} \frac{\varepsilon_1(\omega') - 1}{\omega' - \omega} \, d\omega' \quad (6.5)$$

known as the Kramers–Kronig relations. In these expressions $\mathcal{P}$ in front of the integrals stands for the principal value. In essence, this implies that there is only one unknown function (either ε_1 or ε_2). Assuming the reflectivity R can be measured over a wide range of frequencies, and using the Kramers–Kronig relations, both ε_1 and ε_2 can then be determined. The dielectric constant is related to the conductivity of the solid, and the conductivity can be expressed in terms of the microscopic features, such as the electronic states. Our goal here is first, to derive this connection between dielectric function and conductivity and second, to express the conductivity in terms of all the microscopic properties of the solid as obained from the discussion of the previous chapter.

6.2.1 General Formulation

The conductivity is a response function that describes the response of the system in terms of the induced current density $\mathbf{J}(\mathbf{r}, t)$ to an external electric field $\mathbf{E}(\mathbf{r}, t)$. In general, the induced current at position $\mathbf{r}$ and time t will depend on the values of the field at all points in space $\mathbf{r}'$ and at all times t', so the relation between it and the field will be given by:

$$\mathbf{J}(\mathbf{r}, t) = \int \int \sigma(\mathbf{r} - \mathbf{r}', t - t') \mathbf{E}(\mathbf{r}', t') d\mathbf{r}' dt' \tag{6.6}$$

where we have assumed that the conductivity is translationally and temporally invariant, so it depends only on the position differences $(\mathbf{r} - \mathbf{r}')$ and time differences $(t - t')$. The conductivity relates one vector quantity $\mathbf{J}$ to another vector quantity $\mathbf{E}$, so in principle it is a second-rank tensor. To simplify the notation, we have taken it to be a scalar in Eq. (6.6), which implies that we are dealing with an isotropic solid; this can readily be generalized to take into account all the components of the conductivity tensor in the case of an anisotropic solid, which must also obey all the symmetries of the solid. As we have often done before, we introduce the Fourier transforms of all the quantities involved, which depend on the reciprocal space variable $\mathbf{q}$ (the wave-vector) and the reciprocal time variable ω (the frequency). From the convolution theorem [see Appendix A, Eq. (A.51)], we then obtain:

$$\mathbf{J}(\mathbf{q}, \omega) = \sigma(\mathbf{q}, \omega) \mathbf{E}(\mathbf{q}, \omega)$$

With the use of this relation, the continuity equation connecting the current $\mathbf{J}(\mathbf{r}, t)$ to the induced charge $\rho^{\text{ind}}(\mathbf{r}, t)$, once these quantities have been expressed in terms of their Fourier transforms, leads to:

$$\nabla_{\mathbf{r}} \cdot \mathbf{J} + \frac{\partial \rho^{\text{ind}}}{\partial t} = 0 \Rightarrow \mathbf{q} \cdot \mathbf{J}(\mathbf{q}, \omega) - \omega \rho^{\text{ind}}(\mathbf{q}, \omega) = 0 \Rightarrow \rho^{\text{ind}}(\mathbf{q}, \omega) = \frac{1}{\omega} \mathbf{q} \cdot [\sigma(\mathbf{q}, \omega) \mathbf{E}(\mathbf{q}, \omega)]$$

We also have:

$$\mathbf{E}(\mathbf{r}, t) = -\nabla_{\mathbf{r}} \Phi(\mathbf{r}, t) \Rightarrow \mathbf{E}(\mathbf{q}, \omega) = -i\mathbf{q}\Phi(\mathbf{q}, \omega) \tag{6.7}$$

where $\Phi(\mathbf{r}, t)$ is the electrostatic potential and $\Phi(\mathbf{q}, \omega)$ its Fourier transform [see also the discussion in Appendix B, Eqs (B.42)–(B.53)]. This gives for the induced charge:

$$\rho^{\text{ind}}(\mathbf{q}, \omega) = -\frac{iq^2 \sigma(\mathbf{q}, \omega)}{\omega} \Phi(\mathbf{q}, \omega)$$

For weak external fields we can use the expression $\rho^{\text{ind}}(\mathbf{q}, \omega) = \chi(\mathbf{q}, \omega)\Phi(\mathbf{q}, \omega)$, where $\chi(\mathbf{q}, \omega)$ is the linear response function. This gives, for the conductivity:

$$\sigma(\mathbf{q}, \omega) = \frac{i\omega}{q^2} \chi(\mathbf{q}, \omega)$$

Comparing this result to the general relation between the response function and the dielectric function that we derived in Chapter 4, Eq. (4.110), we obtain the desired relation between the conductivity and the dielectric function:

$$\varepsilon(\mathbf{q}, \omega) = 1 - \frac{4\pi}{i\omega} \sigma(\mathbf{q}, \omega) \tag{6.8}$$

Having established this relation, we next express the conductivity in terms of microscopic properties of the solid (the electronic wavefunctions and their energies and occupation numbers); as a final step, we will use the connection between conductivity and dielectric function to obtain an expression of the latter in terms of the microscopic properties of the solid. This will provide a direct link between the electronic structure at the microscopic level and the experimentally measured dielectric function, which captures the macroscopic response of the solid to an external electromagnetic field.

6.2.2 Drude and Lorentz Models

We discuss first two simple expressions which capture much of the physics. In a classical picture, the electrons will respond to an external electric field $\mathbf{E}$ by accelerating, but there will also be a damping force to their motion due to collisions with other charged particles (electrons and ions). The damping force is expressed as a frictional component of the total force, which is proportional to the velocity and the mass, with the coefficient of friction denoted by η. These considerations give the following expression for the equation of motion of an electron at position $\mathbf{r}$:

$$m_e \frac{d^2 \mathbf{r}}{dt^2} = -e\mathbf{E} - \eta m_e \frac{d\mathbf{r}}{dt} \tag{6.9}$$

with the right-hand side containing all the forces acting on the electron. This equation applies to a free electron. If the electron is bound by a harmonic potential with spring constant κ_0 and corresponding characteristic fequency $\omega_0 = \sqrt{\kappa_0/m_e}$, the above equation contains one more term on the right-hand side:

$$m_e \frac{d^2 \mathbf{r}}{dt^2} = -e\mathbf{E} - \eta m_e \frac{d\mathbf{r}}{dt} - m_e \omega_0^2 \mathbf{r} \tag{6.10}$$

Taking the electric field and the position of the electron to have the same time dependence, $\mathbf{r} = \mathbf{r}_0 \exp(-i\omega t)$, $\mathbf{E} = \mathbf{E}_0 \exp(-i\omega t)$, and substituting in the above equations, we find that the electron velocity is given by:

$$\mathbf{v} = \frac{d\mathbf{r}}{dt} = -i\omega \frac{e}{m_e[(\omega^2 - \omega_0^2) + i\eta\omega]} \mathbf{E}$$

If we have N electrons in volume V, the total current per unit volume $\mathbf{j}$ is:

$$\mathbf{j} = \frac{-eN}{V} \mathbf{v} = \sigma \mathbf{E} \Rightarrow \sigma(\omega) = i\omega \frac{Ne^2}{m_e V} \frac{1}{(\omega^2 - \omega_0^2) + i\eta\omega}$$

where we have introduced the conductivity σ which relates the current to the external field $\mathbf{E}$. Substituting this expression for $\sigma(\omega)$ in Eq. (6.8), we obtain for the dielectric function $\varepsilon(\omega)$:

$$\varepsilon(\omega) = 1 - \left(\frac{4\pi Ne^2}{m_e V} \right) \frac{1}{(\omega^2 - \omega_0^2) + i\eta\omega} \tag{6.11}$$

The quantity in parentheses before the term that contains the frequency dependence is very important, and is defined as the square of the *plasma frequency*, denoted by ω_p:

$$\omega_p \equiv \left(\frac{4\pi e^2}{m_e}n\right)^{1/2}, \quad n = \frac{N}{V} \tag{6.12}$$

This is the characteristic frequency of the response of a uniform electron gas of density n in a uniform background of compensating positive charge (see Problem 1); the modes describing this response are called *plasmons* (mentioned earlier in Chapter 4). The value of ω_p is a measure of the amount of electrons that contribute to this collective mode of response to external fields. We can use the standard expressions for the radius r_s of the sphere corresponding to the average volume per electron, to express the plasma frequency as:

$$\omega_p = 0.716 \left(\frac{1}{r_s/a_0}\right)^{3/2} \times 10^{17} \text{Hz}$$

and since typical values of r_s/a_0 in metals are 2–6, we find that the frequency of plasmon oscillations is in the range of 10^3–10^4 THz, or equivalently, $\hbar\omega_p$ in the range 4–40 eV. We note here that not all valence electrons participate in the plasmon oscillations, because some of those electrons are closely bound to the ions, depending on the type of atom involved. A realistic range of plasma frequencies is 4–16 eV (10^3–4×10^3 THz).

We can now identify from the above expression the contributions of free electrons and bound electrons to the dielectric function and separate the real and imaginary parts in each case. For the bound electron case, $\omega_0 \neq 0$, we obtain:

$$\varepsilon_1^{(b)}(\omega) = 1 - \frac{\omega_p^2(\omega^2 - \omega_0^2)}{(\omega^2 - \omega_0^2)^2 + \eta^2\omega^2}, \quad \varepsilon_2^{(b)}(\omega) = \frac{\omega_p^2\omega\eta}{(\omega^2 - \omega_0^2)^2 + \eta^2\omega^2} \tag{6.13}$$

For the free-electron case, $\omega_0 = 0$, we obtain:

$$\varepsilon_1^{(f)}(\omega) = 1 - \frac{\omega_p^2}{\omega^2 + \eta^2}, \quad \varepsilon_2^{(f)}(\omega) = \frac{\omega_p^2\eta}{\omega(\omega^2 + \eta^2)} \tag{6.14}$$

The general features of these terms are illustrated in Fig. 6.2. Specifically, $\varepsilon_1^{(f)}(\omega)$ changes sign at $\omega = \omega_p$ from negative (for small ω) to positive (for large ω); similarly, $\varepsilon_1^{(b)}(\omega)$ changes sign at $\omega = \omega_0$ from positive (for small ω) to negative (for intermediate ω) and again to positive (for large ω) at some value $\omega > \omega_p$. We also show in Fig. 6.2 the more general case in which there are contributions from both free and bound electrons. These results make it evident that the dielectric function in the limit $\omega \to 0$, called the "static dielectric constant," has divergent real and imaginary parts from the free-electron component while from the bound-charge component the real part is a constant larger than unity and the imaginary part vanishes. We elaborate next on the two limiting cases, the free-electron and bound-electron models.

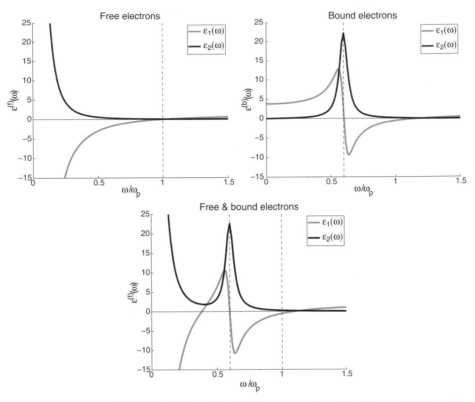

Illustration of the general behavior of the real, $\varepsilon_1(\omega)$ (cyan curves), and imaginary, $\varepsilon_2(\omega)$ (purple curves), parts of the dielectric function with frequency ω in units of the plasma frequency ω_p, for the free-electron case, $\varepsilon^{(f)}(\omega)$ (top-left panel) and the bound-electron case, $\varepsilon^{(b)}(\omega)$ (top-right panel). The bottom figure shows a case where the contributions of the free and bound electrons are added to give the total. In these examples we have taken $\omega_0/\omega_p = 0.6$ for the pole and $\eta/\omega_p = 0.075$.

The free-electron result is known as the "Drude model" or the "longitudinal" component, because it refers to the motion of electrons as a uniform sea of charge along the direction of the external field. This contribution is usually expressed as:

$$\varepsilon^{(f)}(\omega) = 1 - \frac{\omega_p^2}{\omega(\omega + i/\tau)} \tag{6.15}$$

where ω_p is the plasma frequency and τ is the "relaxation time," a quantity with dimensions of time, since it is the inverse of the width η. Since η is associated with the friction that electrons experience as they are moving through the solid, the relaxation time is a measure of the time scale associated with collisions that lead to damping of the oscillatory motion induced by the external field. For a good conductor, the friction is very small or equivalently, on the time scale defined by ω^{-1}, the relaxation time is very large. This situation corresponds to:

$$\omega\tau \gg 1 \Rightarrow \frac{1}{\tau} \ll \omega \Rightarrow \frac{\omega}{\tau} \ll \omega^2$$

and in this limit the dielectric function takes the form

$$\varepsilon^{(f)}(\omega) = 1 - \frac{\omega_p^2}{\omega^2} \tag{6.16}$$

The bound-electron result is known as the "Lorentz model" or "transverse" component, and is usually expressed as:

$$\varepsilon^{(b)}(\omega) = 1 - \sum_i \frac{\lambda_i \omega_p^2}{(\omega^2 - \omega_{0,i}^2) + i\eta_i \omega} \tag{6.17}$$

with ω_p the plasma frequency and $\omega_{0,i}$ the poles in frequency, each associated with a width η_i and a weight λ_i. There are typically many pole values $\omega_{0,i}$, although in experimental measurements a few of the poles dominate because they are associated with a high density of states corresponding to electronic transitions that involve an energy difference equal to this frequency. This is reflected in the value of the weight λ_i.

In the following sections we will explore in more detail the origin of the free-electron and bound-electron behavior in metals and insulators. Before we delve into the detailed calculations, we present some examples of real solids, where the simple ideas discussed so far are already adequate to explain some of the important features of the dielectric function.

In Fig. 6.3 we show the dielectric function of Al, our familiar free-electron-like solid. In this case, the dielectric function is indeed dominated by the free-electron component, as is evident by comparing the measured dielectric function to the simple result expected from the Drude model. The plasma frequency found in this case is $\hbar\omega = 15.2$ eV. If we try to estimate the plasma frequency from the density of electrons in Al, taking into account that there are three valence electrons per atom, we find $\hbar\omega_p = 15.8$ eV, reasonably close to the value obtained by fitting the experimental result to the free-electron model. In addition to the dominant free-electron contribution, there is also a very small bound-electron contribution at $\hbar\omega_0 = 1.4$ eV, which is related to features of the band structure as indicated in Fig. 6.3; we will return to this later and explain its origin.

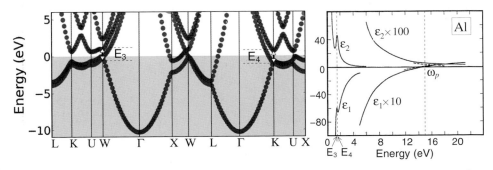

Fig. 6.3 Example of the behavior of the real and imaginary parts of the dielectric function, $\varepsilon_1(\omega)$, $\varepsilon_2(\omega)$, for Al, a representative free-electron-like solid. The value of the plasma frequency $\hbar\omega_p = 15.2$ eV is indicated. There is also a very small bound-electron-like contribution, from the features of the band structure (shown on the left) indicated by the symbols E_3 near the point W and E_4 near the point K. Adapted from H. Ehrenreich, H. R. Philipp, and B. Segall, *Phys. Rev.* **132**, 1918 (1963).

As another example, we consider the case of the noble metals Cu and Ag. In this case, there are 11 valence electrons per atom, which would give plasma frequencies $\hbar\omega_p = 35.9$ eV for Cu and 29.8 eV for Ag. However, we know from studying the electronic structure of these metals that the d-electron bands are far below the Fermi level and they look almost like atomic states with very little dispersion, suggesting that the electrons in these states are not part of the free-electron system; instead, they are closer to bound electrons. Indeed, the real part of the dielectric function $\varepsilon_1(\omega)$ for these two metals can be analyzed into free-electron and bound-electron contributions, as shown in Fig. 6.4. The free-electron contribution obtained from analyzing the experimental results gives a plasma frequency of 9.3 eV for Cu and 9.2 eV for Ag. If we consider only the s valence state as contributing to this behavior, which implies only one valence electron per atom, the estimates for the plasma frequency would be 10.8 eV for Cu and 9.0 eV for Ag, much closer to those obtained from experiment. Moreover, the bound-electron contribution from experimental results suggests a pole at $\hbar\omega_0 \approx 2$ eV for Cu and ≈ 4 eV for Ag. The band structures of these two metals show that the highest d bands are about 2 eV and about 4 eV below the Fermi level for Cu and Ag, respectively, as shown in Fig. 6.4. These observations nicely explain the main features of the experimental results.

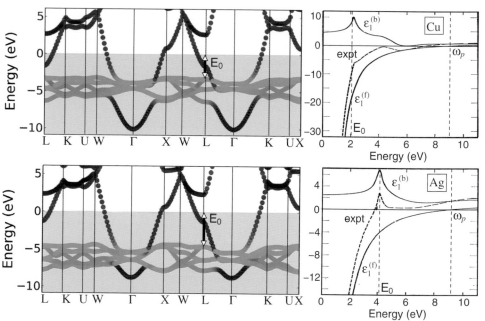

Fig. 6.4 Examples of the behavior of the real part of the dielectric function $\varepsilon_1(\omega)$ for Cu (top) and Ag (bottom): the main features are the pole at $\hbar\omega_0 = 2$ eV for Cu and 4 eV for Ag, and the value of the plasma frequency $\hbar\omega_p = 9.3$ eV for Cu and 9.2 eV for Ag. The corresponding band structures on the left show the feature (transitions from d bands to the Fermi surface, marked as E_0 near the point L) responsible for the behavior near the pole. The experimental results and their analysis into the free-electron $\varepsilon_1^{(f)}$ and the bound-charge $\varepsilon_1^{(b)}$ contributions are adapted from H. Ehrenreich and H. R. Philipp, *Phys. Rev.* **128**, 1622 (1962).

6.2.3 Connection to Microscopic Features

For the calculation of the conductivity, we will rely on the general result for the expectation value of a many-body operator $\hat{O}(\{\mathbf{r}_i\})$, which can be expressed as a sum of single-particle operators $\hat{o}(\mathbf{r}_i)$:

$$\hat{O}(\{\mathbf{r}_i\}) = \sum_i \hat{o}(\mathbf{r}_i)$$

in terms of the matrix elements in the single-particle states, as derived in Appendix C. For simplicity, we will use a single index, the subscript κ, to identify the single-particle states in the crystal, with the understanding that this index represents in shorthand notation both the band index and the wave-vector index. From Eq. (C.21) we find that the expectation value of the many-body operator $\mathcal{O}$ in the single-particle states labeled by κ is:

$$\langle \hat{O} \rangle = \sum_{\kappa,\kappa'} \hat{o}_{\kappa,\kappa'} \gamma_{\kappa',\kappa}$$

where $\hat{o}_{\kappa,\kappa'}$ and $\gamma_{\kappa,\kappa'}$ are the matrix elements of the operator $\hat{o}(\mathbf{r})$ and the single-particle density matrix $\gamma(\mathbf{r}, \mathbf{r}')$. We must therefore identify the appropriate single-particle density matrix and single-particle operator $\hat{o}(\mathbf{r})$ for the calculation of the conductivity. As derived by Ehrenreich and Cohen[3] to first-order perturbation theory, the interaction term of the hamiltonian gives for the single-particle density matrix:

$$\gamma_{\kappa',\kappa}^{\text{int}} = \frac{f_{\kappa'} - f_\kappa}{\epsilon_{\kappa'} - \epsilon_\kappa - \hbar(\omega + i\eta)} \langle \psi_{\kappa'} | \hat{\mathcal{H}}^{\text{int}} | \psi_\kappa \rangle$$

where f_κ is the Fermi occupation number for the state with energy ϵ_κ in the unperturbed system, and η is an infinitesimal positive quantity with the dimensions of frequency.

For the conductivity, the relevant interaction term $\hat{\mathcal{H}}^{\text{int}}(\mathbf{r}, t)$ is that which describes absorption or emission of photons, as mentioned earlier, Eq. (6.2):

$$\hat{\mathcal{H}}^{\text{int}}(\mathbf{r}, t) = \frac{e}{m_e c} \mathbf{A}(\mathbf{r}, t) \cdot \mathbf{p} = \frac{e}{m_e c} \left[e^{i(\mathbf{q} \cdot \mathbf{r} - \omega t)} \mathbf{A}_0 \cdot \mathbf{p} + \text{c.c.} \right] \tag{6.18}$$

where $\hbar\mathbf{q}$ is the momentum and ω the frequency of the electromagnetic field, $\mathbf{A}_0$ is the vector potential, and c.c. stands for complex conjugate. For the transverse component of the field, we have the following relation between the transverse electric field $\mathbf{E}_t$ ($\nabla_\mathbf{r} \cdot \mathbf{E}_t = 0$) and the vector potential $\mathbf{A}_t$:

$$\mathbf{E}_t(\mathbf{r}, t) = -\frac{1}{c} \frac{\partial \mathbf{A}_t}{\partial t} \Rightarrow \mathbf{E}_t(\mathbf{q}, \omega) = \frac{i\omega}{c} \mathbf{A}_t(\mathbf{q}, \omega) \Rightarrow \mathbf{A}_t(\mathbf{q}, \omega) = \frac{c}{i\omega} \mathbf{E}_t(\mathbf{q}, \omega) \tag{6.19}$$

With this, the interaction term takes the form

$$\hat{\mathcal{H}}^{\text{int}}(\mathbf{r}, t) = \frac{-ie}{m_e \omega} \mathbf{E}_t \cdot \mathbf{p} \Rightarrow \langle \psi_{\kappa'} | \hat{\mathcal{H}}^{\text{int}} | \psi_\kappa \rangle = \frac{-ie}{m_e \omega} |\mathbf{E}_t| \langle \psi_{\kappa'} | e^{i\mathbf{q} \cdot \mathbf{r}} \mathbf{e}_t \cdot \mathbf{p} | \psi_\kappa \rangle \tag{6.20}$$

where we denote by $\mathbf{e}_t$ the unit vector along the direction of the field, for which, since we are working with the transverse component, $\mathbf{q} \cdot \mathbf{e}_t = 0$.

---

[3] H. Ehrenreich and M. H. Cohen, *Phys. Rev.* **115**, 786 (1959).

The relevant single-particle operator $\hat{o}(\mathbf{r})$ is the current, since it describes the response of the physical system to the external field. The single-particle current operator is:

$$\hat{\mathbf{j}}(\mathbf{r}) = \frac{-e}{V}\frac{\mathbf{p}}{m_e} \Rightarrow \hat{\mathbf{j}}(\mathbf{q}) = \frac{-e}{m_e V}e^{-i\mathbf{q}\cdot\mathbf{r}}\mathbf{p}$$

Since in the end we want to use the relation $\mathbf{J} = \sigma\mathbf{E}$ to obtain an expression for the conductivity, assuming an isotropic medium, we will consider only the component of the current that is along the external field, that is, its projection in the direction of the field $\mathbf{e}_t$:

$$\mathbf{e}_t \cdot \hat{\mathbf{j}}(\mathbf{q}) = \frac{-e}{m_e V}e^{-i\mathbf{q}\cdot\mathbf{r}}\mathbf{e}_t\cdot\mathbf{p} \Rightarrow j_{\kappa,\kappa'} = \frac{-e}{m_e V}\langle\psi_\kappa|e^{-i\mathbf{q}\cdot\mathbf{r}}\mathbf{e}_t\cdot\mathbf{p}|\psi_{\kappa'}\rangle \quad (6.21)$$

Substituting the expressions from Eqs (6.20) and (6.21) into the expression for the expectation value of the current, and dividing by the magnitude of the external field, we obtain for the conductivity:

$$\sigma_t(\mathbf{q},\omega) = \frac{ie^2}{m_e^2 V}\sum_{\kappa,\kappa'}\frac{1}{\omega}\frac{|\langle\psi_{\kappa'}|e^{i\mathbf{q}\cdot\mathbf{r}}\mathbf{e}_t\cdot\mathbf{p}|\psi_\kappa\rangle|^2}{\epsilon_{\kappa'}-\epsilon_\kappa-\hbar(\omega+i\eta)}\left[f_{\kappa'}-f_\kappa\right], \quad \mathbf{q}\cdot\mathbf{e}_t = 0 \quad (6.22)$$

This expression for the conductivity is known as the Kubo–Greenwood formula.[4] From this result for the conductivity, and using the general relation derived earlier, Eq. (6.8), we find for the dielectric function:

$$\varepsilon_t(\mathbf{q},\omega) = 1 - \frac{4\pi e^2}{m_e^2 V\omega^2}\sum_{\kappa,\kappa'}\frac{|\langle\psi_{\kappa'}|e^{i\mathbf{q}\cdot\mathbf{r}}\mathbf{e}_t\cdot\mathbf{p}|\psi_\kappa\rangle|^2}{\epsilon_{\kappa'}-\epsilon_\kappa-\hbar(\omega+i\eta)}\left[f_{\kappa'}-f_\kappa\right], \quad \mathbf{q}\cdot\mathbf{e}_t = 0 \quad (6.23)$$

In the case of the longitudinal component of the field, the derivation is similar, with the interaction hamiltonian being given by:

$$\hat{\mathcal{H}}^{int}(\mathbf{r},t) = (-e)\Phi(\mathbf{r},t) \Rightarrow (-e)\Phi(\mathbf{q},\omega) = -\frac{ie}{q}[\mathbf{e}_l\cdot\mathbf{E}_l(\mathbf{q},\omega)], \quad \mathbf{q}\cdot\mathbf{e}_l = q$$

as is easily seen from Eq. (6.7). This leads to the following expression for the longitudinal dielectric constant:

$$\varepsilon_l(\mathbf{q},\omega) = 1 - \frac{4\pi e^2}{Vq^2}\sum_{\kappa,\kappa'}\frac{|\langle\psi_{\kappa'}|e^{i\mathbf{q}\cdot\mathbf{r}}|\psi_\kappa\rangle|^2}{\epsilon_{\kappa'}-\epsilon_\kappa-\hbar(\omega+i\eta)}\left[f_{\kappa'}-f_\kappa\right], \quad \mathbf{q}\cdot\mathbf{e}_l = q \quad (6.24)$$

We will write explicitly the real and imaginary parts of the dielectric function, $\varepsilon_1, \varepsilon_2$, as these are measured directly from the reflectivity, and one can be obtained from the other by the Kramers–Kronig relations. In the following we will make use of the mathematical identity

$$\frac{1}{x-i\eta} = \frac{x}{x^2+\eta^2} + i\left[\frac{\eta}{x^2+\eta^2}\right] \Rightarrow \lim_{\eta\to 0^+}\frac{x}{x^2+\eta^2} = \mathcal{P}\frac{1}{x}, \quad \lim_{\eta\to 0^+}\left[\frac{\eta}{x^2+\eta^2}\right] = \pi\delta(x) \quad (6.25)$$

with $\mathcal{P}$ denoting the principal value, that is, the evaluation of the root $x = 0$ by approaching it symmetrically from the left and from the right. Applying Eq. (6.25) to the expression for

---

[4] R. Kubo, *J. Phys. Soc. Japan* **12**, 570 (1957); D. A. Greenwood, *Proc. Phys. Soc. (London)* **A71**, 585 (1958).

the dielectric function, we find:

$$\varepsilon_{1,l}(\mathbf{q}, \omega) = 1 - \frac{4\pi e^2}{V q^2} \sum_{\kappa,\kappa'} \frac{|\langle \psi_{\kappa'} | e^{i\mathbf{q}\cdot\mathbf{r}} | \psi_\kappa \rangle|^2}{\epsilon_{\kappa'} - \epsilon_\kappa - \hbar\omega} \left[f_{\kappa'} - f_\kappa \right] \tag{6.26}$$

$$\varepsilon_{2,l}(\mathbf{q}, \omega) = -\frac{4\pi^2 e^2}{V q^2} \sum_{\kappa,\kappa'} |\langle \psi_{\kappa'} | e^{i\mathbf{q}\cdot\mathbf{r}} | \psi_\kappa \rangle|^2 \delta(\epsilon_{\kappa'} - \epsilon_\kappa - \hbar\omega) \left[f_{\kappa'} - f_\kappa \right] \tag{6.27}$$

for the longitudinal component and similarly for the transverse one:

$$\varepsilon_{1,t}(\mathbf{q}, \omega) = 1 - \frac{4\pi e^2}{m_e^2 V \omega^2} \sum_{\kappa,\kappa'} \frac{|\mathbf{e}_t \cdot \langle \psi_{\kappa'} | e^{i\mathbf{q}\cdot\mathbf{r}} \mathbf{p} | \psi_\kappa \rangle|^2}{\epsilon_{\kappa'} - \epsilon_\kappa - \hbar\omega} \left[f_{\kappa'} - f_\kappa \right] \tag{6.28}$$

$$\varepsilon_{2,t}(\mathbf{q}, \omega) = -\frac{4\pi^2 e^2}{m_e^2 V \omega^2} \sum_{\kappa,\kappa'} |\mathbf{e}_t \cdot \langle \psi_{\kappa'} | e^{i\mathbf{q}\cdot\mathbf{r}} \mathbf{p} | \psi_\kappa \rangle|^2 \delta(\epsilon_{\kappa'} - \epsilon_\kappa - \hbar\omega) \left[f_{\kappa'} - f_\kappa \right] \tag{6.29}$$

Finally, we point out that in the limit of long wavelength, $q \to 0$, which is the physically interesting situation as we shall discuss below, the longitudinal and transverse expressions for *interband* transitions ($n \neq n'$) become essentially the same, the only difference being in the direction of field polarization, embodied in the vectors $\mathbf{e}_l, \mathbf{e}_t$. To see this, we start with the expression for the longitudinal dielectric constant and expand the exponential that appears in the matrix elements to lowest order in q:

$$e^{i\mathbf{q}\cdot\mathbf{r}} \approx 1 + i\mathbf{q} \cdot \mathbf{r} \tag{6.30}$$

which we substitute in the matrix elements. Taking advantage of the facts that $\hat{\mathbf{q}} = \hat{\mathbf{e}}_l$ for the longitudinal component and $\langle \psi_\kappa | \psi_{\kappa'} \rangle = \delta_{\kappa,\kappa'}$, we find:

$$\lim_{q \to 0} [\varepsilon_l(\mathbf{q}, \omega)] = 1 - \frac{4\pi e^2}{V} \sum_{\kappa \neq \kappa'} \frac{|\langle \psi_{\kappa'} | \mathbf{e}_l \cdot \mathbf{r} | \psi_\kappa \rangle|^2}{\epsilon_{\kappa'} - \epsilon_\kappa - \hbar(\omega + i\eta)} \left[f_{\kappa'} - f_\kappa \right]$$

We can now use the relations between momentum and position operators, Eq. (6.3), to rewrite the above expression as:

$$\lim_{q \to 0} [\varepsilon_l(\mathbf{q}, \omega)] = 1 - \frac{4\pi e^2}{m_e^2 V} \sum_{\kappa \neq \kappa'} \frac{|\mathbf{e}_l \cdot \langle \psi_{\kappa'} | \mathbf{p} | \psi_\kappa \rangle|^2}{\epsilon_{\kappa'} - \epsilon_\kappa - \hbar(\omega + i\eta)} \frac{f_{\kappa'} - f_\kappa}{(\epsilon_{\kappa'} - \epsilon_\kappa)^2 / \hbar^2} \tag{6.31}$$

Next we note that because of the denominators that appear under the matrix elements, the only significant contributions come from values of $\hbar\omega$ very close to values of the energy differences ($\epsilon_{\kappa'} - \epsilon_\kappa$). Indeed, if we set ($\epsilon_{\kappa'} - \epsilon_\kappa) \approx \hbar\omega$ in the denominator of the second factor under the summation in Eq. (6.31), we recover an expression which is the same as that for the transverse dielectric function, Eq. (6.23), in the limit $\mathbf{q} \to 0$, with the polarization vector $\mathbf{e}_l$ instead of $\mathbf{e}_t$. For this reason, in the following, we deal only with the expression for the longitudinal dielectric function, which is simpler.

6.2.4 Implications for Crystals

The expressions that we derived above can be used to calculate the dielectric function of crystals, once we have determined the band structure, that is, solved for all the single-particle eigenfunctions $|\psi_\kappa\rangle$ and corresponding eigenvalues ϵ_κ. We will need to introduce again the wave-vector and band indices $\mathbf{k}, \mathbf{k}', n, n'$ explicitly, so that the general expression reads:

$$\varepsilon(\mathbf{q}, \omega) = 1 - \frac{4\pi e^2}{Vq^2} \sum_{\mathbf{k},\mathbf{k}',n,n'} \frac{|\langle \psi_{\mathbf{k}'}^{(n')} | e^{i\mathbf{q}\cdot\mathbf{r}} | \psi_{\mathbf{k}}^{(n)}\rangle|^2}{\epsilon_{\mathbf{k}'}^{(n')} - \epsilon_{\mathbf{k}}^{(n)} - \hbar(\omega + i\eta)} \left[f_{\mathbf{k}'}^{(n')} - f_{\mathbf{k}}^{(n)} \right] \tag{6.32}$$

We note first that all the expressions we derived involve the difference between two Fermi occupation numbers, $\left[f_{\mathbf{k}'}^{(n')} - f_{\mathbf{k}}^{(n)} \right]$. For low enough temperature (which is typically the case), these are either 1 or 0. The only terms that contribute to the sums involved in the expressions for the dielectric functions are those for which:

$$\epsilon_{\mathbf{k}'}^{(n')} > \epsilon_F > \epsilon_{\mathbf{k}}^{(n)} \Rightarrow \left[f_{\mathbf{k}'}^{(n')} = 0, f_{\mathbf{k}}^{(n)} = 1 \right] \quad \text{or} \quad \epsilon_{\mathbf{k}}^{(n)} > \epsilon_F > \epsilon_{\mathbf{k}'}^{(n')} \Rightarrow \left[f_{\mathbf{k}'}^{(n')} = 1, f_{\mathbf{k}}^{(n)} = 0 \right]$$

that is, one of the states involved must be a valence state (occupied) and the other one a conduction state (unoccupied). We can then label the states explicitly by their band index, (c) for conduction and (v) for valence, with $f_{\mathbf{k}}^{(v)} = 1$ and $f_{\mathbf{k}'}^{(c)} = 0$. Taking into account the spin degeneracy of each level, we obtain for the real and imaginary parts:

$$\varepsilon_1(\mathbf{q}, \omega) = 1 + \frac{16\pi e^2}{Vq^2} \sum_{v,\mathbf{k},c,\mathbf{k}'} |\langle \psi_{\mathbf{k}'}^{(c)} | e^{i\mathbf{q}\cdot\mathbf{r}} | \psi_{\mathbf{k}}^{(v)}\rangle|^2 \left[\frac{\epsilon_{\mathbf{k}'}^{(c)} - \epsilon_{\mathbf{k}}^{(v)}}{(\epsilon_{\mathbf{k}'}^{(c)} - \epsilon_{\mathbf{k}}^{(v)})^2 - (\hbar\omega)^2} \right]$$

$$\varepsilon_2(\mathbf{q}, \omega) = \frac{8\pi^2 e^2}{Vq^2} \sum_{v,\mathbf{k},c,\mathbf{k}'} |\langle \psi_{\mathbf{k}'}^{(c)} | e^{i\mathbf{q}\cdot\mathbf{r}} | \psi_{\mathbf{k}}^{(v)}\rangle|^2 \left[\delta(\epsilon_{\mathbf{k}'}^{(c)} - \epsilon_{\mathbf{k}}^{(v)} - \hbar\omega) - \delta(\epsilon_{\mathbf{k}'}^{(c)} - \epsilon_{\mathbf{k}}^{(v)} + \hbar\omega) \right]$$

Next, we take a closer look at the matrix elements involved in the expressions for the dielectric function. Since all electronic states in the crystal can be expressed in our familiar Bloch form:

$$\psi_{\mathbf{k}}^{(n)}(\mathbf{r}) = \frac{e^{i\mathbf{k}\cdot\mathbf{r}}}{\sqrt{V}} u_{\mathbf{k}}^{(n)}(\mathbf{r}) = \frac{e^{i\mathbf{k}\cdot\mathbf{r}}}{\sqrt{V}} \sum_{\mathbf{G}} \alpha_{\mathbf{k}}^{(n)}(\mathbf{G}) e^{i\mathbf{G}\cdot\mathbf{r}} \tag{6.33}$$

we find the following expression for the matrix element:

$$\langle \psi_{\mathbf{k}'}^{(n')} | e^{i\mathbf{q}\cdot\mathbf{r}} | \psi_{\mathbf{k}}^{(n)}\rangle = \sum_{\mathbf{G}\mathbf{G}'} \left[\alpha_{\mathbf{k}'}^{(n')}(\mathbf{G}') \right]^* \alpha_{\mathbf{k}}^{(n)}(\mathbf{G}) \frac{1}{V} \int e^{i(\mathbf{k}-\mathbf{k}'+\mathbf{q}+\mathbf{G}-\mathbf{G}')\cdot\mathbf{r}} d\mathbf{r}$$

The last integral produces a δ-function in reciprocal-space vectors (see Appendix A):

$$\frac{1}{V} \int e^{i(\mathbf{k}-\mathbf{k}'+\mathbf{q}+\mathbf{G})\cdot\mathbf{r}} d\mathbf{r} = \delta(\mathbf{k} - \mathbf{k}' + \mathbf{q} + \mathbf{G}) \Rightarrow \mathbf{k}' = \mathbf{k} + \mathbf{q} + \mathbf{G}$$

However, taking into account the relative magnitudes of the wave-vectors involved in this condition reveals that it boils down to $\mathbf{k}' \approx \mathbf{k}$, or $\mathbf{q} \to 0$. The reason for this is that the momentum of radiation for optical transitions is given by $|\mathbf{q}| = (2\pi/\lambda)$, with λ the wavelength, which in the optical range is $\lambda \sim 10^4$ Å, while the crystal wave-vectors have typical wavelength values of order the interatomic distances, that is, ~ 1 Å. Consequently, the difference between the wave-vectors $\mathbf{k}, \mathbf{k}'$ of the initial and final states due to the photon momentum $\mathbf{q}$ is negligible. Taking the value $\mathbf{q} = 0$ in the above equation, and constraining the summation to the first BZ ($\mathbf{G} = 0$), leads to so-called "direct" transitions, that is, transitions at the same value of $\mathbf{k}$ in the BZ, and therefore between different bands, one occupied and one unoccupied. These are the only allowed optical transitions when no other excitations are present. When other excitations that can carry crystal momentum are present, such as phonons (see Chapter 7), the energy and momentum conservation conditions can be satisfied independently, even for indirect transitions, in which case the initial and final photon states can have different momenta (see Chapter 8). The limit $\mathbf{q} \to 0$ is also useful in analyzing transitions between occupied and unoccupied states across the Fermi level of a metal, for the *same* band; in this case, the limit must be taken carefully for intraband transitions, as we discuss in detail below.

Having established that the long-wavelength (small q) limit is the relevant range for optical transitions, we can now use the results derived earlier, Eq. (6.24), with $\mathbf{q} = \mathbf{k}' - \mathbf{k}$. To analyze the behavior of the dielectric function, we will consider two different situations: we will examine first, transitions at the same wave-vector $\mathbf{k}' = \mathbf{k}$ but between different bands $n \neq n'$ and second, transitions within the same band $n = n'$ but at slightly different values of the wave-vector $\mathbf{k}' = \mathbf{k} + \mathbf{q}$. The first kind correspond to *interband* or *bound-electron* transitions; they correspond to situations where an electron makes a direct transition by absorption or emission of a photon across the band gap in insulators or semiconductors. The second kind are called *intraband* or *free-electron* transitions; they correspond to situations where an electron makes a transition by absorption or emission of a photon across the Fermi level by changing its wave-vector slightly. In both cases, we will be assuming that we are working in the long-wavelength limit $\mathbf{q} \to 0$.

We consider first the effect of direct interband transitions, that is, transitions for which $\mathbf{k} = \mathbf{k}'$ and $n \neq n'$. To ensure direct transitions we introduce a factor $\delta_{\mathbf{k},\mathbf{k}'}$ in the double summation over wave-vectors in the general expression for the dielectric function, Eq. (6.32), from which we obtain, from the transverse component of the dielectric function in the long-wavelength ($\mathbf{q} \to 0$) limit:

$$\varepsilon^{\text{inter}}(\mathbf{q}, \omega) = 1 - \frac{4\pi e^2}{V q^2 \hbar} \sum_{\mathbf{k}, n \neq n'} \left| \langle \psi_{\mathbf{k}}^{(n')} | e^{i\mathbf{q} \cdot \mathbf{r}} | \psi_{\mathbf{k}}^{(n)} \rangle \right|^2 \frac{f_{\mathbf{k}}^{(n')} - f_{\mathbf{k}}^{(n)}}{\omega_{\mathbf{k}}^{(n'n)} - (\omega + i\eta)} \qquad (6.34)$$

where we have defined

$$\hbar \omega_{\mathbf{k}}^{(n'n)} = \epsilon_{\mathbf{k}}^{(n')} - \epsilon_{\mathbf{k}}^{(n)}$$

There are two possible sets of values for the summation indices, giving rise to summation terms

$$[n = v, \ n' = c] \Rightarrow \left[f_{\mathbf{k}}^{(n')} = 0, \ f_{\mathbf{k}}^{(n)} = 1 \right] \to \left| \langle \psi_{\mathbf{k}}^{(c)} | e^{i\mathbf{q} \cdot \mathbf{r}} | \psi_{\mathbf{k}}^{(v)} \rangle \right|^2 \frac{-1}{\omega_{\mathbf{k}}^{(cv)} - (\omega + i\eta)}$$

$$[n = c, \ n' = v] \Rightarrow \left[f_{\mathbf{k}}^{(n')} = 1, \ f_{\mathbf{k}}^{(n)} = 0 \right] \to \left| \langle \psi_{\mathbf{k}}^{(v)} | e^{i\mathbf{q} \cdot \mathbf{r}} | \psi_{\mathbf{k}}^{(c)} \rangle \right|^2 \frac{1}{\omega_{\mathbf{k}}^{(vc)} - (\omega + i\eta)}$$

With these contributions, and taking into account the spin degeneracy of each state, we find:

$$\varepsilon^{\text{inter}}(\mathbf{q},\omega) = 1 + \frac{16\pi e^2}{V\hbar} \sum_{\mathbf{k},c,v} \left| \langle \psi_{\mathbf{k}}^{(c)} | \mathbf{r} | \psi_{\mathbf{k}}^{(v)} \rangle \right|^2 \frac{\omega_{\mathbf{k}}^{(cv)}}{\left(\omega_{\mathbf{k}}^{(cv)} \right)^2 - (\omega + i\eta)^2} \tag{6.35}$$

where we have used the expansion of Eq. (6.30); the summation in the final result is over $\mathbf{k}$ values in the first BZ and (c, v) pairs of values of conduction and valence bands at the same $\mathbf{k}$ value. This last expression is the type of expression we mentioned before, Eq. (6.11), with $\omega_0 \neq 0$. The terms that appear under the summation in the expression for $\varepsilon^{\text{intra}}(\mathbf{q},\omega)$ give large contributions for $\omega \approx \omega_{\mathbf{k}}^{(cv)}$. For these values of ω, the expression reduces to the Lorentz expression, Eq. (6.17), mentioned earlier. In this expression, $\omega_{0,i}$ are the special values for which there is a significant portion of the BZ (many $\mathbf{k}$ values) with $\omega_{\mathbf{k}}^{(cv)} = \omega_{0,i}$. The expression for the imaginary part of the dielectric function in particular is very similar to the expressions that we saw earlier for the DOS. The main difference here is that what appears in $\varepsilon_2^{\text{intra}}(\mathbf{q},\omega)$ is the density of *pairs* of states that have energy difference $\hbar\omega_{0,i}$, rather than the DOS at given energy ϵ. This new quantity is called the "joint density of states" (JDOS), and its calculation is exactly analogous to that of the DOS. This is an important feature of the band structure, which to a large extent determines the behavior of $\varepsilon_2^{\text{intra}}(\mathbf{q},\omega)$. This result corresponds to a material in which only interband transitions are allowed, that is, a material with a band gap. Representative examples of the dielectric function for semiconductors and insulators are shown in Fig. 6.5.

We next work out what happens in the intraband transitions, that is, $n = n'$ and $\mathbf{k}' = \mathbf{k} + \mathbf{q}$. To simplify the notation, we define

$$\hbar\omega_{\mathbf{k}}^{(n)}(\mathbf{q}) = \epsilon_{\mathbf{k}+\mathbf{q}}^{(n)} - \epsilon_{\mathbf{k}}^{(n)} = \mathbf{q} \cdot \nabla_{\mathbf{k}} \epsilon_{\mathbf{k}}^{(n)} + \frac{\hbar^2}{2\overline{m}^{(n)}(\mathbf{k})} q^2, \quad \frac{1}{\overline{m}^{(n)}(\mathbf{k})} \equiv \frac{1}{3\hbar^2} \nabla_{\mathbf{k}}^2 \epsilon_{\mathbf{k}}^{(n)} \tag{6.36}$$

with the Taylor expansion in powers of $\mathbf{q}$ being valid in the limit $\mathbf{q} \to 0$; $\overline{m}^{(n)}(\mathbf{k})$ is an average effective mass,[5] defined by the local ($\mathbf{k}$-dependent) curvature of the band. The expression for the dielectric function now takes the form

$$\varepsilon^{\text{intra}}(\mathbf{q},\omega) = 1 - \frac{4\pi e^2}{Vq^2\hbar} \sum_{\mathbf{k},n} \left| \langle \psi_{\mathbf{k}+\mathbf{q}}^{(n)} | e^{i\mathbf{q}\cdot\mathbf{r}} | \psi_{\mathbf{k}}^{(n)} \rangle \right|^2 \frac{f_{\mathbf{k}+\mathbf{q}}^{(n)} - f_{\mathbf{k}}^{(n)}}{\omega_{\mathbf{k}}^{(n)}(\mathbf{q}) - (\omega + i\eta)} \tag{6.37}$$

The difficulty in setting $\mathbf{q} = 0$ in the present case lies in the fact that in this limit the term $1/q^2$ in front of the summation blows up, but the numerator of the summands also vanishes and in the denominator, $\omega_{\mathbf{k}}^{(n)}(\mathbf{q}) \to 0$, so all these terms need to be considered together when taking the limit. Again, there are two possibilities for the values of the indices and the corresponding terms in the summation:

$$\left[\epsilon_{\mathbf{k}+\mathbf{q}}^{(n)} > \epsilon_{\text{F}}, \; \epsilon_{\mathbf{k}}^{(n)} < \epsilon_{\text{F}} \right] \Rightarrow \left[f_{\mathbf{k}+\mathbf{q}}^{(n)} = 0, \; f_{\mathbf{k}}^{(n)} = 1 \right] \to \left| \langle \psi_{\mathbf{k}+\mathbf{q}}^{(n)} | e^{i\mathbf{q}\cdot\mathbf{r}} | \psi_{\mathbf{k}}^{(n)} \rangle \right|^2 \frac{-1}{\omega_{\mathbf{k}}^{(n)}(\mathbf{q}) - (\omega + i\eta)}$$

---

[5] To obtain this expression we have assumed that the curvature is isotropic, so that we can define its average, and hence a scalar value for the effective mass, which otherwise would be a tensor.

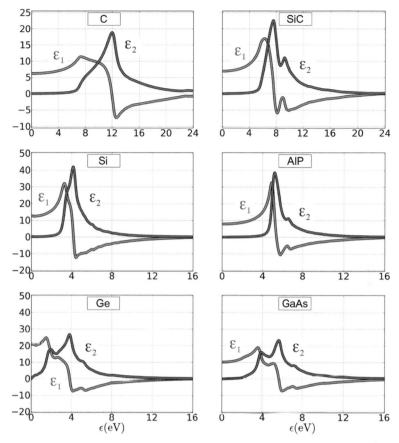

Fig. 6.5 Representative examples of the interband contributions to the dielectric function (real and imaginary parts, $\varepsilon_1^{\text{inter}}$ and $\varepsilon_2^{\text{inter}}$, respectively), for semiconductors (Si, Ge, AlP, GaAs) and insulators (C, SiC), obtained from the band structures calculated by DFT methods (calculations performed by G. A. Tritsaris).

$$\left[\epsilon_{\mathbf{k}+\mathbf{q}}^{(n)} < \epsilon_F, \; \epsilon_{\mathbf{k}}^{(n)} > \epsilon_F\right] \Rightarrow \left[f_{\mathbf{k}+\mathbf{q}}^{(n)} = 1, \; f_{\mathbf{k}}^{(n)} = 0\right] \rightarrow \left|\langle\psi_{\mathbf{k}+\mathbf{q}}^{(n)}|\mathrm{e}^{\mathrm{i}\mathbf{q}\cdot\mathbf{r}}|\psi_{\mathbf{k}}^{(n)}\rangle\right|^2 \; \frac{1}{\omega_{\mathbf{k}}^{(n)}(\mathbf{q}) - (\omega + \mathrm{i}\eta)}$$

and changing the indices of the occupied and unoccupied states, we can rewrite the expression for the dielectric function as:

$$\varepsilon^{\text{intra}}(\mathbf{q}, \omega) = 1 + \frac{4\pi e^2}{Vq^2\hbar} \sum_{\mathbf{k},n} \left|\langle\psi_{\mathbf{k}+\mathbf{q}}^{(n)}|\mathrm{e}^{\mathrm{i}\mathbf{q}\cdot\mathbf{r}}|\psi_{\mathbf{k}}^{(n)}\rangle\right|^2 \left[\frac{2f_{\mathbf{k}}^{(n)}}{\omega_{\mathbf{k}}^{(n)}(\mathbf{q}) - (\omega + \mathrm{i}\eta)} + \frac{2f_{\mathbf{k}}^{(n)}}{\omega_{\mathbf{k}}^{(n)}(\mathbf{q}) + (\omega + \mathrm{i}\eta)}\right]$$

Using the expansion of Eq. (6.30) leads to:

$$\langle\psi_{\mathbf{k}+\mathbf{q}}^{(n)}|\mathrm{e}^{\mathrm{i}\mathbf{q}\cdot\mathbf{r}}|\psi_{\mathbf{k}}^{(n)}\rangle \approx \langle\psi_{\mathbf{k}+\mathbf{q}}^{(n)}|\psi_{\mathbf{k}}^{(n)}\rangle = 1, \quad \text{for } \mathbf{q} \to 0$$

Moreover, for $\mathbf{q} \to 0$ and any finite value of ω, we have:

$$\omega_{\mathbf{k}}^{(n)}(\mathbf{q}) \to 0 \Rightarrow \omega_{\mathbf{k}}^{(n)}(\mathbf{q})/\omega \ll 1$$

and using Taylor expansions in the small quantity $\omega_{\mathbf{k}}^{(n)}(\mathbf{q})/\omega$ for the two terms in the square brackets, we arrive at:

$$\varepsilon^{\text{intra}}(\mathbf{q}, \omega) = 1 + \left(\frac{4\pi e^2}{Vq^2\hbar}\right) \sum_{\mathbf{k},n} 2f_{\mathbf{k}}^{(n)} \left[\frac{-2\omega_{\mathbf{k}}^{(n)}(\mathbf{q})}{\omega(\omega + i\eta)}\right]$$

$$= 1 - \left(\frac{4\pi e^2}{Vq^2\hbar}\right) \left[\frac{4}{\hbar}\mathbf{q} \cdot \left(\sum_{\mathbf{k},n} \nabla_{\mathbf{k}}\epsilon_{\mathbf{k}}^{(n)} f_{\mathbf{k}}^{(n)}\right) + q^2 \left(\sum_{\mathbf{k},n} \frac{2\hbar f_{\mathbf{k}}^{(n)}}{\overline{m}^{(n)}(\mathbf{k})}\right)\right] \frac{1}{\omega(\omega + i\eta)}$$

We notice that, due to Kramers degeneracy, without taking spin into account, $\epsilon_{\mathbf{k}}^{(n)}$ must be an even function of $\mathbf{k}$, since $\epsilon_{-\mathbf{k}}^{(n)} = \epsilon_{\mathbf{k}}^{(n)}$, and therefore its derivative with respect to $\mathbf{k}$ must be an odd function of $\mathbf{k}$; for the same reason, $f_{\mathbf{k}}^{(n)}$ is also an even function of $\mathbf{k}$. These observations imply that the term proportional to $\mathbf{q}$ in the above expression will vanish upon summing over $\mathbf{k}$ in the BZ. Thus, the only term remaining in the summation over $\mathbf{k}$ is the one proportional to $\mathbf{q}^2 = q^2$, which leads to:

$$\varepsilon^{\text{intra}}(\mathbf{q}, \omega) = 1 - \left(\frac{4\pi e^2}{V}\right) \sum_{\mathbf{k},n} \frac{2f_{\mathbf{k}}^{(n)}}{\overline{m}^{(n)}(\mathbf{k})} \frac{1}{\omega(\omega + i\eta)}$$

At this point, we need to specify the values of $\mathbf{k}$ for which the summand does not vanish. The above derivation was based on the assumption that the band labeled n is such that a state with eigenvalue $\epsilon_{\mathbf{k}+\mathbf{q}}^{(n)}$ is occupied (valence state) and another state with eigenvalue $\epsilon_{\mathbf{k}}^{(n)}$ is empty (conduction state), or vice versa, in the limit $\mathbf{q} \to 0$. This means that the band labeled n must be intersected by the Fermi level. We therefore define a surface in reciprocal space, $S_F^{(n)}$, associated with this band and containing all the values of $\mathbf{k}$ at which the Fermi level intersects the band, that is, on this surface $\epsilon_{\mathbf{k}}^{(n)} = \epsilon_F$. The summation that appears in the last equation will then extend over all values of $\mathbf{k}$ on this surface. We define N_{ele}, the number of $\mathbf{k}$ values that belong to all such surfaces of bands that are intersected by the Fermi level:

$$N_{\text{ele}} = \sum_{n} \sum_{\mathbf{k} \in S_F^{(n)}} 2 \tag{6.38}$$

which counts all the electronic states, including spin degeneracy, that belong to surfaces defined by the Fermi level. With this definition, the expression for the dielectric function arising from intraband contributions takes the form

$$\varepsilon^{\text{intra}}(\mathbf{q}, \omega) = 1 - \left(\frac{4\pi e^2 N_{\text{ele}}}{m_e V}\right) \left[\frac{4}{N_{\text{ele}}} \sum_{n} \sum_{\mathbf{k} \in S_F^{(n)}} \frac{m_e}{\overline{m}^{(n)}(\mathbf{k})}\right] \frac{1}{\omega(\omega + i\eta)} \tag{6.39}$$

where we have set $f_{\mathbf{k}}^{(n)} = 1$. The expression for the intraband dielectric function is precisely the type of expression we mentioned before, Eq. (6.11), with $\omega_0 = 0$. In the final result we have also introduced a factor of m_e/N_{ele} inside the square brackets and its reciprocal factor outside, where m_e is the electron mass, to produce more physically transparent expressions. Specifically, the quantity is parentheses is the square of the bare plasma frequency, while the quantity in brackets is a dimensionless number, giving the average of the band curvature on the surface in reciprocal space where the band intersects the Fermi level, for each such band labeled n; this factor modifies the bare plasma frequency to a value appropriate for the particular solid, whose dielectric behavior due to intraband transitions is determined by this set of bands. The remaining term in the expression for the dielectric function, namely $1/\omega(\omega + i\eta)$, determines its dependence on the frequency, a behavior equivalent to the Drude expression for the dielectric function discussed earlier, Eq. (6.15).

The derivations outlined above show that the dielectric function of a metal with several bands will have intraband contributions, described by Eq. (6.39), for small values of ω, as well as interband contributions, described by Eq. (6.35), at higher-frequency values, depending on which pair of bands shows large JDOS, and the corresponding dipole matrix elements do not vanish. In contrast to this, the dielectric function of a semiconductor or insulator will derive only from interband contributions, described by Eq. (6.35), which also depends on large JDOS between pairs of bands and non-vanishing dipole matrix elements. We discuss next representative examples of each type.

6.2.5 Application: Optical Properties of Metals and Semiconductors

We address the behavior of the dielectric function of Al first, shown in Fig. 6.3. We saw there that, in addition to the dominant free-electron contribution, there was a bound-electron contribution at $\hbar\omega_0 = 1.4$ eV. This feature comes from direct transitions between occupied and unoccupied states near the W and K points of the BZ. In these regions, marked E_3 and E_4, there exist bands below the Fermi level and above the Fermi level that are almost parallel, hence there exists a large JDOS, and the energy difference between these pairs of bands is approximately 1.4 eV.

In Fig. 6.6 we show the dielectric function of the noble metals Cu and Ag for a wide range of energies up to 20 eV. In addition to the bound-electron feature discussed earlier (see Fig. 6.4 and related comments), there are many other interesting features arising from transitions between states below and above the Fermi level that have similar dispersion, that is, their energies are nearly parallel. In the low-energy region, a major feature is found around 5 eV, which is manifest as a broad peak in Cu and a broad shoulder in Ag. The transitions responsible for this feature span a range of energies, and are marked as E_1 for those near the L point and as E_2 for those near the X point in the BZ. This behavior closely matches the general behavior of the dielectric function as illustrated in Fig. 6.2. Other features at higher energies can similarly be explained by transitions between parallel bands, of which several examples can be found in different regions of the BZ.

Data for the dielectric functions of Al, Cu, Ag, and various other metals are collected in Table 6.1. In this table we compare actual values of the plasma frequency $\hbar\omega_p$ as extracted from experiments to those obtained from Eq. (6.12), which can be written as:

$$\hbar\omega_p = \left(\frac{16\pi n_{at} n_{el}}{(a/a_0)^3} \right)^{1/2} \times 13.6058 \text{ eV} \tag{6.40}$$

the latter being an expression based on the number of electrons per unit volume available to partipate in plasmon oscillations; in this expression n_{at} is the number of atoms in the

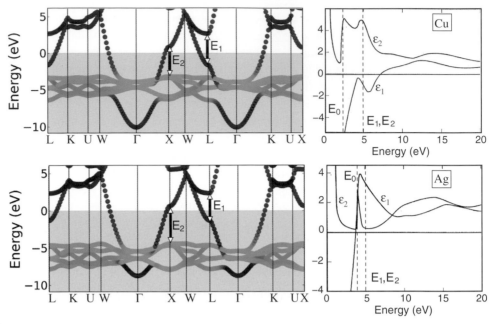

Fig. 6.6 Examples of the behavior of the real and imaginary parts of the dielectric function $\varepsilon_1(\omega)$, $\varepsilon_2(\omega)$ for Cu (top) and Ag (bottom). In addition to the d-band-to-Fermi-surface feature, marked as E_0 and discussed in Fig. 6.4, other low-energy features involve transitions near the L point marked as E_1 and transitions near the X point marked as E_2. The experimental results are adapted from H. Ehrenreich and H. R. Philipp, *Phys. Rev.* **128**, 1622 (1962).

conventional unit cell (for the BCC, FCC, diamond, and zincblende crystals), a is the lattice constant of this unit cell in units of a_0, and n_{el} is the average number of valence electrons per atom. The agreement between these two quantities is quite reasonable when a meaningful value for the average number of valence electrons per atom is used: this means three electrons for Al, one electron for the alkalis (Na, K) and noble metals (Cu, Ag, Au), and just 0.5 electrons for Pt.

For semiconductors, it is often easy to identify the features of the band structure which are responsible for the major peaks of the dielectric function. These features are typically related to transitions between occupied and unoccupied bands which happen to be parallel, that is, they have a constant energy difference over a significant portion of the BZ, because this produces a large JDOS. We show in Fig. 6.7 the examples of Si and Ge. In the case of Si, for instance, there are two such sets of bands, one in which the highest occupied and lowest unoccupied bands differ by a constant energy of ~ 3.5 eV near the Γ point, marked as E_0, and the L point, marked as E_1 in Fig. 6.7, and another where they differ by ~ 4.2 eV near the X point, marked as E_2 in Fig. 6.7. Indeed, a major feature in both the real and imaginary parts of the dielectric function of Si [a node in $\varepsilon_1(\omega)$ and the main peak in $\varepsilon_2(\omega)$] can be seen at $\hbar\omega \approx 4.1$ eV in Fig. 6.5. A second feature at $\hbar\omega \approx 3.8$ eV is seen as a shoulder to the left of the main peak. Similar features appear in the dielectric function of Ge: the peak near $\hbar\omega \approx 2$ eV is due to interband transitions near the L point, marked as E_1, and the peak near $\hbar\omega \approx 4.5$ eV is due to interband transitions near the X point, marked as E_2.

Table 6.1 Plasma frequencies for various metal and semiconductor crystals (the crystal structure and lattice constant a in units of a_0 are also included): ω_p is the estimate from Eq. (6.40), ω_p^{exp} is from fitting the Drude model to experimental data, $\omega_p^{(1)}$ is from the zero of $\varepsilon_1(\omega)$, $\omega_p^{(l)}$ is from the maximum of the energy loss function, Eq. (6.41). All values of the frequencies and inverse relaxation times $(\hbar/\tau)$ are in electron-volts. Data for metals are from M. A. Ordal *et al.*, *Applied Optics* **24**, 4493 (1985) and for semiconductors from H. R. Philipp and H. Ehrenreich, *Phys. Rev.* **129**, 1550 (1963).

Metal	a/a_0	n_{at}	n_{el}	$\hbar\omega_p$	$\hbar\omega_p^{exp}$	$\hbar/\tau$
Na (BCC)	8.11	2	1	5.91	5.71	0.028
K (BCC)	10.07	2	1	4.27	3.72	0.018
Al (FCC)	7.65	4	3	15.77	14.75	0.082
Cu (FCC)	6.82	4	1	10.82	7.39	0.009
Ag (FCC)	7.73	4	1	8.97	9.01	0.018
Au (FCC)	7.71	4	1	9.00	9.03	0.027
Pt (FCC)	7.4	4	0.5	6.76	5.15	0.069

Semiconductor	a/a_0	n_{at}	n_{el}	$\hbar\omega_p$	$\hbar\omega_p^{(1)}$	$\hbar\omega_p^{(l)}$
Si (DIA)	10.26	8	4	16.60	15.0	16.4
Ge (DIA)	10.70	8	4	15.60	13.8	16.0
GaAs (ZBL)	10.68	8	4	15.63	9.7	14.7

As these two examples of semiconductors show, at high enough energies (around 15 eV), the behavior of the dielectric function exhibits the general features expected from both the free-electron and the bound-electron contributions, that is, $\varepsilon_1(\omega)$ crosses from negative to positive values. In the case of semiconductors, where the bound-electron contribution is dominant at low frequencies (see also the computed dielectric functions in Fig. 6.5), it is difficult to assign a value to the plasma frequency ω_p. This is in contrast to the case of metals, where the low-frequency behavior clearly exhibits a free-electron contribution that can be fitted to yield a value for ω_p. For semiconductors there are two alternative ways to estimate ω_p: from the value at which $\varepsilon_1(\omega)$ changes sign, which we denote by $\omega_p^{(1)}$, or from the value at which the so-called "energy loss" function, defined as

$$\frac{\varepsilon_2(\omega)}{|\varepsilon(\omega)|^2} = -\text{Im}\left[\frac{1}{\varepsilon(\omega)}\right] \tag{6.41}$$

exhibits a maximum, which we denote by $\omega_p^{(l)}$; the energy loss function gives a measure of the energy absorbed by the oscillations of the electron gas in response to the external field. In Table 6.1 we show a comparison between the values of ω_p obtained from Eq. (6.40) and those of $\omega_p^{(1)}$ and $\omega_p^{(l)}$ for the representative semiconductors Si, Ge, and GaAs. In general, the value of $\omega_p^{(1)}$ appears to be an underestimate, while the value of $\omega_p^{(l)}$ is higher and closer to ω_p.

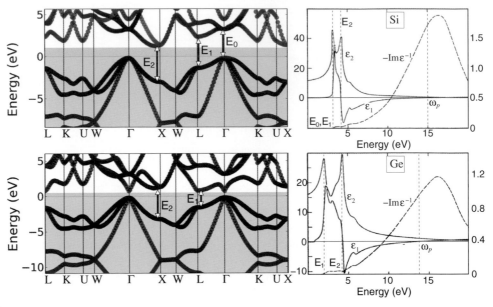

Fig. 6.7 The dielectric function of Si (top) and Ge (bottom), real ε_1 and imaginary ε_2 parts. The band structures on the left show the excitations that produce the main features in the dielectric constant; these are labeled as E_0 for transitions near the Γ point, E_1 for transitions near the L point, and E_2 for transitions near the X point. The value of the plasma frequency marked by ω_p corresponds to $\omega_p^{(1)}$, the root of $\varepsilon_1(\omega)$. The experimental results are adapted from H. R. Philipp and H. Ehrenreich, *Phys. Rev.* **129**, 1550 (1963); the quantity $-\mathrm{Im}\,\varepsilon^{-1}$ is shown on the right axis.

6.3 Excitons

Up to this point we have discussed how crystals can be excited from their ground state by absorption of light quanta, that is, photons; the reverse process, de-excitation of an excited state by emission of photons, is also possible. In treating these effects, we concentrated on dominant features of light absorption of emission, related to portions of the band structure of the crystal where there are many states that can contribute to such processes, namely, there is a large JDOS between occupied and empty bands that can participate in the excitations. In semiconductors and insulators there is another aspect of light absorption that is very important, the low-energy excitations. This refers to energy values close to the band gap, since at least that much energy is needed to induce an excitation by photon absorption in semiconductors and insulators. In this range of energy, the excited electron and the hole left behind are oppositely charged quasiparticles, and as such interact by a screend Coulomb potential. Because their charges are opposite, the interaction is attractive, but the presence of all the other charged particles in the system provides the screening. The attraction between electron and hole quasiparticles leads to formation of bound states. The bound electron–hole pair is called an "exciton." Its presence in the optical absorption spectra of semiconductors and insulators is ubiquitous, and can produce very interesting phenomena, such as condensation of these entities, which act like bosons.

Moreover, the excitons must dissociate in order to produce free carriers (electrons or holes) that can carry current through the material; this process competes with recombination of the electron and the hole, producing emission of a photon. Thus, exciton dissociation is a process of great practical importance in devices like photovoltaics, that depend on electron excitation by light and subsequent motion of the charges (positive and negative) in the device. The creation and behavior of excitons involves both fascinating physics and important consequences for practical applications. In this section we explore the main ideas in the physics of excitons, first through simple heuristic arguments, and then through more detailed models based on the picture of solids that we have developed so far.

6.3.1 General Considerations

Experimentally, the presence of excitons becomes evident when light is absorbed by the solid as individual photons, assuming that each absorbed photon creates an electron–hole pair. The signature of excitons appears in the optical absorption measurements. If there were no excitons, the lowest energy of a photon required to produce an excited electron, while leaving a hole behind, would be equal to the band gap. The presence of the exciton implies that the lowest-energy excitation induced by be absorption of light will correspond to photon energies *smaller* than the band-gap energy, as expected for an electron hole pair that is bound by the Coulomb interaction. Quite often, several exciton peaks can be resolved experimentally, reflecting the features of the valence and conduction bands of the solid, as shown in Fig. 6.8.

There is a wide range of excitons. The limiting cases concern electron–hole pairs that are strongly or weakly bound. In the tightly bound case, the binding energy is of order 1 eV. This is common in insulators, like ionic solids (for example, SiO_2); these are referred to as "Frenkel excitons."[6] In the weakly bound case, the exciton is delocalized over a range of several angstroms, and with a small binding energy of order 0.001–0.1 eV (1–100 meV); these are referred to as Wannier or Mott–Wannier excitons[7] and are common in small band-gap systems, especially semiconductors. These two limiting cases are reasonably well understood, while intermediate cases are more difficult to treat. In Table 6.2 we give some examples of materials that have Frenkel or Mott–Wannier excitons and the corresponding band gaps and exciton binding energies.

The presence of excitons is a genuinely many-body effect, so we need to invoke the many-body hamiltonian to describe the physics. For the purposes of the following treatment, we will find it convenient to cast the problem in the language of Slater determinants composed of single-particle states, so that solving it becomes an exercise in dealing with single-particle wavefunctions. We begin by writing the total many-body hamiltonian $\hat{\mathcal{H}}$ in the form

$$\hat{\mathcal{H}}(\{\mathbf{r}_i\}) = \sum_i \hat{h}(\mathbf{r}_i) + \frac{1}{2}\sum_{i \neq j} \frac{e^2}{|\mathbf{r}_i - \mathbf{r}_j|}, \quad \hat{h}(\mathbf{r}) = -\frac{\hbar^2}{2m_e}\nabla_{\mathbf{r}}^2 + \sum_{\mathbf{R},I} \frac{-Z_I e^2}{|\mathbf{r} - \mathbf{R} - \mathbf{t}_I|} \quad (6.42)$$

[6] J. Frenkel, *Phys. Rev.* **37**, 1276 (1931).
[7] N. Mott, *Trans. Faraday Soc.* **34**, 500 (1938); G. Wannier, *Phys. Rev.* **52**, 191 (1937).

Table 6.2 Examples of Frenkel excitons in ionic solids and Mott–Wannier excitons in semiconductors; the band gap ϵ_{gap} and the binding energies of the first excitonic state, $E_1^{\text{bin}} = \epsilon_{\text{gap}} - E_1^{\text{exc}}$, Eq. (6.48), are given in units of electron-volts. [*Sources*: M. Fox, *Optical Properties of Solids* (Oxford University Press, New York, 2001); K. Takahashi, A. Yoshikawa, and A. Sandhu, (Eds), *Wide Bandgap Semiconductors – Fundamental Properties and Modern Photonic and Electronic Devices* (Springer-Verlag, Berlin, 2007).]

	Frenkel excitons					Mott–Wannier excitons				
	ϵ_{gap}	E_1^{bin}		ϵ_{gap}	E_1^{bin}		ϵ_{gap}	E_1^{bin}		ϵ_{gap}	E_1^{bin}
KI	6.3	0.4	NaI	5.9	0.3	ZnS	3.84	0.037	Si	1.17	0.015
KBr	7.4	0.7	NaBr	7.1	0.4	ZnO	3.37	0.059	Ge	0.74	0.004
KCl	8.7	0.9	NaCl	8.8	0.9	ZnSe	2.83	0.017	AlN	6.28	0.044
KF	10.8	0.9	NaF	11.5	0.8	ZnTe	2.39	0.013	GaN	3.52	0.025
RbCl	8.3	0.4	CsF	9.8	0.5	CdS	2.58	0.027	GaAs	1.52	0.004
RbF	10.3	0.8	LiF	13.7	0.9	CdSe	1.73	0.015	InP	1.42	0.005
						CdTe	1.61	0.012	GaSb	0.80	0.002

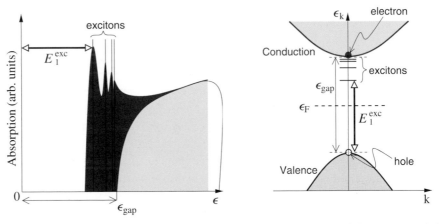

Fig. 6.8   **Left**: Modification of the absorption spectrum in the presence of excitons (solid red curve) relative to the spectrum in the absence of excitons (faint curve); in the latter case absorption begins at exactly the value of the intrinsic band gap, ϵ_{gap}. In this example, we show four bound states for excitons, the sharp peaks below ϵ_{gap}. **Right**: Schematic representation of the conduction and valence bands and the energy levels in the gap corresponding to electron–hole bound states. The energy of the first exciton state is shown as E_1^{exc}; its binding energy is $E_1^{\text{bin}} = \epsilon_{\text{gap}} - E_1^{\text{exc}}$, see Eq. (6.48).

where $\mathbf{R}$ are the lattice vectors and $\mathbf{t}_I$ are the positions of the ions in the unit cell. In this manner, we separate the part that can be dealt with strictly in the single-particle framework, namely $\hat{h}(\mathbf{r})$, which is the non-interacting part, and the part that contains all the complications of electron–electron interactions. For simplicity, in the following we will discuss the case where there is only one atom per unit cell, thus eliminating the index I for the positions of atoms within the unit cell. The many-body wavefunction will have Bloch-like symmetry:

$$\Psi_{\mathbf{K}}(\mathbf{r}_1 + \mathbf{R}, \mathbf{r}_2 + \mathbf{R}, \ldots) = e^{i\mathbf{K}\cdot\mathbf{R}}\Psi_{\mathbf{K}}(\mathbf{r}_1, \mathbf{r}_2, \ldots) \tag{6.43}$$

We will illustrate these issues for the cases of both strongly bound (Frenkel) excitons and weakly bound (Wannier) excitons. For the many-body wavefunctions, we choose to work with Slater determinants in the positions $\{\mathbf{r}\}$ and spins $\{s\}$ of the electrons. To keep the notation simple and the discussion transparent, we will consider the band structure of a solid with a fully occupied valence band and an empty conduction band; generalization to more bands is straightforward. For a solid consisting of N atoms with two electrons per atom (so that we can have a fully occupied valence band), we then have the general many-body wavefunction

$$\Psi_{\mathbf{K}}(\{\mathbf{r}\}, \{s\}) = \frac{1}{\sqrt{(2N)!}}
\begin{vmatrix}
\psi_{\mathbf{k}_1\uparrow}^{(v)}(\mathbf{r}_1) & \psi_{\mathbf{k}_1\uparrow}^{(v)}(\mathbf{r}_2) & \cdots & \psi_{\mathbf{k}_1\uparrow}^{(v)}(\mathbf{r}_{2N}) \\
\psi_{\mathbf{k}_1\downarrow}^{(v)}(\mathbf{r}_1) & \psi_{\mathbf{k}_1\downarrow}^{(v)}(\mathbf{r}_2) & \cdots & \psi_{\mathbf{k}_1\downarrow}^{(v)}(\mathbf{r}_{2N}) \\
\cdot & \cdot & & \cdot \\
\cdot & & \cdot & \cdot \\
\cdot & \cdot & & \cdot \\
\psi_{\mathbf{k}_N\uparrow}^{(v)}(\mathbf{r}_1) & \psi_{\mathbf{k}_N\uparrow}^{(v)}(\mathbf{r}_2) & \cdots & \psi_{\mathbf{k}_N\uparrow}^{(v)}(\mathbf{r}_{2N}) \\
\psi_{\mathbf{k}_N\downarrow}^{(v)}(\mathbf{r}_1) & \psi_{\mathbf{k}_N\downarrow}^{(v)}(\mathbf{r}_2) & \cdots & \psi_{\mathbf{k}_N\downarrow}^{(v)}(\mathbf{r}_{2N})
\end{vmatrix}$$

In the ground state, all the states corresponding to $\mathbf{k}$ values in the first BZ will be occupied, and the total wave-vector will be equal to 0, since for every occupied $\mathbf{k}$-state there is a corresponding $-\mathbf{k}$-state with the same energy. Similarly, the total spin will be 0, because of the equal occupation of single-particle states with up and down spins. Our goal is to create excited states, starting from the ground-state wavefunction, by replacing a single-particle valence wavefunction that is occupied in the ground state with a single-particle conduction wavefunction that is empty in the ground state. When we do this, care must be taken to preserve the important features of the many-body wavefunction that should not be affected by the excitation. Specifically, as argued above, optical excitations involve direct $\mathbf{k}$ transitions, as explained in Section 6.2.4, that is, excitation by absorption (or de-excitation by emission) of a photon produces no change in the crystal momentum of electrons or the total spin.

Because of the difference in the nature of holes and particles, we need to pay special attention to the possible spin states of the entire system. When the electron is removed from a state with spin s_h, the many-body state has a total spin z component $S_z = -s_h$, since the original ground state had spin 0. Therefore, the new state created by adding a particle with spin s_p produces a many-body state with total spin z component $S_z = s_p - s_h$. This reveals that when we deal with hole states, we must take their contribution to the spin as the opposite of what a normal particle would contribute. Taking into consideration the fact that the hole has the opposite wave-vector of a particle in the same state, we conclude that the hole corresponds to the time-reversed particle state, since the effect of the time-reversal operator $\hat{\mathcal{T}}$ on the energy and the wavefunction is:

$$\hat{\mathcal{T}}\epsilon_{\mathbf{k}\uparrow}^{(n)} = \epsilon_{-\mathbf{k}\downarrow}^{(n)}, \quad \hat{\mathcal{T}}|\psi_{\mathbf{k}\uparrow}^{(n)}\rangle = |\psi_{-\mathbf{k}\downarrow}^{(n)}\rangle$$

Table 6.3 Spin configurations for a particle–particle pair and a particle–hole pair for spin-1/2 particles; in the latter case the first spin refers to the particle, the second to the hole.

Spin state		Particle–particle	Particle–hole
S	S_z		
1	1	$\uparrow\uparrow$	$\uparrow\downarrow$
1	0	$\frac{1}{\sqrt{2}}(\uparrow\downarrow + \downarrow\uparrow)$	$\frac{1}{\sqrt{2}}(\uparrow\uparrow - \downarrow\downarrow)$
1	−1	$\downarrow\downarrow$	$\downarrow\uparrow$
0	0	$\frac{1}{\sqrt{2}}(\uparrow\downarrow - \downarrow\uparrow)$	$\frac{1}{\sqrt{2}}(\uparrow\uparrow + \downarrow\downarrow)$

as discussed in Chapter 2. Notice that for the particle–hole system, if we start as usual with the highest S_z state, which in this case is $\uparrow\downarrow$, and proceed to create the rest by applying spin-lowering operators (see Appendix C), there will be an overall minus sign associated with the hole spin-lowering operator due to complex conjugation implied by time reversal. The resulting spin states for the particle–hole system, as identified by the total spin S and its z component S_z, are given in Table 6.3 and contrasted with the particle–particle spin states.

With these considerations, we are now ready to construct specific models to capture the nature and behavior of excitons. But before we discuss such detailed models, we provide a heuristic picture that captures the essence of the problem; we will be able to justify this heuristic picture from the more detailed treatment that follows. We shall assume that we are dealing with two quantum-mechanical entities, a particle with effective mass $\bar{m}_e$ and a hole with effective mass $\bar{m}_h$. These have been created by an excitation relative to the ground state, in which all states below the Fermi level are occupied and separated by an energy gap ϵ_{gap} from the unoccupied states. We think of these as quasiparticles of opposite charge, $-e$ for the particle and $+e$ for the hole, interacting with a screened Coulomb potential:

$$\mathcal{V}(\mathbf{r}_p, \mathbf{r}_h) = -\frac{e^2}{\bar{\varepsilon}_0 |\mathbf{r}_p - \mathbf{r}_h|}$$

where $\bar{\varepsilon}_0$ is the static dielectric constant of the medium in which they move. In this oversimplified picture, the details of the crystal potential that the particle and the hole experience have been subsumed in a few key features: the dielectric constant of the crystal and the effective masses, which are a consequence of the fact that they are derived from electronic states of the crystal. More specifically, these effective masses are determined by the curvature at the extrema of the relevant conduction and valence bands corresponding to the particle and hole states.

The hamiltonian for this simple system is given by:

$$\hat{\mathcal{H}}(\mathbf{r}_p, \mathbf{r}_h) = \epsilon_{\text{gap}} + \frac{\mathbf{p}_p^2}{2\bar{m}_e} + \frac{\mathbf{p}_h^2}{2\bar{m}_h} - \frac{e^2}{\bar{\varepsilon}_0 |\mathbf{r}_p - \mathbf{r}_h|}$$

with $\mathbf{p}_p$ and $\mathbf{p}_h$ the momentum operator for the particle and the hole. Since the energy is measured relative to the gound state, the hamiltonian includes a constant term ϵ_{gap} to reflect the fact that the particle–hole excitation involves the transfer of an electron from an occupied state below the Fermi level to an unoccupied state above the Fermi level, which differ by this much energy. We now make the standard change of coordinates to the center-of-mass $\mathbf{R}$ and the relative coordinate $\mathbf{r}$:

$$\mathbf{R} = \frac{\overline{m}_e \mathbf{r}_p + \overline{m}_h \mathbf{r}_h}{\overline{m}_e + \overline{m}_h}, \quad \mathbf{r} = \mathbf{r}_p - \mathbf{r}_h \tag{6.44}$$

in terms of which we rewrite the hamiltonian as:

$$\hat{\mathcal{H}}(\mathbf{R}, \mathbf{r}) = \epsilon_{\text{gap}} + \frac{\mathbf{P}^2}{2M} + \frac{\mathbf{p}^2}{2\mu} - \frac{e^2}{\bar{\varepsilon}_0 |\mathbf{r}|}$$

with $\mathbf{p}_0$ and $\mathbf{p}$ the momentum operators for the center-of-mass and the relative coordinate, and the two new variables for the masses, M and μ, defined by:

$$M = \overline{m}_e + \overline{m}_h, \quad \mu = \frac{\overline{m}_e \overline{m}_h}{\overline{m}_e + \overline{m}_h} \tag{6.45}$$

Since the last expression of the hamiltonian includes only a kinetic energy term for the center-of-mass coordinate, we can write the wavefunction of the system as:

$$\Psi_{\mathbf{K}}(\mathbf{r}_p, \mathbf{r}_h) = \Psi_{\mathbf{K}}(\mathbf{R}, \mathbf{r}) = \frac{1}{\sqrt{V}} e^{i\mathbf{K}\cdot\mathbf{R}} \psi(\mathbf{r}) \tag{6.46}$$

where we have separated out the center-of-mass motion as a plane wave with wave-vector $\mathbf{K}$. The two-body Schrödinger equation then reads:

$$\hat{\mathcal{H}}(\mathbf{R}, \mathbf{r})\Psi_{\mathbf{K}}(\mathbf{R}, \mathbf{r}) = E_{\mathbf{K}}^{\text{exc}} \Psi_{\mathbf{K}}(\mathbf{R}, \mathbf{r}) \Rightarrow$$

$$\left(-\frac{\hbar^2 \nabla_{\mathbf{r}}^2}{2\mu} - \frac{e^2}{\bar{\varepsilon}_0 |\mathbf{r}|} \right) \psi(\mathbf{r}) = \left(E_{\mathbf{K}}^{\text{exc}} - \epsilon_{\text{gap}} - \frac{\hbar^2 \mathbf{K}^2}{2M} \right) \psi(\mathbf{r}) \tag{6.47}$$

The last expression is simply the Schrödinger equation for a hydrogen-like atom, with effective mass of the "electron" equal to μ and Coulomb potential between the "nucleus" and the "electron" screened by the dielectric constant $\bar{\varepsilon}_0$. We have encountered this model before, in the case of dopant impurities in crystals, Section 5.5; by solving this problem, we recover the solutions corresponding to the hydrogen atom, with both the energy and the length scaled by appropriate factors that involve the dielectric constant $\bar{\varepsilon}_0$ and the effective mass μ. The sequence of energy levels from these solutions corresponds closely to the excitonic peaks found in experimental measurements, discussed in relation to Fig. 6.8. The difference between the value of the gap and the actual energy of the state with the electron–hole pair excitation, a positive quantity, is called the "binding energy" of the exciton, defined as:

$$E_{\mathbf{K},n}^{\text{bin}} = \epsilon_{\text{gap}} - E_{\mathbf{K},n}^{\text{exc}} \tag{6.48}$$

for various states labeled by n. We shall return to these results and provide a detailed discussion of the implications once we have developed a more rigorous justification of the origin of Eq. (6.47).

6.3.2 Strongly Bound (Frenkel) Excitons

For strongly bound (Frenkel) excitons, it is convenient to use a unitary transformation to a new set of basis functions which are localized at the positions of the ions in each unit cell of the lattice. These so-called Wannier functions, discussed in Section 2.7, are defined in terms of the usual band states $\psi_{\mathbf{k},s}^{(v)}(\mathbf{r})$ through the following relation:

$$\phi_s^{(v)}(\mathbf{r} - \mathbf{R}) = \frac{1}{\sqrt{N}} \sum_{\mathbf{k}} e^{-i\mathbf{k}\cdot\mathbf{R}} \psi_{\mathbf{k},s}^{(v)}(\mathbf{r}) \tag{6.49}$$

Using this new basis, we can express the many-body wavefunction as:

$$\Psi_0(\{\mathbf{r}\}) = \frac{1}{\sqrt{(2N)!}} \begin{vmatrix} \phi_{\uparrow}^{(v)}(\mathbf{r}_1 - \mathbf{R}_1) & \cdots & \phi_{\uparrow}^{(v)}(\mathbf{r}_{2N} - \mathbf{R}_1) \\ \phi_{\downarrow}^{(v)}(\mathbf{r}_1 - \mathbf{R}_1) & \cdots & \phi_{\downarrow}^{(v)}(\mathbf{r}_{2N} - \mathbf{R}_1) \\ & \cdot & \\ & \cdot & \\ & \cdot & \\ \phi_{\uparrow}^{(v)}(\mathbf{r}_1 - \mathbf{R}_N) & \cdots & \phi_{\uparrow}^{(v)}(\mathbf{r}_{2N} - \mathbf{R}_N) \\ \phi_{\downarrow}^{(v)}(\mathbf{r}_1 - \mathbf{R}_N) & \cdots & \phi_{\downarrow}^{(v)}(\mathbf{r}_{2N} - \mathbf{R}_N) \end{vmatrix} \tag{6.50}$$

In order to create an exciton wavefunction, we remove a single electron from state $\phi_{s_h}^{(v)}(\mathbf{r} - \mathbf{R}_h)$ and put it in state $\phi_{s_p}^{(c)}(\mathbf{r} - \mathbf{R}_p)$. We next construct Bloch states from the proper basis. Let us denote by

$$\Phi_{\mathbf{R}_i}^{(S,S_z)}(\{\mathbf{r}\}) = \sum_{s_p} \frac{\lambda^{(S,S_z)}(s_p)}{\sqrt{(2N)!}} \begin{vmatrix} \phi_{\uparrow}^{(v)}(\mathbf{r}_1 - \mathbf{R}_1) & \cdots & \phi_{\uparrow}^{(v)}(\mathbf{r}_{2N} - \mathbf{R}_1) \\ \phi_{\downarrow}^{(v)}(\mathbf{r}_1 - \mathbf{R}_1) & \cdots & \phi_{\downarrow}^{(v)}(\mathbf{r}_{2N} - \mathbf{R}_1) \\ & \cdot & \\ \phi_{s_p}^{(c)}(\mathbf{r}_1 - \mathbf{R}_i) & \cdots & \phi_{s_p}^{(c)}(\mathbf{r}_{2N} - \mathbf{R}_i) \\ & \cdot & \\ \phi_{\uparrow}^{(v)}(\mathbf{r}_1 - \mathbf{R}_N) & \cdots & \phi_{\uparrow}^{(v)}(\mathbf{r}_{2N} - \mathbf{R}_N) \\ \phi_{\downarrow}^{(v)}(\mathbf{r}_1 - \mathbf{R}_N) & \cdots & \phi_{\downarrow}^{(v)}(\mathbf{r}_{2N} - \mathbf{R}_N) \end{vmatrix} \tag{6.51}$$

the many-body wavefunction produced by exciting one electron from a valence to a conduction single-particle state at the unit cell identified by the lattice vector $\mathbf{R}_i$. There are two differences between $\Phi_{\mathbf{R}_i}^{(S,S_z)}$ and the ground-state wavefunction Ψ_0 of Eq. (6.50): a valence single-particle state $\phi^{(v)}(\mathbf{r})$ has been changed into a conduction single-particle state $\phi^{(c)}(\mathbf{r})$ at the unit cell identified by $\mathbf{R}_i$, and the spin of this single-particle state is now s_p, which can be different from the spin of the original single-particle state, denoted by s_h. The new many-body state has total spin S and z-projection S_z, produced by the combination of s_p, s_h, in the manner discussed above; in the $\Phi_{\mathbf{R}_i}^{(S,S_z)}$ wavefunction, $\lambda^{(S,S_z)}(s_p)$ are the coefficients corresponding to the (S, S_z) state for each s_p value as given in Table 6.3. Then, the Bloch state obtained by appropriately combining such states is:

$$\Psi_{\mathbf{K}}^{(S,S_z)}(\{\mathbf{r}\}) = \frac{1}{\sqrt{N}} \sum_{i=1}^{N} e^{i\mathbf{K}\cdot\mathbf{R}_i} \Phi_{\mathbf{R}_i}^{(S,S_z)}(\{\mathbf{r}\}) \tag{6.52}$$

The choice we have made in writing down the many-body wavefunction $\Phi_{\mathbf{R}_i}^{(S,S_z)}$, namely that the electron and the hole are localized at the same lattice site $\mathbf{R}_i$, represents the extreme case of a localized Frenkel exciton. This can be generalized to a less restrictive situation, where the electron is created at a lattice site different from the hole position, that is, the particle state is at lattice vector $\mathbf{R}_j, j \neq i$, and appears on the ith row of the Slater determinant (see Problem 10). The energy of the state corresponding to the choice of identical electron and hole positions is then:

$$E_{\mathbf{K}}^{(S,S_z)} = \langle \Psi_{\mathbf{K}}^{(S,S_z)}|\hat{\mathcal{H}}|\Psi_{\mathbf{K}}^{(S,S_z)}\rangle = \frac{1}{N} \sum_{\mathbf{R},\mathbf{R}'} e^{i\mathbf{K}\cdot(\mathbf{R}-\mathbf{R}')}\langle \Phi_{\mathbf{R}'}^{(S,S_z)}|\hat{\mathcal{H}}|\Phi_{\mathbf{R}}^{(S,S_z)}\rangle \tag{6.53}$$

Now we can define the last expectation value as $E_{\mathbf{R}',\mathbf{R}}^{(S,S_z)}$, and obtain for the energy:

$$\begin{aligned} E_{\mathbf{K}}^{(S,S_z)} &= \frac{1}{N} \sum_{\mathbf{R},\mathbf{R}'} e^{i\mathbf{K}\cdot(\mathbf{R}-\mathbf{R}')} E_{\mathbf{R}',\mathbf{R}}^{(S,S_z)} \\ &= \frac{1}{N} \sum_{\mathbf{R}} E_{\mathbf{R},\mathbf{R}}^{(S,S_z)} + \frac{1}{N} \sum_{\mathbf{R}} \sum_{\mathbf{R}' \neq \mathbf{R}} e^{i\mathbf{K}\cdot(\mathbf{R}-\mathbf{R}')} E_{\mathbf{R}',\mathbf{R}}^{(S,S_z)} \\ &= E_0^{(S,S_z)} + \sum_{\mathbf{R}\neq 0} e^{-i\mathbf{K}\cdot\mathbf{R}} E_{\mathbf{R}}^{(S,S_z)} \end{aligned} \tag{6.54}$$

where we have taken advantage of the translational symmetry of the hamiltonian to write $E_{\mathbf{R}',\mathbf{R}}^{(S,S_z)} = E_{\mathbf{R}-\mathbf{R}'}^{(S,S_z)}$; we have also eliminated summations over $\mathbf{R}'$ when the summand does not depend explicitly on $\mathbf{R}$, together with a factor of $(1/N)$.

We can express the quantities $E_0^{(S,S_z)}, E_{\mathbf{R}}^{(S,S_z)}$ in terms of the single-particle states $\phi^{(v)}(\mathbf{r})$ and $\phi^{(c)}(\mathbf{r})$ as follows:

$$E_0^{(S,S_z)} = E_0 + \tilde{E}_0^{(c)} - \tilde{E}_0^{(v)} + U_0^{(S)} \tag{6.55}$$

where E_0 is the ground-state energy, $E_0 = \langle \Psi_0|\hat{\mathcal{H}}|\Psi_0\rangle$, and the other terms that appear in this equation are defined as follows:

$$\tilde{E}_0^{(c)} = \langle \phi_0^{(c)}|\hat{h}|\phi_0^{(c)}\rangle + \tilde{V}_0^{(c)}, \quad \tilde{E}_0^{(v)} = \langle \phi_0^{(v)}|\hat{h}|\phi_0^{(v)}\rangle + \tilde{V}_0^{(v)} \tag{6.56}$$

$$\tilde{V}_0^{(c)} = \sum_{\mathbf{R}} \left[2\langle \phi_0^{(c)}\phi_{\mathbf{R}}^{(v)}|\frac{e^2}{|\mathbf{r}-\mathbf{r}'|}|\phi_0^{(c)}\phi_{\mathbf{R}}^{(v)}\rangle - \langle \phi_0^{(c)}\phi_{\mathbf{R}}^{(v)}|\frac{e^2}{|\mathbf{r}-\mathbf{r}'|}|\phi_{\mathbf{R}}^{(v)}\phi_0^{(c)}\rangle \right] \tag{6.57}$$

$$\tilde{V}_0^{(v)} = \sum_{\mathbf{R}} \left[2\langle \phi_0^{(v)}\phi_{\mathbf{R}}^{(v)}|\frac{e^2}{|\mathbf{r}-\mathbf{r}'|}|\phi_0^{(v)}\phi_{\mathbf{R}}^{(v)}\rangle - \langle \phi_0^{(v)}\phi_{\mathbf{R}}^{(v)}|\frac{e^2}{|\mathbf{r}-\mathbf{r}'|}|\phi_{\mathbf{R}}^{(v)}\phi_0^{(v)}\rangle \right] \tag{6.58}$$

$$U_0^{(S)} = 2\delta_{S,0}\langle \phi_0^{(c)}\phi_0^{(v)}|\frac{e^2}{|\mathbf{r}-\mathbf{r}'|}|\phi_0^{(v)}\phi_0^{(c)}\rangle - \langle \phi_0^{(c)}\phi_0^{(v)}|\frac{e^2}{|\mathbf{r}-\mathbf{r}'|}|\phi_0^{(c)}\phi_0^{(v)}\rangle \tag{6.59}$$

In the expressions above, we have used the shorthand notation $\phi_{\mathbf{R}}^{(n)}(\mathbf{r}) = \phi^{(n)}(\mathbf{r} - \mathbf{R})$ with $n = v$ or c for the valence or conduction states. We have not included spin labels in the single-particle states $\phi^{(n)}(\mathbf{r})_{\mathbf{R}}$, since the spin degrees of freedom have explicitly been taken into account to arrive at these expressions. $\hat{\mathcal{H}}(\{\mathbf{r}_i\}), \hat{h}(\mathbf{r})$ are the many-body and single-particle hamiltonians defined in Eq. (6.42). With the same conventions, the last term appearing under the summation over $\mathbf{R}$ in Eq. (6.54) takes the form

$$E_{\mathbf{R}}^{(S,S_z)} \equiv U_{\mathbf{R}}^{(S)} = \langle \Phi_{\mathbf{R}}^{(S,S_z)} | \hat{\mathcal{H}} | \Phi_0^{(S,S_z)} \rangle$$

$$= 2\delta_{S,0} \langle \phi_{\mathbf{R}}^{(v)} \phi_0^{(c)} | \frac{e^2}{|\mathbf{r} - \mathbf{r}'|} | \phi_{\mathbf{R}}^{(c)} \phi_0^{(v)} \rangle - \langle \phi_{\mathbf{R}}^{(v)} \phi_0^{(c)} | \frac{e^2}{|\mathbf{r} - \mathbf{r}'|} | \phi_0^{(v)} \phi_{\mathbf{R}}^{(c)} \rangle \quad (6.60)$$

With these expressions, the energy of the excited state takes the form

$$E_{\mathbf{K}}^{(S,S_z)} = E_0 + \tilde{E}_0^{(c)} - \tilde{E}_0^{(v)} + U_0^{(0)} + \sum_{\mathbf{R} \neq 0} e^{-i\mathbf{K}\cdot\mathbf{R}} E_{\mathbf{R}}^{(S,S_z)}$$

$$= E_0 + \tilde{E}_0^{(c)} - \tilde{E}_0^{(v)} + \sum_{\mathbf{R}} e^{-i\mathbf{K}\cdot\mathbf{R}} U_{\mathbf{R}}^{(S)} \quad (6.61)$$

The interpretation of these results is the following: the energy $E_{\mathbf{K}}^{(S,S_z)}$ of the system with the exciton in a state of wave-vector $\mathbf{K}$ and spin S with z component S_z is equal to the energy of the ground state E_0, plus the particle energy $\tilde{E}_0^{(c)}$ minus the hole energy $\tilde{E}_0^{(v)}$; the last term, based on the definition of $U_{\mathbf{R}}^{(S)}$ from Eq. (6.59) for $\mathbf{R} = 0$ and Eq. (6.60) for $\mathbf{R} \neq 0$, gives the particle–hole interaction. We note that the particle and hole energies, $\tilde{E}_0^{(c)}$ and $\tilde{E}_0^{(v)}$, represent *quasiparticles* and contain the contribution from the single-particle hamiltonian, $\langle \phi_0^{(n)} | \hat{h} | \phi_0^{(n)} \rangle$, for $n = c$ and $n = v$, as well as the interactions of the particle or the hole with all other electrons in the system, expressed by the terms $\tilde{\mathcal{V}}_0^{(c)}$ and $\tilde{\mathcal{V}}_0^{(v)}$, respectively. All interaction terms are at the Hartree–Fock level and include the direct Coulomb and the exchange contributions, which is a natural consequence of our choice of a Slater determinant for the many-body wavefunction, Eq. (6.50).

To analyze the effects of the presence of excitons, we will assume that $\mathbf{K} = 0, S = S_z = 0$. We then find for the energy of the first excitation:

$$E_1^{\text{exc}} = E_0^{(0,0)} - E_0 = \tilde{E}_0^{(c)} - \tilde{E}_0^{(v)} + \sum_{\mathbf{R}} U_{\mathbf{R}}^{(0)} = \epsilon_{\text{gap}} + \sum_{\mathbf{R}} U_{\mathbf{R}}^{(0)}$$

where we have defined the band-gap energy ϵ_{gap} as the difference between the particle and hole quasiparticle energies:

$$\epsilon_{\text{gap}} = \tilde{E}_0^{(c)} - \tilde{E}_0^{(v)}$$

From our assumption of localized functions for the single-particle states $\phi_{\mathbf{R}}^{(n)}(\mathbf{r})$, we conclude that the interaction term is dominated by $U_0^{(0)}$, that is, the Coulomb and exchange interactions between the particle and the hole situated in the same unit cell, which is given by Eq. (6.59). When the particle and the hole are situated in unit cells a distance $\mathbf{R} \neq 0$ apart, the localized nature of the single-particle states leads to a much weaker interaction, as shown in Fig. 6.9. In this schematic example, which is based on an ionic insulator composed of positive and negative ions, the excitation consists of taking an electron from

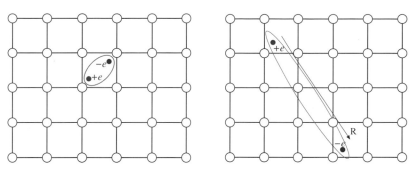

Fig. 6.9 Comparison of Coulomb interaction contributions for excitonic states with the particle and hole situated in the same unit cell (left panel) and in different unit cells a distance **R** apart (right panel); the unit cells are identified by the dashed lines. White circles represent positive ions and black circles negative ions. Red and blue clouds represent the density of particle and hole distributions.

the neighborhood of a positive ion (white circle), where it would normally reside in the ground state, to the neighborhood of a negative ion (black circle), and leaving a hole behind. This involves an excitation energy equal to the band gap of the insulator minus the interaction energy between the two charges. Within the strongly bound picture of the exciton derived above, the final result for the excitation energy is then:

$$E_1^{\text{exc}} \approx \epsilon_{\text{gap}} - \int \int |\phi_0^{(c)}(\mathbf{r})|^2 \frac{e^2}{|\mathbf{r} - \mathbf{r}'|} |\phi_0^{(v)}(\mathbf{r}')|^2 \mathrm{d}\mathbf{r} \, \mathrm{d}\mathbf{r}'$$

$$+ 2 \int \int \phi_0^{(c)*}(\mathbf{r}) \phi_0^{(v)}(\mathbf{r}) \frac{e^2}{|\mathbf{r} - \mathbf{r}'|} \phi_0^{(v)*}(\mathbf{r}') \phi_0^{(c)}(\mathbf{r}') \mathrm{d}\mathbf{r} \, \mathrm{d}\mathbf{r}' \qquad (6.62)$$

Since the states $\phi_0^{(v)}(\mathbf{r})$ and $\phi_0^{(c)}(\mathbf{r})$ are orthogonal, their overlap is very small, and therefore the second integral in the above equation has much smaller value than the first one, which is evidently a positive quantity. This shows that the photon energy $\hbar\omega = E_1^{\text{exc}}$, at which there is the first absorption peak, is *lower* than the energy gap by approximately the value of the first integral, that is, the exciton binding energy defined in Eq. (6.48). Excitations of higher energy can also be obtained, for instance, by creating a hole at some unit cell and a particle at a near-neighbor cell; this will have a lower binding energy, that is, the difference from the value of the gap will be smaller than $E_1^{\text{bin}} = \epsilon_{\text{gap}} - E_1^{\text{exc}}$, which confirms the qualitative picture of excitons shown in Fig. 6.8.

6.3.3 Weakly Bound (Wannier) Excitons

For weakly bound (Wannier) excitons, we will use the Bloch states $\psi_{\mathbf{k},s}^{(n)}(\mathbf{r})$, in terms of which we must express the many-body wavefunction. Our starting point is the expression for the many-body wavefunction of Eq. (6.44), in which we replace a valence state $\psi_{\mathbf{k}_h,s_h}^{(v)}(\mathbf{r})$, which becomes the hole state, with a conduction state $\psi_{\mathbf{k}_p,s_p}^{(c)}(\mathbf{r})$, which becomes the particle state. The value of $\mathbf{k}_h$ can be any of the N values that span the first BZ, while the value of $\mathbf{k}_p$ must be such that $\mathbf{K} = \mathbf{k}_p - \mathbf{k}_h$ to satisfy the total momentum conservation condition; this can be implemented by choosing

$$\text{particle}: \ \mathbf{k}_p = \mathbf{k} + \frac{\mathbf{K}}{2}, \quad \text{hole}: \ \mathbf{k}_h = \mathbf{k} - \frac{\mathbf{K}}{2} \tag{6.63}$$

The value of $\mathbf{k}$, in addition to the value of $\mathbf{K}$, will serve to label the new basis states for the many-body wavefunction, which are now written as $\Psi_{\mathbf{K}}^{(S)}(\mathbf{k}; \{\mathbf{r}\}, \{s\})$. We next construct a linear combination of all such wavefunctions with coefficients $\xi_{\mathbf{K}}(\mathbf{k})$:

$$\Psi_{\mathbf{K}}^{(S)}(\{\mathbf{r}\}, \{s\}) = \sum_{\mathbf{k}} \xi_{\mathbf{K}}^{(S)}(\mathbf{k}) \Psi_{\mathbf{K}}^{(S)}(\mathbf{k}; \{\mathbf{r}\}, \{s\}) \tag{6.64}$$

which is our trial wavefunction for solving the problem. All that remains to be done is to determine the coefficients $\xi_{\mathbf{K}}^{(S)}(\mathbf{k})$. Our goal then will be to derive an equation obeyed by these coefficients, so that we can find their values. This will be accomplished by requiring the state $\Psi_{\mathbf{K}}^{(S)}(\{\mathbf{r}\}, \{s\})$ to be a solution of the Schrödinger equation:

$$\hat{\mathcal{H}} \Psi_{\mathbf{K}}^{(S)}(\{\mathbf{r}\}, \{s\}) = E^{\text{exc}} \Psi_{\mathbf{K}}^{(S)}(\{\mathbf{r}\}, \{s\})$$

with $\hat{\mathcal{H}}$ the many-body hamiltonian for the system.

As we have explained earlier, physically relevant processes for photon absorption or emission do not change the total momentum $\mathbf{K}$ and total spin S state, since the photon momentum is negligible on the scale relevant to crystal electrons. Unless there is some other type of process that could change the total momentum, we must take $\mathbf{K} = 0$ and $S = 0$. As an aside, we note that processes which change the momentum during an excitation are indeed relevant in certain cases, involving, for example, the scattering of a quantum of lattice vibrations, called a phonon; the nature of phonons and their interactions with electrons will be examined later, in Chapter 7. In the following discussion we assume that the proper linear combination of spin states is used to ensure that the total spin $S = 0$, and we consider only the $\mathbf{K} = 0$ case. To simplify the notation, we will drop the superscript $S = 0$ for the total spin and the subscript $\mathbf{K} = 0$ for the total momentum. The many-body Schrödinger equation then takes the form

$$\hat{\mathcal{H}}|\Psi\rangle = E^{\text{exc}}|\Psi\rangle \Rightarrow \sum_{\mathbf{k}'} \xi(\mathbf{k}') \langle \Psi(\mathbf{k})|\hat{\mathcal{H}}|\Psi(\mathbf{k}')\rangle = E^{\text{exc}} \xi(\mathbf{k}) \tag{6.65}$$

where we have taken advantage of the fact that, by construction, the wavefunctions $|\Psi(\mathbf{k})\rangle$ are orthonormal. In order to determine the values of the coefficients $\xi(\mathbf{k})$, we have to diagonalize the hamiltonian matrix in this basis of states. To calculate the matrix elements that appear in Eq (6.65), we go through the same steps as in the case of the Wannier basis, Eq. (6.53). Fortunately, all the work we did to obtain the Wannier basis matrix elements can be adopted here, with the change in notation that the lattice vector we had labeled "0" maps to the wave-vector $\mathbf{k}$ and the lattice vector $\mathbf{R}$ maps to the wave-vector $\mathbf{k}'$; both $\mathbf{k}$ and $\mathbf{k}'$ span the first BZ. We can then simply rewrite the expressions we found above in terms of the Bloch states $|\psi_{\mathbf{k}}\rangle, |\psi_{\mathbf{k}'}\rangle$, to obtain for the diagonal matrix elements ($\mathbf{k} = \mathbf{k}'$):

$$\langle \Psi(\mathbf{k})|\hat{\mathcal{H}}|\Psi(\mathbf{k})\rangle = E_0 + \tilde{E}_{\mathbf{k}}^{(c)} - \tilde{E}_{\mathbf{k}}^{(v)} + U_{\mathbf{k}}$$

with E_0 the ground-state energy and the remaining quantities defined as:

$$\tilde{E}_{\mathbf{k}}^{(c)} = \langle \psi_{\mathbf{k}}^{(c)} | \hat{h} | \psi_{\mathbf{k}}^{(c)} \rangle + \tilde{\mathcal{V}}_{\mathbf{k}}^{(c)}, \quad \tilde{E}_{\mathbf{k}}^{(v)} = \langle \psi_{\mathbf{k}}^{(v)} | \hat{h} | \psi_{\mathbf{k}}^{(v)} \rangle + \tilde{\mathcal{V}}_{\mathbf{k}}^{(v)} \tag{6.66}$$

$$\tilde{\mathcal{V}}_{\mathbf{k}}^{(c)} = \sum_{\mathbf{k}'} 2 \langle \psi_{\mathbf{k}}^{(c)} \psi_{\mathbf{k}'}^{(v)} | \frac{e^2}{|\mathbf{r}-\mathbf{r}'|} | \psi_{\mathbf{k}}^{(c)} \psi_{\mathbf{k}'}^{(v)} \rangle - \langle \psi_{\mathbf{k}}^{(c)} \psi_{\mathbf{k}'}^{(v)} | \frac{e^2}{|\mathbf{r}-\mathbf{r}'|} | \psi_{\mathbf{k}'}^{(v)} \psi_{\mathbf{k}}^{(c)} \rangle \tag{6.67}$$

$$\tilde{\mathcal{V}}_{\mathbf{k}}^{(v)} = \sum_{\mathbf{k}'} 2 \langle \psi_{\mathbf{k}}^{(v)} \psi_{\mathbf{k}'}^{(v)} | \frac{e^2}{|\mathbf{r}-\mathbf{r}'|} | \psi_{\mathbf{k}}^{(v)} \psi_{\mathbf{k}'}^{(v)} \rangle - \langle \psi_{\mathbf{k}}^{(v)} \psi_{\mathbf{k}'}^{(v)} | \frac{e^2}{|\mathbf{r}-\mathbf{r}'|} | \psi_{\mathbf{k}'}^{(v)} \psi_{\mathbf{k}}^{(v)} \rangle \tag{6.68}$$

$$U_{\mathbf{k}} = 2 \langle \psi_{\mathbf{k}}^{(c)} \psi_{\mathbf{k}}^{(v)} | \frac{e^2}{|\mathbf{r}-\mathbf{r}'|} | \psi_{\mathbf{k}}^{(v)} \psi_{\mathbf{k}}^{(c)} \rangle - \langle \psi_{\mathbf{k}}^{(c)} \psi_{\mathbf{k}}^{(v)} | \frac{e^2}{|\mathbf{r}-\mathbf{r}'|} | \psi_{\mathbf{k}}^{(c)} \psi_{\mathbf{k}}^{(v)} \rangle \tag{6.69}$$

while for the off-diagonal matrix elements, $\mathbf{k}' \neq \mathbf{k}$, we find:

$$\langle \Psi(\mathbf{k}) | \hat{\mathcal{H}} | \Psi(\mathbf{k}') \rangle = 2 \langle \psi_{\mathbf{k}}^{(c)} \psi_{\mathbf{k}'}^{(v)} | \frac{e^2}{|\mathbf{r}-\mathbf{r}'|} | \psi_{\mathbf{k}}^{(v)} \psi_{\mathbf{k}'}^{(c)} \rangle - \langle \psi_{\mathbf{k}}^{(c)} \psi_{\mathbf{k}'}^{(v)} | \frac{e^2}{|\mathbf{r}-\mathbf{r}'|} | \psi_{\mathbf{k}'}^{(c)} \psi_{\mathbf{k}}^{(v)} \rangle$$

With these expressions, we can start evaluating the relevant integrals so that we can solve for the coefficients $\xi(\mathbf{k})$ in Eq. (6.65). These integrals can in principle be evaluated numerically, since we assume that the single-particle states $\psi_{\mathbf{k}}^{(n)}(\mathbf{r})$ are known. For a more transparent solution, we will make some simplifying assumptions. The first is that we will treat the quasiparticle energies $\tilde{E}_{\mathbf{k}}^{(n)}$ as known, without having to evaluate either the single-particle terms or the two-particle contributions, $\tilde{\mathcal{V}}_{\mathbf{k}}^{(n)}$. In fact, we have seen in the previous chapter that reasonable quasiparticle energies can be obtained from first-principles calculations based on density functional theory or other similar approaches. We then need to concern ourselves only with the terms $U_{\mathbf{k}}$ and the off-diagonal matrix elements of the hamiltonian, both of which involve two types of integrals, the Coulomb and the exchange interactions. We will use the familiar expression for the Bloch states, Eq. (6.33), with the assumption that the periodic part, $u_{\mathbf{k}}^{(n)}(\mathbf{r})$, is a slowly varying function of $\mathbf{k}$. We then obtain, for the Coulomb integral:

$$\langle \psi_{\mathbf{k}}^{(c)} \psi_{\mathbf{k}'}^{(v)} | \frac{e^2}{|\mathbf{r}-\mathbf{r}'|} | \psi_{\mathbf{k}'}^{(c)} \psi_{\mathbf{k}}^{(v)} \rangle =$$

$$\frac{1}{V^2} \int \int u_{\mathbf{k}}^{(c)*}(\mathbf{r}) u_{\mathbf{k}'}^{(v)*}(\mathbf{r}') \frac{e^2}{|\mathbf{r}-\mathbf{r}'|} u_{\mathbf{k}'}^{(c)}(\mathbf{r}) u_{\mathbf{k}}^{(v)}(\mathbf{r}') e^{-i(\mathbf{k}-\mathbf{k}')\cdot(\mathbf{r}-\mathbf{r}')} d\mathbf{r} d\mathbf{r}'$$

The fast-varying part of this integral involves the complex exponential multiplied by the Coulomb potential, with the other parts being slowly varying. We can approximate the integral by taking out the slowly varying parts and replacing them by their averages, denoted by $\langle \cdots \rangle$. The expression for the integral then becomes:

$$\langle \psi_{\mathbf{k}}^{(c)} \psi_{\mathbf{k}'}^{(v)} | \frac{e^2}{|\mathbf{r}-\mathbf{r}'|} | \psi_{\mathbf{k}'}^{(c)} \psi_{\mathbf{k}}^{(v)} \rangle \approx \langle u_{\mathbf{k}}^{(c)*} u_{\mathbf{k}'}^{(c)} \rangle \langle u_{\mathbf{k}'}^{(v)*} u_{\mathbf{k}}^{(v)} \rangle \mathcal{V}_C(\mathbf{k}-\mathbf{k}')$$

with the last term that appears in this result being the Fourier transform of the Coulomb potential:

$$\mathcal{V}_C(\mathbf{k}) \equiv \frac{1}{V} \int e^{-i\mathbf{k}\cdot\mathbf{r}} \frac{e^2}{|\mathbf{r}|} d\mathbf{r}$$

The exchange integral is:

$$\langle\psi_{\mathbf{k}}^{(c)}\psi_{\mathbf{k}'}^{(v)}|\frac{e^2}{|\mathbf{r}-\mathbf{r}'|}|\psi_{\mathbf{k}}^{(v)}\psi_{\mathbf{k}'}^{(c)}\rangle = \frac{1}{V^2}\int\int u_{\mathbf{k}}^{(c)*}(\mathbf{r})u_{\mathbf{k}'}^{(v)*}(\mathbf{r}')\frac{e^2}{|\mathbf{r}-\mathbf{r}'|}u_{\mathbf{k}}^{(v)}(\mathbf{r})u_{\mathbf{k}'}^{(c)}(\mathbf{r}')d\mathbf{r}d\mathbf{r}'$$

which, with the same approximations as above, takes the form

$$\left\langle u_{\mathbf{k}}^{(c)*}u_{\mathbf{k}}^{(v)}\right\rangle\left\langle u_{\mathbf{k}'}^{(v)*}u_{\mathbf{k}'}^{(c)}\right\rangle\frac{1}{V^2}\int\int\frac{e^2}{|\mathbf{r}-\mathbf{r}'|}d\mathbf{r}d\mathbf{r}' = \left\langle u_{\mathbf{k}}^{(c)*}u_{\mathbf{k}}^{(v)}\right\rangle\left\langle u_{\mathbf{k}'}^{(v)*}u_{\mathbf{k}'}^{(c)}\right\rangle\mathcal{V}_C(0)$$

but the orthogonality of the periodic parts of the Bloch wavefunctions gives:

$$\left\langle u_{\mathbf{k}}^{(c)*}u_{\mathbf{k}}^{(v)}\right\rangle = \int u_{\mathbf{k}}^{(c)*}(\mathbf{r})u_{\mathbf{k}}^{(v)}(\mathbf{r})d\mathbf{r} = 0, \quad \left\langle u_{\mathbf{k}'}^{(v)*}u_{\mathbf{k}'}^{(c)}\right\rangle = \int u_{\mathbf{k}'}^{(v)*}(\mathbf{r})u_{\mathbf{k}'}^{(c)}(\mathbf{r})d\mathbf{r} = 0$$

that is, the exchange integral does not contribute. Moreover, since the periodic part of the Bloch wavefunction is slowly varying with $\mathbf{k}$, we can assume that for $\mathbf{k}'$ close to $\mathbf{k}$ the functions with the same superscipt index are approximately equal, which leads to:

$$\left\langle u_{\mathbf{k}}^{(c)*}u_{\mathbf{k}'}^{(c)}\right\rangle \approx \int u_{\mathbf{k}}^{(c)*}(\mathbf{r})u_{\mathbf{k}}^{(c)}(\mathbf{r})d\mathbf{r} = 1, \quad \left\langle u_{\mathbf{k}'}^{(v)*}u_{\mathbf{k}}^{(v)}\right\rangle \approx \int u_{\mathbf{k}}^{(v)*}(\mathbf{r})u_{\mathbf{k}'}^{(v)}(\mathbf{r})d\mathbf{r} = 1$$

Putting all these results together, we find for the hamiltonian matrix elements:

$$\langle\Psi(\mathbf{k})|\hat{\mathcal{H}}|\Psi(\mathbf{k})\rangle = E_0 + \tilde{E}_{\mathbf{k}}^{(c)} - \tilde{E}_{\mathbf{k}}^{(v)} - \mathcal{V}_C(0), \quad \langle\Psi(\mathbf{k})|\hat{\mathcal{H}}|\Psi(\mathbf{k}')\rangle = -\mathcal{V}_C(\mathbf{k}-\mathbf{k}')$$

We can neglect the ground-state energy E_0 and the average Coulomb potential $\mathcal{V}_C(0)$, since these are just constants that shift the spectrum. Also, a more careful treatment of many-body effects reveals that the Coulomb integral is modified by the dielectric function,[8] which itself is energy and wave-vector dependent, as shown in previous sections of this chapter. For the purposes of the present qualitative discussion, it suffices to include just the static part of the dielectric function $\bar{\varepsilon}_0$, as an overall factor modifying the Coulomb integral, that is, we take

$$\overline{\mathcal{V}}_C(\mathbf{k}-\mathbf{k}') = \frac{1}{\bar{\varepsilon}_0 V}\int\frac{e^2}{|\mathbf{r}|}e^{-i(\mathbf{k}-\mathbf{k}')\cdot\mathbf{r}}d\mathbf{r}$$

The matrix equation then becomes:

$$\left[\tilde{E}_{\mathbf{k}}^{(c)} - \tilde{E}_{\mathbf{k}}^{(v)}\right]\xi(\mathbf{k}) - \frac{1}{\bar{\varepsilon}_0 V}\sum_{\mathbf{k}'}\int\frac{e^2}{|\mathbf{r}|}e^{-i(\mathbf{k}-\mathbf{k}')\cdot\mathbf{r}}d\mathbf{r}\,\xi(\mathbf{k}') = E^{\mathrm{exc}}\xi(\mathbf{k})$$

and turning as usual the summation over $\mathbf{k}'$ values into an integral over the BZ, we arrive at:

$$\left[\tilde{E}_{\mathbf{k}}^{(c)} - \tilde{E}_{\mathbf{k}}^{(v)}\right]\xi(\mathbf{k}) - \frac{1}{\bar{\varepsilon}_0(2\pi)^3}\int\int\frac{e^2}{|\mathbf{r}|}e^{-i(\mathbf{k}-\mathbf{k}')\cdot\mathbf{r}}d\mathbf{r}\,\xi(\mathbf{k}')\,d\mathbf{k}' = E^{\mathrm{exc}}\xi(\mathbf{k}) \qquad (6.70)$$

Next, we assume that the excitation is taking place across the band gap, between the two band extrema, as shown schematically in Fig. 6.8. In this case, we can write the

[8] M. Rohlfing and S. G. Louie, *Phys. Rev. B* **62**, 4927 (2000).

quasiparticle energies near the band extrema as quadratic approximations in $\mathbf{k}$, with the effective masses of the electron and the hole:

$$\tilde{E}_{\mathbf{k}}^{(c)} = \epsilon_F + \frac{\epsilon_{\text{gap}}}{2} + \frac{\hbar^2 \mathbf{k}^2}{2\bar{m}_e}, \quad \tilde{E}_{\mathbf{k}}^{(v)} = \epsilon_F - \frac{\epsilon_{\text{gap}}}{2} + \frac{\hbar^2 \mathbf{k}^2}{2\bar{m}_h} \quad (6.71)$$

where we have assumed the Fermi level ϵ_F to be in the middle of the band gap, as expected for an intrinsic semiconductor or insulator. The two expressions lead to:

$$\left[\tilde{E}_{\mathbf{k}}^{(c)} - \tilde{E}_{\mathbf{k}}^{(v)} \right] = \epsilon_{\text{gap}} + \frac{\hbar^2 \mathbf{k}^2}{2\mu}$$

Finally, we introduce the Fourier transform of $\xi(\mathbf{k})$, which is a function of the real-space position $\mathbf{r}$ that we call $\psi(\mathbf{r})$:

$$\xi(\mathbf{k}) = \frac{1}{V} \int \psi(\mathbf{r}) e^{-i\mathbf{k}\cdot\mathbf{r}} d\mathbf{r} \Rightarrow \psi(\mathbf{r}) = \frac{1}{(2\pi)^3} \int \xi(\mathbf{k}) e^{i\mathbf{k}\cdot\mathbf{r}} d\mathbf{k} \quad (6.72)$$

and substituting this in the equation for $\xi(\mathbf{k})$, we find:

$$\left[-\frac{\hbar^2 \nabla_{\mathbf{r}}^2}{2\mu} - \frac{e^2}{\bar{\varepsilon}_0 |\mathbf{r}|} \right] \psi(\mathbf{r}) = (E^{\text{exc}} - \epsilon_{\text{gap}}) \psi(\mathbf{r}) \quad (6.73)$$

with μ the reduced mass of the particle–hole pair, defined in Eq. (6.45).

The last equation is precisely what we had anticipated based on a simple heuristic picture, Eq. (6.47), with $\mathbf{K} = 0$, the case we have been analyzing here. The energy levels obtained from solving this equation are given by:

$$E_n^{\text{exc}} = \epsilon_{\text{gap}} - \frac{\mu e^4}{2\hbar^2 \bar{\varepsilon}_0^2 n^2} = \epsilon_{\text{gap}} - \left(\frac{m_e e^4}{2\hbar^2} \right) \frac{\mu/m_e}{\bar{\varepsilon}_0^2} \frac{1}{n^2} \quad n = 1, 2, \ldots$$

where the quantity in parentheses is equal to 1 Ry = 13.6058 eV, the energy scale for the actual hydrogen atom. This sequence of values is very similar to typical experimental results, as shown schematically in Fig. 6.8; it is the same sequence of energy values as in the hydrogen atom, scaled by the factor $(\mu/m_e)/\bar{\varepsilon}_0^2$. Figure 6.10 shows a remarkable experimental demonstration of how reasonable this description is, for the case of the insulator Cu_2O, with the exciton states of index up to $n = 25$ clearly resolved. In this case, the factor $(\mu/m_e)/\bar{\varepsilon}_0^2 = 0.00676$, giving an "effective Rydberg" value of 92 meV; the static dielectric constant of Cu_2O is $\bar{\varepsilon}_0 = 7.5$, and the effective mass $\mu = 0.38\,m_e$.

The eigenfunctions are also the same as in the case of a hydrogen-like atom, with effective electron mass μ and effective nuclear charge $+e/\bar{\varepsilon}_0$; for instance, the lowest eigenfunction, for $n = 1$, is:

$$\psi_1(\mathbf{r}) = A_1 e^{-r/\bar{a}}, \quad \bar{a} = \frac{\hbar^2 \bar{\varepsilon}_0}{\mu e^2} = \left(\frac{\hbar^2}{m_e e^2} \right) \frac{\bar{\varepsilon}_0}{\mu/m_e} = a_0 \frac{\bar{\varepsilon}_0}{\mu/m_e} \quad (6.74)$$

where A_1 is the normalization constant and the quantity in parentheses is the Bohr radius, a_0, which sets the length scale for the actual hydrogen atom. From our discussion of the heuristic model, it follows that this is the wavefunction that describes the *relative* motion of the particle and the hole, while the center-of-mass coordinate does not enter the picture

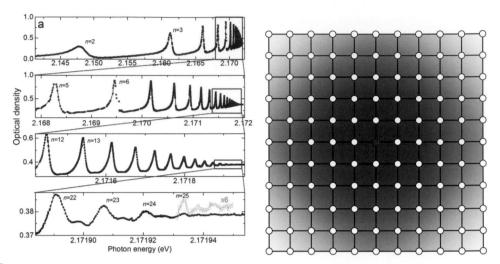

Fig. 6.10   **Left**: Excitons in Cu_2O, an insulator with a direct band gap $\epsilon_{gap} = 2.17208$ eV [adapted from "Giant Rydberg excitons in the copper oxide Cu_2O," T. Kazimierczuk, D. Fröhlich, S. Scheel, H. Stolz, and M. Bayer, *Nature* **514**, 343 (2014)]. **Right**: Schematic illustration of the wavefunction of an *s*-state of the exciton, for a fixed position of the hole (blue circle) and the distribution of the electron (red cloud) around it, according to the wavefunction of Eq. (6.74).

because we chose $\mathbf{K} = 0$. Had we considered the case $\mathbf{K} \neq 0$, we would have obtained a plane-wave-like behavior for the center-of-mass coordinate; this case can be developed by following the same steps as above and taking the particle and hole wave-vectors to be given by Eq. (6.63). The picture that emerges then is that the weakly bound particle–hole pair moves as a neutral entity with center-of-mass momentum $\hbar\mathbf{K}$, and the distribution of one charge relative to the other is described by hydrogen-like wavefunctions with the characteristic length scaled by a factor $\bar{\varepsilon}_0/(\mu/m_e)$ relative to the Bohr radius a_0, which, for example, in the case of Cu_2O mentioned above, is approximately 20. To visualize this behavior we show schematically in Fig. 6.10 the wavefunction for the electron–hole pair in the lowest-energy *s*-like state described by Eq. (6.74), with the position of the hole fixed at some point, for a fictitious 2D square lattice. To obtain this picture, we start with the many-body wavefunction of Eq. (6.64), integrate out all the coordinates other than those of the particle $\mathbf{r}_p$ and the hole $\mathbf{r}_h$, whence the expression of Eq. (6.46) emerges, and then plot the values of this effective two-particle state as a function of $\mathbf{r}_p$ for a fixed value of $\mathbf{r}_h$, supressing the wave-like contribution from the dependence on the center-of-mass coordinate $\mathbf{R}$. The reverse picture as a function of $\mathbf{r}_h$, with the electron fixed at some position $\mathbf{r}_p$, also holds.

Further Reading

1. *Electronic Structure and Optical Properties of Semiconductors*, M. L. Cohen and J. R. Chelikowsky (Springer-Verlag, Berlin, 1988).

2. *Electronic States and Optical Transitions in Solids*, F. Bassani and G. Pastori Parravicini (Pergamon Press, Oxford, 1975).

3. H. Ehrenreich, "Electromagnetic transport in solids: Optical properties and plasma effects," in J. Tauc (Ed.), *Optical Properties of Solids* (Course 34 of E. Fermi International Summer School, Varenna, Italy, 1966). This article presents a fundamental theory for solids and includes several examples, some of which were discussed in the text.

4. H. R. Philipp and H. Ehrenreich, "Ultraviolet optical properties," in R. K. Willardson and A. C. Beer (Eds), *Semiconductors and Semimetals*, Vol. 3, p. 93 (Academic Press, New York, 1967). This is a general review article, including the theoretical background within the one-electron framework, but with emphasis on the experimental side.

5. *Optical Properties of Solids*, F. Wooten (Academic Press, New York, 1972). This book contains a comprehensive treatment of optical properties of solids.

Problems

1. Prove that if a uniform electron gas is displaced by a small amount in one direction relative to the uniform background of compensating positive charge and then is allowed to relax, it will oscillate with the plasma frequency, Eq. (6.12).

2. Derive the expressions for the real and imaginary parts of the dielectric function as a function of frequency due to interband transitions, starting with the expression of Eq. (6.35).

3. Derive the expressions for the real and imaginary parts of the dielectric function as a function of frequency due to intraband transitions, starting with the expression of Eq. (6.39).

4. Compare the expression for the dielectric function due to interband transitions, Eq. (6.35), to the Lorentz result, Eq. (6.17), and identify the constants in the latter in terms of fundamental constants and the microscopic properties of the solid. Provide a physical interpretation for the constants in the Lorentz expression. Apply this analysis to the case of Si, whose band structure is shown in Fig. 6.7 and real and imaginary parts of the dielectric function are given in Fig. 6.5.

5. Compare the expression for the dielectric function due to intraband transitions, Eq. (6.39), to the Drude result, Eq. (6.15), and identify the constants in the latter in terms of fundamental constants and the microscopic properties of the solid. Provide a physical interpretation for the constants in the Drude expression.

6. From the general expression for the dielectric function due to interband transitions, Eq. (6.35), show that the static dielectric constant for semiconductors and insulators, $\varepsilon^{\text{inter}}(\omega = 0)$, is a real constant greater than 1.

7. An important check on the calculation of the dielectric function values is the so-called "sum rule":

$$\int_0^\infty \omega \, \varepsilon_2(\mathbf{q}, \omega) \, d\omega = \frac{\pi}{2}\omega_p^2 \tag{6.75}$$

where ω_p is the plasma frequency, Eq. (6.12). We can prove this relation by the following steps. First, show that the double commutator of the hamiltonian $\hat{\mathcal{H}}$ with $e^{\pm i\mathbf{q}\cdot\mathbf{r}}$ gives:

$$[\hat{\mathcal{H}}, e^{i\mathbf{q}\cdot\mathbf{r}}] = e^{i\mathbf{q}\cdot\mathbf{r}}\left(\frac{\hbar^2 q^2}{2m_e} - i\frac{\hbar^2}{m_e}\mathbf{q}\cdot\nabla_{\mathbf{r}}\right) \Rightarrow \left[e^{-i\mathbf{q}\cdot\mathbf{r}}, [\hat{\mathcal{H}}, e^{i\mathbf{q}\cdot\mathbf{r}}]\right] = \frac{\hbar^2 q^2}{m_e}$$

We can take the expectation value of the left-hand side in the last relation with respect to a state $|\psi_\kappa\rangle$, and then insert a complete set of states $1 = \sum_{\kappa'}|\psi_{\kappa'}\rangle\langle\psi_{\kappa'}|$ to obtain:

$$\sum_{\kappa'}\left[|\langle\psi_{\kappa'}|e^{i\mathbf{q}\cdot\mathbf{r}}|\psi_\kappa\rangle|^2 + |\langle\psi_\kappa|e^{i\mathbf{q}\cdot\mathbf{r}}|\psi_{\kappa'}\rangle|^2\right](\epsilon_{\kappa'} - \epsilon_\kappa) = \frac{\hbar^2 q^2}{m_e}$$

Using this result, show that:

$$\frac{1}{V}\sum_{\kappa,\kappa'}\left[f_\kappa|\langle\psi_{\kappa'}|e^{i\mathbf{q}\cdot\mathbf{r}}|\psi_\kappa\rangle|^2 + f_{\kappa'}|\langle\psi_\kappa|e^{i\mathbf{q}\cdot\mathbf{r}}|\psi_{\kappa'}\rangle|^2\right](\epsilon_{\kappa'} - \epsilon_\kappa) = \frac{\hbar^2 q^2}{m_e}\frac{N}{2V}$$

where N is the number of electrons in the solid, V is the total volume, and f_κ are Fermi occupation numbers. By comparing this expression to that for the imaginary part of the dielectric function, Eq. (6.27), and integrating over ω, we can establish the sum rule, Eq. (6.75).

8. Prove the expressions for the various terms in the energy of a Frenkel exciton represented by a Slater determinant of Wannier functions, given in Eqs (6.55)–(6.60).

9. Using the Fourier transform of the coefficients $\xi(\mathbf{k})$, Eq. (6.72), in the exciton equation, Eq. (6.70), and the result of Eq. (6.71), prove the final equation for the exciton wavefunction, Eq. (6.73).

10. Write the many-body wavefunction $\Phi_{\mathbf{R}_i,\mathbf{R}_j}^{(S,S_z)}(\{\mathbf{r}\}, \{s\})$ for a Frenkel exciton with the electron and hole at different lattice sites $\mathbf{R}_i$ and $\mathbf{R}_j$. Using this expression, construct a many-body wavefunction with the proper Bloch character for a periodic solid as an expansion over $\Phi_{\mathbf{R}_i,\mathbf{R}_j}^{(S,S_z)}(\{\mathbf{r}\}, \{s\})$ states with appropriate coefficients. Compare this wavefunction with the expression of Eq. (6.52) and discuss the choice of expansion coefficients that make the two many-body wavefunctions identical.

11. Using data from Tables 5.2 and 5.3 for the static dielectric constants and effective masses of electrons and holes, estimate the binding energies of excitons for the semiconductors Si, Ge, GaAs, and compare with the experimental values given in Table 6.2.

7 Lattice Vibrations and Deformations

Up to now we have focused mainly on the behavior of electrons in a solid, given the fixed positions of all the ions. This is justified in terms of the Born–Oppenheimer approximation, and allowed us to study a number of interesting phenomena, including the interaction of solids with light. We next move on to a detailed account of what happens when the ions in a solid move away from their positions at equilibrium, that is, the lowest-energy structure of the entire system. We will again employ the Born–Oppenheimer approximation which is valid for most situations since the typical time scale for motion of ions, 10^{-15} sec (femtoseconds), is much slower than the relaxation time of electrons, 10^{-18} sec (attoseconds). We will examine two broad types of motion: the vibrations of atoms with small amplitude around their equilibrium positions, known as "phonon" modes, and the elastic deformation of a solid, that is, distortion by relatively small changes of atomic positions relative to equilibrium of a very large number of atoms that do not permanently change the shape of the solid. These topics will allow us to access some of the thermal properties of solids, which have to do with transport of mechanical energy by lattice vibrations, and some features of solids that characterize their response to external pressure, like bulk and elastic moduli.

7.1 Lattice Vibrations: Phonon Modes

At a non-zero temperature the atoms that form a crystalline lattice vibrate about their equilibrium positions, with an amplitude that depends on the temperature. Because a crystalline solid has symmetries, these thermal vibrations can be analyzed in terms of collective modes of motion of the ions. These modes correspond to collective excitations, which are typically called "phonons,"[1] although strictly speaking phonons are the quanta of these excitations. Unlike electrons, phonons are bosons: their total number is not fixed, nor is there a Pauli exclusion principle governing the occupation of any particular phonon state. This is easily rationalized, if we consider the real nature of phonons, that is, collective vibrations of the atoms in a crystalline solid which can be excited arbitrarily by heating (or hitting) the solid. We next discuss the physics of phonons and how they can be used to describe the thermal properties of solids.

The vibrational motions of the atoms are typically discussed in the context of classical dynamics, while the quantum mechanics of the electrons is hidden in the interaction

---

[1] The word phonon derives from the Greek noun $\phi\omega\nu\eta$, "phoni," for voice.

potential between the ions, as discussed in detail in Chapter 4, Sections 4.1 and 4.10. Suppose that the positions of the ions in the crystalline solid at zero temperature are determined by the vectors

$$\mathbf{R}_{ni} = \mathbf{R}_n + \mathbf{t}_i \tag{7.1}$$

where the $\mathbf{R}_n$ are the Bravais lattice vectors and the $\mathbf{t}_i$ are the positions of ions in one primitive unit cell, with the convention that $|\mathbf{t}_i| < |\mathbf{R}_n|$ for all non-zero lattice vectors. Then the deviation of each ionic position at non-zero temperature from its zero-temperature position can be denoted as:

$$\mathbf{S}_{ni} = \delta \mathbf{R}_{ni} \tag{7.2}$$

with n running over all the PUCs of the crystal and i running over all ions in the PUC. In terms of these vectors, the kinetic energy K of the ions will be:

$$K = \sum_{n,i} \frac{1}{2} M_i \left[\frac{d\mathbf{S}_{ni}}{dt} \right]^2 = \sum_{ni\alpha} \frac{1}{2} M_i \left(\frac{dS_{ni\alpha}}{dt} \right)^2 \tag{7.3}$$

where M_i is the mass of ion i and α labels the cartesian coordinates of the vectors $\mathbf{S}_{ni}$ ($\alpha = x, y, z$ in three dimensions).

When ions move, the system has a potential energy in excess of the frozen-ion potential energy U^{ion}, discussed earlier in Chapter 4. We denote this excess potential energy as ΔU, and refer to it in the following as simply the "potential energy," since this is the term of interest when atoms in the solid deviate from the ideal positions. ΔU can be written as a Taylor-series expansion in powers of $\mathbf{S}_{ni}$. The harmonic approximation consists of keeping only the second-order terms in this Taylor series, with the zeroth-order term being a constant (which we can set to zero for convenience). We will also take the first-order terms to be zero, since the system is expected to be in an equilibrium configuration at zero temperature, which represents a minimum of the total energy. Higher-order terms in the expansion are considered negligible. In this approximation, the potential energy ΔU is then given by:

$$\Delta U = \frac{1}{2} \sum_{n,i,\alpha;m,j,\beta} S_{ni\alpha} \frac{\partial^2 E^{\text{tot}}}{\partial R_{ni\alpha} \partial R_{mj\beta}} S_{mj\beta} \tag{7.4}$$

where the total energy E^{tot} depends on all the atomic coordinates $\mathbf{R}_{ni}$. We define the so-called force-constant matrix by:

$$F_{ni\alpha,mj\beta} = \frac{\partial^2 E^{\text{tot}}}{\partial R_{ni\alpha} \partial R_{mj\beta}} \tag{7.5}$$

which has the dimensions of a spring constant, [energy]/[length]^2. In terms of this expression, the potential energy becomes:

$$\Delta U = \frac{1}{2} \sum_{n,i,\alpha;m,j,\beta} S_{ni\alpha} F_{ni\alpha,mj\beta} S_{mj\beta} \tag{7.6}$$

The size of the force-constant matrix is $d \times n_{\text{at}} \times N$, where d is the dimensionality of space (the number of values for α), n_{at} is the number of atoms in the PUC, and N is the number of PUCs in the crystal. We notice that the following relations hold:

$$\frac{\partial^2 \Delta U}{\partial S_{ni\alpha} \partial S_{mj\beta}} = F_{ni\alpha,mj\beta} = \frac{\partial^2 E^{\mathrm{tot}}}{\partial R_{ni\alpha} \partial R_{mj\beta}} \tag{7.7}$$

$$\frac{\partial \Delta U}{\partial S_{ni\alpha}} = \sum_{m,j,\beta} F_{ni\alpha,mj\beta} S_{mj\beta} = \frac{\partial E^{\mathrm{tot}}}{\partial R_{ni\alpha}} \tag{7.8}$$

where the first set of equations is a direct consequence of the definition of the force-constant matrix and the harmonic approximation for the energy, while the second set of equations is a consequence of Newton's third law, since the left-hand side represents the negative of the α component of the total force on ion i in the unit cell labeled by $\mathbf{R}_n$.

The motion of the ions will be governed by the following equations:

$$M_i \frac{\mathrm{d}^2 S_{ni\alpha}}{\mathrm{d}t^2} = -\frac{\partial E^{\mathrm{tot}}}{\partial R_{ni\alpha}} = -\sum_{m,j,\beta} F_{ni\alpha,mj\beta} S_{mj\beta} \tag{7.9}$$

where we have used Eq. (7.8). We can try to solve the equations of motion by assuming sinusoidal expressions for the time dependence of their displacements:

$$S_{ni\alpha}(t) = \frac{1}{\sqrt{M_i}} \tilde{u}_{ni\alpha} e^{-i\omega t} \tag{7.10}$$

where ω is the frequency of oscillation and we have explicitly introduced the mass of the ions in the definition of the new variables $\tilde{u}_{ni\alpha}$. This gives, when substituted in the equations of motion:

$$\omega^2 \tilde{u}_{ni\alpha} = \sum_{m,j,\beta} F_{ni\alpha,mj\beta} \frac{1}{\sqrt{M_i M_j}} \tilde{u}_{mj\beta} \tag{7.11}$$

We define a new matrix, which we will call the dynamical matrix, through

$$\tilde{D}_{ni\alpha,mj\beta} = \frac{1}{\sqrt{M_i M_j}} F_{ni\alpha,mj\beta} \tag{7.12}$$

In terms of this matrix, the equations of motion can be written as:

$$\sum_{m,j,\beta} \tilde{D}_{ni\alpha,mj\beta} \tilde{u}_{mj\beta} = \omega^2 \tilde{u}_{ni\alpha} \Rightarrow \underline{\mathbf{D}} \cdot \tilde{\mathbf{u}} = \omega^2 \tilde{\mathbf{u}} \tag{7.13}$$

where we have used bold symbols for the dynamical matrix and the vector of ionic displacements in the last expression. This is an eigenvalue equation, the solution of which gives the values of the frequency and the vectors that describe the corresponding ionic displacements. The size of the dynamical matrix is the same as the size of the force-constant matrix (i.e. $d \times n_{\mathrm{at}} \times N$). Obviously, it is impossible to diagonalize such a matrix for a crystal in order to find the eigenvalues and eigenfunctions when $N \to \infty$.

We need to reduce this eigenvalue equation to a manageable size, so that it can be solved. To this end, we note that from the definition of the dynamical matrix:

$$\tilde{D}_{ni\alpha,mj\beta} = \frac{1}{\sqrt{M_i M_j}} F_{ni\alpha,mj\beta} = \frac{1}{\sqrt{M_i M_j}} \frac{\partial^2 E^{\mathrm{tot}}}{\partial R_{ni\alpha} \partial R_{mj\beta}} \tag{7.14}$$

If both positions $R_{ni\alpha}, R_{mj\beta}$ were to be shifted by the same lattice vector $\mathbf{R}'$, the result of differentiation of the energy with ionic positions must be the same because of the translational invariance of the hamiltonian. This leads to the conclusion that the dynamical matrix can only depend on the distance $\mathbf{R}_n - \mathbf{R}_m$ and not on the specific values of n and m, that is:

$$\tilde{D}_{ni\alpha,mj\beta} = \tilde{D}_{i\alpha,j\beta}(\mathbf{R}_n - \mathbf{R}_m) \tag{7.15}$$

Accordingly, we can define the ionic displacements as follows:

$$\tilde{u}_{ni\alpha}(\mathbf{k}) = u_{i\alpha}(\mathbf{k}) \, e^{i\mathbf{k}\cdot\mathbf{R}_n} \tag{7.16}$$

which gives for the eigenvalue equation:

$$\sum_{j,\beta}\sum_{m} \tilde{D}_{i\alpha,j\beta}(\mathbf{R}_n - \mathbf{R}_m) \, e^{-i\mathbf{k}\cdot(\mathbf{R}_n - \mathbf{R}_m)} \, u_{j\beta}(\mathbf{k}) = \omega^2 \, u_{i\alpha}(\mathbf{k}) \tag{7.17}$$

and with the definition

$$D_{i\alpha,j\beta}(\mathbf{k}) = \sum_{\mathbf{R}} \tilde{D}_{i\alpha,j\beta}(\mathbf{R}) \, e^{-i\mathbf{k}\cdot\mathbf{R}} = \sum_{n} e^{-i\mathbf{k}\cdot\mathbf{R}_n} \, \frac{1}{\sqrt{M_i M_j}} \, \frac{\partial^2 \Delta U}{\partial S_{ni\alpha} \partial S_{0j\beta}} \tag{7.18}$$

where the last expression is obtained with the use of Eq. (7.7), the eigenvalue equation takes the form

$$\sum_{j,\beta} D_{i\alpha,j\beta}(\mathbf{k}) \, u_{j\beta}(\mathbf{k}) = \omega^2 \, u_{i\alpha}(\mathbf{k}) \Rightarrow \underline{\mathbf{D}}(\mathbf{k}) \cdot \mathbf{u}_{\mathbf{k}}^{(l)} = \left(\omega_{\mathbf{k}}^{(l)}\right)^2 \mathbf{u}_{\mathbf{k}}^{(l)} \tag{7.19}$$

Since $D_{i\alpha,j\beta}(\mathbf{k})$ has a dependence on the wave-vector $\mathbf{k}$, so will the eigenvalues ω and the eigenvectors $u_{i\alpha}$, both of which have now been labeled by the value of $\mathbf{k}$. We have also introduced an additional "band" index for the eigenvalues and corresponding eigenvectors, needed for the reason explained below. The size of the new matrix is $d \times n_{at}$, which is a manageable size for crystals that typically contain few atoms per PUC. In transforming the problem to a manageable size, we realize that we now have to solve this eigenvalue problem for all the allowed values of $\mathbf{k}$, which of course are all the values in the first BZ. There are N distinct values of $\mathbf{k}$ in the first BZ, where N is the number of PUCs in the crystal, that is, no information has been lost in the transformation. Arguments similar to those applied to the solution of the single-particle hamiltonian for the electronic states can also be used here, to reduce the reciprocal-space volume where we need to obtain a solution down to the IBZ only. The solutions to the eigenvalue equation will need to be labeled by the additional "band" index l, which takes $d \times n_{at}$ values at each $\mathbf{k}$ value, to take into account all the different ions in the PUC and the dimensionality of space in which they move. The solution for the displacement of ion j in the PUC at lattice vector $\mathbf{R}_n$ will then be given by:

$$\mathbf{S}_{nj}^{(l)}(\mathbf{k}, t) = \frac{1}{\sqrt{M_j}} \, \hat{\mathbf{e}}_{\mathbf{k}j}^{(l)} \, e^{i\mathbf{k}\cdot\mathbf{R}_n} e^{-i\omega_{\mathbf{k}}^{(l)} t} \tag{7.20}$$

where $\hat{\mathbf{e}}_{\mathbf{k}j}^{(l)}$ is the set of d components of the eigenvector that denote the displacement of ion j in d dimensions. The eigenvectors can be chosen to be orthonormal:

$$\sum_{j} \left[\hat{\mathbf{e}}_{\mathbf{k}j}^{(l)}\right]^* \cdot \hat{\mathbf{e}}_{\mathbf{k}j}^{(l')} = \delta_{ll'} \tag{7.21}$$

In terms of these displacements, the most general ionic motion of the crystal can be expressed as:

$$\mathbf{S}_{n,j}(t) = \sum_{l,\mathbf{k}} c_{\mathbf{k}}^{(l)} \frac{1}{\sqrt{M_j}} \, \hat{\mathbf{e}}_{\mathbf{k}j}^{(l)} \, e^{i\mathbf{k}\cdot\mathbf{R}_n} e^{-i\omega_{\mathbf{k}}^{(l)}t} \tag{7.22}$$

where the coefficients $c_{\mathbf{k}}^{(l)}$ correspond to the amplitude of oscillation of the mode with frequency $\omega_{\mathbf{k}}^{(l)}$. We note that the eigenvalues of the frequency $\omega_{\mathbf{k}}^{(l)}$ obey the symmetry $\omega_{\mathbf{k}+\mathbf{G}}^{(l)} = \omega_{\mathbf{k}}^{(l)}$, where $\mathbf{G}$ is any reciprocal lattice vector. This symmetry is a direct consequence of the property $D_{i\alpha,j\beta}(\mathbf{k}) = D_{i\alpha,j\beta}(\mathbf{k}+\mathbf{G})$, which is evident from the definition of the dynamical matrix, Eq. (7.18). The symmetry of the eigenvalue spectrum allows us to solve the eigenvalue equations for $\omega_{\mathbf{k}}^{(l)}$ in the first BZ only, just like we did for the electronic energies. In both cases the underlying symmetry of the crystal that leads to this simplification is the translational periodicity. These analogies are summarized below.

Analogies between phonons and electrons in crystals

	Phonons	Electrons

$$\frac{d^2}{dt^2}S_{ni\alpha} = -\sum_{m,j,\beta} \sqrt{\frac{M_j}{M_i}} \tilde{D}_{ni\alpha,mj\beta} S_{mj\beta} \quad (1) \qquad i\hbar\frac{\partial}{\partial t}\psi(\mathbf{r},t) = \hat{\mathcal{H}}^{\mathrm{sp}}\psi(\mathbf{r},t)$$

$$S_{ni\alpha}(t) = \frac{1}{\sqrt{M_i}}\tilde{u}_{ni\alpha}e^{-i\omega t} \quad (2) \qquad \psi(\mathbf{r},t) = \psi(\mathbf{r})e^{-i\epsilon t/\hbar}$$

$$\sum_{m,j,\beta} \tilde{D}_{ni\alpha,mj\beta}\tilde{u}_{mj\beta} = \omega^2\tilde{u}_{ni\alpha} \quad (3) \qquad \hat{\mathcal{H}}^{\mathrm{sp}}\psi(\mathbf{r}) = \epsilon\psi(\mathbf{r})$$

$$\tilde{D}_{ni\alpha,mj\beta} = \tilde{D}_{i\alpha,j\beta}(\mathbf{R}_n - \mathbf{R}_m) \quad (4) \qquad \mathcal{V}(\mathbf{r}+\mathbf{R}) = \mathcal{V}(\mathbf{r})$$

$$\tilde{u}_{ni\alpha}(\mathbf{k}) = u_{i\alpha}(\mathbf{k})e^{i\mathbf{k}\cdot\mathbf{R}_n} \quad (5) \qquad \psi_{\mathbf{k}}(\mathbf{r}) = u_{\mathbf{k}}(\mathbf{r})e^{i\mathbf{k}\cdot\mathbf{r}}, \ u_{\mathbf{k}}(\mathbf{r}+\mathbf{R}) = u_{\mathbf{k}}(\mathbf{r})$$

$$D_{i\alpha,j\beta}(\mathbf{k}) = \sum_{\mathbf{R}} \tilde{D}_{i\alpha,j\beta}(\mathbf{R})\, e^{-i\mathbf{k}\cdot\mathbf{R}} \quad (6) \qquad \hat{\mathcal{H}}_{\mathbf{k}}^{\mathrm{sp}} = e^{-i\mathbf{k}\cdot\mathbf{r}}\hat{\mathcal{H}}^{\mathrm{sp}}e^{i\mathbf{k}\cdot\mathbf{r}}$$

$$\underline{\mathbf{D}}(\mathbf{k})\mathbf{u}_{\mathbf{k}}^{(l)} = \left(\omega_{\mathbf{k}}^{(l)}\right)^2 \mathbf{u}_{\mathbf{k}}^{(l)} \quad (7) \qquad \hat{\mathcal{H}}_{\mathbf{k}}^{\mathrm{sp}}u_{\mathbf{k}}^{(n)}(\mathbf{r}) = \epsilon_{\mathbf{k}}^{(n)}u_{\mathbf{k}}^{(n)}(\mathbf{r})$$

$$\mathbf{S}_{nj}^{(l)}(\mathbf{k},t) = \frac{1}{\sqrt{M_j}} \, \hat{\mathbf{e}}_{\mathbf{k}j}^{(l)} \, e^{i\mathbf{k}\cdot\mathbf{R}_n} e^{-i\omega_{\mathbf{k}}^{(l)}t} \quad (8) \qquad \psi_{\mathbf{k}}^{(n)}(\mathbf{r},t) = u_{\mathbf{k}}^{(n)}(\mathbf{r})e^{i\mathbf{k}\cdot\mathbf{r}}e^{-i\epsilon_{\mathbf{k}}^{(n)}t/\hbar}$$

(1)　Fundamental equation of motion.
(2)　Separation of time-dependent and time-independent components.
(3)　Eigenvalue equation for time-independent component.
(4)　Fundamental symmetry of operator acting on time-independent component.
(5)　Consequences of fundamental symmetry (Bloch states).
(6)　Operator acting on periodic component of Bloch states.
(7)　Eigenvalue equation for periodic component of Bloch states.
(8)　General form of eigenvectors/eigenfunctions, including time-dependent part.

Finally, we can calculate a phonon density of states, just like we did for the electronic eigenvalues in a crystal. The expression for the phonon DOS, by analogy to that for electronic states, for a system in d dimensions is:

$$g(\omega) = \frac{1}{(2\pi)^d} \sum_{k,l} \int_{\omega_k^{(l)}=\omega} \frac{1}{|\nabla_k \omega_k^{(l)}|} dS_k \tag{7.23}$$

where S_k is a surface in reciprocal space on which the condition $\omega_k^{(l)} = \omega$ is satisfied and the summation is over values of k in the first BZ and over values of the index l of phonon branches at a given value of k.

7.2 The Born Force-Constant Model

We present next a very simple model for calculating the dynamical matrix. In its ideal, zero-temperature configuration, we can think of the crystal as consisting of atoms with "bonds" between them. These bonds can have covalent, ionic, or metallic character, and in the context of the very simple model we are considering, all the bonds are localized between pairs of atoms. For situations where this picture is reasonable, it is appropriate to consider an effective potential $u(R_i, R_j)$ between each pair of atoms at positions R_i, R_j (for simplicity of notation, we use here a single index, i or j, to denote the atomic position vectors, with the understanding that this index contains both the unit cell index and the basis index). The total potential energy ΔU includes the contributions from all pairs of atoms:

$$\Delta U = \sum_{\langle ij \rangle} u(R_i, R_j)$$

where the notation $\langle ij \rangle$ stands for summation over all the distinct nearest-neighbor pairs. We shall assume that there are two types of distortions of the bonds relative to their ideal positions and orientations:

1. bond stretching, that is, elongation or contraction of the length of bonds, expressed as $\Delta |R_i - R_j|$, and
2. bond bending, that is, change of the orientation of bonds relative to their original position, expressed as $\Delta\theta_{ij}$, where θ_{ij} is the angle describing the orientation of the bond with respect to a convenient direction in the undistorted, ideal crystal.

For the bond-stretching term, $u_r(R_i, R_j)$, we will take the contribution to the potential energy to be proportional to the square of the amount of stretching:

$$u_r(R_i, R_j) \sim (\Delta |R_i - R_j|)^2 \tag{7.24}$$

For the bond-bending term, $u_\theta(\mathbf{R}_i, \mathbf{R}_j)$, we will take the contribution to the potential energy to be proportional to the square of the angle that describes the change in bond orientation:

$$u_\theta(\mathbf{R}_i, \mathbf{R}_j) \sim (\Delta\theta_{ij})^2 \tag{7.25}$$

It is important to justify on physical grounds our choices for the behavior of the bond-stretching and bond-bending terms. First, we note that these choices are consistent with the harmonic approximation on which the entire model is based, as discussed in the previous section. The behavior of the bond-stretching term is easy to rationalize: we think of bonds as simple harmonic springs, with a spring constant κ_r, implying that the force due to changes in the equilibrium bond length is proportional to the amount of extension or contraction, $\Delta|\mathbf{R}_i - \mathbf{R}_j|$, and hence the potential energy is quadratic in this variable, as expressed by Eq. (7.24). The behavior of the bond-bending term is a little trickier to explain. This term is related to the preference of atoms to have certain angles between the bonds they form to nearest neighbors. For example, in tetrahedrally bonded solids, with bonds formed by interaction between sp^3 hybrid orbitals, the preferred bond angle is 109.47°; in planar structures with bonds formed by interaction between sp^2 hybrid oribtals, the preferred bond angle is 120° (for examples of such solids, see Chapter 1). Any deviation from the preferred bond angle will lead to an increase in the potential energy, even if there are no changes in the bond lengths. The reason is that a deviation in angular orientation leads to changes in the nature of the hybrid orbitals that are involved in the bonding, which produces a less favorable (higher potential energy) structure. While these examples were picked from covalently bonded solids, similar arguments can be invoked in the general case. Thus, the bond-bending contribution is really a three-body term, since we need three different atomic positions to define an angle between the two bonds.

To see the connection of this three-body term to the bond-bending model, we define the angle between the bonds of atom pairs ij and jk as θ_{ijk}, with the apex at the middle atom j; similarly, the angle between the bonds of atom pairs ji and il is θ_{jil}, with the apex at the middle atom i. When atoms i and j move from their ideal crystal positions, it is these two bond angles that will be affected. We take, as is common, the energy cost for changing a bond angle from its preferred value, θ_0, in the ideal, undistorted crystal to be proportional to a trigonometric function of the new angle minus the value of this function at θ_0; this difference is squared, to take into account the equal contributions to the energy for positive and negative changes in the value of the function relative to the ideal one. We will assume that the displacements of the atoms i and j and their neighbors are such that only the two bond angles with apex at those two atoms are affected, and take the energy cost to be:

$$u_\theta(\mathbf{R}_i, \mathbf{R}_j) \sim \left\{ \left[\cos(\theta_{ijk}) - \cos(\theta_0)\right]^2 + \left[\cos(\theta_{jil}) - \cos(\theta_0)\right]^2 \right\}$$

As shown in the simple diagram of Fig. 7.1, we can write the angle θ_{ijk} as $\theta_{ijk} = \theta_0 - \Delta\theta_{ij}$, where $\Delta\theta_{ij}$ is the change in orientation of the ij pair bond relative to its orientation in the

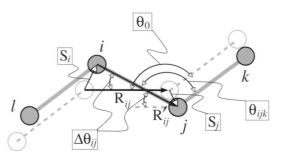

The origin of the bond-bending term in the force-constant model: the open circles denote the positions of atoms in the ideal configuration and the filled circles their displaced positions; $\mathbf{S}_i, \mathbf{S}_j$ are the atomic displacements; $\mathbf{R}_{ij}$ (blue arrow) and $\mathbf{R}'_{ij}$ (red arrow) are the original and the distorted relative distances between atoms labeled i and j; $\Delta\theta_{ij}$ is the change in orientation of the bond between these two atoms, equal to the change in the bond angle between pair ij and pair jk, which has value θ_0 in the undistorted configuration.

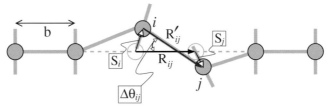

Bond stretching and bond bending in the force-constant model: open circles denote the positions of atoms in the ideal configuration and filled circles their displaced positions; only the atoms labeled i and j have been displaced by $\mathbf{S}_i, \mathbf{S}_j$; $\mathbf{R}_{ij}$ (blue arrow) is the original relative distance ($|\mathbf{R}_{ij}| = b$), $\mathbf{R}'_{ij}$ (red arrow) is the distorted relative distance, and $\Delta\theta_{ij}$ is the change in orientation of the bond (the angular distortion). The thick blue lines indicate bonds, some of which are stretched or bent due to the displacement of atoms i and j.

ideal crystal. Expanding to lowest order in the small angle $\Delta\theta_{ij}$, we find that the above expression for the potential energy cost takes the form

$$\left[\cos(\theta_{ijk}) - \cos(\theta_0)\right]^2 = \left[\cos(\theta_0 - \Delta\theta_{ij}) - \cos(\theta_0)\right]^2 \approx \sin^2(\theta_0)(\Delta\theta_{ij})^2$$

and the same result is found for the term involving the angle θ_{jil}. This result justifies our choice for the dependence of the potential energy on the bond-bending contribution, Eq. (7.25). A significant advantage of this formulation is that now we have to deal with two-body terms, that is, terms that involve only *pairs* of atoms as opposed to *triplets* of atoms, needed to specify the bond angles $\theta_{ijk}, \theta_{jil}$; this greatly simplifies calculations, both analytical and numerical. One caveat is that this model is *not* invariant under zero-frequency global rotations as a crystal should be. Nevertheless, it can still provide a reasonable description at non-zero frequencies for small displacements of the atoms from their equilibrium positions.

To determine the contributions of bond stretching and bond bending to the potential energy, we will use the diagram shown in Fig. 7.2. The bond-stretching energy is given by:

$$\Delta U_r = \frac{1}{2} \kappa_r \sum_{\langle ij \rangle} \left(\Delta |\mathbf{R}_{ij}|\right)^2 \tag{7.26}$$

where κ_r is the force constant for bond stretching, and $\Delta|\mathbf{R}_{ij}|$ is the change in the length of the bond between atoms i and j. If we assume that all the bonds are equal to b (this can easily be generalized for systems with several types of bonds), then the change in bond length between atoms i and j is given by $\Delta|\mathbf{R}_{ij}| = |\mathbf{R}'_{ij}| - b$, where $\mathbf{R}'_{ij}$ is the new distance between atoms i and j. We will denote by $\mathbf{S}_i$ the departure of ion i from its equilibrium ideal position, which gives for the new distance between displaced atoms i and j:

$$\mathbf{R}'_{ij} = \mathbf{S}_j - \mathbf{S}_i + \mathbf{R}_{ij} \Rightarrow |\mathbf{R}'_{ij}| = \left[\left(\mathbf{S}_j - \mathbf{S}_i + \mathbf{R}_{ij}\right)^2\right]^{1/2} \tag{7.27}$$

The displacements $\mathbf{S}_i$ will be taken to be always much smaller in magnitude than the bond distances $\mathbf{R}_{ij}$, consistent with the assumption of small deviations from equilibrium, in which case the harmonic approximation makes sense. Using this fact, we can expand in powers of the small quantity $|\mathbf{S}_i - \mathbf{S}_j|/b$, which gives to lowest order:

$$|\mathbf{R}'_{ij}| = b\left(1 + \frac{\mathbf{R}_{ij} \cdot (\mathbf{S}_i - \mathbf{S}_j)}{b^2}\right) \Rightarrow \Delta|\mathbf{R}_{ij}| = |\mathbf{R}'_{ij}| - b = \hat{\mathbf{R}}_{ij} \cdot (\mathbf{S}_i - \mathbf{S}_j) \tag{7.28}$$

which in turn gives for the bond-stretching energy:

$$\Delta U_r = \frac{1}{2}\kappa_r \sum_{\langle ij \rangle} \left[(\mathbf{S}_j - \mathbf{S}_i) \cdot \hat{\mathbf{R}}_{ij}\right]^2 \tag{7.29}$$

By similar arguments we can obtain an expression in terms of the variables $\mathbf{S}_i, \mathbf{S}_j$ for the bond-bending energy, defined to be:

$$\Delta U_\theta = \frac{1}{2}\kappa_\theta \sum_{\langle ij \rangle} b^2 \left(\Delta\theta_{ij}\right)^2 \tag{7.30}$$

with κ_θ the bond-bending force constant and $\Delta\theta_{ij}$ the change in the orientation of the bond between atoms i and j. Using the same notation for the new positions of the ions (see Fig. 7.2), and the assumption that the displacements $\mathbf{S}_i$ are much smaller than the bond distances $\mathbf{R}_{ij}$, we obtain for the bond-bending term:

$$\mathbf{R}'_{ij} \cdot \mathbf{R}_{ij} = |\mathbf{R}'_{ij}|b\cos(\Delta\theta_{ij}) \approx \left[1 - \frac{1}{2}(\Delta\theta_{ij})^2\right]|\mathbf{R}'_{ij}|b \Rightarrow \frac{\mathbf{R}'_{ij} \cdot \mathbf{R}_{ij}}{|\mathbf{R}'_{ij}|b} = 1 - \frac{1}{2}(\Delta\theta_{ij})^2 \tag{7.31}$$

For the left-hand side of this last equation, we find:

$$\frac{\mathbf{R}'_{ij} \cdot \mathbf{R}_{ij}}{|\mathbf{R}'_{ij}|b} = \left[(\mathbf{S}_j - \mathbf{S}_i + \mathbf{R}_{ij}) \cdot \mathbf{R}_{ij}\right]\frac{1}{b^2}\left[1 + \frac{(\mathbf{S}_i - \mathbf{S}_j)^2}{b^2} - \frac{2\mathbf{R}_{ij} \cdot (\mathbf{S}_i - \mathbf{S}_j)}{b^2}\right]^{-1/2}$$

$$= \left[1 + \frac{\mathbf{R}_{ij} \cdot (\mathbf{S}_j - \mathbf{S}_i)}{b^2}\right]\left[1 + \frac{\mathbf{R}_{ij} \cdot (\mathbf{S}_i - \mathbf{S}_j)}{b^2} - \frac{1}{2}\frac{(\mathbf{S}_i - \mathbf{S}_j)^2}{b^2} + \frac{3}{2}\frac{|\mathbf{R}_{ij} \cdot (\mathbf{S}_i - \mathbf{S}_j)|^2}{b^4}\right]$$

$$= 1 - \frac{|\mathbf{S}_j - \mathbf{S}_i|^2}{2b^2} + \frac{\left[(\mathbf{S}_j - \mathbf{S}_i) \cdot \mathbf{R}_{ij}\right]^2}{2b^4}$$

where we have used the Taylor expansion of $(1 + x)^{-1/2}$ (see Appendix A) and kept only terms up to second order in the small quantities $|\mathbf{S}_i - \mathbf{S}_j|/b$. This result, when compared to the right-hand side of the previous equation, gives:

$$(\Delta\theta_{ij})^2 b^2 = |\mathbf{S}_j - \mathbf{S}_i|^2 - \left[(\mathbf{S}_j - \mathbf{S}_i) \cdot \hat{\mathbf{R}}_{ij}\right]^2 \qquad (7.32)$$

which leads to the following expression for the bond-bending energy:

$$\Delta U_\theta = \frac{1}{2}\kappa_\theta \sum_{\langle ij \rangle} \left[|\mathbf{S}_j - \mathbf{S}_i|^2 - \left[(\mathbf{S}_j - \mathbf{S}_i) \cdot \hat{\mathbf{R}}_{ij}\right]^2\right] \qquad (7.33)$$

Combining the two contributions, Eqs (7.29) and (7.33), we obtain for the total potential energy:

$$\Delta U = \frac{1}{2} \sum_{\langle ij \rangle} \left[(\kappa_r - \kappa_\theta)\left[(\mathbf{S}_j - \mathbf{S}_i) \cdot \hat{\mathbf{R}}_{ij}\right]^2 + \kappa_\theta |\mathbf{S}_j - \mathbf{S}_i|^2\right] \qquad (7.34)$$

where the sum runs over all pairs of atoms in the crystal that are bonded, that is, over all pairs of nearest neighbors.

7.3 Applications of the Force-Constant Model

We will apply next the force-constant model to a few simple examples that include all the essential features to demonstrate the behavior of phonons, in 1D, 2D, and 3D systems. In each case we will need to set up and diagonalize the dynamical matrix $D_{i\alpha,j\beta}(\mathbf{k})$ as defined in Eq. (7.18). We take the unit cell of interest to correspond to the zero Bravais lattice vector, and consider nearest-neighbor interactions only, in the spirit of the Born force-constant model. The potential energy ΔU then has contributions from all the pairs of atoms in the crystal that are nearest neighbors. We focus on contributions to ΔU that involve the atoms of the unit cell corresponding to the zero lattice vector, and employ the full notation for the displacement of ions $S_{ni\alpha}, S_{0j\beta}$: n labels the unit cells which contain atoms that are nearest neighbors to the atoms of the unit cell with zero lattice vector; i, j label the atoms in the unit cells, and α, β label the spatial coordinates ($\alpha, \beta = x$ in 1D, $\alpha, \beta = x$ or y in 2D, $\alpha, \beta = x, y$, or z in 3D).

For crystals with n_{at} atoms per unit cell, there are $3 \times n_{at}$ modes for each value of $\mathbf{k}$. When $n_{at} > 1$, there are two types of phonon modes: modes in which the motion of atoms within one unit cell is in phase are referred to as "acoustic," and are lower-frequency modes; modes in which the motion of atoms within one unit cell is π out of phase are referred to as "optical," and are higher-frequency modes. For example, in three dimensions there are three acoustic and three optical modes for a crystal with two atoms per unit cell, and their frequencies in general are not degenerate, except when this is required by symmetry. In cases where there is only one atom per unit cell, there are only acoustic modes. In all cases, the acoustic modes have a frequency that is linear in the wave-vector $\mathbf{k}$ near the center of the BZ ($\mathbf{k} \to 0$); the limit $\mathbf{k} = 0$ corresponds to the translation of the entire crystal in one direction, so naturally the frequency of this motion is zero. The slope of the linear behavior of the frequency $\omega_{\mathbf{k}}^{(n)}$ near $\mathbf{k} \to 0$ with respect to the wave-vector gives the velocity of acoustic (very long wavelength) waves in the crystal, which can be different along different crystallographic directions.

Example 7.1   Periodic chain in one dimension

We first apply the force-constant model to investigate phonon modes in a 1D infinite chain with one atom per unit cell and lattice constant a, as shown in Fig. 7.3. We assume first that all motion is confined in the direction of the chain, taken as $\hat{x}$. Since we are dealing with one atom per unit cell with mass M, we do not need to use the atom index i, which we omit for simplicity. The potential energy in this case, with $\mathbf{S}_n, \mathbf{S}_m (n = 0, m = \pm 1)$ having components only in the $\hat{x}$ direction along the chain, is given by:

$$\Delta U_x = \frac{\kappa_r}{2} \left[(S_{\bar{1}x} - S_{0x})^2 + (S_{0x} - S_{1x})^2 \right]$$

Taking derivatives of this potential energy with respect to the atomic displacements gives the dynamical matrix $D_{x,x}(k_x)$ and its solution $\omega_{k_x}^{(1)}$:

$$D_{x,x}(k_x) = \frac{1}{M} \left[2\kappa_r - \kappa_r(e^{ik_x a} + e^{-ik_x a}) \right] \Rightarrow \omega_{k_x}^{(1)} = \sqrt{\frac{2\kappa_r}{M}} [1 - \cos(k_x a)]^{1/2}$$

A plot of this expression is shown in Fig. 7.3.

Next, we consider the situation in which motion of the atoms in the direction perpendicular to the chain, $\hat{y}$, is also allowed. In this case, the potential energy includes another contribution from the displacements in this direction, given by:

$$\Delta U_y = \frac{\kappa_\theta}{2} \left[(S_{\bar{1}y} - S_{0y})^2 + (S_{0y} - S_{1y})^2 \right]$$

Now we have to solve a dynamical matrix of size 2×2, since we have $n_{at} = 2$ independent directions and still one atom per unit cell, but the off-diagonal matrix elements of this matrix are zero because there are no terms in the potential energy that involve the variables x and y simultaneously. The new matrix element is:

$$D_{y,y}(k_x) = \frac{1}{M} \left[2\kappa_\theta - \kappa_\theta(e^{ik_x a} + e^{-ik_x a}) \right]$$

There are therefore two solutions, one $\omega_{k_x}^{(1)}$ exactly the same as before, and the second given by

$$\omega_{k_x}^{(2)} = \sqrt{\frac{2\kappa_\theta}{M}} [1 - \cos(k_x a)]^{1/2}$$

which is also shown in Fig. 7.3, with $\kappa_\theta < \kappa_r$.

We calculate the DOS for the phonon modes as a function of the frequency ω, using Eq. (7.23). For the first type of phonons corresponding to frequencies $\omega_{k_x}^{(1)}$, the DOS, which we will call $g^{(1)}(\omega)$, is given by:

$$g^{(1)}(\omega) = \frac{1}{L} \sum_{\mathbf{k}} \delta(\omega - \omega_{\mathbf{k}}^{(1)}) = \frac{1}{2\pi} \int_{\omega_{\mathbf{k}}^{(1)} = \omega} \left| \frac{d\omega_{k_x}^{(1)}}{dk_x} \right|^{-1} dk_x$$

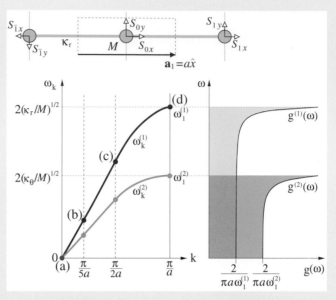

Fig. 7.3

Illustration of the 1D chain with lattice constant a and one atom of mass M per unit cell, and spring constant κ_r for bond-length distortions along the chain and κ_θ for bond-angle distortions. The frequency of the phonon modes for displacements along the chain, $\hat{x}$ direction, labeled $\omega_{\mathbf{k}}^{(1)}$ (red curve) and perpendicular to it, $\hat{y}$ direction, labeled $\omega_{\mathbf{k}}^{(2)}$ (green curve) are shown as functions of $\mathbf{k} = k_x\hat{x}$. The corresponding DOS are shown as functions of ω, in a red-shaded curve for $g^{(1)}(\omega)$ and a green-shaded curve for $g^{(2)}(\omega)$. The atomic displacements of the modes with frequencies marked (a), (b), (c), (d) are shown in Fig. 7.4.

with the integral restricted to values of k_x that satisfy the condition $\omega_{\mathbf{k}}^{(1)} = \omega$. Differentiating the expression of $\omega_{k_x}^{(1)}$ with respect to k_x, we obtain:

$$\frac{d\omega_{k_x}^{(1)}}{dk_x} = \frac{1}{\omega_{k_x}^{(1)}} \left(\frac{\kappa_r a}{M} \right) \sin(k_x a)$$

We solve for $\sin(k_x a)$ under the condition $\omega_{k_x}^{(1)} = \omega$:

$$\cos(k_x a) = 1 - \frac{\omega^2}{(2\kappa_r/M)} \Rightarrow$$

$$\sin(k_x a) = \left(1 - \cos^2(k_x a) \right)^{1/2} = \frac{\omega}{(2\kappa_r/M)^{1/2}} \left[2 - \frac{\omega^2}{(2\kappa_r/M)} \right]^{1/2}$$

and with this result we find for the density of states:

$$g^{(1)}(\omega) = \frac{2}{\pi a} \frac{1}{(2\kappa_r/M)^{1/2}} \left[2 - \frac{\omega^2}{(2\kappa_r/M)} \right]^{-1/2}$$

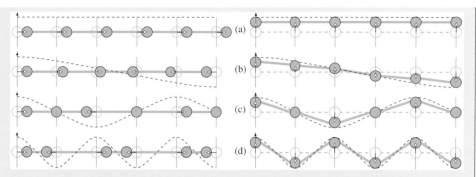

Fig. 7.4

The atomic displacements corresponding to phonon modes in the 1D chain with one atom per unit cell, with frequencies indicated by (a), (b), (c), (d) in Fig. 7.3 for motion along the chain direction (left) and perpendicular to the chain direction (right). The black dashed lines are plots of the function $\cos(k_x a)$ for the corresponding values of the wave-vector: (a) $k_x = 0$, (b) $k_x = \pi/5a$, (c) $k_x = \pi/2a$, (d) $k_x = \pi/a$.

where an extra factor of 2 comes from the fact that there are two values of k_x in the BZ for which the condition $\omega_{k_x}^{(1)} = \omega$ is satisfied for every value of ω in the range $[0, \omega_1^{(1)}]$. The result is plotted in Fig. 7.3. Similarly, we can calculate the DOS $g^{(2)}(\omega)$ for phonon modes of the second type, also shown in Fig. 7.3. In both cases the DOS is constant, equal to $2/\pi a \omega_1^{(i)}$ for $i = 1, 2$, for the range of values where the phonon frequency is linear in k_x, and it diverges near the maximum value of the frequency.

Finally, it is instructive to analyze the actual motion of atoms that corresponds to different modes. This is done by solving for the eigenvectors of the dynamical matrix

$$D_{x,x}(k_x)\mathbf{S}_{k_x} = \frac{2\kappa_r}{M}\left[1 - \cos(k_x a)\right]\mathbf{S}_{k_x}$$

which gives the trivial solution that $\mathbf{S}_{k_x}$ is a constant (which we will take to be unity) and the corresponding displacements of the ions at unit cells labeled by n at lattice vectors $\mathbf{R}_n = na\hat{x}$ are given by:

$$\mathbf{S}_n(k_x) = \frac{1}{\sqrt{M}}e^{i\mathbf{k}\cdot\mathbf{R}_n}\,\hat{x} = \frac{1}{\sqrt{M}}e^{ink_x a}\,\hat{x}$$

These are shown in Fig. 7.4 for selected values of the k_x wave-vector. Allowing for motion in the y direction produces similar phonon modes with displacements in that direction, also shown in Fig. 7.4 for the same selected values of k_x.

Example 7.2   Periodic chain with two different masses in one dimension

An interesting variation of this model is the case when there are two ions per unit cell with different masses, $M_1 \neq M_2$. Now we need to use all three indices for the potential: n, m for the lattice vectors; i, j for the two ions; and x, y for the two directions of motion. We only discuss the case of motion along the x direction. The expression for the potential energy becomes:

$$\Delta U_x = \frac{\kappa_r}{2}\left[(S_{\bar{1}2x} - S_{01x})^2 + (S_{01x} - S_{02x})^2 + (S_{02x} - S_{11x})^2\right]$$

This expression for the potential leads to a dynamical matrix with diagonal elements:

$$D_{1x,1x}(k_x) = \frac{2\kappa_r}{M_1}, \quad D_{2x,2x}(k_x) = \frac{2\kappa_r}{M_2}$$

and off-diagonal elements:

$$D_{2x,1x}(k_x) = -\frac{\kappa_r}{\sqrt{M_1 M_2}}\left(1 + e^{ik_x a}\right), \quad D_{1x,2x}(k_x) = D_{2x,1x}^*(k_x)$$

We define the reduced mass μ and the ratio of the masses ζ:

$$\mu = \frac{M_1 M_2}{M_1 + M_2}, \quad \zeta = \frac{M_2}{M_1}$$

and solve for the eigenvalues of this matrix to find the phonon frequencies:

$$\omega_{k_x}^{(\pm)} = \sqrt{\frac{\kappa_r}{\mu}} \left[1 \pm \left\{1 - \frac{4\zeta}{(1+\zeta)^2}\sin^2\left(\frac{k_x a}{2}\right)\right\}^{1/2}\right]^{1/2} \tag{7.35}$$

The phonon modes now have two branches, labeled with the superscripts $(-)$, corresponding to lower frequencies and $(+)$, corresponding to higher frequencies; these are the "acoustic" and "optical" branches, respectively. The plot for these two phonon branches as functions of the wave-vector k_x is shown in Fig. 7.3. By following the same procedure as in the previous example, we can also calculate the DOS for each branch as a function of ω, which is included in Fig. 7.3.

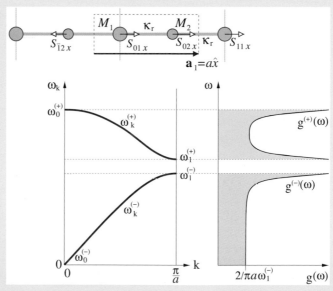

Fig. 7.5 Illustration of the 1D chain with lattice constant a and two atoms per unit cell with masses M_1, M_2 and spring constants κ_r for bond-length distortions along the chain $\hat{x}$ between pairs of nearest neighbors. The corresponding acoustic, $\omega_{k_x}^{(-)}$ (red curve) and optical, $\omega_{k_x}^{(+)}$ (blue curve) phonon mode dispersions are also shown as functions of $\mathbf{k} = k_x \hat{x}$. The DOS of the two branches is shown as a function of ω on the right.

The frequencies near the center and near the edge of the BZ are of particular interest. We will use the subscript 0 for values of $|\mathbf{k}| \approx 0$ and the subscript 1 for values of $|\mathbf{k}| \approx \pi/a$. Near $k_x \approx 0$, the acoustic and optical branch frequencies become:

$$\omega_0^{(-)} \approx \sqrt{\frac{2\kappa_r}{\mu}} \left[\frac{\sqrt{\zeta}}{2(1+\zeta)} \right] ak_x, \quad \omega_0^{(+)} \approx \sqrt{\frac{2\kappa_r}{\mu}} \left[1 - \frac{\zeta a^2 k_x^2}{8(1+\zeta)^2} \right]$$

that is, the lower branch is linear in k_x and the higher branch approaches a constant value as $k_x \to 0$. At $k_x = \pi/a$, the frequencies of the two branches become:

$$\omega_1^{(-)} = \sqrt{\frac{2\kappa_r}{\mu}} \left(\frac{\zeta}{1+\zeta} \right)^{1/2}, \quad \omega_1^{(+)} = \sqrt{\frac{2\kappa_r}{\mu}} \left(\frac{1}{1+\zeta} \right)^{1/2}$$

where we have assumed $\zeta < 1$. There is a gap between these two values equal to:

$$\omega_1^{(+)} - \omega_1^{(-)} = \sqrt{\frac{2\kappa_r}{\mu}} \left(\frac{1 - \zeta^{1/2}}{1 + \zeta^{1/2}} \right)^{1/2}$$

which is also evident in the DOS for the two branches, as shown in Fig. 7.5.

Example 7.3   Periodic chain in two dimensions

Our next example consists of a system that is periodic in one dimension with lattice vector $\mathbf{a}_1$ but exists in 2D space, and has two atoms per unit cell at positions $\mathbf{t}_1, \mathbf{t}_2$ with equal masses for the ions, $M_1 = M_2 = M$:

$$\mathbf{a}_1 = \frac{a}{\sqrt{2}}[\hat{x} + \hat{y}], \quad \mathbf{t}_1 = \frac{a}{2\sqrt{2}}\hat{x}, \quad \mathbf{t}_2 = -\frac{a}{2\sqrt{2}}\hat{x}$$

The physical system is shown in Fig. 7.6. Since there are two atoms per unit cell ($n_{\text{at}} = 2$) and the system exists in two dimensions ($d = 2$), the size of the dynamical matrix will be $2 \times 2 = 4$. In this case we only need the following three terms in the expression for the potential energy:

$$\Delta U_{01,02} = \frac{1}{2}(\kappa_r - \kappa_\theta)(S_{01x} - S_{02x})^2 + \frac{1}{2}\kappa_\theta \left[(S_{01x} - S_{02x})^2 + (S_{01y} - S_{02y})^2 \right]$$

$$\Delta U_{01,12} = \frac{1}{2}(\kappa_r - \kappa_\theta)(S_{01y} - S_{12y})^2 + \frac{1}{2}\kappa_\theta \left[(S_{01x} - S_{12x})^2 + (S_{01y} - S_{12y})^2 \right]$$

$$\Delta U_{\bar{1}1,02} = \frac{1}{2}(\kappa_r - \kappa_\theta)(S_{\bar{1}1y} - S_{02y})^2 + \frac{1}{2}\kappa_\theta \left[(S_{\bar{1}1x} - S_{02x})^2 + (S_{\bar{1}1y} - S_{02y})^2 \right]$$

The diagonal elements of the dynamical matrix $D_{i\alpha,j\beta}$ are:

$$D_{1x,1x}(\mathbf{k}) = D_{2x,2x}(\mathbf{k}) = D_{1y,1y}(\mathbf{k}) = D_{2y,2y}(\mathbf{k}) = \kappa_r + \kappa_\theta$$

and the off-diagonal ones are:

$$D_{1x,2x}(\mathbf{k}) = D_{1y,2y}(\mathbf{k}) = -\kappa_r - \kappa_\theta e^{i\mathbf{k}\cdot\mathbf{a}_1}$$

$$D_{2x,1x}(\mathbf{k}) = D_{1x,2x}^*(\mathbf{k}), \quad D_{2y,1y}(\mathbf{k}) = D_{1y,2y}^*(\mathbf{k})$$

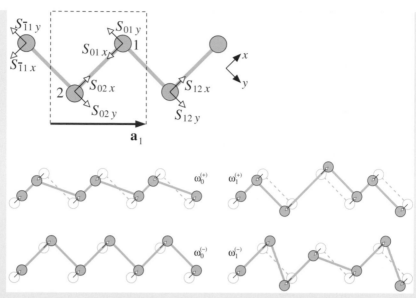

Fig. 7.6

Top: Definition of the model of the periodic chain in two dimensions, with lattice constant $\mathbf{a}_1 = (a/\sqrt{2})(\hat{x} + \hat{y})$ and two atoms per unit cell (labeled "1" and "2") at positions $\mathbf{t}_1 = (a/2\sqrt{2})\hat{x}, \mathbf{t}_2 = -(a/2\sqrt{2})\hat{x}$. The displacements $S_{ni\alpha}$ are shown for the two atoms in the unit cell and their nearest neighbors, $n = 0, \pm 1; i = 1, 2; \alpha = x, y$. **Bottom**: The atomic displacements of the phonon modes with frequencies marked as $\omega_0^{(\pm)}$ and $\omega_1^{(\pm)}$ in Fig. 7.3 for the periodic chain with lattice constant $\mathbf{a}_1 = a\hat{x}$ and two atoms per unit cell at positions $\mathbf{t}_1, \mathbf{t}_2$ in two dimensions.

with the rest equal to zero. The diagonalization of this matrix gives the frequency eigenvalues

$$\omega_{\mathbf{k}}^{(\pm)} = \frac{1}{\sqrt{M}} \left[(\kappa_r + \kappa_\theta) \pm \left(\kappa_r^2 + \kappa_\theta^2 + 2\kappa_r\kappa_\theta \cos(\mathbf{k} \cdot \mathbf{a}_1) \right)^{1/2} \right]^{1/2} \tag{7.36}$$

with each value of $\omega_{\mathbf{k}}^{(\pm)}$ doubly degenerate due to the equivalence of the x and y directions. We can define the new variable

$$\zeta = \frac{\kappa_\theta}{\kappa_r}$$

and rewrite the above expression for the frequencies as:

$$\omega_{\mathbf{k}}^{(\pm)} = \sqrt{\frac{\kappa_r + \kappa_\theta}{M}} \left[1 \pm \left\{ 1 - \frac{4\zeta}{(1+\zeta)^2} \sin^2\left(\frac{\mathbf{k} \cdot \mathbf{a}_1}{2} \right) \right\}^{1/2} \right]^{1/2}$$

which is the same expression as Eq. (7.35), since the system has periodicity only in one dimension, namely along $\mathbf{a}_1$; the only thing that differs is the overall factor in front, and the value of the variable ζ. We conclude from this that the *behavior* of the different eigenvalues is the same as the 1D chain with two atoms in the unit cell, shown in Fig. 7.5, with appropriate changes in the values since the relevant parameters here are the new factor in front, $\sqrt{(\kappa_r + \kappa_\theta)/M}$ and the value of $\zeta = \kappa_\theta/\kappa_r$. We note that in general, $\kappa_r > \kappa_\theta$ or equivalently $\zeta < 1$, since the cost of stretching bonds is typically greater than the cost of bending bonds. The relevant direction in reciprocal space is the vector $\mathbf{k}_1 = (k/\sqrt{2})(\hat{x} + \hat{y})$, with $0 \leq k \leq \pi/a$; in the perpendicular direction, $\mathbf{k}_2 = (k/\sqrt{2})(\hat{x} - \hat{y})$, the eigenvalues have no dispersion.

We analyze the behavior of the eigenvalues and the corresponding eigenvectors near the center, $\mathbf{k} = k(\hat{x} + \hat{y})/\sqrt{2}$, and near the edge of the BZ, $\mathbf{k} = [(\pi/a) - k](\hat{x} + \hat{y})/\sqrt{2}$. This gives the following results, to leading order in the small quantity $k \ll \pi/a$:

$$\text{at } \mathbf{k} = \frac{k}{\sqrt{2}}(\hat{x} + \hat{y}): \quad \omega_0^{(-)} = \sqrt{\frac{\zeta\kappa_r}{2M(1+\zeta)}}\, ak, \quad \omega_0^{(+)} = \sqrt{\frac{2(\kappa_r + \kappa_\theta)}{M}}\left[1 - \frac{\zeta a^2 k^2}{8(1+\zeta)^2}\right]$$

$$\text{at } \mathbf{k} = \frac{(\pi/a) - k}{\sqrt{2}}(\hat{x} + \hat{y}): \quad \omega_1^{(-)} = \sqrt{\frac{2\kappa_\theta}{M}}, \quad \omega_1^{(+)} = \sqrt{\frac{2\kappa_r}{M}}$$

We see that the lowest frequency at $\mathbf{k} \approx 0$, $\omega_0^{(-)}$, is linear in $k = |\mathbf{k}|$. From the eigenvector of this eigenvalue in the limit $k \to 0$, we find that this mode corresponds to motion of all the atoms in the same direction and by the same amount, that is, uniform translation of the crystal. This motion does not involve bond bending or bond stretching distortions. The eigenvector of the lower branch eigenvalue $\omega_1^{(-)}$ at $|\mathbf{k}| = (\pi/a)$ corresponds to motion of the two ions which is in phase within one unit cell, but π out of phase in neighboring unit cells. This motion involves bond bending distortions only, hence the frequency eigenvalue $\omega_1^{(-)} \sim \sqrt{\kappa_\theta}$. For the eigenvalue $\omega_0^{(+)}$, we find an eigenvector which corresponds to motion of the ions against each other within one unit cell, while the motion of equivalent ions in neighboring unit cells is in phase. This mode involves both bond stretching and bond bending, hence the frequency eigenvalue $\omega_0^{(+)} \sim \sqrt{\kappa_r + \kappa_\theta}$. Finally, for the eigenvalue $\omega_1^{(+)}$, we find an eigenvector which corresponds to motion of the ions against each other in the same unit cell, while the motion of equivalent ions in neighboring unit cells is π out of phase. This distortion involves bond stretching only, hence the frequency eigenvalue $\omega_1^{(+)} \sim \sqrt{\kappa_r}$. The motion of ions corresponding to these four modes is illustrated in Fig. 7.6. There are four other modes which are degenerate to these and involve similar motion of the ions in the $\hat{y}$ direction. In this example we get two branches because we have two atoms per unit cell, so the in-phase and out-of-phase motion of atoms in the unit cell produce two distinct modes. Moreover, there are doubly degenerate modes at each value of $\mathbf{k}$ because we are dealing with a 2D crystal which has symmetric couplings in the $\hat{x}$ and $\hat{y}$ directions. This degeneracy would be broken if the couplings between atoms were not symmetric in the different directions.

Example 7.4   2D square crystal
As another example we consider the 2D square lattice with lattice constant a in the x and y directions, and one atom of mass M per unit cell. This is illustrated in Fig. 7.7.

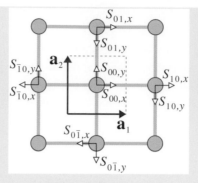

Fig. 7.7 The 2D square lattice with lattice constants $\mathbf{a}_1 = a\hat{x}$, $\mathbf{a}_2 = a\hat{y}$, with one atom per unit cell. The displacements of one atom at the center and its four nearest neighbors, that enter in the expressions for the dynamical matrix, are shown by arrows.

Since we now have two independent lattice vectors, $\mathbf{a}_1 = a\hat{x}$, $\mathbf{a}_2 = a\hat{y}$, we will employ a slightly modified notation to exemplify the setup: we denote the atomic displacements as $S_{nm,\alpha}$, where n and m are the indices identifying the multiples of vectors $\mathbf{a}_1$ and $\mathbf{a}_2$, respectively, that determine the unit cells, and $\alpha = x$ or y. For the unit cell at the origin, the atomic displacement has components $S_{00,x}$, $S_{00,y}$, while for its four nearest neighbors, the components are $S_{nm,\alpha}$ with $n = \pm 1$ and $m = \pm 1$ (the values -1 are denoted as $\bar{1}$ in each case). The potential energy is given by:

$$\Delta U = \frac{1}{2}(\kappa_r - \kappa_\theta)[(S_{00,x} - S_{10,x})^2 + (S_{00,x} - S_{\bar{1}0,x})^2 +$$
$$(S_{00,y} - S_{01,y})^2 + (S_{00,y} - S_{0\bar{1},y})^2]$$
$$+ \frac{1}{2}\kappa_\theta[(S_{00,x} - S_{10,x})^2 + (S_{00,x} - S_{\bar{1}0,x})^2 + (S_{00,y} - S_{10,y})^2 + (S_{00,y} - S_{\bar{1}0,y})^2]$$
$$+ \frac{1}{2}\kappa_\theta[(S_{00,x} - S_{01,x})^2 + (S_{00,x} - S_{0\bar{1},x})^2 + (S_{00,y} - S_{01,y})^2 + (S_{00,y} - S_{0\bar{1},y})^2]$$

Differentiating the potential energy, we obtain the matrix elements of the dynamical matrix:

$$D_{x,x}(\mathbf{k}) = 2\kappa_r + 2\kappa_\theta - \kappa_r(e^{ik_x a} + e^{-ik_x a}) - \kappa_\theta(e^{ik_y a} + e^{-ik_y a})$$

$$D_{y,y}(\mathbf{k}) = 2\kappa_r + 2\kappa_\theta - \kappa_r(e^{ik_y a} + e^{-ik_y a}) - \kappa_\theta(e^{ik_x a} + e^{-ik_x a})$$

with the off-diagonal matrix elements zero. Diagonalizing this matrix we find the eigenvalues of the two frequencies:

$$\omega_{\mathbf{k}}^{(1)} = \left[\frac{2\kappa_r}{M}(1 - \cos(k_x a)) + \frac{2\kappa_\theta}{M}(1 - \cos(k_y a))\right]^{(1/2)}$$

$$\omega_{\mathbf{k}}^{(2)} = \left[\frac{2\kappa_r}{M}(1 - \cos(k_y a)) + \frac{2\kappa_\theta}{M}(1 - \cos(k_x a))\right]^{(1/2)}$$

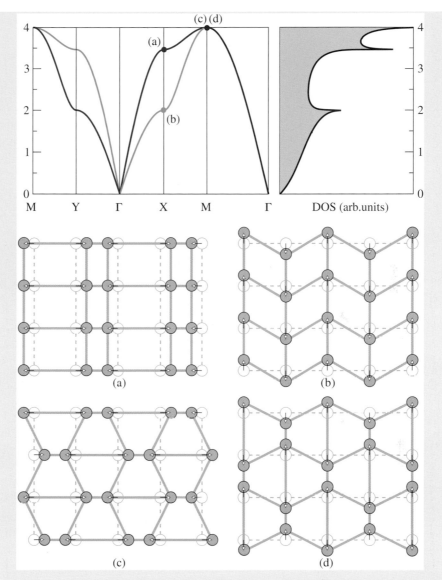

Fig. 7.8

Top: The frequencies of the phonon modes $\omega_{\mathbf{k}}^{(1)}$ (red curve) and $\omega_{\mathbf{k}}^{(2)}$ (green curve) of the 2D square lattice along different high-symmetry directions of the BZ, M–Y–Γ–X–M–Γ, where the points Γ, X, Y, M are defined in terms of (k_x, k_y) components as: $\Gamma = (0, 0)$, $X = (\pi/a, 0)$, $Y = (0, \pi/a)$, $M = (\pi/a, \pi/a)$. The two modes are degenerate (purple curve) along the Γ–M direction. The DOS is also shown on the right. **Bottom**: Atomic displacements of the phonon modes with frequencies marked as (a), (b), (c), and (d).

These are shown in Fig. 7.8 along the high-symmetry directions of the BZ. We also show the corresponding DOS of all phonon modes, which exhibits the characteristic features of a 2D system near the extrema (minimum at Γ and maximum at M) and saddle points (X and Y). As expected, since there is only one atom per unit cell, in this case we have only acoustic modes, both having linear behavior with k near the origin. However, there are two different slopes for the two different frequencies along the Γ–X and Γ–Y directions, due to different ways of displacing the atoms for a given **k** value.

In Fig. 7.8 we also show the eigenvectors of the atomic displacements corresponding to the eigenmodes with frequencies labeled (a), (b), (c), and (d). We recognize in (a) a mode at the X point of the BZ that involves only bond stretching of half of the crystal bonds (those that lie in the $\hat{x}$ direction) with the displacement of the atoms exclusively in the $\hat{x}$ direction, and in (b) a mode also at the X point that involves only bond bending of half of the crystal bonds (those that originally lie in the $\hat{x}$ direction) with the displacement of the atoms exclusively in the $\hat{y}$ direction. There are corresponding modes with equal frequencies at the Y point of the BZ which involve motion of the atoms in the direction perpendicular to that of modes (a) and (b). The two modes labeled (c) and (d) at the M point of the BZ involve both bond stretching/compressing of half of the bonds and bond bending of the other half of the bonds. These two modes have the same frequency, since they are equivalent by a rotation of the crystal by $\pi/2$.

Example 7.5   Phonons in a 3D crystal
As an illustration of the above ideas in a more realistic example, we discuss briefly the calculation of phonon modes in Si. Si is the prototypical covalent solid, where both bond-bending and bond-stretching terms are very important. If we wanted to obtain the phonon spectrum of Si by a force-constant model, we would need parameters that can describe accurately the relevant contributions of the bond-bending and bond-stretching forces. For example, fitting the values of κ_r and κ_θ to reproduce the experimental values for the highest-frequency optical mode at Γ, labeled LTO(Γ),[a] and the lowest-frequency acoustic mode at X, labeled TA(X), gives the spectrum shown in Fig. 7.9 (see also Problem 1).

Although such a procedure gives estimates for the bond-stretching and bond-bending parameters, the model is not very accurate. For instance, the values of the other two phonon modes at X are off by +5% for the mode labeled TO(X) and −12% for the mode labeled LOA(X), compared to experimental values. Interactions with several neighbors beyond the nearest ones are required in order to obtain frequencies closer to the experimentally measured spectrum throughout the BZ. However, establishing the values of force constants for interactions beyond nearest neighbors is rather complicated and must rely on experimental input. Moreover, with such extensions the model loses its simplicity and the transparent physical meaning of the parameters.

Alternatively, one can use the formalism discussed in Chapter 4 (DFT/LDA and its refinements) for calculating total energies or forces, as follows. The atomic displacements that correspond to various phonon modes can be established either by symmetry (for instance, using group theory arguments) or in conjunction with simple

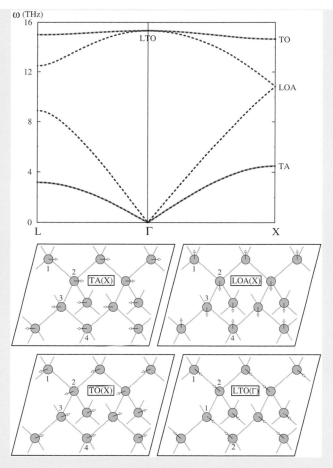

Fig. 7.9

Top: The phonon spectrum of Si along the high-symmetry directions L−Γ−X, calculated within the force-constant model. **Bottom**: Atomic displacements associated with the phonon modes in Si at the Γ and X points in the BZ; inequivalent atoms are labeled 1–4.

calculations based on a force-constant model, as in the example discussed above. For instance, we show in Fig. 7.9 the atomic displacements corresponding to certain high-symmetry phonon modes for Si, as obtained from the force-constant model. For a given atomic displacement, the energy of the system relative to its equilibrium structure, or the restoring forces on the atoms, can be calculated as a function of the phonon amplitude. These energy differences are then fitted by a second or higher-order polynomial in the phonon amplitude, and the coefficient of the second-order term gives the phonon frequency (in the case of a force calculation the fit starts at first order in the phonon amplitude). Higher-order terms give the anharmonic contributions which correspond to phonon–phonon interactions. This straightforward method, called the "frozen phonon" approach, gives results that are remarkably close to experimental values. Its limitation is that only phonons of relatively high symmetry can be calculated, such as at the center, the boundaries, and a few special points

Table 7.1 Frequencies of four high-symmetry phonon modes in Si (THz), at the center (Γ) and the boundary (X) of the BZ, as obtained by theoretical calculations and by experimental measurements. DFT results are from: J. R. Chelikowsky and M. L. Cohen, "Ab initio pseudopotentials and the structural properties of semiconductors," in P. T. Landsberg (Ed.), *Handbook on Semiconductors*, Vol. 1, p. 59 (North Holland, Amsterdam, 1992); M. T. Yin and M. L. Cohen, *Phys. Rev. B* **26**, 5668 (1982). The force-constant model is based on nearest-neighbor interactions only, with $\sqrt{\kappa_r/M_{Si}} = 8.828$ THz, $\sqrt{\kappa_\theta/M_{Si}} = 2.245$ THz, values that were chosen to reproduce exactly the experimental frequencies marked by asterisks. The atomic displacements corresponding to these modes are shown in Fig. 7.9.

	LTO(Γ)	LOA(X)	TO(X)	TA(X)
DFT energy calculation	15.16	12.16	13.48	4.45
DFT force calculation	15.14	11.98	13.51	4.37
Force-constant model	15.53	10.83	14.65	4.49
Experiment	15.53*	12.32	13.90	4.49*

within the BZ. The reason is that phonon modes with low symmetry involve the coherent motion of atoms across many unit cells of the ideal crystal; in order to represent this motion, a large supercell (a multiple of the ideal crystal unit cell) that contains all the inequivalent atoms in a particular phonon mode must be used. For instance, the unit cell for the LTO(Γ) mode in Fig. 7.9 involves only two atoms, the same as in the PUC of the perfect crystal, while the supercell for all the modes at X involves four atoms, that is, twice as many as in the PUC of the perfect crystal. The computational cost of the quantum-mechanical calculations of energy and forces increases sharply with the size of the unit cell, making computations for large unit cells with many atoms intractable. However, the information from the few high-symmetry points can usually be interpolated to yield highly accurate phonon frequencies throughout the entire BZ. In Table 7.9 we compare the calculated frequencies[b] for the phonon modes of Fig. 7.9 to experimental values.

---

[a] "L" stands for longitudinal, "T" for transverse, "O" for optical, and "A" for acoustic.

[b] M. T. Yin and M. L. Cohen, *Phys. Rev. B* **26**, 3259 (1982).

7.4 Phonons as Harmonic Oscillators

We next draw an analogy between the phonon hamiltonian and that of a collection of independent harmonic oscillators. We begin with the most general expression for the displacement of ions, Eq. (7.22). In that equation, we combine the amplitude $c_{\mathbf{k}}^{(l)}$ with the time-dependent part of the exponential $\exp[-i\omega_{\mathbf{k}}^{(l)}t]$ into a new variable $Q_{\mathbf{k}}^{(l)}(t)$, and express the most general displacement of ions as:

$$\mathbf{S}_{nj}(t) = \sum_{l,\mathbf{k}} Q_{\mathbf{k}}^{(l)}(t) \frac{1}{\sqrt{M_j}} \hat{\mathbf{e}}_{kj}^{(l)} e^{i\mathbf{k}\cdot\mathbf{R}_n} \tag{7.37}$$

These quantities must be real, since they describe ionic displacements in real space. Therefore, the coefficients of factors $\exp[i\mathbf{k}\cdot\mathbf{R}_n]$ and $\exp[i(-\mathbf{k})\cdot\mathbf{R}_n]$ that appear in the expansion must be complex conjugate, leading to the relation

$$\left[Q_{\mathbf{k}}^{(l)}(t)\hat{\mathbf{e}}_{kj}^{(l)} \right]^* = Q_{-\mathbf{k}}^{(l)}(t)\hat{\mathbf{e}}_{-kj}^{(l)} \tag{7.38}$$

and since this relation must hold for any time t, we deduce that:

$$\left[Q_{\mathbf{k}}^{(l)}(t) \right]^* = Q_{-\mathbf{k}}^{(l)}(t), \quad \left[\hat{\mathbf{e}}_{kj}^{(l)} \right]^* = \hat{\mathbf{e}}_{-kj}^{(l)} \tag{7.39}$$

The kinetic energy of the system of ions, per unit cell of the crystal, will be given in terms of the displacements $\mathbf{S}_{nj}(t)$ as:

$$\frac{K}{N} = \sum_{n,j} \frac{M_j}{2N} \left(\frac{\mathrm{d}\mathbf{S}_{nj}}{\mathrm{d}t} \right)^2 = \frac{1}{2N} \sum_{n,j} \sum_{\mathbf{k}l,\mathbf{k}'l'} \frac{\mathrm{d}Q_{\mathbf{k}}^{(l)}}{\mathrm{d}t} \frac{\mathrm{d}Q_{\mathbf{k}'}^{(l')}}{\mathrm{d}t} \hat{\mathbf{e}}_{kj}^{(l)} \cdot \hat{\mathbf{e}}_{k'j}^{(l')} e^{i(\mathbf{k}+\mathbf{k}')\cdot\mathbf{R}_n} \tag{7.40}$$

where N is the number of unit cells. Using the relations of Eq. (7.39) and the fact that the sum over the Bravais lattice vectors $\mathbf{R}_n$ represents a δ-function in the $\mathbf{k}$ argument [see Appendix A, Eq. (A.70)], we obtain for the kinetic energy:

$$\frac{K}{N} = \frac{1}{2} \sum_{\mathbf{k},l,l',j} \frac{\mathrm{d}Q_{\mathbf{k}}^{(l)}}{\mathrm{d}t} \frac{\mathrm{d}Q_{\mathbf{k}}^{(l')*}}{\mathrm{d}t} \hat{\mathbf{e}}_{kj}^{(l)} \cdot \hat{\mathbf{e}}_{kj}^{(l')*} = \frac{1}{2} \sum_{\mathbf{k},l} \frac{\mathrm{d}Q_{\mathbf{k}}^{(l)}}{\mathrm{d}t} \frac{\mathrm{d}Q_{\mathbf{k}}^{(l)*}}{\mathrm{d}t} \tag{7.41}$$

where we have set $\mathbf{G} = 0$ in the argument of the δ-function since the wave-vectors $\mathbf{k}, \mathbf{k}'$ lie in the first BZ, and the last equality was obtained with the help of Eq. (7.21). The potential energy is also expressed in terms of the atomic displacements and the force-constant matrix:

$$\frac{\Delta U}{N} = \frac{1}{N} \sum_{n,i,\alpha;m,j,\beta} \frac{1}{2\sqrt{M_i M_j}} \sum_{\mathbf{k},l;\mathbf{k}',l'} Q_{\mathbf{k}}^{(l)} \hat{e}_{ki\alpha}^{(l)} e^{i\mathbf{k}\cdot\mathbf{R}_n} F_{ni\alpha,mj\beta} Q_{\mathbf{k}'}^{(l')} \hat{e}_{k'j\beta}^{(l')} e^{i\mathbf{k}'\cdot\mathbf{R}_m}$$

We now follow the same steps as in the derivation of the kinetic energy result, after using the definition of the $\mathbf{k}$-dependent dynamical matrix in terms of the force-constant matrix, and the fact that $\hat{e}_{k'j\beta}^{(l')}$ are its eigenvectors with eigenvalues $(\omega_{\mathbf{k}}^{(l')})^2$, to find:

$$\frac{\Delta U}{N} = \frac{1}{2N} \sum_{n,i,\alpha;m,j,\beta} \sum_{\mathbf{k},l;\mathbf{k}',l'} Q_{\mathbf{k}}^{(l)} \hat{e}_{ki\alpha}^{(l)} e^{i(\mathbf{k}+\mathbf{k}')\cdot\mathbf{R}_n} D_{ni\alpha,mj\beta} Q_{\mathbf{k}'}^{(l')} \hat{e}_{k'j\beta}^{(l')} e^{i\mathbf{k}'\cdot(\mathbf{R}_m-\mathbf{R}_n)}$$

$$= \frac{1}{2N} \sum_{n} \sum_{i,\alpha;j,\beta} \sum_{\mathbf{k},l;\mathbf{k}',l'} Q_{\mathbf{k}}^{(l)} \hat{e}_{ki\alpha}^{(l)} e^{i(\mathbf{k}+\mathbf{k}')\cdot\mathbf{R}_n} Q_{\mathbf{k}'}^{(l')} D_{i\alpha,j\beta}(\mathbf{k}') \hat{e}_{k'j\beta}^{(l')}$$

$$= \frac{1}{2} \sum_{\mathbf{k},l;l'} \sum_{i,\alpha} Q_{\mathbf{k}}^{(l)} \hat{e}_{ki\alpha}^{(l)} Q_{-\mathbf{k}}^{(l')} \left(\omega_{-\mathbf{k}}^{(l')} \right)^2 \hat{e}_{-ki\alpha}^{(l')} = \frac{1}{2} \sum_{\mathbf{k},l} Q_{\mathbf{k}}^{(l)} Q_{-\mathbf{k}}^{(l)*} \left(\omega_{\mathbf{k}}^{(l)} \right)^2$$

where we have also used the orthogonality relations of the eigenvectors, Eq. (7.21), and the relations of Eq. (7.39). Combining the expressions for the kinetic and potential energies, we obtain the total energy of a system of phonons that represents the vibrational state of the ions in the solid:

$$\frac{E^{\text{vib}}}{N} = \frac{1}{2} \sum_{\mathbf{k},l} \left[\left| \frac{dQ_{\mathbf{k}}^{(l)}}{dt} \right|^2 + \left(\omega_{\mathbf{k}}^{(l)} \right)^2 \left| Q_{\mathbf{k}}^{(l)} \right|^2 \right] \tag{7.42}$$

which is formally identical to the total energy of a collection of independent harmonic oscillators with frequencies $\omega_{\mathbf{k}}^{(l)}$, where $Q_{\mathbf{k}}^{(l)}$ is the free variable describing the motion of the harmonic oscillator. This expression also makes it easy to show that the kinetic and potential contribution to the total energy are equal, as expected for harmonic oscillators (see Problem 2).

This analysis shows that a solid in which atoms are moving is equivalent to a number of phonon modes that are excited, with the atomic motion given as a superposition of the harmonic modes corresponding to the excited phonons. The total energy of the excitation is that of a collection of independent harmonic oscillators with the proper phonon frequencies. Notice that the $Q_{\mathbf{k}}^{(l)}$s contain the amplitude of the vibration, as seen from Eq. (7.37), and consequently, the total energy involves the absolute value squared of the amplitude of every phonon mode that is excited. Since the atomic motion in the solid must be quantized, the harmonic oscillators describing this motion should be treated as quantum-mechanical ones. Indeed, in Eq. (7.42), $Q_{\mathbf{k}}^{(l)}$ can be thought of as the quantum-mechanical position variable and $dQ_{\mathbf{k}}^{(l)}/dt$ as the conjugate momentum variable of a harmonic oscillator identified by index l and wave-vector $\mathbf{k}$. The amplitude of the vibration, contained in $Q_{\mathbf{k}}^{(l)}$, can be interpreted as the number of excited phonons of this particular mode. In principle, there is no limit for this amplitude, therefore an arbitrary number of phonons of each frequency can be excited; all these phonons contribute to the internal energy of the solid. The fact that an arbitrary number of phonons of each mode can be excited indicates that phonons must be treated as bosons. With this interpretation, $Q_{\mathbf{k}}^{(l)}$ and its conjugate variable $dQ_{\mathbf{k}}^{(l)}/dt$ obey the proper commutation relations for bosons. This implies that the vibrational total energy per unit cell, due to the phonon excitations, must be given by:

$$E_s^{\text{vib}} = \sum_{\mathbf{k},l} \left(n_{\mathbf{k},s}^{(l)} + \frac{1}{2} \right) \hbar \omega_{\mathbf{k}}^{(l)} \tag{7.43}$$

where $n_{\mathbf{k},s}^{(l)}$ is the number of phonons of frequency $\omega_{\mathbf{k}}^{(l)}$ that have been excited in a particular state (denoted by s) of the system. This expression is appropriate for quantum harmonic oscillators, with $n_{\mathbf{k},s}^{(l)}$ allowed to take any non-negative integer value. One interesting aspect of this expression is that, even in the ground state of the system when none of the phonon modes is excited, that is, $n_{\mathbf{k}0}^{(l)} = 0$, there is a certain amount of energy in the system due to the so-called zero-point motion associated with quantum harmonic oscillators; this arises from the factors $\frac{1}{2}$ which are added to the phonon occupation numbers in Eq. (7.43). We will see below that this has measurable consequences (see Chapter 8).

In practice, if the atomic displacements are not too large, the harmonic approximation to phonon excitations is reasonable. For large displacements, anharmonic terms become increasingly important, and a more elaborate description is necessary which takes into account phonon–phonon interactions arising from the anharmonic terms. Evidently, this places a limit on the number of phonons that can be excited before the harmonic approximation breaks down.

7.5 Application: Specific Heat of Crystals

We can use the concept of phonons to determine the thermal properties of crystals, and in particular their specific heat. This is especially interesting at low temperatures, where the quantum nature of excitations becomes important, and gives behavior drastically different from the classical result. We discuss this topic next, beginning with a brief review of the result of the classical theory.

7.5.1 The Classical Picture

The internal energy per unit volume associated with the motion of a collection of classical particles at inverse temperature $\beta = 1/k_B T$ (where k_B is Boltzmann's constant) contained in a volume V is given by:

$$\frac{E^{vib}}{V} = \frac{1}{V} \frac{\int \{d\mathbf{R}\}\{d\mathbf{p}\} E^{vib}(\{\mathbf{r}\}, \{\mathbf{p}\}) e^{-\beta E^{vib}(\{\mathbf{r}\},\{\mathbf{p}\})}}{\int \{d\mathbf{r}\}\{d\mathbf{p}\} e^{-\beta E^{vib}(\{\mathbf{r}\},\{\mathbf{p}\})}}$$

$$= -\frac{1}{V}\frac{\partial}{\partial \beta} \ln\left[\int \{d\mathbf{r}\}\{d\mathbf{p}\} e^{-\beta E^{vib}(\{\mathbf{r}\},\{\mathbf{p}\})}\right] \tag{7.44}$$

where $\{\mathbf{r}\}$ and $\{\mathbf{p}\}$ are the coordinates and momenta of the particles. In the harmonic approximation, the energy is given in terms of the displacements $\delta \mathbf{R}_n$ from ideal positions as:

$$E^{vib} = \frac{1}{2}\sum_{n,n'} \delta \mathbf{R}_n \cdot \mathbf{F}(\mathbf{R}_n - \mathbf{R}_{n'}) \cdot \delta \mathbf{R}_{n'} + \frac{1}{2}\sum_n \frac{\mathbf{p}_n^2}{M_n} \tag{7.45}$$

where $\mathbf{F}(\mathbf{R}_n - \mathbf{R}_{n'})$ is the force-constant matrix, defined in Eq. (7.5); the indices n, n' run over all the particles in the system. Since both the potential and the kinetic parts involve quadratic expressions, we can rescale all coordinates by $\beta^{1/2}$, in which case the integral in the last expression in Eq. (7.44) is multiplied by a factor of $\beta^{-3N_{at}}$, where N_{at} is the total number of atoms in 3D space. What remains is independent of β, giving for the internal energy per unit volume:

$$\frac{E^{vib}}{V} = \frac{3N_{at}}{V} k_B T \tag{7.46}$$

from which the specific heat per unit volume can be calculated:

$$c(T) = \frac{\partial}{\partial T}\left[\frac{E^{vib}}{V}\right] = 3k_B(N_{at}/V) \tag{7.47}$$

This turns out to be a constant independent of temperature. This behavior is referred to as the Dulong–Petit law, which is valid at high temperatures.

7.5.2 The Quantum-Mechanical Picture

The quantum-mechanical calculation of the internal energy in terms of phonons gives the following expression:

$$\frac{E^{\text{vib}}}{V} = \frac{1}{V} \frac{\sum_s E_s^{\text{vib}} e^{-\beta E_s^{\text{vib}}}}{\sum_s e^{-\beta E_s^{\text{vib}}}} \tag{7.48}$$

where E_s^{vib} is the vibrational energy per unit cell corresponding to a particular state of the system, which involves a certain number of excited phonons. We will take E_s^{vib} from Eq. (7.43) that we derived earlier. Just like in the classical discussion, we express the total internal energy as:

$$\frac{E^{\text{vib}}}{V} = -\frac{1}{V} \frac{\partial}{\partial \beta} \ln\left(\sum_s e^{-\beta E_s^{\text{vib}}} \right) \tag{7.49}$$

There is a neat mathematical trick that allows us to express this in a more convenient form. Consider the expression

$$\sum_s e^{-\beta E_s^{\text{vib}}} = \sum_s \exp\left[-\sum_{\mathbf{k},l} \beta \left(n_{\mathbf{k},s}^{(l)} + \frac{1}{2} \right) \hbar \omega_{\mathbf{k}}^{(l)} \right] \tag{7.50}$$

which involves a sum of all exponentials containing terms $\beta(n_{\mathbf{k}}^{(l)} + \frac{1}{2})\hbar\omega_{\mathbf{k}}^{(l)}$ with all possible non-negative integer values of $n_{\mathbf{k}}^{(l)}$. Now consider the expression

$$\prod_{\mathbf{k},l} \sum_{n=0}^{\infty} e^{-\beta(n+\frac{1}{2})\hbar\omega_{\mathbf{k}}^{(l)}} = \left(e^{-\beta\hbar\frac{1}{2}\omega_{\mathbf{k}_1}^{(l_1)}} + e^{-\beta\hbar\frac{3}{2}\omega_{\mathbf{k}_1}^{(l_1)}} + \cdots \right)\left(e^{-\beta\hbar\frac{1}{2}\omega_{\mathbf{k}_2}^{(l_2)}} + e^{-\beta\hbar\frac{3}{2}\omega_{\mathbf{k}_2}^{(l_2)}} + \cdots \right)\cdots$$

It is evident that when these products are expanded out we obtain exponentials with exponents $(n + \frac{1}{2})\beta\hbar\omega_{\mathbf{k}}^{(l)}$ with all possible non-negative integer values of n. Therefore, the expressions in these two equations are the same, and we can substitute the second expression for the first one in Eq. (7.49). Notice further that the geometric series summation gives:

$$\sum_{n=0}^{\infty} e^{-\beta(n+\frac{1}{2})\hbar\omega_{\mathbf{k}}^{(l)}} = \frac{e^{-\beta\hbar\omega_{\mathbf{k}}^{(l)}/2}}{1 - e^{-\beta\hbar\omega_{\mathbf{k}}^{(l)}}} \tag{7.51}$$

which leads to, for the total internal energy per unit volume:

$$\frac{E^{\text{vib}}}{V} = -\frac{1}{V} \frac{\partial}{\partial \beta} \ln\left[\prod_{\mathbf{k},l} \frac{e^{-\beta\hbar\omega_{\mathbf{k}}^{(l)}/2}}{1 - e^{-\beta\hbar\omega_{\mathbf{k}}^{(l)}}} \right] = \frac{1}{V} \sum_{\mathbf{k},l} \hbar\omega_{\mathbf{k}}^{(l)} \left(\bar{n}_{\mathbf{k}}^{(l)} + \frac{1}{2} \right) \tag{7.52}$$

where we have defined $\bar{n}_{\mathbf{k}}^{(l)}$ as:

$$\bar{n}_{\mathbf{k}}^{(l)}(T) = \frac{1}{e^{\beta \hbar \omega_{\mathbf{k}}^{(l)}} - 1} \tag{7.53}$$

This quantity represents the average occupation of the phonon state with frequency $\omega_{\mathbf{k}}^{(l)}$, at temperature T, and is appropriate for bosons (see Appendix D). From Eq. (7.52) we can now calculate the specific heat per unit volume:

$$c(T) = \frac{\partial}{\partial T} \left[\frac{E^{\mathrm{vib}}}{V} \right] = \frac{1}{V} \sum_{\mathbf{k},l} \hbar \omega_{\mathbf{k}}^{(l)} \frac{\partial}{\partial T} \bar{n}_{\mathbf{k}}^{(l)}(T) \tag{7.54}$$

We examine first the behavior of this expression in the low-temperature limit. In this limit, $\beta \hbar \omega_{\mathbf{k}}^{(l)}$ becomes very large, and therefore $\bar{n}_{\mathbf{k}}^{(l)}(T)$ becomes negligibly small, except when $\omega_{\mathbf{k}}^{(l)}$ happens to be very small. We saw in earlier discussion that $\omega_{\mathbf{k}}^{(l)}$ goes to zero linearly near the center of the BZ ($\mathbf{k} \approx 0$), for the acoustic branches. We can then write near the center of the BZ $\omega_{\mathbf{k}}^{(l)} = v_{\hat{\mathbf{k}}}^{(l)} k$, where $k = |\mathbf{k}|$ is the wave-vector magnitude and $v_{\hat{\mathbf{k}}}^{(l)}$ is the sound velocity which depends on the direction of the wave-vector $\hat{\mathbf{k}}$ and the acoustic branch label l. For large β we can then approximate all frequencies as $\omega_{\mathbf{k}}^{(l)} = v_{\hat{\mathbf{k}}}^{(l)} k$ over the entire BZ, since for large values of k the contributions are negligible anyway, because of the factor $\exp[\beta \hbar v_{\hat{\mathbf{k}}}^{(l)} k]$ which appears in the denominator. Turning the sum into an integral as usual, we obtain for the specific heat at low temperature:

$$c(T) = \frac{\partial}{\partial T} \sum_{l} \int \frac{d\mathbf{k}}{(2\pi)^3} \frac{\hbar v_{\hat{\mathbf{k}}}^{(l)} k}{e^{\beta \hbar v_{\hat{\mathbf{k}}}^{(l)} k} - 1} \tag{7.55}$$

We will change variables as in the case of the classical calculation to make the integrand independent of inverse temperature β, by taking $\beta \hbar v_{\hat{\mathbf{k}}}^{(l)} k = t_{\hat{\mathbf{k}}}^{(l)}$, which gives a factor of β^{-4} in front of the integral, while the integral has now become independent of temperature. In fact, we can use the same argument about the negligible contributions of large values of k to extend the integration to infinity in the variable k, so that the result is independent of the BZ shape. Denoting by c_0 the value of the constant which is obtained by performing the integration and encompasses all other constants in front of the integral, we then find:

$$c(T) = c_0 \frac{\partial}{\partial T} T^4 = 4c_0 T^3 \tag{7.56}$$

that is, the behavior of the specific heat at low temperature is cubic in the temperature and not constant as the Dulong–Petit law suggests from the classical calculation.

In the high-temperature limit, that is, when $k_B T \gg \hbar \omega_{\mathbf{k}}^{(l)}$, the Bose occupation factors take the form

$$\frac{1}{e^{\beta \hbar \omega_{\mathbf{k}}^{(l)}} - 1} \approx \frac{k_B T}{\hbar \omega_{\mathbf{k}}^{(l)}} \tag{7.57}$$

and with this, the expression for the specific heat becomes:

$$c(T) = \frac{1}{V} \sum_{\mathbf{k},l} k_B = 3k_B (N_{\mathrm{at}}/V) \tag{7.58}$$

which is the same result as in the classical calculation. Thus, at sufficiently high temperatures the Dulong–Petit law is recovered.

7.5.3 The Debye Model

A somewhat more quantitative discussion of the behavior of the specific heat as a function of the temperature is afforded by the Debye model. In this model we assume all frequencies to be linear in the wave-vector magnitude, as acoustic modes near $\mathbf{k} \approx 0$ are. For simplicity we assume we are dealing with an isotropic solid, so we can take $\omega = vk$ for all the different acoustic branches, with v (the sound velocity) being the same in all directions. The total number of phonon modes is equal to $3N_{at}$, where N_{at} is the total number of atoms in the crystal, and for each value of $\mathbf{k}$ there are $3n_{at}$ normal modes in three dimensions (n_{at} being the number of atoms in the PUC). Strictly speaking, the Debye model makes sense only for crystals with one atom per PUC, because it assumes all modes to be acoustic in nature (they behave like $\omega \sim k$ for $k \to 0$). Just like in the case of electronic states, we can use a normalization argument to relate the density of the crystal, $n = N_{at}/V$, to the highest value of the wave-vector that phonons can assume (denoted by k_D):

$$\sum_{\mathbf{k}} 3n_{at} \to 3V \int_0^{k_D} \frac{4\pi k^2 \mathrm{d}k}{(2\pi)^3} = 3N_{at} \Rightarrow n = \frac{N_{at}}{V} = \frac{k_D^3}{6\pi^2} \tag{7.59}$$

from which the value of k_D is determined in terms of n. Notice that k_D is determined by considering the total number of *phonon modes*, not the actual number of phonons present in some excited state of the system, which of course can be anything since phonons are bosons. We can also define the Debye frequency ω_D and Debye temperature Θ_D, through:

$$\omega_D = vk_D, \quad \Theta_D = \frac{\hbar\omega_D}{k_B} \tag{7.60}$$

With these definitions, we obtain the following expression for the specific heat at any temperature:

$$c(T) = 9nk_B \left(\frac{T}{\Theta_D}\right)^3 \int_0^{\Theta_D/T} \frac{t^4 e^t}{(e^t - 1)^2} \mathrm{d}t \tag{7.61}$$

In the low-temperature limit, with Θ_D much larger than T, the upper limit of the integral is approximated by ∞, and there is no temperature dependence left in the integral, which gives a constant when evaluated explicitly. This reduces to the expression we discussed above, Eq. (7.56). What happens for higher temperatures? Notice first that the integrand is always a positive quantity. As the temperature increases, the upper limit of the integral becomes smaller and smaller, and the value obtained from integration is lower than the value obtained in the low-temperature limit. Therefore, as the temperature increases, the value of the specific heat increases slower than T^3, as shown in Fig. 7.10. As we saw above, at sufficiently high temperature the specific heat eventually approaches the constant value of the Dulong–Petit law.

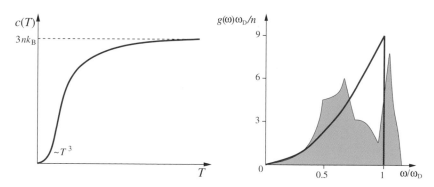

Fig. 7.10 **Left**: Behavior of the specific heat c as a function of temperature T in the Debye model. The asymptotic value for large T corresponds to the Dulong–Petit law. **Right**: Comparison of the density of states in the Debye model as given by Eq. (7.64), shown by the red line and a realistic calculation [adapted from R. Stedman, L. Almqvist, and G. Nilsson, *Phys. Rev.* **162**, 549 (1967)], shown by the shaded curve, for the Al crystal.

To explore the physical meaning of the Debye temperature, we note that from its definition:

$$n_{\mathbf{k}}^{(l)}(T) = \frac{1}{e^{(\omega_{\mathbf{k}}^{(l)}/\omega_D)(\Theta_D/T)} - 1}$$

For $T > \Theta_D$ all phonon modes are excited, that is, $\bar{n}_{\mathbf{k}}^{(l)}(T) > 0$ for all frequencies, while for $T < \Theta_D$ the high-frequency phonon modes are frozen, that is, $\bar{n}_{\mathbf{k}}^{(l)}(T) \approx 0$ for those modes. The Debye model is oversimplified, because it treats the frequency spectrum as linearly dependent on the wave-vectors, $\omega = vk$ for all modes. For a realistic calculation we need to include the actual frequency spectrum, which produces a density of states with features that depend on the structure of the solid, analogous to what we discussed for the electronic energy spectrum. An example, the Al crystal, is shown in Fig. 7.10.

Within the Debye model, the density of phonon modes at frequency ω per unit volume of the crystal, $g(\omega)$, can easily be obtained starting with the general expression

$$g(\omega) = \frac{1}{V} \sum_{\mathbf{k},l} \delta(\omega - \omega_{\mathbf{k}}^{(l)}) = \sum_{l} \int \frac{d\mathbf{k}}{(2\pi)^3} \delta(\omega - \omega_{\mathbf{k}}^{(l)}) = \sum_{l} \int_{\omega = \omega_{\mathbf{k}}^{(l)}} \frac{dS_{\mathbf{k}}}{(2\pi)^3} \frac{1}{|\nabla_{\mathbf{k}} \omega_{\mathbf{k}}^{(l)}|}$$

where the last expression involves an integral over a surface in $\mathbf{k}$-space on which $\omega_{\mathbf{k}}^{(l)} = \omega$. In the Debye model we have $\omega_{\mathbf{k}}^{(l)} = v|\mathbf{k}|$, which gives $|\nabla_{\mathbf{k}} \omega_{\mathbf{k}}^{(l)}| = v$, with v the average sound velocity. For fixed value of $\omega = \omega_{\mathbf{k}}^{(l)} = v|\mathbf{k}|$, the integral over a surface in $\mathbf{k}$-space that corresponds to this value of ω is equal to the surface of a sphere of radius $|\mathbf{k}| = \omega/v$, and since there are three phonon branches in 3D space, all having the same average sound velocity within the Debye model, we find for the density of states:

$$g(\omega) = \frac{3}{(2\pi)^3} \frac{1}{v} 4\pi \left(\frac{\omega}{v}\right)^2 = \frac{3}{2\pi^2} \frac{\omega^2}{v^3} \tag{7.62}$$

If we calculate the total number of available phonon modes per unit volume N_{ph}/V up to the Debye frequency ω_D, we find:

$$\frac{N_{ph}}{V} = \int_0^{\omega_D} g(\omega)d\omega = \frac{1}{2\pi^2}\frac{\omega_D^3}{v^3} = 3\frac{1}{6\pi^2}k_D^3 \tag{7.63}$$

which, from Eq. (7.59), is equal to $3(N_{at}/V)$, exactly as we would expect for a 3D crystal containing N_{at} atoms. Using the definition of the Debye frequency, we can express the density of states as:

$$\frac{g(\omega)\omega_D}{n} = 9\left(\frac{\omega}{\omega_D}\right)^2 \tag{7.64}$$

which is a convenient relation between dimensionless quantities. In Fig. 7.10 we show a comparison of the density of states in the Debye model as given by Eq. (7.64), and as obtained by a realistic calculation for the Al crystal. This comparison illustrates how simple the Debye model is relative to real phonon modes. Nevertheless, the model is useful in that it provides a simple justification for the behavior of the specific heat as a function of temperature as well as a rough measure of the highest phonon frequency, ω_D. In fact, the form of the specific heat predicted by the Debye model, Eq. (7.61), is often used to determine the Debye temperature, and through it the Debye frequency. Since this relation involves temperature dependence, a common practice is to fit the observed specific heat at given T to one half the value of the Dulong–Petit law as obtained from Eq. (7.61), which determines Θ_D. Results of this fitting approach for the Debye temperatures of several elemental solids are given in Table 7.2.

Table 7.2 Debye temperatures Θ_D (K) and frequencies ω_D (THz) for several elemental solids. Values of Θ_D and ω_D were determined by fitting the observed value of the specific heat at a certain temperature to half the value of the Dulong–Petit law through Eq. (7.61). Source: J. de Launay, Solid State Physics, Vol. 2 (Academic Press, Boston, MA, 1956). * In the diamond phase.

Element	Θ_D	ω_D	Element	Θ_D	ω_D	Element	Θ_D	ω_D
Li	400	8.33	Na	150	3.12	K	100	2.08
Be	1000	20.84	Mg	318	6.63	Ca	230	4.79
B	1250	26.05	Al	394	8.21	Ga	240	5.00
C*	1860	38.76	Si	625	13.02	Ge	350	7.29
As	285	5.94	Sb	200	4.17	Bi	120	2.50
Cu	315	6.56	Ag	215	4.48	Au	170	3.54
Zn	234	4.88	Cd	120	2.50	Hg	100	2.08
Cr	460	9.58	Mo	380	7.92	W	310	6.46
Mn	400	8.33	Fe	420	8.75	Co	385	8.02
Ni	375	7.81	Pd	275	5.73	Pt	230	4.79

7.5.4 Thermal Expansion Coefficient

An important physical effect which is related to phonon modes and to the specific heat is the thermal expansion of a solid. The thermal expansion coefficient α is defined as the rate of change of the linear dimension L of the solid with temperature T, normalized by this linear dimension, at constant pressure P:

$$\alpha \equiv \frac{1}{L} \left(\frac{\partial L}{\partial T} \right)_P \tag{7.65}$$

The linear dimension of the solid L is related to its volume V through $L = V^{1/3}$, which gives for the thermal expansion coefficient in terms of the volume:

$$\alpha = \frac{1}{3V} \left(\frac{\partial V}{\partial T} \right)_P \tag{7.66}$$

This can also be expressed in terms of the bulk modulus B, the negative inverse of which is the rate of change of the volume with pressure, normalized by the volume, at constant temperature:

$$B^{-1} \equiv -\frac{1}{V} \left(\frac{\partial V}{\partial P} \right)_T \tag{7.67}$$

[see also the equivalent definition, Eq. (7.78)]. With the help of standard thermodynamic relations (see Appendix D), we obtain for the thermal expansion coefficient in terms of the bulk modulus and the pressure:

$$\alpha = \frac{1}{3B} \left(\frac{\partial P}{\partial T} \right)_V \tag{7.68}$$

The pressure is given by the negative rate of change of the Helmholtz free energy F with volume, at constant temperature:

$$P = - \left(\frac{\partial F}{\partial V} \right)_T \tag{7.69}$$

where $F = E - TS$, with E the internal energy and S the entropy (see Appendix D). When changes in the internal energy E are due exclusively to phonon excitations, we can use E from Eq. (7.52), and with this derive the appropriate expression for S (see Problem 5), to obtain for the thermal expansion coefficient:

$$\alpha = \frac{1}{3B} \sum_{\mathbf{k},l} \left(-\frac{\partial \hbar \omega_{\mathbf{k}}^{(l)}}{\partial V} \right)_T \left(\frac{\partial \bar{n}_{\mathbf{k}}^{(l)}}{\partial T} \right)_V \tag{7.70}$$

This involves the same quantities as the expression for the specific heat c, Eq. (7.54), that is, the phonon frequencies $\omega_{\mathbf{k}}^{(l)}$ and the corresponding average occupation numbers $\bar{n}_{\mathbf{k}}^{(l)}$ at temperature T. In fact, it is customary to express the thermal expansion coefficient in terms of the specific heat as:

$$\alpha = \frac{\gamma c(T)}{3B} \tag{7.71}$$

where the coefficient γ is known as the Grüneisen parameter. This quantity is given by:

$$\gamma = \frac{1}{c(T)} \sum_{\mathbf{k},l} -\frac{\partial \hbar \omega_{\mathbf{k}}^{(l)}}{\partial V} \frac{\partial \bar{n}_{\mathbf{k}}^{(l)}}{\partial T} = -V \left[\sum_{\mathbf{k},l} \frac{\partial \hbar \omega_{\mathbf{k}}^{(l)}}{\partial V} \frac{\partial \bar{n}_{\mathbf{k}}^{(l)}}{\partial T} \right] \left[\sum_{\mathbf{k},l} \hbar \omega_{\mathbf{k}}^{(l)} \frac{\partial \bar{n}_{\mathbf{k}}^{(l)}}{\partial T} \right]^{-1} \qquad (7.72)$$

We can simplify the notation in this equation by defining the contributions to the specific heat from each phonon mode as:

$$c_{\mathbf{k}}^{(l)} = \hbar \omega_{\mathbf{k}}^{(l)} \frac{\partial \bar{n}_{\mathbf{k}}^{(l)}}{\partial T} \qquad (7.73)$$

and the mode-specific Grüneisen parameters:

$$\gamma_{\mathbf{k}}^{(l)} = -\frac{V}{\omega_{\mathbf{k}}^{(l)}} \frac{\partial \omega_{\mathbf{k}}^{(l)}}{\partial V} = -\frac{\partial (\ln \omega_{\mathbf{k}}^{(l)})}{\partial (\ln V)} \qquad (7.74)$$

in terms of which the Grüneisen parameter takes the form

$$\gamma = \left[\sum_{\mathbf{k},l} \gamma_{\mathbf{k}}^{(l)} c_{\mathbf{k}}^{(l)} \right] \left[\sum_{\mathbf{k},l} c_{\mathbf{k}}^{(l)} \right]^{-1} \qquad (7.75)$$

The Grüneisen parameter is a quantity that can be measured directly by experiment.

Important warning. We should alert the reader to the fact that so far we have taken into account only the excitation of phonons as contributing to changes in the internal energy of the solid. This is appropriate for semiconductors and insulators where electron excitation across the band gap is negligible for usual temperatures. For these solids, for temperatures as high as their melting point, the thermal energy is still much lower than the band-gap energy, the minimum energy required to excite electrons from their ground state. For metals, on the other hand, the excitation of electrons is as important as that of phonons; when it is included in the picture, it gives different behavior of the specific heat and the thermal expansion coefficient at low temperature. These effects can be treated explicitly by modeling the electrons as a Fermi liquid, which requires a many-body picture; a detailed description of the thermodynamics of the Fermi liquid is given Fetter and Walecka.[2]

7.6 Application: Mössbauer Effect

Another interesting manifestation of the quantum nature of phonons is the emission of recoil-less radiation from crystals, known as the Mössbauer effect. Consider a nucleus that can emit γ-rays due to nuclear transitions. In free space, the energy and momentum conservation requirements produce a shift in the γ-ray frequency and a broadening of the spectrum. Both of these changes obscure the nature of the nuclear transition. We can estimate the amount of broadening by the following argument: the momentum of the

---

[2] *Quantum Theory of Many-Particle Systems*, A. L. Fetter and J. D. Walecka (McGraw-Hill, New York, 1971).

emitted photon of wave-vector $\mathbf{K}$ is $\hbar\mathbf{K}$, which must be equal to the recoil of the nucleus $M\mathbf{v}$, with M the mass of the nucleus (assuming that the nucleus is initially at rest). The recoil energy of the nucleus is $E_R = Mv^2/2 = \hbar^2\mathbf{K}^2/2M$, and the photon frequency is given by $\hbar\omega_0/c = \hbar|\mathbf{K}|$, so that $E_R = \hbar^2\omega_0^2/2Mc^2$. The broadening is then obtained by $E_R = \hbar\Delta\omega$, which gives:

$$\frac{\Delta\omega}{\omega_0} = \left(\frac{E_R}{2Mc^2}\right)^{1/2}$$

This can be much greater than the natural broadening which comes from the finite lifetime of nuclear processes that give rise to the γ-ray emission in the first place. When the emission of γ-rays takes place within a solid, the presence of phonons can change things significantly. Specifically, much of the photon momentum can be carried by phonons with low energy, near the center of the BZ, and the emission can be recoil-less, in which case the γ-rays will have a frequency and width equal to their natural values. We next develop this picture by treating phonons as quantum oscillators.

The transition probability between initial ($|i\rangle$) and final ($\langle f|$) states is given by:

$$P_{i\to f} = \left|\langle f|\hat{O}(\mathbf{K}, \{p\})|i\rangle\right|^2$$

where $\hat{O}(\mathbf{K}, \{p\})$ is the operator for the transition, $\mathbf{K}$ is the wave-vector of the radiation, and $\{p\}$ are nuclear variables that describe the internal state of the nucleus. Due to translational and galilean invariance, we can write the operator for the transition as:

$$\hat{O}(\mathbf{K}, \{p\}) = \exp[-i\mathbf{K}\cdot\mathbf{r}]\hat{O}(\{p\})$$

where the last operator involves nuclear degrees of freedom only. We exploit this separation to write:

$$P_{i\to f} = \left|\langle f|\hat{O}(\{p\})|i\rangle\right|^2 P_{0\to 0}$$

where the states $\langle f|, |i\rangle$ now refer to nuclear states, and $P_{0\to 0}$ is the probability that the crystal will be left in its ground state, with only phonons of low frequency $\omega_{\mathbf{q}\to 0}$ emitted.

We next calculate $P_{0\to 0}$ using the quantum picture of phonons. The hamiltonian for the crystal degrees of freedom involved in the transition is simply the phonon energy, as derived in Eq. (7.42). For simplicity, in the following we assume that there is only one type of atom in the crystal and one atom per unit cell, so that all atoms have the same mass $M_i = M$; this can easily be extended to the general case. This allows us to exclude the factor of $1/\sqrt{M}$ from the definition of the coordinates $Q_{\mathbf{q}}^{(l)}(t)$ (since the mass is now the same for all phonon modes), distinct from what we did above, see Eq. (7.37); with this new definition, the quantities Q_s have the dimension of length. To simplify the notation, we also combine the $\mathbf{q}$ and l indices into a single index s. With these definitions, the total energy of the phonons takes the form

$$E^{\text{vib}} = \sum_s \left[\frac{1}{2M}\left(M\frac{dQ_s}{dt}\right)^2 + \frac{1}{2}M\omega_s^2 Q_s^2\right] = \sum_s\left[\frac{1}{2M}P_s^2 + \frac{1}{2}M\omega_s^2 Q_s^2\right], \quad P_s = M\frac{dQ_s}{dt}$$

which, as discussed before, corresponds to a set of independent harmonic oscillators, with P_s the canonical momenta. The harmonic oscillator coordinates Q_s are related to the atomic displacements $\mathbf{S}$ by:

$$\mathbf{S} = \sum_s Q_s \hat{\mathbf{e}}_s$$

where $\hat{\mathbf{e}}_s$ is the eigenvector corresponding to the phonon mode with frequency ω_s. Each of the oscillators obeys the hamiltonian

$$\hat{\mathcal{H}}^{\text{vib}} \phi_n^{(s)}(Q) = \left[-\frac{\hbar^2 \nabla_Q^2}{2M} + \frac{1}{2} M \omega_s^2 Q^2 \right] \phi_n^{(s)}(Q) = \epsilon_n^{(s)} \phi_n^{(s)}(Q)$$

The wavefunction of each independent harmonic oscillator in its ground state is:

$$\phi_0^{(s)}(Q_s) = \left(\frac{1}{\pi r_s^2} \right)^{1/4} e^{-Q_s^2/2r_s^2}$$

where the constant $1/r_s = M\omega_s/\hbar$ (see Appendix C).

With all these expressions we are now in a position to calculate the probability of recoilless emission in the ground state of the crystal, at $T = 0$, so that all oscillators are in their ground state before and after the transition. This probability is given by:

$$P_{0 \to 0} \sim \left| \langle 0| e^{i\mathbf{K} \cdot \mathbf{S}} |0 \rangle \right|^2 = \left| \prod_s \langle \phi_0^{(s)} | e^{i\mathbf{K} \cdot \hat{\mathbf{e}}_s Q_s} | \phi_0^{(s)} \rangle \right|^2$$

$$= \left| \prod_s \left(\frac{1}{\pi r_s^2} \right)^{1/2} \int e^{-Q_s^2/r_s^2} e^{i\mathbf{K} \cdot \hat{\mathbf{e}}_s Q_s} dQ_s \right|^2 = \prod_s e^{-(\mathbf{K} \cdot \hat{\mathbf{e}}_s)^2 r_s^2/2}$$

But there is a simple relation between $r_s^2/2$ and the average value of Q_s^2 in the ground state of the harmonic oscillator:

$$\frac{r_s^2}{2} = \int \left(\frac{1}{\pi r_s^2} \right)^{1/2} Q_s^2 e^{-Q_s^2/r_s^2} dQ_s = \langle \phi_0^{(s)} | Q_s^2 | \phi_0^{(s)} \rangle = \langle Q_s^2 \rangle$$

with the help of which we obtain:

$$\left| \langle 0| e^{i\mathbf{K} \cdot \mathbf{S}} |0 \rangle \right|^2 = \exp \left[-\sum_s (\mathbf{K} \cdot \hat{\mathbf{e}}_s)^2 \langle Q_s^2 \rangle \right]$$

Now we will assume that all the phonon modes l have the same average $(Q_\mathbf{q}^{(l)})^2$, which, following the same steps as in Eq. (8.10), leads to:

$$\left| \langle 0| e^{i\mathbf{K} \cdot \mathbf{S}} |0 \rangle \right|^2 = \exp \left[-\sum_\mathbf{q} \mathbf{K}^2 \langle Q_\mathbf{q}^2 \rangle \right]$$

By a similar argument, we obtain:

$$\langle \mathbf{S}^2 \rangle = \left\langle \left| \sum_{\mathbf{q},l} Q_\mathbf{q}^{(l)} \hat{\mathbf{e}}_\mathbf{q}^{(l)} \right|^2 \right\rangle = 3 \sum_\mathbf{q} \langle Q_\mathbf{q}^2 \rangle \Rightarrow P_{0 \to 0} \sim e^{-\mathbf{K}^2 \langle \mathbf{S}^2 \rangle/3}$$

For harmonic oscillators the average kinetic and potential energies are equal and each is half of the total energy (see Problem 2). This gives:

$$M\omega_s^2 \left\langle |Q_s|^2 \right\rangle = \left(n_s + \frac{1}{2} \right) \hbar\omega_s \Rightarrow \left\langle |Q_s|^2 \right\rangle = \frac{\hbar}{M\omega_s} \left(n_s + \frac{1}{2} \right) \tag{7.76}$$

with n_s the phonon occupation numbers. This result holds in general for a canonical distribution, from which we can obtain:

$$\frac{1}{3}\mathbf{K}^2 \left\langle \mathbf{S}^2 \right\rangle = \frac{\hbar^2 \mathbf{K}^2}{2M} \frac{2}{3} \sum_s \frac{1}{\hbar\omega_s} \left(n_s + \frac{1}{2} \right) = \frac{2}{3} E_R \sum_s \frac{1}{\hbar\omega_s} \left(n_s + \frac{1}{2} \right)$$

with E_R the recoil energy as before, which is valid for a system with a canonical distribution at temperature T. At $T = 0$, n_s vanishes and if we use the Debye model with $\omega_s = vk$ (that is, we take an average velocity $v_s = v$), we obtain:

$$\frac{1}{3}\mathbf{K}^2 \left\langle \mathbf{S}^2 \right\rangle = \frac{1}{3} E_R \sum_{\mathbf{q}} \frac{1}{\hbar v k}$$

from which we can calculate explicitly the sum over $\mathbf{q}$ by turning it into an integral with upper limit k_D (the Debye wave-vector), to obtain:

$$P_{0 \to 0} \sim \exp\left[-\frac{3}{2} \frac{E_R}{k_B \Theta_D} \right] \tag{7.77}$$

with Θ_D the Debye temperature.

As we have seen before, energy conservation gives $E_R = \hbar\omega_0^2 / 2Mc^2$, which is a fixed value since the value of ω_0 is the natural frequency of the emitted γ-rays. This last expression shows that if the Debye temperature is high enough ($k_B \Theta_D \gg E_R$), then $P_{0 \to 0} \sim 1$, in other words, the probability of the transition involving *only* nuclear degrees of freedom, without having to balance the recoil energy, will be close to 1. This corresponds to recoil-less emission.

7.7 Elastic Deformations of Solids

As we have seen in Chapter 4, total-energy calculations allow the determination of the optimal lattice constant of any crystalline phase of a material, its cohesive energy, and its bulk modulus. The total-energy vs. volume curves for different phases of a material, like those for Si shown in Fig. 4.8, exhibit a quadratic behavior of the energy near the minimum. A simple polynomial fit to the calculated energy as a function of volume near the minimum readily yields estimates for the equilibrium volume V_0, and hence the lattice constant $a_0 \sim V_0^{1/3}$, that corresponds to the lowest-energy structure. The standard way to fit the total-energy curve near its minimum is by a polynomial in powers of $V^{1/3}$, where V is the volume. The bulk modulus B of a solid is defined as the change in the pressure P when the volume V changes, normalized by the volume (in the present calculation we are assuming zero-temperature conditions):

$$B \equiv -V \frac{\partial P}{\partial V}$$
(7.78)

Using the relation between the pressure and the total energy at zero temperature, we obtain for the bulk modulus:

$$P = -\frac{\partial E}{\partial V} \Rightarrow B = V \frac{\partial^2 E}{\partial V^2}$$
(7.79)

which is easily obtained from the total-energy vs. volume curves, as the curvature at its minimum.

A quadratic behavior of the energy as a function of the lattice constant is referred to as "harmonic" or "elastic" behavior. It implies simple harmonic-spring-like forces that restore the volume to its equilibrium value for small deviations from it. This is behavior found in all materials. It implies that the solid returns to its original equilibrium state once the agent (external force) that produced the deviation is removed. If the distortion is large, then the energy is no longer quadratic in the amount of deviation from equilibrium and the solid cannot return to its equilibrium state; it is permanently deformed. This is referred to as "plastic" deformation. The elastic (harmonic) deformation can be described in much the same terms as the small deviation of individual atoms from their equilibrium positions on which we based the theory of lattice vibrations. In the case of elastic deformation a great number of atoms are moved from their equilibrium positions in the same direction in different parts of the solid. The amount of deformation is called strain and the resulting forces per unit area are called stress. These are both tensors because their values depend on the direction in which atoms are displaced and crystals are not necessarily isotropic. The theory that describes the effects of elastic deformation is called "elasticity theory."

7.7.1 Phenomenological Models of Solid Deformation

Before we develop the basic concepts of elasticity theory and how they apply to simple cases of elastic deformation of solids, we discuss two phenomenological models that predict with remarkable accuracy some important quantities related to the energetics of solid deformations. The first, developed by M. L. Cohen[3] and based on the arguments of J. C. Phillips[4] about the origin of energy bands in semiconductors, deals with the bulk modulus of covalently bonded solids. The second, developed by Rose, Smith, Guinea, and Ferrante,[5] concerns the energy-vs.-volume relation and asserts that, when scaled appropriately, it is a universal relation for all solids; this is referred to as the universal binding energy relation (UBER). We discuss these phenomenological theories in some detail, because they represent important tools, as they distill the results of elaborate calculations into simple, practical expressions.

[3] M. L. Cohen, *Phys. Rev. B* **32**, 7988 (1985).
[4] *Bonds and Bands in Semiconductors*, J. C. Phillips (Academic Press, New York, 1973).
[5] J. H. Rose, J. R. Smith, F. Guinea, and J. Ferrante, *Phys. Rev. B* **29**, 2963 (1984).

Cohen's theory gives the bulk modulus of a solid which has average coordination $\langle N_c \rangle$, as:

$$B = \frac{\langle N_c \rangle}{4}(1971 - 220I)d^{-3.5}\,\text{GPa} \tag{7.80}$$

where d is the bond length (Å) and I is a dimensionless number which describes the ionicity: $I = 0$ for the homopolar crystals C, Si, Ge, Sn; $I = 1$ for III–V compounds (where the valence of each element differs by 1 from the average valence of 4); and $I = 2$ for II–VI compounds (where the valence of each element differs by 2 from the average valence of 4). The basic physics behind this expression is that the bulk modulus is intimately related to the average electron density in a solid where the electrons are uniformly distributed, as discussed in Chapter 4, Problem 5. For covalently bonded solids, the electronic density is concentrated between nearest-neighbor sites, forming the covalent bonds. Therefore, in this case the crucial connection is between the bond length, which provides a measure of the electron density relevant to bonding strength in the solid, and the bulk modulus. Cohen's argument produced the relation $B \sim d^{-3.5}$, which, dressed with the empirical constants that give good fits for a few representative solids, leads to Eq. (7.80). Later in this section we give actual values of bulk moduli B of several representative semiconductors (see Table 7.4); as can be checked from these values, the bulk moduli of semiconductors are captured with an accuracy which is remarkable, given the simplicity of the expression in Eq. (7.80). In fact, this expression has been used to *predict* theoretically solids with very high bulk moduli,[6] possibly exceeding those of naturally available hard materials like diamond.

The UBER theory gives the energy-vs.-volume relation as a two-parameter expression:

$$\begin{aligned}
E(a) &= E^{\text{coh}}\overline{E}(\overline{a}) \\
\overline{E}(\overline{a}) &= -[1 + \overline{a} + 0.05\overline{a}^3]\exp(-\overline{a}) \\
\overline{a} &= (a - a_0)/l
\end{aligned} \tag{7.81}$$

where a is the lattice constant (which has a very simple relation to the volume per atom, depending on the lattice), a_0 is its equilibrium value, l is a length scale, and E^{coh} is the cohesive energy; the latter two quantities are the two parameters that are specific to each solid. This curve has a single minimum and goes asymptotically to zero for large distances; the value of its minimum corresponds to the cohesive energy. For example, we show in Fig. 7.11 the fit of this expression to DFT/GGA values for a number of representative elemental solids, including metals with s (Li, K, Ca) and sp (Al) electrons, sd electron metals with magnetic (Fe, Ni) or non-magnetic (Cu, Mo) ground states, and semiconductors (Si) and insulators (C). As seen from this example, the fit to this rather wide range of solids is quite satisfactory; the values of the lattice constant and cohesive energy a_0 and E^{coh}, extracted from this fit, differ somewhat from the experimental values but this is probably related more to limitations of the DFT/GGA formalism than the UBER theory. The degree to which this simple expression fits theoretically calculated values for a wide range of solids, as well as the energy of adhesion, molecular binding, and chemisorption of

---

[6] A. Y. Liu and M. L. Cohen, *Phys. Rev. B* **41**, 10727 (1990).

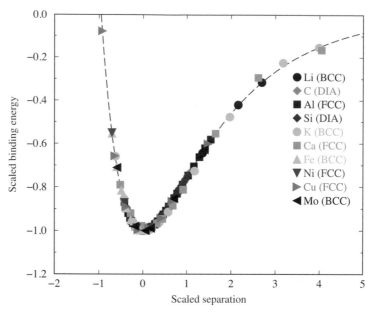

Fig. 7.11 Fit of calculated total energies of various elemental solids to the universal binding energy relation; the calculations are based on DFT/GGA.

gas molecules on surfaces of metals, is remarkable. The UBER expression is very useful for determining, with a few electronic structure calculations of the total energy, a number of important quantities like the cohesive energy (or absorption, chemisorption, adhesion energies, when applied to surface phenomena), bulk modulus, critical stresses, etc.

7.7.2 Elasticity Theory: The Strain and Stress Tensors

Elasticity theory considers solids from a macroscopic point of view, and deals with them as an elastic continuum. The basic assumption is that a solid can be divided into small elements, each of which is considered to be of macroscopic size, that is, very large on the atomic scale. Moreover, it is also assumed that we are dealing with small changes in the state of the solid with respect to a reference configuration, so that the response of the solid is well within the elastic regime, in other words, the amount of deformation is proportional to the applied force just like in a spring. Although these assumptions may seem very restrictive, limiting the applicability of elasticity theory to very large scales, in most cases this theory is essentially correct all the way down to scales of order a few atomic distances. This is due to the fact that it takes a lot of energy to distort solids far from their equilibrium reference state, and for small deviations from that state solids behave very similarly to elastic springs.

We begin with the definitions of the strain and stress tensors in a solid. The reference configuration is usually taken to be the equilibrium structure of the solid, on which there are no external forces. We define strain as the amount by which a small element of the solid is distorted with respect to the reference configuration The arbitrary point in the solid at

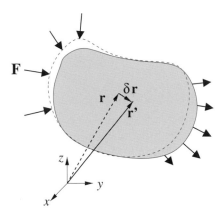

Fig. 7.12 Definition of the strain tensor for the cartesian (x, y, z) coordinate system: the arbitrary point in the solid at $\mathbf{r} = x\hat{x} + y\hat{y} + z\hat{z}$ in the reference configuration moves to $\mathbf{r}' = \mathbf{r} + \delta\mathbf{r} = (x + u)\hat{x} + (y + v)\hat{y} + (z + w)\hat{z}$ under the action of external forces, $\mathbf{F}$, indicated by blue arrows, on the solid (in this example the left part is under compression and the right part is under tension). The undistorted shape and position are shown as dashed lines.

(x, y, z) cartesian coordinates moves to $(x + u, y + v, z + w)$ when the solid is strained, as illustrated in Fig. 7.12. Each of the displacement fields u, v, w is a function of the position in the solid: $u(x, y, z), v(x, y, z), w(x, y, z)$.

The normal components of the strain are defined to be:

$$\epsilon_{xx} = \frac{\partial u}{\partial x}, \quad \epsilon_{yy} = \frac{\partial v}{\partial y}, \quad \epsilon_{zz} = \frac{\partial w}{\partial z} \tag{7.82}$$

These quantities represent the simple stretch of the solid in one direction in response to a force acting in this direction, if there were no deformation in the other two directions. A solid, however, can be deformed in various ways, beyond the simple stretch along the chosen axes x, y, z. In particular, it can be sheared as well as stretched. Stretching and shearing are usually coupled, that is, when a solid is pulled (or pushed) not only does it elongate (or contract) by stretching, but it also changes size in the direction perpendicular to the pulling (or pushing). The shear itself will induce strain in directions other than the direction in which the shearing force acts. To describe this type of deformation we introduce the shear-strain components given by:

$$\epsilon_{xy} = \frac{1}{2}\left(\frac{\partial u}{\partial y} + \frac{\partial v}{\partial x}\right), \quad \epsilon_{yz} = \frac{1}{2}\left(\frac{\partial v}{\partial z} + \frac{\partial w}{\partial y}\right), \quad \epsilon_{zx} = \frac{1}{2}\left(\frac{\partial w}{\partial x} + \frac{\partial u}{\partial z}\right) \tag{7.83}$$

which are symmetric, $\epsilon_{yx} = \epsilon_{xy}, \epsilon_{zy} = \epsilon_{yz}, \epsilon_{xz} = \epsilon_{zx}$. Thus, we need a second-rank tensor to describe the state of strain, which is a symmetric tensor with diagonal elements ϵ_{ii} ($i = x, y, z$) and off-diagonal elements ϵ_{ij} ($i, j = x, y, z$). We also introduce, for future use, the rotation tensor

$$\xi_{xy} = \frac{1}{2}\left(\frac{\partial u}{\partial y} - \frac{\partial v}{\partial x}\right), \quad \xi_{yz} = \frac{1}{2}\left(\frac{\partial v}{\partial z} - \frac{\partial w}{\partial y}\right), \quad \xi_{zx} = \frac{1}{2}\left(\frac{\partial w}{\partial x} - \frac{\partial u}{\partial z}\right) \tag{7.84}$$

which describes the infinitesimal rotation of the volume element at (x, y, z). This is an antisymmetric tensor, $\xi_{ji} = -\xi_{ij}$ ($i, j = x, y, z$).

There is a different notation which makes the symmetric or antisymmetric nature of the strain and rotation tensors more evident: we define the three cartesian axes as x_1, x_2, x_3 (instead of x, y, z) and the three displacement fields as u_1, u_2, u_3 (instead of u, v, w). Then the strain tensor takes the form

$$\epsilon_{ij} = \frac{1}{2}\left(\frac{\partial u_i}{\partial x_j} + \frac{\partial u_j}{\partial x_i}\right) \tag{7.85}$$

which holds for both the diagonal and off-diagonal components. This form makes the symmetric character of the strain tensor obvious. Similarly, the rotation tensor with this alternative notation takes the form

$$\xi_{ij} = \frac{1}{2}\left(\frac{\partial u_i}{\partial x_j} - \frac{\partial u_j}{\partial x_i}\right) \tag{7.86}$$

which is obviously an antisymmetric tensor. This notation becomes particularly useful in the calculation of physical quantities when other multiple-index tensors are also involved, as we will see below. For the moment, we will use the original notation in terms of x, y, z and u, v, w, which makes it easier to associate the symbols in 3D space with the corresponding physical quantities.

It is a straightforward geometrical exercise to show that, if the coordinate axes x and y are rotated around the axis z by an angle Θ, then the strain components in the new coordinate frame $(x'y'z')$ are given by:

$$\epsilon_{x'x'} = \frac{1}{2}(\epsilon_{xx} + \epsilon_{yy}) + \frac{1}{2}(\epsilon_{xx} - \epsilon_{yy})\cos 2\Theta + \epsilon_{xy}\sin 2\Theta$$

$$\epsilon_{y'y'} = \frac{1}{2}(\epsilon_{xx} + \epsilon_{yy}) - \frac{1}{2}(\epsilon_{xx} - \epsilon_{yy})\cos 2\Theta - \epsilon_{xy}\sin 2\Theta$$

$$\epsilon_{x'y'} = \frac{1}{2}(\epsilon_{yy} - \epsilon_{xx})\sin 2\Theta + \epsilon_{xy}\cos 2\Theta \tag{7.87}$$

The last expression is useful, because it can be turned into the following equation:

$$\tan 2\Theta = \frac{\epsilon_{xy}}{\epsilon_{xx} - \epsilon_{yy}} \tag{7.88}$$

through which we can identify the rotation around the z-axis which will make the shear components of the strain ϵ_{xy} vanish. For this rotation, only normal strains will survive; the rotated axes are called the principal strain axes. Using this rotation angle, we find that the normal strains along the principal strain axes are given by:

$$\epsilon_{x'x'} = \frac{\epsilon_{xx} + \epsilon_{yy}}{2} + \frac{\sqrt{(\epsilon_{xx} - \epsilon_{yy})^2 + 4\epsilon_{xy}^2}}{2}, \quad \epsilon_{y'y'} = \frac{\epsilon_{xx} + \epsilon_{yy}}{2} - \frac{\sqrt{(\epsilon_{xx} - \epsilon_{yy})^2 + 4\epsilon_{xy}^2}}{2}$$

Next we define the stress tensor. The definition of stress is a generalization of the definition of pressure (i.e. force per unit area). We take the small element of the solid on which the external forces act to be an orthogonal parallelepiped with the directions normal to its surfaces defining the axis directions $\hat{x}, \hat{y}, \hat{z}$. The forces that act on these surfaces can have arbitrary directions: for example, the force on the surface identified by $+\hat{x}$ can have F_x, F_y, F_z components. Consequently, we need a second-rank tensor to describe stress, just like for strain. We define the diagonal elements of the stress tensor

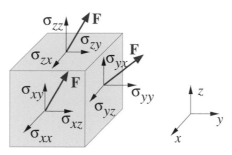

Fig. 7.13 Definition of the stress tensor for the cartesian (x, y, z) coordinate system. The blue arrows labeled **F** denote the forces acting on the three faces of the cube. The resulting stress components are denoted as $\sigma_{ij} = F_j/A_i$, with F_j the force components and A_i the area elements, $i, j = x, y, z$.

as $\sigma_{ii} = F_i/A_i$ $(i = x, y, z)$, where F_i is the component of the force that acts on the surface in the $\hat{\mathbf{i}}$ direction and A_i is the area of this surface of the volume element (see Fig. 7.13). Similarly, the off-diagonal components of the stress are denoted by σ_{ij} and are given by $\sigma_{ij} = F_j/A_i$. Care must be taken in defining the sign of the stress components: the normal directions to the surfaces of the volume element are taken to always point outward from the volume enclosed by the parallelepiped. Thus, if the force applied to a surface points outward, the stress is positive (called tensile stress), while if it is pointing inward, the stress is negative (called compressive stress). The stress tensor is also symmetric, $\sigma_{ji} = \sigma_{ij}$ $(i, j = x, y, z)$, which is a consequence of torque equilibrium. Similar definitions and relations hold for the cylindrical polar coordinate system (r, θ, z), which is related to the cartesian coordinate system (x, y, z) through $r = \sqrt{x^2 + y^2}$ and $\theta = \tan^{-1}(y/x)$.

Just like in the analysis of strains, a rotation of the x, y coordinate axes by an angle θ around the z-axis gives the corresponding relations between stresses in the old $(\sigma_{xx}, \sigma_{yy}, \sigma_{xy})$ and new $(\sigma_{x'x'}, \sigma_{y'y'}, \sigma_{x'y'})$ coordinate systems:

$$\sigma_{x'x'} = \frac{1}{2}(\sigma_{xx} + \sigma_{yy}) + \frac{1}{2}(\sigma_{xx} - \sigma_{yy})\cos 2\Theta + \sigma_{xy}\sin 2\Theta$$

$$\sigma_{y'y'} = \frac{1}{2}(\sigma_{xx} + \sigma_{yy}) - \frac{1}{2}(\sigma_{xx} - \sigma_{yy})\cos 2\Theta - \sigma_{xy}\sin 2\Theta$$

$$\sigma_{x'y'} = \frac{1}{2}(\sigma_{yy} - \sigma_{xx})\sin 2\Theta + \sigma_{xy}\cos 2\Theta \tag{7.89}$$

From these relations we can determine the angle Θ that will give vanishing shear stress $\sigma_{x'y'}$ in the new coordinate system, through the relation

$$\tan 2\Theta = \frac{2\sigma_{xy}}{\sigma_{xx} - \sigma_{yy}} \tag{7.90}$$

which determines the principal stress axes. On the planes that are perpendicular to the principal stress axes, the shear stress is zero, and only the normal component of the stress survives. There are two possible values of the angle Θ that can satisfy the equation for the principal stress, which differ by $\pi/2$. The two values of the normal stress from Eq. (7.89) are given by:

$$\sigma_{x'x'} = \frac{\sigma_{xx} + \sigma_{yy}}{2} + \frac{\sqrt{(\sigma_{xx} - \sigma_{yy})^2 + 4\sigma_{xy}^2}}{2}, \quad \sigma_{y'y'} = \frac{\sigma_{xx} + \sigma_{yy}}{2} - \frac{\sqrt{(\sigma_{xx} - \sigma_{yy})^2 + 4\sigma_{xy}^2}}{2} \tag{7.91}$$

which turn out to be the maximum and minimum values of the stress.

The condition of equilibrium, for a volume element on which an arbitrary force $\mathbf{f}$ is exerted, is:

$$\mathbf{f} = \left(-\sum_{j=x,y,z} \frac{\partial \sigma_{xj}}{\partial j} \right) \hat{x} + \left(-\sum_{j=x,y,z} \frac{\partial \sigma_{yj}}{\partial j} \right) \hat{y} + \left(-\sum_{j=x,y,z} \frac{\partial \sigma_{zj}}{\partial j} \right) \hat{z} \tag{7.92}$$

where the vector $\mathbf{f}$ has dimensions of force/volume. The equilibrium in terms of torque is assured by the fact that the stress tensor is symmetric, $\sigma_{ij} = \sigma_{ji}$ $(i, j = x, y, z)$.

The stress and strain tensors are useful in describing the response of a solid to an external load. In the simplest expression, we assume that only a normal stress is acting in the x direction, in response to which the solid is deformed by:

$$\epsilon_{xx} = \frac{1}{Y} \sigma_{xx} \tag{7.93}$$

that is, if the stress is compressive (negative), the solid is compressed in the x direction and if the stress is tensile (positive), the solid is elongated. The amount of elongation or contraction is determined by the positive coefficient $1/Y$, where the constant Y is called Young's modulus. The above equation is the familiar Hooke's law for elastic strain. For an arbitrary stress applied to the solid, we need to invoke all of the strain components to describe its state of deformation; the corresponding general form of Hooke's law is referred to as a constitutive relation. The traditional way of doing this is through the elastic constants of the solid. These are represented by a tensor of rank 4, C_{ijkl}, with each index taking three values corresponding to x, y, z, but not all of the combinations of the four subscripts give independent elastic constant values. For the independent values, a contracted version of this tensor is used, which is a matrix with only two indices C_{ij}, each of which takes six values (this is referred to as Voigt notation). The connection between the two sets of elastic constants is made through the following identification of indices:

$$1 \rightarrow xx, \quad 2 \rightarrow yy, \quad 3 \rightarrow zz, \quad 4 \rightarrow yz, \quad 5 \rightarrow zx, \quad 6 \rightarrow xy$$

The constitutive law relating stress to strain, in terms of the contracted elastic constants, is:

$$\begin{pmatrix} \sigma_{xx} \\ \sigma_{yy} \\ \sigma_{zz} \\ \sigma_{yz} \\ \sigma_{zx} \\ \sigma_{xy} \end{pmatrix} = \begin{bmatrix} C_{11} & C_{12} & C_{13} & C_{14} & C_{15} & C_{16} \\ C_{12} & C_{22} & C_{23} & C_{24} & C_{25} & C_{26} \\ C_{13} & C_{23} & C_{33} & C_{34} & C_{35} & C_{36} \\ C_{14} & C_{24} & C_{34} & C_{44} & C_{45} & C_{46} \\ C_{15} & C_{25} & C_{35} & C_{45} & C_{55} & C_{56} \\ C_{16} & C_{26} & C_{36} & C_{46} & C_{56} & C_{66} \end{bmatrix} \begin{pmatrix} \epsilon_{xx} \\ \epsilon_{yy} \\ \epsilon_{zz} \\ \epsilon_{yz} \\ \epsilon_{zx} \\ \epsilon_{xy} \end{pmatrix} \tag{7.94}$$

The symmetry of the strain tensor further constrains the 36 components of the matrix C_{ij}, leaving only 21 independent values in the most general case: these are the six diagonal elements (corresponding to $i = j$) and the 15 off-diagonal ones on the *upper* triangle only (corresponding to $j < i$), with those on the *lower* triangle (corresponding to $j > i$) related by symmetry, $C_{ij} = C_{ji}$, as indicated explicitly in Eq. (7.94). These equations imply that the solid is linear elastic, that is, its deformation is linear in the applied stress. This is valid for many physical situations if the deformation is sufficiently small.

The use of the more general tensor of rank 4, C_{ijkl}, simplifies the constitutive equations when the Einstein repeated-index summation convention is employed; this convention consists of summing over all the allowed values of indices which appear more than once on one side of an equation. For example, the above-mentioned linear elastic constitutive relations are written as:

$$\sigma_{ij} = C_{ijkl}\epsilon_{kl} \tag{7.95}$$

where summation over the values of the indices k and l on the right-hand side is implied because they appear twice (they are repeated) on the same side of the equation. Of course, not all the values of C_{ijkl} are independent, because of symmetry; specifically, the symmetries we have already mentioned, $\epsilon_{ij} = \epsilon_{ji}$ and $\sigma_{ij} = \sigma_{ji}$, imply that $C_{ijkl} = C_{jikl} = C_{ijlk} = C_{jilk}$, which makes many of the components of the four-index tensor C_{ijkl} identical. In the most general case, there are 21 independent components, just like for the contracted notation with two indices. Additional symmetries of the solid further reduce the number of independent values of the elastic constants. For example, in an isotropic solid, many of the components vanish, and of the non-vanishing ones only two are independent. Similarly, in a solid of cubic symmetry, there are only three independent elastic constants, as we show in the examples below.

7.7.3 Strain Energy Density

Next, we make the connection between energy, stress, and strain. We will calculate the energy per unit volume in terms of the applied stress and the corresponding strain, which is called the strain energy density (the repeated index summation is implied in the following derivation). We assume that the energy is at its lowest value, taken to be zero, at the initial state of the system before stress is applied. The rate of change of the energy will be given by:

$$\frac{dE}{dt} = \int_V f_i \frac{\partial u_i}{\partial t} dV + \int_S \sigma_{ij} n_j \frac{\partial u_i}{\partial t} dS \tag{7.96}$$

where we have used the alternative notation $u_1 = u, u_2 = v, u_3 = w$ and $x_1 = x, x_2 = y, x_3 = z$, employed in Eqs (7.85) and (7.86). The first term in Eq. (7.96) is due to volume distortions, with f_i the i component of the force on the volume element dV. The second term in Eq. (7.96) is due to surface distortions: the surface contribution involves the vector **n** with components n_i ($i = 1, 2, 3$), which is the unit vector normal to a surface element dS; the product $\sigma_{ij} n_j$ represents the i component of the force on the surface element. By turning the surface integral to a volume one through the divergence theorem, we obtain:

$$\frac{dE}{dt} = \int_V \left[f_i \frac{\partial u_i}{\partial t} + \frac{\partial}{\partial x_j} \left(\sigma_{ij} \frac{\partial u_i}{\partial t} \right) \right] dV \tag{7.97}$$

Newton's law of motion $\mathbf{F} = m(\mathrm{d}^2\mathbf{r}/\mathrm{d}t^2)$ applied to the volume element takes the form

$$\rho\frac{\partial^2 u_i}{\partial t^2} = f_i + \frac{\partial \sigma_{ij}}{\partial x_j} \tag{7.98}$$

where ρ is the density. In this equation f_i is the component of the external force exerted on the volume element and $\partial \sigma_{ij}/\partial x_j$ is the contribution of the stress, which also amounts to a force on the volume element [see also Eq. (7.92), where the corresponding relation for equilibrium conditions is given]. From the definitions of the strain and rotation tensors, Eqs (7.85) and (7.86), we have:

$$\frac{\partial}{\partial t}\left(\frac{\partial u_i}{\partial x_j}\right) = \frac{\partial}{\partial t}\left(\epsilon_{ij} + \xi_{ij}\right) \tag{7.99}$$

Using Eqs (7.98) and (7.99) to simplify the expression for the change in the energy, Eq. (7.97), we obtain:

$$\frac{\mathrm{d}E}{\mathrm{d}t} = \frac{\mathrm{d}}{\mathrm{d}t}(K+U) = \int_V\left[\rho\frac{\partial^2 u_i}{\partial t^2}\frac{\partial u_i}{\partial t}\right]\mathrm{d}V + \int_V\left[\sigma_{ij}\frac{\partial \epsilon_{ij}}{\partial t}\right]\mathrm{d}V \tag{7.100}$$

where we have taken advantage of the symmetric and antisymmetric nature of the stress and the rotation tensors to set $\sigma_{ij}\xi_{ij} = 0$. In Eq. (7.100), the first term on the right-hand side represents the kinetic energy K and the second the potential energy U. To obtain the kinetic energy K, we integrate over time the first term on the right-hand side of the last equation:

$$K = \frac{1}{2}\int_V \rho\left[\frac{\partial u_i}{\partial t}\right]^2 \mathrm{d}V \tag{7.101}$$

The strain energy density, denoted by Σ, is defined as the potential energy per unit volume;[7] from the expression of the second term on the right-hand side of Eq. (7.100), Σ takes the form

$$U = \int_V \Sigma \mathrm{d}V \Rightarrow \frac{\partial \Sigma}{\partial t} = \sigma_{ij}\frac{\partial \epsilon_{ij}}{\partial t} \Rightarrow \sigma_{ij} = \frac{\partial \Sigma}{\partial \epsilon_{ij}} \tag{7.102}$$

Now we can use the Taylor expansion for the strain energy density in terms of small strains:

$$\Sigma = \Sigma_0 + \frac{1}{2}\left[\frac{\partial^2\Sigma}{\partial\epsilon_{ij}\partial\epsilon_{kl}}\right]_0 \epsilon_{ij}\epsilon_{kl} \tag{7.103}$$

where the first-order term is omitted on the assumption that we are referring to changes from an equilibrium state. Taking a derivative of this expression with respect to ϵ_{mn}, we obtain:

$$\sigma_{mn} = \frac{\partial\Sigma}{\partial\epsilon_{mn}} = \frac{1}{2}\left[\frac{\partial^2\Sigma}{\partial\epsilon_{ij}\partial\epsilon_{kl}}\right]_0 (\delta_{im}\delta_{jn}\epsilon_{kl} + \delta_{km}\delta_{ln}\epsilon_{ij})$$

$$= \frac{1}{2}\left[\frac{\partial^2\Sigma}{\partial\epsilon_{mn}\partial\epsilon_{kl}}\right]_0 \epsilon_{kl} + \frac{1}{2}\left[\frac{\partial^2\Sigma}{\partial\epsilon_{mn}\partial\epsilon_{ij}}\right]_0 \epsilon_{ij} = \left[\frac{\partial^2\Sigma}{\partial\epsilon_{mn}\partial\epsilon_{ij}}\right]_0 \epsilon_{ij} \tag{7.104}$$

[7] Depending on the conditions, we can identify U as the internal energy if the loading process is adiabatic, that is, without exchange of heat, or as the Helmholtz free energy if the loading process is isothermal, that is, quasistatic reversible under constant temperature.

which, by comparison to Eq. (7.95), yields:

$$C_{ijkl} = \left[\frac{\partial^2 \Sigma}{\partial \epsilon_{ij} \partial \epsilon_{kl}} \right]_0 \tag{7.105}$$

This last equation demonstrates that the elastic constant tensor can be obtained by calculating the variations to second order in the potential energy density with respect to small strains. Using this expression, we can write the strain energy density as:

$$\Sigma = \Sigma_0 + \frac{1}{2} C_{ijkl} \epsilon_{ij} \epsilon_{kl} = \Sigma_0 + \frac{1}{2} \sigma_{ij} \epsilon_{ij} \tag{7.106}$$

which shows that the strain energy density due to elastic deformation is obtained by multiplying the stress with the strain tensors. Another interesting result that this expression reveals is that the strain energy density is quadratic in the strain tensor ϵ_{ij}. The relations we have just described within continuum elasticity and the corresponding general relations between forces $\mathbf{F}$ and the time derivatives of the position $\mathbf{r}$ or spatial derivatives of the potential energy $U(\mathbf{r})$ are summarized below.

Correspondence between classical mechanics and continuum elasticity

Mechanics   Elasticity

(1) $\mathbf{F} = m \dfrac{d^2 \mathbf{r}}{dt^2}$ $f_i + \dfrac{\partial \sigma_{ij}}{\partial x_j} = \rho \dfrac{\partial^2 u_i}{\partial t^2}$

(2) $\mathbf{F} = -\nabla_{\mathbf{r}} U(\mathbf{r})$ $\sigma_{ij} = \dfrac{\partial \Sigma}{\partial \epsilon_{ij}}$

(3) $\mathbf{F} = -\kappa(\mathbf{r} - \mathbf{r}_0)$ $\sigma_{ij} = C_{ijkl} \epsilon_{kl}$

(4) $U(\mathbf{r}) = U_0 + \dfrac{1}{2}\kappa(\mathbf{r} - \mathbf{r}_0)^2$ $\Sigma = \Sigma_0 + \dfrac{1}{2} C_{ijkl} \epsilon_{ij} \epsilon_{kl}$

(1) Equation of motion.
(2) Force from potential energy gradient/stress from strain energy density.
(3) Hooke's law: linear spring force/linear stress–strain relation.
(4) Harmonic approximation: potential energy/strain energy density.

7.7.4 Isotropic Solid

We next examine the case of an isotropic, elastic solid, that is, a solid in which the elastic response in different directions is the same, and hence the corresponding elastic constants are equivalent. First, when a normal strain σ_{xx} is applied in the x direction, the solid not only deforms according to Eq. (7.93), but also deforms in the y, z directions according to:

$$\epsilon_{yy} = \epsilon_{zz} = -\nu \epsilon_{xx} = -\nu \frac{\sigma_{xx}}{Y} \tag{7.107}$$

where ν is called Poisson's ratio. Poisson's ratio is a positive quantity: when a material is pulled in one direction its cross-section perpendicular to the pulling direction becomes smaller (a rod gets thinner when pulled along its length).[8] From this we conclude that if three normal stresses $\sigma_{xx}, \sigma_{yy}, \sigma_{zz}$ are applied to an element which is an orthogonal parallelepiped with faces normal to the x, y, z axes, its state of strain will be given by:

$$\epsilon_{xx} = \frac{1}{Y}[\sigma_{xx} - \nu(\sigma_{yy} + \sigma_{zz})], \quad \epsilon_{yy} = \frac{1}{Y}[\sigma_{yy} - \nu(\sigma_{xx} + \sigma_{zz})], \quad \epsilon_{zz} = \frac{1}{Y}[\sigma_{zz} - \nu(\sigma_{xx} + \sigma_{yy})]$$

$$(7.108)$$

An analogous expression holds for shear stresses. Namely, if a shear stress σ_{xy} (or σ_{yz}, σ_{zx}) is applied to a solid, the corresponding shear strains are given by:

$$\epsilon_{xy} = \frac{1}{2\mu}\sigma_{xy}, \quad \epsilon_{yz} = \frac{1}{2\mu}\sigma_{yz}, \quad \epsilon_{zx} = \frac{1}{2\mu}\sigma_{zx} \qquad (7.109)$$

In these expressions the shear modulus μ is used to relate strain to stress for the off-diagonal (shear) components. Of the three elastic constants introduced so far, Young's modulus Y, Poisson's ratio ν, and shear modulus μ, only two are independent, because they obey the relation

$$Y = 2\mu(1 + \nu) \Rightarrow \mu = \frac{Y}{2(1 + \nu)} \qquad (7.110)$$

The relations for the strain can be written compactly with use of the Kronecker δ and the Einstein summation convention, as:

$$\epsilon_{ij} = \frac{1 + \nu}{Y}\sigma_{ij} - \delta_{ij}\frac{\nu}{Y}(\sigma_{xx} + \sigma_{yy} + \sigma_{zz}), \quad i, j = x, y, z \qquad (7.111)$$

These equations can be inverted to give the stresses in terms of the strains:

$$\sigma_{ij} = 2\mu\epsilon_{ij} + \delta_{ij}\lambda(\epsilon_{xx} + \epsilon_{yy} + \epsilon_{zz}), \quad i, j = x, y, z \qquad (7.112)$$

where the quantity λ is defined as:

$$\lambda = \frac{Y\nu}{(1 + \nu)(1 - 2\nu)} = \frac{2\mu\nu}{1 - 2\nu} \qquad (7.113)$$

and is known as Lamé's constant. In cartesian coordinates, the above equation represents the following set of relations:

$$\sigma_{xx} = 2\mu\epsilon_{xx} + \lambda(\epsilon_{xx} + \epsilon_{yy} + \epsilon_{zz}) = \frac{Y}{(1 + \nu)(1 - 2\nu)}\left[(1 - \nu)\epsilon_{xx} + \nu(\epsilon_{yy} + \epsilon_{zz})\right]$$

$$\sigma_{yy} = 2\mu\epsilon_{yy} + \lambda(\epsilon_{xx} + \epsilon_{yy} + \epsilon_{zz}) = \frac{Y}{(1 + \nu)(1 - 2\nu)}\left[(1 - \nu)\epsilon_{yy} + \nu(\epsilon_{xx} + \epsilon_{zz})\right]$$

$$\sigma_{zz} = 2\mu\epsilon_{zz} + \lambda(\epsilon_{xx} + \epsilon_{yy} + \epsilon_{zz}) = \frac{Y}{(1 + \nu)(1 - 2\nu)}\left[(1 - \nu)\epsilon_{zz} + \nu(\epsilon_{xx} + \epsilon_{yy})\right]$$

$$\sigma_{xy} = 2\mu\epsilon_{xy}, \quad \sigma_{yz} = 2\mu\epsilon_{yz}, \quad \sigma_{zx} = 2\mu\epsilon_{yz}$$

---

[8] It is often mentioned that in certain exceptional solids, like cork, Poisson's ratio can be negative. Cork actually has a very small (near zero) Poisson ratio, but recently some materials have been synthesized that do have negative Poisson ratio.

Table 7.3 Values of the shear modulus μ, Lamé's constant λ, and Poisson's ratio ν for representative covalent and ionic solids (left column) and metals (right column). λ and ν are given in megabars (1 Mbar = 100 GPa); ν is dimensionless. The crystals in each category are ordered by their shear modulus value in decreasing order. *Source:* Hirth and Lothe, mentioned in Further Reading.

Covalent and ionic solids				Metals			
Crystal	μ	λ	ν	Crystal	μ	λ	ν
				W	1.6	2.01	0.278
C	5.36	0.85	0.068	Mo	1.23	1.89	0.305
Si	0.681	0.524	0.218	Cr	1.21	0.778	0.130
Ge	0.564	0.376	0.200	Ni	0.947	1.17	0.276
				Fe	0.860	1.21	0.291
MgO	1.29	0.68	0.173	Cu	0.546	1.006	0.324
LiF	0.515	0.307	0.187	Ag	0.338	0.811	0.354
PbS	0.343	0.393	0.267	Au	0.310	1.46	0.412
NaCl	0.148	0.146	0.248	Al	0.265	0.593	0.347
KCl	0.105	0.104	0.250	Pb	0.101	0.348	0.387
AgBr	0.087	0.345	0.401	Na	0.038	0.025	0.201
				K	0.017	0.029	0.312

Values of the shear modulus, Lamé's constant, and Poisson's ratio for representative solids are given in Table 7.3.

Finally, we use these expressions to obtain an equation for the strain energy density, starting from Eq. (7.106), and setting the constant term $\Sigma_0 = 0$. Inserting the expressions we found for the stresses σ_{ij} in terms of the strains ϵ_{ij} above, we arrive at:

$$\Sigma_{\text{iso}} = \frac{\mu(1-\nu)}{1-2\nu}(\epsilon_{xx}^2 + \epsilon_{yy}^2 + \epsilon_{zz}^2) + 2\mu(\epsilon_{xy}^2 + \epsilon_{yz}^2 + \epsilon_{zx}^2) + \lambda(\epsilon_{xx}\epsilon_{yy} + \epsilon_{yy}\epsilon_{zz} + \epsilon_{zz}\epsilon_{xx}) \quad (7.114)$$

an expression that will prove useful in the following derivations.

Isotropic Solid under Hydrostatic Pressure

We use the expressions derived so far to analyze the effects of hydrostatic pressure applied to an isotropic solid. We consider an orthogonal parallelepiped element in the solid which experiences stresses:

$$\sigma_{xx} = \sigma_{yy} = \sigma_{zz} = -p, \quad \sigma_{xy} = \sigma_{yz} = \sigma_{zx} = 0 \quad (7.115)$$

with the change in hydrostatic pressure $p > 0$. The ensuing strains are then given by:

$$\epsilon_{xx} = \epsilon_{yy} = \epsilon_{zz} = -\frac{1-2\nu}{Y}p, \quad \epsilon_{xy} = \epsilon_{yz} = \epsilon_{zx} = 0 \quad (7.116)$$

The change in volume of the element will be given by:

$$dx(1 + \epsilon_{xx}) \times dy(1 + \epsilon_{yy}) \times dz(1 + \epsilon_{zz}) = (1 + \epsilon_{xx} + \epsilon_{yy} + \epsilon_{zz})dxdydz \quad (7.117)$$

where we neglect terms higher than first order in the strains. We define the total fractional change in volume $\bar{\epsilon} = \epsilon_{xx} + \epsilon_{yy} + \epsilon_{zz}$, and find from the above equation that:

$$\bar{\epsilon} = -\frac{3}{Y}(1 - 2\nu)p$$

But the total fractional change in volume is also given as:

$$\bar{\epsilon} = \frac{\Delta V}{V} = -\frac{1}{B}p$$

where B is the bulk modulus of the solid, as defined in Eq. (7.78). This relationship reveals that the bulk modulus is expressed in terms of Young's modulus and Poisson's ratio as:

$$B = \frac{Y}{3(1 - 2\nu)} = \frac{2\mu(1 + \nu)}{3(1 - 2\nu)} = \lambda + \frac{2}{3}\mu \tag{7.118}$$

where we have also used the definition of Lamé's constant, Eq. (7.113).

We can take advantage of this result to express various quantities in terms of the bulk modulus. Specifically, for the stress–strain relations we find:

$$\sigma_{ij} = B\epsilon_{kk}\delta_{ij} + 2\mu \left(\epsilon_{ij} - \frac{1}{3}\epsilon_{kk} \right) = B\bar{\epsilon}\delta_{ij} + 2\mu \left(\epsilon_{ij} - \frac{\bar{\epsilon}}{3} \right)$$

which for the diagonal terms becomes:

$$\sigma_{xx} = B\bar{\epsilon} + 2\mu \left(\epsilon_{xx} - \frac{\bar{\epsilon}}{3} \right), \quad \sigma_{yy} = B\bar{\epsilon} + 2\mu \left(\epsilon_{yy} - \frac{\bar{\epsilon}}{3} \right), \quad \sigma_{zz} = B\bar{\epsilon} + 2\mu \left(\epsilon_{zz} - \frac{\bar{\epsilon}}{3} \right)$$

By adding these three expressions and defining the volume stress $\bar{\sigma} = (\sigma_{xx} + \sigma_{yy} + \sigma_{zz})/3$ (which is also called the hydrostatic component of stress), we obtain:

$$\bar{\epsilon} = \frac{1}{B}\bar{\sigma} \tag{7.119}$$

The result of Eq. (7.119) applies to any state of stress, and the quantities $\bar{\epsilon}, \bar{\sigma}$ are both invariant with respect to orthogonal transformations of the coordinate axes. The reason for this simple relation is that the shear strains, which in the present example are zero, do not contribute to any changes in the volume.

We can also use the expressions derived so far to express the strain energy density in a form that provides some insight into the different contributions. We start with our expression from Eq. (7.114) and use the relations between B, μ, λ from Eq. (7.118) to write the strain energy density as:

$$\Sigma_{\text{iso}} = \left(\frac{3}{2}B - \lambda \right)(\epsilon_{xx}^2 + \epsilon_{yy}^2 + \epsilon_{zz}^2) + 2\mu(\epsilon_{xy}^2 + \epsilon_{yz}^2 + \epsilon_{zx}^2) + \lambda(\epsilon_{xx}\epsilon_{yy} + \epsilon_{yy}\epsilon_{zz} + \epsilon_{zz}\epsilon_{xx})$$

but we also have from Eq. (7.118):

$$\frac{3}{2}B - \lambda = \frac{1}{2}\lambda + \mu \quad \text{and} \quad \frac{1}{2}\lambda = \frac{1}{2}B - \frac{1}{3}\mu$$

and combining terms to form a perfect square, we find

$$\Sigma_{\text{iso}} = \left(\frac{1}{2}B - \frac{1}{3}\mu \right)(\epsilon_{xx} + \epsilon_{yy} + \epsilon_{zz})^2 + \mu(\epsilon_{xx}^2 + \epsilon_{yy}^2 + \epsilon_{zz}^2) + 2\mu(\epsilon_{xy}^2 + \epsilon_{yz}^2 + \epsilon_{zx}^2)$$

which can be written in a more compact and elegant form using Einstein notation:

$$\Sigma_{iso} = \frac{1}{2}B(\epsilon_{kk})^2 + \mu\left[\epsilon_{ij}\epsilon_{ij} - \frac{1}{3}(\epsilon_{kk})^2\right] \tag{7.120}$$

This is an interesting expression because it clearly shows that the first term, which is quadratic in the volume change captured by $\bar{\epsilon} = (\epsilon_{xx} + \epsilon_{yy} + \epsilon_{zz})$, is multiplied by the coefficient $(1/2)B$ and corresponds to the energy cost for uniform volume expansion or contraction, while the rest of the contribution is related to shear, and contains all the off-diagonal terms of the strain.

Stress Waves in Isotropic Solid

Finally, we calculate the speed of different types of stress waves through an isotropic solid. To this end, we use similar manipulations as above for the strain energy density to rewrite the general equilibrium condition of Eq. (7.98), for the case of an isotropic solid with no external forces, as:

$$(\lambda + \mu)\frac{\partial^2 u_j}{\partial x_i \partial x_j} + \mu\frac{\partial^2 u_i}{\partial x_j \partial x_j} = \rho\frac{\partial^2 u_i}{\partial t^2}$$

or equivalently, in vector notation:

$$(\lambda + \mu)\nabla_{\mathbf{r}}(\nabla_{\mathbf{r}} \cdot \mathbf{u}) + \mu\nabla_{\mathbf{r}}^2\mathbf{u} = \rho\frac{\partial^2 \mathbf{u}}{\partial t^2} \tag{7.121}$$

This last expression allows us to find the speed with which waves can propagate in an isotropic solid. We assume that the displacement field $\mathbf{u}$ has a wave form

$$\mathbf{u}(\mathbf{r}, t) = \mathbf{e}f(\mathbf{q} \cdot \mathbf{r} - ct) \tag{7.122}$$

where $\mathbf{e}$ is the polarization of the wave, $\mathbf{q}$ is the direction of propagation, and c is its speed. Using this expression for $\mathbf{u}(\mathbf{r}, t)$ in Eq. (7.121), we obtain

$$(\lambda + \mu)f''(\mathbf{q} \cdot \mathbf{r} - ct)(\mathbf{e} \cdot \mathbf{q})\mathbf{q} + \mu\mathbf{e}f''(\mathbf{q} \cdot \mathbf{r} - ct)|\mathbf{q}|^2 = c^2\rho\mathbf{e}f''(\mathbf{q} \cdot \mathbf{r} - ct)$$

where $f''(x)$ is the second derivative of $f(x)$ with respect to its argument. For dilatational waves, $\mathbf{e} \parallel \mathbf{q}$, we then find that the speed of the wave is given by:

$$c_d = \sqrt{\frac{\lambda + 2\mu}{\rho}} = \sqrt{\left(B + \frac{4}{3}\mu\right)\frac{1}{\rho}} \tag{7.123}$$

while for shear waves, $\mathbf{e} \perp \mathbf{q}$, the speed of the wave is given by:

$$c_s = \sqrt{\frac{\mu}{\rho}} \tag{7.124}$$

Note that in order to make the expression given above an identity for any function $f(x)$, for either dilatational or shear waves, we must have $|\mathbf{q}|^2 = 1$; this implies that there is no length scale for the wave, that is, no natural scale to set the wavelength. We also note that on the surface of the solid there can only exist shear waves; the speed of propagation of these waves, called the Rayleigh speed c_R, is typically $c_R \approx 0.9c_s$.

7.7.5 Solid with Cubic Symmetry

We examine next the case of a solid which has cubic symmetry, which is quite common in crystals. The minimal symmetry in such a case consists of axes of threefold rotation, which are the major axes of the cube, as illustrated in Fig. 3.5(a). The rotations by $2\pi/3$ around these symmetry axes lead to the following permutations of the x, y, z axes:

$$1 : x \to y \to z \to x$$
$$2 : x \to z \to -y \to x$$
$$3 : x \to -z \to y \to x$$
$$4 : x \to -y \to -z \to x \tag{7.125}$$

as described in Chapter 3, Table 3.3. Now we can use these general symmetries of the solid to deduce relations between the elastic constants. The simplest way to derive these relations is to consider the strain energy density expression of Eq. (7.106), which must be invariant under any of the symmetry operations. This expression for the energy is quadratic in the strains, so that the terms appearing in it can have one of the following forms:

$$\epsilon_{ii}^2, \quad \epsilon_{ii}\epsilon_{jj}, \quad \epsilon_{ij}^2, \quad \epsilon_{ii}\epsilon_{ij}, \quad \epsilon_{ii}\epsilon_{jk}, \quad \epsilon_{ij}\epsilon_{jk} \tag{7.126}$$

with $(i, j, k) = (x, y, z)$. Each type of the first three terms in Eq. (7.126) gives a contribution to the energy that must be the same for all possible values of the indices, since the three axes x, y, z are equivalent due to the cubic symmetry. Specifically, the contribution of ϵ_{xx}^2 must be the same as that of ϵ_{yy}^2 and that of ϵ_{zz}^2, and similarly for the terms ϵ_{ij}^2 and $\epsilon_{ii}\epsilon_{jj}$. These considerations imply:

$$C_{11} = C_{22} = C_{33}, \quad C_{12} = C_{13} = C_{23}, \quad C_{44} = C_{55} = C_{66}$$

so there are at least three independent elastic constants, as claimed earlier. Moreover, the first three terms in Eq. (7.126), when summed over the values of the indices, are unchanged under the symmetry operations of Eq. (7.125), while the last three terms change sign. For example, the term

$$\epsilon_{xx}^2 + \epsilon_{yy}^2 + \epsilon_{zz}^2$$

[which corresponds to the first term in Eq. (7.126), since $C_{11} = C_{22} = C_{33}$] is unchanged under any of the operations in Eq. (7.125). In contrast to this, the term $\epsilon_{xx}\epsilon_{xy}$ under the third operation in Eq. (7.125) is transformed into:

$$\epsilon_{xx}\epsilon_{xy} \to \epsilon_{\bar{z}\bar{z}}\epsilon_{\bar{z}x} = -\epsilon_{zz}\epsilon_{zx}$$

(where we have used the notation $\bar{z} = -z$ for the subscripts of the strain), which is another term of the same character but opposite sign. The coefficients of terms that change sign must vanish, since otherwise, for an arbitrary state of strain, the energy would not be invariant under the symmetry operations. This implies vanishing of the following coefficients:

$$C_{14} = C_{15} = C_{16} = 0, \quad C_{24} = C_{25} = C_{26} = 0, \quad C_{34} = C_{35} = C_{36} = 0$$

and similar considerations lead to:

$$C_{45} = C_{46} = C_{56} = 0$$

leaving as the only independent coefficients C_{11}, C_{12}, C_{44}. This result leads to the following expression for the strain energy density of a solid with cubic symmetry:

$$\Sigma_{\text{cubic}} = \frac{1}{2} C_{11} \left(\epsilon_{xx}^2 + \epsilon_{yy}^2 + \epsilon_{zz}^2 \right) + 2C_{44} \left(\epsilon_{xy}^2 + \epsilon_{yz}^2 + \epsilon_{zx}^2 \right)$$
$$+ C_{12} \left(\epsilon_{xx}\epsilon_{yy} + \epsilon_{yy}\epsilon_{zz} + \epsilon_{zz}\epsilon_{xx} \right) \tag{7.127}$$

where we have taken the constant term Σ_0 to be zero for simplicity. This expression is also useful for calculating the elastic constants C_{11}, C_{12}, C_{44}, using total-energy calculations of the type discussed in detail in Chapter 4. Specifically, for C_{11} it suffices to distort the solid so that of all the strain components, only ϵ_{xx} is non-zero, and then calculate the total energy as a function of the magnitude of ϵ_{xx} and fit it with a second-order term in this variable: the coefficient of this term is $\frac{1}{2}C_{11}$. In a similar fashion, one can obtain values for C_{12} and C_{44}.

The diagonal components of the strain correspond to the change in volume, so ϵ_{xx} corresponds to a volume change due to elongation or contraction of the solid in one direction only. In this sense, the elastic constant C_{11} is related to the response of the solid to elongation or contraction in one direction only. The strains $\epsilon_{ij}, i \neq j$ describe pure shear distortions of the solid without changes in the volume, so that the elastic constant C_{44} is related to the response of the solid to shear forces. The elastic constant C_{12} involves two different diagonal components of the strain, so that it is related to simultaneous distortions (elongation or contraction but not shear) along two different orthogonal axes. From this information, we can also deduce that the bulk modulus of a solid with cubic symmetry is related to the elastic constants C_{11} and C_{12} only (see Problem 12). In Table 7.4 we give the values of the elastic constants of some representative cubic solids.

7.7.6 Thin Plate Equilibrium

We consider next the case of a thin plate of thickness h in the z direction, which is very small compared to the size of the plate in the other two, x and y directions. This situation is relevant to graphene and various other materials that can be produced in the form of 2D, atomically thin layers. The thin plate can be analyzed as an isotropic solid in the x, y dimensions. We imagine that the plate is slightly bent, so that it deviates locally by a very small amount from being planar. Thus, the displacements u, v, w along the x, y, z axes, respectively, will be small, and specifically we will take the displacement along z to depend only on the position along the plane of the plate, $w(x, y)$. Referring back to the definition of the various components of the stress in Fig. 7.13 in the limit where the solid has infinitesimal extent along the z-axis, we realize that we must set:

$$\sigma_{xz} = \sigma_{yz} = \sigma_{zz} = 0$$

Then, using the stress–strain relation for an isotropic solid, Eq. (7.112), we deduce that for this situation we have: $\epsilon_{xz} = \epsilon_{yz} = 0$. We will next use the definition of the shear strains in

Table 7.4 Bulk moduli B (GPa) and elastic constants, C_{11}, C_{12}, C_{44} (Mbar) for representative cubic crystals at room temperature. *Sources*: D. R. Lide, (Ed.), *CRC Handbook of Chemistry and Physics* (CRC Press, Boca Raton, FL, 1999-2000); S. Haussühl, *Z. Phys.* **159**, 223 (1960). The names of crystal structures are DIA = diamond, ZBL = zincblende, CUB = cubic, FCC = face centered cubic, BCC = body centered cubic (see also Chapter 1).

Crystal	B	C_{11}	C_{12}	C_{44}	Crystal	B	C_{11}	C_{12}	C_{44}
C (DIA)	443	10.76	1.250	5.760	Ir (FCC)	320	5.800	2.420	2.560
Si (DIA)	98.8	1.658	0.639	0.796	W (BCC)	310	5.224	2.044	1.608
Ge (DIA)	77.2	1.284	0.482	0.667	Mo (BCC)	273	4.637	1.578	1.092
					Pt (FCC)	230	3.467	2.507	0.765
AlSb (ZBL)	58.2	0.894	0.443	0.416	Au (FCC)	220	1.924	1.630	0.420
GaP (ZBL)	88.7	1.412	0.625	0.705	Ta (BCC)	200	2.602	1.545	0.826
GaAs (ZBL)	74.8	1.188	0.537	0.594	Cr (BCC)	190	3.398	0.586	0.990
GaSb (ZBL)	57.0	0.884	0.403	0.432	Ni (FCC)	186	2.481	1.549	1.242
InP (ZBL)	71.0	1.022	0.576	0.460	Pd (FCC)	181	2.271	1.760	0.717
InAs (ZBL)	60.0	0.833	0.453	0.396	Nb (BCC)	170	2.465	1.345	0.287
InSb (ZBL)	47.4	0.672	0.367	0.302	Fe (BCC)	168	2.260	1.400	1.160
					V (BCC)	162	2.287	1.190	0.432
ZnS (ZBL)	77.1	1.046	0.653	0.461	Cu (FCC)	137	1.683	1.221	0.757
ZnSe (ZBL)	62.4	0.807	0.488	0.441	Ag (FCC)	101	1.240	0.937	0.461
ZnTe (ZBL)	51.0	0.713	0.408	0.312	Al (FCC)	72.2	1.067	0.604	0.283
					Pb (FCC)	46.0	0.497	0.423	0.150
LiCl (CUB)	36.9	0.493	0.231	0.250	Li (BCC)	11.6	0.135	0.114	0.088
NaCl (CUB)	28.6	0.495	0.129	0.129	Na (BCC)	6.8	0.074	0.062	0.042
KCl (CUB)	20.8	0.407	0.071	0.063	K (BCC)	3.2	0.037	0.031	0.019
RbCl (CUB)		0.361	0.062	0.047	Rb (BCC)	3.1	0.030	0.025	0.017
CsCl (CUB)		0.364	0.088	0.080	Cs (BCC)	1.6	0.025	0.021	0.015

terms of the displacements, Eq. (7.83), and integrate once to obtain:

$$\epsilon_{xz} = 0 \Rightarrow \frac{\partial w}{\partial x} = -\frac{\partial u}{\partial z} \Rightarrow u = \left(-\frac{\partial w}{\partial x}\right)z, \quad \epsilon_{yz} = 0 \Rightarrow \frac{\partial w}{\partial y} = -\frac{\partial v}{\partial z} \Rightarrow v = \left(-\frac{\partial w}{\partial y}\right)z$$

where we have set the constants of integration to zero, to reflect the fact that without any deformation the strains are zero and the plate is flat, lying on the x, y plane. From these results we can calculate the different components of the strain, finding:

$$\epsilon_{xx} = \frac{\partial u}{\partial x} = \left(-\frac{\partial^2 w}{\partial x^2}\right)z, \quad \epsilon_{yy} = \frac{\partial v}{\partial y} = \left(-\frac{\partial^2 w}{\partial y^2}\right)z, \quad \epsilon_{xy} = \frac{1}{2}\left(\frac{\partial u}{\partial y} + \frac{\partial v}{\partial x}\right) = \left(-\frac{\partial^2 w}{\partial x \partial y}\right)z$$

Similarly, the condition for σ_{zz} vanishing, from Eq. (7.83), leads to:

$$\epsilon_{zz}(1 - v) + v(\epsilon_{xx} + \epsilon_{yy}) \Rightarrow \epsilon_{zz} = \frac{v}{1 - v}\left(\frac{\partial^2 w}{\partial x^2} + \frac{\partial^2 w}{\partial y^2}\right)z$$

Having calculated all the components of the strain for this case, we can use them in the expression for the strain energy density, Eq. (7.114), which then becomes, after grouping terms:

$$\Sigma_{\text{plt}} = \frac{\mu}{1-\nu} z^2 \left[\left(\frac{\partial^2 w}{\partial x^2} + \frac{\partial^2 w}{\partial y^2} \right)^2 + 2(1-\nu) \left\{ \left(\frac{\partial^2 w}{\partial x \partial y} \right)^2 - \frac{\partial^2 w}{\partial x^2} \frac{\partial^2 w}{\partial y^2} \right\} \right] \tag{7.128}$$

To obtain the total free energy of the plate we must integrate this expression over the entire volume of the plate. This volume extends from $z = -h/2$ to $z = h/2$ in the z direction at each point (x, y), and over the entire x, y plane. Thus, the free energy per unit area of the x, y plane will be given by:

$$\gamma = \frac{Yh^3}{24(1-\nu^2)} \left[\left(\frac{\partial^2 w}{\partial x^2} + \frac{\partial^2 w}{\partial y^2} \right)^2 + 2(1-\nu) \left\{ \left(\frac{\partial^2 w}{\partial x \partial y} \right)^2 - \frac{\partial^2 w}{\partial x^2} \frac{\partial^2 w}{\partial y^2} \right\} \right] \tag{7.129}$$

where we have used the relations of Eq. (7.110) to express μ in terms of Y, ν. To derive the equation of equilibrium of the plate, we need to perform a variational calculation to ensure that the energy is at a minimum, and set the expression obtained in this way to the sum of the external forces f_z, acting per unit area of the plate in the direction perpendicular to it. This is a rather elaborate calculation, performed in detail in the book by L. D. Landau and E. M. Lifshitz, *Theory of Elasticity* (see Further Reading). The result is:

$$f_z - \frac{Yh^3}{12(1-\nu^2)} \left(\frac{\partial^2 w}{\partial x^2} + \frac{\partial^2 w}{\partial y^2} \right)^2 = 0 \tag{7.130}$$

where w, f_z are both functions of the in-plane variables x, y.

7.8 Application: Phonons of Graphene

As a case study in applying the concepts developed above, we examine the phonon spectrum of graphene, a 2D layer of C atoms forming a honeycomb lattice. The structure is illustrated in Fig. 7.14, along with first-neighbor and second-neighbor interactions associated with the two basis atoms of the primitive unit cell, labeled A and B, in the Born force-constant model. The lattice constants and atomic positions are defined as:

$$\mathbf{a}_1 = \frac{a}{2} \left(\sqrt{3}\hat{x} + \hat{y} \right), \quad \mathbf{a}_2 = \frac{a}{2} \left(\sqrt{3}\hat{x} - \hat{y} \right), \quad \mathbf{t}_A = \frac{a}{\sqrt{3}}\hat{x}, \quad \mathbf{t}_B = 0$$

Even though all atoms lie on the x, y plane in the lowest-energy configuration, we will consider their motion in all three dimensions.

From the first-neighbor interactions, which couple atoms at different sites (A with B), with bond-stretching and bond-bending constants denoted as κ_r, κ_θ, the contributions to

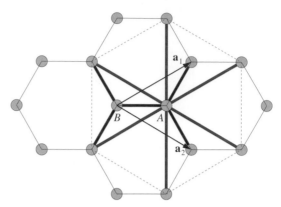

Fig. 7.14 The Born force-constant model for graphene. Gray circles represent the C atoms on the plane and black lines the bonds between them. The red arrows indicate the primitive lattice vectors $\mathbf{a}_1, \mathbf{a}_2$ with the two basis atoms in the unit cell labeled A and B, with first-neighbor interactions indicated by blue lines. Second-neighbor interactions are also shown, by magenta lines, only for site A situated at the center of a hexagon outlined by thin dashed lines (see Problem 13).

the potential energy from all terms that involve at least one site within the primitive unit cell are:

$$\Delta U_{AB} = \frac{\kappa_r - \kappa_\theta}{2} \left\{ \left[\left(\mathbf{S}_{00}^A - \mathbf{S}_{00}^B \right) \cdot \hat{\mathbf{R}}_{00}^{AB} \right]^2 + \left[\left(\mathbf{S}_{00}^A - \mathbf{S}_{10}^B \right) \cdot \hat{\mathbf{R}}_{10}^{AB} \right]^2 + \right.$$

$$\left. \left[\left(\mathbf{S}_{00}^A - \mathbf{S}_{01}^B \right) \cdot \hat{\mathbf{R}}_{01}^{AB} \right]^2 + \left[\left(\mathbf{S}_{00}^B - \mathbf{S}_{\bar{1}0}^A \right) \cdot \hat{\mathbf{R}}_{\bar{1}0}^{BA} \right]^2 + \left[\left(\mathbf{S}_{00}^B - \mathbf{S}_{0\bar{1}}^A \right) \cdot \hat{\mathbf{R}}_{0\bar{1}}^{BA} \right]^2 \right\} +$$

$$\frac{\kappa_\theta}{2} \left\{ \left| \mathbf{S}_{00}^A - \mathbf{S}_{00}^B \right|^2 + \left| \mathbf{S}_{00}^A - \mathbf{S}_{10}^B \right|^2 + \left| \mathbf{S}_{00}^A - \mathbf{S}_{01}^B \right|^2 + \left| \mathbf{S}_{00}^B - \mathbf{S}_{\bar{1}0}^A \right|^2 + \left| \mathbf{S}_{00}^B - \mathbf{S}_{0\bar{1}}^A \right|^2 \right\}$$

where $\mathbf{S}_{nm}^A$ stands for the displacement of an atom at site A in the unit cell displaced relative to the origin by $n\mathbf{a}_1 + m\mathbf{a}_2$ (and similarly for $\mathbf{S}_{nm}^B$) and the unit vectors along the bonds are given by:

$$\hat{\mathbf{R}}_{nm}^{AB} = \frac{\mathbf{t}_A - (\mathbf{t}_B + n\mathbf{a}_1 + m\mathbf{a}_2)}{|\mathbf{t}_A - (\mathbf{t}_B + n\mathbf{a}_1 + m\mathbf{a}_2)|}$$

with $n, m = 0, \pm 1$ and $\bar{1}$ stands for -1. Taking derivates of this expression, according to Eq. (7.18), to derive the dynamical matrix, we find the matrix shown in Table 7.5. First, we note that the 6×6 dynamical matrix, coupling the motion of atoms at sites A and B in all three dimensions, splits into two smaller diagonal matrices, one being a 4×4 matrix that couples the x, y components of the displacements, and another being a 2×2 matrix that couples the z components of the displacements. This implies that within the Born force-constant model, the motion out of the plane is completely decoupled from the in-plane motion, which has important consequences, as discussed below. The various entries in this table are calculated as follows. The diagonal matrix elements for the first-neighbor interactions are:

$$D_{xy} = \frac{3}{2} (\kappa_r + \kappa_\theta), \quad D_z = 3\kappa_\theta$$

Table 7.5 Dynamical matrix for graphene in the Born force-constant model, including contributions from first nearest neighbors only. The various entries are given explicitly in the text.

	A_x	A_y	A_z	B_x	B_y	B_z
A_x	D_{xy}	0	0	$C(\mathbf{k})$	$E(\mathbf{k})$	0
A_y	0	D_{xy}	0	$E(\mathbf{k})$	$F(\mathbf{k})$	0
A_z	0	0	D_z	0	0	$G(\mathbf{k})$
B_x	$C^*(\mathbf{k})$	$E^*(\mathbf{k})$	0	D_{xy}	0	0
B_y	$E^*(\mathbf{k})$	$F^*(\mathbf{k})$	0	0	D_{xy}	0
B_z	0	0	$G^*(\mathbf{k})$	0	0	D_z

while the off-diagonal matrix elements are:

$$C(\mathbf{k}) = -\kappa_r - \frac{\kappa_r + 3\kappa_\theta}{4} \left(e^{i\mathbf{k}\cdot\mathbf{a}_1} + e^{i\mathbf{k}\cdot\mathbf{a}_2} \right) \quad F(\mathbf{k}) = -\kappa_\theta - \frac{3\kappa_r + \kappa_\theta}{4} \left(e^{i\mathbf{k}\cdot\mathbf{a}_1} + e^{i\mathbf{k}\cdot\mathbf{a}_2} \right)$$

$$E(\mathbf{k}) = -\frac{\sqrt{3}\,(\kappa_r - \kappa_\theta)}{4} \left(e^{i\mathbf{k}\cdot\mathbf{a}_1} - e^{i\mathbf{k}\cdot\mathbf{a}_2} \right), \quad G(\mathbf{k}) = -\kappa_\theta \left(1 + e^{i\mathbf{k}\cdot\mathbf{a}_1} + e^{i\mathbf{k}\cdot\mathbf{a}_2} \right)$$

The results from diagonalizing the matrix to obtain the frequency eigenvalues are shown in Fig. 7.15, along high-symmetry directions of the 2D Brillouin zone. We also show in this figure results from first-principles calculations based on DFT for comparison. The values of the force constants κ_r, κ_θ were chosen to reproduce approximately the value of the first-principles results for the highest optical frequency at Γ and the lowest acoustic frequency at M; we have also allowed the value of κ_θ that appears in the matrix for the z displacements to be different from the one for the x, y displacements, and adjusted it to give approximately the same value as the first-principles calculations for the highest mode at Γ.

One glaring difference, which *cannot* be fixed for any choice of parameters in the Born force-constant model, is the behavior of the out-of-plane acoustic modes. Indeed, this behavior is peculiar. All three acoustic frequencies in the force-constant model are linear in the wave-vector magnitude $|\mathbf{k}|$ for small values, as for all other examples we have discussed in this chapter. In contrast to this, the lowest acoustic frequency from the first-principles calculations is *quadratic* in the wave-vector magnitude $|\mathbf{k}|$. This unusual behavior is due to the special nature of the graphene 2D sheet. Specifically, in the real material the out-of-plane motion of atoms is coupled to the in-plane motion; in the force-constant model, no such coupling is allowed, as a consequence of the fact that there are no out-of-plane bonds. Any attempt to capture this behavior within the Born force-constant model, for example, by including next-nearest-neighbor interactions as if there were effective in-plane bonds between them, is doomed to failure.[9] But because there are no atoms to inhibit the displacement in the direction perpendicular to the plane, the effect of the coupling between

---

[9] To explore this issue, the reader is guided through the steps of including second-neighbor interactions in the force-constant model (shown schematically in Fig. 7.14) in Problem 13

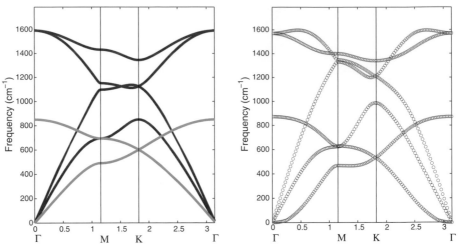

Fig. 7.15   The phonon spectrum of the honeycomb lattice (graphene) in 3D space. **Left**: Born force-constant model with first-neighbor interactions only. The blue bands involve motion on the plane of the lattice (x, y), the green bands involve motion perpendicular to the plane (z-axis). **Right**: The phonon bands of graphene (red dots) as obtained from first-principles calculations based on DFT (performed by D. T. Larson).

in-plane and out-of-plane motion is second order, and therefore the dependence of the frequency on the wave-vector is higher order than linear.

The quadratic behavior of the lowest acoustic mode is nicely captured by the analysis of the thin plate equilibrium within elasticity theory, discussed in the previous section. From Eq. (7.130), to describe the dynamical behavior of the plate we set the external forces per unit area f_z equal to zero and let the total force per unit area be equal to the mass per unit area times acceleration in the z direction:

$$-\frac{Yh^3}{12(1 - v^2)}\left(\frac{\partial^2 w}{\partial x^2} + \frac{\partial^2 w}{\partial y^2}\right)^2 = \rho h \frac{\partial^2 w}{\partial t^2}$$

where ρ is the mass density per unit volume. We define the constant

$$\Lambda = \left(\frac{Yh^2}{12\rho(1 - v^2)}\right)^{1/2}$$

and introduce wave-like harmonic behavior for the variable w in both space and time:

$$w(x, y, t) = w_0 e^{i\mathbf{k}\cdot\mathbf{r}} e^{-i\omega t}$$

with $\mathbf{k}$ and $\mathbf{r}$ being 2D vectors with x, y components only:

$$\mathbf{r} = x\hat{x} + y\hat{y}, \quad \mathbf{k} = k_x\hat{x} + k_y\hat{y}$$

Substituting the expression for $w(x, y, t)$ in the equation for the dynamical response of the plate, we find that the frequency and the wave-vector must satisfy the relation

$$\omega^2 = \Lambda^2|\mathbf{k}|^4 \Rightarrow \omega = \pm\Lambda|\mathbf{k}|^2$$

This result explains the quadratic behavior of the frequency related to the out-of-plane motion for small wave-vector magnitude. This type of motion is referred to as the "flexural mode" of graphene. The curvature of ω for small $|\mathbf{k}|$ is determined by the in-plane Young's modulus Y and Poisson's ratio ν, as well as the thickness of graphene h and its mass density ρ, all of which enter in the constant Λ.

Further Reading

1. *The Mathematical Theory of Elasticity*, A. E. H. Love (Cambridge University Press, Cambridge, 1927). This is a classic book on the theory of elasticity, with advanced treatments of many important topics.
2. *Elasticity: Tensor, Dyadic and Engineering Approaches*, P. C. Cou and N. J. Pagano (Van Nostrand, Princeton, NJ, 1967). This is an accessible and concise treatment of the theory of elasticity.
3. *Theory of Elasticity*, 3rd English edition, L. D. Landau and E. M. Lifshitz, translated by J. B. Sykes and W. H. Reid (Pergamon Press, New York, 1986). A classic text with in-depth treatment of many advanced topics.
4. *Modeling Materials: Continuum, Atomistic and Multiscale Techniques*, E. B. Tadmor and R. E. Miller (Cambridge University Press, Cambridge, 2011). This is a modern treatment of many important topics in modeling materials, with a very concise account of elasticity theory.
5. *Theory of Dislocations*, J. P. Hirth and J. Lothe (Krieger, Malabar, 1992). This is a standard reference for the physics of dislocations, which are defects responsible for the behavior of solids in the plastic regime, that is, beyond the elastic range.

Problems

1. Use the Born force-constant model to calculate the phonon frequencies for silicon along the $L-\Gamma-X$ directions, where $L = (\pi/a)(1, 1, 1)$ and $X = (2\pi/a)(1, 0, 0)$. Take the ratio of the bond-stretching and bond-bending force constants to be $\kappa_r/\kappa_\theta = 16$. Fit the value of κ_r to reproduce the experimental value for the highest optical mode at Γ, which is 15.53 THz, and use this value to obtain the frequencies of the various modes at X; compare these to the values given in Table 7.9. Determine the atomic displacements for the normal modes at Γ, the lowest acoustic branch at X and L, and the highest optical branch at X and L.

2. Show that the kinetic and potential energy contribution to the energy of a simple harmonic oscillator are equal. Show explicitly (that is, not by invoking the analogy to a set of independent harmonic oscillators) that the same holds for the kinetic and potential energy contributions to the energy of a collection of phonons, as given in Eq. (7.42).

coupling A_x to A_x, A_y to A_y, and A_z to A_z, respectively (and similarly for the B sites), and the off-diagonal matrix element

$$H(\mathbf{k}) = -\frac{\sqrt{3}}{2}\left(\kappa_r^{(2)} - \kappa_\theta^{(2)}\right)[\cos(\mathbf{k}\cdot\mathbf{a}_1) - \cos(\mathbf{k}\cdot\mathbf{a}_2)]$$

coupling A_x to A_y and B_x to B_y. Find values of the new parameters $\kappa_r^{(2)}$, $\kappa_\theta^{(2)}$ to obtain as close a match as possible to the DFT results shown in Fig. 7.15. Comment on which features of the phonon dispersion can be affected by the inclusion of second-neighbor couplings and which cannot.

8 Phonon Interactions

In the previous chapter we considered the quantized vibrations of ions in a crystalline solid, the phonons, as independent "particles." We used this concept to investigate some of the properties of solids related to ionic vibrations, like the specific heat and the thermal expansion coefficient. In this chapter we look at the interaction of phonons with other quasiparticles, like electrons, and other particles, like neutrons, as well as with photons, the quanta of the electromagnetic field.

8.1 Phonon Scattering Processes

The presence of phonons in a crystal is manifest, and the phonon properties can be measured, by the scattering of particles or waves when they interact with the excited phonons. A common method for measuring phonon frequencies is by scattering of neutrons, which are electrically neutral particles of mass comparable to that of ions. We discuss first the example of phonon scattering by particles, and then describe in more detail the theoretical framework for calculating scattering processes and their effects.

The hamiltonian for the system of ions in an infinite crystal and a particle being scattered by those ions is given by:

$$\hat{\mathcal{H}} = \sum_n \frac{\hat{\mathbf{P}}_n^2}{2M_n} + \mathcal{V}(\{\mathbf{R}_n + \mathbf{S}_n\}) + \frac{\hat{\mathbf{p}}^2}{2m} + + \sum_n v(\mathbf{r} - (\mathbf{R}_n + \mathbf{S}_n)) \qquad (8.1)$$

where $\mathbf{R}_n$ are the positions of ions in the ideal crystal, $\mathbf{S}_n \equiv \delta\mathbf{R}_n$ are their displacements due to phonons, as expressed in Eq. (7.37), and $\hat{\mathbf{P}}_n$ the momentum operators for the ionic coordinates; $\mathbf{r}$ and $\hat{\mathbf{p}}$ are the position and momentum operator for the scattering particle, m its mass, and $v(\mathbf{r})$ the potential through which it interacts with the ions of the solid. The potential energy term $\mathcal{V}(\{\mathbf{R}_n + \mathbf{S}_n\})$ depends on all the ionic coordinates and represents the interactions among ions. The problem now can be posed as follows: given the initial and final states of the scattering particles, characterized by energy E_1, E_2 and momentum $\mathbf{p}_1$, $\mathbf{p}_2$, and the initial state of the crystal, characterized by the set of phonon occupation numbers $\{n_{\mathbf{k},1}^{(l)}\}$, what are the possible final states of the crystal, which will be characterized by the new set of phonon occupation numbers $\{n_{\mathbf{k},2}^{(l)}\}$? Each $n_{\mathbf{k},s}^{(l)}$ represents the number of phonons excited in the state with wave-vector $\mathbf{k}$ of the lth phonon band, for the state of the entire system labeled s. We define the change in occupation of each phonon state as:

$$\Delta n_{\mathbf{k}}^{(l)} \equiv n_{\mathbf{k},2}^{(l)} - n_{\mathbf{k},1}^{(l)} \qquad (8.2)$$

Conservation of energy in the system gives:

$$E_1 + \sum_{\mathbf{k},l} \hbar\omega_{\mathbf{k}}^{(l)} \left(n_{\mathbf{k},1}^{(l)} + \frac{1}{2} \right) = E_2 + \sum_{\mathbf{k},l} \hbar\omega_{\mathbf{k}}^{(l)} \left(n_{\mathbf{k},2}^{(l)} + \frac{1}{2} \right)$$

and using the definition of $\Delta n_{\mathbf{k}}^{(l)}$, Eq. (8.2), we express the change in energy as:

$$E_2 - E_1 = -\sum_{\mathbf{k},l} \hbar\omega_{\mathbf{k}}^{(l)} \Delta n_{\mathbf{k}}^{(l)} \tag{8.3}$$

It takes a slightly more elaborate treatment to establish the proper expression for the change in momentum of the particle, because we have to take into account the translational symmetries of the system that are related to momentum conservation. Specifically, by following similar steps as in the proof of Bloch's theorem (see Chapter 2), we can establish that the hamiltonian of Eq. (8.1) commutes with the translational operator for changing the particle position by a lattice vector $\mathbf{R}$, because the particle–crystal interaction is invariant under such a change. This implies that the corresponding translational operator is given by:

$$\hat{\mathcal{R}}_{\mathbf{R}} = e^{-i\mathbf{R}\cdot\hat{\mathbf{p}}/\hbar}$$

An additional symmetry comes from permutations between ionic positions, due to the indistinguishability of these positions within the crystal. The corresponding operator is:

$$\hat{\mathcal{P}}_{\mathbf{R}} = e^{-i\mathbf{R}\cdot\mathbf{K}_s}$$

where the new quantity introduced, $\mathbf{K}_s$, is the wave-vector for the entire crystal in the phonon state labeled s, which is determined by the phonon occupation numbers $n_{\mathbf{k},s}^{(l)}$ and is given by:

$$\mathbf{K}_s = \sum_{\mathbf{k},l} \mathbf{k}\, n_{\mathbf{k},s}^{(l)}$$

This is in close analogy to the wave-vector of individual electrons that experience the effects of lattice periodicity, as discussed in detail in Chapter 2. The total translational symmetry of the system is then described by the combined operator

$$\hat{\mathcal{R}}_{\mathbf{R}}\hat{\mathcal{P}}_{\mathbf{R}} = e^{-i\mathbf{R}\cdot(\hat{\mathbf{p}}/\hbar + \mathbf{K}_s)}$$

Thus, we can use the eigenfunctions of this operator as the basis for the eigenfunctions of the hamiltonian of the system, which ensures that the eigenvalues of this operator, $\mathbf{p}/\hbar + \mathbf{K}_s$, will be conserved. This leads to the following relation for the momentum in the system at the initial and final states:

$$\frac{\mathbf{p}_1}{\hbar} + \sum_{\mathbf{k},l} \mathbf{k}\, n_{\mathbf{k},1}^{(l)} + \mathbf{G}_1 = \frac{\mathbf{p}_2}{\hbar} + \sum_{\mathbf{k},l} \mathbf{k}\, n_{\mathbf{k},2}^{(l)} + \mathbf{G}_2 \Rightarrow \frac{\mathbf{p}_1}{\hbar} + \sum_{\mathbf{k},l} \mathbf{k}\, n_{\mathbf{k},1}^{(l)}$$

$$= \frac{\mathbf{p}_2}{\hbar} + \sum_{\mathbf{k},l} \mathbf{k}\, n_{\mathbf{k},2}^{(l)} - \mathbf{G}$$

where we have taken advantage of the relation $\exp(i\mathbf{R}\cdot\mathbf{G}) = 1$, which holds for any Bravais lattice vector $\mathbf{R}$ and any reciprocal-space lattice vector $\mathbf{G}$, to introduce an arbitrary lattice vector $\mathbf{G}_1$, $\mathbf{G}_2$ to each side of the equation, but their difference is simply another reciprocal-space lattice vector: $\mathbf{G} = \mathbf{G}_1 - \mathbf{G}_2$. This result then leads to the following expression for the momentum difference between initial and final states:

$$\mathbf{p}_2 - \mathbf{p}_1 = -\sum_{\mathbf{k},l} \hbar\mathbf{k} \, \Delta n_{\mathbf{k}}^{(l)} + \hbar\mathbf{G} \tag{8.4}$$

Example 8.1   Neutron scattering by phonons

We apply the general expressions derived above to analyze what happens in neutron-scattering experiments that probe the phonon properties of solids. Neutrons are typically scattered by atomic nuclei in the solid through inelastic collisions that involve changes in the energy and momentum of the neutron, which imply a change in the number of phonons that are excited in the solid; this change number of phonons is determined by Eqs (8.3) and (8.4). If we assume that experimental conditions can be tuned to involve a single phonon, corresponding to $\Delta n_{\mathbf{k}}^{(l)} = \pm 1$, then we will have for the energy and momentum before ($E_1, \mathbf{p}_1$) and after ($E_2, \mathbf{p}_2$) a scattering event:

$$\text{phonon absorption}: \quad E_2 = E_1 + \hbar\omega_{\mathbf{k}}^{(l)}, \quad \mathbf{p}_2 = \mathbf{p}_1 + \hbar(\mathbf{k} + \mathbf{G})$$

$$\text{phonon emission}: \quad E_2 = E_1 - \hbar\omega_{\mathbf{k}}^{(l)}, \quad \mathbf{p}_2 = \mathbf{p}_1 - \hbar(\mathbf{k} + \mathbf{G})$$

Using the fact that $\omega_{\mathbf{k}\pm\mathbf{G}}^{(l)} = \omega_{\mathbf{k}}^{(l)}$, we obtain the following relation from conservation of energy:

$$\frac{\mathbf{p}_2^2}{2m_{\mathrm{n}}} = \frac{\mathbf{p}_1^2}{2m_{\mathrm{n}}} \pm \hbar\omega_{(\mathbf{p}_2-\mathbf{p}_1)/\hbar}^{(l)} \tag{8.5}$$

with $(+)$ corresponding to absorption and $(-)$ to emission of a phonon, and m_{n} being the neutron mass. In experiment, $\mathbf{p}_1$ is the momentum of the incident neutron beam and $\mathbf{p}_2$ the momentum of the scattered beam. In the situation we are considering, Eq. (8.5) has solutions only if the second term on the right-hand side corresponds to a phonon frequency at some value of $\mathbf{k}$ in the BZ:

$$\omega_{(\mathbf{p}_2-\mathbf{p}_1)/\hbar}^{(l)} = \omega_{\mathbf{k}}^{(l)}$$

In reality it is impossible to separate the single-phonon events from those that involve many phonons. Thus, for every value of the energy and momentum of neutrons there will be a broad background, corresponding to multi-phonon processes. However, the flux of scattered neutrons as a function of their energy, $\mathbf{p}_2^2/2m_{\mathrm{n}}$, will exhibit sharp peaks in certain directions $\hat{\mathbf{p}}_2$ which correspond to phonon frequencies that satisfy Eq. (8.5). From these peaks, one determines the phonon spectrum by scanning the energy of the scattered neutron beam along different directions.

8.1.1 Scattering Formalism

To describe phonon interactions in a more general context, we consider the scattering of a quantum-mechanical particle, modeled by a plane-wave, from a solid in the presence of phonons. This particle can be an electron, a photon, or even another phonon; we explore the scattering of these three types of particles in more detail in subsequent sections.

We denote the incident wave-vector of the particle by $\mathbf{k}$ and the scattered wave-vector by $\mathbf{k}'$. The matrix element $\mathcal{M}_{\mathbf{k},\mathbf{k}'}$ for the scattering process will be given by the expectation value of the scattering potential, that is, the potential of all the ions in the crystal, $\tilde{\mathcal{V}}_{\text{xtl}}(\mathbf{r})$, between initial and final states:

$$\mathcal{M}_{\mathbf{k},\mathbf{k}'} = \frac{1}{N V_{\text{PUC}}} \int e^{-i\mathbf{k}'\cdot\mathbf{r}} \tilde{\mathcal{V}}_{\text{xtl}}(\mathbf{r}) e^{i\mathbf{k}\cdot\mathbf{r}} d\mathbf{r} , \quad \tilde{\mathcal{V}}_{\text{xtl}}(\mathbf{r}) = \sum_{nj} \mathcal{V}_{\text{ion}}(\mathbf{r} - \mathbf{t}_j - \mathbf{R}_n)$$

where $\tilde{\mathcal{V}}_{\text{ion}}(\mathbf{r} - \mathbf{t}_j - \mathbf{R}_n)$ is the potential of an ion situated at $\mathbf{t}_j$ in the unit cell of the crystal at the lattice vector $\mathbf{R}_n$; N is the total number of unit cells in the crystal, each of volume V_{PUC}. For simplicity, we will assume that the solid contains only one type of ions, so there is only one type of ionic potential $\tilde{\mathcal{V}}_{\text{ion}}(\mathbf{r})$. Inserting the expression for the crystal potential in the scattering matrix element, and defining $\mathbf{q} = \mathbf{k}' - \mathbf{k}$, we obtain:

$$\mathcal{M}_{\mathbf{k},\mathbf{k}+\mathbf{q}} = \frac{1}{N} \sum_n \tilde{\mathcal{V}}_{\text{ion}}(\mathbf{q}) \sum_j e^{-i\mathbf{q}\cdot(\mathbf{t}_j+\mathbf{R}_n)} , \quad \tilde{\mathcal{V}}_{\text{ion}}(\mathbf{q}) = \frac{1}{V_{\text{PUC}}} \int \tilde{\mathcal{V}}_{\text{ion}}(\mathbf{r}) e^{-i\mathbf{q}\cdot\mathbf{r}} d\mathbf{r}$$

with $\tilde{\mathcal{V}}_{\text{ion}}(\mathbf{q})$ the Fourier transform of the ionic potential $\tilde{\mathcal{V}}_{\text{ion}}(\mathbf{r})$. If $\mathbf{q}$ happens to be equal to a reciprocal lattice vector $\mathbf{G}$, then the scattering matrix element takes the form

$$\mathcal{M}_{\mathbf{k},\mathbf{k}+\mathbf{G}} = \frac{1}{N} \sum_n \tilde{\mathcal{V}}_{\text{ion}}(\mathbf{G}) \sum_j e^{-i\mathbf{G}\cdot\mathbf{t}_j} = \tilde{\mathcal{V}}_{\text{ion}}(\mathbf{G}) \mathcal{S}(\mathbf{G})$$

where we identified the sum over the ions in the unit cell j as the structure factor $\mathcal{S}(\mathbf{G})$, the quantity we had defined in Chapter 2, Eq. (2.106); the other sum over the crystal cells n is canceled in this case by the factor $1/N$. If $\mathbf{q} \neq \mathbf{G}$, we can generalize the definition of the structure factor as follows:

$$\mathcal{S}(\mathbf{q}) \equiv \frac{1}{N} \sum_{nj} e^{-i\mathbf{q}\cdot(\mathbf{t}_j+\mathbf{R}_n)}$$

in which case the scattering matrix element takes the form

$$\mathcal{M}_{\mathbf{k},\mathbf{k}+\mathbf{q}} = \tilde{\mathcal{V}}_{\text{ion}}(\mathbf{q}) \mathcal{S}(\mathbf{q})$$

We are now interested in determining the behavior of this matrix element when, due to thermal motion, the ions are not at their ideal crystalline positions. Obviously, only the structure factor $\mathcal{S}(\mathbf{q})$ is affected by this departure of the ions from their ideal positions, so we examine its behavior. The thermal motion leads to deviations from the ideal crystal positions, which we denote by $\mathbf{s}_{nj}$. With these deviations, the structure factor takes the form

$$\mathcal{S}(\mathbf{q}) = \frac{1}{N} \sum_{n,j} e^{-i\mathbf{q}\cdot(\mathbf{t}_j+\mathbf{R}_n)} e^{-i\mathbf{q}\cdot\mathbf{s}_{nj}}$$

For simplicity, we will assume from now on that there is only one ion in each unit cell, which allows us to eliminate the summation over the index j. We can express the deviations of the ions from their crystalline positions $\mathbf{S}_n$ in terms of the amplitudes of phonons $Q_{\mathbf{q}'}^{(l)}$

and the corresponding eigenvectors $\hat{\mathbf{e}}_{\mathbf{q}'}^{(l)}$, with l denoting the phonon branch, which gives for the structure factor:

$$\mathbf{S}_n = \sum_{l,\mathbf{q}'} Q_{\mathbf{q}'}^{(l)} \hat{\mathbf{e}}_{\mathbf{q}'}^{(l)} e^{i\mathbf{q}'\cdot\mathbf{R}_n} \Rightarrow \mathcal{S}(\mathbf{q}) = \frac{1}{N} \sum_n e^{-i\mathbf{q}\cdot\mathbf{R}_n} \exp\left[-i\mathbf{q}\cdot\sum_{l,\mathbf{q}'}(Q_{\mathbf{q}'}^{(l)}\hat{\mathbf{e}}_{\mathbf{q}'}^{(l)})e^{i\mathbf{q}'\cdot\mathbf{R}_n}\right] \quad (8.6)$$

We will take the deviations from ideal positions to be small quantities and the corresponding phonon amplitudes to be small. To simplify the calculations, we define the vectors $\mathbf{f}_{\mathbf{q}'n}^{(l)}$ as:

$$\mathbf{f}_{n\mathbf{q}'}^{(l)} = Q_{\mathbf{q}'}^{(l)}\hat{\mathbf{e}}_{\mathbf{q}'}^{(l)} e^{-i\mathbf{q}'\cdot\mathbf{R}_n} \quad (8.7)$$

which, by our assumption above, will have small magnitude. Using this fact, we can expand the exponential with square brackets in Eq. (8.6) as follows:

$$\exp\left[-i\mathbf{q}\cdot\sum_{l,\mathbf{q}'}\mathbf{f}_{n\mathbf{q}'}^{(l)}\right] = \prod_{l,\mathbf{q}'}\exp\left[-i\mathbf{q}\cdot\mathbf{f}_{n\mathbf{q}'}^{(l)}\right] = \prod_{l,\mathbf{q}'} = \left[1 - i\mathbf{q}\cdot\mathbf{f}_{n\mathbf{q}'}^{(l)} - \frac{1}{2}\left(\mathbf{q}\cdot\mathbf{f}_{n\mathbf{q}'}^{(l)}\right)^2 + \cdots\right]$$

Keeping only terms up to second order in $|\mathbf{f}_{n\mathbf{q}'}^{(l)}|$ in the last expression gives the following result:

$$\left[1 - i\mathbf{q}\cdot\sum_{l,\mathbf{q}'}\mathbf{f}_{n\mathbf{q}'}^{(l)} - \frac{1}{2}\sum_{l,\mathbf{q}'}\left(\mathbf{q}\cdot\mathbf{f}_{n\mathbf{q}'}^{(l)}\right)^2 - \sum_{l\mathbf{q}'<l'\mathbf{q}''}\left(\mathbf{q}\cdot\mathbf{f}_{n\mathbf{q}'}^{(l)}\right)\left(\mathbf{q}\cdot\mathbf{f}_{n\mathbf{q}''}^{(l')}\right)\right]$$

$$= \left[1 - i\mathbf{q}\cdot\sum_{l,\mathbf{q}'}\mathbf{f}_{n\mathbf{q}'}^{(l)} - \frac{1}{2}\sum_{ll',\mathbf{q}'\mathbf{q}''}\left(\mathbf{q}\cdot\mathbf{f}_{n\mathbf{q}'}^{(l)}\right)\left(\mathbf{q}\cdot\mathbf{f}_{n\mathbf{q}''}^{(l')}\right)\right] \quad (8.8)$$

Let us consider the physical meaning of these terms by order, when the above expression is substituted in Eq. (8.6).

(0) The zeroth-order term in Eq. (8.8) gives:

$$\mathcal{S}_0(\mathbf{q}) = \frac{1}{N}\sum_n e^{-i\mathbf{q}\cdot\mathbf{R}_n}$$

which is the structure factor for a crystal with the ionic positions frozen at the ideal crystal sites, that is, in the absence of any phonon excitations.

(1) The first-order term in Eq. (8.8), omitting the overall minus sign, gives:

$$i\mathbf{q}\cdot\sum_{\mathbf{q}'l}\frac{1}{N}\sum_n \mathbf{f}_{n\mathbf{q}'}^{(l)} e^{-i\mathbf{q}\cdot\mathbf{R}_n} = i\mathbf{q}\cdot\sum_{\mathbf{q}'l}\frac{1}{N}\sum_n Q_{\mathbf{q}'}^{(l)}\hat{\mathbf{e}}_{\mathbf{q}'}^{(l)} e^{i(\mathbf{q}'-\mathbf{q})\cdot\mathbf{R}_n}$$

$$= i\mathbf{q}\cdot\sum_{\mathbf{q}'l} Q_{\mathbf{q}'}^{(l)}\hat{\mathbf{e}}_{\mathbf{q}'}^{(l)}\delta(\mathbf{q}'-\mathbf{q}-\mathbf{G}) = i\sum_l(\mathbf{q}\cdot\hat{\mathbf{e}}_{\mathbf{q}+\mathbf{G}}^{(l)})Q_{\mathbf{q}+\mathbf{G}}^{(l)}$$

This is an interesting result: it corresponds to the scattering from a single phonon mode, with wave-vector $\mathbf{q}$ and amplitude $Q_{\mathbf{q}+\mathbf{G}}^{(l)}$, and involves the projection of the scattering wave-vector $\mathbf{q}$ onto the polarization $\hat{\mathbf{e}}_{\mathbf{q}+\mathbf{G}}^{(l)}$ of this phonon. If $\mathbf{G} = 0$, these processes are called normal single-phonon scattering processes; if $\mathbf{G} \neq 0$, they are called Umklapp processes. The latter usually contribute less to scattering, so we will ignore them in the following.

(2) The second-order term in Eq. (8.8), again omitting the overall minus sign, gives:

$$\sum_{ll'\mathbf{q}'\mathbf{q}''}\sum_n (\mathbf{q}\cdot\mathbf{f}_{n\mathbf{q}'}^{(l)})(\mathbf{q}\cdot\mathbf{f}_{n\mathbf{q}''}^{(l')})\frac{e^{-i\mathbf{q}\cdot\mathbf{R}_n}}{2N} = \sum_{ll'\mathbf{q}'\mathbf{q}''}(\mathbf{q}\cdot Q_{\mathbf{q}'}^{(l)}\hat{\mathbf{e}}_{\mathbf{q}'}^{(l)})(\mathbf{q}\cdot Q_{\mathbf{q}''}^{(l')}\hat{\mathbf{e}}_{\mathbf{q}''}^{(l')})\sum_n \frac{e^{i(\mathbf{q}'+\mathbf{q}''-\mathbf{q})\cdot\mathbf{R}_n}}{2N}$$

$$= \frac{1}{2}\sum_{ll'\mathbf{q}'\mathbf{q}''}(\mathbf{q}\cdot Q_{\mathbf{q}'}^{(l)}\hat{\mathbf{e}}_{\mathbf{q}'}^{(l)})(\mathbf{q}\cdot Q_{\mathbf{q}''}^{(l')}\hat{\mathbf{e}}_{\mathbf{q}''}^{(l')})\delta(\mathbf{q}'+\mathbf{q}''-\mathbf{q}) = \frac{1}{2}\sum_{ll'\mathbf{q}'}(\mathbf{q}\cdot Q_{\mathbf{q}'}^{(l)}\hat{\mathbf{e}}_{\mathbf{q}'}^{(l)})(\mathbf{q}\cdot Q_{\mathbf{q}-\mathbf{q}'}^{(l')}\hat{\mathbf{e}}_{\mathbf{q}-\mathbf{q}'}^{(l')})$$

This expression can be interpreted as the scattering from two phonons, one of wave-vector $\mathbf{q}'$, the other of wave-vector $\mathbf{q}-\mathbf{q}'$; the corresponding phonon amplitudes and projections of the scattering wave-vector onto the phonon polarizations are also involved.

This set of processes can be represented in a more graphical manner in terms of diagrams, as illustrated in Fig. 8.1. The incident and scattered wave-vectors are represented by normal vectors, while the wave-vectors of the phonons are represented by wavy lines of the proper direction and magnitude to satisfy momentum conservation by vector addition. Simple rules can be devised to make the connection between such diagrams and the expressions in the equations above that give their contribution to the structure factor, and hence the scattering cross-section. For example, the following simple rules would generate the terms we calculated above from the diagrams shown in Fig. 8.1:

(i) Every vertex where a phonon wave-vector $\mathbf{q}'$ intersects other wave-vectors introduces a factor of $(-i)\sum_l Q_{\mathbf{q}'}^{(l)}\mathbf{q}\cdot\hat{\mathbf{e}}_{\mathbf{q}'}^{(l)}$, with $\mathbf{q}$ the total scattering wave-vector.
(ii) A factor of $(1/m!)$ accompanies the term of order m in $\mathbf{q}$.

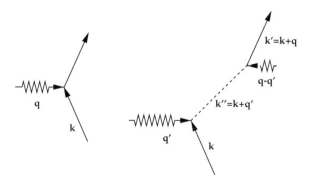

Fig. 8.1   Diagrams for the one and two-phonon scattering processes: $\mathbf{k}$ is the wave-vector of the incoming particle, $\mathbf{k}'$ the wave-vector of the outgoing particle, and $\mathbf{q},\mathbf{q}',\mathbf{q}''=\mathbf{q}-\mathbf{q}'$ are the wave-vectors of the phonons (represented by the zigzag lines) that scatter from the particle. The solid lines represent actual particle states, the dashed line labeled as $\mathbf{k}''$ represents virtual states. At each vertex of the interaction, momentum is conserved, leading to the relations between wave-vectors that are indicated.

(iii) When an intermediate phonon wave-vector $\mathbf{q}'$, other than the final scattering phonon wave-vector $\mathbf{q}$, appears in a diagram, it is accompanied by a sum over all allowed values.

These rules can be applied to generate diagrams of higher order, which we have not discussed here. The use of diagrammatic techniques simplifies many calculations of the type outlined above. For perturbative calculations in many-body interacting systems, such techniques are indispensable.

8.2 Application: The Debye–Waller Factor

In neutron-scattering experiments, the width of peaks in the elastic scattering at any value of $\mathbf{G}$ does not broaden with increasing temperature, but the peak heights decrease by a factor $\exp[-2W]$, where W is a quantity related to the phonon properties. This quantity is known as the Debye–Waller factor. It is a measure of the effect of thermal motion in reducing the apparent periodicity of the lattice.

For an arbitrary state of the system we are not interested in individual phonon-scattering processes, but rather in the thermal average over such events. We are also typically interested in the absolute value squared of the scattering matrix element, which enters in the expression for the cross-section, the latter being the experimentally measurable quantity. When we average over the sums that appear in the absolute value squared of the structure factor, the contribution that survives is:

$$\left\langle |\mathcal{S}(\mathbf{q})|^2 \right\rangle = |\mathcal{S}_0(\mathbf{q})|^2 \left[1 - \frac{1}{N} \sum_{l\mathbf{q}'} \left\langle \left| \mathbf{q} \cdot \mathbf{f}_{n\mathbf{q}'}^{(l)} \right|^2 \right\rangle \right] \tag{8.9}$$

Taking advantage again of the fact that the magnitude of $\mathbf{f}_{n\mathbf{q}}^{(l)}$ vectors is small, we can approximate the factor in square brackets by:

$$\mathrm{e}^{-2W} \approx \left[1 - \frac{1}{N} \sum_{l\mathbf{q}'} \left\langle \left| \mathbf{q} \cdot \mathbf{f}_{n\mathbf{q}'}^{(l)} \right|^2 \right\rangle \right] \Rightarrow W = \frac{1}{2N} \sum_{l\mathbf{q}'} \left\langle \left| \mathbf{q} \cdot \mathbf{f}_{n\mathbf{q}'}^{(l)} \right|^2 \right\rangle$$

This relation defines the Debye–Waller factor W. In this expression, the index n of $\mathbf{f}_{n\mathbf{q}'}^{(l)}$ is irrelevant, since it only appears in the complex exponential $\exp[-i\mathbf{q}' \cdot \mathbf{R}_n]$, which is eliminated by the absolute value; thus, W contains no dependence on n. Substituting the expression of $\mathbf{f}_{n\mathbf{q}}^{(l)}$ from Eq. (8.7), we find:

$$W = \frac{1}{2N} \sum_{l\mathbf{q}'} \left\langle \left| \mathbf{q} \cdot Q_{\mathbf{q}'}^{(l)} \hat{\mathbf{e}}_{\mathbf{q}'}^{(l)} \right|^2 \right\rangle$$

We will assume next that the amplitudes $|Q_{\mathbf{q}'}^{(l)}|^2$ are independent of the phonon mode l, which leads to:

$$\sum_{l\mathbf{q}'}\left\langle \left| \mathbf{q} \cdot Q_{\mathbf{q}'}^{(l)}\hat{\mathbf{e}}_{\mathbf{q}'}^{(l)} \right|^2 \right\rangle = \sum_{\mathbf{q}'}\left\langle |Q_{\mathbf{q}'}|^2 \right\rangle \sum_{l}|\mathbf{q} \cdot \hat{\mathbf{e}}_{\mathbf{q}'}^{(l)}|^2 = |\mathbf{q}|^2 \sum_{\mathbf{q}'}\left\langle |Q_{\mathbf{q}'}|^2 \right\rangle \tag{8.10}$$

because in our simple example of one ion per unit cell the polarization vectors $\hat{\mathbf{e}}_{\mathbf{k}}^{(l)}$ cover the same space as the cartesian coordinates, therefore:

$$\sum_{l}|\mathbf{q} \cdot \hat{\mathbf{e}}_{\mathbf{k}}^{(l)}|^2 = |\mathbf{q}|^2$$

With these simplifications, W takes the form

$$W = \frac{1}{2N}|\mathbf{q}|^2 \sum_{\mathbf{q}'}\left\langle |Q_{\mathbf{q}'}|^2 \right\rangle$$

We invoke the same argument as in Eq. (7.76) to express $\left\langle Q_{\mathbf{q}'}^2 \right\rangle$ in terms of the corresponding occupation number $n_{\mathbf{q}'}$ and frequency $\omega_{\mathbf{q}'}$:

$$M\omega_{\mathbf{q}'}^2 \left\langle |Q_{\mathbf{q}'}|^2 \right\rangle = \left(n_{\mathbf{q}'} + \frac{1}{2} \right)\hbar\omega_{\mathbf{q}'} \Rightarrow \left\langle |Q_{\mathbf{q}'}|^2 \right\rangle = \frac{\hbar}{M\omega_{\mathbf{q}'}}\left(n_{\mathbf{q}'} + \frac{1}{2} \right)$$

with $\omega_{\mathbf{q}}$ the phonon frequencies, M the mass of the ions, and $n_{\mathbf{q}}$ the phonon occupation numbers. When this is substituted in the expression for W, it leads to:

$$W = \frac{1}{2N}|\mathbf{q}|^2 \sum_{\mathbf{q}'}\frac{\hbar}{M\omega_{\mathbf{q}'}}\left(n_{\mathbf{q}'} + \frac{1}{2} \right) = \frac{\hbar^2|\mathbf{q}|^2}{2MN}\sum_{\mathbf{q}'}\left(n_{\mathbf{q}'} + \frac{1}{2} \right)\frac{1}{\hbar\omega_{\mathbf{q}'}} \tag{8.11}$$

As usual, we will turn the sum over $\mathbf{q}'$ into an integral over the first BZ. We will also employ the Debye approximation in which all phonon frequencies are given by $\omega_{\mathbf{q}} = vk$ with the same average v; there is a maximum frequency ω_{D}, and related to it is the Debye temperature $k_{\mathrm{B}}\Theta_{\mathrm{D}} = \hbar\omega_{\mathrm{D}}$. This approximation leads to the following expression for the Debye–Waller factor:

$$W = \frac{3\hbar^2|\mathbf{q}|^2 T^2}{2Mk_{\mathrm{B}}\Theta_{\mathrm{D}}^3}\int_0^{\Theta_{\mathrm{D}}/T}\left(\frac{1}{e^t - 1} + \frac{1}{2} \right)t\,dt \tag{8.12}$$

with T the temperature. The limits of high temperature and low temperature are interesting. At high temperature, the upper limit of the integral in Eq. (8.12) is a very small quantity, so we can expand the exponential in the integrand in powers of t and evaluate the integral to obtain:

$$W_\infty \equiv \lim_{T \gg \Theta_{\mathrm{D}}} W = \frac{3\hbar^2|\mathbf{q}|^2 T}{2Mk_{\mathrm{B}}\Theta_{\mathrm{D}}^2}$$

which has a strong linear dependence on the temperature. Thus, at high enough temperature, when all the phonon modes are excited, the effect of the thermal motion will be manifest strongly in the structure factor which, in absolute value squared, will be multiplied by a factor

$$e^{-2W_\infty} = \exp\left[-\frac{3\hbar^2|\mathbf{q}|^2 T}{Mk_B\Theta_D^2}\right]$$

For low temperatures, the upper limit in the integral in Eq. (8.12) is a very large number; the exact result of the integration contains a complicated dependence on temperature, but we see that the term $\frac{1}{2}$ inside the bracket in the integrand is now significant. For $T \to 0$, this term provides the dominant contribution to the integral:

$$W_0 \equiv \lim_{T\to 0} W = \frac{3\hbar^2|\mathbf{q}|^2}{8Mk_B\Theta_D}$$

which is comparable to the value of W_∞ at $T = \Theta_D$. This suggests that even in the limit of zero temperature there will be a significant modification of the structure factor by $\exp[-2W_0]$ relative to its value for a frozen lattice of ions, due to the phonon degrees of freedom; this modification arises from the zero-point motion associated with phonons, that is, the $\frac{1}{2}$ term added to the phonon occupation numbers. This effect is another manifestation of the quantum-mechanical nature of phonon excitations in a crystal.

8.3 Phonon–Photon Interactions

The interaction of phonons with electromagnetic waves takes two different forms. The first couples the motion of ions directly to photons and leads to transitions between vibrational states of the lattice; the energy ($\hbar\omega$) of the electromagnetic radiation that couples directly to phonon modes is of order 0.5 eV, which corresponds to the infrared regions of the spectrum (see Fig. 8.2), so this process, typically absorption of radiation, is referred to as the "IR" signal. In a different form, light couples to phonon excitations indirectly, by first creating electronic excitations which then interact with (scatter from) phonon modes; for additional introductory discussion, see the sources mentioned in Further Reading. In the following we shall refer to the frequency of the incident radiation as ω_I and the frequency of the scattered radiation as ω_S. There are three possibilities when such a process takes place:

(a) the scattering is *elastic*, that is, the incident and scattered radiation has the same frequency, which is often referred to as "Rayleigh scattering" (originally defined for scattering of particles much smaller than the wavelength of light);
(b) the scattering is *inelastic*, and the scattered radiation has lower energy than the incoming one, which is called the "Stokes" excitation mechanism;
(c) the inelastic scattering leads to radiation with frequency higher than the incident one, which is called the "anti-Stokes" excitation mechanism.

The three types of processes are shown schematically in Fig. 8.2. Rayleigh scattering is the dominant process, but does not provide any useful information about the internal modes (phonons) of the system. For the Stokes mechanism, the system usually starts out in its

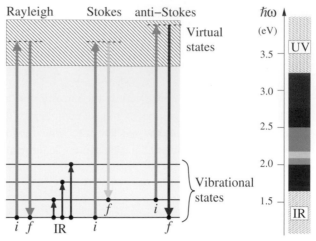

Fig. 8.2 Illustration of excitations involved in Rayleigh, Stokes, and anti-Stokes scattering, and infrared absorption; in each case the initial (i) and final (f) states are marked by small dots. The diagram on the right illustrates the range of photon energies $\hbar\omega$ (eV) for the visible, infrared (IR), and ultraviolet (UV) spectra.

lowest-energy state, from which it can make transitions to higher-energy virtual states, and then fall back into states that are higher in energy than the initial one; the difference in energy between final and initial vibrational states is a *positive* quantity, and is equal to the energy of the *emitted* phonon, $\hbar\omega_{\mathbf{q}}$, that makes the transition between the two vibrational states possible:

$$\text{Stokes}: \quad E_f - E_i = \hbar\omega_{\mathbf{q}} = \hbar(\omega_I - \omega_S) > 0$$

For the anti-Stokes mechanism, the system must start at a state with energy *higher* than the lowest one, so that after excitation to a virtual state it can return to a state of *lower* energy than the initial one; the difference in energy between final and initial states, in this case a *negative* quantity, is equal to the energy of the *absorbed* phonon that makes the transition between the two vibrational states possible:

$$\text{anti-Stokes}: \quad E_f - E_i = -\hbar\omega_{\mathbf{q}} = \hbar(\omega_I - \omega_S) < 0$$

In both cases, conservation of energy gives:

$$E_f + \hbar\omega_S = E_i + \hbar\omega_I \tag{8.13}$$

The virtual states involved in the inelastic processes will be discussed in more detail below. The inelastic processes are referred to as "Raman scattering"; they depend on the phonon modes of the solid and reveal information about these modes. The difference in frequency between incident and scattered radiation is referred to as the "Raman shift," and as the expressions above make clear, it is equal to the frequency of either an emitted or an absorbed phonon:

$$\Delta\omega = |\omega_I - \omega_S| = \omega_{\mathbf{q}}$$

8.3.1 Infrared Absorption

A quantitative description of IR absorption relies on the same concepts as the absorption of light by electronic excitations. The underlying physical mechanism is the coupling of the electromagnetic field to charged particles, which in this case are the ions of the solid that are undergoing collective motion described by the phonon modes. By analogy to our derivation of electron excitations (see Chapter 6), if a phonon mode has the characteristics of a charge distribution it can couple to the electromagnetic field the familiar interaction term in the hamiltonian, Eq. (6.2), which results in transition probabilities that involve matrix elements of the dipole operator [see Eq. (6.4)] between states of the system. In the present case, these are vibrational states of the system and the transitions between them are due to phonons. We will use the notation $\{Q\} = \{Q_1, Q_2, \ldots, Q_{3N}\}$ to represent all the phonon modes of the crystal with N ions. Generally, we are dealing with the system in its ground electronic state, which we denote with the superscript (0); for this case, we denote the initial and final vibrational states as $\phi_f^{(0)}(\{Q\})$, $\phi_i^{(0)}(\{Q\})$ and write the expectation value of the dipole operator between them as:

$$[\alpha_{fi}]_\kappa = \langle \phi_f^{(0)}(\{Q\})|\hat{d}_\kappa(\{Q\})|\phi_i^{(0)}(\{Q\})\rangle, \quad \kappa = x, y, z$$

In terms of this, the IR absorption intensity will be proportional to:

$$I_{fi} \sim \sum_\kappa |\langle \phi_f^{(0)}(\{Q\})|\hat{d}_\kappa(\{Q\})|\phi_i^{(0)}(\{Q\})\rangle|^2, \quad \kappa = x, y, z$$

We next expand the dipole moment in a Taylor expansion in terms of the variables Q_s which correspond to the various phonon modes, using a single index s to denote both the wave-vector $\mathbf{k}$ and the branch l of a phonon mode, and keeping only the lowest-order terms (up to linear in Q_s):

$$\hat{d}_\kappa(\{Q\}) = \hat{d}_\kappa^{(0)} + \sum_s \hat{d}_\kappa^{(s)} Q_s, \quad \hat{d}_\kappa^{(s)} = \left[\frac{\partial \hat{d}_\kappa(\{Q\})}{\partial Q_s} \right]_{\{Q\}=0}$$

which leads to:

$$\langle \phi_f^{(0)}(\{Q\})|\hat{d}_\kappa(\{Q\})|\phi_i^{(0)}(\{Q\})\rangle = \hat{d}_\kappa^{(0)} \langle \phi_f^{(0)}(\{Q\})|\phi_i^{(0)}(\{Q\})\rangle$$
$$+ \sum_s \hat{d}_\kappa^{(s)} \langle \phi_f^{(0)}(\{Q\})|Q_s|\phi_i^{(0)}(\{Q\})\rangle \qquad (8.14)$$

This expression allows us to figure out the contributions to the IR signal. For a transition between two different states, the first term vanishes since they must be orthogonal, $\langle \phi_f^{(0)}(\{Q\})|\phi_i^{(0)}(\{Q\})\rangle = 0$. In order to have a non-vanishing contribution, at least one of the $\kappa = x, y, z$ components must have both non-vanishing $\hat{d}_\kappa^{(s)}$ and non-vanishing matrix element $\langle \phi_f^{(0)}(\{Q\})|Q_s|\phi_i^{(0)}(\{Q\})\rangle$. In particular, the first condition tells us that the phonon mode described by Q_s must be related to *a change in the dipole moment*, such that $\hat{d}_\kappa^{(s)} \neq 0$, which is the most important characteristic. If, in addition, the corresponding matrix element $\langle \phi_f^{(0)}(\{Q\})|Q_s|\phi_i^{(0)}(\{Q\})\rangle \neq 0$, then we have identified what is referred to as an "IR active" mode; the value of the matrix element determines how strongly the IR active mode contributes to the signal.

8.3.2 Raman Scattering

Inelastic scattering of radiation by phonons was discovered experimentally by C. V. Raman in 1928. The theory of Raman scattering is based on the dispersion equation, referred to as the Kramers–Heisenberg–Dirac (KHD) theory. A first formulation that gave insight into the selection rules that govern Raman scattering was developed by J. H. Van Vleck.[1] The modern formulation of the theory was given by A. Albrecht,[2] which we follow here.

From KHD theory, the scattering cross-section σ_{fi} between an initial and a final state denoted by i and f, as a function of the incident radiation frequency ω_I and normalized by the incident intensity, is given by:

$$\sigma_{fi}(\omega_I) = \frac{8\pi}{9c^4}(\omega_I \pm \omega_\mathbf{q})^4 \sum_{\kappa,\lambda} \left|[\alpha_{fi}]_{\kappa\lambda}(\omega_I)\right|^2, \quad \{\kappa,\lambda\} = \{x, y, z\}$$

where $\omega_\mathbf{q}$ is the frequency of the phonon involved in the scattering and $\boldsymbol{\alpha}_{fi}(\omega_I)$ is the transition amplitude (a second-rank tensor). The two different signs of the $\omega_\mathbf{q}$ term correspond to the Stokes and anti-Stokes contributions, mentioned earlier. The elements of the transition amplitude are:

$$[\alpha_{fi}]_{\kappa\lambda}(\omega_I) = \sum_j \sum_n \frac{[\mathcal{M}_{fn,0j}]_\kappa [\mathcal{M}_{ni,j0}]_\lambda}{E_{nj} - E_{i0} - \hbar\omega_I + i\eta} + \frac{[\mathcal{M}_{fn,0j}]_\lambda [\mathcal{M}_{ni,j0}]_\kappa}{E_{nj} - E_{f0} + \hbar\omega_I + i\eta} \tag{8.15}$$

where the matrix elements that appear in the numerators under the double summation are:

$$[\mathcal{M}_{ni,j0}]_\lambda = \langle \phi_n^{(j)}(\{Q\})\psi_j(\{Q\};\{\mathbf{r}\}) | \hat{d}_\lambda | \phi_i^{(0)}(\{Q\})\psi_0(\{Q\};\{\mathbf{r}\}) \rangle \tag{8.16}$$

$$[\mathcal{M}_{fn,0j}]_\kappa = \langle \phi_f^{(0)}(\{Q\})\psi_0(\{Q\};\{\mathbf{r}\}) | \hat{d}_\kappa | \phi_n^{(j)}(\{Q\})\psi_j(\{Q\};\{\mathbf{r}\}) \rangle \tag{8.17}$$

with the following meaning: $\phi_i^{(0)}(\{Q\}), \phi_f^{(0)}(\{Q\})$ are the initial and final vibrational states, both in the ground state of the electronic system, and $\phi_n^{(j)}(\{Q\})$ is an intermediate state, associated with the electronic system state j (all three are functions of the phonon degrees of freedom denoted collectively by $\{Q\}$), and $\psi_0(\{Q\};\{\mathbf{r}\}), \psi_j(\{Q\};\{\mathbf{r}\})$ are wavefunctions of the electronic system, in the ground state and the excited state j, respectively, which depend both on the phonic state, indicated by $\{Q\}$, and the electronic degrees of freedom, the latter denoted collectively by $\{\mathbf{r}\}$. E_{f0}, E_{nj}, E_{i0} are the corresponding energies of the system in the particular electronic and vibrational states.

These expressions help us visualize what happens "behind the scenes" when the incident radiation of frequency ω_I is scattered into the radiation with frequency ω_S in the presence of a phonon mode in the solid. This is shown schematically in Fig. 8.3. Specifically, the term in Eq. (8.16) describes the process of starting with the electronic system in the ground state $\psi_0(\{Q\},\{\mathbf{r}\})$ at a given vibrational state labeled i, which is then excited to an electronic state labeled $\psi_j(\{Q\};\{\mathbf{r}\})$ and a new vibrational state labeled n within this manifold of possible states; the term in Eq. (8.17) describes the subsequent decay back to the electronic ground state $\psi_0(\{Q\};\{\mathbf{r}\})$ and a different vibrational state labeled f. The exitation takes place by

[1] J. H. Van Vleck, *Proc. Nat. Acad. Sci. (US)* **15**, 754 (1929).
[2] A. C. Albrecht, *J. Chem. Phys.* **34**, 1476 (1961).

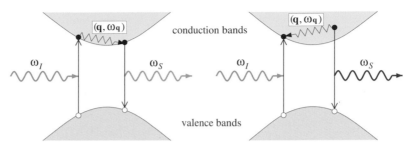

Illustration of excitations involved in Raman scattering processes: the wavy colored lines represent photons (absorbed or emitted) and the zigzag lines represent phonons of wave-vector $\mathbf{q}$ and frequency $\omega_\mathbf{q}$. **Left**: Excitation of an electron–hole pair by absorption of a photon of frequency ω_I, followed by *emission* of a phonon by the electron, which is scattered to a new state and recombines with a hole to emit a photon of frequency $\omega_S < \omega_I$. **Right**: Excitation of an electron–hole pair by absorption of a photon of frequency ω_I, followed by *absorption* of a phonon by the electron, which is scattered to a new state and recombines with a hole to emit a photon of frequency $\omega_S > \omega_I$.

the creation of an electron–hole pair from the absorption of a photon of energy $\hbar\omega_I$, with the electron emitting a phonon of wave-vector $\mathbf{q}$ and frequency $\omega_\mathbf{q}$ to end up in a different excited state, from which it decays to the ground by the emission of a photon of energy $\hbar\omega_S$. At each stage energy and momentum are conserved, but the emission of the phonon of frequency $\omega_\mathbf{q}$ is what prevents the electron from returning to its original state. The phonon emission is manifest by the change in vibrational state of the system, which changes from the one labeled i to the one labeled f. The transition amplitude involves a summation over all possible phonon modes of wave-vector $\mathbf{q}$ and frequency $\omega_\mathbf{q}$, collectively represented by the index n, as well as over all excited electronic states labeled j that satisfy the energy and momentum conservation rules. A similar description applies to the second term in Eq. (8.15). In this case, the excited electron absorbs a phonon of wave-vector $\mathbf{q}$ and frequency $\omega_\mathbf{q}$, and is scattered into a new state of higher energy, from which it can decay to the ground state by emitting a photon. The Stokes mechanism gives a stronger contribution to the signal, since it is easier for the excited electron to emit phonons and lose some of its energy, rather than to have to wait to absorb a particular phonon before it can decay to the ground state.

The processes that make the most significant contributions are those for which the denominators are very small, and the matrix elements do not vanish. Typically, the energy of the incident photon $\hbar\omega_I$ is much larger than that of emitted phonons. If the photon energy is close to that required for an electronic transition between states $\psi_0(\{Q\}; \{\mathbf{r}\})$ and $\psi_j(\{Q\}; \{\mathbf{r}\})$, then it is likely to find phonon modes that satisfy the energy and momentum conservation rules, which leads to resonance and a large signal in the scattering cross-section. By spanning the values of the incident radiation frequency, such phonon modes can be identified. The term $i\eta$ in the denominators in Eq. (8.15) serves to describe the finite lifetime of the excited electronic state and gives a finite width to the resonance.

The final result will depend on which matrix elements, for frequencies ω_I close to resonance, have non-vanishing values. These are determined by symmetry considerations, similar to what we discussed for the case of electronic excitations (see Chapter 6), only now we need to take into consideration both electronic excitations and matrix elements that involve vibrational states of the system. To explore this further, we define:

$$[\hat{d}_{0j}(\{Q\})]_\kappa \equiv \langle \psi_0(\{Q\}; \{\mathbf{r}\}) | \hat{d}_\kappa | \psi_j(\{Q\}; \{\mathbf{r}\}) \rangle \tag{8.18}$$

that is, integrate over the electronic degrees of freedom first, which gives the following expressions for the matrix elements:

$$
\begin{aligned}
\left[\mathcal{M}_{ni,j0} \right]_\lambda &= \langle \phi_n^{(j)}(\{Q\}) | [\hat{d}_{j0}(\{Q\})]_\lambda | \phi_i^{(0)}(\{Q\}) \rangle \\
\left[\mathcal{M}_{fn,0j} \right]_\kappa &= \langle \phi_f^{(0)}(\{Q\}) | [\hat{d}_{0j}(\{Q\})]_\kappa | \phi_n^{(j)}(\{Q\}) \rangle
\end{aligned}
\tag{8.19}
$$

Next, we expand the dipole expectation value as we did for the case of IR absorption in a Taylor series in the variables of the phonon modes Q_s:

$$[\hat{d}_{0j}(\{Q\})]_\kappa = [\hat{d}_{0j}^{(0)}]_\kappa + \sum_s [\hat{d}_{0j}^{(s)}(\{Q\})]_\kappa Q_s, \quad [\hat{d}_{0j}^{(s)}(\{Q\})]_\kappa = \left[\frac{\partial [\hat{d}_{0j}(\{Q\})]_\kappa}{\partial Q_s} \right]_{\{Q\}=0} \tag{8.20}$$

which gives, for the matrix elements to lowest order in Q_s:

$$\left[\mathcal{M}_{ni,j0}^{(0)} \right]_\lambda = [\hat{d}_{j0}^{(0)}]_\lambda \, \langle \phi_n^{(j)}(\{Q\}) | \phi_i^{(0)}(\{Q\}) \rangle, \quad \left[\mathcal{M}_{fn,0j}^{(0)} \right]_\kappa = [\hat{d}_{0j}^{(0)}]_\kappa \, \langle \phi_f^{(0)}(\{Q\}) | \phi_n^{(j)}(\{Q\}) \rangle$$

Now, in contrast to the case of IR absorption, these matrix elements do not automatically vanish, because in general we will have:

$$\langle \phi_n^{(j)}(\{Q\}) | \phi_i^{(0)}(\{Q\}) \rangle \neq 0 \quad \text{and} \quad \langle \phi_f^{(0)}(\{Q\}) | \phi_n^{(j)}(\{Q\}) \rangle \neq 0$$

since the two vibrational states involved in each product belong to different electronic states, the ground state (labeled 0) and the excited state (labeled j). Moreover, the amplitude in Eq. (8.15) involves a summation over n, that is, all the vibrational states of the excited electronic state, which guarantees that there will be some vibrational states of the excited electronic state with large overlap with the vibrational states of the electronic ground state. Consequently, the leading (lowest)-order terms will be proportional to $[\hat{d}_{j0}^{(0)}]_\lambda$, which implies that Raman scattering is qualitatively different from the case of IR absorption, because this term involves no change in the dipole moment. The conclusion that we draw from this analysis is that Raman active modes are essentially opposite in nature to IR active modes: a mode that is IR active involves a change in the dipole moment and is therefore *not active* in Raman scattering, and vice versa (see also Problem 2 for further discussion).

8.4 Phonon–Electron Interactions: Superconductivity

One of the most striking manifestations of the interaction between electrons and phonons is the phenomenon of conventional superconductivity, which was discovered in 1911 by Kamerling Onnes.[3] Superconductivity is an essentially many-body phenomenon which cannot be described simply within the single-particle picture. Because of its fascinating

---

[3] K. Onnes, *Leiden Commun.* **120** (1911).

nature and many applications, superconductivity has been the focus of intense theoretical and experimental investigations ever since its discovery. The hallmark of superconductivity is a vanishing electrical resistance below the critical temperature T_c. There is no measurable DC resistance in a superconductor below this temperature, and if a current is set up it will flow without dissipation practically forever: experiments trying to detect changes in the magnetic field associated with current in a superconductor give estimates that it is constant for 10^6–10^9 years! Thus, the superconducting state is not a state of merely very low resistance, but one with a truly zero resistance. This is different from the case of very good conductors. In fact, materials which are very good conductors in their normal state typically do not exhibit superconductivity. The reason is that in very good conductors there is little coupling between phonons and electrons, since it is scattering by phonons which gives rise to the temperature-dependent resistance in a conductor, whereas electron–phonon coupling is crucial for superconductivity. The drop in resistance from its normal value above T_c to zero takes place over a range of temperatures of order 10^{-2}–$10^{-3}T_c$, that is, the transition is rather sharp.

For conventional superconductors, T_c is in the range of a few kelvin, which has made it difficult to take advantage of this extraordinary behavior in practical applications, because cooling the specimen to within a few degrees of absolute zero is quite difficult (it requires the use of liquid helium, the only non-solid substance at such low temperatures, as the coolant, which is expensive and cumbersome). We give examples of conventional superconductors in Table 8.1. A new class of superconducting materials was discovered in 1986, with T_c that can exceed 130 K and thus can be cooled to the superconducting state with nitrogen which is liquid below 77 K; we mention a few important facts about these superconductors at the end of this section.

The theory of superconductivity involved a long search for a proper microscopic picture of how electrons create the very special superconducting state. In addition to the microscopic picture, there are many aspects of superconductivity that can be described

Table 8.1 Critical temperature T_c (K), critical field H_0 (Oe), and Debye frequency $\hbar\omega_D$ (meV) of elemental conventional superconductors. *Source*: D. R. Lide (Ed.), *CRC Handbook of Chemistry and Physics* (CRC Press, Boca Raton, FL, 1999–2000).

Element	T_c	H_0	$\hbar\omega_D$	Element	T_c	H_0	$\hbar\omega_D$
Al	1.175	105	36.2	Ru	0.49	69	50.0
Cd	0.517	28	18.0	Nb	9.25	2060	23.8
Gd	1.083	58	28.0	Ta	4.47	829	22.2
Hg	3.949	339	8.0	Tc	7.80	1410	35.4
In	3.408	282	9.4	Th	1.38	1.6	14.2
Ir	0.113	16	36.6	Ti	0.40	56	35.8
Mo	0.915	96	39.6	Tl	2.38	178	6.8
Sn	3.722	305	50.0	V	5.40	1408	33.0
W	0.015	1.2	33.0	Os	0.66	70	43.1
Pb	7.196	803	8.3	Zn	0.85	54	26.7
Re	1.697	200	0.4	Zr	0.61	47	25.0

by phenomenological theories. A full treatment of the subject is beyond the scope of the present book (we give some suggestions for texts that cover these topics in the Further Reading). Instead, we concentrate here on the microscopic theory of superconductivity as it was developed in the seminal paper of J. Bardeen, L. N. Cooper, and J. R. Schriefer[4] (BCS theory); this theory relies on treating explicitly the interaction of phonons with electrons, the theme of the present section.

8.4.1 BCS Theory of Superconductivity

There are two main ingredients in the BCS theory of superconductivity. The first is an effective attractive interaction between two electrons that have opposite momenta (larger in magnitude than the Fermi momentum) and opposite spins, which leads to the formation of the so-called "Cooper pairs." The second is the condensation of the Cooper pairs into a single coherent quantum state, which is called the "superconducting condensate"; this is the state responsible for all the manifestations of superconducting behavior. We discuss both ingredients in some detail.

The Attractive Potential Between Electrons Due to Phonons

Our starting point is the system of electrons described by the hamiltonian

$$\hat{\mathcal{H}} = \hat{\mathcal{H}}_0^{\text{sp}} + \hat{\mathcal{H}}^{\text{int}}$$

with the interaction term describing the electron–electron as well as the electron–ion interactions when ions are allowed to move. The single-particle wavefunctions $|\psi_{\mathbf{k}}\rangle$ are eigenfunctions of the first term, the kinetic energy, with eigenvalues $\epsilon_{\mathbf{k}}$:

$$\hat{\mathcal{H}}_0^{\text{sp}}|\psi_{\mathbf{k}}\rangle = \epsilon_{\mathbf{k}}|\psi_{\mathbf{k}}\rangle$$

and form a complete orthonormal set. We shall examine in more detail the nature of the interaction when ions are allowed to move: it contains a new term relative to the case of fixed ions, which represents electron–electron interactions mediated by the exchange of phonons. The physical origin of this new term is shown schematically in Fig. 8.4: an electron moving through the solid attracts the positively charged ions which come closer to it as it approaches and then return slowly to their equilibrium positions once the electron has passed by. We can describe this motion in terms of phonons emitted by the traveling electron. It is natural to assume that the other electrons will be affected by this distortion of the ionic positions; since the electrons themselves are attracted to the ions, the collective motion of ions toward one electron will translate into an effective attraction of other electrons toward the first one. Frölich[5] and Bardeen and Pines[6] showed that the effective

[4] J. Bardeen, L. N. Cooper, and J. R. Schriefer, *Phys. Rev.* **108**, 1175 (1957).
[5] H. Frölich, *Proc. Roy. Soc. A* **215**, 291 (1952).
[6] J. Bardeen and D. Pines, *Phys. Rev.* **99**, 1140 (1955).

Fig. 8.4   Illustration of attractive effective interaction between two electrons mediated by phonons: on the left is the distortion that the first electron (blue dot) induces to the lattice of ions (red spheres) as it passes through; on the right is the second electron with opposite momentum near the original position, at a later time: the lattice distortion, which decays on a time scale much longer than electronic motion, favors the presence of the second electron in that position, leading to an effective attraction.

interaction between electrons due to exchange of a phonon takes a form that explains and quantifies this attraction.

While the full proof of the effective interaction between electrons in the presence of phonons requires a treatment based on field theory, which is beyond the scope of the present discussion, we provide here a heuristic argument to justify the form of this interaction. We begin by noting that when the positively charged ions move, they can produce their own plasmon modes with characteristic plasma frequency given by:

$$\Omega_p = \left(\frac{4\pi N_{\mathrm{ion}} Z^2 e^2}{MV} \right)^{1/2}$$

where M is the mass of the ions, Ze their valence charge, and N_{ion} is the total number of ions in the solid of volume V (we have assumed here for simplicity that there is only one type of ions in the solid). We saw earlier (Chapter 6) that free electrons produce a dielectric function that involves the plasma frequency of the electron gas, see Eq. (6.16). We would therefore expect similar screening to be produced by the motion of the ions treated as free charges, expressed in terms of a dielectric function with exactly the same form as that of free electrons. However, in the corresponding expression for the ions, we should include not the bare plasma frequency for the ion gas, Ω_p, as written above, but a screened one, due to screening of the ionic motion by the presence of electrons, which is embodied in the electronic dielectric constant $\varepsilon(\mathbf{k})$. Since the restoring force of the harmonic potential between ions is proportional to the square of the frequency, and the dielectric constant screens the potential itself, we conclude that the screened plasma frequency of the ions must be given by:

$$\omega_{\mathbf{k}}^2 = \frac{\Omega_p^2}{\varepsilon(\mathbf{k})}$$

with $\omega_{\mathbf{k}}$ representing a phonon frequency. With this modification, the analog of Eq. (6.16) for the dielectric constant of ions, including screening from electrons, will be:

$$\varepsilon_{\mathrm{ion}}(\omega) = 1 - \frac{\Omega_p^2/\varepsilon(\mathbf{k})}{\omega^2} = 1 - \frac{\omega_{\mathbf{k}}^2}{\omega^2}$$

The combined effect of screening on the electron–electron interaction will be given by an effective dielectric function which is the product of the electronic and ionic contributions to screening, with the two components treated as independent of each other. Consistent with the qualitative nature of the argument, we will take the electronic states to be those of free electrons in a Hartree–Fock picture. We recall that the bare (unscreened) Coulomb interaction in this basis is:

$$V_{\mathbf{kk'}}^{\text{bare}} = \langle \psi_{\mathbf{k'}} | \frac{e^2}{|\mathbf{r}-\mathbf{r'}|} | \psi_{\mathbf{k}} \rangle = \frac{4\pi e^2}{V|\mathbf{k}-\mathbf{k'}|^2}$$

[see Appendix A, Eq. (A.74)], where we have specialized the interaction to two electron states with wave-vectors $\mathbf{k}$ and $\mathbf{k'}$; in the following we define $\mathbf{q} = \mathbf{k}-\mathbf{k'}$. For the frequency-independent part, we can take the electronic contribution to the dielectric function to be that of the Thomas–Fermi model, that is:

$$\varepsilon(\mathbf{q}) = 1 + \frac{k_s^2}{|\mathbf{q}|^2}$$

with k_s the screening inverse length [see Eq. (4.115)]. Now, using the argument outlined above, we can write the effective screened interaction between the electrons as:

$$V_{\mathbf{kk'}}^{\text{eff}} = \frac{1}{\varepsilon_{\text{ion}}(\omega)} \frac{1}{\varepsilon(\mathbf{q})} V_{\mathbf{kk'}}^{\text{bare}} = \frac{1}{\varepsilon_{\text{ion}}(\omega)} \frac{4\pi e^2}{V(k_s^2 + |\mathbf{q}|^2)} = \frac{4\pi e^2}{V(k_s^2 + |\mathbf{q}|^2)} \left(\frac{\omega^2}{\omega^2 - \omega_{\mathbf{q}}^2} \right)$$

Finally, we associate the frequency ω with the energy difference between the electronic states, $\omega = (\epsilon_{\mathbf{k'}} - \epsilon_{\mathbf{k}})/\hbar$, which allows us to rewrite the effective interaction as:

$$V_{\mathbf{kk'}}^{\text{eff}} = \frac{4\pi e^2}{V(k_s^2 + |\mathbf{q}|^2)} \left[1 + \frac{(\hbar\omega_{\mathbf{q}})^2}{(\epsilon_{\mathbf{k'}} - \epsilon_{\mathbf{k}})^2 - (\hbar\omega_{\mathbf{q}})^2} \right]$$

In the above expression for the effective interaction between electrons there are two terms: the first term (corresponding to the factor 1 in the square brackets) is the screened electron interaction in the absence of ionic motion; therefore, we can associate the second term with the contribution of the phonons:

$$V_{\mathbf{kk'}}^{\text{phon}} = \frac{4\pi e^2}{V(k_s^2 + |\mathbf{q}|^2)} \frac{(\hbar\omega_{\mathbf{q}})^2}{(\epsilon_{\mathbf{k'}} - \epsilon_{\mathbf{k}})^2 - (\hbar\omega_{\mathbf{q}})^2}$$

From the form of this term, it is evident that if the magnitude of the energy difference $|\epsilon_{\mathbf{k'}} - \epsilon_{\mathbf{k}}|$ is smaller than the phonon energy $\hbar\omega_{\mathbf{q}}$, the phonon contribution to the interaction is *attractive*; in other words, if the energies of the two electrons $\epsilon_{\mathbf{k}}, \epsilon_{\mathbf{k'}}$ lie within a very narrow range, so that their difference can be made very small, then the exchange of a phonon of frequency $\omega_{\mathbf{q}}$ between them will lead to effective attraction. The presence of the Fermi sea is what provides the conditions for realizing this situation.

The Cooper Pair Problem

To show that an attractive interaction due to phonons can actually produce binding between electron pairs, Cooper[7] investigated the energy of the Fermi sea in the normal state and compared this with a state in which two electrons are removed from the sea and allowed to interact. This led to the conclusion that for an *arbitrarily* weak attractive interaction

---

[7] L. N. Cooper, *Phys. Rev.* **104**, 1189 (1956).

between the two electrons, the new state has a lower energy, implying that the Fermi sea is unstable with respect to the formation of pairs if there is an attractive interaction. Note that this is not yet a theory of the superconducting state; for the full theory we must also treat the many-body aspect of the problem, that is, what happens when the electrons have formed pairs. Note also that the binding between electrons from an arbitrarily weak attraction arises from the restriction of the phase space due to the presence of the Fermi sea; normally binding in 3D solids would require a finite threshold value of the attraction.

To demonstrate these notions, consider two electrons described by their positions, momenta, and spins $(\mathbf{r}_i, \mathbf{k}_i, s_i)$, $i = 1, 2$ in single-particle plane-wave states:

$$\psi_{\mathbf{k}_i}(\mathbf{r}) = \frac{1}{\sqrt{V}} e^{i\mathbf{k}_i \cdot \mathbf{r}}$$

Anticipating an attractive interaction, we choose the spatial state to be symmetric and thus lower in energy. This means that the two-electron system will be in an antisymmetric total-spin state, that is, described by the following two-particle wavefunction:

$$\Phi(\mathbf{k}_1, \mathbf{k}_2, s_1, s_2 | \mathbf{r}_1, \mathbf{r}_2) = \frac{1}{\sqrt{2V}} \left[e^{i(\mathbf{k}_1 \cdot \mathbf{r}_1 + \mathbf{k}_2 \cdot \mathbf{r}_2)} + e^{i(\mathbf{k}_1 \cdot \mathbf{r}_2 + \mathbf{k}_2 \cdot \mathbf{r}_1)} \right] \frac{1}{\sqrt{2}} (\uparrow_1 \downarrow_2 - \downarrow_1 \uparrow_2)$$

We introduce coordinates for the center-of-mass $(\mathbf{R}, \mathbf{K})$ and relative $(\mathbf{r}, \mathbf{k})$ motion:

$$\mathbf{R} = \frac{\mathbf{r}_1 + \mathbf{r}_2}{2}, \quad \mathbf{K} = \mathbf{k}_1 + \mathbf{k}_2, \quad \mathbf{r} = \mathbf{r}_1 - \mathbf{r}_2, \quad \mathbf{k} = \frac{\mathbf{k}_1 - \mathbf{k}_2}{2}$$

and write the spatial part of the wavefunction as:

$$\Phi(\mathbf{k}, \mathbf{K} | \mathbf{r}, \mathbf{R}) = \frac{e^{i\mathbf{K} \cdot \mathbf{R}}}{\sqrt{2V}} \left[e^{i\mathbf{k} \cdot \mathbf{r}} + e^{-i\mathbf{k} \cdot \mathbf{r}} \right]$$

We will concentrate on states with total momentum $\mathbf{K} = 0$, since we expect the lowest-energy state to have zero total momentum. In this case the individual momenta take values $\mathbf{k}_1 = \mathbf{k}, \mathbf{k}_2 = -\mathbf{k}$, that is, we are dealing with a pair of particles with *opposite* momenta. The hamiltonian for this system is:

$$\hat{\mathcal{H}} = \frac{1}{2m_e}(\hat{\mathbf{p}}_1^2 + \hat{\mathbf{p}}_2^2) + \hat{\mathcal{H}}^{\text{int}} = \frac{1}{2m_e} \left(\frac{1}{2}\hat{\mathbf{P}}^2 + 2\hat{\mathbf{p}}^2 \right) + \hat{\mathcal{H}}^{\text{int}}$$

We can ignore the total-momentum term $\hat{\mathbf{P}}$, because we are only considering $\mathbf{K} = 0$ states. We express the wavefunction in the relative coordinate $\mathbf{r}$ as:

$$\phi(\mathbf{r}) = \sum_{|\mathbf{k}'| > k_{\text{F}}} \alpha(\mathbf{k}') e^{i\mathbf{k}' \cdot \mathbf{r}}$$

with k_{F} the Fermi momentum, and require it to be an eigenfunction of the hamiltonian with eigenvalue E:

$$\hat{\mathcal{H}}\phi(\mathbf{r}) = E\phi(\mathbf{r})$$

To determine the coefficients $\alpha(\mathbf{k}')$, we take matrix elements of the hamiltonian with individual factors $\exp(-i\mathbf{k} \cdot \mathbf{r})$ and use the Schrödinger equation for $\phi(\mathbf{r})$ to find:

$$\int e^{-i\mathbf{k} \cdot \mathbf{r}} (\hat{\mathcal{H}} - E) \sum_{|\mathbf{k}'| > k_{\text{F}}} \alpha(\mathbf{k}') e^{i\mathbf{k}' \cdot \mathbf{r}} \, d\mathbf{r} = 0 \Rightarrow (2\epsilon_{\mathbf{k}} - E)\,\alpha(\mathbf{k}) + \sum_{|\mathbf{k}'| > k_{\text{F}}} \alpha(\mathbf{k}') \langle \mathbf{k} | \hat{\mathcal{H}}^{\text{int}} | \mathbf{k}' \rangle = 0$$

We next rewrite the last equation in terms of the density of states $g(\epsilon)$ and switch to the energy $\epsilon_\mathbf{k}$ as the variable instead of the wave-vector $\mathbf{k}$, to obtain:

$$(2\epsilon_\mathbf{k} - E)\alpha(\epsilon_\mathbf{k}) + \int_{\epsilon_{\mathbf{k}'} > \epsilon_F} g(\epsilon_{\mathbf{k}'})\langle\mathbf{k}|\hat{\mathcal{H}}^{\text{int}}|\mathbf{k}'\rangle\alpha(\epsilon_{\mathbf{k}'})\mathrm{d}\epsilon_{\mathbf{k}'} = 0 \qquad (8.21)$$

where ϵ_F is the Fermi energy. Following Cooper (see footnote 7), we consider a simple model:

$$\langle\mathbf{k}|\hat{\mathcal{H}}^{\text{int}}|\mathbf{k}'\rangle = -\mathcal{V} < 0, \quad \text{for } \epsilon_F < \epsilon_\mathbf{k}, \epsilon_{\mathbf{k}'} < \epsilon_F + \delta\epsilon \qquad (8.22)$$
$$= 0 \ \text{ otherwise}$$

that is, the interaction term is taken to be a constant independent of the energy for energies $\epsilon_{\mathbf{k}'}, \epsilon_\mathbf{k}$ within a narrow shell of width $\delta\epsilon$ above the Fermi level, and 0 for other energy values.

We provide a rough argument to justify this choice for the interaction term; the detailed justification has to do with subtle issues related to the optimal choice for the superconducting ground state, which lie beyond the scope of the present treatment. As indicated in Fig. 8.4, the distortion of the lattice induced by an electron would lead to an attractive interaction with any other electron put in the same position at a later time. The delay must be restricted to times of order $1/\omega_D$, where ω_D is the Debye frequency, or the distortion will decay away. These considerations lead to the estimate of the value of ϵ_0, the thickness of the shell above ϵ_F where the energies $\epsilon_\mathbf{k}$ lie:

$$\delta\epsilon \approx \hbar\omega_D \qquad (8.23)$$

There is, however, no restriction on the momentum of the second electron from these considerations. It turns out that the way to maximize the effect of the interaction is to take the electrons in pairs with opposite momenta, because this ensures that no single-particle state is double counted or left out of the many-body ground state of the system built from the electron pairs.

Substituting the choice for the interaction term, Eq. (8.22), in the last equation for the coefficients $\alpha(\epsilon_\mathbf{k})$, Eq. (8.21), after defining the constant C, we get:

$$C \equiv \mathcal{V} \int_{\epsilon_F}^{\epsilon_F + \hbar\omega_D} g(\epsilon_{\mathbf{k}'})\alpha(\epsilon_{\mathbf{k}'})\mathrm{d}\epsilon_{\mathbf{k}'} \ \Rightarrow \ \alpha(\epsilon_\mathbf{k}) = \frac{C}{2\epsilon_\mathbf{k} - E} \qquad (8.24)$$

and substituting this value for $\alpha(\epsilon_\mathbf{k})$ in the definition of C, we arrive at:

$$1 = \mathcal{V} \int_{\epsilon_F}^{\epsilon_F + \hbar\omega_D} \frac{g(\epsilon')}{2\epsilon' - E}\mathrm{d}\epsilon'$$

We approximate the density of states in the above expression with its value at the Fermi energy, since the range of integration is a thin shell of values near ϵ_F:

$$\frac{1}{g(\epsilon_F)\mathcal{V}} = \int_{\epsilon_F}^{\epsilon_F + \hbar\omega_D} \frac{1}{2\epsilon' - E}\mathrm{d}\epsilon' = \frac{1}{2}\ln\left(\frac{2\epsilon_F + 2\hbar\omega_D - E}{2\epsilon_F - E}\right)$$

Next, we define the binding energy of the pair Δ, the amount by which the energy is lowered upon formation of the pair relative to the energy of the two independent electrons:

$$E = 2\epsilon_F - \Delta \tag{8.25}$$

and use it in the above expression to arrive at:

$$\Delta = \frac{2\hbar\omega_D}{\exp[2/\mathcal{V}g(\epsilon_F)] - 1} \tag{8.26}$$

If the quantity $\mathcal{V}g(\epsilon_F)$ is small, then to a good approximation the value of Δ is given by:

$$\Delta \approx 2\hbar\omega_D \, \exp[-2/\mathcal{V}g(\epsilon_F)] = 2\hbar\omega_D \, \exp[-1/\mathcal{V}g_2(2\epsilon_F)] \tag{8.27}$$

where we defined the pair density of states $g_2(\epsilon)$ in terms of the regular density of states $g(\epsilon)$ as:

$$g_2(\epsilon) = \frac{1}{2}g(\epsilon/2)$$

This result shows that the Fermi sea is indeed unstable, since we can lower the energy of the system by the amount Δ, no matter how small the interaction $\mathcal{V}$ is.

With the expression we obtained for the coefficients $\alpha(\epsilon_k)$, Eq. (8.24), we can write the pair wavefunction as:

$$\phi(\mathbf{r}) = \frac{C}{\sqrt{V}} \sum_{|\mathbf{k}|>k_F} \frac{1}{2\epsilon_k - E} \cos(\mathbf{k} \cdot \mathbf{r}) = \frac{C}{\sqrt{V}} \sum_{|\mathbf{k}|>k_F} \frac{1}{2\xi_k + \Delta} \cos(\mathbf{k} \cdot \mathbf{r})$$

where we have defined the quantities $\xi_k = \epsilon_k - \epsilon_F$, which are positive since $\epsilon_k > \epsilon_F$; we have also used the definition of the binding energy Δ from Eq. (8.25), which is a positive quantity. The coefficients of the $\cos(\mathbf{k} \cdot \mathbf{r})$ terms in the last expansion are always positive; they take their largest value for $\xi_k = 0$, that is, for $\epsilon_k = \epsilon_F$ or $|\mathbf{k}| = k_F$. For $\xi_k \sim \Delta$, the value has already fallen from its maximum value by a factor of $1/3$. These considerations show that only electrons with energy within Δ of ϵ_F contribute significantly to the summation. Since $\Delta \ll \hbar\omega_D$ for weak coupling, this is a very small energy range, and the details of the behavior of the matrix element $\langle \mathbf{k}|\hat{\mathcal{H}}^{int}|\mathbf{k}'\rangle$ should not matter much in this energy range, consistent with the assumption that its value is a constant, Eq. (8.22). The energy range of interest turns out to be $\sim k_B T_c$ around ϵ_F.

Note that if the pair momentum $\mathbf{K}$ does not vanish as we assumed in the derivation above, but is small, then the result is a binding energy given by:

$$\Delta(\mathbf{K}) = \Delta - \frac{v_F\hbar|\mathbf{K}|}{2}$$

where v_F is the Fermi velocity. Cooper also showed that the radius R of the bound electron pair is:

$$R \sim \frac{\hbar^2 k_F}{m_e \Delta}$$

For typical values of k_F and Δ, this radius is $R \sim 10^4 \text{Å}$, which is a very large distance on the atomic scale.

subject to the conditions

$$v_{\mathbf{k}}^2 + u_{\mathbf{k}}^2 = 1 \quad \text{and} \quad \langle \Psi_{\text{BCS}}^{(s)} | \hat{N} | \Psi_{\text{BCS}}^{(s)} \rangle = 2 \sum_{\mathbf{k}} v_{\mathbf{k}}^2 = N \tag{8.40}$$

To perform this minimization, we first define the angle $\theta_{\mathbf{k}}$ using

$$u_{\mathbf{k}} \equiv \sin \frac{\theta_{\mathbf{k}}}{2}, \quad v_{\mathbf{k}} \equiv \cos \frac{\theta_{\mathbf{k}}}{2} \tag{8.41}$$

from which we obtain:

$$v_{\mathbf{k}}^2 = \cos^2 \frac{\theta_{\mathbf{k}}}{2} = \frac{1}{2}(1 + \cos \theta_{\mathbf{k}}), \quad u_{\mathbf{k}} v_{\mathbf{k}} = \cos \frac{\theta_{\mathbf{k}}}{2} \sin \frac{\theta_{\mathbf{k}}}{2} = \frac{1}{2} \sin \theta_{\mathbf{k}}$$

and with these, the expression for the ground-state energy becomes:

$$E_{\text{BCS}} = \sum_{\mathbf{k}} (\epsilon_{\mathbf{k}} - \mu)(1 + \cos \theta_{\mathbf{k}}) + \frac{1}{4} \sum_{\mathbf{k},\mathbf{k}'} V_{\mathbf{k}\mathbf{k}'} \sin \theta_{\mathbf{k}} \sin \theta_{\mathbf{k}'}$$

Now the only variational parameter is $\theta_{\mathbf{k}}$. Differentiating with respect to this parameter and setting the derivative of E_{BCS} to zero yields:

$$\tan \theta_{\mathbf{k}} = \frac{1}{2(\epsilon_{\mathbf{k}} - \mu)} \sum_{\mathbf{k}'} V_{\mathbf{k}\mathbf{k}'} \sin \theta_{\mathbf{k}'} \tag{8.42}$$

We can rewrite this equation, with some hindsight, by defining the following two quantities that will later be shown to be the energy-gap parameter and quasiparticle excitation energy, respectively:

$$\Delta_{\mathbf{k}} \equiv -\sum_{\mathbf{k}'} V_{\mathbf{k}\mathbf{k}'} u_{\mathbf{k}'} v_{\mathbf{k}'} = -\frac{1}{2} \sum_{\mathbf{k}'} V_{\mathbf{k}\mathbf{k}'} \sin \theta_{\mathbf{k}'} \tag{8.43}$$

$$E_{\mathbf{k}} \equiv \left(\Delta_{\mathbf{k}}^2 + (\epsilon_{\mathbf{k}} - \mu)^2 \right)^{1/2} \tag{8.44}$$

Substituting $\Delta_{\mathbf{k}}$ in Eq. (8.42) gives:

$$\tan \theta_{\mathbf{k}} = -\frac{\Delta_{\mathbf{k}}}{(\epsilon_{\mathbf{k}} - \mu)} \Rightarrow \cos \theta_{\mathbf{k}} = -\frac{(\epsilon_{\mathbf{k}} - \mu)}{\Delta_{\mathbf{k}}} \sin \theta_{\mathbf{k}}$$

from which we obtain:

$$\sin^2 \theta_{\mathbf{k}} + \cos^2 \theta_{\mathbf{k}} = 1 = \sin^2 \theta_{\mathbf{k}} \left(1 + \frac{(\epsilon_{\mathbf{k}} - \mu)^2}{\Delta_{\mathbf{k}}^2} \right) \Rightarrow \sin \theta_{\mathbf{k}} = \frac{\Delta_{\mathbf{k}}}{E_{\mathbf{k}}}$$

Using this value in Eq. (8.42) leads to:

$$\Delta_{\mathbf{k}} = -\frac{1}{2} \sum_{\mathbf{k},\mathbf{k}'} V_{\mathbf{k}\mathbf{k}'} \frac{\Delta_{\mathbf{k}'}}{E_{\mathbf{k}'}} = -\frac{1}{2} \sum_{\mathbf{k},\mathbf{k}'} V_{\mathbf{k}\mathbf{k}'} \frac{\Delta_{\mathbf{k}'}}{\sqrt{(\Delta_{\mathbf{k}'}^2 + (\epsilon_{\mathbf{k}'} - \mu)^2}} \tag{8.45}$$

This integral equation is known as the BCS gap equation and must be solved for $\Delta_{\mathbf{k}}$ and μ under the second constraint of Eq. (8.40), because the first constraint is automatically satisfied by our choice of $u_{\mathbf{k}}$ and $v_{\mathbf{k}}$ in terms of the single variable $\theta_{\mathbf{k}}$, Eq. (8.41). Since we are working with single-particle states that in the normal system at $T = 0$ are filled up to the Fermi level, with energy ϵ_{F}, the chemical potential can effectively be taken to be $\mu = \epsilon_{\text{F}}$.

The BCS Model

To make further progress in describing the state of the system within the BCS hamiltonian, we need to find explicit solutions to Eq. (8.45). This can easily be done in a limiting case (trivial solution), or by introducing a model that simplifies the equation enough to obtain an analytic result, the BCS model. We investigate these two cases next.

The trivial solution is $\Delta_{\mathbf{k}} = 0$, which then gives $v_{\mathbf{k}} = 1$ for $\epsilon_{\mathbf{k}} < \epsilon_F$ and $v_{\mathbf{k}} = 0$ for $\epsilon_{\mathbf{k}} > \epsilon_F$. This is just the normal state of the system at $T = 0$, described by a single Slater determinant with single-particle states occupied up to $\epsilon_{\mathbf{k}} = \epsilon_F$.

A non-trivial solution can be obtained by approximating $\mathcal{V}_{\mathbf{kk}'}$ to be similar to the interaction used in the Cooper pair problem; this is known as the BCS model. Specifically, we take:

$$\mathcal{V}_{\mathbf{kk}'} = -\mathcal{V}, \quad \text{for } \epsilon_F - \hbar\omega_D < \epsilon_{\mathbf{k}}, \epsilon_{\mathbf{k}'} < \epsilon_F + \hbar\omega_D$$
$$= 0, \quad \text{otherwise} \tag{8.46}$$

This form of $\mathcal{V}_{\mathbf{kk}'}$ implies that:

$$\Delta_{\mathbf{k}} = \Delta, \quad \text{for } \epsilon_F - \hbar\omega_D < \epsilon_{\mathbf{k}}, \epsilon_{\mathbf{k}'} < \epsilon_F + \hbar\omega_D$$
$$= 0, \quad \text{otherwise} \tag{8.47}$$

The gap equation, Eq. (8.45), then simplifies to the solution of Δ from

$$1 = \frac{1}{2}\mathcal{V}\sum_{\mathbf{k}} \frac{1}{\sqrt{\Delta^2 + (\epsilon_{\mathbf{k}'}^2 - \epsilon_F)^2}} \Rightarrow 1 = \frac{\mathcal{V}}{2}\int_{\epsilon_F-\hbar\omega_D}^{\epsilon_F+\hbar\omega_D} \frac{g(\epsilon)}{\sqrt{\Delta^2 + (\epsilon - \epsilon_F)^2}}d\epsilon$$

$$\Rightarrow 1 \approx \mathcal{V}g(\epsilon_F)\sinh^{-1}\left(\frac{\hbar\omega_D}{\Delta}\right)$$

where, in the last two steps, we introduced the density of states $g(\epsilon)$ and assumed that $\hbar\omega_D \ll \epsilon_F$. For weak coupling, that is, if $g(\epsilon_F)\mathcal{V} < 1$, the result simplifies further to give:

$$\Delta \approx 2\hbar\omega_D e^{-1/g(\epsilon_F)\mathcal{V}} \tag{8.48}$$

which is an expression similar to the binding energy of a Cooper pair, Eq. (8.27). The magnitude of Δ is much smaller than $\hbar\omega_D$, which in turn is much smaller than the Fermi energy, as discussed in Chapter 7:

$$\Delta \ll \hbar\omega_D \ll \epsilon_F$$

which shows that the superconducting gap is a minuscule fraction of the Fermi energy.

As yet, however, we still have not determined what the physical meaning of Δ is. This will be discussed in the next subsection. Nevertheless, given Δ, we can obtain the associated optimized $u_{\mathbf{k}}, v_{\mathbf{k}}$ from

$$v_{\mathbf{k}}^2 = \frac{1}{2}\left(1 - \frac{\epsilon_{\mathbf{k}} - \epsilon_F}{E_{\mathbf{k}}}\right) = \frac{1}{2}\left(1 - \frac{\xi_{\mathbf{k}}}{E_{\mathbf{k}}}\right), \quad u_{\mathbf{k}}^2 = \frac{1}{2}\left(1 + \frac{\epsilon_{\mathbf{k}} - \epsilon_F}{E_{\mathbf{k}}}\right) = \frac{1}{2}\left(1 + \frac{\xi_{\mathbf{k}}}{E_{\mathbf{k}}}\right)$$

where we have also defined for convenience $\xi_{\mathbf{k}} \equiv \epsilon_{\mathbf{k}} - \epsilon_F$; $u_{\mathbf{k}}$ and $v_{\mathbf{k}}$ are plotted in Fig. 8.6 as functions of ξ. Note that even though these results are for $T = 0$, the plot

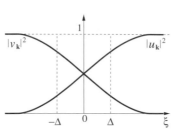
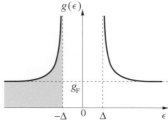

Fig. 8.6 Features of the BCS model. **Left**: Cooper pair occupation variables $|u_\mathbf{k}|^2$ and $|v_\mathbf{k}|^2$ as a function of the energy $\xi = \epsilon - \epsilon_F$. **Right**: Superconducting density of states $g(\epsilon)$ as a function of the energy ϵ; g_F is the density of states in the normal state at the Fermi level. The filled states are shown shaded in pink.

for $v_\mathbf{k}^2$ is reminiscent of the Fermi distribution function for a normal system at a non-zero temperature of order Δ. As we shall see later, the transition temperature T_c from normal to superconducting behavior is also of order Δ. Thus, one could argue that as the system is cooled below T_c, it is the relative phases of the many-body states describing the normal system that get locked into a single quantum state, rather than changes in the occupation numbers of the single-particle states. Using the above expressions for $v_\mathbf{k}$ and $u_\mathbf{k}$, the optimized ground-state energy becomes:

$$E_{\text{BCS}} = \sum_\mathbf{k} \xi_\mathbf{k} \left(1 - \frac{\xi_\mathbf{k}}{E_\mathbf{k}} \right) - \frac{\Delta^2}{\mathcal{V}}$$

Now recall that $\Delta = 0$ corresponds to the normal state at $T = 0$. Therefore, the difference in energy between the superconducting and normal states given by $\delta E = E_{\text{BCS}} - E_N$ is:

$$\delta E = \sum_\mathbf{k} \xi_\mathbf{k} \left(1 - \frac{\xi_\mathbf{k}}{E_\mathbf{k}} \right) - \frac{\Delta^2}{\mathcal{V}} - \sum_\mathbf{k} \xi_\mathbf{k} \left(1 - \frac{\xi_\mathbf{k}}{|\xi_\mathbf{k}|} \right)$$

Since $\xi_\mathbf{k}/|\xi_\mathbf{k}| = 1$ for $|\mathbf{k}| > k_F$ and -1 for $|\mathbf{k}| < k_F$, the summation over $\mathbf{k}$ in the third term of the expression above will only involve terms for $|\mathbf{k}| < k_F$. This summation can then be included in the first term to give:

$$\delta E = \sum_{|\mathbf{k}|<k_F} \left(-\xi_\mathbf{k} - \frac{\xi_\mathbf{k}^2}{E_\mathbf{k}} \right) + \sum_{|\mathbf{k}|>k_F} \left(\xi_\mathbf{k} - \frac{\xi_\mathbf{k}^2}{E_\mathbf{k}} \right) - \frac{\Delta^2}{\mathcal{V}}$$

Using $E_\mathbf{k} = \sqrt{\Delta^2 + \xi_\mathbf{k}^2}$ and $E_\mathbf{k} = |\xi_\mathbf{k}|$ when $|\xi_\mathbf{k}| > \hbar\omega_D$, the summations above will give zero unless $|\xi_\mathbf{k}| < \hbar\omega_D$. This produces:

$$\delta E = g(\epsilon_F) \int_{-\hbar\omega_D}^{\hbar\omega_D} \left(\xi - \frac{\xi^2}{\sqrt{\Delta^2 + \xi^2}} \right) d\xi - \frac{\Delta^2}{\mathcal{V}}$$

and in the limit of small $g(\epsilon_F)\mathcal{V}$ we obtain:

$$\delta E = -\frac{1}{2} g(\epsilon_F)\Delta^2 = -2(\hbar\omega_D)^2 g(\epsilon_F) e^{-2/g(\epsilon_F)\mathcal{V}}$$

which proves that the superconducting-state energy is lower than that of the normal state.

Quasiparticle Excitations in BCS Theory

We consider next what the excited states of the BCS hamiltonian might look like. The quasiparticle excitation energy is simply defined to be the total excitation energy when an extra electron is added to the system. Suppose we add this extra electron in the single-particle state $\psi_{\mathbf{k},\uparrow}$, with its pair state $\psi_{-\mathbf{k},\downarrow}$ empty. The only effect of this on the BCS hamiltonian is to block the pair $(\mathbf{k},\uparrow;-\mathbf{k},\downarrow)$ from participating in the pairing. Thus, the energy of the total system will be increased by the energy of the added electron $\xi_{\mathbf{k}}$ and removal of terms from the total energy, Eq. (8.39), involving $\mathbf{k}$. This gives a quasiparticle energy $E_{\mathbf{k}}^{\mathrm{QP}}$ equal to:

$$E_{\mathbf{k}}^{\mathrm{QP}} = \xi_{\mathbf{k}} - 2v_{\mathbf{k}}^2 \xi_{\mathbf{k}} - 2u_{\mathbf{k}} v_{\mathbf{k}} \sum_{\mathbf{k}'} V_{\mathbf{k}\mathbf{k}'} u_{\mathbf{k}'} v_{\mathbf{k}'} \tag{8.49}$$

Using our earlier definitions of $u_{\mathbf{k}}, v_{\mathbf{k}}$ in terms of the variable $\theta_{\mathbf{k}}$, and the definitions of $\Delta_{\mathbf{k}}, E_{\mathbf{k}}$, Eqs (8.43) and (8.44), we find:

$$E_{\mathbf{k}}^{\mathrm{QP}} = \xi_{\mathbf{k}} - \left(1 - \frac{\xi_{\mathbf{k}}}{E_{\mathbf{k}}}\right)\xi_{\mathbf{k}} + \frac{\Delta_{\mathbf{k}}}{E_{\mathbf{k}}}\Delta_{\mathbf{k}} = \frac{\xi_{\mathbf{k}}^2}{E_{\mathbf{k}}} + \frac{\Delta_{\mathbf{k}}^2}{E_{\mathbf{k}}} = E_{\mathbf{k}}$$

that is, the quasiparticle excitation energy is just $E_{\mathbf{k}}$, the quantity defined in Eq. (8.44). Moreover, since the minimum energy to create one quasiparticle is $E_{\mathbf{k}} = \Delta_{\mathbf{k}}$, for $\xi_{\mathbf{k}} = 0$, this validates our earlier suggestion that $\Delta_{\mathbf{k}}$ plays the role of the quasiparticle energy gap.

Interestingly, $E_{\mathbf{k}}$ is also the energy to remove a particle from the state $\psi_{-\mathbf{k},\downarrow}$ in the system. Again, the only energies left are $\xi_{\mathbf{k}}$ and the fact that the pair $(\mathbf{k},\uparrow;-\mathbf{k},\downarrow)$ no longer contributes. Since the minimum energy to remove a particle from the system is $\Delta_{\mathbf{k}}$, and to then add the particle back into the system is also $\Delta_{\mathbf{k}}$, any electronic transition of the system must have a minimum energy of $2\Delta_{\mathbf{k}}$. This effect can be directly observed experimentally. To explain this observation, we consider contact between a superconductor and a metal. Typically, the superconducting gap is small enough to allow us to approximate the density of states in the metal as a constant over a range of energies at least equal to 2Δ around the Fermi level. This situation is equivalent to metal–semiconductor contact, as discussed in detail in Chapter 2. We can therefore apply the results of that discussion directly to the metal-superconductor contact. The measured differential conductance at $T = 0$ will be given by Eq. (2.118):

$$\left[\frac{\mathrm{d}I}{\mathrm{d}V_{\mathrm{b}}}\right]_{T=0} = e|\mathcal{T}|^2 g^{(\mathrm{S})}(\epsilon_{\mathrm{F}} + eV_{\mathrm{b}}) g_{\mathrm{F}}^{(\mathrm{M})}$$

with $\mathcal{T}$ the tunneling matrix element and $g^{(\mathrm{S})}(\epsilon), g^{(\mathrm{M})}(\epsilon)$ the density of states of the superconductor and the metal, the latter evaluated at the Fermi level, $g_{\mathrm{F}}^{(\mathrm{M})} = g^{(\mathrm{M})}(\epsilon_{\mathrm{F}})$. This shows that by scanning the bias voltage V_{b}, one samples the density of states of the superconductor. Thus, the measured differential conductance will reflect all the features of the superconductor density of states, including the superconducting gap. However, making measurements at $T = 0$ is not physically possible. At non-zero temperature $T > 0$ the occupation number $n^{(\mathrm{M})}(\epsilon)$ is not a sharp step-function but a smooth one whose derivative is an analytic representation of the δ-function (see Appendix A). In this case, when the voltage V_{b} is scanned, the measured differential conductance is also a smooth function representing a smoothed version of the superconductor density of states: the superconducting gap is still clearly evident, although not as an infinitely sharp feature.

BCS Theory at Non-zero Temperature

Finally, we consider the situation at non-zero temperature T. It is straightforward to generalize the $T = 0$ result in the following way: instead of the BCS wavefunction at $T = 0$, we will have some typical excited state $|\Psi_{BCS}(T)\rangle$ which will include the presence of quasiparticle excitations that will conspire to reduce the pairing interactions and consequently the superconducting gap. If we define the probability that a quasiparticle is excited at temperature T with energy $E_{\mathbf{k}\uparrow}(T)$ to be $f_{\mathbf{k},\uparrow}(T)$, then since the quasiparticles are to a good approximation independent fermions, we have:

$$f_{\mathbf{k},\uparrow}(T) = \frac{1}{1 + \exp[\beta E_{\mathbf{k},\uparrow}(T)]}$$

where $\beta = 1/k_B T$. Let us now consider adding a quasiparticle in the state $\mathbf{k}$ to the system at $T > 0$. This will be the same as the result in Eq. (8.49) for $T = 0$, except that we now need to multiply the third term in that equation by the probabilities that the quasiparticle states $\psi_{\mathbf{k}',\uparrow}$ and $\psi_{-\mathbf{k}',\downarrow}$ are empty at $T > 0$. For low T, and small values of $f_{\mathbf{k}',\uparrow} = f_{-\mathbf{k}',\downarrow}$, this gives:

$$E_{\mathbf{k}}^{QP}(T) = \xi_{\mathbf{k}}\left(1 - 2v_{\mathbf{k}}^2\right) - 2u_{\mathbf{k}}v_{\mathbf{k}}\sum_{\mathbf{k}'} V_{\mathbf{k}\mathbf{k}'} u_{\mathbf{k}'} v_{\mathbf{k}'} \left(1 - f_{\mathbf{k}',\uparrow} - f_{-\mathbf{k}',\downarrow}\right)$$

$$= \xi_{\mathbf{k}}\left(1 - 2v_{\mathbf{k}}^2\right) - 2u_{\mathbf{k}}v_{\mathbf{k}}\sum_{\mathbf{k}'} V_{\mathbf{k}\mathbf{k}'} u_{\mathbf{k}'} v_{\mathbf{k}'} \tanh\left(\frac{E_{\mathbf{k}'}(T)}{2k_B T}\right)$$

If we now define a new gap

$$\Delta_{\mathbf{k}}(T) \equiv -\sum_{\mathbf{k}'} V_{\mathbf{k}\mathbf{k}'} u_{\mathbf{k}'} v_{\mathbf{k}'} \tanh\left(\frac{E_{\mathbf{k}'}(T)}{2k_B T}\right)$$

then the gap equation takes the form

$$\Delta_{\mathbf{k}}(T) = -\frac{1}{2}\sum_{\mathbf{k}'} V_{\mathbf{k}\mathbf{k}'} \frac{\Delta_{\mathbf{k}'}(T)}{\sqrt{\xi_{\mathbf{k}'}^2 + \Delta_{\mathbf{k}'}^2(T)}} \tanh\left(\frac{E_{\mathbf{k}'}(T)}{2k_B T}\right) \tag{8.50}$$

with

$$E_{\mathbf{k}}(T) = \sqrt{\xi_{\mathbf{k}'}^2 + \Delta_{\mathbf{k}'}^2(T)}$$

Note that when $T = 0$, we get $f_{\mathbf{k},s} = 0$:

$$\tanh\left(\frac{E_{\mathbf{k}'}(T)}{2k_B T}\right) = 1$$

and we recover the corresponding expression as in Eq. (8.45). The new equation for the temperature-dependent gap can be used with the BCS model to calculate the transition temperature T_c at which the superconductor becomes normal, that is, $\Delta(T_c) = 0$. In the weak coupling limit, this becomes (see Problem 5):

$$k_B T_c = 1.14 \hbar \omega_D e^{-1/g_F V_0} \tag{8.51}$$

This relation provides an explanation for the so-called "isotope effect," that is, the dependence of T_c on the mass of the ions involved in the crystal. From our discussion of phonons in Chapter 7, we can take the Debye frequency to be:

$$\omega_D \sim \left(\frac{\kappa}{M} \right)^{1/2}$$

where κ is the relevant force constant and M the mass of the ions; this leads to $T_c \sim M^{-1/2}$, the relation that quantifies the isotope effect. Moreover, combining our earlier result for Δ at zero temperature, Eq. (8.48), with Eq. (8.51), we find:

$$2\Delta = 3.52 k_B T_c \tag{8.52}$$

where 2Δ is the zero-temperature value of the superconducting gap. This relation, referred to as the "law of corresponding states," is obeyed quite accurately by a wide range of conventional superconductors, confirming the BCS theory.

8.4.2 The McMillan Formula for T_c

We close this section with a brief discussion of how the T_c in *conventional* superconductors, arising from electron–phonon coupling as described by the BCS theory, can be calculated. McMillan[9] proposed a formula to evaluate T_c as:

$$T_c = \frac{\Theta_D}{1.45} \exp\left[-\frac{1.04(1 + \lambda)}{\lambda - \mu^* - 0.62\lambda\mu^*} \right] \tag{8.53}$$

where Θ_D is the Debye temperature, λ is a constant describing electron–phonon coupling strength, and μ^* is another constant describing the repulsive Coulomb interaction strength. This expression is valid for $\lambda < 1.25$. The value of μ^* is difficult to obtain from calculations, but in any case the value of this constant tends to be small. For *sp* metals like Pb, its value has been estimated from tunneling measurements to be $\mu^* \sim 0.1$; for other cases this value is scaled by the density of states at the Fermi level, g_F, taking the value of g_F for Pb as the norm. The exponential dependence of T_c on the other parameter, λ, necessitates a more accurate estimate of its value. It turns out that λ can actually be obtained from electronic structure calculations. We outline the calculation of λ following the treatment of Chelikowsky and Cohen.[10] λ is expressed as the average over all phonon modes of the constants $\lambda_{\mathbf{k}}^{(l)}$:

$$\lambda = \sum_l \int \lambda_{\mathbf{k}}^{(l)} d\mathbf{k}$$

[9] W. G. McMillan, *Phys. Rev.* **167**, 331 (1968).
[10] J. R. Chelikowsky and M. L. Cohen, "Ab initio pseudopotentials and the structural properties of semiconductors," in P. T. Landsberg (Ed.), *Handbook on Semiconductors*, Vol. 1, p. 59 (North Holland, Amsterdam, 1992).

which describes the coupling of an electron to a particular phonon mode identified by the index l and the wave-vector $\mathbf{k}$ (see Chapter 7):

$$\lambda_{\mathbf{k}}^{(l)} = \frac{2g_F}{\hbar \omega_{\mathbf{k}}^{(l)}} \left\langle |\mathcal{M}(n, \mathbf{q}, n', \mathbf{q}'; l, \mathbf{k})|^2 \right\rangle_F$$

where $\mathcal{M}(n, \mathbf{q}, n', \mathbf{q}'; l, \mathbf{k})$ is the electron–phonon matrix element and $\langle \cdots \rangle_F$ denotes an average over the Fermi surface. The electron–phonon matrix elements are given by:

$$\mathcal{M}(n, \mathbf{q}, n', \mathbf{q}'; l, \mathbf{k}) = \sum_j \sqrt{\frac{\hbar}{2M_j \omega_{\mathbf{k}}^{(l)}}} \langle \psi_{\mathbf{q}}^{(n)} | \left[\hat{\mathbf{e}}_{\mathbf{k}j}^{(l)} \cdot \frac{\delta \mathcal{V}}{\delta \mathbf{t}_j} \right] | \psi_{\mathbf{q}'}^{(n')} \rangle \delta(\mathbf{q} - \mathbf{q}' - \mathbf{k}) \qquad (8.54)$$

where the summation on j is over the ions of mass M_j at positions $\mathbf{t}_j$ in the unit cell, $\psi_{\mathbf{q}}^{(n)}, \psi_{\mathbf{q}'}^{(n')}$ are electronic wavefunctions in the ideal crystal, and $\hat{\mathbf{e}}_{\mathbf{k}j}^{(l)}$ is the phonon polarization vector (more precisely, the part of the polarization vector corresponding to the position of ion j). The term $\delta \mathcal{V}/\delta \mathbf{t}_j$ is the change in crystal potential due to the presence of the phonon. This term can be evaluated as:

$$\sum_l \hat{\mathbf{e}}_{\mathbf{k}j}^{(l)} \cdot \frac{\delta \mathcal{V}}{\delta \mathbf{t}_j} = \frac{\mathcal{V}_{\mathbf{k}}^{(l)} - \mathcal{V}_0}{u_{\mathbf{k}}^{(l)}} \qquad (8.55)$$

where $\mathcal{V}_0$ is the ideal crystal potential, $\mathcal{V}_{\mathbf{k}}^{(l)}$ is the crystal potential in the presence of the phonon mode identified by $l, \mathbf{k}$, and $u_{\mathbf{k}}^{(l)}$ is the average ionic displacement corresponding to this phonon mode. The terms in Eq. (8.55) can readily be evaluated through the computational methods discussed in Chapter 6, by introducing ionic displacements $\mathbf{u}_{\mathbf{k}j}^{(l)}$ corresponding to a phonon mode and evaluating the crystal potential difference resulting from this distortion, with:

$$u_{\mathbf{k}}^{(l)} = \frac{1}{N_{\text{ion}}} \left(\sum_j |\mathbf{u}_{\mathbf{k}j}^{(l)}|^2 \right)^{1/2}$$

the ionic displacement averaged over the N_{ion} ions in the unit cell. Using this formalism, it was predicted and later verified experimentally that Si under high pressure would be a superconductor.[11]

8.4.3 High-Temperature Superconductors

Finally, we discuss very briefly the new class of high-temperature superconductors, discovered in 1986 by Bednorz and Müller, in which the T_c is much higher than in conventional superconductors: in general it is in the range of ~ 90 K, but in certain compounds it can exceed 130 K (see Table 8.2). This is well above the freezing point of N_2 (77 K), so this much more abundant and cheap substance can be used as the coolant to bring the superconducting materials below their critical point. The new superconductors have complex crystal structures characterized by Cu–O octahedra arranged in various ways, as in

[11] K. J. Chang and M. L. Cohen, *Phys. Rev. B* **31**, 7819 (1985).

Table 8.2 Critical temperature T_c (K) of representative high-temperature superconductors. In several cases there is fractional occupation of the dopant atoms, denoted by x (with $0 < x < 1$), or there are equivalent structures with two different elements, denoted by (X, Y); in such cases, the value of the highest T_c is given. *Source:* D. R. Lide (Ed.), *CRC Handbook of Chemistry and Physics* (CRC Press, Boca Raton, FL, 1999–2000).

Material	T_c	Material	T_c
$La_2CuO_{4+\delta}$	39	$SmBaSrCu_3O_7$	84
$La_{2-x}(Sr,Ba)_xCuO_4$	35	$EuBaSrCu_3O_7$	88
$La_2Ca_{1-x}Sr_xCu_2O_6$	60	$GdBaSrCu_3O_7$	86
$YBa_2Cu_3O_7$	93	$DyBaSrCu_3O_7$	90
$YBa_2Cu_4O_8$	80	$HoBaSrCu_3O_7$	87
$Y_2Ba_4Cu_7O_{15}$	93	$YBaSrCu_3O_7$	84
$Tl_2Ba_2CuO_6$	92	$ErBaSrCu_3O_7$	82
$Tl_2CaBa_2Cu_2O_8$	119	$TmBaSrCu_3O_7$	88
$Tl_2Ca_2Ba_2Cu_3O_{10}$	128	$HgBa_2CuO_4$	94
$TlCaBa_2Cu_2O_7$	103	$HgBa_2CaCu_2O_6$	127
$TlSr_2Y_{0.5}Ca_{0.5}Cu_2O_7$	90	$HgBa_2Ca_2Cu_3O_8$	133
$TlCa_2Ba_2Cu_3O_8$	110	$HgBa_2Ca_3Cu_4O_{10}$	126
$Bi_2Sr_2CuO_6$	10	$Pb_2Sr_2La_{0.5}Ca_{0.5}Cu_3O_8$	70
$Bi_2CaSr_2Cu_2O_8$	92	$Pb_2(Sr,La)_2Cu_2O_6$	32
$Bi_2Ca_2Sr_2Cu_3O_{10}$	110	$(Pb,Cu)Sr_2(La,Ca)Cu_2O_7$	50

the perovskites (see discussion of these types of structures in Chapter 1) and decorated by various other elements. In the following we will refer to the conventional, low-temperature superconductors as the conventional superconductors, to distinguish them from the high-temperature kind. We give examples of high-temperature superconductors in Table 8.2.

The discovery of high-temperature superconductors has opened the possibility of many practical applications, but these materials are ceramics and therefore more difficult to utilize than the conventional superconductors, which are typically elemental metals or simple metallic alloys. It also reinvigorated theoretical interest in superconductivity, since it seemed doubtful that the microscopic mechanisms responsible for low-temperature superconductivity could also explain its occurrence at such high temperatures. J. G. Bednorz and K. A. Müller were awarded the 1987 Nobel Prize for Physics for their discovery, which has sparked an extraordinary amount of activity, both in experiment and in theory, to understand the physics of the high-temperature superconductors.

We give here a short discussion of the physics of high-temperature superconductors, mostly in order to bring out their differences from conventional ones. We first review the main points of the theory that explain the physics of conventional superconductors, as derived in the previous section. These are:

(a) Electrons form pairs, called Cooper pairs, due to an attractive interaction mediated by the exchange of phonons. In a Cooper pair, the electrons have opposite momenta and opposite spins in a spin-singlet configuration.

(b) Cooper pairs combine to form a many-body wavefunction which has lower energy than the normal, non-superconducting state, and represents a coherent state of the entire electron gas. Excitations above this ground state are represented by quasiparticles which correspond to broken Cooper pairs.

(c) In order to derive simple expressions relating the superconducting gap Δ to the critical temperature (or other experimentally accessible quantities), we assume that there is no dependence of the gap, the quasiparticle energies, or the attractive interaction potential on the electron wave-vectors $\mathbf{k}$. This assumption implies an isotropic solid and a spherical Fermi surface. This model is referred to as "s-wave" pairing, due to the lack of any spatial features in the physical quantities of interest (similar to the behavior of an s-like atomic wavefunction, see Appendix C).

(d) We also assume that we are in the limit in which the product of the density of states at the Fermi level with the strength of the interaction potential, $g_F V_0$, which is a dimensionless quantity, is much smaller than 1; we call this the weak-coupling limit.

The high-temperature superconductors based on copper oxides (see Table 8.2) conform to some, but not all, of these points. Points (a) and (b) apparently apply to those systems as well, with one notable exception: there seems to exist grounds for doubting that electron–phonon interactions alone can be responsible for the pairing of electrons. Magnetic order plays an important role in the physics of these materials, and the presence of strong antiferromagnetic interactions may be intimately related to the mechanism(s) of electron pairing. On the other hand, some variation of the basic theme of electron–phonon interaction, also taking into account the strong anisotropy in these systems (which is discussed next), may be able to capture the reason for electron pairing. Point (c) seems to be violated by these systems in several important ways. The copper-oxide superconductors are strongly anisotropic, having as a main structural feature planes of linked Cu–O octahedra which are decorated by other elements, as illustrated in Fig. 8.7. In addition to this anisotropy of the crystal in real space, there exist strong indications that important physical

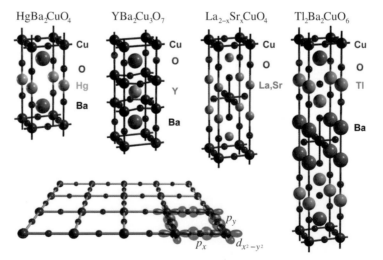

Fig. 8.7 Examples of the structure of high-T_c compounds (see Table 8.2). The diagram on the lower left shows the Cu–O planes that are common to all structures, with the p_x, p_y orbitals of O atoms (red clouds) and $d_{x^2-y^2}$ orbitals of Cu atoms (blue clouds) involved in forming the in-plane bonds [adapted from N. Barisic *et al.*, *PNAS* **110**, 12235 (2013)].

quantities like the superconducting gap are not featureless, but possess structure and hence have a strong dependence on the electron wave-vectors. These indications point to what is called "d-wave" pairing, that is, dependence of the gap on $\mathbf{k}$, similar to what is exhibited by a d-like atomic wavefunction. Finally, point (d) also appears to be strongly violated by the high-temperature superconductors, which seem to be in the "strong-coupling" limit. Indications to this effect come from the very short coherence length in these systems, which is of order a few lattice constants as opposed to $\sim 10^4 \text{Å}$, and from the fact that the ratio $2\Delta/k_B T_c$ is not 3.52, as the weak-coupling limit of BCS theory, Eq. (8.52), predicts but is in the range 4–7. These departures from the behavior of conventional superconductors have sparked much theoretical debate on what are the microscopic mechanisms responsible for this exotic behavior.

Further Reading

1. *Vibrational Spectroscopy in Life Science*, F. Siebert and P. Hildenbrandt (Wiley-VCH, Weinheim, 2008). This book offers a pedagogical introduction to the theory of infrared absorption and Raman spectroscopy and its applications to biomolecules.

2. *Theory of Superconductivity*, J. R. Schrieffer (Benjamin/Cummings, Reading, MA, 1964). This is a classic account of the BCS theory of superconductivity.

3. *Introduction to Superconductivity*, M. Tinkham (2nd edn, McGraw-Hill, New York, 1996). This is a standard reference, with comprehensive coverage of all aspects of superconductivity, including the high-temperature copper-oxide superconductors.

4. *Superconductivity in Metals and Alloys*, P. G. de Gennes (Addison-Wesley, Reading, MA, 1966).

5. R. Beyers and T. M. Shaw, "The structure of YBCO and its derivatives," in H. Ehrenreich and D. Turnbull (Eds), *Solid State Physics*, Vol. 42, pp. 135–212 (Academic Press, Boston, MA, 1989). This article reviews the structure of representative high-temperature superconductor ceramic crystals.

6. K. C. Hass, "Electronic structure of copper-oxide semiconductors," in H. Ehrenreich and D. Turnbull (Eds), *Solid State Physics*, Vol. 42, pp. 213–270 (Academic Press, Boston, MA, 1989). This article reviews the electronic properties of HTSC-related ceramic crystals.

7. "Electronic structure of the high-temperature oxide superconductors," W. E. Pickett, *Rev. Mod. Phys.*, **61**, 433 (1989). This is a thorough review of experimental and theoretical studies of the electronic properties of the oxide superconductors.

Problems

1. Provide the steps in the derivation of the Debye–Waller factor, Eq. (8.12), starting with Eq. (8.11), the general expression for W that appears in the average of the absolute value squared of the structure factor $\langle |\mathcal{S}(\mathbf{q})|^2 \rangle$.

2.   Calculate the amplitude for Raman scattering to zeroth and first order in the phonon-mode amplitudes Q_s, using the Taylor expansion of the expectation of the dipole moment, Eq. (8.20), and the matrix elements of Eq. (8.19). These are known as the A and B Albrecht terms; explain what types of processes these terms describe.

3.   Prove the commutator relations between pair creation–destruction operators, Eq. (8.34), using their definition in terms of the single-particle creation–destruction operators, Eq. (8.33), and their commutator relations, Eq. (8.32).

4.   Prove the expression for N_{RMS}^2 given in Eq. (8.37) through steps similar to what we used to derive the expression for $\overline{N}$, Eq. (8.36).

5.   Starting from the non-zero temperature BCS gap equation, Eq. (8.50), and using a BCS model similar to the $T = 0$ case but with the gap being a temperature-dependent quantity $\Delta(T)$, show that:

     (a) The gap equation at non-zero temperature gives:

     $$\int_0^{\hbar\omega_D} \frac{1}{\sqrt{\xi^2 + \Delta^2(T)}} \tanh\left(\frac{\sqrt{\xi^2 + \Delta^2(T)}}{2k_B T}\right) d\xi = \frac{1}{g_F V_0}$$

     From this expression prove that the gap $\Delta(T)$ is a monotonically decreasing function of T, which becomes 0 at some temperature that we identify as the transition temperature T_c.

     (b) Show that for $T = T_c$, in the weak coupling limit, the above equation yields Eq. (8.51). By comparing this with the result at zero temperature, Eq. (8.48), prove the law of corresponding states, Eq. (8.52).

Dynamics and Topological Constraints

In previous chapters we examined the behavior of electrons in solids when they are in their ground state, or when they are excited within the single-particle picture, from a state of energy below to one above the Fermi level. In all these discussions we did not explicitly consider the motion of electrons. In the present chapter, we shall study the *dynamics* of electrons in solids, that is, how they move as a function of time under the influence of external fields. We do this in three steps: we first construct a very simple picture which is intuitive but incomplete, yet gives a general sense of how we approach these issues; we next describe fundamental properties of the electron wavefunctions when they evolve as functions of time, in which the phase of each wavefunction becomes of central importance – this is the idea of the "geometric" or Berry's phase; finally, we take a close look at the physics of topological constraints that can arise and their effect on the dynamics of electrons, especially those related to edges and surfaces of solids.

As before, we will rely on the picture of electrons in crystals, whose behavior captures much of the behavior of common materials and can be used as the basis for more general treatments. Before we treat the dynamics of crystal electrons in detail, we shall review what one might expect for the motion of electrons in external electromagnetic fields, using a simplified picture where the effects of crystal periodicity are absent. The experimentally observed motion of electrons confined in two dimensions is richer and more interesting than might be expected from simple considerations. Striking examples are the quantized Hall effects (integer and fractional), which can be explained without taking into account the effect of the crystal potential, and topological insulators, where the Bloch character of states plays an important role.

9.1 Electrons in External Electromagnetic Fields

The response of electrons to external electromagnetic fields is a standard means of probing the properties of solids. It is also a common way to create useful devices that exploit the properties of materials. This includes electronic, optical, and magnetic devices, but also devices that take advantage of the coupling of mechanical forces and electrical polarization in solids. The response of electrons to electromagnetic fields is particularly interesting when the electrons are confined in two dimensions. Therefore, we shall focus on that configuration. This may seem to be a rather artificial constraint, but it is not. Recent advances in materials synthesis have made it possible to produce physical realizations

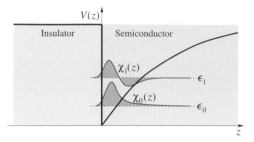

Fig. 9.1 The confining potential $V(z)$ in the z direction (thick black line) and the energies ϵ_0, ϵ_1 and wavefunctions $\chi_0(z), \chi_1(z)$ of the two lowest states, at an insulator–semiconductor junction. The Fermi level lies between ϵ_0 and ϵ_1.

of the 2D sheet of electrons, either in atomically thin layers like graphene, or through the creation of quantum wells. We review briefly the justification for 2D confinement of electrons in the latter case.

Through advanced methods of crystal growth, it is possible to form an atomically flat interface between two crystals. We will take x, y to be the coordinate axes on the interface plane and z the direction perpendicular to it; the magnetic field will be $\mathbf{B} = B\hat{z}$. By combining an insulator on one side and a doped semiconductor on the other side of the interface, it is possible to produce a potential well in the direction perpendicular to the interface plane which quantizes the motion of electrons in this direction. This is illustrated in Fig. 9.1: the confining potential $V(z)$ can be approximated as nearly linear for small values of z on the doped semiconductor side and a hard wall on the insulator side. In this simple case, the electron wavefunctions $\psi(\mathbf{r})$ can be factored into two parts:

$$\psi(\mathbf{r}) = \bar{\psi}(x, y)\chi_n(z)$$

where the first part $\bar{\psi}(x, y)$ describes the motion on the interface plane and the second part $\chi_n(z)$ describes motion perpendicular to it. The latter part experiences the confining potential $V(z)$, which gives rise to discrete levels. When these levels are well separated in energy and the temperature is lower than the separation between levels, only the lowest level is occupied. When all electrons are at this lowest level, corresponding to the wavefunction $\chi_0(z)$, their motion in the z direction is confined, making the system of electrons essentially a 2D one. This arrangement is called a quantum well and the system of electrons confined in two dimensions in the semiconductor is called an inversion layer. A more detailed discussion of electronic states in doped semiconductors and the inversion layer is given in Chapter 5.

In the discussion of this section, we will momentarily ignore the crystal momentum $\mathbf{k}$ and band index normally associated with electronic states in the crystal, because the density of confined electrons in the inversion layer is such that only a very small fraction of one band is occupied. Specifically, the density of confined electrons is of order $n = 10^{12}$ cm^{-2}, whereas a full band corresponding to a crystal plane can accommodate of order $[10^{24}$ cm$^{-3}]^{2/3} = 10^{16}$ cm^{-2} states. Therefore, the confined electrons occupy a small fraction of the band near its minimum. It should be noted that the phenomena we will describe below can also be observed for a system of holes at similar density. For both the electron and the hole systems, the presence of the crystal interface is crucial in achieving confinement in two dimensions, but is otherwise not important in the behavior of the charge carriers (electrons or holes) under the conditions considered here. In the case of atomically thin 2D crystals, like graphene and other layered materials, it becomes important to include the in-plane $\mathbf{k}$ vector of the electronic states; we return to this issue and give a more complete treatment in following sections of this chapter.

9.1.1 Classical Hall Effect

We consider a classical treatment of what happens when electrons confined in a 2D plane, (x, y), are subject to an electric field in the x direction, $E\hat{x}$, and a magnetic field in the z direction, $B\hat{z}$, perpendicular to the plane, as shown in Fig. 9.2 (in the following discussion the components without superscripts are the externally applied fields). The applied electric field produces a force on the electrons:

$$\mathbf{F}^{\text{ele}} = -e\mathbf{E}$$

which will induce a current density:

$$\mathbf{j} = \frac{-e}{A} \sum_{i-1}^{N_e} \mathbf{v}_i$$

where A is the area of the sample of length L and width W, $A = LW$, and the sum runs over all electrons N_e in the system. Instead of following the course of each electron, we define the drift velocity:

$$\mathbf{v} = \frac{1}{N_e} \sum_{i=1}^{N_e} \mathbf{v}_i \Rightarrow \mathbf{j} = \frac{-e}{A} N_e \mathbf{v} = n(-e)\mathbf{v}$$

where we have introduced the density of electrons on the plane, $n = N_e/A$; note that the electron drift velocity is opposite to the current flow, as shown in Fig. 9.2. Due to the

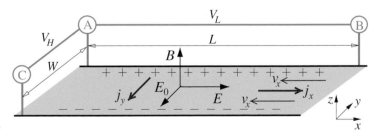

Fig. 9.2 Geometry of the 2D electrons in the Hall effect, with external magnetic field $\mathbf{B} = B\hat{z}$ and the x, y components of the electric field $\mathbf{E} = E\hat{x} - E_0\hat{y}$. The blue arrows indicate the initial current in the x direction, j_x, and the current induced by the Lorentz force, j_y; the red arrows indicate the electron velocity, v_x, due to the potential difference V_L.

presence of the magnetic field, the moving electrons will also experience a Lorentz force perpendicular to their velocity:

$$\mathbf{F}^{\mathrm{mag}} = -\frac{e}{c}\mathbf{v} \times \mathbf{B}$$

Because of the limits in electron motion imposed by the geometry of the sample, which has a width $W \ll L$, the electrons pushed by the Lorentz force will end up accumulating on one side of the sample along the y direction, and will be depleted from the other side. This imbalance in charge distribution will result in an effective electric field, denoted as $E_0(-\hat{y})$ in Fig. 9.2, which acts to oppose the further accumulation of electrons on one side of the sample. When equilibrium is established, the effect of this field will be to cancel the Lorentz force, and the resulting motion will only be along the x direction.[1] The electrons will also experience a drag force due to scattering, which is typically proportional to their velocity, with $\overline{m}\eta$ the drag coefficient and $\overline{m}$ the effective mass:

$$\mathbf{F}^{\mathrm{drag}} = -\overline{m}\eta\mathbf{v}$$

The total force will be given by the expression

$$\overline{m}\frac{d\mathbf{v}}{dt} = \mathbf{F}^{\mathrm{tot}} = \mathbf{F}^{\mathrm{ele}} + \mathbf{F}^{\mathrm{mag}} + \mathbf{F}^{\mathrm{drag}} = -e\mathbf{E} - \frac{e}{c}\mathbf{v} \times \mathbf{B} - \overline{m}\eta\mathbf{v}$$

where we have used $\overline{m}$ as the effective mass for electrons, allowing for the fact that they are interacting with other electrons and their crystal environment. Under steady-state conditions, the drift velocity will not change, which gives:

$$-e\mathbf{E} - \frac{e}{c}\mathbf{v} \times \mathbf{B} - \overline{m}\eta\mathbf{v} = 0 \Rightarrow \mathbf{v} = -\frac{1}{\overline{m}\eta}\left[e\mathbf{E} + \frac{e}{c}\mathbf{v} \times \mathbf{B}\right]$$

and from this we can calculate the steady-state current:

$$\mathbf{j} = n(-e)\mathbf{v} = \frac{ne^2}{\overline{m}\eta}\mathbf{E} + \frac{ne^2}{\overline{m}\eta c}\mathbf{v} \times \mathbf{B}$$

We can rewrite this last expression in terms of the current as:

$$\left[\mathbf{j} - \frac{e}{\overline{m}\eta c}\mathbf{B} \times \mathbf{j}\right] = \frac{ne^2}{\overline{m}\eta}\mathbf{E} \tag{9.1}$$

with $\mathbf{E}$ the total electric field, consisting of two components: the externally applied field $E_x = E$ and the field generated by the steady-state distribution of charges, $E_y = -E_0$, which opposes the magnetic force. The current is expressed in terms of the resistivity tensor $\underline{\rho}$ and the total electric field $\mathbf{E} = E_x\hat{x} + E_y\hat{y}$ as:

$$\underline{\rho} \cdot \mathbf{j} = \mathbf{E} \Rightarrow \begin{pmatrix} \rho_{xx} & \rho_{xy} \\ \rho_{yx} & \rho_{yy} \end{pmatrix}\begin{pmatrix} j_x \\ j_y \end{pmatrix} = \begin{pmatrix} E_x \\ E_y \end{pmatrix}$$

---

[1] We caution the reader that this is actually an approximation because in two dimensions the Coulomb interaction that falls as $1/r$ is not completely screened at small momenta and therefore the long-range tail can never be screened by a localized charge distribution. This is resolved by some power-law dependence of the distribution of the screening charge.

which, taking into account that $\mathbf{B} = B\hat{z}$ and the above expression relating the current and electric and magnetic fields, Eq. (9.1), leads to the following results for the diagonal and off-diagonal components of the resistivity tensor:

$$\rho_{xx} = \rho_{yy} = \rho_0 = \frac{\overline{m}\eta}{ne^2}, \quad \rho_{xy} = -\rho_{yx} = \frac{B}{nec} \tag{9.2}$$

Often, the conductivity tensor $\underline{\sigma}$ is also used to express the relationship between the current and the electric field as $\mathbf{j} = \underline{\sigma} \cdot \mathbf{E}$. The conductivity tensor is the inverse of the resistivity tensor, $\underline{\sigma} = \underline{\rho}^{-1}$, and can easily be derived from the expressions given above (see Problem 1).

With the above definitions, we are now in a position to describe quantitatively the classical picture of the Hall effect for the geometry shown in Fig. 9.2. In experiments, one measures the voltage V_L along the x direction (the direction of the external electric field), over the length L of the sample between leads A and B, and along the y direction, V_H, along the width W of the sample, between leads A and C. The conditions for the measurement of V_L and V_H are under constant current I in the x direction. The current density is $j_x = I/W$, since W is the cross-section of the sample through which the current in the x direction flows. With these definitions, the electric field and the voltage in the longitudinal (x) and transverse (y) directions are related by $V_L = E_x L$ and $V_H = E_y W$, which lead to the expressions for the resistance in the two directions:

$$R_L = \frac{V_L}{I} = \frac{E_x L}{j_x W} = \rho_{xx}\frac{L}{W} = \rho_0 \frac{L}{W}, \quad R_W = \frac{V_H}{I} = \frac{E_y W}{j_x W} = \rho_{yx} = -\frac{B}{nec}$$

These relations give, for the behavior of the longitudinal and transverse (Hall) voltages as functions of the magnetic field:

$$V_L \sim \text{constant}, \quad V_H \sim \text{linear in B}$$

This is indeed what experiments show, for relatively small values of the field, $B < 0.5\,\text{T}$ (see Fig. 9.3). In 1980, pioneering work by von Klitzing and coworkers[2] showed that for high magnetic fields and low temperatures, the Hall resistivity ρ_{xy} as a function of B has wide plateaus that correspond to quantized values of $(h/e^2)/(l + 1)$ with l an integer; moreover, when ρ_{xy} has these quantized values, the diagonal resistivity ρ_{xx} vanishes (see Fig. 9.2). This fascinating observation was called the integer quantum Hall effect (IQHE); K. von Klitzing was awarded the 1985 Nobel Prize for Physics for its discovery. To explain this behavior, we have to provide a proper quantum-mechanical description of the motion of electrons on the 2D plane in the presence of the external magnetic field.

9.1.2 Landau Levels

Up to this point we have treated the electrons as a set of classical charge carriers, except for the confinement in the z direction which introduces quantization of the motion along z, and localization in the x, y plane. Now we introduce the effects of quantization in the remaining two dimensions. We begin with the Schrödinger equation that describes the

[2] K. von Klitzing, G. Dorda, and M. Pepper, *Phys. Rev. Lett.* **45**, 494 (1980).

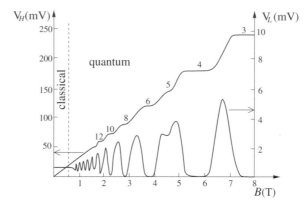

Fig. 9.3 Schematic representation of experimental measurements in the Hall effect For small values of the magnetic field, $B < 0.5$ T), the behavior of the measured voltages V_H. (blue curve, left axis) and V_L (red curve, right axis) is consistent with the classical picture. For $B > 0.5$ T, the voltages show quantized values, which become more pronounced as B increases: The Hall voltage V_H exhibits plateaus for certain regions of B values, which correspond to quantized values of the resistivity ρ_{xy} given by $(h/e^2)/(l+1)$ with l an integer, and for the same regions the longitudinal voltage V_L, and hence the diagonal conductivity ρ_{xx}, vanishes.

motion of electrons with effective mass $\overline{m}$ in the x, y plane. The presence of the magnetic field is represented by the vector potential $\mathbf{A}(\mathbf{r})$, which we choose as:

$$\mathbf{A}(\mathbf{r}) = -By\hat{x} \Rightarrow \mathbf{B} = \nabla_{\mathbf{r}} \times \mathbf{A}(\mathbf{r}) = B\hat{z} \qquad (9.3)$$

a choice known as the "Landau gauge." As far as the presence of the electric field is concerned, an estimate of the relative magnitude of the components E_x and E_y, for typical setups where $L \gg W$ in the geometry for the Hall measurements shown in Fig. 9.2, indicates that E_y is larger than E_x by at least two orders of magnitude. In other words, other than setting up the current in the x direction, the electric field component E_x can be ignored compared to the E_y component. We let the value of the latter, at steady state, be equal to $-E_0$, so that the potential energy of electrons associated with this field, which appears in the hamiltonian, is $\mathcal{V}(\mathbf{r}) = eE_0y$. The full hamiltonian is then:

$$\hat{\mathcal{H}}_L = \frac{1}{2\overline{m}}\left(\mathbf{p} + \frac{e}{c}\mathbf{A}(\mathbf{r})\right)^2 + eE_0y = \frac{1}{2\overline{m}}\left[\left(p_x - \frac{eBy}{c}\right)^2 + p_y^2\right] + eE_0y \qquad (9.4)$$

The solutions to this hamiltonian are of the form

$$\overline{\psi}(x, y) = \frac{1}{\sqrt{L}}e^{ik_x x}\phi_{k_x}(y), \quad k_x = \frac{2\pi k}{L}, \quad k : \text{integer}$$

where we have imposed Born–von Karman boundary conditions in the x direction through the choice of the wave-vector k_x, a process with which we are familiar from the discussion of Chapter 2. We next substitute this expression for the wavefunction in the Schrödinger equation to obtain:

$$\left\{\frac{1}{2\overline{m}}\left[\left(\hbar k_x - \frac{eBy}{c}\right)^2 + p_y^2\right] + eE_0y\right\}\phi_{k_x}(y) = \epsilon_{k_x}\phi_{k_x}(y)$$

To solve for the eigenvalues and eigenfunctions of this equation, we add to and subtract from the hamiltonian the term

$$\frac{\overline{m}}{2}\left(\frac{cE_0}{B}\right)^2$$

which allows us to complete the square in the y variable, and we arrive at:

$$\left[\frac{p_y^2}{2\overline{m}} + \frac{1}{2}\overline{m}\left(\frac{eB}{\overline{m}c}\right)^2(y-\bar{y})^2\right]\phi_{k_x}(y) = \left[\epsilon_{k_x} - \frac{\overline{m}c^2}{2}\left(\frac{E_0}{B}\right)^2 + eE_0\bar{y}\right]\phi_{k_x}(y) \qquad (9.5)$$

where we have introduced the variable $\bar{y}$, defined as:

$$\bar{y} = \frac{\hbar k_x c}{eB} + \frac{\overline{m}c^2 E_0}{eB^2} \qquad (9.6)$$

The equation for the eigenfunctions and eigenvalues of the y variable has now become that of the harmonic oscillator potential with the origin shifted to $\bar{y}$, and the energy levels of the harmonic oscillator are shifted by the two terms that appear in the square bracket of the right-hand side. The spring constant κ for the harmonic potential and the corresponding characteristic frequency ω_c are given by:

$$\kappa = \overline{m}\left(\frac{eB}{\overline{m}c}\right)^2 \Rightarrow \omega_c = \left(\frac{\kappa}{\overline{m}}\right)^{1/2} = \frac{eB}{\overline{m}c} \qquad (9.7)$$

with the latter referred to as the "cyclotron frequency"; in terms of this, the energy levels of the harmonic oscillator are given by:

$$\bar{\epsilon}^{(j)} = \left(j + \frac{1}{2}\right)\hbar\omega_c, \quad j = 0, 1, 2, \ldots$$

(see Appendix C). The original eigenvalues of the energy will then be given by:

$$\epsilon_{k_x}^{(j)} = \left(j + \frac{1}{2}\right)\hbar\omega_c + \frac{\overline{m}c^2}{2}\left(\frac{E_0}{B}\right)^2 - eE_0\bar{y} = \left(j + \frac{1}{2}\right)\hbar\omega_c - \frac{\overline{m}c^2}{2}\left(\frac{E_0}{B}\right)^2 - \hbar k_x c\frac{E_0}{B} \qquad (9.8)$$

These energy levels are known as "Landau levels"; the full wavefunctions $\bar{\psi}(x,y)$ consist of a component $\phi_{k_x}^{(j)}(y)$ coming from the harmonic potential in the y variable, centered at $\bar{y}$, and a plane-wave component in the x variable with wave-vector k_x.

How many electrons can each Landau level accommodate? To answer this question, we consider the different possible values of $\bar{y}$, the position of the center of the harmonic oscillator wavefunction. From the definition of the value of $\bar{y}$, Eq. (9.6), we have:

$$\delta\bar{y} = \frac{\hbar c}{eB}\delta k_x = \frac{\hbar c}{eB}\frac{2\pi\,\delta k}{L} = \left(\frac{hc}{e}\right)\frac{1}{BL} \qquad (9.9)$$

since k is an integer (recall that $\hbar = h/2\pi$). $\delta\bar{y}$ are the steps in the value of $\bar{y}$, and since the total width of the sample in the y direction is W, the number of possible positions of the center is given by:

$$N_{\mathrm{L}} = \frac{W}{\delta\bar{y}} = \frac{e}{hc}(WL)B = \frac{e}{hc}AB$$

where we have used $A = WL$ for the area of the sample. This is a very interesting result, because $\Phi = AB$ is the total magnetic flux through the sample and the quantity $\phi_0 = (hc/e)$ is the magnetic flux quantum, so this result shows that the number of states accommodated by the Landau level is the total number of flux quanta:

$$\phi_0 = \frac{hc}{e} \Rightarrow N_L = \frac{AB}{(hc/e)} = \frac{\Phi}{\phi_0} \tag{9.10}$$

9.1.3 Quantum Hall Effect

The quantization of electron states in Landau levels plays a central role in the description of the quantized version of the Hall effect. The basic picture for explaining the quantum Hall effect (QHE) was constructed by R. B. Laughlin.[3] An important role in the manifestation of this phenomenon is played by states at the edge of the sample, an issue that was elucidated by the elegant treatment of these states by B. I. Halperin.[4] The edge states of 2D systems have become the focus of much attention recently as they underlie very interesting physics phenomena, to which we return in later sections of this chapter. We present here the main ideas of Laughlin and Halperin that led to the explanation of the QHE.

We begin by considering the system in a loop, rather than a planar, ribbon geometry, which will facilitate the justification of how various terms arise from changes in the external magnetic field. In this geometry, the x direction becomes the tangential direction along the loop of total circumference L, the y direction is perpendicular to x and lies on the loop surface which had width W, and the z direction becomes the radial direction perpendicular to the loop surface, as illustrated in Fig. 9.4. This is consistent with our assumption of Born–von Karman periodic boundary conditions in the x direction, with the

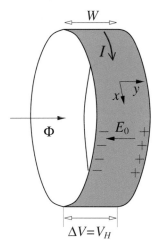

Fig. 9.4　Loop geometry for establishing the quantum Hall effect (after R. B. Laughlin, see footnote 3).

[3] R. B. Laughlin, *Phys. Rev. B* **23**, 5632 (1981).
[4] B. I. Halperin, *Phys. Rev. B* **25**, 2185 (1982).

sample length L in this direction being much larger than the sample width W; in the limit $L \to \infty$ we recover the planar ribbon geometry discussed above. We will assume that there is a magnetic flux Φ through the torus, and any changes in it will produce an electromotive force $\mathcal{E}$ according to Faraday's law:

$$\mathcal{E} = -\frac{1}{c}\frac{d\Phi}{dt}$$

that drives a current along the x direction of the loop. A small adiabatic change in the flux, $\delta\Phi$, introduced over a small time interval δt will produce a current I, and in order to maintain this current work δU must be produced at the rate

$$\frac{\delta U}{\delta t} = -\mathcal{E}I = \frac{1}{c}I\frac{\delta\Phi}{\delta t} \Rightarrow I = c\frac{\delta U}{\delta\Phi}$$

Now suppose the change in flux $\delta\Phi$ is associated with changes in the magnetic field $\mathbf{B} = \nabla \times \mathbf{A}$:

$$\delta\Phi = \delta\left(\int_S \nabla \times \mathbf{A} \cdot d\mathbf{S}\right) = \delta\left(\oint_L \mathbf{A} \cdot d\mathbf{l}\right)$$

where $d\mathbf{S}$ is the surface element of the surface S that the field $\mathbf{B}$ pierces, and $d\mathbf{l} = dx\,\hat{x}$ is the length element along the circumference of the loop L that encloses the surface S. If we choose the vector potential to be $\mathbf{A} = A_x\hat{x}$, consistent with the Landau gauge, Eq. (9.3), we obtain:

$$\delta\Phi = L\delta A_x$$

since the geometry of the loop is fixed. This result, together with the above expression for the current I, produces the expression

$$I = \frac{c}{L}\frac{\delta U}{\delta A_x}$$

Next we calculate the change in internal energy of the system δU when the vector potential is adiabatically changed by a small amount $\delta\mathbf{A} = \delta A_x\hat{x}$. This change introduces a change in the hamiltonian of Eq. (9.4):

$$\mathbf{p} \to \mathbf{p} + \frac{e}{c}\delta A_x\hat{x}$$

Since the wavefunctions $\bar{\psi}(x,y)$ have been factorized into products of functions that depend only on the x variable, or only on the y variable, we examine how this change in the x component of the momentum operator affects the x-dependent part of the wavefunction only, which is a plane wave with wavenumber $k_x = 2\pi k/L$ with k an integer. The change in the momentum operator produces gauge transformation of the wavefunction:

$$\frac{p_x^2}{2m}f(x) = \epsilon f(x) \Rightarrow \frac{1}{2m}\left(p_x + \frac{e}{c}\delta A_x\right)^2\bar{f}(x) = \epsilon\bar{f}(x)$$

that is, the original solution $f(x)$ to the plane-wave equation is replaced by $\bar{f}(x)$, which is given by:

$$\bar{f}(x) = f(x)\exp\{-i[(e/\hbar c)\delta A_x]x\}$$

Generally, there are two classes of states, localized and extended. For a localized state, which obeys the Schrödinger equation

$$\left[\frac{1}{2m} \left(\mathbf{p} + \frac{e}{c} \delta A_x \, \hat{x} \right)^2 + \tilde{V}(x, y) \right] \tilde{\psi}(x, y) = \tilde{\epsilon} \tilde{\psi}(x, y)$$

the above gauge transformation is still valid, and because the state is confined to a small region of (x, y) values, it amounts to a change in phase which does not lead to a change in the energy eigenvalue $\tilde{\epsilon}$. For an extended state, in order to have a valid gauge transformation, we must have:

$$\exp\{-i[(e/\hbar c)\delta A_x]x\} = \exp\{-i[(e/\hbar c)\delta A_x](x + L)\} \Rightarrow \delta A_x = \frac{hc}{e} \frac{i}{L}, \quad i : \text{integer}$$

but this leaves all the states unchanged due to periodic boundary conditions, and therefore their energies do not change. This suggests that any change in the total energy of the system must come from changes in the *occupancy* of the levels, without any change in the energy levels themselves. Connecting all this to our analysis of the Landau levels, we see that a change in the vector potential from $\mathbf{A}$ to $\mathbf{A}'$ by $\delta A_x \hat{x}$ implies a change

$$\mathbf{A}' = (-By + \delta A_x)\hat{x} = -B\left(y - \frac{\delta A_x}{B} \right)\hat{x} = -By'\hat{x} \Rightarrow y' = y - \frac{\delta A_x}{B}$$

in other words, the change in the vector potential is equivalent to shifting all values of the y variable, and hence the centers of the harmonic oscillator orbits, by $-(\delta A_x / B)$. Using the expression that we derived above for δA_x in the case of extended states, we find that the shift in the position of the harmonic oscillator orbits is then given by:

$$\delta \bar{y} = -\frac{\delta A_x}{B} = -\left(\frac{hc}{e} \right) \frac{i}{LB}, \quad i : \text{integer}$$

which is an integer multiple of the *spacing* between positions of the centers in the y coordinate, as we found earlier, Eq. (9.9). This already suggest that, since all the orbits are contained within the width W of the ribbon in the y direction, some of them must be transferred from one edge to the opposite. To simplify the discussion, let us assume that $i = -1$ in the above expression, that is, the change in vector potential is:

$$\delta A_x = \left(\frac{hc}{e} \right) \frac{1}{L} = \phi_0 \frac{1}{L}$$

which is equivalent to changing the flux by one flux quantum, ϕ_0. We had also found earlier [see Eq. (9.9)] that the change in $\bar{y}$ corresponds to a change in the k_x wave-vector:

$$\delta k_x = \frac{eB}{\hbar c} \delta \bar{y} = \frac{eB}{\hbar c} \left(\frac{hc}{e} \right) \frac{1}{BL} = \frac{2\pi}{L} \quad \text{and} \quad \delta k_x = \frac{2\pi}{L} \delta k \Rightarrow \delta k = 1$$

that is, under this change in vector potential all the values of the index k are shifted by unity, which gives:

$$\epsilon_{k_x}^{(j)}(\delta A_x) = \epsilon_{k'_x}^{(j)}(0), \quad k_x = \frac{2\pi k}{L}, \quad k'_x = \frac{2\pi(k + 1)}{L}, \quad k : \text{integer}$$

With these considerations, we can now calculate the change in internal energy of the system induced by the change in flux:

$$\delta U = \sum_{j,k} \left[\epsilon_{k_x}^{(j)}(\delta A_x) - \epsilon_{k_x}^{(j)}(0) \right] = \left[\sum_{j,k+1} \epsilon_{k_x}^{(j)} \right] - \left[\sum_{j,k} \epsilon_{k_x}^{(j)} \right] = \sum_{j} \left[\epsilon_{k_{\max}}^{(j)} - \epsilon_{k_{\min}}^{(j)} \right]$$

where $k_{\min}, k_{\max}$ are the minimum and maximum values of the wave-vector k_x. From the expression for the Landau-level energies, Eq. (9.8), we find that the energy difference between the levels that correspond to these two values is given by:

$$\epsilon_{k_{\max}}^{(j)} - \epsilon_{k_{\min}}^{(j)} = -eE_0 \Delta \bar{y} = -eV_H$$

with $\Delta \bar{y}$ the total difference in values of the variable $\bar{y}$, that is, $\Delta \bar{y} = W$, and the Hall voltage being given by $V_H = E_0 W$. We can now sum over the index $j = 0, 1, \ldots, l$ with l the index value of the highest occupied Landau level, to obtain:

$$\delta U = -eV_H(l+1) \Rightarrow I = \frac{c}{L}\frac{\delta U}{\delta A_x} = -\frac{ec}{\phi_0}(l+1)V_H \Rightarrow \rho_{xy} = -\frac{V_H}{I} = \frac{\phi_0}{ec(l+1)} = \frac{(h/e^2)}{l+1}$$

This is an important result, since it establishes that the resistivity is quantized and takes values that are integer fractions of h/e^2, the quantum of resistivity:

$$\rho_{xy} = \frac{B}{nec} = \frac{(h/e^2)}{l+1} \tag{9.11}$$

The entire argument presented above to establish the quantization of the Hall resistivity rests on the assumption that it is possible to change the occupancy of the Landau levels by transferring edge states from one side of the sample to the other, in the y direction. Now the states in the interior of the ribbon are separated in energy by a finite amount, $\hbar\omega_c$, and hence no change in their occupancy can be achieved by a small change in flux $\delta\Phi$. This leaves as the only option that edge states corresponding to a Landau level are degenerate with a level of different index value. This type of degeneracy was established through an elegant argument by B. I. Halperin. We shall briefly present parts of this argument. For the purposes of this discussion, we focus on the harmonic oscillator equation that the y-dependent part of the wavefunctions obeys, Eq. (9.5), and put aside the extra terms on the right-hand side that depend on the field E_0 (we will return to them later); accordingly, we also drop the subscript k_x of the wavefunctions, which obey:

$$\left[\frac{p_y^2}{2m} + \frac{1}{2}\bar{m}\omega_c^2 (y - \bar{y})^2 \right] \bar{\phi}^{(j)}(y; \bar{y}) = \bar{\epsilon}^{(j)}(\bar{y})\bar{\phi}^{(j)}(y; \bar{y})$$

As long as $\bar{y}$ is far from the edges of the ribbon, $y = \pm W/2$, the solution to this equation is the usual set of eigenfunctions and eigenvalues of the harmonic oscillator potential (see Appendix C):

$$\bar{\epsilon}^{(j)}(\bar{y}) = \left(j + \frac{1}{2} \right)\hbar\omega_c, \quad |\bar{y}| \ll W/2$$

If $\bar{y}$ approaches one of the edges, the wavefunctions must be distorted to ensure that they vanish at the edge, $y = \pm W/2$, while they preserve the number of nodes that the harmonic oscillator solutions have. We examine what happens for $\bar{y} = W/2$ (analogous results are obtained for the other case, $\bar{y} = -W/2$). In this case, the wavefunction must obey

$$\bar{\phi}^{(j)}(y, W/2) \to 0 \quad \text{for} \quad y \to -\infty \quad \text{and} \quad y = W/2$$

and has j nodes, including the one at $y = W/2$, but this is the same as the wavefunction of the harmonic oscillator with index $2j + 1$, centered at $W/2$. The consequence of this result is that the energy of the state with index j, when $\bar{y} = W/2$, will be equal to that of the harmonic oscillator state with index $2j + 1$ far from the edge, that is:

$$\bar{\epsilon}^{(j)}(W/2) = \left((2j+1) + \frac{1}{2}\right)\hbar\omega_c = \left(2j + \frac{3}{2}\right)\hbar\omega_c$$

The values of $\bar{y}$ can also lie outside the interval $[-W/2, W/2]$, but then the energy becomes even higher, and rises eventually as $\overline{m}\omega_c^2(|\bar{y}| - W/2)^2/2$, for $|\bar{y}| > W/2 + r_c$, where r_c is the characteristic length scale of the harmonic oscillator wavefunction (see Appendix C). The energy levels of the states as a function of the distance of their center $\bar{y}$ from the edges of the ribbon at $y = \pm W/2$ are shown in Fig. 9.5. This establishes the degeneracy of edge states connected to a Landau level to states of a different index value, making the change of occupancy possible. The resulting density of states exhibits similar features with that of the system with disorder, see Fig. 9.7 later.

This treatment allows us to make an analogy with the classical picture of electrons in a 2D ribbon on the x, y plane, on which a magnetic field $\mathbf{B} = B\hat{z}$ is applied perpendicular to the ribbon plane, as illustrated in Fig. 9.6. For classical charged particles of charge e, the magnetic field will induce circular orbits in the interior of the ribbon, of radius $r_c = \overline{m}cv/eB$ (the cyclotron radius), where v is the magnitude of the original velocity of the

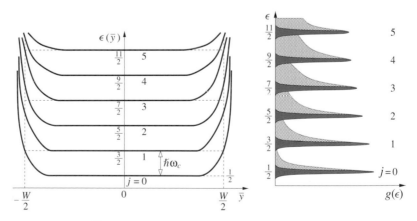

Fig. 9.5 **Left**: The energy levels $\bar{\epsilon}^{(j)}(\bar{y})$ of states in the ribbon as a function of the position of their centers $\bar{y}$ from the edges at $y = \pm W/2$; the red lines correspond to the ideal harmonic oscillator levels, separated by $\hbar\omega_c$ (after B. I. Halperin, see footnote 4). **Right**: The density of states $g(\epsilon)$ for the states that include edge effects (blue-shaded regions) and the ideal harmonic oscillator levels, represented by sharp δ-function-like peaks in red.

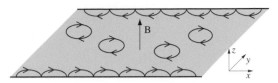

Fig. 9.6 Schematic representation of classical orbits of charged particles, in the bulk (red circles) and at the edges (blue skipping orbits), of a ribbon on the x, y plane with external magnetic field $\mathbf{B} = B\hat{z}$ perpendicular to it.

particles.[5] At the edges, there is not enough space for circular orbits, so skipping edge orbits are induced. These are chiral states, that is, they travel in opposite directions at the two opposite edges of the ribbon. The bulk circular orbits do not contribute to any current along the ribbon length. The only states that can carry current along the ribbon are edge states. If the number of occupied states at the two opposite edges of the ribbon is the same, the net current is zero. Therefore, the occupancy of the states at opposite edges must be different. This is consistent with our earlier discussion, which established that the energy levels of the opposite edge states differ by $-eV_H$, that is, their distance from the Fermi level will differ by this amount and their occupancy will indeed be different, leading to the net current I. Finally, Halperin also showed that weak disorder will not affect this picture. Since the edge states are not destroyed by disorder or by scattering from bulk states, the quantization of the Hall resistance is a robust effect, as manifest by the plateaus in the experimental measurements, see Fig. 9.3.

What leads to the existence of plateaus in the value of ρ_{xy} is the fact that not all states in the system are extended Bloch states. In a real system, the Landau levels are not δ-functions in energy but are broadened due to the presence of impurities, as indicated in Fig. 9.7. Only those states with energy very close to the ideal Landau levels are extended and participate in the current, while the states lying between ideal Landau levels are localized and do not participate in the current. As the magnetic field is varied and localized states with energy away from an ideal Landau level are being filled, there can be no change in the conductivity of the system, which remains stuck to its quantized value corresponding to the ideal Landau level. This produces the plateaus in the Hall resistivity ρ_{xy}. The plateaus in Hall resistivity can be measured with unprecedented precision, reaching 1 part in 10^7. This has led to advances in metrology, since the ratio e^2/h appears in several fundamental units. For example, the fine structure constant is given by:

$$\alpha = \frac{e^2}{\hbar c} = \left(\frac{e^2}{h}\right)\left(\frac{2\pi}{c}\right)$$

and the speed of light c is known to be 1 part in 10^9 from independent measurements. Thus, using phenomena related to the behavior of electrons in a solid, one can determine the value of a fundamental constant that plays a central role in high-energy physics!

---

[5] The use of the same symbol, r_c, for the cyclotron radius and the characteristic length of the harmonic oscillator is intentional, because for the physical system discussed here the two quantities are of the same order of magnitude (see footnote 4).

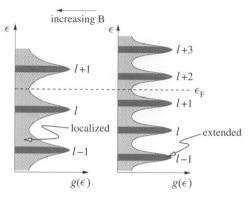

Fig. 9.7 Density of states $g(\epsilon)$ and filling of Landau levels, including the effects of disorder, at different values of the magnetic field B. When the magnetic field takes the values $B = nch/e(l + 1)$, with $l = 0, 1, 2, \ldots$, exactly $l + 1$ Landau levels are filled and the Fermi level, denoted by a horizontal dashed line, lies between levels with index l and $l + 1$ in the figure on the left. In the figure on the right, corresponding to a smaller value of the magnetic field $B = nch/e(l + 2)$ and $l + 2$ Landau levels filled, the Fermi level lies between levels with index $l + 1$ and $l + 2$. The spacing between levels is $\hbar\omega_c = \hbar(eB/\overline{m}c)$. The disorder broadens the sharp δ-function-like DOS of the ideal system and creates states (blue-shaded regions) between those of the ideal system.

Fascinating though the discovery of the IQHE may have seemed, an even more remarkable observation was in store: a couple of years after the experiments of von Klitzing and coworkers, experiments on samples of very high purity at high magnetic fields and low temperatures by Tsui, Störmer, and Gossard[6] revealed that the plateaus in the Hall resistivity can also occur for values *larger* than (h/e^2). In particular, they occur at values $3, 3/2, 5, 5/2, 5/3, \ldots$, in units of (h/e^2). This discovery was called the fractional quantum Hall effect (FQHE). The explanation of the FQHE involves very interesting many-body physics, which lies beyond the scope of the present treatment. Suffice to say that the FQHE has generated enormous interest in the condensed matter physics community. A hierarchy of states at fractions

$$\nu = \frac{p}{2qp + 1}$$

with p, q integers, that lead to plateaus in the Hall conductivity, has been predicted theoretically[7] and observed experimentally; all these fractions have odd denominators. The states corresponding to the partial fillings are referred to as "incompressible quantum fluids." Laughlin originally postulated a many-body wavefunction[8] to describe the states with $\nu = 1/(2q + 1)$, where q is a positive integer; this wavefunction for N_e electrons has the form

$$\Psi_q(\mathbf{r}_1, \ldots, \mathbf{r}_{N_e}) = \prod_{j<k}^{N_e} (\zeta_j - \zeta_k)^{(2q+1)} \exp\left(-\sum_{k=1}^{N_e} |\zeta_k|^2\right), \quad \zeta_k = x_k + iy_k$$

[6] D. C. Tsui, H. L. Störmer, and A. C. Gossard, *Phys. Rev. Lett.* **48**, 1559 (1982); *Phys. Rev. B* **25**, 1405 (1982).
[7] J. K. Jain, *Phys. Rev. Lett.* **63**, 199 (1989); *Phys. Rev. B* **40**, 8079 (1989).
[8] R. B. Laughlin, *Phys. Rev. Lett.* **50**, 1395 (1983).

Evidently, this is an antisymmetric wavefunction upon exchange of any pair of electron positions. The magnitude squared of this wavefunction can be written as:

$$| \Psi_q(\mathbf{r}_1, \ldots, \mathbf{r}_{N_e}) |^2 = \exp\left[\Phi_q(\zeta_1, \ldots, \zeta_{N_e})/(q + 1/2) \right]$$

$$\Phi_q(\zeta_1, \ldots, \zeta_{N_e}) = (2q + 1)^2 \sum_{j<k}^{N_e} \ln |\zeta_j - \zeta_k| - (2q + 1) \sum_{k=1}^{N_e} |\zeta_k|^2$$

where $\Phi_q(\zeta_1, \ldots, \zeta_{N_e})$ has the form of an electrostatic potential energy between charged particles in two dimension: the first term represents the mutual repulsion between particles of charge $(2q + 1)$ and the second term the attraction to a uniform background of opposite charge. This analogy can be used to infer the relation of this wavefunction to a system of quasiparticles with fractional charge.

Many interesting theoretical ideas have been developed to account for the behavior of the incompressible quantum fluids which involves the highly correlated motion of all the electrons in the system. Theories have also been developed to address the properties of partially filled Landau levels with even denominator fractions (such as $\nu = 1/2$), at which there is no quantized Hall effect; it is intriguing that those states, despite the highly correlated motion of the underlying physical particles, seem to be described well in terms of *independent fermions*, each of which carries an even number of fictitious magnetic flux quanta[9]. R. B. Laughlin, D. C. Tsui, and H. L. Störmer were awarded the 1998 Nobel Prize for Physics for their work on the FQHE.

9.1.4 de Haas–van Alphen Effect

The analysis of the behavior of the system in terms of Landau levels is also helpful in explaining an interesting phenomenon, known as the "de Haas–van Alphen" effect. This effect is useful in measuring experimentally the Fermi surface of metals, an important feature of their electronic structure, which to a large extent determines how electrons in the crystal respond to external perturbations. We describe this effect briefly here.

The basic picture can be understood by the following argument. Let's assume that in order to accommodate the total number of electrons N_e we need to fill exactly all the Landau levels up to index value l, that is, we need exactly $l + 1$ Landau levels starting with level $j = 0$. Then we will have:

$$N_e = (l + 1)N_L = (l + 1)\frac{AB}{\phi_0} \Rightarrow l + 1 = \frac{N_e}{A}\frac{\phi_0}{B} = n\frac{ch}{eB} \Rightarrow$$

$$B = n\frac{ch}{e(l + 1)} = n\frac{\phi_0}{(l + 1)} \tag{9.12}$$

with $n = N_e/A$ the density of electrons in the plane. If the magnetic field takes the special value $B_1 = nch/e(l + 1)$, exactly $l + 1$ Landau levels will be filled with index numbers $j = 0, 1, \ldots, l$. If the electrons cannot be accommodated exactly by an integer number of

---

[9] B. I. Halperin, P. A. Lee, and N. Read, *Phys. Rev. B* **47**, 7312 (1993).

Landau levels, then levels up to index l will be completely filled and the level with index $l + 1$ will be partially filled. In this case we will have:

$$(l+1)N_L < N_e < (l+2)N_L \Rightarrow \frac{1}{l+2} < b < \frac{1}{l+1}, \quad b \equiv \frac{B}{n\phi_0}$$

where we have defined the variable b for convenience. In this case, the total energy of the system is given by:

$$U(B) = \sum_{j=0}^{l} N_L \hbar\omega_c \left(j + \frac{1}{2}\right) + [N_e - (l+1)N_L]\,\hbar\omega_c \left(l + \frac{3}{2}\right), \quad \text{for } \frac{1}{l+2} < b < \frac{1}{l+1}$$

Using the definition of the frequency ω_c, the Bohr magneton μ_B, and the variable b, we obtain for the total energy per particle:

$$\frac{U(B)}{N} = \mu_B \left(\frac{N_e\phi_0}{A}\right) \left[b(2l+3) - b^2(l^2 + 3l + 2)\right], \quad \text{for } \frac{1}{l+2} < b < \frac{1}{l+1}, \; l \geq 0$$

For a value of the field larger than the critical value $B_0 = N_e\phi_0/A$, which corresponds to the critical value $b_0 = 1$ of the variable b, only the lowest Landau level ($l = 0$) will be filled, and the corresponding value of b will satisfy $b > b_0 = 1$, giving for the energy per particle:

$$\frac{U(B)}{N} = \mu_B \left(\frac{N_e\phi_0}{A}\right) b, \quad \text{for } b > b_0 = 1, \; l = 0$$

We can now calculate the magnetization M, defined as:

$$M \equiv -\frac{1}{A}\frac{\partial U}{\partial B} = -\frac{N_e}{A}\mu_B, \quad \text{for } b > 1, \; l = 0$$

$$= \frac{N_e}{A}\mu_B \left[2(l^2 + 3l + 2)b - (2l+3)\right], \quad \text{for } \frac{1}{l+2} < b < \frac{1}{l+1}, \; l \geq 0$$

A plot of the magnetization in units of $n\mu_B$ as a function of the variable b, which is the magnetic field in units of $n\phi_0$, is shown in Fig. 9.8: the magnetization shows oscillations between positive and negative values as a function of the magnetic field! Although we showed how it arises in a 2D system, the effect is observed in 3D systems as well, in which case the discontinuities in the magnetization are rounded off. A number of other physical quantities exhibit the same type of oscillations as a function of the magnetic field B; examples include the conductivity (this is known as the Shubnikov–de Haas effect and is usually easier to measure than the de Haas–van Alphen effect), the sound attenuation, the magnetoresistance, and the magnetostriction (strain induced on the sample by the magnetic field).

The oscillations of the magnetization as a function of the magnetic field can be used to map out the Fermi surface of metals, through the following argument, due to Onsager.[10] When a 3D metal is in an external magnetic field, the plane perpendicular to the field will intersect the Fermi surface of the metal, producing a cross-sectional area that depends on the shape of the Fermi surface. Of course, there are many cross-sections of a 3D Fermi

[10] L. Onsager, *Phil. Mag.* **43**, 1006 (1952).

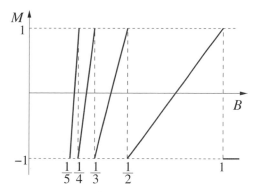

Fig. 9.8   Illustration of the de Haas–van Alphen effect in a 2D electron gas: the magnetization M, measured in units of $(N_e/A)\mu_B$, exhibits oscillations as a function of the magnetic field B, measured in units of $N_e\phi_0/A = n\phi_0$.

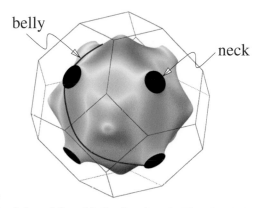

Fig. 9.9   The Fermi surface of Au [adapted from M. Gradhand *et al.*, *Phys. Rev. B* **80**, 224413 (2009)]. The thin gray lines show the first Brillouin zone of the FCC crystal. The smallest and largest cross-sections in the (111) direction are indicated as "neck" (black-filled circles on the hexagonal faces of the BZ) and "belly" (blue circles).

surface along a certain direction, but the relevant ones are those with the largest and smallest areas (the extremal values), as illustrated in Fig. 9.9. These cross-sectional areas will play a role analogous to the area of the plane on which the electrons were confined in the above example. Thus, our simple treatment of electrons confined in two dimensions becomes relevant to the behavior of electrons with wave-vectors lying on a certain plane that intersects the Fermi surface. Although additional complications arise from the band structure, in principle, we can use the oscillations of the magnetization to determine the extremal cross-sectional areas enclosed by the Fermi surface on this plane. By changing the orientation of the metal with respect to the direction of the field, different cross-sections of the Fermi surface come into play, which makes it possible to determine the entire 3D Fermi surface by reconstructing it from its various cross-sections. Usually, a band-structure calculation of the type discussed in detail in Chapter 5 is indispensable in reconstructing the exact Fermi surface, in conjunction with magnetic measurements which give the precise values of the extremal areas on selected planes.

9.2 Dynamics of Crystal Electrons: Single-Band Picture

Up to this point we have treated electrons as essentially free particles, confined only by an external potential that restricts their motion on the x, y plane, and examined what happens when we subject them to external electromagnetic fields; we have paid no attention to the effects of atomic-scale potentials, such as that provided by the ions in a solid. To complete the picture of the dynamics of electrons in real solids, we will need to introduce again the features of the atomic structure of solids. As we have done in previous chapters, we will focus on the crystalline case, which can also form the basis for treating a wider range of systems (see the discussion in Chapter 1). We will describe the dynamics of crystal electrons in two stages: first, in this section, we will discuss a picture that establishes a connection between the forces exerted on electrons by external electromagnetic fields and the intrinsic features of electronic motion due to the crystal potential, that is, the Bloch character of electronic states. Specifically, we will derive the intuitively pleasing results that the expectation values of the velocity and of the acceleration of crystal electrons are related to derivatives of the crystal energy $\epsilon_{\mathbf{k}}^{(n)}$ with respect to the crystal momentum $\hbar\mathbf{k}$. This picture, however, does not capture the full range of the electronic response. Then, in later sections of this chapter, we will explore more subtle effects of how crystal electrons respond to external forces; these include important changes in the *phase* of electronic wavefunctions (Berry's phase) that are manifest in experimentally measurable quantities.

We start by deriving expressions for the time evolution of the position and velocity of a crystal electron in the state $\psi_{\mathbf{k}}^{(n)}$. To this end, we will need to allow the crystal momentum to acquire a time dependence, $\mathbf{k} = \mathbf{k}(t)$, and include its time derivatives where appropriate. Since we are dealing with a particular band, we will omit the band index n for simplicity. Considering the time-dependent position of a crystal electron $\mathbf{r}(t)$ as a quantum-mechanical operator, we have from the usual formulation in the Heisenberg picture (see Appendix C):

$$\frac{d\mathbf{r}(t)}{dt} = \frac{i}{\hbar}\left[\hat{\mathcal{H}}, \mathbf{r}\right] \Rightarrow \langle\psi_{\mathbf{k}}^{(n)}|\frac{d\mathbf{r}(t)}{dt}|\psi_{\mathbf{k}}^{(n)}\rangle = \frac{i}{\hbar}\langle\psi_{\mathbf{k}}^{(n)}|[\hat{\mathcal{H}}, \mathbf{r}]|\psi_{\mathbf{k}}^{(n)}\rangle \tag{9.13}$$

with $[\hat{\mathcal{H}}, \mathbf{r}]$ the commutator of the hamiltonian with the position operator. Now we can take advantage of the following identity:

$$\nabla_{\mathbf{k}}\left(e^{-i\mathbf{k}\cdot\mathbf{r}}\hat{\mathcal{H}}e^{i\mathbf{k}\cdot\mathbf{r}}\right) = ie^{-i\mathbf{k}\cdot\mathbf{r}}[\hat{\mathcal{H}}, \mathbf{r}]e^{i\mathbf{k}\cdot\mathbf{r}}$$

whose proof involves simple differentiations of the exponentials with respect to $\mathbf{k}$, and we recall the relation derived in Eq. (2.24), to rewrite the right-hand side of the previous equation as:

$$\frac{i}{\hbar}\langle\psi_{\mathbf{k}}^{(n)}|[\hat{\mathcal{H}}, \mathbf{r}]|\psi_{\mathbf{k}}^{(n)}\rangle \equiv \frac{i}{\hbar}\int u_{\mathbf{k}}^{(n)*}(\mathbf{r})e^{-i\mathbf{k}\cdot\mathbf{r}}[\hat{\mathcal{H}}, \mathbf{r}]e^{i\mathbf{k}\cdot\mathbf{r}}u_{\mathbf{k}}^{(n)}(\mathbf{r})d\mathbf{r}$$

$$= \frac{1}{\hbar}\int u_{\mathbf{k}}^{(n)*}(\mathbf{r})\left[\nabla_{\mathbf{k}}\left(e^{-i\mathbf{k}\cdot\mathbf{r}}\hat{\mathcal{H}}e^{i\mathbf{k}\cdot\mathbf{r}}\right)\right]u_{\mathbf{k}}^{(n)}(\mathbf{r})d\mathbf{r}$$

$$= \frac{1}{\hbar}\int u_{\mathbf{k}}^{(n)*}(\mathbf{r})\left[\nabla_{\mathbf{k}}\hat{\mathcal{H}}(\mathbf{p} + \hbar\mathbf{k}, \mathbf{r})\right]u_{\mathbf{k}}^{(n)}(\mathbf{r})d\mathbf{r}$$

By comparing the last expression to the original one, Eq. (9.13), we derive the interesting result that the operator that gives the average velocity in a band, with the basis $|u_{\mathbf{k}}^{(n)}\rangle$, is:

$$\mathbf{v}_{\mathbf{k}}^{(n)} = \frac{1}{\hbar}\nabla_{\mathbf{k}}\hat{\mathcal{H}}(\mathbf{p}+\hbar\mathbf{k},\mathbf{r}) \Rightarrow \langle\mathbf{v}_{\mathbf{k}}^{(n)}\rangle = \frac{1}{\hbar}\langle u_{\mathbf{k}}^{(n)}|\left[\nabla_{\mathbf{k}}\hat{\mathcal{H}}(\mathbf{p}+\hbar\mathbf{k},\mathbf{r})\right]|u_{\mathbf{k}}^{(n)}\rangle \qquad (9.14)$$

a result that will prove useful in the following section. Next, we move the differentiation with respect to $\mathbf{k}$, $\nabla_{\mathbf{k}}$, outside the integral and subtract the additional terms produced by this change, which leads to:

$$\frac{i}{\hbar}\langle\psi_{\mathbf{k}}^{(n)}|[\hat{\mathcal{H}},\mathbf{r}]|\psi_{\mathbf{k}}^{(n)}\rangle = \frac{1}{\hbar}\nabla_{\mathbf{k}}\int u_{\mathbf{k}}^{(n)*}(\mathbf{r})\hat{\mathcal{H}}(\mathbf{p}+\hbar\mathbf{k},\mathbf{r})u_{\mathbf{k}}^{(n)}(\mathbf{r})d\mathbf{r}$$
$$-\frac{1}{\hbar}\left[\int\left(\nabla_{\mathbf{k}}u_{\mathbf{k}}^{(n)*}(\mathbf{r})\right)\hat{\mathcal{H}}(\mathbf{p}+\hbar\mathbf{k},\mathbf{r})_{\mathbf{k}}^{(n)}(\mathbf{r})d\mathbf{r} + \int u_{\mathbf{k}}^{(n)*}(\mathbf{r})\hat{\mathcal{H}}(\mathbf{p}+\hbar\mathbf{k},\mathbf{r})\left(\nabla_{\mathbf{k}}u_{\mathbf{k}}^{(n)}(\mathbf{r})\right)d\mathbf{r}\right]$$

We deal with the last two terms in the above expression separately. Recalling that $u_{\mathbf{k}}^{(n)}(\mathbf{r})$ is an eigenfunction of the hamiltonian $\hat{\mathcal{H}}(\mathbf{p}+\hbar\mathbf{k},\mathbf{r})$ with eigenvalue $\epsilon_{\mathbf{k}}$, we obtain for these two terms:

$$\langle\nabla_{\mathbf{k}}u_{\mathbf{k}}^{(n)}|\hat{\mathcal{H}}(\mathbf{p}+\hbar\mathbf{k})|u_{\mathbf{k}}^{(n)}\rangle + \langle u_{\mathbf{k}}^{(n)}|\hat{\mathcal{H}}(\mathbf{p}+\hbar\mathbf{k})|\nabla_{\mathbf{k}}u_{\mathbf{k}}^{(n)}\rangle =$$
$$\epsilon_{\mathbf{k}}\left(\langle\nabla_{\mathbf{k}}u_{\mathbf{k}}^{(n)}|u_{\mathbf{k}}^{(n)}\rangle + \langle u_{\mathbf{k}}^{(n)}|\nabla_{\mathbf{k}}u_{\mathbf{k}}^{(n)}\rangle\right) = \epsilon_{\mathbf{k}}\nabla_{\mathbf{k}}\langle u_{\mathbf{k}}^{(n)}|u_{\mathbf{k}}^{(n)}\rangle = 0$$

since $\langle u_{\mathbf{k}}^{(n)}|u_{\mathbf{k}}^{(n)}\rangle = 1$ for properly normalized wavefunctions. This is a special case of the more general Hellmann–Feynman theorem, which we encountered in Chapter 4. This leaves the following result:

$$\langle\psi_{\mathbf{k}}^{(n)}|\frac{d\mathbf{r}(t)}{dt}|\psi_{\mathbf{k}}^{(n)}\rangle = \frac{1}{\hbar}\nabla_{\mathbf{k}}\int\psi_{\mathbf{k}}^{(n)*}(\mathbf{r})\hat{\mathcal{H}}\psi_{\mathbf{k}}^{(n)}(\mathbf{r})d\mathbf{r} \Rightarrow \langle\mathbf{v}_{\mathbf{k}}^{(n)}\rangle = \frac{1}{\hbar}\nabla_{\mathbf{k}}\epsilon_{\mathbf{k}}^{(n)} \qquad (9.15)$$

where we have identified the velocity $\langle\mathbf{v}_{\mathbf{k}}^{(n)}\rangle$ of a crystal electron in state $\psi_{\mathbf{k}}^{(n)}$ with the expectation value of the time derivative of the operator $\mathbf{r}(t)$ in that state. This result is equivalent to Eq. (2.50), which we derived earlier for the momentum of a crystal electron in state $\psi_{\mathbf{k}}^{(n)}$.

The results derived so far are correct when dealing with electronic states describing the ground state of the system without any external perturbations. Using these results to describe the response of the system to external fields would therefore be incomplete. Nevertheless, we will proceed here to do just this for the purpose of discussing the dynamics of electrons in external fields in a simple and intuitive, even if incomplete, picture. We will then derive a more complete description in the following section.

To continue the discussion of electron dynamics in this simple picture, we take a derivative of the velocity in state $\psi_{\mathbf{k}}^{(n)}$ with respect to time, and using the chain rule for differentiation with respect to $\mathbf{k}$, we find:

$$\frac{d\langle\mathbf{v}_{\mathbf{k}}^{(n)}\rangle}{dt} = \frac{1}{\hbar}\left(\frac{d\mathbf{k}}{dt}\cdot\nabla_{\mathbf{k}}\right)\nabla_{\mathbf{k}}\epsilon_{\mathbf{k}}^{(n)}$$

which we can write in terms of cartesian components ($i, j = x, y, z$) as:

$$\frac{d\langle v_{\mathbf{k},i}^{(n)}\rangle}{dt} = \sum_j \left(\hbar\frac{dk_j}{dt}\right)\left[\frac{1}{\hbar^2}\frac{\partial^2 \epsilon_{\mathbf{k}}^{(n)}}{\partial k_j \partial k_i}\right] = \sum_j \left(\hbar\frac{dk_j}{dt}\right)\frac{1}{m_{ji}^{(n)}(\mathbf{k})} \tag{9.16}$$

where we have identified the term in square brackets as the inverse effective mass tensor derived in Eq. (2.51). With this identification, this last equation has the form

$$\text{acceleration} = \text{force} \times \frac{1}{\text{mass}}$$

as might be expected, but the mass is not a simple scalar quantity as it would be in the case of free electrons; instead, the mass is now a second-rank tensor corresponding to the behavior of crystal electrons. The form of Eq. (9.16) compels us to identify the quantities in parentheses on the right-hand side as the components of the external force acting on the crystal electron:

$$\hbar\frac{d\mathbf{k}}{dt} = \mathbf{F} \tag{9.17}$$

We note two important points. First, the identification of the time derivative of $\hbar\mathbf{k}$ with the external force is not a proper proof but only an *inference* from dimensional analysis; external forces could actually change the wavefunctions and the eigenvalues of the single-particle hamiltonian, an issue to which we return below. Second, this relation is a state-independent equation, implying that the wave-vectors evolve in the same manner for all states. For instance, if crystal electrons were subject to a constant external electric field $\mathbf{E}$, they would experience a force $\mathbf{F} = -e\mathbf{E}$ and according to Eq. (9.17), this would lead to:

$$\hbar\frac{d\mathbf{k}}{dt} = -e\mathbf{E} \tag{9.18}$$

We discuss below how this picture applies to a simple 1D example.

Example 9.1   Motion of electrons in one dimension, one-band model

We consider the simplest case, a 1D crystal with a single energy band, given by:

$$\epsilon_k = \epsilon + 2\tau \cos(ka)$$

with ϵ a reference energy and τ a negative constant; since we are dealing with a single band, we are going to dispense with the band index n, to simplify the notation. This is shown in Fig. 9.10: the lattice constant is a, so the first BZ for this crystal extends from $-\pi/a$ to π/a, and the energy ranges from a minimum of $\epsilon + 2\tau$ (at $k = 0$) to a maximum of $\epsilon - 2\tau$ (at $k = \pm\pi/a$). The momentum for this state is given by:

$$\mathbf{p}(k) = p(k)\hat{\mathbf{x}} = \frac{m_e}{\hbar}\frac{d\epsilon_k}{dk}\hat{\mathbf{x}} = -\frac{2m_e\tau a}{\hbar}\sin(ka)\hat{\mathbf{x}}$$

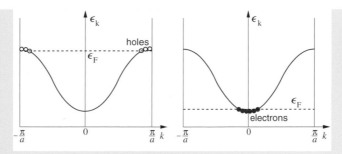

Fig. 9.10 Single-particle energy eigenvalues ϵ_k for the simple example that illustrates the motion of electrons and holes in a one-band, 1D crystal.

When the system is in an external electric field $\mathbf{E} = -E_0\hat{\mathbf{x}}$ with E_0 a positive constant, the time evolution of the wave-vector will be:

$$\frac{d\mathbf{k}}{dt} = -\frac{e}{\hbar}\mathbf{E} \Rightarrow \frac{dk}{dt} = \frac{e}{\hbar}E_0$$

We will consider some limiting cases for this idealized system.

(1) A single electron in this band would start at $k = 0$ at $t = 0$ (before the application of the external field it occupies the lowest-energy state), then its k would increase at a constant rate $(eE_0/\hbar)$ until the value π/a is reached, and then it would re-enter the first BZ at $k = -\pi/a$ and continue this cycle. The same picture would hold for a few electrons initially occupying the bottom of the band.

(2) If the band were completely full, then all the wave-vectors would be changing in the same way, and all states would remain completely full. Since this creates no change in the system, we conclude that no current would flow: a full band cannot contribute to current flow in an ideal crystal!

(3) If the band were mostly full, with only a few states empty at wave-vectors near the boundaries of the first BZ (at $k = \pm\pi/a$) where the energy is a maximum, then the total current would be:

$$\mathbf{I} = -e\sum_{k \leq k_F} \mathbf{v}(k) = -e\sum_{k \leq k_F} \frac{p(k)}{m_e}\hat{\mathbf{x}} = \left[-e\sum_{k \in BZ} \frac{p(k)}{m_e} + e\sum_{k > k_F} \frac{p(k)}{m_e}\right]\hat{\mathbf{x}} = e\sum_{k > k_F} \frac{p(k)}{m_e}\hat{\mathbf{x}}$$

where, in the first two sums, the summation is over k values in the first BZ, restricted to $k \leq k_F$, that is, over occupied states only, whereas in the last sum it is restricted over $k > k_F$, that is, over unoccupied states only. The last equality follows from the fact that the sum over all values of $k \in BZ$ corresponds to the full band, which as explained above does not contribute to the current. Thus, in this case the system behaves like a set of positively charged particles (holes) corresponding to the unoccupied states near the top of the band. In our simple 1D example we can use the general expression for the effective mass derived in Eq. (2.51), to find:

$$\frac{1}{\overline{m}} = \frac{1}{\hbar^2}\frac{\mathrm{d}^2\epsilon_k}{\mathrm{d}k^2} \Rightarrow \overline{m} = \hbar^2\left(\left[\frac{\mathrm{d}^2\epsilon_k}{\mathrm{d}k^2}\right]_{k=\pm\pi/a}\right)^{-1} = \frac{\hbar^2}{2\tau a^2} \tag{9.19}$$

which is a *negative* quantity (recall that $\tau < 0$). Using the Taylor expansion of the cosine function near $k_0 = \pm\pi/a$, we can write the energy near the top of the band as:

$$\epsilon_{\pm(\pi/a)+k} = \epsilon + 2\tau\left[-1 + \frac{1}{2}(ka)^2 + \cdots\right] = (\epsilon - 2\tau) + \tau a^2 k^2 = (\epsilon - 2\tau) + \frac{\hbar^2 k^2}{2\overline{m}} \tag{9.20}$$

with the effective mass $\overline{m}$ being the negative quantity found in Eq. (9.19). From the general expression Eq. (9.19), we find the time derivative of the current in our 1D example to be:

$$\frac{\mathrm{d}\mathbf{I}}{\mathrm{d}t} = e\sum_{k>k_F}\frac{1}{m_e}\frac{\mathrm{d}p(k)}{\mathrm{d}t}\hat{\mathbf{x}} = e\sum_{k>k_F}\frac{\mathrm{d}}{\mathrm{d}t}\left(\frac{1}{\hbar}\frac{\mathrm{d}}{\mathrm{d}k}\epsilon_k\right)\hat{\mathbf{x}} = \frac{e}{\overline{m}}\sum_{k>k_F}\hbar\frac{\mathrm{d}k}{\mathrm{d}t}\hat{\mathbf{x}}$$

where we have used Eq. (2.50) to obtain the second equality and Eq. (9.20) to obtain the third equality. Now, assuming that we are dealing with a single-hole state at the top of the band, and using the general result of Eq. (9.17), we obtain

$$\frac{\mathrm{d}\mathbf{I}}{\mathrm{d}t} = e\frac{\mathbf{F}}{\overline{m}}$$

which is simply Newton's equation of motion for a *positively* charged particle of charge $+e$ with *negative* effective mass $\overline{m}$ to an external force $\mathbf{F}$.

9.3 Time-Reversal Invariance

So far we have treated time as a new variable, separate from the other degrees of freedom of the hamiltonian. In the most general case, the hamiltonian can also contain a time dependence, usually due to the time dependence of external fields (like the electromagnetic field). When the hamiltonian is time independent, there is a deep and important symmetry that governs the behavior of electrons, the time-reversal symmetry. We briefly touched upon this in our discussion of symmetries of the band structure for crystal electrons, when the spin degree of freedom is not explicitly taken into account (see Chapter 3). Since time is a variable of central importance in dynamics, we revisit the issue of time-reversal symmetry in a more careful manner. This is accomplished by working with the Dirac equation that treats the wavefunctions as spinors, that is, it includes the spin degree of freedom explicitly. We consider the state of a particle with spin s, described by its wavefunction $\psi_s(\mathbf{r}, t)$, where s is the spin. We will specialize the discussion to particles with spin $s = 1/2$, since this is the case relevant to electrons. A more detailed discussion of the Dirac equation for spin-1/2 particles is provided in Appendix C (see Section C.8), including the definition of Pauli matrices, Eq. (C.43), that are used in the following discussion.

We seek to define the time-reversal operator $\hat{\mathcal{T}}$ such that when it is applied to an eigenstate $\psi_s(\mathbf{r}, t)$ of the hamiltonian $\hat{\mathcal{H}}$ it produces a state $\tilde{\psi}(\mathbf{r}, -t)$ with the time direction inverted ($t \rightarrow -t$) and a new spin value denoted $\tilde{s}$, which is also a solution of the Dirac equation. The fact that time reversal of the state $\psi_s(\mathbf{r}, t)$ produces another solution of the Dirac equation is a consequence of the fact that the hamiltonian appearing in the Dirac equation, Eq. (C.69), has no explicit time dependence. Our strategy will be to first work out the conditions that this operator must satisfy, then guess its form and prove that this form works. From the definition of the operator we have:

$$\hat{\mathcal{T}}\psi_s(\mathbf{r}, t) = \tilde{\psi}_{\tilde{s}}(\mathbf{r}, -t) \quad \text{such that} \quad i\hbar \frac{\partial \psi_s}{\partial t} = \hat{\mathcal{H}}\psi_s \Rightarrow i\hbar \frac{\partial \tilde{\psi}_{\tilde{s}}}{\partial t} = \hat{\mathcal{H}}\tilde{\psi}_{\tilde{s}} \tag{9.21}$$

We next consider the inverse of the time-reversal operator $\hat{\mathcal{T}}^{-1}$, and use it to undo the effect of time reversal, that is, to restore the direction of time:

$$\hat{\mathcal{T}}^{-1}\tilde{\psi}_{\tilde{s}}(\mathbf{r}, -t) = \hat{\mathcal{T}}^{-1}\hat{\mathcal{T}}\psi_s(\mathbf{r}, t) = \psi_s(\mathbf{r}, t) \quad \text{and} \quad \hat{\mathcal{T}}^{-1}\frac{\partial}{\partial t}\tilde{\psi}_{\tilde{s}} = -\frac{\partial}{\partial t}\hat{\mathcal{T}}^{-1}\tilde{\psi}_{\tilde{s}} = -\frac{\partial \psi_s}{\partial t}$$

We will use these relations in figuring out how the inverse of the time-reversal operator should affect the Dirac equation for the state $\tilde{\psi}_{\tilde{s}}$: we apply $\hat{\mathcal{T}}^{-1}$ to the equation for $\tilde{\psi}_{\tilde{s}}$ and insert the product $\hat{\mathcal{T}}\hat{\mathcal{T}}^{-1} = 1$ between the factor ($i\hbar$) and the time derivative of $\tilde{\psi}_{\tilde{s}}$ on the left-hand side of the Dirac equation, and between the hamiltonian and $\tilde{\psi}$ on the right-hand side. The result is:

$$\hat{\mathcal{T}}^{-1}(i\hbar)\hat{\mathcal{T}}\hat{\mathcal{T}}^{-1}\frac{\partial \tilde{\psi}_{\tilde{s}}}{\partial t} = \hat{\mathcal{T}}^{-1}\hat{\mathcal{H}}\hat{\mathcal{T}}\hat{\mathcal{T}}^{-1}\tilde{\psi}_{\tilde{s}} \Rightarrow -\hat{\mathcal{T}}^{-1}(i\hbar)\hat{\mathcal{T}}\frac{\partial \psi_s}{\partial t} = \hat{\mathcal{T}}^{-1}\hat{\mathcal{H}}\hat{\mathcal{T}}\psi_s \tag{9.22}$$

where we have used the relations derived earlier. If $\hat{\mathcal{T}}$ is indeed the right operator for time reversal, then Eqs (9.21) and (9.22) must be compatible, that is, we must have:

$$-\hat{\mathcal{T}}^{-1}(i\hbar)\hat{\mathcal{T}} = i\hbar \quad \text{and} \quad \hat{\mathcal{T}}^{-1}\hat{\mathcal{H}}\hat{\mathcal{T}} = \hat{\mathcal{H}} \tag{9.23}$$

An expression that satisfies these conditions is given by:

$$\hat{\mathcal{T}} = \underline{\mathbf{S}}\hat{\mathcal{C}}, \quad \underline{\mathbf{S}} = i\begin{pmatrix} \sigma_y & 0 \\ 0 & \sigma_y \end{pmatrix}, \quad \hat{\mathcal{C}} = \text{complex conjugation} \tag{9.24}$$

where σ_y is one of the Pauli spin matrices [see Appendix C, Eq. (C.43)]. This expression for $\hat{\mathcal{T}}$ and the properties of the Pauli matrices prove that the two conditions above are satisfied (the reader is guided through this proof by the steps of Problem 2). We emphasize that this choice is not unique, because multiplying the operator by an arbitrary phase factor $\exp(i\theta)$ gives a valid time-reversal operator. For example, another convenient choice for the time-reversal operator is:

$$\hat{\mathcal{T}} = \exp\left[i\frac{\pi}{2}\underline{\mathbf{S}}\hat{\mathcal{C}}\right] \tag{9.25}$$

(see Problem 3).

9.3.1 Kramers Degeneracy

We will next consider the implications of time reversal for a state which is composed of a stationary and a time-dependent part, with the stationary part being identified by its momentum $\mathbf{p} = \hbar\mathbf{k}$ and its spin s. The proof that $\hat{\mathcal{T}}$ is a proper time-reversal operator also shows that it commutes with the hamiltonian:

$$\hat{\mathcal{T}}^{-1}\hat{\mathcal{H}}\hat{\mathcal{T}} = \hat{\mathcal{H}} \Rightarrow \hat{\mathcal{H}}\hat{\mathcal{T}} = \hat{\mathcal{T}}\hat{\mathcal{H}}$$

which means that if the state $\psi_{\mathbf{k},s}(\mathbf{r})$ is a stationary eigenstate of the hamiltonian with energy eigenvalue $\epsilon_{\mathbf{k},s}$ then its time-reversed partner $\hat{\mathcal{T}}\psi_{\mathbf{k},s}(\mathbf{r},t)$ will also be an eigenstate with the same eigenvalue:

$$\hat{\mathcal{H}}\psi_{\mathbf{k},s} = \epsilon_{\mathbf{k},s}\psi_{\mathbf{k},s} \Rightarrow \hat{\mathcal{T}}\hat{\mathcal{H}}\psi_{\mathbf{k},s} = \epsilon_{\mathbf{k},s}\hat{\mathcal{T}}\psi_{\mathbf{k},s} \Rightarrow \hat{\mathcal{H}}(\hat{\mathcal{T}}\psi_{\mathbf{k},s}) = \epsilon_{\mathbf{k},s}(\hat{\mathcal{T}}\psi_{\mathbf{k},s})$$

Now we denote as $\tilde{\mathbf{k}}$ and $\tilde{s}$ the momentum and spin labels of the stationary part in the time-reversed state $\tilde{\psi}$; these are quantities whose values we will need to figure out. With these definitions, the above relation takes the form

$$\hat{\mathcal{H}}\tilde{\psi}_{\tilde{\mathbf{k}},\tilde{s}} = \epsilon_{\mathbf{k},s}\tilde{\psi}_{\tilde{\mathbf{k}},\tilde{s}}$$

Since we know the form of the operator $\hat{\mathcal{T}}$, we can determine how the stationary part of the state is affected by time reversal. Specifically, the time-reversed state will have the opposite momentum $\tilde{\mathbf{k}} = -\mathbf{k}$, since the momentum operator changes sign under $\hat{\mathcal{T}}$. Moreover, from the form of the matrix $\underline{\mathbf{S}}$ in the expression for $\hat{\mathcal{T}}$, we conclude that the spin of the original state must be flipped in the time-reversed state (recall that our assumption at the beginning of the section was that we are dealing with spin-1/2 particles). In the following we will use the notation $\uparrow$ and $\downarrow$ to denote states whose expectation value $\langle\sigma_z\rangle$ of the Pauli spin matrix σ_z is either positive or negative, respectively. We emphasize that in the presence of terms in the hamiltonian that include the spin operator, like the spin–orbit coupling term, the eigenfunctions can no longer be described in terms of the up- or down-spin states. These considerations establish that if the original stationary state has labels $\mathbf{k}$ and $\uparrow$ for the momentum and the spin, the corresponding stationary part in the time-reversed state will have labels $-\mathbf{k}$ and $\downarrow$, that is, the energies and stationary parts of the two wavefunctions will be related by:

$$\epsilon_{-\mathbf{k},\downarrow} = \epsilon_{\mathbf{k},\uparrow}, \quad \psi_{-\mathbf{k},\downarrow}(\mathbf{r}) = i\sigma_y\psi^*_{\mathbf{k},\uparrow}(\mathbf{r}) \tag{9.26}$$

These relations are known as "Kramers degeneracy" for particles with spin 1/2.

9.4 Berry's Phase

The discussion so far is missing a key ingredient, which is the arbitrary phase φ that can multiply a system's wavefunction $|\psi\rangle$, without changing the expectation value of any operator $\hat{\mathcal{O}}$ that corresponds to a physical observable:

$$|\psi\rangle \rightarrow |\tilde{\psi}\rangle = e^{i\varphi}|\psi\rangle \Rightarrow \langle\psi|\hat{\mathcal{O}}|\psi\rangle = \langle\tilde{\psi}|\hat{\mathcal{O}}|\tilde{\psi}\rangle$$

This holds because the action of the operator $\hat{O}$ does not affect the phase factor $e^{i\varphi}$, which is just a constant complex number of magnitude 1. It turns out that there exist physical observables that cannot be described by the expectation value of any operator; the arbitrary phase can play a key role in the description of such observables. This can occur when the system has a *parametric dependence* in its hamiltonian. In this case, the phase is often referred to as the "geometric" phase, since it describes the dependence of the wavefunction on parameters related to geometric features of the system. We encountered one such general example in the discussion of the Born–Oppenheimer approximation in Chapter 4. We note that in a truly isolated system, where the hamiltonian is not a function of some parametric variable, Berry's phase is not obtained.

9.4.1 General Formulation

To define Berry's phase in a general context, let us define a generic parametric hamiltonian such that:

$$\hat{\mathcal{H}}(\mathbf{q})|\psi_{\mathbf{q}}\rangle = \epsilon_{\mathbf{q}}|\psi_{\mathbf{q}}\rangle$$

where $\mathbf{q}$ is some parameter. At this point we leave the nature of the parameter $\mathbf{q}$ intentionally undefined, because as we will see later, it can be associated with different features of the system.[11] When solving the Schrödinger equation for the wavefunction of a system, either analytically or numerically, an arbitrary phase φ appears in the form $e^{i\varphi}$; we assume we are dealing with the system in its ground state. When finding an analytic solution, we may choose the simplest mathematical expression for the wavefunction, but this implies choosing the value of this phase to be zero; when calculating a numerical solution, the algorithm we use typically selects a random value for this phase, which is not set by the differential equation. Suppose that this phase for two different ground states of the system, identified by different values of the geometric parameter $\mathbf{q}$, takes the values φ_1 and φ_2 for the values of the parameter $\mathbf{q}_1$ and $\mathbf{q}_2$. Next, suppose that we wish to explicitly remove the phase φ for each case, so we would multiply the first wavefunction by $e^{-i\varphi_1}$ and the second by $e^{-i\varphi_2}$:

$$|\psi_{\mathbf{q}_1}\rangle \rightarrow |\tilde{\psi}_{\mathbf{q}_1}\rangle = e^{-i\varphi_1}|\psi_{\mathbf{q}_1}\rangle, \quad |\psi_{\mathbf{q}_2}\rangle \rightarrow |\tilde{\psi}_{\mathbf{q}_2}\rangle = e^{-i\varphi_2}|\psi_{\mathbf{q}_2}\rangle$$

a process which is called "choosing a gauge." Then the overlap of the two phase-corrected wavefunctions would be:

$$\langle \tilde{\psi}_{\mathbf{q}_1}|\tilde{\psi}_{\mathbf{q}_2}\rangle = e^{i(\varphi_1-\varphi_2)}\langle \psi_{\mathbf{q}_1}|\psi_{\mathbf{q}_2}\rangle$$

But since in this last expression we have removed all arbitrary phase dependence, the left-hand side of the equation must be the *magnitude* of the overlap, $|\langle \psi_{\mathbf{q}_1}|\psi_{\mathbf{q}_2}\rangle|$, which gives a useful way of defining the *relative* phase $\Delta\varphi_{12} = (\varphi_1 - \varphi_2)$ as:

$$e^{-i\Delta\varphi_{12}} = \frac{\langle \psi_{\mathbf{q}_1}|\psi_{\mathbf{q}_2}\rangle}{|\langle \psi_{\mathbf{q}_1}|\psi_{\mathbf{q}_2}\rangle|} \qquad (9.27)$$

---

[11] The discussion in this section follows closely the treatment in the review articles by R. Resta and by D. Xiao, M.-C. Chang, and Q. Niu (see Further Reading).

By separating real and imaginary parts in the above definition, we can obtain an alternative way for expressing the relative phase as:

$$\Delta\varphi_{12} = -\text{Im}\left[\ln\left(\langle\psi_{\mathbf{q}_1}|\psi_{\mathbf{q}_2}\rangle\right)\right] \tag{9.28}$$

In the limit of small difference $\Delta\mathbf{q}$ between $\mathbf{q}_1$ and $\mathbf{q}_2$, which corresponds to a change $\Delta\varphi$ in phase, the above expressions become:

$$e^{-i\Delta\varphi} = \frac{\langle\psi_{\mathbf{q}}|\psi_{\mathbf{q}+\Delta\mathbf{q}}\rangle}{|\langle\psi_{\mathbf{q}}|\psi_{\mathbf{q}+\Delta\mathbf{q}}\rangle|} \Rightarrow -i\Delta\varphi \approx \langle\psi_{\mathbf{q}}|\nabla_{\mathbf{q}}\psi_{\mathbf{q}}\rangle \cdot \Delta\mathbf{q} \tag{9.29}$$

where we have assumed that our choice of gauge ensures a differentiable phase and we have kept the leading-order terms in the Taylor expansion of both sides. For infinitesimal changes $d\mathbf{q}$, the infinitesimal change in the phase becomes:

$$d\varphi = i\langle\psi_{\mathbf{q}}|\nabla_{\mathbf{q}}\psi_{\mathbf{q}}\rangle \cdot d\mathbf{q} \tag{9.30}$$

With this last expression, we can now calculate the total change in phase, that is, Berry's phase γ, along a closed path C in the space of the parameter $\mathbf{q}$:

$$\gamma = \oint_C d\varphi = i\oint_C \langle\psi_{\mathbf{q}}|\nabla_{\mathbf{q}}\psi_{\mathbf{q}}\rangle \cdot d\mathbf{q} \tag{9.31}$$

The interesting property of γ is that it is *gauge invariant*. To show this explicitly, we consider a new wavefunction $|\tilde{\psi}_{\mathbf{q}}\rangle$ that involves a gauge transformation by a phase $\varphi(\mathbf{q})$ which is a continuous differentiable function of the variable $\mathbf{q}$:

$$|\tilde{\psi}_{\mathbf{q}}\rangle = e^{-i\varphi(\mathbf{q})}|\psi_{\mathbf{q}}\rangle \tag{9.32}$$

Differentiating with $\mathbf{q}$ we obtain

$$\nabla_{\mathbf{q}}|\tilde{\psi}_{\mathbf{q}}\rangle = \left[-i\nabla_{\mathbf{q}}\varphi(\mathbf{q})\right]|\tilde{\psi}_{\mathbf{q}}\rangle + e^{-i\varphi(\mathbf{q})}\left[\nabla_{\mathbf{q}}|\psi_{\mathbf{q}}\rangle\right]$$

and using this result to calculate the Berry phase for the new wavefunction we find:

$$\tilde{\gamma} = i\oint_C \langle\tilde{\psi}_{\mathbf{q}}|\nabla_{\mathbf{q}}\tilde{\psi}_{\mathbf{q}}\rangle \cdot d\mathbf{q} = i\oint_C \langle\tilde{\psi}_{\mathbf{q}}|e^{-i\varphi(\mathbf{q})}\nabla_{\mathbf{q}}|\psi_{\mathbf{q}}\rangle \cdot d\mathbf{q} + \oint_C \nabla_{\mathbf{q}}\varphi(\mathbf{q}) \cdot d\mathbf{q} = \gamma$$

since from the definition of $|\tilde{\psi}_{\mathbf{q}}\rangle$ we have:

$$\langle\tilde{\psi}_{\mathbf{q}}| = \langle\psi_{\mathbf{q}}|e^{i\varphi(\mathbf{q})} \Rightarrow \langle\tilde{\psi}_{\mathbf{q}}|e^{-i\varphi(\mathbf{q})}\nabla_{\mathbf{q}}|\psi_{\mathbf{q}}\rangle = \langle\psi_{\mathbf{q}}|\nabla_{\mathbf{q}}\psi_{\mathbf{q}}\rangle$$

and the integral of $\nabla_{\mathbf{q}}\varphi(\mathbf{q})$ over a closed loop vanishes identically.

A different way of establishing gauge invariance is by using the expression of Eq. (9.28), and splitting the closed path into N infinitesimal intervals of size $d\mathbf{q}$ starting at $\mathbf{q}_0$ and ending at $\mathbf{q}_{N+1} = \mathbf{q}_0$, with $\mathbf{q}_{i+1} = \mathbf{q}_i + d\mathbf{q}$ (we are assuming the limit $N \to \infty$ is properly defined). The result becomes:

$$\Delta\varphi = -\text{Im}\left[\sum_{i=0}^{N}\ln\left(\langle\psi_{\mathbf{q}_i}|\psi_{\mathbf{q}_{i+1}}\rangle\right)\right] = -\text{Im}\left[\ln\left(\prod_{i=0}^{N}\langle\psi_{\mathbf{q}_i}|\psi_{\mathbf{q}_{i+1}}\rangle\right)\right]$$

In the last expression, if we were to introduce a gauge transformation of the type given in Eq. (9.32), the phases cancel between successive steps because they arise from bras and kets with the same index, and the phases at the initial and final points of the path are the same so they also cancel; thus, the whole expression is gauge invariant. This is in contrast to the value of the *integrand* in Eq. (9.31), which *is* gauge dependent. The fact that γ is gauge invariant is crucial; otherwise, there would be no hope of relating it to a physical observable. The value of γ actually depends on the path we choose for the contour integration, as is evident from the discussion of the Aharonov–Bohm effect presented in Section 9.5.

In the following, we shall encounter several cases where the parameter $\mathbf{q}$ is a real (as opposed to complex) 3D vector, as this is the most usual situation for application of these concepts. In this case, the expressions above naturally lead to two new quantities. The first is the integrand of Eq. (9.31) itself, called "Berry connection":

$$\mathcal{A}(\mathbf{q}) = i\langle \psi_{\mathbf{q}} | \nabla_{\mathbf{q}} \psi_{\mathbf{q}} \rangle \tag{9.33}$$

which is a three-component vector. The second is an antisymmetric second-rank tensor, called "Berry curvature":

$$\Omega_{\alpha\beta}(\mathbf{q}) = \frac{\partial \mathcal{A}_\beta(\mathbf{q})}{\partial q_\alpha} - \frac{\partial \mathcal{A}_\alpha(\mathbf{q})}{\partial q_\beta} \tag{9.34}$$
$$- i \left[\langle \frac{\partial \psi_{\mathbf{q}}}{\partial q_\alpha} | \frac{\partial \psi_{\mathbf{q}}}{\partial q_\beta} \rangle - \langle \frac{\partial \psi_{\mathbf{q}}}{\partial q_\beta} | \frac{\partial \psi_{\mathbf{q}}}{\partial q_\alpha} \rangle \right] = -2\, \text{Im}\, \langle \frac{\partial \psi_{\mathbf{q}}}{\partial q_\alpha} | \frac{\partial \psi_{\mathbf{q}}}{\partial q_\beta} \rangle$$

Since we are working with three-component vectors, we can associate the elements of the second-rank tensor $\Omega_{\alpha\beta}$ with the components of a vector $\mathbf{\Omega}$ by the cyclic permutations of the indices, for instance, in cartesian coordinates Ω_{xy} would be the $\hat{z}$ component of $\mathbf{\Omega}$, and so on (this is formally known as "using the Levi–Civita antisymmetric tensor $\epsilon_{\alpha\beta\gamma}$" in the relation $\Omega_{\alpha\beta} = \epsilon_{\alpha\beta\gamma} \Omega_\gamma$). This allows us to write the vector $\mathbf{\Omega}$ as the curl of the vector $\mathcal{A}$, leading to the expression

$$\mathbf{\Omega}(\mathbf{q}) = \nabla_{\mathbf{q}} \times \mathcal{A}(\mathbf{q}) \tag{9.35}$$

These definitions simplify the expression for Berry's phase, which becomes:

$$\gamma = \oint_C \mathcal{A}(\mathbf{q}) \cdot d\mathbf{q} = \int_S \nabla_{\mathbf{q}} \times \mathcal{A}(\mathbf{q}) \cdot d\mathcal{S} = \int_S \mathbf{\Omega}(\mathbf{q}) \cdot d\mathcal{S} \tag{9.36}$$

where S is the surface enclosed by the curve C and $d\mathcal{S}$ is the infinitesimal element on it with the surface unit vector pointing in the direction determined by the right-hand rule, given the direction in which C is traversed. The second expression for γ in Eq. (9.36) is obtained from the first using Stokes' theorem [see Appendix A, Eq. (A.23)] and justifies the name "Berry curvature" for $\mathbf{\Omega}$. These expressions are interesting not only for their formal simplicity and elegance, but also because they are related to familiar physical quantities and have an important manifestation, as we describe next.

An alternate way for calculating Berry's phase, expressed as a surface integral of the Berry curvature, Eq. (9.36), is the following. From perturbation theory, and introducing superscripts to identify the ground state. $\psi_{\mathbf{q}}^{(0)}, \epsilon_{\mathbf{q}}^{(0)}$ and excited states $\psi_{\mathbf{q}}^{(n)}, \epsilon_{\mathbf{q}}^{(n)}$ of the

$$A_\theta^{(1)}(\theta,\phi) = i\langle\psi_\omega^{(1)}|\frac{\partial}{\partial\theta}|\psi_\omega^{(1)}\rangle = i\begin{pmatrix} \sin(\theta/2)e^{-i\phi} \\ -\cos(\theta/2) \end{pmatrix}^\dagger \cdot \begin{pmatrix} \frac{1}{2}\cos(\theta/2)e^{-i\phi} \\ \frac{1}{2}\sin(\theta/2) \end{pmatrix} = 0$$

$$A_\phi^{(1)}(\theta,\phi) = i\langle\psi_\omega^{(1)}|\frac{\partial}{\partial\phi}|\psi_\omega^{(1)}\rangle = i\begin{pmatrix} \sin(\theta/2)e^{-i\phi} \\ -\cos(\theta/2) \end{pmatrix}^\dagger \cdot \begin{pmatrix} -i\sin(\theta/2)e^{-i\phi} \\ 0 \end{pmatrix}$$

$$= \frac{1-\cos(\theta)}{2}$$

From these we can also calculate the Berry curvature:

$$\Omega_{\theta\phi}^{(1)}(\theta,\phi) = \frac{\partial A_\phi^{(1)}}{\partial\theta} - \frac{\partial A_\theta^{(1)}}{\partial\phi} = \frac{1}{2}\sin(\theta) \tag{9.41}$$

We can calculate Berry's phase in two ways. First, we use the Berry connection integrated over the closed path around the horizontal circle C (see Fig. 9.11), with the variable of integration denoted by $\omega' = (\theta,\phi')$, so that the path C is described by $d\omega' = \hat{\phi}d\phi'$, with $\phi' \in [0,2\pi)$:

$$\gamma^{(1)}(\theta) = \oint_C \mathbf{A}^{(1)}(\omega') \cdot d\omega' = \int_0^{2\pi} A_\phi^{(1)}(\theta,\phi')d\phi' = \pi(1-\cos(\theta))$$

Second, we the Berry curvature, integrated over the surface enclosed by the circle C, that is, the part of the sphere north of the horizontal circle in which the wavefunction is well defined. This gives:

$$\gamma^{(1)}(\theta) = \int_0^{2\pi} d\phi' \int_0^\theta \Omega_{\theta\phi}^{(1)}(\theta',\phi')d\theta' = 2\pi\int_0^\theta \frac{1}{2}\sin(\theta')d\theta' = \pi(1-\cos(\theta))$$

If we let $\theta \to \pi$ (without actually acquiring this value because the wavefunction there is not properly defined), we find that the integral of the Berry curvature over the entire sphere becomes:

$$\int_0^{2\pi} d\phi' \int_0^\pi \Omega_{\theta\phi}^{(1)}(\theta',\phi')d\theta' = \lim_{\theta\to\pi}[\pi(1-\cos(\theta))] = 2\pi$$

that is, it becomes an integral multiple of 2π. This is related to the first Chern number for this state, a topic we discuss in more detail below (see also Problem 4).

9.4.2 Berry's Phase for Electrons in Crystals

We turn next to electrons in a crystal and the manifestations of Berry's phase in this physical system. We have seen in Section 9.2 that it is convenient to deal with the system in terms of the wavefunctions $|u_\mathbf{k}^{(n)}\rangle$ which have full translational periodicity, rather than the true Bloch states $|\psi_\mathbf{k}^{(n)}\rangle$. The hamiltonian that applies to the states $|u_\mathbf{k}^{(n)}\rangle$ is:

$$\hat{\mathcal{H}}_\mathbf{k}(\mathbf{p},\mathbf{r}) = \frac{1}{2m_e}[-i\hbar\nabla_\mathbf{r} + \hbar\mathbf{k}]^2 + \mathcal{V}(\mathbf{r})$$

to which we will later add the electric field as an external perturbation. We calculate first the Berry curvature for this system. The parameter that characterizes the hamiltonian and the states is evidently $\mathbf{k}$, the wave-vector. With the definitions introduced above, the Berry connection and Berry curvature for the nth band are given by:

$$
\mathcal{A}^{(n)}(\mathbf{k}) = i\langle u_{\mathbf{k}}^{(n)} | \nabla_{\mathbf{k}} u_{\mathbf{k}}^{(n)}\rangle \Rightarrow \Omega_{\alpha\beta}^{(n)}(\mathbf{k}) = i\left[\langle \frac{\partial u_{\mathbf{k}}^{(n)}}{\partial k_\alpha} | \frac{\partial u_{\mathbf{k}}^{(n)}}{\partial k_\beta}\rangle - \langle \frac{\partial u_{\mathbf{k}}^{(n)}}{\partial k_\beta} | \frac{\partial u_{\mathbf{k}}^{(n)}}{\partial k_\alpha}\rangle \right] \tag{9.42}
$$

$$
\mathbf{\Omega}^{(n)}(\mathbf{k}) = i\langle \nabla_{\mathbf{k}} u_{\mathbf{k}}^{(n)}| \times |\nabla_{\mathbf{k}} u_{\mathbf{k}}^{(n)}\rangle \tag{9.43}
$$

and the expression for Berry's phase takes the form

$$
\gamma^{(n)} = \oint_C i\langle u_{\mathbf{k}}^{(n)} | \nabla_{\mathbf{k}} u_{\mathbf{k}}^{(n)}\rangle \cdot d\mathbf{k} \tag{9.44}
$$

where the closed path C is in the space of wave-vectors $\mathbf{k}$, that is, the BZ of the crystal. A closed path can also be realized by moving from one point in $\mathbf{k}$ space to the point $\mathbf{k} + \mathbf{G}$, where $\mathbf{G}$ is a reciprocal lattice vector, since the two points are then equivalent, as we have seen earlier (see the discussion in Chapter 2). In this case, γ is known as "Zak's phase."[12] All the quantities defined so far are intrinsic properties of the band structure as they depend exclusively on the band wavefunctions. A closed path C is required to obtain Berry's phase $\gamma^{(n)}$, but is not necessary in the definition of the Berry curvature $\mathbf{\Omega}^{(n)}(\mathbf{k})$, which is a gauge-invariant quantity. We will see some manifestations of this in the following.

A related important issue is what happens when the hamiltonian acquires a time dependence due to a term that does not destroy the translational periodicity of the crystal. We seek a general result that will not depend on the nature of the perturbation, so it can be applied in any situation. For a sufficiently weak perturbation, we can still use the original wavefunctions as the basis to express the new states that are perturbed by a small amount. We use a Taylor expansion for the perturbed states:

$$
|\tilde{u}_{\mathbf{k}}^{(n)}\rangle = |u_{\mathbf{k}}^{(n)}\rangle + \delta |u_{\mathbf{k}}^{(n)}\rangle = |u_{\mathbf{k}}^{(n)}\rangle + \delta t | \frac{\partial u_{\mathbf{k}}^{(n)}}{\partial t}\rangle
$$

and then develop a way to express them in terms of the unperturbed ones. One perspective is to use perturbation theory (see Appendix C) and consider the time derivative of the hamiltonian as the perturbation itself. Another perspective is to start with the time-dependent Schrödinger equation (see Appendix C) and take the time derivative to be the ratio of the small differences $\delta\psi = \psi(t) - \psi(t_0)$ and $\delta t = t - t_0$, which leads to the identification of the small time interval δt with the inverse of the hamiltonian operator, shifted by the energy of the unperturbed state, and multiplied by $i\hbar$. From either approach we arrive at the general expression for the perturbation of the wavefunction that involves an expansion over the unperturbed states:

$$
\delta |u_{\mathbf{k}}^{(n)}\rangle = -i\hbar \sum_{m\neq n} \frac{|u_{\mathbf{k}}^{(m)}\rangle \langle u_{\mathbf{k}}^{(m)}|}{\epsilon_{\mathbf{k}}^{(n)} - \epsilon_{\mathbf{k}}^{(m)}} | \frac{\partial u_{\mathbf{k}}^{(n)}}{\partial t}\rangle
$$

---

[12] J. Zak, *Phys. Rev. Lett.* **62**, 2747 (1989).

With this, we calculate the expectation of the velocity operator, Eq. (9.14), in the *perturbed* states to obtain, for the velocity:

$$\langle \mathbf{v}_{\mathbf{k}}^{(n)} \rangle = \frac{1}{\hbar} \langle u_{\mathbf{k}}^{(n)} | \nabla_{\mathbf{k}} \hat{\mathcal{H}}_{\mathbf{k}} | u_{\mathbf{k}}^{(n)} \rangle - i \sum_{m \neq n} \left[\frac{\langle u_{\mathbf{k}}^{(n)} | \nabla_{\mathbf{k}} \hat{\mathcal{H}}_{\mathbf{k}} | u_{\mathbf{k}}^{(m)} \rangle \langle u_{\mathbf{k}}^{(m)} | \partial u_{\mathbf{k}}^{(n)} / \partial t \rangle}{\epsilon_{\mathbf{k}}^{(n)} - \epsilon_{\mathbf{k}}^{(m)}} - \text{c.c.} \right] \quad (9.45)$$

with c.c. standing for the complex conjugate. The first term in this expression was worked out in the previous section, where we found that it is equal to the result of Eq. (9.15). For the second term, we follow the same steps of the derivation that led to Eq. (9.15) together with the orthogonality of the wavefunctions $|u_{\mathbf{k}}^{(n)}\rangle$ and the completeness of this basis (see Problem 5), and we write the derivative with respect to $\mathbf{k}$ in the equivalent notation $\nabla_{\mathbf{k}} = \partial/\partial\mathbf{k}$ to arrive at the following expression for the velocity:

$$\langle \mathbf{v}_{\mathbf{k}}^{(n)} \rangle = \frac{1}{\hbar} \nabla_{\mathbf{k}} \epsilon_{\mathbf{k}}^{(n)} - i \left[\langle \frac{\partial u_{\mathbf{k}}^{(n)}}{\partial \mathbf{k}} | \frac{\partial u_{\mathbf{k}}^{(n)}}{\partial t} \rangle - \langle \frac{\partial u_{\mathbf{k}}^{(n)}}{\partial t} | \frac{\partial u_{\mathbf{k}}^{(n)}}{\partial \mathbf{k}} \rangle \right] \quad (9.46)$$

In the second term of this expression we recognize a quantity that has the same structure as the Berry curvature of the band, but is now expressed with respect to the two independent parameters $\mathbf{k}$ and t. This is an important result with several consequences that we explore further in the next section.

Example 9.3   Berry's phase for the 2D honeycomb lattice

We revisit the 2D honeycomb lattice, discussed in Chapter 2, to apply the concepts developed so far; recall that this example is actually relevant to a real physical system, graphene (that is, a single plane of threefold sp^2-bonded C atoms). In the 2D honeycomb lattice, there are two sites per unit cell, labeled A and B, each forming a 2D hexagonal sublattice, with real-space lattice vectors $\mathbf{a}_1, \mathbf{a}_2$:

$$\mathbf{a}_1 = (a/2)\left(\sqrt{3}\hat{x} + \hat{y}\right), \quad \mathbf{a}_2 = (a/2)\left(\sqrt{3}\hat{x} - \hat{y}\right) \quad (9.47)$$

The two sites can be equivalent as in graphene, corresponding to two identical C atoms, or inequivalent as in hBN, corresponding to the B and N atoms. We already investigated the model for equivalent A and B sites, at the level of the tight-binding hamiltonian with nearest-neighbor hopping matrix elements in Chapter 2. To include inequivalent sites in the model, at the same level of tight-binding treatment, we need to make the on-site energies different, say $\epsilon_A \neq \epsilon_B$. We can then define:

$$\bar{\epsilon} = \frac{\epsilon_A + \epsilon_B}{2}, \quad \mathcal{V}_0 = \frac{\epsilon_A - \epsilon_B}{2} \Rightarrow \epsilon_A = \bar{\epsilon} + \mathcal{V}_0, \quad \epsilon_B = \bar{\epsilon} - \mathcal{V}_0$$

and set $\bar{\epsilon} = 0$, since this is the same as a constant shift of all the energy values which does not affect the physics. This leaves a hamiltonian with diagonal matrix elements $\mathcal{V}_0$ and $-\mathcal{V}_0$, a variation of the graphene model (this model was introduced by M. J. Rice and E. J. Mele[a]). This model amounts to breaking the inversion symmetry,

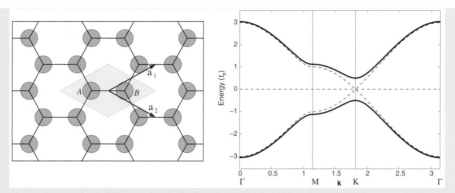

Fig. 9.12

Left: The honeycomb lattice with the primitive lattice vectors $\mathbf{a}_1, \mathbf{a}_2$ (red arrows) and the primitive unit cell highlighted in blue; the sites A and B can be inequivalent, when the on-site energies $\epsilon_A \neq \epsilon_B$. The black lines indicate the nearest-neighbor hopping matrix elements of strength t_π. **Right**: The corresponding band structure, in solid lines, along the high-symmetry directions of the BZ; the model parameters are $\mathcal{V}_0 = 0.5|t_\pi|$, leading to $\epsilon_{\mathrm{gap}}(\mathbf{K}) = 2\mathcal{V}_0 = |t_\pi|$. The band structure of the lattice *with* inversion symmetry is shown in dashed lines for comparison.

also referred to as P-symmetry, of the honeycomb lattice for $\mathcal{V}_0 \neq 0$, as illustrated in Fig. 9.12. For this case, we have chosen the unit cell to be a rhombus, and the origin of the coordinate system to be at the mid-point of the nearest-neighbor distance between sites A and B. This choice of origin demonstrates the inversion symmetry of the lattice if the two sites are equivalent, which is broken for the case of inequivalent sites. (a different choice of origin that demonstrates the inversion symmetry is the center of the hexagon formed by the atoms, that is, a corner of the blue-shaded rhombus in Fig. 9.12). We had found that the tight-binding hamiltonian for graphene is given by [see Eq. (2.86)]:

$$\hat{\mathcal{H}}_{\mathbf{k}} = \begin{pmatrix} \epsilon_p & t_\pi f_{\mathbf{k}}^* \\ t_\pi f_{\mathbf{k}} & \epsilon_p \end{pmatrix}, \quad f_{\mathbf{k}} = \left[1 + e^{i\mathbf{k}\cdot\mathbf{a}_1} + e^{i\mathbf{k}\cdot\mathbf{a}_2}\right]$$

With the addition of the symmetry-breaking diagonal terms, after setting $\epsilon_p = 0$, the hamiltonian can be written in the more compact form

$$\hat{\mathcal{H}}_{\mathbf{k}} = \begin{pmatrix} \mathcal{V}_0 & t_\pi f_{\mathbf{k}}^* \\ t_\pi f_{\mathbf{k}} & -\mathcal{V}_0 \end{pmatrix} = \mathbf{h}(\mathbf{k})\cdot\boldsymbol{\sigma} = h_x(\mathbf{k})\sigma_x + h_y(\mathbf{k})\sigma_y + h_z(\mathbf{k})\sigma_z \quad (9.48)$$

where $\boldsymbol{\sigma} = \sigma_x\hat{x} + \sigma_y\hat{y} + \sigma_z\hat{z}$ with $\sigma_x, \sigma_y, \sigma_z$ the Pauli spin matrices, and we have defined the vector operator $\mathbf{h}(\mathbf{k})$ with components

$$h_x(\mathbf{k}) \equiv t_\pi[1 + \cos(\mathbf{k}\cdot\mathbf{a}_1) + \cos(\mathbf{k}\cdot\mathbf{a}_2)],$$

$$h_y(\mathbf{k}) \equiv t_\pi[\sin(\mathbf{k}\cdot\mathbf{a}_1) + \sin(\mathbf{k}\cdot\mathbf{a}_2)], \quad h_z(\mathbf{k}) \equiv \mathcal{V}_0$$

The band structures for both the case with inversion symmetry ($\mathcal{V}_0 = 0$) and with broken inversion symmetry ($\mathcal{V}_0 \neq 0$) are shown in Fig. 9.12.

We investigate first the solution of this hamiltonian for $\mathcal{V}_0 = 0$. For this case, the eigenvalues and corresponding eigenfunctions are given by:

$$\epsilon_{\mathbf{k}}^{(1)} = -t_\pi |f_{\mathbf{k}}|, \quad |u_{\mathbf{k}}^{(1)}\rangle = \frac{1}{\sqrt{2}} \begin{pmatrix} f_{\mathbf{k}}^*/|f_{\mathbf{k}}| \\ -1 \end{pmatrix}$$

$$\epsilon_{\mathbf{k}}^{(2)} = t_\pi |f_{\mathbf{k}}|, \quad |u_{\mathbf{k}}^{(2)}\rangle = \frac{1}{\sqrt{2}} \begin{pmatrix} 1 \\ f_{\mathbf{k}}/|f_{\mathbf{k}}| \end{pmatrix}$$

Using these, we can calculate the Berry curvature for this model. The components of the Berry connection for the first state are given by:

$$\mathcal{A}_j^{(1)}(\mathbf{k}) = \mathrm{i}\langle u_{\mathbf{k}}^{(1)}| \frac{\partial}{\partial k_j} |u_{\mathbf{k}}^{(1)}\rangle = -\frac{1}{2|f_{\mathbf{k}}|^2} \left(A_{\mathbf{k},j} f_{\mathbf{k}}^* + A_{\mathbf{k},j}^* f_{\mathbf{k}} \right), \quad j = x, y \qquad (9.49)$$

with the definitions

$$A_{\mathbf{k},j} \equiv \left(a_{1,j} e^{\mathrm{i}\mathbf{k}\cdot\mathbf{a}_1} + a_{2,j} e^{\mathrm{i}\mathbf{k}\cdot\mathbf{a}_2} \right), \quad j = x, y$$

From these, we find that the Berry curvature

$$\Omega_{k_x k_y}^{(1)}(\mathbf{k}) = \frac{\partial \mathcal{A}_y^{(1)}(\mathbf{k})}{\partial k_x} - \frac{\partial \mathcal{A}_x^{(1)}(\mathbf{k})}{\partial k_y}$$

vanishes identically as long as the expressions for the Berry connection are analytic. These expressions are analytic everywhere except when $f_{\mathbf{k}} = 0$, which happens at the six corners of the BZ, labeled K, K', see Fig. 9.13. When these points are excluded from a closed path of integration around the BZ boundary, the contribution of remaining straight lines along the boundary is zero, because the portions on opposite sides of the hexagon are traversed in opposite directions and are equivalent by symmetry (they are related by reciprocal-space vectors). Thus, we only need to calculate the contribution of these points. We define the coordinates of the representative points K and K' as:

$$\mathbf{K} = \frac{2\pi}{a} \left(\frac{1}{\sqrt{3}}\hat{x} + \frac{1}{3}\hat{y} \right), \quad \mathbf{K}' = -\mathbf{K}$$

with each one having two other equivalent points by symmetry (related by reciprocal-space vectors), as shown in Fig. 9.13. Next, we expand in Taylor series around these points, $\mathbf{k} = \mathbf{K} + \mathbf{q}$ or $\mathbf{k}' = \mathbf{K}' - \mathbf{q} = -(\mathbf{K} + \mathbf{q}) = -\mathbf{k}$ with $|\mathbf{q}| \to 0$, identified by the blue arrows in Fig. 9.13. We treat the case of $\mathbf{k}$ near $\mathbf{K}$ first, which leads to the hamiltonian

$$\hat{\mathcal{H}}_{\mathbf{K}+\mathbf{q}} = \mathbf{h}(\mathbf{q}) \cdot \boldsymbol{\sigma}, \quad \mathbf{h}(\mathbf{q}) = e^{\mathrm{i}\tilde{\phi}} \frac{\sqrt{3}a t_\pi}{2\hbar} (\hbar\mathbf{q})$$

where we have explicitly included the hopping matrix element t_π and factors of $\hbar$, to make the connection to the Dirac equation. We have also included a phase factor, $\exp[\mathrm{i}\tilde{\phi}]$, where $\tilde{\phi}$ depends on the arbitrary choice of cartesian coordinate axes, used to express the lattice vectors. However, this phase does not affect the physics: it does not enter into the expression for the energy eigenvalues and only introduces

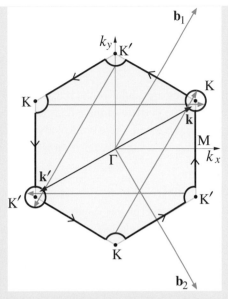

Fig. 9.13 Brillouin zone of the 2D honeycomb lattice, with the high-symmetry points Γ, K, K', M identified. The thick black lines and small circular arcs subtending an angle of $2\pi/3$ each indicate the integration path around the BZ boundary that avoids the singularities at the corners. The green arrows indicate reciprocal lattice vectors $\pm\mathbf{b}_1$ and $\pm(\mathbf{b}_1 + \mathbf{b}_2)$ (horizontal arrows) for displacement of the circular arcs to complete a circle around a single K or K' point. The arbitrary point in the neighborhood of K or K' is denoted by the vectors $\mathbf{k}, \mathbf{k}'$.

a constant phase in the wavefunctions, so we can safely ignore it in the rest of the discussion; this is simply equivalent to rotating the cartesian axes by an angle that sets $\tilde{\phi} = 0$, which leaves all dot products invariant. Since $\hbar\mathbf{q}$ has the dimensions of momentum, the above hamiltonian is exactly equivalent to the Dirac hamiltonian for massless particles (see Appendix C, section C.8), the only difference being that here the equivalent of the "speed of light" is the expression

$$v_{\mathrm{F}} = \frac{\sqrt{3}at_\pi}{2\hbar}$$

that is, the Fermi velocity of the particles at these points in the BZ. Near the point K, the quantities that appear in the wavefunctions $|u_{\mathbf{k}}^{(i)}\rangle, i = 1, 2$, take the form

$$f_{\mathbf{K}+\mathbf{q}} = \frac{\sqrt{3}a}{2}(q_x + iq_y), \quad |f_{\mathbf{K}+\mathbf{q}}| = \frac{\sqrt{3}a}{2}|\mathbf{q}|$$

We switch to spherical coordinates for the vector $\mathbf{q}$:

$$\mathbf{q} = |\mathbf{q}| \left[\sin(\theta)\cos(\phi)\hat{x} + \sin(\theta)\sin(\phi)\hat{y} + \cos(\theta)\hat{z}\right]$$

in which we must set $\theta = \pi/2$, since $\mathbf{q}$ is a vector on the (q_x, q_y) plane. This leads to the following expressions for the eigenvalues and eigenfunctions near the corners of the BZ:

$$\epsilon^{(1)}_{\mathbf{K+q}} = -\hbar v_F |\mathbf{q}|, \quad |u^{(1)}_{\mathbf{K+q}}\rangle = \frac{1}{\sqrt{2}} \begin{pmatrix} e^{-i\phi} \\ -1 \end{pmatrix}$$

$$\epsilon^{(2)}_{\mathbf{K+q}} = \hbar v_F |\mathbf{q}|, \quad |u^{(2)}_{\mathbf{K+q}}\rangle = \frac{1}{\sqrt{2}} \begin{pmatrix} 1 \\ e^{i\phi} \end{pmatrix}$$

Both eigenvalues are linear in the magnitude of $|\mathbf{q}|$, that is, they correspond to the Dirac cones at the corners of the BZ, as discussed in Chapter 2. As far as the eigenfunctions are concerned, these are exactly the solutions for the spin in the monopole field that we found in the previous example, when we set in those expressions

$$\theta = \frac{\pi}{2} \Rightarrow \sin(\theta/2) = \cos(\theta/2) = \frac{1}{\sqrt{2}}$$

due to the 2D nature of the vector $\mathbf{q}$. Thus, for the case of $\mathcal{V}_0 = 0$, the energy–momentum dispersion relation of the two solutions is linear in $|\mathbf{q}|$ and the eigenfunctions correspond to the solutions of the spin in a monopole field constrained to the equator of the sphere. With this approach we have solved the problem of analyticity of the wavefunctions near the points K and K$'$. We can now perform the integration around the whole path shown in Fig. 9.13, including the six circular arcs that avoid the corners of the BZ and take the limit $|\mathbf{q}| \to 0$. By translational periodicity in reciprocal space, the six circular arcs combine to make a full circle around one of the three equivalent K points and another full circle around one of the three equivalent K$'$ points, since each arc subtends an angle of $2\pi/3$ (see Fig. 9.13). We can also directly borrow the results we derived for the Berry connection of a spin in a monopole field:

$$\mathcal{A}^{(1)}_\theta(\pi/2, \phi) = 0, \quad \mathcal{A}^{(1)}_\phi(\pi/2, \phi) = \sin^2(\pi/4) = \frac{1}{2}$$

where we have used $\theta = \pi/2$, and then integrate over a full circle of radius $|\mathbf{q}|$ in reciprocal space around the point K, denoted as $C(\mathrm{K})$, to obtain:

$$\oint_{C(\mathrm{K})} \mathcal{A}^{(1)}(\boldsymbol{\omega}) \cdot d\boldsymbol{\omega} = \int_0^{2\pi} \mathcal{A}^{(1)}_\phi(\pi/2, \phi) d\phi = \pi$$

The result for the integration around a full circle centered at K$'$, denoted as $C(\mathrm{K}')$, gives the opposite result:

$$\oint_{C(\mathrm{K}')} \mathcal{A}^{(1)}(\boldsymbol{\omega}) \cdot d\boldsymbol{\omega} = -\int_0^{2\pi} \mathcal{A}^{(1)}_\phi(\pi/2, \phi) d\phi = -\pi$$

so adding these two contributions we get as final answer a zero value for Berry's phase.

We next solve the model in which $\mathcal{V}_0$ is not zero, starting again with the hamiltonian of Eq. (9.48); we will take $\mathcal{V}_0 > 0$ for the moment, and mention the case $\mathcal{V}_0 < 0$ at the end of the discussion. The two eigenvalues of the hamiltonian are:

$$\epsilon^{(1)}_{\mathbf{k}} = -\sqrt{t_\pi^2 |f_{\mathbf{k}}|^2 + \mathcal{V}_0^2}, \quad \epsilon^{(2)}_{\mathbf{k}} = \sqrt{t_\pi^2 |f_{\mathbf{k}}|^2 + \mathcal{V}_0^2}$$

and the new band structure is shown in Fig. 9.13, for the choice $\mathcal{V}_0 = 0.5|t_\pi|$. In the neighborhood of the six corners of the BZ, with $|\mathbf{q}| \to 0$, we define the new variable $\kappa_z = \mathcal{V}_0/\hbar v_F$, so that the hamiltonian becomes:

$$\hat{\mathcal{H}}_{\mathbf{K}+\mathbf{q}} = \mathbf{h}(\mathbf{q}) \cdot \boldsymbol{\sigma}, \quad \mathbf{h}(\mathbf{q}) \equiv \hbar v_F \left[q_x \hat{x} + q_y \hat{y} + \kappa_z \hat{z} \right]$$

For a fixed value of $\mathbf{q}$, we recognize that this hamiltonian is equivalent to the model of a spin interacting with the field of a monopole, discussed in the previous example; here, the relevant vector which determines the position of the spin has components (q_x, q_y, κ_z). Solving for the eigenvalues of this hamiltonian in the limit $\kappa_z \gg |\mathbf{q}| \to 0$, we obtain:

$$\epsilon_{\mathbf{K}+\mathbf{q}}^{(1)} = -\hbar v_F \sqrt{|\mathbf{q}|^2 + \kappa_z^2} \approx -\mathcal{V}_0 - \hbar v_F \frac{|\mathbf{q}|^2}{2\kappa_z}$$

$$\epsilon_{\mathbf{K}+\mathbf{q}}^{(2)} = \hbar v_F \sqrt{|\mathbf{q}|^2 + \kappa_z^2} \approx \mathcal{V}_0 + \hbar v_F \frac{|\mathbf{q}|^2}{2\kappa_z}$$

This is a spectrum of energy values qualitatively different from the case $\mathcal{V}_0 = 0$ because there is a gap between the two eigenvalues for $|\mathbf{q}| = 0$, given by:

$$\epsilon_{\text{gap}}(\mathbf{K}) = \lim_{|\mathbf{q}| \to 0} \left[\epsilon_{\mathbf{K}+\mathbf{q}}^{(2)} - \epsilon_{\mathbf{K}+\mathbf{q}}^{(1)} \right] = 2\mathcal{V}_0$$

For both eigenvalues, the dependence of the energy on $|\mathbf{q}|$ is quadratic, with an effective mass $\overline{m}$ given by:

$$\pm \hbar v_F \frac{|\mathbf{q}|^2}{2\kappa_z} = \frac{\hbar^2 |\mathbf{q}|^2}{2\overline{m}} \Rightarrow \overline{m} = \pm \frac{\hbar \kappa_z}{v_F} = \pm \frac{\mathcal{V}_0}{v_F^2}$$

with the positive value corresponding to $\epsilon_{\mathbf{K}+\mathbf{q}}^{(2)}$ (electrons) and the negative to $\epsilon_{\mathbf{K}+\mathbf{q}}^{(1)}$ (holes). Because the effective mass $\overline{m}$ is proportional to the symmetry-breaking term $\mathcal{V}_0$, the latter is usually referred to as simply the "mass" in the literature, as it leads to massive quasiparticles at the valence and conduction band edges. These features are clearly seen in the band structure of the model in Fig. 9.13.

In order to obtain the two eigenfunctions corresponding to the eigenvalues $\epsilon_{\mathbf{K}+\mathbf{q}}^{(1)}$, $\epsilon_{\mathbf{K}+\mathbf{q}}^{(2)}$, we define an angle θ which is determined by the values of its cosine and sine:

$$\cos(\theta/2) \equiv \frac{2\kappa_z}{\sqrt{|\mathbf{q}|^2 + (2\kappa_z)^2}}, \quad \sin(\theta/2) \equiv \frac{|\mathbf{q}|}{\sqrt{|\mathbf{q}|^2 + (2\kappa_z)^2}} \tag{9.50}$$

With the use of these quantities, we find that the eigenfunctions corresponding to the eigenvalues $\epsilon_{\mathbf{K}+\mathbf{q}}^{(1)}, \epsilon_{\mathbf{K}+\mathbf{q}}^{(2)}$ turn out to be exactly the same as the eigenfunctions of the spin in a monopole field discussed in the previous example! In the limit $|\mathbf{q}| \to 0$ we have $\theta = 0$, that is, the solutions correspond to the spin being at the North pole. For small but finite values of $\mathbf{q}$, the spin is at some point in the northern hemisphere; the latitude is determined by the relative values of $\kappa_z = \mathcal{V}_0/\hbar v_F$ and $|\mathbf{q}|$. If $\mathcal{V}_0 < 0$ the role of the states labeled (1) and (2) is interchanged, and the spin is in the southern hemisphere. Having made these connections to the model of a spin interacting with the monopole field, we can then import all the results obtained for that model, for the Berry curvature and Berry's phase of the present model; we leave this as an exercise for the reader (see Problem 9).

[a] M. J. Rice and E. J. Mele, *Phys. Rev. Lett.* **49**, 1455 (1982).

9.5 Applications of Berry's Phase

We consider next some important manifestations of Berry's phase, which include the Aharonov–Bohm effect, the proper definition of the polarization in crystals, and the motion of electrons in crystals under the influence of a uniform external electric field; this last case will complete the picture that we presented in Section 9.2.

9.5.1 Aharonov–Bohm Effect

An interesting application of the concepts introduced so far is the Aharonov–Bohm effect. Our quantum-mechanical system consists of an electron state $\psi(\mathbf{r})$ localized in space by a confining potential $\mathcal{V}(\mathbf{r})$, which obeys the Schrödinger equation

$$\hat{\mathcal{H}}\psi(\mathbf{r}) = \left[-\frac{\hbar^2 \nabla_{\mathbf{r}}^2}{2m_e} + \mathcal{V}(\mathbf{r}) \right] \psi(\mathbf{r}) = \epsilon \psi(\mathbf{r})$$

in free space. Now we imagine that the origin of the confining potential is shifted by $\mathbf{R}$, so that the hamiltonian acquires a parametric dependence on $\mathbf{R}$:

$$\hat{\mathcal{H}}_{\mathbf{R}}\psi_{\mathbf{R}}(\mathbf{r}) = \left[-\frac{\hbar^2 \nabla_{\mathbf{r}}^2}{2m_e} + \mathcal{V}(\mathbf{r} - \mathbf{R}) \right] \psi_{\mathbf{R}}(\mathbf{r}) = \epsilon_{\mathbf{R}} \psi_{\mathbf{R}}(\mathbf{r})$$

but due to the translational invariance of the system, the wavefunction is simply shifted by $\mathbf{R}$ and the energy of this state remains the same:

$$\psi_{\mathbf{R}}(\mathbf{r}) = \psi(\mathbf{r} - \mathbf{R}), \quad \epsilon_{\mathbf{R}} = \epsilon$$

We can allow the parameter $\mathbf{R}$ to span a closed path C along which the above relations will hold. Next we consider that there is a constant magnetic field $\mathbf{B}$ generated by a long solenoid within a small region of the area enclosed by the path C, and this region is very far from the path, as shown in Fig. 9.14. The new hamiltonian for the system is (see Appendix C):

$$\hat{\mathcal{H}}_{\mathbf{R}} = \frac{1}{2m_e} \left[-i\hbar\nabla_{\mathbf{r}} + \frac{e}{c}\mathbf{A}(\mathbf{r}) \right]^2 + \mathcal{V}(\mathbf{r} - \mathbf{R})$$

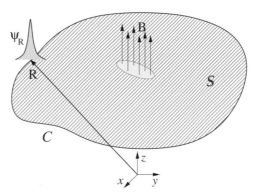

Fig. 9.14 Illustration of Aharonov–Bohm effect description in terms of Berry's phase: the localized wavefunction $\psi_{\mathbf{R}}$ is transported around a closed path C which is far from the region where a uniform magnetic field exists; the wavefunction acquires a Berry's phase γ during this process.

where $\mathbf{A}(\mathbf{r})$ is the vector potential of the magnetic field, $\mathbf{B} = \nabla_{\mathbf{r}} \times \mathbf{A}(\mathbf{r})$. The new wavefunction for this hamiltonian is given by:

$$\psi_{\mathbf{R}}(\mathbf{r}) = \exp\left(-\frac{ie}{\hbar c} \int_{\mathbf{R}}^{\mathbf{r}} \mathbf{A}(\mathbf{r}') \cdot d\mathbf{r}'\right) \psi(\mathbf{r} - \mathbf{R})$$

that is, the wavefunction has acquired a phase $\varphi_{\mathbf{R}}$:

$$\varphi_{\mathbf{R}}(\mathbf{r}) = -\frac{e}{\hbar c} \int_{\mathbf{R}}^{\mathbf{r}} \mathbf{A}(\mathbf{r}') \cdot d\mathbf{r}'$$

which evidently depends on the path which makes the wavefunction ill-defined (it is not single-valued). For the situation illustrated in Fig. 9.14, we can restore single-valuedness by carefully choosing the path of integration: the vector potential is zero for values of the integration variable $\mathbf{r}'$ away from the region where the magnetic field exists, so as long as the wavefunction is localized and the magnetic field is far from the path of $\mathbf{R}$, the wavefunction can be chosen to be well-behaved (single-valued). This does not mean that the wavefunction does not change when it is displaced *around* the closed path: it acquires a Berry's phase which can be calculated from the theory outlined above. The Berry connection for this example is:

$$\mathcal{A}(\mathbf{R}) = i\langle\psi_{\mathbf{R}}|\nabla_{\mathbf{R}}\psi_{\mathbf{R}}\rangle = \frac{-e}{\hbar c}\mathbf{A}(\mathbf{R}) + i\int \psi^*(\mathbf{r} - \mathbf{R})\nabla_{\mathbf{R}}\psi(\mathbf{r} - \mathbf{R}) \cdot d\mathbf{r}$$

We can show that the last term is identically zero by the following argument. We define $\mathbf{r}' = \mathbf{r} - \mathbf{R}$ and use the chain rule to switch the variable of differentiation to $\mathbf{r}'$:

$$\int \psi^*(\mathbf{r}')\nabla_{\mathbf{r}'}\psi(\mathbf{r}') \cdot d\mathbf{r}' \left(\frac{\partial\mathbf{r}'}{\partial\mathbf{R}}\right) = -\int \psi^*(\mathbf{r}')\nabla_{\mathbf{r}'}\psi(\mathbf{r}') \cdot d\mathbf{r}'$$

but this last expression is proportional to the expectation value of momentum in the state $|\psi\rangle$, which for a bound state is zero. Berry's phase is then given by:

$$\gamma = \frac{-e}{\hbar c}\oint_C \mathbf{A}(\mathbf{R}) \cdot d\mathbf{R} = \frac{-e}{\hbar c}\int_S (\nabla_{\mathbf{r}} \times \mathbf{A}(\mathbf{r})) \cdot d\mathcal{S} = \frac{-e}{\hbar c}\int_S \mathbf{B} \cdot d\mathcal{S} = -\frac{2\pi}{\phi_0}\Phi$$

with S the surface enclosed by the path C and the first equality being the result of Stokes' theorem; $\phi_0 = (hc/e)$ is the flux quantum. We conclude that the value of γ is simply proportional to the total flux Φ of the $\mathbf{B}$ field through the surface enclosed by the path C. Note that this flux is entirely due to the small region in which $\mathbf{B}$ exists, and this region is not sampled directly by the localized wavefunction $\psi_{\mathbf{R}}(\mathbf{r})$ as it is transported around the closed path. This surprising result, predicted by Aharonov and Bohm,[16] is counterintuitive in terms of classical physics because the path of the particle's displacement is far from the region where the magnetic field exists, so it is a manifestation of the quantum nature of the system. The fact that this phase has been measured in interference experiments[17] is a remarkable validation of quantum theory.

9.5.2 Polarization of Crystals

An important application of the results we have derived above is the ability to calculate the polarization of crystals. As we have seen, there are positive and negative charges in crystals and their distribution in space can give rise to a net dipole moment per unit volume, which we call the polarization $\mathbf{P}$ and define as:

$$\mathbf{P} = \frac{1}{V_{\mathrm{PUC}}} \int_{V_{\mathrm{PUC}}} \mathbf{r}\rho(\mathbf{r})d\mathbf{r} \tag{9.51}$$

where $\rho(\mathbf{r})$ is the total charge density within the primitive unit cell and V_{PUC} is its volume; the polarization is proportional to the net electric dipole moment of the system (see also Appendix B); equivalently, we can integrate over the volume of the entire crystal and divide by it. The calculation of $\mathbf{P}$ was a vexing problem, because the expression in Eq. (9.51) depends on the choice of the origin of the unit cell, a problem inherent to the periodic crystal, as illustrated in Fig. 9.15, but absent in a finite system (see Problem 6). In several seminal papers that established the modern theory of polarization, R. Resta[18] and independently D. Vanderbilt and coworkers[19] took a different approach which solved the problem of properly calculating the polarization of crystals and established the connection to the Berry curvature. The key insight was to use the *current* $\mathbf{j}(\mathbf{r}, t)$ rather than the charge density $\rho(\mathbf{r}, t)$, because the current is periodic in the crystal, while $\mathbf{r}\rho(\mathbf{r}, t)$ is not.

To establish the connection between polarization and current, we consider what happens when small changes are introduced in the charge distribution of a crystal by adiabatically turning on an external field. The change in the α-component of the polarization, ΔP_α, the only experimentally measurable quantity, can be expressed as:

$$\Delta P_\alpha = \int_{\lambda_1}^{\lambda_2} \frac{\partial P_\alpha}{\partial \lambda} d\lambda \tag{9.52}$$

[16] Y. Aharonov and D. Bohm, *Phys. Rev.* **115**, 485 (1959).
[17] See M. Peshkin and A. Tonomura, *The Aharonov–Bohm Effect* (Springer, Berlin, 1989).
[18] R. Resta, *Ferroelectrics*, **136**, 51 (1992).
[19] R. D. King-Smith and D. Vanderbilt, *Phys. Rev. B* **47**, 1651 (1993); D. Vanderbilt and R. D. King-Smith, *Phys. Rev. B* **48**, 4442 (1993).

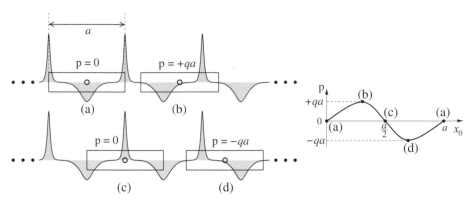

Fig. 9.15 Illustration of the ambiguity in determining the dipole moment **p** in the unit cell of a 1D crystal with period a, which contains localized positive (red curves) and negative (blue curves) charges in each unit cell; the diagram on the right gives the value of **p** as a function of the position of the unit cell origin x_0 (denoted by a black circle).

where λ is a parameter used to express the adiabatic change in the hamiltonian that produces the polarization. We can actually relate λ to the time t by considering a time-dependent hamiltonian whose change over a period from $t = 0$ to $t = T$ produces the polarization. This can be established from the relation between the dipole moment (or, equivalently, the polarization) per unit volume **P**, the charge density $\rho(\mathbf{r})$, derived from Eq. (9.51), and the continuity equation (see Appendix B for details), which together give:

$$\nabla_\mathbf{r} \cdot \mathbf{P}(\mathbf{r}) = -\rho(\mathbf{r}), \quad \frac{\partial \rho}{\partial t} + \nabla_\mathbf{r} \cdot \mathbf{j} = 0 \rightarrow \nabla_\mathbf{r} \cdot \left(\frac{\partial \mathbf{P}}{\partial t} - \mathbf{j} \right) = 0$$

Up to a divergence-free component of the current, which is usually related to the magnetization current that vanishes in the bulk solid (for more details, see the review article by Xiao *et al.* in Further Reading), we can integrate this last expression over time to get:

$$\Delta P_\alpha = \int_0^T \frac{\partial P_\alpha}{\partial t} \, \mathrm{d}t = \int_0^T j_\alpha \, \mathrm{d}t$$

We can now employ Eq. (9.46) for the velocity, multiplied by the charge $-e$ and divided by the volume V, to get the current per unit volume:

$$j_\alpha^{(n)}(\mathbf{k}, t) = -\frac{e}{\hbar V} \frac{\partial \epsilon_\mathbf{k}^{(n)}}{\partial k_\alpha} + \mathrm{i} \frac{e}{V} \left[\langle \frac{\partial u_\mathbf{k}^{(n)}}{\partial k_\alpha} | \frac{\partial u_\mathbf{k}^{(n)}}{\partial t} \rangle - \langle \frac{\partial u_\mathbf{k}^{(n)}}{\partial t} | \frac{\partial u_\mathbf{k}^{(n)}}{\partial k_\alpha} \rangle \right]$$

for the contribution of band n at $\mathbf{k}$. For the total current, we need to integrate over the entire BZ and sum over the filled bands, which produces the total change in polarization. Upon integration over $\mathbf{k}$, the first term of the current vanishes because of the symmetry of the bands, Eq. (2.41), while in the second term we recognize the Berry curvature of each band:

$$\Omega_{k_\alpha t}^{(n)}(\mathbf{k}, t) = \mathrm{i} \left[\langle \frac{\partial u_\mathbf{k}^{(n)}}{\partial k_\alpha} | \frac{\partial u_\mathbf{k}^{(n)}}{\partial t} \rangle - \langle \frac{\partial u_\mathbf{k}^{(n)}}{\partial t} | \frac{\partial u_\mathbf{k}^{(n)}}{\partial k_\alpha} \rangle \right]$$

in the parameters k_α and t. Putting these results together produces the following expression for the change in polarization:

$$\Delta P_\alpha = e \sum_{n:\text{filled}} \int_{\text{BZ}} \int_0^T \Omega_{k_\alpha t}^{(n)}(\mathbf{k}, t) dt \, \frac{d\mathbf{k}}{(2\pi)^3}$$

In the same manner that we have assumed the polarization to be induced by an adiabatic change in a time-dependent hamiltonian, we could also allow the changes to be induced by variation of another physical parameter λ; this could for instance, be, a smooth, adiabatic deformation of the crystal lattice, which we will take to vary between 0 and 1, leading to the more general expression

$$\Delta P_\alpha = e \sum_{n:\text{filled}} \int_{\text{BZ}} \int_0^1 \Omega_{k_\alpha \lambda}^{(n)}(\mathbf{k}, \lambda) d\lambda \, \frac{d\mathbf{k}}{(2\pi)^3}$$

In this case, we can take advantage of the relation between the Berry curvature and Berry connection:

$$\Omega_{k_\alpha \lambda}^{(n)}(\mathbf{k}, \lambda) = \frac{\partial \mathcal{A}_\lambda^{(n)}}{\partial k_\alpha} - \frac{\partial \mathcal{A}_{k_\alpha}^{(n)}}{\partial \lambda}$$

to perform the integrals. For the term $\partial \mathcal{A}_\lambda^{(n)}/\partial k_\alpha$, we perform first the integral over the BZ, which vanishes as long as we choose the parameter λ to make sure that $\mathcal{A}_\lambda^{(n)}$ is periodic in reciprocal space, in the same sense that the wavefunctions $u_{\mathbf{k}}^{(n)}(\mathbf{r})$ are periodic, see Eq. (2.42). For the term $\partial \mathcal{A}_{k_\alpha}^{(n)}/\partial \lambda$, we perform first the integral over λ, which gives simply the integrand evaluated at the end points, leading to:

$$\Delta P_\alpha = -e \sum_{n:\text{filled}} \int_{\text{BZ}} \left[\mathcal{A}_{k_\alpha}^{(n)} \right]_{\lambda=0}^{\lambda=1} \frac{d\mathbf{k}}{(2\pi)^3} = \frac{-ie}{(2\pi)^3} \sum_{n:\text{filled}} \int_{\text{BZ}} \left[\langle u_{\mathbf{k},\lambda}^{(n)} | \frac{\partial}{\partial k_\alpha} | u_{\mathbf{k},\lambda}^{(n)} \rangle \right]_{\lambda=0}^{\lambda=1} d\mathbf{k}$$

where in the last expression we have written explicitly the Berry connection in terms of the wavefunctions which contain the λ dependence from the hamiltonian; this is the expression derived by King-Smith and Vanderbilt for the polarization of crystals that has made possible unambiguous calculations of this important property. Note that by integrating over λ we lose any information about the actual path in λ-space. Therefore, the value of the polarization is given up to a multiple of the quantum of polarization $(e\mathbf{R}/V_{\text{PUC}})$, with $\mathbf{R}$ being one of the Bravais lattice vectors with smallest magnitude. This means that the path of integration may have passed several times from one end of the PUC to its opposite end, separated by $\mathbf{R}$, a fact lost when only the end points of the path are considered.

9.5.3 Crystal Electrons in Uniform Electric Field

We next consider the effects of an external electric field. If we were to express the electric field in terms of the electrostatic potential $\mathbf{E}(\mathbf{r}) = -\nabla_{\mathbf{r}}\Phi(\mathbf{r})$, a constant (smooth over large distances in the crystal) field would require a linearly increasing potential $\Phi(\mathbf{r})$ which would destroy the crystal periodicity. Instead, we can use a time-dependent vector potential $\mathbf{A}(\mathbf{r}, t)$ to express the electric field as $\mathbf{E}(\mathbf{r}) = (-1/c)\partial \mathbf{A}(\mathbf{r}, t)/\partial t$ (see Appendix B). This choice, at the cost of introducing the independent variable t, allows us to maintain the Bloch states as the basis for describing the physics. We also know that the introduction of such a term changes the hamiltonian to (see Appendix C):

$$\hat{\mathcal{H}}_{\mathbf{k}}(\mathbf{p}, \mathbf{r}, t) = \frac{1}{2m_e} \left[\mathbf{p} + \hbar\mathbf{k} + \frac{e}{c}\mathbf{A}(\mathbf{r}, t) \right]^2 + \mathcal{V}(\mathbf{r})$$

We can now use the result of Eq. (9.46) in this situation where the origin of the adiabatic change of the wavefunctions with time is the presence of the external electric field. First, we define a new variable $\mathbf{q}$:

$$\hbar\mathbf{q} = \hbar\mathbf{k} + \frac{e}{c}\mathbf{A}(\mathbf{r}, t) \tag{9.53}$$

which contains both the $\mathbf{k}$ dependence and the t dependence of the hamiltonian. Since we assumed that $\mathbf{A}(\mathbf{r}, t)$ preserves the translational symmetry of the hamiltonian, $\mathbf{k}$ is still a good quantum number, and is therefore a constant of motion, which implies:

$$\frac{\partial \mathbf{k}}{\partial t} = 0 \Rightarrow \hbar\frac{\partial \mathbf{q}}{\partial t} = \frac{e}{c}\frac{\partial \mathbf{A}}{\partial t} = -e\mathbf{E}$$

In Eqs (9.17) and (9.18) we *assumed* that a similar expression held for the wave-vector $\mathbf{k}$. Here we have *proved* that it holds instead for the variable $\mathbf{q}$. We can then use these expressions in Eq. (9.46) for the velocity (see Problem 7), which becomes:

$$\langle \mathbf{v}_{\mathbf{k}}^{(n)} \rangle = \frac{1}{\hbar}\nabla_{\mathbf{k}}\epsilon_{\mathbf{k}}^{(n)} + \frac{e}{\hbar}\mathbf{E} \times \mathbf{\Omega}^{(n)}(\mathbf{k}) \tag{9.54}$$

with $\mathbf{\Omega}^{(n)}(\mathbf{k})$ the Berry curvature of the band, Eq. (9.43). The first term in the above expression is the contribution from band dispersion, the result we had derived from the treatment of the pure crystal that led to Eq. (9.15). The second term is an additional contribution to the velocity, which was referred to as "anomalous velocity" before its relation to the Berry curvature was recognized (for original references, see the review article by Xiao *et al.* in Further Reading). This term is transverse to the electric field and will therefore give rise to a Hall current. This observation makes evident that the Berry curvature and Berry's phase have an important relation to Hall physics. As a case in point, we show in Fig. 9.16 the calculated Berry curvature $\Omega_{xy}^{(n)}(\mathbf{k})$ for the highest occupied (valence) band of an interesting 2D solid, MoS_2 (see also the discussion in Chapter 5). The Berry curvature takes opposite values in different regions of the BZ, indicating that electrons of different wave-vectors will experience opposite forces, leading to interesting behavior. It is evident from this plot that the strongest contribution will be at the corners of the hexagonal BZ, referred to as K and K′ valleys. Manipulating these effects for applications in electronic devices has been termed "valleytronics."

9.6 Chern Numbers

We will next address some interesting implications of the Berry curvature and Berry connection. We will motivate these implications through a physical argument on a simple

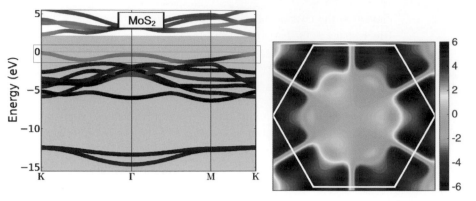

Fig. 9.16 **Left**: The band structure of a single layer of MoS_2. The highest valence band is shown enclosed in a red rectangle near the Fermi level (zero value on the energy axis). **Right**: The calculated Berry curvature ($Å^{-2}$) for the highest valence band; the white hexagon shows the first Brillouin zone [from S. Fang and E. Kaxiras, *Phys. Rev. B* **93**, 235153 (2016)].

example, and then present a more general argument. Along the way, we will introduce the notion of the Chern number, an important quantity that characterizes the features of the Berry curvature and is intimately related to Berry's phase.

Building on the last result we derived for electrons in a uniform electric field, we assume that the carriers are confined in a 2D sheet (the x, y plane) and that the electric field is in the $\hat{x}$ direction, $\mathbf{E} = E_x\hat{x}$, as illustrated in Fig. 9.2. Consistent with these assumptions and the geometry shown in Fig. 9.2, the electron wavefunctions have no dependence on the k_z component of the wave-vector, and the velocity of electrons from Eq. (9.54) will be given by:

$$\langle \mathbf{v}_{\mathbf{k}}^{(n)} \rangle = \frac{1}{\hbar}\nabla_{\mathbf{k}}\epsilon_{\mathbf{k}}^{(n)} - \frac{e}{\hbar}E_x\Omega_{xy}^{(n)}(\mathbf{k})\hat{y}$$

This is the contribution from band n at the $\mathbf{k}$ value of the wave-vector. To get the total current per unit volume, we need to multiply by the charge $-e$ and integrate over the whole BZ. As we have argued in the previous section, integration over the entire BZ of the first term gives a zero contribution due to the symmetry of the bands, Eq. (2.41), while the second term gives:

$$j_y = \frac{e^2}{\hbar}E_x\frac{1}{V}\sum_{\mathbf{k}\in\text{BZ}}\Omega_{xy}^{(n)}(\mathbf{k}) \Rightarrow \frac{j_y}{E_x} = \frac{e^2}{\hbar}\int_{\text{BZ}}\Omega_{xy}^{(n)}(\mathbf{k})\frac{\mathrm{d}k_x\mathrm{d}k_y}{(2\pi)^2}$$

since we are dealing with a purely 2D situation. The last expression, identified as the Hall conductivity $\sigma_{xy} = j_y/E_x$, can also be written as:

$$\sigma_{xy} = \frac{e^2}{h}C_1^{(n)}, \quad C_1^{(n)} \equiv \frac{1}{2\pi}\int_{\text{BZ}}\Omega_{xy}^{(n)}(\mathbf{k})\mathrm{d}k_x\mathrm{d}k_y \qquad (9.55)$$

where we have defined the quantity $C_1^{(n)}$, called the "first Chern number" of the band, and expressed the conductivity in units of the quantum of conductance, e^2/h. For this simple example, the Hall conductivity must be equal to an integer number of quanta of conductance, because its value is determined by the number of electrons transferred from one side of the sample to the opposite side in the $\hat{x}$ direction. In other words, at least for the simple case considered here, we have shown that the first Chern number of the band must be an integer. We note that, due to the 2D nature of the system, the Berry curvature takes the form

$$\boldsymbol{\Omega}^{(n)}(\mathbf{k}) = i\left[\langle\frac{\partial u_{\mathbf{k}}^{(n)}}{\partial k_x}|\frac{\partial u_{\mathbf{k}}^{(n)}}{\partial k_y}\rangle - \langle\frac{\partial u_{\mathbf{k}}^{(n)}}{\partial k_y}|\frac{\partial u_{\mathbf{k}}^{(n)}}{\partial k_x}\rangle\right]\hat{z} = \Omega_{xy}^{(n)}(\mathbf{k})\hat{z}$$

and $dk_x dk_y$ is the surface element $d\boldsymbol{S}_{\mathbf{k}} = dk_x dk_y\hat{z}$ in reciprocal space for a surface contained within the boundary of the BZ. These considerations compel us to write the first Chern number as:

$$2\pi C_1^{(n)} \equiv \int_{BZ} \boldsymbol{\Omega}^{(n)}(\mathbf{k}) \cdot d\boldsymbol{S}_{\mathbf{k}} \qquad (9.56)$$

and in this expression we recognize Berry's phase on the right-hand side. The conclusion is that, if the first Chern number is an integer, then Berry's phase for a path that goes around the BZ is an integer multiple of 2π, which produces an invariant of the wavefunction.

The reason for introducing the notion of the Chern number is that it constitutes a mathematically precise way of classifying the properties of the surface described by the Berry curvature. Higher Chern numbers can also be defined by proper generalization of the surface curvature.[20] What is relevant for the purposes of our discussion is that the integer value of the first Chern number describes the torus in 2D space that the symmetric nature of the BZ forms: recall that the functions $u_{\mathbf{k}}^{(n)}(\mathbf{r})$ are periodic in $\mathbf{k}$, with the BZ being the unit cell of this periodic pattern, see Eq. (2.42).

We next give another argument, beyond the simple motivational example discussed so far, why the first Chern number is an integer. We consider a 2D case again, and assume a square lattice which implies that the first BZ is given by the region shown in Fig. 9.17. We perform the integral that defines the first Chern number for this case, Eq. (9.56). Instead of actually doing the surface integral, we use Stokes' theorem to turn it into a line integral of the Berry connection along the boundary of the BZ, the four segments AB–BC–CD–DA (to simplify the notation we use the symbols 1 for the values $k_x = \pi/a$ and $k_y = \pi/a$ and $\bar{1}$ for $k_x = -\pi/a$ and $k_y = -\pi/a$):

$$2\pi C_1^{(n)} = \oint_{ABCD} \boldsymbol{\mathcal{A}}^{(n)}(k_x, k_y) \cdot d\mathbf{k}$$

$$= \int_{AB} \mathcal{A}_x^{(n)}(k_x, \bar{1})dk_x + \int_{BC} \mathcal{A}_y^{(n)}(1, k_y)dk_y + \int_{CD} \mathcal{A}_x^{(n)}(k_x, 1)dk_x + \int_{DA} \mathcal{A}_y^{(n)}(\bar{1}, k_y)dk_y$$

$$= \int_{\bar{1}}^{1} \left[\mathcal{A}_x(k_x, \bar{1}) - \mathcal{A}_x(k_x, 1)\right] dk_x + \int_{\bar{1}}^{1} \left[\mathcal{A}_y(1, k_y) - \mathcal{A}_y(\bar{1}, k_y)\right] dk_y$$

Now we recall that

$$\mathcal{A}_j^{(n)}(k_x, k_y) = i\langle u_{\mathbf{k}}^{(n)}|\frac{\partial}{\partial k_j}|u_{\mathbf{k}}^{(n)}\rangle, \quad j = x, y$$

[20] S. S. Chern, "Characteristic classes of hermitian manifolds," *Ann. Math.* **47**, 85 (1946).

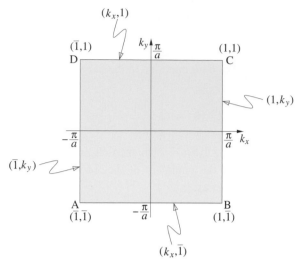

Fig. 9.17 The first Brillouin zone of the 2D square lattice. The corners are labeled $A = (-\pi/a, -\pi/a) \equiv (\bar{1}, \bar{1})$, $B = (\pi/a, -\pi/a) \equiv (1, \bar{1})$, $C = (\pi/a, \pi/a) \equiv (1, 1)$, $D = (-\pi/a, \pi/a) \equiv (\bar{1}, 1)$, and the points between the corners are labeled $(k_x, \bar{1})$, $(1, k_y)$, $(k_x, 1)$, $(\bar{1}, k_y)$, respectively, with $-1 \leq k_x, k_y \leq 1$, where k_x, k_y are expressed in units of π/a.

and that the wavefunctions at opposite ends of the BZ represent physically equivalent states, so they must be related by a phase factor:

$$|u_{k_x,1}^{(n)}\rangle = e^{-i\varphi_x(k_x)}|u_{k_x,\bar{1}}^{(n)}\rangle, \quad |u_{1,k_y}^{(n)}\rangle = e^{-i\varphi_y(k_y)}|u_{\bar{1},k_y}^{(n)}\rangle \tag{9.57}$$

Substituting these expressions in the integrals we need to evaluate along the path, we find:

$$2\pi C_1^{(n)} = \varphi_x(1) - \varphi_x(\bar{1}) - \varphi_y(1) + \varphi_y(\bar{1}) \tag{9.58}$$

We can also use the phase relations to connect the states at the corners of the closed path, that is, we make the connections:

$$|u_{k_A}^{(n)}\rangle \rightarrow |u_{k_B}^{(n)}\rangle \rightarrow |u_{k_C}^{(n)}\rangle \rightarrow |u_{k_D}^{(n)}\rangle \rightarrow |u_{k_A}^{(n)}\rangle$$

by employing the phase differences between these states from the definition of the phase factors, Eq. (9.57):

$$|u_{\bar{1},\bar{1}}^{(n)}\rangle = e^{i\varphi_y(\bar{1})}|u_{1,\bar{1}}^{(n)}\rangle = e^{i[\varphi_y(\bar{1})+\varphi_x(1)]}|u_{1,1}^{(n)}\rangle = e^{i[\varphi_y(\bar{1})+\varphi_x(1)-\varphi_y(1)]}|u_{\bar{1},1}^{(n)}\rangle$$
$$= e^{i[\varphi_y(\bar{1})+\varphi_x(1)-\varphi_y(1)-\varphi_x(\bar{1})]}|u_{\bar{1},\bar{1}}^{(n)}\rangle$$

and because we end up with the same wavefunction, which is single-valued, we must have:

$$e^{i[\varphi_y(\bar{1})+\varphi_x(1)-\varphi_y(1)-\varphi_x(\bar{1})]} = 1 \Rightarrow \varphi_y(\bar{1}) + \varphi_x(1) - \varphi_y(1) - \varphi_x(\bar{1}) = 2\pi l, \quad l : \text{integer}$$

Since the left-hand side of the last equation is the same as the right-hand side of Eq. (9.58), it proves that the first Chern number $C_1^{(n)}$ must be an integer.

9.7 Broken Symmetry and Edge States

The concepts we have addressed so far are also relevant in crystals exhibiting broken symmetries. Two important symmetries are inversion symmetry in real space and time-reversal symmetry. Inversion symmetry applies to a crystal for which the change $\mathbf{r} \to -\mathbf{r}$ leaves it invariant. Time-reversal symmetry applies to a hamiltonian with no time dependence and leads to the fundamental symmetry $\mathbf{k} \to -\mathbf{k}$ in reciprocal space for spinless particles, as discussed earlier (see Section 9.3). Which symmetry is broken or retained can have important effects on the behavior of the crystal, such as the nature of edge states, which can lead to new types of electronic behavior. We give examples of these notions next. Throughout this discussion, we will use the example of the 2D honeycomb lattice to illustrate these ideas. This is convenient, as we have already explored several aspects of this model, see Example 9.4.2. It is also relevant, beyond its usefulness as a "toy model," because it can describe much of the physics of several real 2D solids, like graphene, hexagonal boron nitride (hBN), and related materials.

9.7.1 Broken Symmetry in Honeycomb Lattice

As we have seen already, the breaking of inversion symmetry (P-symmetry) in real space in the honeycomb lattice produces a gap in the band structure. If the energy difference of the A and B sites is $\mathcal{V}_0$, the gap turns out to be $\epsilon_{\text{gap}} = 2|\mathcal{V}_0|$. Notice that, even though P-symmetry is broken in this model, inversion symmetry in *reciprocal* space still holds, because of Kramers degeneracy, $\epsilon_{-\mathbf{k}} = \epsilon_{\mathbf{k}}$ (for spinless fermions), which is a consequence of time-reversal symmetry.

An important generalization of the honeycomb lattice model, in the tight-binding approximation, described by the hamiltonian $\hat{\mathcal{H}}_{\mathbf{k}}$ of Eq. (9.48), was introduced by F. D. M. Haldane.[21] For this and related work, Haldane shared the 2016 Nobel Prize in Physics with D. J. Thouless and J. M. Kosterlitz (see also the discussion in Section 9.8). This model includes the following additional terms, both of which contribute to the diagonal elements of the hamiltonian matrix:

$$h_2^{(+)}(\mathbf{k}) = 2t_2 \cos(\varphi)[\cos(\mathbf{k} \cdot \mathbf{a}_1) + \cos(\mathbf{k} \cdot \mathbf{a}_2) + \cos(\mathbf{k} \cdot \mathbf{a}_3)]I_2$$
$$h_2^{(-)}(\mathbf{k}) = 2t_2 \sin(\varphi)[\sin(\mathbf{k} \cdot \mathbf{a}_1) - \sin(\mathbf{k} \cdot \mathbf{a}_2) - \sin(\mathbf{k} \cdot \mathbf{a}_3)]\sigma_z \qquad (9.59)$$

where I_2 is the 2×2 identity matrix and $\mathbf{a}_3 = \mathbf{a}_1 - \mathbf{a}_2$. The meaning of these terms is the following: $h_2^{(+)}(\mathbf{k})$ describes the hopping to second-nearest neighbors (hence the subscript 2) with hopping matrix element t_2; these neighbors in the honeycomb lattice are situated at $\pm\mathbf{a}_1, \pm\mathbf{a}_2, \pm\mathbf{a}_3$. Note that sites of each type (A or B) have second neighbors of the same type, that is, the second neighbors of an A site are all A sites, and similarly for the B sites. This term is obtained in the usual way, according to the rules we described in Chapter 2, the only difference being the extra factor $\cos(\varphi)$, which we will address shortly. The second

---

[21] F. D. M. Haldane, *Phys. Rev. Lett.* **61**, 2015 (1988).

term, $h_2^{(-)}(\mathbf{k})$, is unusual in the context of tight-binding theory, because it is diagonal, that is, it couples sites that are equivalent by lattice translations (and therefore of the same type, A to A and B to B); however, it is *not* invariant under inversion in reciprocal space, $\mathbf{k} \to -\mathbf{k}$, which violates time-reversal invariance. Thus, assuming $t_2 \sin(\varphi) \neq 0$, the model introduced by Haldane breaks time-reversal symmetry, also referred to as *T*-symmetry.

To obtain a physical picture of what these terms imply, we imagine that the hopping between second-neighbor sites within a single unit cell takes place in circular fashion and in the same direction, as illustrated in Fig. 9.18: from any given site, the successive hops in a clockwise direction from site A would be $\mathbf{a}_1, -\mathbf{a}_2, -\mathbf{a}_3$, and similarly for site B. If a magnetic flux density $\mathbf{B} = B(\mathbf{r})\hat{z}$ threads the lattice, with zero total flux through the unit cell, then we can choose the corresponding vector potential $\mathbf{A}(\mathbf{r})$ to be periodic, and the hopping matrix elements will be multiplied by the factor

$$\exp\left[i\frac{e}{\hbar c}\int \mathbf{A}(\mathbf{r}) \cdot d\mathbf{r}\right]$$

where the integral is over the hopping path. We can choose these phases in a consistent way, so that the total phase around a closed path is equal to the flux enclosed by the path, in units of the flux quantum $\phi_0 = hc/e$. For the nearest-neighbor hops, the path involves the entire perimeter of the unit cell, and the total flux enclosed is zero due to our assumption that there is zero total flux through the cell, which means we do not need to include a phase in the matrix elements. For the path traced by the second-nearest-neighbor hops (SNN-hop), the total phase in going around the unit cell is given by:

$$\varphi = \frac{e}{\hbar c}\oint_{\text{SNN-hop}} \mathbf{A}(\mathbf{r}) \cdot d\mathbf{r} = -\frac{6\pi}{\phi_0}(2\Phi_a + \Phi_b)$$

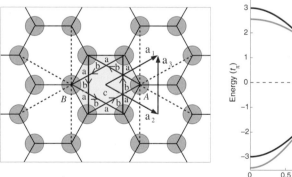
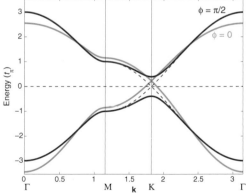

Fig. 9.18 Honeycomb lattice with broken time-reversal symmetry. **Left**: The lattice with the primitive lattice vectors $\mathbf{a}_1, \mathbf{a}_2$ (red arrows) and the primitive unit cell highlighted in blue; the sites A and B are inequivalent for $\mathcal{V}_0 \neq 0$. The black lines indicate the nearest-neighbor hopping matrix elements of strength t_π and the solid and dashed purple lines represent the next-nearest-neighbor hopping matrix elements, of strength t_2. **Right**: The corresponding band structure, for two values of the parameter ϕ, along the high-symmetry directions of the BZ; the model parameters are $t_2 = 0.075t_\pi$, $\mathcal{V}_0 = 0$. Green lines: $\phi = 0$, no band gap; red lines: $\phi = \pi/2$, with gap $\epsilon_{\text{gap}}(\mathbf{K}) = 6\sqrt{3}|t_2|$. The band structure of the lattice *with* time-reversal and inversion symmetry is shown in dashed red lines for comparison.

where Φ_a, Φ_b are the fluxes through the small triangles labeled "a" and "b" in Fig. 9.18. These considerations lead to the two terms in the hamiltonian introduced by Haldane, in the presence of the magnetic field $\mathbf{B}(\mathbf{r})$, with the phase φ given by the expression above. The band structure for $\varphi = 0$ and $\pi/2$ is shown in Fig. 9.18: the first case corresponds to the Rice–Mele model, with the added feature of second-neighbor-hopping in the usual manner, which shifts the energy level but maintains the essential character of a semiconductor with gap $\epsilon_{gap} = 2|\mathcal{V}_0|$; the second case is a qualitatively different situation, in which the band gap is a result of breaking the time-reversal invariance and is equal to $\epsilon_{gap} = 2|\mathcal{V}_0| + 6\sqrt{3}|t_2|$. Haldane showed that this new hamiltonian, for certain values of the parameters $\mathcal{V}_0, t_2, \varphi$ (t_π sets the energy scale), leads to quantized values of the conductivity in units of e^2/h, just like in the case of the quantum Hall effect. Since this effect is produced not by a macroscopic external field but by a local field with special properties, namely zero total flux through the unit cell, it is sometimes called the "anomalous quantum Hall effect" (AQHE).

9.7.2 Edge States of Honeycomb Lattice

As we saw in Section 9.1, the edge states of a sample can play a central role in how electrons respond to external fields. To explore how they may affect the properties of the system, we consider the nature of edge states in the honeycomb lattice. In order to create edges in the honeycomb lattice, we have to cut a ribbon out of the infinite 2D lattice. The ribbon will have finite width in one direction and essentially infinite length in the orthogonal direction. Even though the width is finite, we generally think of it as being very large on the scale of atomic distances. There are many ways to construct such structures, starting with the ideal 2D lattice. For simplicity, we will consider only ribbons that have edges of the highest symmetry, that is, they have the shortest periodic lattice vector along the edge, and have the same type of edge on their two sides. For the honeycomb lattice, the two shortest vectors that define straight edges are given by $\mathbf{a}_3 = \mathbf{a}_1 - \mathbf{a}_2$ and $\mathbf{a}_4 = \mathbf{a}_1 + \mathbf{a}_2$ in terms of the original 2D lattice constants defined earlier, Eq. (9.47). These vectors in cartesian coordinates are given by:

$$\mathbf{a}_{ZZ} = \mathbf{a}_3 = \mathbf{a}_1 - \mathbf{a}_2 = a\hat{\mathbf{y}}, \quad \mathbf{a}_{AC} = \mathbf{a}_4 = \mathbf{a}_1 + \mathbf{a}_2 = \sqrt{3}a\hat{\mathbf{x}} \tag{9.60}$$

The two types of edges, referred to as "zigzag" (ZZ) and "armchair" (AC), are shown in Fig. 9.19. In order to properly define a ribbon, we also need to specify its width. A zigzag ribbon can be constructed by replicating the two-atom basis at the edge, labeled A, B, along one of the original 2D lattice vectors, $\mathbf{a}_1$ or $\mathbf{a}_2$. Alternatively, it can be constructed by repeating the four-atom unit, labeled B, A, B, A, a finite number N of times along the $\mathbf{a}_4 = \mathbf{a}_1 + \mathbf{a}_2$ vector (this larger repeat vector corresponds to a doubling of the cell), as shown in Fig. 9.19. Similarly, an armchair ribbon can be constructed by replicating the four-atom basis at the edge, labeled B, A, B, A, a finite number N of times along the lattice vector $\mathbf{a}_3$, as shown in Fig. 9.19. In the zigzag ribbon, the edge contains only one type of site (A or B) on one side and only the opposite type on the other side. In this type of ribbon, the interaction between second-nearest-neighbor atoms involves sites belonging to the *same* sublattice, each of which has only two bonds to the interior of the ribbon. By contrast, in the armchair-edged ribbon, the edge contains both types of sites, A and B, each

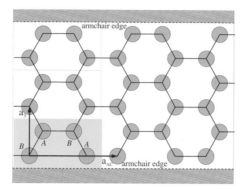 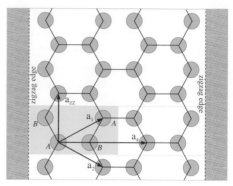

Fig. 9.19 The structure of ribbons with armchair (left panel) and zigzag (right panel) edges. The ribbons shown are of small finite width to illustrate their atomic structure. The blue-shaded regions indicate the unit that produces the full ribbon, when repeated periodically along $\mathbf{a}_{ZZ}$ (infinite length) and a finite number N of times along $\mathbf{a}_4$ (finite width) for the zigzag ribbon, or along $\mathbf{a}_{AC}$ (infinite length) and a finite number N of times along $\mathbf{a}_3$ (finite width) for the armchair ribbon; for illustration purposes, in these examples only ribbons for very small width are shown, $N = 2$ for the zigzag ribbon and $N = 3$ for the armchair ribbon.

of which has a bond to the ribbon interior and a bond to another edge site. In this case, there are no second-nearest-neighbor sites of the same type along the ribbon edge. The types of neighbors are important in determining the nature of edge states.

To have a unified picture, we adopt the $\mathbf{a}_3$ and $\mathbf{a}_4$ vectors as the basis for forming either the zigzag or the armchair ribbon; this does not impose any restrictions on the limit of large N, the number of units that form the width of the ribbon. The true periodicity along each ribbon edge is described by the two vectors defined in Eq. (9.60); since these vectors describe the 1D periodicity of the ribbon along its (in principle infinite) length, the corresponding reciprocal lattice vectors are $\mathbf{b}_3, \mathbf{b}_4$:

$$\mathbf{b}_{ZZ} = \mathbf{b}_3 = \frac{1}{2}(\mathbf{b}_1 - \mathbf{b}_2) = \frac{2\pi}{a}\hat{\mathbf{y}}, \quad \mathbf{b}_{AC} = \mathbf{b}_4 = \frac{1}{2}(\mathbf{b}_1 + \mathbf{b}_2) = \frac{2\pi}{\sqrt{3}a}\hat{\mathbf{x}}$$

With this convention, we can now examine how the electronic states of the original honeycomb lattice are affected when ribbons are formed. First, we consider the BZ of the ribbons: it is defined by the new reciprocal lattice vectors, $\mathbf{b}_3$ and $\mathbf{b}_4$, and their fractions, when a number N of units are included in the total ribbon width, either along $\mathbf{a}_3$ for the armchair-edged ribbon, or along $\mathbf{a}_4$ for the zigzag-edged ribbon. The new reciprocal lattice vectors produce a shrinking of the original BZ, such that the new BZ is half the size, as shown in Fig. 9.20(a), (b). The shrinking results in the three K points on the boundary of the original BZ being mapped onto a single point in the interior of the new BZ, and similarly the three K′ points on the original BZ boundary are mapped onto a single interior point. Each of these points is at a distance of 2/3 from the BZ center to the edge, the position of the latter being defined by the midpoint of the vectors $\pm\mathbf{b}_3$.

For an armchair-edged ribbon with $N\mathbf{a}_3$ units along its width, the corresponding BZ has a width of $\mathbf{b}_3/N$, as shown in Fig. 9.20(c). Since N is large, we can assume that it is a multiple of 3, which means that the K and K′ points are mapped by the shrinking along the $\mathbf{b}_3$ direction to the BZ center, Γ. This allows us to understand how the bands of the

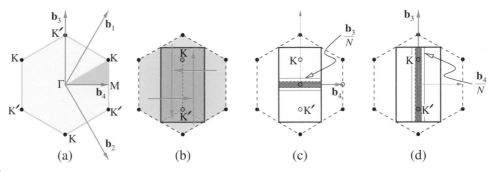

Fig. 9.20 (a) The BZ of the original honeycomb lattice, shown in yellow and corresponding to the $\mathbf{b}_1, \mathbf{b}_2$ reciprocal lattice vectors. When zigzag-edged ribbons with vector $\mathbf{b}_3$, or armchair ribbons with vector $\mathbf{b}_4$, are formed, the size of the BZ shrinks to half the original size. (b) The green arrows indicate how the light blue and light red sections are displaced by the reciprocal lattice vectors $\pm\mathbf{b}_3$ and $\pm\mathbf{b}_4$ respectively, to be mapped into the new first BZ, which is outlined in purple. (c) The armchair-edged ribbon BZ, which corresponds to shrinking by a factor of N along the $\mathbf{b}_3$ vector, shown as the purple-shaded rectangle. (d) The zigzag-edged ribbon BZ, which corresponds to shrinking by a factor of N along the $\mathbf{b}_4$ vector, shown as the purple-shaded rectangle.

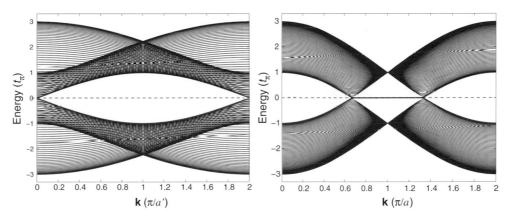

Fig. 9.21 Energy bands of ribbons cut from the 2D honeycomb lattice with armchair (left) or zigzag (right) edges; the edge states are shown in blue. In each case, the reciprocal-space vector $\mathbf{k}$ goes from 0 to $2\pi/a_{\text{edge}}$. In terms of a, the lattice constant of the 2D honeycomb lattice, the armchair-edged ribbon has $a_{\text{edge}} = \sqrt{3}a = a'$ while the zigzag-edged ribbon has $a_{\text{edge}} = a$.

armchair-edged ribbon will look: in the original 2D solid, the bands have a Dirac point with no gap at the K and K′ points, all of which have now been mapped at Γ. Therefore, we expect the Dirac point to appear at the center of the ribbon BZ, and of course at its periodic images in reciprocal space. We expect that at all other points of the BZ there is a gap separating the π and π^* states. The lattice constant of this ribbon along its edge is $|\mathbf{a}_{\text{AC}}| = a' = \sqrt{3}a$ and the length of the reciprocal space vector is $|\mathbf{b}_4| = 2\pi/a'$. A plot of the calculated bands for a wide ribbon is shown in Fig. 9.21, for k values from 0 to $2\pi/a'$, with the bands exhibiting the expected features.

For a zigzag-edged ribbon with $N\mathbf{a}_4$ units along its width, the corresponding BZ has a width of $\mathbf{b}_4/N$, as shown in Fig. 9.20(d). The K and K′ points appear at a distance 2/3 from the center to the edge of the BZ, no matter what the width is. This allows us to understand how the bands of the zigzag-edged ribbon will look: in the original 2D solid, the bands have a Dirac point with no gap at the K and K′ points, so we expect the Dirac point to appear at 2/3 of the distance from the center to the edge of the BZ. We expect that at all other points of the BZ there is a gap separating the π and π^* states. The lattice constant of this ribbon along its edge is $|\mathbf{a}_{ZZ}| = a$ and the length of the reciprocal space vector is $|\mathbf{b}_3| = 2\pi/a$. A plot of the calculated bands for a wide ribbon is shown in Fig. 9.21, for k values from 0 to $2\pi/a$, with the bands exhibiting the expected features. There is, however, a new feature here, which is exclusive to the zigzag ribbon: a flat band appears, which is twofold degenerate with constant energy, connecting the points K and K′ across the boundary of the BZ, through the midpoint of the vector $\mathbf{b}_3$; we note that without any symmetry-breaking terms in the hamiltonian, the two points are equivalent by inversion symmetry in reciprocal space, a consequence of time-reversal invariance (Kramers degeneracy, discussed in Section 9.3).

We next consider what happens to the edge states of the zigzag ribbon when we introduce symmetry-breaking terms in the hamiltonian. Edge states can always be identified by examining the wavefunction character. When dealing with an edge state, the wavefunction is mostly composed of orbitals associated with the edge sites. We plot in Fig. 9.22 the energy bands for the zigzag-edged ribbon obtained from the hamiltonian with broken P-symmetry (the Rice–Mele model) or with broken T-symmetry (the Haldane model). We see that the effect of breaking P-symmetry is the same as in the case of the bulk 2D lattice, that is, the system is a conventional insulator with a band gap $\epsilon_{\text{gap}} = 2\mathcal{V}_0$; the edge states exist, but are split into a valence (occupied) and a conduction (unoccupied) state, each of which has no dispersion for the range of wave-vector values between K and K′. This is a consequence of the fact that, even though inversion symmetry is broken in the real-space lattice, the reciprocal space still has inversion symmetry due to time-reversal symmetry, which requires the eigenvalues of a given band to be equal at these two points. By contrast, the breaking of T-symmetry in the hamiltonian (Haldane model)

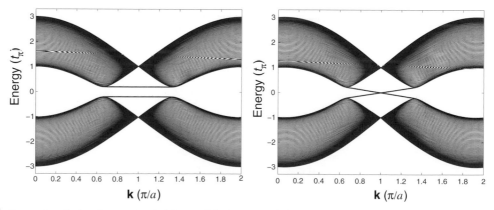

Fig. 9.22   Energy bands for zigzag-edged ribbons of the honeycomb lattice, with broken symmetry; the edge states are shown in blue. **Left**: Hamiltonian with broken P-symmetry (Rice–Mele model), with $\mathcal{V}_0 = 0.2|t_\pi|$. **Right**: Hamiltonian with broken T-symmetry (Haldane model), with $\mathcal{V}_0 = 0$, $\varphi = \pi/2$, $t_2 = 0.075t_\pi$.

removes the $\mathbf{k} \to -\mathbf{k}$ symmetry in reciprocal space, so it is no longer necessary to have the edge-band eigenvalues degenerate at these two points. Indeed, in this model, the eigenvalues of the two edge states cross the band gap when the wave-vector spans the values between K and K′, which are no longer equivalent points! This is a very intriguing feature, as it implies that, even though the bulk is an insulator with a clear gap, the edges introduce states that span the entire gap so they must be metallic. This unusual behavior has been termed "topological insulator."

9.8 Topological Constraints

In a seminal paper, Thouless and coworkers (TKNN)[22] showed that, by using the Kubo formula to calculate the 2D conductivity, the off-diagonal component or Hall conductivity takes the form

$$\sigma_{xy} = \frac{e^2}{\hbar} \int_{\text{BZ}} \Omega_{xy}(\mathbf{k}) \frac{\mathrm{d}k_x \mathrm{d}k_y}{(2\pi)^2} = \frac{e^2}{h} C_1$$

a result we discussed earlier, using the anomalous velocity expression, Eq. (9.55) (here we have dropped the band index of the Berry curvature for simplicity). They also proved that the conductivity is quantized in units of (e^2/h):

$$\sigma_{xy} = N\frac{e^2}{h}, \quad N : \text{integer} \Rightarrow C_1 = \frac{1}{2\pi} \int_{\text{BZ}} \Omega_{xy}(\mathbf{k})\mathrm{d}k_x\mathrm{d}k_y = N, \quad N : \text{integer}$$

which evidently implies that the first Chern number is equal to the integer N. These connections have important implications.

The Chern number, or Chern invariant, can be used to characterize the behavior of a material. This quantity is obtained directly from the crystal wavefunctions $|u_{\mathbf{k}}^{(n)}\rangle$, and is related to Berry's phase, as we saw in Section 9.4. Band structures characterized by a band gap belong to the same class as the Chern invariant, in the sense that they can be continuously deformed into one another without closing the band gap. Thus, the Chern number cannot change if the hamiltonian varies smoothly as a function of a parameter. Because $\Omega(\mathbf{k})$ represents a curvature in the space of this parameter, which in the case of crystals is the crystal wave-vector $\mathbf{k}$, we can classify the behavior of crystalline materials by the same mathematical notions that characterize surfaces. A key quantity in this respect is the so-called "genus" that counts the number of holes in a surface. To make the analogy more transparent, we concentrate on 2D surfaces which are easy to visualize and are relevant to the preceding discussion of 2D solids like graphene. The simplest 2D surface is that of a sphere, which has genus $g = 0$ (no holes). This corresponds to Chern number $C_1 = 0$, the case of a conventional (or trivial) insulator. Any surface which can be continuously deformed to that of a sphere has the same genus, hence any material with a zero value of the Chern invariant behaves in the same manner, that is, as a **trivial** insulator. The next simplest surface is that of a torus (think of a donut in terms of an everyday life

[22] D. J. Thouless, M. Kohmoto, M. P. Nightingale, and M. den Nijs, *Phys. Rev. Lett.* **49**, 405 (1982).

object), which has one hole and hence genus $g = 1$. This corresponds to a **topological** insulator, which has insulating behavior in the bulk but with edge states that are metallic: the energy eigenvalues corresponding to edge states span the bulk band gap so that the Fermi level necessarily intersects it. Any surface that can be continuously deformed to that of a torus (think, for example, of a coffee cup with a handle which can be continuously deformed into a donut) belongs to the same class, hence any material with Chern invariant equal to $C_1 = 1$ has similar behavior. We have already encountered such a case, the system described by the Haldane model of the honeycomb lattice with broken T-symmetry, which is a quantum Hall state.

Halperin's proof that the edge states are omnipresent and responsible for conduction in the quantum Hall state, and Haldane's proof that the bulk insulator with broken T-symmetry is a quantum Hall state, coupled with the TKNN result connecting the Hall conductivity and the Berry curvature, established the important link between the Chern invariant and the behavior of edge states: for a system in which the Chern invariant is 1, the Fermi level must intersect the energy of edge states in the gap of a topological insulator, as illustrated in Fig. 9.23. If there is more than one edge state, as in the case of the honeycomb ribbon with zigzag edges, the Fermi level will intersect each of these states in the bulk gap. This intersection *cannot* be avoided by moving the Fermi level in the gap through doping or gating the material. Thus, the presence of edge states with metallic character is a robust feature of a topological insulator.

Another important development was the realization that spin–orbit interaction in materials can introduce additional features in the topological behavior. Kane and Mele showed that, when spin–orbit interaction terms are included, topological constraints can exist for systems even without breaking T-symmetry, that is, without invoking the presence of a magnetic field.[23] This has to do with the effects of Kramers degeneracy, see Eq. (9.26). Specifically, Kramers degeneracy requires that $\epsilon_{-\mathbf{k},\downarrow} = \epsilon_{\mathbf{k},\uparrow}$. Now, consider a 1D system with periodicity a in the x direction, so that at the center of the BZ, $k_x = 0$, and at the edges of the BZ, $k_x = \pm\pi/a$, the "spin-up" and "spin-down" states must be degenerate since at those points $\mathbf{k} = -\mathbf{k}$ (the equality holds up to a reciprocal lattice vector, in the

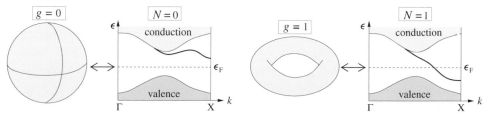

Fig. 9.23 Illustration of the analogy between the genus g of a surface ($g = 0$ for sphere, $g = 1$ for torus) and the topological or Chern invariant N for a band structure ($N = 0$ for a trivial insulator, $N = 1$ for a topological insulator); blue lines represent the energy bands corresponding to an edge state. Γ is the center and X is the edge of the 1D first Brillouin zone.

[23] C. L. Kane and E. J. Mele, *Phys. Rev. Lett.* **95**, 226801 (2005); **95**, 146802 (2005).

present case $2\pi/a$, see discussion in Chapter 2). We note here again the cursory use of the terms "spin up" and "spin down" for the expectation values of the σ_z spin operator; strictly speaking, this terminology is meaningful only when there are no spin-interaction terms in the hamiltonian. As in the case of the honeycomb lattice ribbons discussed earlier, the 1D edges with periodic structure can exhibit states in the band gap of the bulk 2D system, that we identify as the states of interest here. Then, it is possible to connect the degenerate states at the BZ boundaries in different ways. For instance, assuming there are two such degenerate states at the BZ center, $k_x = 0$, coming from the opposite edges of the ribbon, they can connect to degenerate states at the BZ edge in such a way that the Fermi level in the gap will inersect a number of bands given by:

$$\mu_0 = 2(2l), \quad l : \text{positive integer}$$

as shown in Fig. 9.24; when there are no crossings, $\mu_0 = 0$, this situation will produce a conventional insulator. Alternatively, the edge states may connect to other bulk bands through dispersion and mixing, so that the Fermi level will intersect a number of bands given by:

$$\mu_1 = 2(2l + 1), \quad l : \text{positive integer}$$

which cannot be zero, no matter where the Fermi level is placed in the gap. Therefore, this scenario will lead to metallic behavior of spin-polarized bands. To characterize this situation, an additional invariant is needed, since the Chern invariant is zero because none of the symmetries that can lead to non-zero Chern values are broken. The new invariant was called the "Z_2 topological invariant"; its mathematical formulation is more complicated than the definition of the Chern invariant and lies beyond the scope of the present treatment. The types of behavior can then be classified according to the values of different symmetry

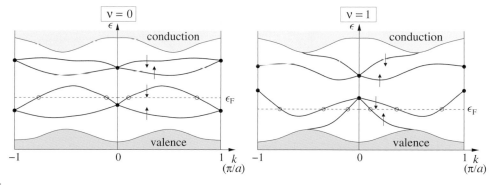

Fig. 9.24 Illustration of the behavior of edge states in the case of different values of the topological invariant Z_2, labeled $\nu = 0$ and $\nu = 1$. Shaded red and blue regions represent the bands of the 2D bulk projected along the edge direction, while blue and red curves correspond to the edge states with up and down spin, which are degenerate at the BZ center ($k_x = 0$) and edges ($k_x = \pm\pi/a$) due to Kramers degeneracy. For $\nu = 0$, the number of allowed band crossings by the Fermi level is $\mu_0 = 0, 4, 8, \ldots$ (the case $\mu_0 = 4$ is shown); for $\nu = 1$, the number of allowed band crossings is $\mu_1 = 2, 6, 10, \ldots$ (the case $\mu_1 = 6$ is shown).

operators, like T-symmetry, and the values of the various topological invariants, like the Chern number or the Z_2 invariant. The topological arguments can be extended to a manifold of occupied states, as discussed in more detail in the three review articles given in Further Reading (items 5, 6, and 7) and the references cited therein.

The notions we presented in relation to 2D systems can be generalized to 3D solids, in which case the edge states become surface states. Of special interest are states with linear energy–momentum dispersion relation in three dimensions; systems where this occurs are referred to as "Weyl semimetals." In this situation, some interesting features appear, related to important topological constraints. We discuss an example of a Weyl semimetal in general terms; a comprehensive recent review of this topic has been provided by N. P. Armitage, E. J. Mele, and A. Vishwanath (see Further Reading). To fix ideas, we adopt a simple model, the tight-binding solution of which yields the essential physics. Our model consists of a 3D cubic lattice with two types of atoms, one with an s-like orbital (A site) and the other with a p-like orbital (B site). We first illustrate and solve this model in its 2D version, a square lattice, with a two-atom basis, which is easier to visualize and is shown in Fig. 9.25. We emphasize that the orbitals associated with each atom in this model are not necessarily the usual s and p atomic orbitals, but only have the *symmetry* of these types of orbitals (even and odd under inversion, respectively, also referred to as "even-parity" and "odd-parity" orbitals). Indeed, the p-like state we chose in Fig. 9.25 has four lobes, as a hybrid of atomic p orbitals would, for instance the symmetric combination $(p_x + p_y)/\sqrt{2}$. We allow the following hopping matrix elements in this model: between nearest-neighbor sites A and B with value $t_1 > 0$ and between second-neighbor A–A sites, with value $t_s < 0$ and B–B sites, with value $t_p > 0$; the set of hamiltonian matrix elements is completed

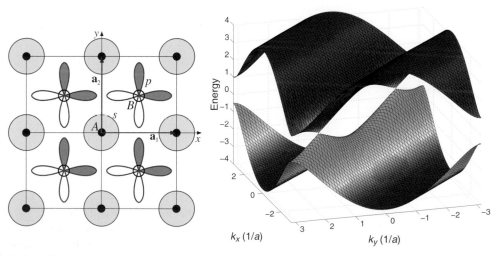

Fig. 9.25 **Left**: A 2D model square lattice with two atoms per unit cell represented by the filled and open black circles, with s (red circle) and p-like hybrid (purple lobes) orbitals. **Right**: The corresponding energy (on the vertical axis) bands, as functions of the k_x, k_y (horizontal plane axes) components of the wave-vector, in the range $\pm\pi/a$; parameter values: $\epsilon_p = 0.5, t_1 = 1, t_s = -0.5, t_p = 0.5$. The two Dirac points that appear along the $k_x = -k_y$ line are evident.

by the on-site energies $\epsilon_A = \epsilon_s, \epsilon_B = \epsilon_p$, the first of which we define as the reference zero-energy value. The hamiltonian for this model has the form

$$\hat{\mathcal{H}} = \begin{pmatrix} \epsilon_A + t_s\, B(\mathbf{k}) & t_1\, C(\mathbf{k}) \\ t_1\, C^*(\mathbf{k}) & \epsilon_B + t_p\, B(\mathbf{k}) \end{pmatrix} \tag{9.61}$$

$$B(\mathbf{k}) = 2[\cos(\mathbf{k} \cdot \mathbf{a}_1) + \cos(\mathbf{k} \cdot \mathbf{a}_2)]$$

$$C(\mathbf{k}) = (1 - e^{-i\mathbf{k} \cdot \mathbf{a}_1})(1 + e^{-i\mathbf{k} \cdot \mathbf{a}_2}) + (1 + e^{-i\mathbf{k} \cdot \mathbf{a}_1})(1 - e^{-i\mathbf{k} \cdot \mathbf{a}_2})$$

and the corresponding energy bands are shown in Fig. 9.25. The interesting feature here is the presence of two Dirac points along the $k_x = -k_y$ line, with linear energy–momentum relations along all directions in reciprocal space but with different slopes (see Problem 11).

We now extend this model to the case of the 3D cubic lattice, assuming the same type of atomic and orbital sites, but without the second-neighbor hopping matrix elements.[24] Instead, we consider the spin-up and spin-down states for each electron, and introduce an external magnetic field in the same direction as the spin orientation (the z direction), the presence of which endows the spins with energy $\pm \mathcal{V}_z$. We will also take the on-site energies of the two orbitals to differ by $2\mathcal{V}_0$. Finally, we will work with a Taylor expansion of the energy near the Dirac point, taking the slope to be $\hbar v_F$, where v_F is the Fermi velocity, which we assume for simplicity to be uniform in $\mathbf{k}$-space. Under these assumptions, the hamiltonian is a 4×4 matrix with the form

$$\hat{\mathcal{H}} = \begin{pmatrix} \mathcal{V}_0 I_2 + \mathcal{V}_z \sigma_z & \hbar v_F \mathbf{k} \cdot \boldsymbol{\sigma} \\ \hbar v_F \mathbf{k} \cdot \boldsymbol{\sigma} & -\mathcal{V}_0 I_2 + \mathcal{V}_z \sigma_z \end{pmatrix} \tag{9.62}$$

where $\boldsymbol{\sigma} = \sigma_x \hat{x} + \sigma_y \hat{y} + \sigma_z \hat{z}$ with $\sigma_x, \sigma_y, \sigma_z$ the Pauli spin matrices and I_2 is the 2×2 identity matrix. For $\mathcal{V}_0 = 0$ and $\mathcal{V}_z = 0$, this hamiltonian reduces to two copies of the graphene hamiltonian, Eq. (9.48), which we studied in the previous section; the bands are two degenerate Dirac cones at $\mathbf{k} = 0$, the degeneracy coming from the two spin states. When we allow values $\mathcal{V}_0 \neq 0$ and $\mathcal{V}_z \neq 0$, the bands split and the four energy eigenvalues are given by the expression

$$\epsilon_{\mathbf{k}}^{(\alpha,\beta)} = \alpha \left[\mathcal{V}_0^2 + \mathcal{V}_z^2 + \hbar^2 v_F^2 |\mathbf{k}|^2 + 2\beta \mathcal{V}_z \left(\hbar^2 v_F^2 k_z^2 + \mathcal{V}_0^2 \right)^{1/2} \right]^{1/2} , \quad \alpha, \beta = \pm 1 \tag{9.63}$$

For the case $|\mathcal{V}_0| > |\mathcal{V}_z|$, the four solutions are split into two pairs, one pair with valence-like (negative effective mass) and the other pair with conduction-like (positive effective mass) character, which are separated by gaps throughout the BZ, as in a conventional semiconductor. But for $|\mathcal{V}_0| < |\mathcal{V}_z|$, a more interesting feature arises: a plot of the four bands for this case is shown in Fig. 9.26. Two of the bands have a gap throughout the BZ, while the other two touch with Dirac cones at the points $\pm \mathbf{k}_0$:

$$\mathbf{k}_0 = \left(0, 0, k_z^{(W)} \right), \quad k_z^{(W)} = \frac{\sqrt{\mathcal{V}_z^2 - \mathcal{V}_0^2}}{\hbar v_F} \tag{9.64}$$

---

[24] This model was introduced by A. Burkov, M. Hook, and L. Balents, *Phys. Rev. B* **84**, 235126 (2011); see also the discussion in the review article by Armitage, Mele, and Vishwanath, listed in Further Reading.

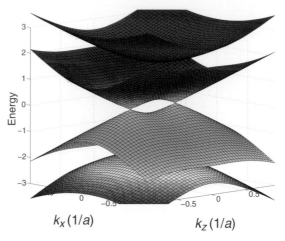

$k_x\,(1/a)$ $k_z\,(1/a)$

Fig. 9.26 The four energy bands of the 3D cubic lattice model with two atoms per unit cell having an s-like (even parity) and a p-like (odd parity) orbital, with on-site energies separated by $2\mathcal{V}_0$, and including spin degrees of freedom in an external magnetic field in the z direction, which splits the energy of the spin states by $2\mathcal{V}_z$ (see footnote 24). The energy bands are plotted for $k_y = 0$; the two Weyl nodes at $\pm\mathbf{k}_0$, Eq. (9.64), are evident.

These points are called "Weyl nodes"; assuming that the Fermi level is at the zero value of the energy, the presence of the Weyl nodes produces the semimetallic character of the system. The reason for this terminology is that the hamiltonian of Eq. (9.62) is formally equivalent to the model first introduced by H. Weyl,[25] which describes two massless fermions with definite chirality, that is, spin pointing along (positive chirality or right-handed particle) or opposite to (negative chirality or left-handed particle) the direction of the momentum. The Weyl particles come in pairs, since they are the two parts of the same solution of the Dirac equation (for more details see Appendix C, Section C.8). The interesting feature of the Weyl nodes is the following: a sphere in reciprocal space surrounding a Weyl node is pierced by a total amount of Berry flux equal to $2\pi\chi$, where $\chi = \pm 1$ is the chirality of this point. According to our definition of the first Chern number, this implies that each Weyl node has a non-vanishing Chern number equal to its chirality, $C_1 = \chi$. When the sphere is extended to cover the entire BZ, the net value of the Chern number must be zero, since the Weyl nodes come in pairs of opposite chirality, just like the Weyl massless particles. Therefore, Weyl nodes must appear or disappear in pairs, as the parameters of the hamiltonian change value.

In our example of a 3D system, we assumed that an external magnetic field splits the two spin states to produce the Weyl nodes at $\pm\mathbf{k}_0$. It turns out that the magnitude of the external field required to produce a splitting $2|\mathcal{V}_z|$ large enough to make the separation of the Weyl points noticeable on the scale of the BZ size far exceeds available experimental fields, so it is not likely to produce this behavior on the basis of the simple model considered

[25] H. Weyl, *Proc. Natl. Acad. Sci. USA* **15**, 323 (1929).

here. Thus, the field we introduced should be considered as an effective internal field. As we saw in the simpler example of the 2D square lattice discussed before, the terms in the hamiltonian that produce similar behavior can arise from hopping matrix elements under the right circumstances. The search for such behavior may then involve materials described by effective hamiltonians where interactions between different neighbors conspire to produce the Weyl semimetal behavior. A characteristic feature that reveals the presence of Weyl points is the "Fermi arc," the band dispersion along a direction of the BZ that connects two Weyl nodes at the Fermi level, as is easy to imagine by considering a plane that cuts through the Dirac cones and passes through the two Weyl nodes. This feature can be generated by surfaces, the presence of which implies a projection of the 3D reciprocal space onto a lower-dimensional space. Using angle-resolved photoemission spectroscopy (ARPES), it is possible to detect such features in the band structure of solids that possess Weyl nodes.[26] Interestingly, Weyl nodes can also be produced in photonic crystals, where their presence is detected by tuning the incident microwave frequency and performing angle-resolved transmission measurements.[27] More examples and possible realizations are discussed in the review by Armitage, Mele, and Vishwanath (see Further Reading).

In addition to their theoretical interest, the materials whose behavior is characterized by topological constraints, often referred to as "Dirac materials," could be important for practical applications such as quantum computing. What makes the topological materials particularly interesting is that the edge or surface states are robust, that is, they are not eliminated by disorder. In general, disorder can lead to localization of electronic states, a phenomenon of great importance for real materials, known as "Anderson localization." The physics behind this is that disorder, when it is sufficiently strong, prevents electrons from forming extended, band-like states, so that the available states are localized. However, as Kane and Mele showed for the topological spin-polarized states, and Halperin showed earlier in the context of the quantum Hall state, edge states of this type are not affected by even strong disorder, as long as the T-symmetry is not broken. This ensures that these states can result in conductance under a wide range of circumstances. Edge states of opposite spin move in opposite directions, which leads to "filtering" of the spin current, providing an additional knob to control the properties of the electron current. Even more complicated situations can be envisioned, including topological superconductors and topological metals, as well as exotic features such as quasiparticles that are half an ordinary Dirac fermion, called "Majorana fermions," which are their own anti-particles. The field exploring the consequences of this type of behavior, and prediciting how it can be realized in real material systems, has blossomed into a most exciting area of condensed matter physics and remains very active at the time of writing this book (see review articles in Further Reading).

---

[26] S.-Y. Xu *et al.*, *Science* **347**, 294 (2015).
[27] L. Lu, Z. Wang, D. Ye, J. D. Joannopoulos, and M. Soljacic, *Science* **349**, 622 (2015).

Further Reading

1. *The Quantum Hall Effect*, R. E. Prange and S. M. Girvin (Eds) (2nd edn, Springer-Verlag, New York, 1990).

2. T. Chakraborty, "The quantum Hall effect," in P. T. Landsbery (Ed.), *Handbook on Semiconductors*, Vol. 1, pp. 977–1038 (North Holland, Amsterdam, 1992).

3. R. Resta, "Manifestations of Berry's phase in molecules and condensed matter," *J. Phys.: Condens. Matter* **12**, R107 (2000). This is an excellent review of Berry's phase and its applications.

4. D. Xiao, M.-C. Chang, and Q. Niu, "Berry phase effects on electronic properties," *Rev. Mod. Phys.* **82**, 1959 (2010). This is another excellent review article on Berry's phase and topological constraints.

5. M. Z. Hasan and C. L. Kane, "Colloquium: Topological insulators," *Rev. Mod. Phys.* **82**, 3045 (2010). This is a wonderfully pedagogical review article on topological insulators with an extensive list of references.

6. X.-L. Qi and S.-C. Zhang, "Topological insulators and superconductors," *Rev. Mod. Phys.* **83**, 1057 (2011).

7. N. P. Armitage, E. J. Mele, and A. Vishwanath, "Weyl and Dirac semimetals in three-dimensional solids," *Rev. Mod. Phys.* **90**, 015001 (2018).

Problems

1.   Calculate the diagonal σ_{xx}, σ_{yy} and off-diagonal σ_{xy}, σ_{yx} components of the resistivity tensor for the case of the 2D electron system, as shown in Fig. 9.2. Define the scattering time τ through $\gamma = \overline{m}/\tau$ and express the conductivity components in terms of τ, the cyclotron frequency $\omega_c = eB/\overline{m}c$, and $\sigma_0 = ne^2\tau/\overline{m}$. Show that when the diagonal resistivity $\rho_{xx} = 0$, the diagonal conductivities $\sigma_{xx} = \sigma_{yy} = 0$ and the effective conductivity, defined through $j_x = \sigma_{\text{eff}}E_x$, diverges, $\sigma_{\text{eff}} \to \infty$.

2.   We wish to prove that the time-reversal operator defined in Eq. (9.24) satisfies the proper conditions, Eq. (9.23) (for this problem it may be useful to briefly review the Dirac equation, see Appendix C, Section C.8). We note first that $\hat{C}^2 = 1$, which means that $\hat{C}$ is its own inverse operator.

      (a) Show that $\underline{\mathbf{S}}$ is a real matrix and that $\underline{\mathbf{S}}^{-1} = -\underline{\mathbf{S}}$. Using this, show that the first of the conditions in Eq. (9.23) is satisfied:

$$-\hat{\mathcal{T}}^{-1}(i\hbar)\hat{\mathcal{T}} = i\hbar$$

      (b) Show that

$$\hat{\mathcal{T}}^{-1}\hat{\mathcal{H}}\hat{\mathcal{T}} = \underline{\mathbf{S}}^{-1}\hat{\mathcal{H}}^*\underline{\mathbf{S}}$$

(c) From the last expression, we conclude that all that remains to be shown for the second condition to be satisfied is that:

$$\underline{S}^{-1}\hat{\mathcal{H}}^*\underline{S} = \hat{\mathcal{H}}$$

Show that this is automatically satisfied for the last two terms in the hamiltonian.

(d) For the momentum term, $\mathbf{p}^* = -\mathbf{p}$, so all that remains to be shown is that $\underline{S}^{-1}\alpha_j^*\underline{S} = -\alpha_j$ for all three components $j = x, y, z$. Show that for each of these components:

$$\underline{S}^{-1}\alpha_j^*\underline{S} = \begin{pmatrix} 0 & \sigma_y\sigma_j^*\sigma_y \\ \sigma_y\sigma_j^*\sigma_y & 0 \end{pmatrix}$$

(e) We will work out the three different components separately, using relations that can be proven from the definitions of the Pauli matrices. First, show that

$$\sigma_y^* = -\sigma_y \text{ and } \sigma_y^2 = 1 \Rightarrow \sigma_y\sigma_y^*\sigma_y = -\sigma_y$$

Then, show that for $j = x, z$:

$$\sigma_j^* = \sigma_j \text{ and } \sigma_j\sigma_y = -\sigma_y\sigma_z \Rightarrow \sigma_y\sigma_j^*\sigma_y = -\sigma_j.$$

Thus, for each component separately we have established that $\sigma_y\sigma_j^*\sigma_y = -\sigma_j$. Show that this leads to the desired result, namely:

$$\underline{S}^{-1}\alpha_j^*\underline{S} = -\alpha_j$$

which completes the proof that the operator $\hat{\mathcal{T}}$ defined in Eq. (9.24) satisfies the necessary conditions to be a time-reversal operator, Eq. (9.23).

3   Use the properties of the Pauli matrix $\sigma_y^2 = 1$ [see definition in Eq. (C.43)] and the series expansions of the exponential, the cosine, and the sine functions to prove that the choice of the time-reversal operator given in Eq. (9.25) is equivalent to that of Eq. (9.24).

4   Calculate the Berry curvature of the higher-energy state in the model of the spin in a magnetic monopole field, $\Omega^{(2)}(\omega)$, discussed in Example 9.2. Calculate the integral of the Berry curvature, summed over both states of the system, over the space spanned by the two angular variables, $\omega = (\theta, \phi)$.

5   Derive the expression of Eq. (9.46) starting with the expression of Eq. (9.45). As a first step, pull the differentiation with respect to $\mathbf{k}$ outside the first expectation value in the numerator to obtain:

$$\langle u_{\mathbf{k}}^{(n)}|\nabla_{\mathbf{k}}\hat{\mathcal{H}}_{\mathbf{k}}|u_{\mathbf{k}}^{(m)}\rangle = \nabla_{\mathbf{k}}\left(\langle u_{\mathbf{k}}^{(n)}|\hat{\mathcal{H}}_{\mathbf{k}}|u_{\mathbf{k}}^{(m)}\rangle\right) - \langle\nabla_{\mathbf{k}}u_{\mathbf{k}}^{(n)}|\hat{\mathcal{H}}_{\mathbf{k}}|u_{\mathbf{k}}^{(m)}\rangle - \langle u_{\mathbf{k}}^{(n)}|\hat{\mathcal{H}}_{\mathbf{k}}|\nabla_{\mathbf{k}}u_{\mathbf{k}}^{(m)}\rangle$$

Then use the orthogonality of the wavefunctions $\langle u_{\mathbf{k}}^{(n)}|u_{\mathbf{k}}^{(m)}\rangle = \delta_{nm}$ and the fact that they are eigenfunctions of $\hat{\mathcal{H}}_{\mathbf{k}}$ with eigenvalues $\epsilon_{\mathbf{k}}^{(n)}, \epsilon_{\mathbf{k}}^{(m)}$ to show that the above expression becomes:

$$\langle u_{\mathbf{k}}^{(n)}|\nabla_{\mathbf{k}}\hat{\mathcal{H}}_{\mathbf{k}}|u_{\mathbf{k}}^{(m)}\rangle = \left(\epsilon_{\mathbf{k}}^{(n)} - \epsilon_{\mathbf{k}}^{(m)}\right)\langle\nabla_{\mathbf{k}}u_{\mathbf{k}}^{(n)}|u_{\mathbf{k}}^{(m)}\rangle$$

Next, show that when this last expression is inserted in Eq. (9.45) and combined with the completeness of the basis:

$$\sum_m |u_{\mathbf{k}}^{(m)}\rangle\langle u_{\mathbf{k}}^{(m)}| = 1$$

it leads to the desired result, Eq. (9.46).

6   Show that for a finite system, the polarization $\mathbf{P}$ as defined in Eq. (9.51) is independent of the position of the origin if the system is neutral (total charge zero); assume that the origin is displaced by an arbitrary vector $\mathbf{R}_0$ and show that the value of $\mathbf{P}$ is not affected.

7   Prove the expression for the velocity $\langle v_{\mathbf{k}}^{(n)}\rangle$ given in Eq. (9.54). To this end, you will need to use the chain-rule relations that are direct consequences of Eq. (9.53), namely:

$$\frac{\partial}{\partial \mathbf{k}} = \frac{\partial \mathbf{q}}{\partial \mathbf{k}}\frac{\partial}{\partial \mathbf{q}} = \frac{\partial}{\partial \mathbf{q}} \quad \text{and} \quad \frac{\partial}{\partial t} = \frac{\partial \mathbf{q}}{\partial t}\cdot\frac{\partial}{\partial \mathbf{q}} = -\frac{e}{\hbar}\mathbf{E}\cdot\frac{\partial}{\partial \mathbf{q}}$$

to substitute in Eq. (9.46). Then, use the definition of the Berry curvature of the band, Eq. (9.43), to arrive at the desired result.

8   For the 2D honeycomb lattice with identical atoms at A and B sites, Eq. (9.48) with $\mathcal{V}_0 = 0$, using the expressions for the Berry connection from Eq. (9.49), show explicitly that the Berry curvature $\Omega_{xy}(\mathbf{k})$ vanishes for $\mathbf{k}$ values different from those at the corners of the BZ, $\mathbf{K}, \mathbf{K}'$.

9   We consider the model of the 2D honeycomb lattice with broken P-symmetry (Rice–Mele model), Eq. (9.48), and examine the solutions near the high-symmetry points K and K′ of the BZ.

(a) Calculate explicitly the eigenfunctions at $\mathbf{K}+\mathbf{q}$ for $|\mathbf{q}| \to 0$.

(b) Use the definition of the angle θ through the relations Eq. (9.50) and show that the eigenfunctions are identical to those of the model of a spin in a monopole field, Eq. (9.40).

(c) For the lower-energy band, evaluate the integral of the Berry curvature over the entirety of the BZ: do this by using Stokes' theorem to relate this integral to the line integral of the Berry connection, and then use Figure 9.13 to show that the line integral reduces to just loops around the K and K′ points, where the analogy with the spin in a monopole field holds. What do you conclude for this model?

10   We consider the model of the honeycomb lattice with broken T-symmetry (Haldane model), Eq. (9.59), for the case $\varphi = \pi/2, \mathcal{V}_0 = 0$. We let $t_2 = t_\pi/\lambda$ and assume $\lambda > 0$.

(a) Calculate the energy eigenvalues near the high-symmetry point K, $\mathbf{k} = \mathbf{K} + \mathbf{q}, |\mathbf{q}| \to 0$, as functions of $\mathbf{q}$: show that they are quadratic in $|\mathbf{q}|$, and that the energy gap at K is given by $\epsilon_{\text{gap}}(\mathrm{K}) = 6\sqrt{3}|t_2|$.

(b) Calculate the effective mass of electrons and holes at K, as functions of the parameters t_π and $\lambda = t_\pi/t_2$. Show that the effective mass becomes infinite (which implies no dispersion in the neighborhood of K) for the value $\lambda = \sqrt{18}$.

11   Solve the 2D square model with two atoms per unit cell, one with an s-like orbital
     and the other with a p-like orbital, described in Section 9.8. Specifically:

(a) Calculate the bands in terms of the model parameters $\epsilon_p, t_1, t_s, t_p$. Take $t_s = 0, t_p = 0$, that is, turn off the second-neighbor hopping and take the two orbitals
     to be degenerate, $\epsilon_p = 0$. Plot the bands for this case along high-symmetry
     directions of the BZ and comment on their behavior.

(b) Next, turn on the second-neighbor hopping by taking $t_s = -t_p \neq 0$, take the two
     orbitals to be non-degenerate, $\epsilon_p > 0$, and find the position of the Dirac points
     in terms of these parameters. Calculate the Fermi velocity at the Dirac points as
     a function of direction in reciprocal space.

(c) Calculate the integral of the Berry curvature and from it the first Chern number,
     using Eq. (9.56). What do you conclude for this model?

10 Magnetic Behavior of Solids

In this chapter we examine the magnetic behavior in different types of solids. Magnetic order in a solid, induced either by an external field or by inherent properties of the structure, may be destroyed by thermal effects, that is, the tendency to increase the entropy by randomizing the direction of microscopic magnetic moments. Thus, magnetic phenomena are typically observed at low temperatures where the effect of entropy is not strong enough to destroy magnetic ordering.

We begin by defining the terms applied to describe the various types of magnetic behavior. A system is called **paramagnetic** if it has no inherent magnetization, but when subject to an external field develops magnetization which is aligned with the field; this corresponds to situations where the microscopic magnetic moments tend to be oriented in the same direction as the external magnetic field. A system is called **diamagnetic** if it has no inherent magnetization, but when subject to an external field develops magnetization that is opposite to the field; this corresponds to situations where the induced microscopic magnetic moments tend to shield the external magnetic field. Finally, a system may exhibit magnetic order even in the absence of an external field. If the microscopic magnetic moments tend to be oriented in the same direction, the system is described as **ferromagnetic**. A variation on this theme is the situation in which microscopic magnetic moments tend to have parallel orientation but are not necessarily equal at neighboring sites, which is described as **ferrimagnetic** behavior. If magnetic moments at neighboring sites tend to point in opposite directions, the system is described as **antiferromagnetic**. In the latter case there is inherent magnetic order due to the orientation of the microscopic magnetic moments, but the net macroscopic magnetization is zero.

A natural way to classify magnetic phenomena is by the origin of the microscopic magnetic moments. This origin can be ascribed to two factors: the intrinsic magnetic moment (spin) of quantum-mechanical particles and the magnetic moment arising from the motion of charged particles in an external electromagnetic field (orbital moment). In real systems the two sources of magnetic moment are actually coupled. In particular, the motion of electrons in the electric field of the nuclei leads to spin–orbit coupling. This coupling produces typically a net moment (total spin) which characterizes the entire atom or ion. In solids, there is another effect that leads to magnetic behavior, namely the motion in an external field of itinerant (metallic) electrons, which are not bound to any particular nucleus. We consider first the behavior of systems in which the microscopic magnetic moments are individual spins, whether these are due to single electrons (spin 1/2) or to compound objects like atoms or ions (which can have total spin of various values). Such phenomena include the behavior of magnetic atoms or ions as non-interacting spins in insulators, the behavior of non-interacting electron spins in metals, and the behavior of interacting classical and quantum spins on a lattice. We will also study the behavior of crystal electrons in an external magnetic field, that is, orbital moments, which gives rise to interesting classical and quantum phenomena.

10.1 Overview of Magnetic Behavior of Insulators

The magnetic behavior of insulators can usually be discussed in terms of single atoms or ions which have a magnetic moment. In such situations the most important consideration is how the spin and angular momentum of electrons in the electronic shells combine to produce the total spin of the atom or ion. The rules that govern this behavior are the usual quantum-mechanical rules for combining spin and angular momentum vectors, supplemented by Hund's rules that specify which states are preferred energetically. There are three quantum numbers that identify the state of an atom or ion: the total spin S, the orbital angular momentum L, and the total angular momentum J (this assumes that spin–orbit coupling is weak, so that S and L can be used independently as good quantum numbers). Hund's rules determine the values of S, L, J for a given electronic shell and a given number of electrons. The first rule states that the spin state is such that the total spin S is maximized. The second rule states that the occupation of angular momentum states l_z in the shell (there are $2l + 1$ of them for a shell with nominal angular momentum l) is such that the orbital angular momentum L is maximized, for those situations consistent with the first rule. The third rule states that $J = |L - S|$ when the electronic shell is less than half filled ($n \leq 2l+1$) and $J = L+S$ when it is more than half filled ($n \geq 2l+1$), where n is the number of electrons in the shell; when the shell is exactly half filled the two expressions happen to give the same result for J. Of course, the application of all rules must also be consistent with the Pauli exclusion principle.

To illustrate how Hund's rules work, consider the elements in the second row of the Periodic Table, columns I-A through VII-A, as shown in Fig. 1.2. In Li, the single-valence electron is in the $2s$ shell in a total-spin $S = 1/2$ state, this being the only possibility; the corresponding orbital angular momentum is $L = 0$ and the total angular momentum is $J = |L - S| = |L + S| = 1/2$. In Be, the two valence electrons fill the $2s$ shell with opposite spins in a total-spin $S = 0$ state; the corresponding orbital angular momentum is $L = 0$ and the total angular momentum is $J = 0$. In B, two of the three valence electrons are in a filled $2s$ shell and one in the $2p$ shell, in a total-spin $S = 1/2$ state, this being the only possibility for the spin; the angular momentum state $l_z = 1$ of the $2p$ shell is singly occupied, resulting in a state with $L = 1$ and $J = |L - S| = 1/2$. In C, two of the four valence electrons are in a filled $2s$ shell and the other two are in the $2p$ shell, both with the same spin in a total-spin $S = 1$ state; the angular momentum states $l_z = 1$ and $l_z = 0$ of the $2p$ shell are singly occupied, resulting in a state with $L = 1$ and $J = |L - S| = 0$. In N, two of the five valence electrons are in a filled $2s$ shell and the other three are in the $2p$ shell, all of them with the same spin in a total-spin $S = 3/2$ state; the angular momentum states $l_z = 1, l_z = 0$, and $l_z = -1$ of the $2p$ shell are singly occupied, resulting in a state with $L = 0$ and $J = |L - S| = L + S = 3/2$. In O, two of the six valence electrons are in a filled $2s$ shell and the other four are in the $2p$ shell, a pair with opposite spins and the remaining two with the same spin in a total-spin $S = 1$ state; the angular momentum states $l_z = 0$ and $l_z = -1$ of the $2p$ shell are singly occupied and the state $l_z = +1$ is doubly occupied, resulting in a state with $L = 1$ and $J = L + S = 2$. In F, two of the seven valence electrons are in a filled $2s$ shell and the other five are in the $2p$ shell, in two pairs with opposite spins and the remaining one giving rise to a total-spin $S = 1/2$ state; the angular momentum states $l_z = +1$ and $l_z = 0$ of the $2p$ shell are doubly occupied and the state $l_z = -1$ is singly occupied, resulting in a state with $L = 1$ and $J = L + S = 3/2$. Ne, with closed $2s$ and $2p$ electronic shells, has $S = 0$, $L = 0$, and therefore also $J = 0$. It is straightforward to derive the corresponding results for the d and f shells, which are given in Table 10.1. We also provide in this table the standard symbols for the various states, as traditionally denoted in spectroscopy: $^{(2S+1)}X_J$, with $X = S, P, D, F, G, H, I$ for $L = 0, 1, 2, 3, 4, 5, 6$, by analogy to the usual notation for the nominal angular momentum of the shell, s, p, d, f, g, h, i for $l = 0, 1, 2, 3, 4, 5, 6$.

From the above discussion we conclude that the B, C, N, and O isolated atoms would have non-zero magnetic moment due to electronic spin states alone. However, when they are close to other atoms their valence electrons form hybrid orbitals from which bonding and anti-bonding states are produced which are filled by pairs of electrons of opposite spin in the energetically favorable configurations (see the discussion in Chapter 1 of how covalent bonds are formed between such atoms); Li can donate its sole valence electron to become a positive ion with closed electronic shell and F can gain an electron to become a negative ion with closed electronic shell, while Be is already in a closed electronic shell configuration. Thus, none of these atoms would lead to magnetic behavior in solids due to electronic spin alone.

Insulators in which atoms have completely filled electronic shells, like the noble elements or the purely ionic alkali halides (composed of atoms from columns I and VII of the Periodic Table) are actually the simplest cases: the atoms or ions in these solids differ very little from isolated atoms or ions, because of the stability of the closed electronic shells. The presence of the crystal produces a very minor perturbation to the atomic configuration of electronic shells. The magnetic behavior of noble element solids and alkali halides predicted by the analysis at the individual atom or ion level is in excellent agreement with experimental measurements.

There also exist insulating solids that contain atoms with partially filled $4f$ electronic shells (the lanthanides series, see Chapter 1). The electrons in these shells are largely unaffected by the presence of the crystal, because they are relatively tightly bound to the nucleus: for the lanthanides there are no core f electrons and therefore the wavefunctions of these electrons penetrate close to the nucleus. Consequently, insulators that contain such atoms or ions will exhibit the magnetic response of a collection of individual atoms or ions. It is tempting to extend this description to the atoms with partially filled $3d$ electronic shells (the fourth row of the Periodic Table, containing the magnetic elements Fe, Co, and Ni); these d electrons can also penetrate close to the nucleus since there are no d electrons in the core. However, these electrons mix more strongly with the other valence electrons and therefore the presence of the crystal has a significant influence on their behavior. The predictions of the magnetic behavior of these solids based on the behavior of the constituent

Table 10.1 Total spin S, orbital angular momentum L, and total angular momentum J numbers for the $l = 2$ (d shell) and $l = 3$ (f shell) as they are being filled by n electrons, where $1 \leq n \leq 2(2l + 1)$, according to Hund's rules; of the two expressions for J, the $-$ sign applies for $n \leq (2l + 1)$ and the $+$ sign applies for $n \geq (2l + 1)$. The standard spectroscopic symbols $^{(2S+1)}X_J$ are also given with $X = S, P, D, F, G, H, I$ for $L = 0, 1, 2, 3, 4, 5, 6$. For the $3d$-shell transition metals and the $4f$-shell rare earth elements we give the calculated (p_{th}) and experimental (p_{exp}) values of the effective Bohr magneton number; note the ionization of the various elements (right superscript) which makes them correspond to the indicated state. All p_{th} values for $l = 2$ are obtained from the quenched total angular momentum expression ($L = 0 \Rightarrow J = S$) and are in much better agreement with experiment than those from the general expression, Eq. (10.8). *Source*: D.R. Lide (Ed.), *CRC Handbook of Chemistry and Physics* (CRC Press, Boca Raton, FL, 1999–2000).

l	n	S $= \left\vert\sum s_z\right\vert$	L $= \left\vert\sum l_z\right\vert$	J $= \vert L \mp S\vert$	$^{(2S+1)}X_J$		p_{th}	p_{exp}
2	1	1/2	2	3/2	$^2D_{3/2}$	Ti^{3+}	1.73	1.8
2	2	1	3	2	$^3F_2$	V^{3+}	2.83	2.8
2	3	3/2	3	3/2	$^4F_{3/2}$	Cr^{3+}	3.87	3.7
2	4	2	2	0	$^5D_0$	Mn^{3+}	4.90	5.0
2	5	5/2	0	5/2	$^6S_{5/2}$	Fe^{3+}	5.92	5.9
2	6	2	2	4	$^5D_4$	Fe^{2+}	4.90	5.4
2	7	3/2	3	9/2	$^4F_{9/2}$	Co^{2+}	3.87	4.8
2	8	1	3	4	$^3F_4$	Ni^{2+}	2.83	3.2
2	9	1/2	2	5/2	$^2D_{5/2}$	Cu^{2+}	1.73	1.9
2	10	0	0	0	$^1S_0$	Zn^{2+}	0.00	
3	1	1/2	3	5/2	$^2F_{5/2}$	Ce^{3+}	2.54	2.4
3	2	1	5	4	$^3H_4$	Pr^{3+}	3.58	3.5
3	3	3/2	6	9/2	$^4I_{9/2}$	Nd^{3+}	3.62	3.5
3	4	2	6	4	$^5I_4$	Pm^{3+}	2.68	
3	5	5/2	5	5/2	$^6H_{5/2}$	Sm^{3+}	0.84	1.5
3	6	3	3	0	$^7F_0$	Eu^{3+}	0.00	3.4
3	7	7/2	0	7/2	$^8S_{7/2}$	Gd^{3+}	7.94	8.0
3	8	3	3	6	$^7F_6$	Tb^{3+}	9.72	9.5
3	9	5/2	5	15/2	$^6H_{15/2}$	Dy^{3+}	10.63	10.6
3	10	2	6	8	$^5I_8$	Ho^{3+}	10.60	10.4
3	11	3/2	6	15/2	$^4I_{15/2}$	Er^{3+}	9.59	9.5
3	12	1	5	6	$^3H_6$	Tm^{3+}	7.57	7.3
3	13	1/2	3	7/2	$^2F_{7/2}$	Yb^{3+}	4.54	4.5
3	14	0	0	8	$^1S_0$	Lu^{3+}	0.00	

atoms is not very close to experimental observations. A more elaborate description, along the lines presented in the following sections, is required for these solids.

In insulating solids whose magnetic behavior arises from individual ions or atoms, there is a common feature in the response to an external magnetic field H, known as the Curie law. This response is measured through the magnetic susceptibility, which according to the Curie law is inversely proportional to the temperature. The magnetic susceptibility is defined as:

$$\chi = \frac{\partial M}{\partial H}, \quad M = -\frac{\partial F}{\partial H}$$

where M is the magnetization and F the free energy. In order to prove the Curie law we use a simple model. We assume that the ion or atom responsible for the magnetic behavior has a total angular momentum J, so there are $(2J + 1)$ values of its J_z component, $J_z = -J, \ldots, +J$, with the z-axis defined by the direction of the external magnetic field, which give rise to the following energy levels for the system:

$$E_{J_z} = m_0 J_z H, \quad J_z = -J, \ldots, +J \tag{10.1}$$

with m_0 the magnetic moment of the atoms or ions. The canonical partition function for this system, consisting of N atoms or ions, is given by:

$$Z_N = \left(\sum_{J_z=-J}^{J} e^{-\beta m_0 J_z H} \right)^N = \left(\frac{e^{\beta m_0 (J+\frac{1}{2})H} - e^{-\beta m_0 (J+\frac{1}{2})H}}{e^{\beta m_0 \frac{1}{2} H} - e^{-\beta m_0 \frac{1}{2} H}} \right)^N$$

where $\beta = 1/k_B T$ is the inverse temperature. We define the variable

$$w = \frac{\beta m_0 H}{2}$$

in terms of which the free energy becomes:

$$F = -\frac{1}{\beta} \ln Z_N = -\frac{N}{\beta} \ln \left[\frac{\sinh(2J + 1)w}{\sinh w} \right]$$

which gives, for the magnetization M and the susceptibility χ per unit volume:

$$M = \frac{Nm_0}{2V} \left[(2J + 1) \coth(2J + 1)w - \coth w \right]$$

$$\chi = \frac{Nm_0^2}{4V} \beta \left[\frac{1}{\sinh^2 w} - \frac{(2J + 1)^2}{\sinh^2(2J + 1)w} \right]$$

with V the total volume containing the N magnetic ions or atoms. In the limit of small w, which we justify below, the lowest few terms in the series expansions of $\coth x$ and $\sinh x$ (see Appendix A) give:

$$M = \frac{N}{V} \frac{J(J + 1)m_0^2}{3} \frac{1}{k_B T} H \tag{10.2}$$

$$\chi = \frac{N}{V} \frac{J(J + 1)m_0^2}{3} \frac{1}{k_B T} \tag{10.3}$$

This is exactly the form of Curie law, that is, the susceptibility is inversely proportional to the temperature. In order for this derivation to be valid, we have to make sure that w is much smaller than 1. From the definition of w, we obtain the condition

$$m_0 H \ll 2k_B T \tag{10.4}$$

Typical values of the magnetic moment m_0 (see the following discussion) are of order the *Bohr magneton*, which is defined as:

$$\mu_B \equiv \frac{e\hbar}{2m_e c} = 0.579 \times 10^{-8} \text{ eV/G}$$

The Bohr magneton may also be written as:

$$\mu_B = \frac{e\hbar}{2m_e c} = \frac{e}{2\pi\hbar c} \frac{\pi\hbar^2}{m_e} = \frac{2\pi a_0^2}{\phi_0}\text{Ry}, \quad \phi_0 \equiv \frac{hc}{e} \tag{10.5}$$

where $h = 2\pi\hbar$ is Planck's constant and ϕ_0 is the value of the flux quantum which involves only the fundamental constants h, c, e. From the units involved in the Bohr magneton we see that even for very large magnetic fields of order 10^4 G, the product $m_0 H$ is of order 10^{-4} eV $\sim 10^0$ K, so the condition of Eq. (10.4) is satisfied reasonably well except at very low temperatures (below 10^0 K) and very large magnetic fields (larger than 10^4 G).

The values of the total spin S, orbital angular momentum L, and total angular momentum J for the state of the atom or ion determine the exact value of the magnetic moment m_0. The interaction energy of an electron in a state of total spin $\mathbf{S}$ and orbital angular momentum $\mathbf{L}$, with an external magnetic field $\mathbf{H}$, to lowest order in the field is given by:

$$\langle \mu_B(\mathbf{L} + g_0\mathbf{S}) \cdot \mathbf{H} \rangle \tag{10.6}$$

where g_0 is the *electronic g-factor*

$$g_0 = 2\left(1 + \frac{\alpha}{2\pi} + \mathcal{O}(\alpha^2) + \cdots\right) = 2.0023$$

with $\alpha = e^2/\hbar c = 1/137$ the fine structure constant. The angular brackets in Eq. (10.6) denote the expectation value of the operator in the electron state. Choosing the direction of the magnetic field to be the z-axis, we conclude that in order to evaluate the effect of the magnetic field we must calculate the expectation values of the operator $(L_z + g_0 S_z)$, which, in the basis of states with definite J, L, S, and J_z quantum numbers, are given by:

$$\mu_B H \langle (L_z + g_0 S_z) \rangle = \mu_B H g(JLS) J_z \tag{10.7}$$

with $g(JLS)$ the Landé g-factors

$$g(JLS) = \frac{1}{2}(g_0 + 1) + \frac{1}{2}(g_0 - 1)\frac{S(S+1) - L(L+1)}{J(J+1)}$$

[see Appendix C, Eq. (C.46) with $\lambda = g_0$]. Comparing Eq. (10.7) to Eq. (10.1), we conclude that the magnetic moment is given by:

$$m_0 = g(JLS)\mu_B \approx \left[\frac{3}{2} + \frac{S(S+1) - L(L+1)}{2J(J+1)}\right]\mu_B \tag{10.8}$$

where we have used $g_0 \approx 2$, a very good approximation. When this expression is used for the theoretical prediction of the Curie susceptibility, Eq. (10.3), the results compare very well to experimental measurements for rare earth $4f$ ions. Typically, the comparison is made through the so-called "effective Bohr magneton number," defined in terms of the magnetic moment m_0 as:

$$p = \frac{m_0\sqrt{J(J+1)}}{\mu_B} = g(JLS)\sqrt{J(J+1)}$$

Values of p calculated from the expression of Eq. (10.8) and obtained from experiment by fitting the measured susceptibility to the expression of Eq. (10.3) are given in Table 10.1. There is only one blatant disagreement between theoretical and experimental values for these rare earth ions, namely Eu, for which $J = 0$; in this case, the vanishing value of the total angular momentum means that the linear (lowest-order) contribution of the magnetic field to the energy assumed in Eq. (10.6) is not adequate, and higher-order contributions are necessary to capture the physics.

When it comes to the effective Bohr magneton number for the $3d$ transition metal ions, the expression of Eq. (10.8) is not successful in reproducing the values measured experimentally. In that case, much better agreement between theory and experiment is obtained if instead we use for the magnetic moment the expression $m_0 = g_0\mu_B \approx 2\mu_B$, and take $J = S$ in the expression for the magnetic susceptibility, Eq. (10.3), that is, if we set $L = 0$. The reason is that in the $3d$ transition metals the presence of the crystal environment strongly affects the valence electrons due to the nature of their wavefunctions, as discussed above, and consequently their total angular momentum quantum numbers are not those determined by Hund's rules which apply to spherical atoms or ions. Apparently, the total spin quantum number is still good, that is, Hund's first rule still holds, and it can be used to describe the magnetic behavior of ions in the crystalline environment, but the orbital angular momentum number is no longer a good quantum number. This phenomenon is referred to as "quenching" of the orbital angular momentum due to crystal field splitting of the levels.

10.2 Overview of Magnetic Behavior of Metals

We consider next the magnetic behavior of metallic electrons. We provide first the general picture and then examine in some detail specific models that put this picture on a quantitative basis. In the simplest possible picture, we can view the electrons in metals as free fermions of spin 1/2 in an external magnetic field. This leads to paramagnetic behavior, which is referred to as **Pauli paramagnetism**.

The orbital motion of electrons in an external electromagnetic field always leads to a diamagnetic response because it tries to shield the external field. This behavior is referred to as **Landau diamagnetism**; for a treatment of this effect, see the book by Peierls.[1] Pauli paramagnetism and Landau diamagnetism are effects encountered in solids that are very good conductors, that is, the behavior of their valence electrons is close to that of the free-electron gas. The two effects are of similar magnitude but opposite sign, as mentioned

[1] R. E. Peierls, *Quantum Theory of Solids* (Oxford University Press, Oxford, 1955).

above, and they are both much weaker than the response of individual atoms or ions. Thus, in order to measure these effects one must be able to separate the effect that comes from the ions in the solid, which is not always easy. In the following we discuss in more detail the theory of the magnetic response of metallic solids in certain situations where it is feasible to identify its source.

10.2.1 Free Fermions in Magnetic Field: Pauli Paramagnetism

We will derive the behavior of a system of fermions with magnetic moment m_0; to apply these results to the electron gas, we simply need to set $m_0 = g_0 \mu_B$. The energy in an external magnetic field H will be given by:

$$\epsilon_{\mathbf{k}\uparrow} = \frac{\hbar^2 \mathbf{k}^2}{2m} - m_0 H, \quad \epsilon_{\mathbf{k}\downarrow} = \frac{\hbar^2 \mathbf{k}^2}{2m} + m_0 H$$

where we have assigned the lowest energy to the spin pointing along the direction of the field (spin $\uparrow$) and the highest energy to the spin pointing against the direction of the field (spin $\downarrow$). The occupation of a state with wave-vector $\mathbf{k}$ and spin $\uparrow$ will be denoted by $n_{\mathbf{k}\uparrow}$ and the total number of particles with spin up by $N_\uparrow$; $n_{\mathbf{k}\downarrow}$ and $N_\downarrow$ are the corresponding quantities for spin-down particles. The total energy of the system in a configuration with $N_\uparrow$ spins pointing up and $N_\downarrow$ spins pointing down is given by:

$$E(N_\uparrow, N_\downarrow) = \sum_{\mathbf{k}} (n_{\mathbf{k}\uparrow} + n_{\mathbf{k}\downarrow}) \frac{\hbar^2 \mathbf{k}^2}{2m} - m_0 H (N_\uparrow - N_\downarrow)$$

$$\sum_{\mathbf{k}} n_{\mathbf{k}\uparrow} = N_\uparrow, \quad \sum_{\mathbf{k}} n_{\mathbf{k}\downarrow} = N_\downarrow, \quad \sum_{\mathbf{k}} (n_{\mathbf{k}\uparrow} + n_{\mathbf{k}\downarrow}) = N$$

The canonical partition function will be given as the sum over all configurations consistent with the above expressions, which is equivalent to summing over all values of $N_\uparrow$ from 0 to N, with $N_\downarrow = N - N_\uparrow$:

$$Z_N(H,T) = \sum_{N_\uparrow=0}^{N} e^{\beta m_0 H(2N_\uparrow - N)} \left[\sum_{\{n_{\mathbf{k}\uparrow}\}} e^{-\beta \sum_{\mathbf{k}} n_{\mathbf{k}\uparrow} \hbar^2 \mathbf{k}^2 / 2m} \sum_{\{n_{\mathbf{k}\downarrow}\}} e^{-\beta \sum_{\mathbf{k}} n_{\mathbf{k}\downarrow} \hbar^2 \mathbf{k}^2 / 2m} \right]$$

We can take advantage of the definition of the partition function of spinless fermions:

$$Z_N^{(0)} = \sum_{\{n_{\mathbf{k}}\}} e^{-\beta \sum_{\mathbf{k}} n_{\mathbf{k}} \hbar^2 \mathbf{k}^2 / 2m}, \quad \sum_{\mathbf{k}} n_{\mathbf{k}} = N$$

to rewrite the partition function $Z_N(H,T)$ as:

$$Z_N(H,T) = e^{-\beta m_0 H N} \sum_{N_\uparrow=0}^{N} e^{2\beta m_0 H N_\uparrow} Z_{N_\uparrow}^{(0)} Z_{N-N_\uparrow}^{(0)}$$

and from this we obtain for the expression $(\ln Z_N)/N$, which appears in thermodynamic averages:

$$\frac{1}{N} Z_N(H, T) = -\beta m_0 H + \frac{1}{N} \ln \left[\sum_{N_\uparrow=0}^{N} e^{2\beta m_0 H N_\uparrow + \ln\left(Z_{N_\uparrow}^{(0)}\right) + \ln\left(Z_{N-N_\uparrow}^{(0)}\right)} \right]$$

We will approximate the above summation over $N_\uparrow$ by the largest term in it, with the usual assumption that in the thermodynamic limit this term dominates; to find the value of $N_\uparrow$ that corresponds to this term, we take the derivative of the exponent with respect to $N_\uparrow$ and set it equal to zero, obtaining:

$$2m_0 H + \frac{1}{\beta} \frac{\partial}{\partial N_\uparrow} \ln\left(Z_{N_\uparrow}^{(0)}\right) - \frac{1}{\beta} \frac{\partial}{\partial (N - N_\uparrow)} \ln\left(Z_{N-N_\uparrow}^{(0)}\right) = 0$$

where, in the second partial derivative, we have changed variables from $N_\uparrow$ to $N - N_\uparrow$. From Eqs (D.62), we can identify the two partial derivatives in the above expression with the negative chemical potentials, thus obtaining the relation

$$\mu^{(0)}(\bar{N}_\uparrow) - \mu^{(0)}(N - \bar{N}_\uparrow) = 2m_0 H$$

where we have denoted by $\bar{N}_\uparrow$ the value corresponding to the largest term in the partition function. We define the average magnetization per particle m as:

$$m(T) = \frac{M}{N} = \frac{m_0(\bar{N}_\uparrow - \bar{N}_\downarrow)}{N}$$

in terms of which we can express $N_\uparrow$ and $N_\downarrow$ as:

$$\bar{N}_\uparrow = \frac{1 + m/m_0}{2} N, \quad \bar{N}_\downarrow = \frac{1 - m/m_0}{2} N$$

With these expressions, the relation we had derived above for the chemical potentials becomes:

$$\mu^{(0)}\left(\frac{1 + m/m_0}{2} N\right) - \mu^{(0)}\left(\frac{1 - m/m_0}{2} N\right) = 2m_0 H$$

For small external field, we expect $N_\uparrow$ and $N_\downarrow$ to be close to $N/2$ and the average magnetization m to be very small, therefore we can use Taylor expansions around $N/2$ to obtain:

$$\left[\frac{\partial \mu^{(0)}(qN)}{\partial q}\right]_{q=1/2} \left(\frac{m}{2m_0}\right) - \left[\frac{\partial \mu^{(0)}(qN)}{\partial q}\right]_{q=1/2} \left(-\frac{m}{2m_0}\right) = 2m_0 H$$

$$\Rightarrow m(T) = 2m_0^2 H \left[\frac{\partial \mu^{(0)}(qN)}{\partial q}\right]_{q=1/2}^{-1}$$

With this expression, the magnetic susceptibility per particle χ takes the form

$$\chi(T) \equiv \frac{\partial m}{\partial H} = 2m_0^2 \left[\frac{\partial \mu^{(0)}(qN)}{\partial q}\right]_{q=1/2}^{-1}$$

while the magnetic susceptibility per volume is simply $\chi N/V = n\chi$.

To analyze the behavior of the system at various temperature regimes we will employ the behavior of the ideal Fermi gas (the results used here are derived for convenience in Appendix D). Specifically, in the low-temperature limit we use Eq. (D.91) and the first term in the corresponding expansion of $h_3(z)$:

$$h_3(z) = \frac{4}{3\sqrt{\pi}} \left[(\ln z)^{3/2} + \frac{\pi^2}{8} (\ln z)^{-1/2} + \cdots \right]$$

as well as the relation between the fugacity z and the chemical potential $\mu^{(0)}$, that is, $z = \exp[\beta\mu^{(0)}]$ to obtain:

$$\mu^{(0)}(N) = k_B T \left(\frac{3\sqrt{\pi}}{4} \frac{N}{V} \lambda^3 \right)^{2/3} \Rightarrow \left[\frac{\partial \mu^{(0)}(qN)}{\partial q} \right]_{q=1/2} = \frac{4}{3} \frac{h^2}{2m} \left(\frac{3N}{8\pi V} \right)^{2/3}$$

with λ the thermal wavelength. For spin-1/2 fermions, the Fermi energy is given by:

$$\epsilon_F = \frac{h^2}{2m} \left(\frac{3N}{8\pi V} \right)^{2/3}$$

which, together with the previous result, gives for the magnetic susceptibility per particle at zero temperature:

$$\chi(T = 0) = \frac{3m_0^2}{2\epsilon_F}$$

Keeping the first two terms in the low-temperature expansion of $h_3(z)$, we obtain the next term in the susceptibility per particle:

$$\chi(T) = \frac{3m_0^2}{2\epsilon_F} \left[1 - \frac{\pi^2}{12} \left(\frac{k_B T}{\epsilon_F} \right)^2 \right], \quad \text{low } T \text{ limit}$$

In the high-temperature limit we use the corresponding expansion of $h_3(z)$, which is:

$$h_3(z) = z - \frac{z^2}{2^{3/2}} + \cdots$$

With the lowest term only, we obtain:

$$\mu^{(0)}(N) = k_B T \ln \left(\frac{N}{V} \lambda^3 \right) \Rightarrow \left[\frac{\partial \mu^{(0)}(qN)}{\partial q} \right]_{q=1/2} = 2k_B T$$

which gives for the susceptibility per particle:

$$\chi(T \to \infty) = \frac{m_0^2}{k_B T}$$

Keeping the first two terms in the high-temperature expansion of $h_3(z)$ we obtain the next term in the susceptibility per particle:

$$\chi(T) = \frac{m_0^2}{k_B T} \left[1 - \left(\frac{N\lambda^3}{V 2^{5/2}} \right) \right], \quad \text{high } T \text{ limit}$$

Turning our attention to the electron gas, the high-temperature and low-temperature limits of the susceptibility per unit volume are given by:

$$\chi(T \to \infty) = \frac{g_0^2 \mu_B^2}{k_B T} n, \quad \chi(T = 0) = \frac{3}{2} \frac{g_0^2 \mu_B^2}{\epsilon_F} n$$

with $n = N/V$ the density of particles. We see that in the high-temperature limit the susceptibility exhibits behavior consistent with the Curie law. In this case, the relevant scale for the temperature is the Fermi energy, which gives for the condition at which the Curie law should be observed in the free-electron model:

$$\epsilon_F = \left(\frac{9\pi}{4} \right)^{2/3} \frac{1}{(r_s/a_0)^2} \mathrm{Ry} < k_B T$$

For typical values of r_s in metals, $(r_s/a_0) \sim 2$–6, we find that the temperature must be of order 10^4–10^5 K in order to satisfy this condition, which is too high to be relevant for any real solids. In the opposite extreme, at $T = 0$, it is natural to expect that the only important quantities are those related to the filling of the Fermi sphere, that is, the density n and Fermi energy ϵ_F. Using the relation between the density and the Fermi momentum $k_F^3 = 3\pi^2 n$, which applies to the uniform electron gas in three dimensions, we can rewrite the susceptibility of the free-electron gas at $T = 0$ as:

$$\chi = g_0^2 \mu_B^2 \frac{m_e k_F}{\pi^2 \hbar^2} = g_0^2 \mu_B^2 g_F \tag{10.9}$$

where $g_F = g(\epsilon_F)$ is the density of states evaluated at the Fermi level [recall that the density of states in the case of free fermions of spin 1/2 in three dimensions is given by:

$$g(\epsilon) = \frac{\sqrt{2}}{\pi^2} \left(\frac{m_e}{\hbar^2} \right)^{3/2} \epsilon^{1/2}$$

see Eq. (2.121)]. The expression of Eq. (10.9) is a general form of the susceptibility, useful in several contexts (see, for example, the following discussion on magnetization of band electrons).

In the following, we specialize the discussion to models that provide a more detailed account of the magnetic behavior of metals. The first step in making the free-electron picture more realistic is to include exchange effects explicitly, that is, to invoke a Hartree–Fock picture, but without the added complications imposed by band-structure effects. We next analyze a model that takes into account band-structure effects in an approximate manner. These models are adequate to introduce the important physics of magnetic behavior in metals without going into detailed specialized discussions.

10.2.2 Magnetization in Hartree–Fock Free-Electron Model

In our discussion of the free-electron model in Chapter 1 we assumed that there are equal numbers of up and down spins, and therefore we took the ground state of the system to have non-magnetic character. This is not necessarily the case. We show here that, depending on the density of the electron gas, the ground state can have magnetic character. This arises from considering explicitly the exchange interaction of electrons due to their fermionic nature. The exchange interaction is taken into account by assuming a Slater determinant of free-particle states for the many-body wavefunction, as we did in the Hartree–Fock approximation (see Chapter 4). In that case, after averaging over the spin states, we found for the ground-state energy per particle, Eq. (4.61):

$$\frac{E^{\mathrm{HF}}}{N} = \frac{3}{5}\frac{\hbar^2 k_{\mathrm{F}}^2}{2m_e} - \frac{3e^2}{4\pi}k_{\mathrm{F}} = \left[\frac{3}{5}(k_{\mathrm{F}}a_0)^2 - \frac{3}{2\pi}(k_{\mathrm{F}}a_0)\right] \mathrm{Ry} \qquad (10.10)$$

where k_{F} is the Fermi momentum; in the last expression we have used rydbergs as the unit of energy and the Bohr radius a_0 as the unit of length. In the following discussion we will express all energies in units of rydbergs, and for simplicity we do not include that unit in the equations explicitly. The first term in the total energy E^{HF} is the kinetic energy, while the second represents the effect of the exchange interactions. When the two spin states of the electrons are averaged, the Fermi momentum is related to the total number of fermions N in volume V by:

$$n = \frac{N}{V} = \frac{k_{\mathrm{F}}^3}{3\pi^2}$$

as shown in Chapter 1. By a similar calculation as the one that led to the above result, we find that if we had $N_\uparrow$ ($N_\downarrow$) electrons with spin up (down), occupying states up to Fermi momentum $k_{F\uparrow}$ ($k_{F\downarrow}$), the relation between this Fermi momentum and the number of fermions in volume V would be:

$$N_\uparrow = V\frac{k_{F\uparrow}^3}{6\pi^2}, \quad N_\downarrow = V\frac{k_{F\downarrow}^3}{6\pi^2}$$

Notice that for $N_\uparrow = N_\downarrow - N/2$, we would obtain:

$$k_{F\uparrow} = k_{F\downarrow} = k_{\mathrm{F}}$$

as expected for the non-magnetic case. For each set of electrons with spin up or spin down considered separately, we would have a total energy given by an equation analogous to the spin-averaged case:

$$\frac{E_\uparrow^{\mathrm{HF}}}{N_\uparrow} = \frac{3}{5}(k_{F\uparrow}a_0)^2 - \frac{3}{2\pi}(k_{F\uparrow}a_0), \quad \frac{E_\downarrow^{\mathrm{HF}}}{N_\downarrow} = \frac{3}{5}(k_{F\downarrow}a_0)^2 - \frac{3}{2\pi}(k_{F\downarrow}a_0)$$

The coefficients for the kinetic-energy and exchange-energy terms in the spin-polarized cases are exactly the same as for the spin-averaged case, because the factors of 2 that enter into the calculation of the spin-averaged results are compensated for by the factor of 2 difference in the definition of $k_{F\uparrow}, k_{F\downarrow}$ relative to that of k_{F}.

We next consider a combined system consisting of a total number of N electrons, with $N_\uparrow$ of them in the spin-up state and $N_\downarrow$ in the spin-down state. We define the magnetization number M and the magnetization per particle m as:

$$M = N_\uparrow - N_\downarrow, \quad m = \frac{M}{N} \qquad (10.11)$$

From these definitions and from the fact that $N = N_\uparrow + N_\downarrow$, we find:

$$N_\uparrow = (1+m)\frac{N}{2}, \quad N_\downarrow = (1-m)\frac{N}{2} \qquad (10.12)$$

Using these expressions in the total energies $E_\uparrow^{HF}, E_\downarrow^{HF}$ and employing the definition of the average volume per particle through the variable r_s, the radius of the sphere that encloses this average volume:

$$\frac{4\pi}{3} r_s^3 = \frac{N}{V} \Rightarrow \left(\frac{3\pi^2 N}{V}\right)^{1/3} = (9\pi/4)^{1/3} r_s$$

we obtain for the total energy per particle:

$$\frac{E_\uparrow^{HF} + E_\downarrow^{HF}}{N} = \frac{1.10495}{(r_s a_0)^2}\left[(1+m)^{5/3} + (1-m)^{5/3}\right] - \frac{0.45816}{(r_s a_0)}\left[(1+m)^{4/3} + (1-m)^{4/3}\right]$$

Notice that in the simple approximation we are considering, the free-electron model, there are no other terms involved in the total energy since we neglect the Coulomb interactions and there is no exchange interaction between spin-up and spin-down electrons.

Using this expression for the total energy we must determine what value of the magnetization corresponds to the lowest-energy state for a given value of r_s. From the powers of r_s involved in the two terms of the total energy, it is obvious that for low r_s the kinetic energy dominates while for high r_s the exchange energy dominates. We expect that when exchange dominates it is more likely to get a magnetized ground state. Indeed, if $m \neq 0$ there will be more than half of the particles in one of the two spin states and that will give larger exchange energy than the $m = 0$ case with exactly half of the electrons in each spin state. The extreme case of this is $m = \pm 1$, when all of the electrons are in one spin state. Therefore, we expect that for low enough r_s (high-density limit) the energy will be dominated by the kinetic part and the magnetization will be 0, while for large enough r_s (low-density limit) the energy will be dominated by the exchange part and the magnetization will be $m = \pm 1$. This is exactly what is found by scanning the values of r_s and determining for each value the lowest-energy state as a function of m: the ground state has $m = 0$ for $r_s \leq 5.450$ and $m = \pm 1$ for $r_s > 5.450$, as shown in Fig. 10.1. It is interesting that the transition from the unpolarized to the polarized state is abrupt and complete (from zero polarization, $m = 0$ to full polarization, $m = \pm 1$). It is also interesting that the kinetic-energy and exchange-energy contributions to the total energy are discontinuous at the transition point, but the total energy itself is continuous. The transition point is a local maximum in the total energy, with two local minima of equal energy on either side of it, the one on the left (at $r_s = 4.824$) corresponding to a non-magnetic ground state and the one on the right (at $r_s = 6.077$) to a ferromagnetic ground state. The presence of the two minima of equal energy is indicative of how difficult it would be to stabilize the ferromagnetic ground state based on exchange interactions, as we discuss in more detail next.

While the above discussion is instructive on how exchange interactions alone can lead to non-zero magnetization, this simple model is not adequate to describe realistic systems. There are several reasons for this, having to do with the limitations of both the Hartree–Fock approximation and the uniform electron gas approximation. Firstly, the Hartree–Fock approximation is not realistic because of screening and correlation effects (see the discussion in Chapter 4); attempts to fix the Hartree–Fock approximation destroy

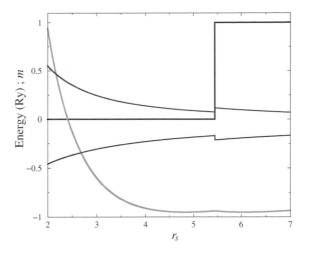

Fig. 10.1 The total energy multiplied by a factor of 10 (green line) and the magnetization m (red line) of the polarized electron gas, as functions of r_s (in units of a_0). The blue lines indicate the contributions of the kinetic part (positive) and the exchange part (negative). For low values of r_s the kinetic energy dominates and the total energy is positive ($r_s \leq 2.411$), while for larger values of r_s the exchange energy dominates and the total energy is negative. Notice that both contributions are discontinuous at $r_s = 5.450$, the transition point at which the magnetization jumps from a value of $m = 0$ to a value of $m = 1$, but the total energy is continuous.

the transition to the non-zero magnetization state. Moreover, there are actually other solutions beyond the uniform density solution which underlies the preceding discussion, that have lower energy; these are referred to as spin density waves. Finally, for very low densities where the state with non-zero magnetization is supposed to be stable, there is a different state in which electrons maximally avoid each other to minimize the total energy: this state is known as the "Wigner crystal," with the electrons localized at crystal sites to optimize the Coulomb repulsion. Thus, it is not surprising that a state with non-zero magnetization arising purely from exchange interactions in a uniform electron gas is not observed in real solids. The model is nevertheless useful in motivating the origin of interactions that can produce ferromagnetic order in metals: the essential idea is that, neglecting all other contributions, the exchange effect can lead to polarization of the spins at low enough density, resulting in a ferromagnetic state.

10.2.3 Magnetization of Band Electrons

We next develop a model, in some sense a direct extension of the preceding analysis, which includes the important aspects of the band structure in magnetic phenomena. The starting point for this theory is the density of states derived from the band structure, $g(\epsilon)$. We assume that the electronic states can be filled up to the Fermi level ϵ_F for a spin-averaged configuration with $N_\uparrow = N_\downarrow = N/2$. We then perturb this state of the system by moving a few electrons from the spin-down state to the spin-up state, so that $N_\uparrow > N/2, N_\downarrow < N/2$. The energy of the electrons that have been transferred to the spin-up state will be slightly higher than the Fermi level, while the highest energy of electrons in the depleted spin state

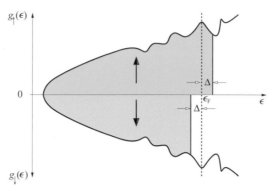

Fig. 10.2 Illustration of the occupation of spin-up electron states $g_\uparrow(\epsilon)$ and spin-down electron states $g_\downarrow(\epsilon)$, with corresponding Fermi levels $\epsilon_F + \Delta$ and $\epsilon_F - \Delta$, where ϵ_F is the Fermi energy for the spin-averaged state. In this example we have chosen $g(\epsilon)$ to have a symmetric local maximum over a range of 2Δ around $\epsilon = \epsilon_F$.

will be somewhat lower than the Fermi energy. We will take the number of electrons with spin up and spin down to be given by:

$$N_\uparrow = \int_{-\infty}^{\epsilon_F+\Delta} g(\epsilon)d\epsilon, \quad N_\downarrow = \int_{-\infty}^{\epsilon_F-\Delta} g(\epsilon)d\epsilon \quad (10.13)$$

where we have introduced Δ to denote the deviation of the highest occupied state in each case from the Fermi level (a positive deviation in the case of spin-up electrons and a negative one in the case of spin-down electrons, as shown in Fig. 10.2). Notice that, in the small Δ limit, in order to conserve the total number of electrons when we transfer some from the spin-down state to the spin-up state, the density of states must be symmetric around its value at ϵ_F, which means that $g(\epsilon)$ is constant in the neighborhood of ϵ_F or it is a local extremum at $\epsilon = \epsilon_F$; in either case, the derivative of the density of states $g'(\epsilon)$ must vanish at $\epsilon = \epsilon_F$.

Expanding the integrals over the density of states as Taylor series in Δ through the expressions given in Eqs (A.24), (A.40) and using the definition of the magnetization number from Eq. (10.11), we obtain:

$$M = 2\Delta g_F \left[1 + \mathcal{O}(\Delta^2)\right] \quad (10.14)$$

where we have used the symbol $g_F = g(\epsilon_F)$ for the density of states at the Fermi level. We can also define the band energy of the system, which in this case can be identified as the Hartree energy E^H, that is, the energy due to electron–electron interactions other than those from exchange and correlation. This contribution comes from summing the energy of electronic levels associated with the two spin states:

$$E^H = E_\uparrow + E_\downarrow = \int_{-\infty}^{\epsilon_F+\Delta} \epsilon g(\epsilon)d\epsilon + \int_{-\infty}^{\epsilon_F-\Delta} \epsilon g(\epsilon)d\epsilon \quad (10.15)$$

Through the same procedure of expanding the integrals as Taylor series in Δ, this expression gives for the band energy:

$$E^{\mathrm{H}} = 2 \int_{-\infty}^{\epsilon_{\mathrm{F}}} \epsilon g(\epsilon) d\epsilon + \Delta^2 \left[g_{\mathrm{F}} + \epsilon_{\mathrm{F}} g'(\epsilon_{\mathrm{F}}) \right] \tag{10.16}$$

where we have kept terms only up to second order in Δ. Consistent with our earlier assumptions about the behavior of the density of states near the Fermi level, we will take $g'(\epsilon_{\mathrm{F}}) = 0$. The integral in the above equation represents the usual band energy for the spin-averaged system, with the factor of 2 coming from the two spin states associated with each electronic state with energy ϵ. The remaining term is the change in the band energy due to spin polarization:

$$\delta E^{\mathrm{H}} = \Delta^2 g_{\mathrm{F}} \tag{10.17}$$

Our goal is to compare this term, which is always positive, to the change in the exchange energy due to spin polarization. We will define the exchange energy in terms of a parameter $-J$, to which we refer as the "exchange integral": this is the contribution to the energy due to exchange of a pair of particles. We will take $J > 0$, that is, we assume that the system gains energy due to exchange interactions, as we proved explicitly for the Hartree–Fock free-electron model in Chapter 4. Since exchange applies only to particles of the same spin, the total contribution to the exchange energy from the spin-up and the spin-down sets of particles will be given by:

$$E^{\mathrm{X}} = -J \left[\frac{N_{\uparrow}(N_{\uparrow} - 1)}{2} + \frac{N_{\downarrow}(N_{\downarrow} - 1)}{2} \right] \tag{10.18}$$

Using the expressions for $N_{\uparrow}$ and $N_{\downarrow}$ in terms of the magnetization, Eqs (10.11) and (10.12), we find that the exchange energy is:

$$E^{\mathrm{X}} = -\frac{J}{2} \left(\frac{N}{2} \right)^2 \left[(1 + m)(1 + m - \frac{2}{N}) + (1 - m)(1 - m - \frac{2}{N}) \right]$$

If we subtract from this expression the value of the exchange energy for $m = 0$, corresponding to the spin-averaged state with $N_{\uparrow} = N_{\downarrow} = N/2$, we find that the change in the exchange energy due to spin polarization is:

$$\delta E^{\mathrm{X}} = -J \frac{M^2}{4} \tag{10.19}$$

which is always negative (recall that $J > 0$). The question then is, under what conditions will the gain in exchange energy due to spin polarization be larger than the cost in band energy, Eq. (10.17)? Using our result for the magnetization from Eq. (10.14), and keeping only terms to second order in Δ, we find that the two changes in the energy become equal for $J = 1/g_{\mathrm{F}}$ and the gain in exchange energy dominates over the cost of band energy for

$$J > \frac{1}{g_{\mathrm{F}}}$$

which is known as the "Stoner criterion" for spontaneous magnetization.[2] This treatment applies to zero temperature; for non-zero temperature, in addition to the density of states in Eq. (10.13), we must also take into account the Fermi occupation numbers (Problem 3).

To complete the discussion, we examine the magnetic susceptibility χ of a system in which spontaneous magnetization arises when the Stoner criterion is met. We will assume that the magnetization develops adiabatically, producing an effective magnetic field H whose change is proportional to the magnetization change

$$dH = \chi^{-1} dM \mu_B$$

where we have also included the Bohr magneton μ_B to give the proper units for the field, and the constant of proportionality is by definition the inverse of the susceptibility. From the preceding discussion we find that the change in total energy due to the spontaneous magnetization is given by:

$$\delta E_{tot} = \delta E^H + \delta E^X = \frac{1 - g_F J}{4 g_F} M^2$$

and is a negative quantity for $J > 1/g_F$ (the system gains energy due to the magnetization). This change can also be calculated from the magnetization in the presence of the induced field H', which ranges from 0 to some final value H when the magnetization M' ranges from 0 to its final value M:

$$\delta E_{tot} = -\int_0^H (M' \mu_B) dH' = -\chi^{-1} \mu_B^2 \int_0^M M' dM' = -\frac{1}{2} \chi^{-1} M^2 \mu_B^2$$

where we have included an overall minus sign to indicate that the spontaneous magnetization lowers the total energy. From the two expressions for the change in total energy we find:

$$\chi = \frac{2 g_F}{|1 - g_F J|} \mu_B^2 \tag{10.20}$$

which should be compared to the Pauli susceptibility, Eq. (10.9): the susceptibility of band electrons is enhanced by the factor $|1 - g_F J|^{-1}$, which involves explicitly the density of states at the Fermi energy g_F and the exchange integral J. For example, the value of this factor for Pd, extracted from measurements of the specific heat, is $|1 - g_F J| \sim 13$ (for more details see Chapter 4, Volume 1 of Jones and March mentioned in Further Reading).

The preceding discussion of magnetic effects in band electrons is based on the assumption of a particularly simple behavior of the band structure near the Fermi level. When these assumptions are not valid, a more elaborate theory which takes into account the realistic band structure is needed.[3]

[2] E .C. Stoner, *Proc. Roy. Soc. A* **165**, 372 (1938); *Proc. Roy. Soc. A* **169**, 339 (1939).
[3] H. Vosko and J. J. Perdew, *Can. J. Phys.* **53**, 1385 (1975); O. Gunnarsson, *J. Phys. F* **6**, 587 (1976); J. F. Janak, *Phys. Rev. B* **16**, 255 (1977).

10.3 Classical Spins: Simple Models on a Lattice

A standard model for physical systems consisting of many atomic-scale particles with magnetic moment, which can respond to an external magnetic field, is that of spins on a lattice. The value of the spin depends on the constituents of the physical system. The simplest models deal with spins that can point either in the direction along the external field (they are aligned with or parallel to the field) or in the opposite direction (anti-aligned or antiparallel). In terms of a microscopic description of the system, this corresponds to spin-1/2 particles with their z component taken along the direction of the field, and having as possible values $s_z = +1/2$ (spins aligned with the field) or $s_z = -1/2$ (spins pointing in the direction opposite to the field). The interaction between the spins and the field can be dominant, with spin–spin interactions vanishingly small, or the interaction among the spins can be dominant, with the spin–external field interaction of minor importance. We discuss these two extreme cases, and give examples of their physical realization.

10.3.1 Non-interacting Spins on a Lattice: Negative Temperature

We consider the following model: a number N of magnetic dipoles is subject to an external magnetic field H. The dipoles have magnetic moment m_0 and are taken to be oriented either along or against the magnetic field. Moreover, we assume that the dipoles do not interact among themselves and are distinguishable, by virtue of being situated at fixed positions in space, for example at the sites of a crystalline lattice. Since each dipole can be oriented either parallel or antiparallel to the external field, its energy can have only two values. We assign to the lower energy of the two states (which by convention corresponds to the dipole pointing along the external field) the value $-\epsilon = -m_0 H$ and to the higher energy of the two states the value $\epsilon = m_0 H$.

  We assume that in a given configuration of the system there are n dipoles oriented along the direction of the field and therefore $N - n$ dipoles are pointing against the direction of the field. The total internal energy of the system $\mathcal{E}$ and its magnetization $\mathcal{M}$ are given by:

$$\mathcal{E} = -n\epsilon + (N - n)\epsilon = (N - 2n)m_0 H$$
$$\mathcal{M} = nm_0 - (N - n)m_0 = -(N - 2n)m_0 \qquad (10.21)$$

with the direction of the magnetic field H taken to define the positive direction of magnetization (for instance, the $+\hat{\mathbf{z}}$ axis). The partition function for this system in the canonical ensemble is given by:

$$Z_N = \sum_{n, N-n} \frac{N!}{n! \, (N - n)!} e^{-\beta(-n\epsilon + (N-n)\epsilon)}$$

$$= \sum_{n=0}^{N} \frac{(N - n + 1)(N - n + 2) \cdots N}{n!} \left(e^{-\beta\epsilon}\right)^n \left(e^{\beta\epsilon}\right)^{N-n} \qquad (10.22)$$

where the first sum implies summation over all possible values of n and $N - n$, which is equivalent to summing over all values of n from 0 to N. In the last expression we recognize the familiar binomial expansion [see Appendix A, Eq. (A.39)], so with $a = \exp[\beta\epsilon], b = \exp[-\beta\epsilon], p = N$ we obtain:

$$Z_N = (e^{\beta\epsilon} + e^{-\beta\epsilon})^N = (2\cosh(\beta\epsilon))^N \tag{10.23}$$

Having calculated the partition function, we can use it to obtain the thermodynamics of the system with the standard expressions we have derived in the context of ensemble theory. We first calculate the free energy F from the usual definition:

$$F = -k_B T \ln Z_N = -N k_B T \ln\left(2\cosh(\beta\epsilon)\right) \tag{10.24}$$

and from that the entropy through

$$S = -\left(\frac{\partial F}{\partial T}\right)_H = N k_B \left[\ln\left(2\cosh(\beta\epsilon)\right) - \beta\epsilon\tanh(\beta\epsilon)\right] \tag{10.25}$$

We can use the calculated free energy and entropy to obtain the average internal energy E through

$$E = F + TS = -N\epsilon\tanh(\beta\epsilon) \tag{10.26}$$

The same result is obtained by calculating the average of $\mathcal{E}$ with the use of the partition function

$$E = \langle\mathcal{E}\rangle = \frac{1}{Z_N}\sum_{n,N-n}\frac{N!}{n!(N-n)!}(-n\epsilon + (N-n)\epsilon)e^{-\beta(-n\epsilon+(N-n)\epsilon)}$$

$$= -\left(\frac{\partial}{\partial\beta}\right)_H \ln Z_N = -N\epsilon\tanh(\beta\epsilon) \tag{10.27}$$

By the same method we can calculate the average magnetization M as the average of $\mathcal{M}$:

$$M = \langle\mathcal{M}\rangle = \frac{1}{Z_N}\sum_{n,N-n}\frac{N!}{n!(N-n)!}(nm_0 - (N-n)m_0)e^{-\beta(-nm_0H+(N-n)m_0H)}$$

$$= \frac{1}{\beta}\left(\frac{\partial}{\partial H}\right)_T \ln Z_N = Nm_0\tanh(\beta\epsilon) \tag{10.28}$$

Notice that the last equality can also be written in terms of a partial derivative of the free energy defined in Eq. (10.24), as:

$$M = \frac{1}{\beta}\left(\frac{\partial\ln Z_N}{\partial H}\right)_T = k_B T\left(\frac{\partial\ln Z_N}{\partial H}\right)_T = -\left(\frac{\partial F}{\partial H}\right)_T \tag{10.29}$$

We can now determine the relation between average magnetization and internal energy, using $\epsilon = m_0 H$:

$$E = -N\epsilon\tanh(\beta\epsilon) = -Nm_0 H\tanh(\beta\epsilon) = -HM \tag{10.30}$$

which is what we would expect, because of the relation between the internal energy and magnetization $\mathcal{E} = -H\mathcal{M}$, as is evident from Eq. (10.21).

If we desired to complete the description of the system in terms of thermodynamic potentials and their derivatives, then the above results imply the following relations:

$$E = -MH \Rightarrow dE = -MdH - HdM$$

$$F = E - TS \Rightarrow dF = dE - TdS - SdT = -MdH - HdM - TdS - SdT$$

We have also shown that

$$M = -\left(\frac{\partial F}{\partial H}\right)_T, \quad S = -\left(\frac{\partial F}{\partial T}\right)_H$$

which together require that

$$dF = -MdH - SdT \Rightarrow -HdM - TdS = 0 \Rightarrow TdS = -HdM$$

This last result tells us that the usual expression for the change in internal energy as the difference between the heat absorbed and the work done by the system must be expressed as:

$$dE = TdS - MdH$$

which compels us to identify H as the equivalent of the volume of the system and M as the equivalent of the pressure, so that MdH becomes the equivalent of the mechanical work PdV done by the system. This equivalence is counterintuitive, because the magnetic field is the intensive variable and the magnetization is the extensive variable, so the reverse equivalence (H with P and M with V) would seem more natural. What is peculiar about this system is that its internal energy is directly related to the work done on the system $E = -MH$, while in the standard example of the ideal gas we have seen that the internal energy is a function of the temperature alone, $E = C_V T$, and the work done on the system is an independent term PV. Although it is counterintuitive, upon some reflection the correspondence between the magnetic field and volume and between the magnetization and pressure appears more reasonable: the magnetic field is the actual constraint that we impose on the system, just like confining a gas of atoms within the volume of a container; the magnetization is the response of the system to the external constraint, just like the exertion of pressure on the walls of the container is the response of the gas to its confinement within the volume of the container. An additional clue that H should be identified with the volume in this case is the fact that the partition function depends only on N and H, the two variables that we consider fixed in the canonical ensemble; in general, the canonical ensemble consists of fixing the number of particles and the volume of the system. Taking these analogies one step further, we might wish to define the Gibbs free energy for this system as:

$$G = F + HM = E - TS + HM = -TS$$

which in turn gives:

$$dG = -SdT - TdS = -SdT + HdM$$

and from this we can obtain the entropy and the magnetic field as partial derivatives of G with respect to T and M:

$$S = \left(\frac{\partial G}{\partial T}\right)_M, \quad H = \left(\frac{\partial G}{\partial M}\right)_T$$

which completes the description of the system in terms of thermodynamic potentials, fields, and variables. As a final step, we can calculate the specific heat at constant magnetic field which, as we argued above, should correspond to the specific heat at constant volume in the general case:

$$C_H = \left(\frac{\partial E}{\partial T}\right)_H = \frac{\partial \beta}{\partial T}\left(\frac{\partial E}{\partial T}\right)_H = Nk_B\beta^2\epsilon^2\frac{1}{\cosh^2(\beta\epsilon)}$$

In order to discuss the behavior of this system in more detail, it is convenient to define the variable

$$u = \frac{E}{N\epsilon}$$

and express the entropy S and the inverse temperature multiplied by the energy per dipole, that is, $\beta\epsilon$, as functions of u. We find from Eqs (10.25) and (10.26):

$$\beta\epsilon = \tanh^{-1}(-u) = \frac{1}{2}\ln\left(\frac{1-u}{1+u}\right)$$

$$\frac{S}{Nk_B} = \beta\epsilon u + \ln 2 - \frac{1}{2}\ln\left(1 - u^2\right)$$

$$= \ln 2 - \frac{1}{2}[(1-u)\ln(1-u) + (1+u)\ln(1+u)]$$

It is obvious from its definition and from the fact that the lowest and highest possible values for the internal energy are $-N\epsilon$ and $N\epsilon$, that u lies in the interval $u \in [-1, 1]$. The plots of $\beta\epsilon$ and S/Nk_B as functions of u are given in Fig. 10.3. These plots suggest the intriguing possibility that in this system it is possible to observe *negative* temperatures, while the entropy has the unusual behavior that it rises and then falls with increasing u, taking its

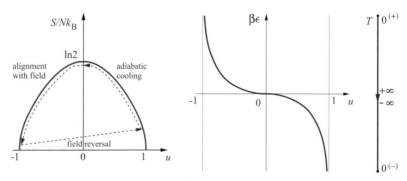

Fig. 10.3 The entropy per dipole S/Nk_B and the inverse temperature $\beta\epsilon$ as functions of $u = E/N\epsilon$ for the model of dipoles in external magnetic field H. The arrow on the temperature scale points from lower to higher values.

maximum value for $u = 0$. Let us elaborate on this, at first sight perplexing, behavior. We consider the behavior of the entropy first. Near the extreme values of u we have:

$$u \to -1^{(+)} \Rightarrow S = 0^{(+)}, \quad u \to +1^{(-)} \Rightarrow S = 0^{(+)}$$

where the $(-)$ and $(+)$ superscripts denote approach from below or from above, respectively. This is reassuring, because it is at least consistent with the third law of thermodynamics: the entropy goes to zero from above, when the energy of the system reaches its lowest or its highest value. Moreover, in the limit of very high temperature, or very low external field, u must be zero: in either limit, the natural state of the system is to have half of the dipoles oriented along the direction of the field and the other half against it, giving a total energy equal to zero. This is also the state of maximum entropy, as we would expect. Indeed, in this state $n = N/2 \Rightarrow N - n = N/2$ and the number of configurations associated with it is:

$$\frac{N!}{n! \, (N - n)!} = \frac{N!}{(N/2)! \, (N/2)!}$$

which is larger than for any other value of n and $N - n$ different from $N/2$. Using Stirling's formula we find that, in the microcanonical ensemble at fixed energy $E = 0$, the entropy of this state is:

$$S = k_B \ln \left[\frac{N!}{(N/2)! \, (N/2)!} \right] \approx k_B \left[N \ln N - N - 2 \left(\frac{N}{2} \ln \left(\frac{N}{2} \right) - \frac{N}{2} \right) \right] = N k_B \ln 2$$

as expected. From this state, which also has zero magnetization, the system can be brought to a state of higher magnetization or lower energy, either by increasing the magnetic field or by lowering the temperature. In either case it will be moving from the maximum of the entropy value at $u = 0$ toward lower values of the entropy in the range of $u \in [-1, 0]$, since the value of u is decreasing as the energy of the system decreases. This is consistent with the second law of thermodynamics, with the entropy decreasing because work is being done on the system in order to align the dipoles to the external field.

As far as the temperature is concerned, we have:

$$\text{for } u \to -1^{(+)} : \beta \epsilon = \frac{1}{2} \ln \left(\frac{2^{(-)}}{0^{(+)}} \right) \to +\infty \Rightarrow T \to 0^{(+)}, \quad T > 0$$

$$\text{for } u \to +1^{(-)} : \beta \epsilon = \frac{1}{2} \ln \left(\frac{0^{(+)}}{2^{(-)}} \right) \to -\infty \Rightarrow T \to 0^{(-)}, \quad T < 0$$

that is, for $-1 \leq u \leq 0$ the temperature is *positive*, ranging from $0^{(+)}$ to $+\infty$, while for $0 \leq u \leq +1$ the temperature is *negative*, ranging from $-\infty$ to $0^{(-)}$. The question now is, what is the meaning of a negative temperature? In the microcanonical ensemble picture, the relation between temperature and entropy is given by:

$$\frac{1}{T} = \frac{\partial S}{\partial E}$$

If the system is in the state of maximum entropy discussed above, for which $E = 0$, increasing the energy by a small amount will decrease the entropy and will produce a large negative temperature. Indeed, if an infinitesimal excess of dipoles point in the direction against the external magnetic field, which corresponds to an increase in the energy ($\delta E > 0$, $u > 0$), the number of states available to the system is smaller than in the case of zero energy, and the accompanying infinitesimal change in the entropy ($\delta S < 0$) will yield a temperature $T \to -\infty$. Since this situation involves an increase of the energy, it must be interpreted as corresponding to a temperature *higher* than $T \to +\infty$, which represented the case of maximum entropy and zero energy:

$$T \to +\infty : \quad S = Nk_B \ln 2, \quad E = 0$$

As the energy of the system increases further, $u \in [0, 1]$, the temperature will continue to *increase*, from $-\infty$ to $0^{(-)}$. Thus, the proper way to interpret the temperature scale is by considering $T \to 0^{(+)}$ to lie at the beginning of the scale (lowest values), $T \to -\infty$ to lie in the middle of the scale and be immediately followed by $T \to +\infty$, and $T \to 0^{(-)}$ to be at the end of the scale (highest values), as illustrated in Fig. 10.3. The reason for this seemingly peculiar scale is that the natural variable is actually the inverse temperature $\beta = 1/k_B T$, which on the same scale is monotonically decreasing: it starts at $+\infty$ and goes to 0 from positive values, then continues to decrease by assuming negative values just below 0 and going all the way to $-\infty$.

Finally, we note that even though the situation described above may seem strange, it can actually be realized in physical systems. This was demonstrated first by Purcell and Pound for nuclear spins (represented by the dipoles in our model) in an insulator,[4] and more recently for nuclear spins in metals.[5] The way these experiments work is to first induce a large magnetization to the system by aligning the nuclear spins with the external field and then to suddenly reverse the external field direction, as indicated in Fig. 10.3. When the direction of the field is reversed, the system is in the range of negative temperatures and in a state of very high energy and low entropy. The success of these experiments relies on weak interactions between the dipoles, so that the sudden reversal of the external field does not change their state; in the simple model discussed here, the interactions between the dipoles are neglected altogether. Once the external field has been reversed, the system can be allowed to cool adiabatically, along the negative temperature scale from $0^{(-)}$ to $-\infty$. When the temperature has cooled to $-\infty$ the system is in the zero-energy, zero-magnetization, maximum-entropy state, from which it can again be magnetized by applying an external magnetic field. In all stages of the process, the second law of thermodynamics is obeyed, with entropy increasing when no work is done on the system.

10.3.2 Interacting Spins on a Lattice: Ising Model

The Ising model is an important paradigm because it captures the physics of several physical systems with short-range interactions. The standard is a system of spins on a lattice with nearest-neighbor interactions. The Ising model is also relevant to the lattice gas model and the binary alloy model. The lattice gas is a system of particles that can be situated at discrete positions only, but exhibit otherwise the behavior of a gas in the sense

---

[4] E. M. Purcell and R. V. Pound, *Phys. Rev.* **81**, 279 (1951).
[5] P. Hakonen and O. V. Lounasmaa, *Science* **265**, 1821 (1994).

that they move freely between positions which are not already occupied by other particles; each particle experiences the presence of other particles only if they are nearest neighbors at the discrete sites which they can occupy, which can be taken as points on a lattice. The binary alloy model assumes that there exists a regular lattice whose sites are occupied by one of two types of atoms, and the interactions between atoms at nearest-neighbor sites depend on the type of atoms. Both models can be formally mapped onto the Ising model.

The Ising model in its simplest form is described by the hamiltonian

$$\hat{\mathcal{H}} = -J \sum_{\langle ij \rangle} s_i s_j - H \sum_i s_i \tag{10.31}$$

where J and H are constants, and the s_is are spins situated at regular lattice sites. The notation $\langle ij \rangle$ implies summation over nearest-neighbor pairs only. The constant J, called the coupling constant, represents the interaction between nearest neighbors. If J is positive it tends to make nearest-neighbor spins parallel, that is, $s_i s_j = |s_i||s_j|$ because this gives the lowest-energy contribution and thus corresponds to ferromagnetic behavior. If J is negative, the lowest-energy contribution comes from configurations in which nearest neighbors are antiparallel, that is, $s_i s_j = -|s_i||s_j|$ and this corresponds to antiferromagnetic behavior. H represents an external field which applies to all spins on the lattice; the second term in the hamiltonian tends to align spins parallel to the field, since this lowers its contribution to the energy. The values of s_i depend on the spin of the particles. For spin-1/2 particles, $s_i = \pm(1/2)$ but a factor of $(1/4)$ can be incorporated in the value of the coupling constant and a factor of $(1/2)$ in the value of the external field, leaving as the relevant values $s_i = \pm 1$. Similarly, for spin-1 particles, $s_i = \pm 1, 0$. The simple version of the model described above can be extended to considerably more complicated forms. For instance, the coupling constant may depend on the sites it couples, in which case J_{ij} assumes a different value for each pair of indices ij. The range of interactions can be extended beyond nearest neighbors. The external field can be non-homogeneous, that is, it can have different values at different sites, H_i. Finally, the lattice on which the spins are assumed to reside plays an important role: the dimensionality of the lattice and its connectivity (number of nearest neighbors of a site) are essential features of the model.

In the present discussion we concentrate on the simplest version of the model, with positive coupling constant J between nearest neighbors only, and homogeneous external field H. We also consider spins that take values $s_i = \pm 1$, that is, we deal with spin-1/2 particles. Our objective is to determine the stable phases of the system in the thermodynamic limit, that is, when the number of spins $N \to \infty$ and the external field $H \to 0$. In this sense, the physics of the model is dominated by the interactions between nearest neighbors rather than by the presence of the external field; the latter only breaks the symmetry between spins pointing up (along the field) or down (opposite to the field). The phases of the system are characterized by the average value M of the total magnetization $\mathcal{M}$, at a given temperature:

$$M(T) = \langle \mathcal{M} \rangle, \quad \mathcal{M} = \sum_i s_i$$

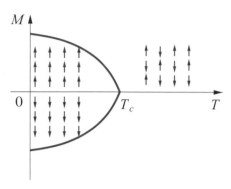

Fig. 10.4 The average magnetization M as a function of temperature for the Ising model. The sets of up–down arrows illustrate the state of the system in the different regions of temperature.

From the expression for the free energy of the system at temperature T:

$$F = E - TS = \langle \hat{\mathcal{H}} \rangle - TS$$

we can argue that there are two possibilities. First, the stable phase of the system has average magnetization zero, which we call the disordered phase; this phase is favored by the entropy term. Second, the stable phase of the system has non-zero average magnetization, which we call the ordered phase; this phase is favored by the internal energy term. The behavior of the system as a function of temperature reflects a competition between these two terms. At low enough temperature we expect the internal energy to win, and the system to be ordered; at zero temperature the obvious stable phase is one with all the spins parallel to each other. At high enough temperature we expect the entropy term to win, and the system to be disordered. There are then two important questions:

 (i) Is there a transition between the two phases at non-zero temperature?
(ii) If there is a transition, what is its character?

If there is a transition at non-zero temperature, we would expect the magnetization to behave in the manner illustrated in Fig. 10.4: at the critical temperature T_c order starts to develop as reflected in the magnetization acquiring a non-zero value, which increases steadily as the temperature is lowered and assumes its highest value at $T = 0$, when all spins have been aligned. The role of the weak external magnetic field, $H \to 0$, is to choose one of the two possible orientations for alignment of the spins, up (positive magnetization) or down (negative magnetization); otherwise it does not affect the behavior of the system.

In one dimension, the nearest-neighbor Ising model does not exhibit a phase transition at non-zero temperature. This can be established by a simple argument: consider the lowest energy configuration in which all spins point in the same direction. We choose some arbitrary point between a pair of spins in the chain of N spins (with $N \to \infty$) and flip all the spins to the right; we call this point a domain wall since it splits the chain into two domains with all spins parallel within each domain. The change in the free energy introduced by the domain wall is given by:

$$\Delta F = \Delta E - T\Delta S = 2J - Tk_B \ln(N - 1)$$

where the first term comes from the change in internal energy and the second from the change in the entropy: the entropy of the initial configuration is zero, because there is only one state with all spins pointing up; the entropy of a configuration with all the spins flipped to the right is $k_B \ln(N - 1)$, because there are $N - 1$ choices of where the domain wall can be introduced. It is evident that for any non-zero temperature, in the thermodynamic limit the entropy term wins. Since this argument can be repeated for each domain, the stable phase of the system cannot be ordered because the introduction of domain walls always lowers the free energy due to the entropy term. For the 1D system it is also possible to find an analytical solution to the problem, which again shows the absence of a phase transition for non-zero temperature (see Problem 4).

In two or higher dimensions, the Ising model does exhibit a phase transition at non-zero temperature. The existence of a phase transition at non-zero temperature was first established by Peierls. The problem has been solved analytically by L. Onsager[6] in two dimensions, but remains unsolved in more than two dimensions. In the latter case one resorts to approximate or numerical methods to obtain a solution.

An instructive approximate method for finding the critical temperature is the so-called mean field (also known as the Bragg–Williams) approximation. Since this is a general approach that can be applied to many situations, we discuss it in some detail. The starting point is to consider a system of N spins on a lattice with q nearest neighbors per site, and assume that it has average magnetization M, the result of $N_\uparrow$ spins pointing up and $N_\downarrow$ spins pointing down. Our goal is to analyze this system with the usual tools of statistical mechanics, that is, by calculating the partition function in the canonical ensemble, since we are dealing with a fixed number of particles. We introduce the average magnetization per spin $m = M/N$ and express the various quantities that appear in the partition function in terms of m and N:

$$m = \frac{M}{N} = \frac{N_\uparrow - N_\downarrow}{N} \text{ and } N_\uparrow + N_\downarrow = N \Rightarrow N_\uparrow = \frac{N}{2}(1 + m), \quad N_\downarrow = \frac{N}{2}(1 - m)$$

The energy of the configuration with average magnetization per spin m is given by:

$$\mathcal{E}(m) = -\left(\frac{q}{2}Jm^2 + Hm\right)N$$

where the first term is the contribution, on average, of the $N/2$ pairs of spins. The degeneracy of this configuration is:

$$Q(m) = \frac{N!}{N_\uparrow! N_\downarrow!} = \frac{N!}{(N(1 + m)/2)!\,(N(1 - m)/2)!}$$

Using these results we can express the canonical partition function as:

$$Z_N = \sum_m Q(m) e^{-\beta \mathcal{E}(m)} \tag{10.32}$$

---

[6] L. Onsager, *Phil. Mag.* **43**, 1006 (1952).

This sum, with m ranging from 0 to 1 in increments of $2/N$, is difficult to evaluate, so we replace it by the largest term and assume that its contribution is dominant.[7] We can then use Stirling's formula, Eq. (D.32), to obtain:

$$\frac{1}{N}\ln Z_N = \beta\left(\frac{q}{2}J(m^*)^2 + Hm^*\right) - \frac{1+m^*}{2}\ln\left(\frac{1+m^*}{2}\right) - \frac{1-m^*}{2}\ln\left(\frac{1-m^*}{2}\right)$$

where we have used m^* to denote the value of the average magnetization per spin that gives the largest term in the sum of the canonical partition function. In order to determine the value of m^* which maximizes the summand in Eq. (10.32), we take its derivative with respect to m and evaluate it at $m = m^*$. It is actually more convenient to deal with the logarithm of this term, since the logarithm is a monotonic function; therefore we apply this procedure to the logarithm of the summand and obtain

$$m^* = \tanh\left[\beta(qJm^* + H)\right] \tag{10.33}$$

This equation can be solved graphically by finding the intersection of the function $y_1(m) = \tanh(\beta qJm)$ with the straight line $y_0(m) = m$, as shown in Fig. 10.5. For high enough T (small enough β), the equation has only one root, $m^* = 0$, which corresponds to the disordered phase. For low enough T (large enough β), the equation has three roots, $m^* = 0, \pm m_0$. The value of the temperature that separates the two cases, called the critical temperature, can be determined by requiring that the slopes of the two functions $y_1(m)$ and $y_0(m)$ are equal at the origin:

$$\frac{d}{dm}\tanh(\beta qJm) = 1 \Rightarrow k_BT_c = Jq \tag{10.34}$$

This, of course, is an approximate result, based on the mean-field approach; for instance, for the 2D square lattice this result gives $k_BT_c = 4J$, whereas the exact solution due to Onsager gives $k_BT_c = 2.27J$ (see footnote 6). Of the three roots of Eq. (10.33) for $T < T_c$, we must choose the one that minimizes the free energy per particle F_N/N, obtained in the usual way from the partition function:

$$\frac{F_N}{N} = -\frac{1}{\beta N}\ln Z_N = \frac{q}{2}J(m^*)^2 + \frac{1}{2\beta}\ln\left[1 - (m^*)^2\right] - \frac{1}{\beta}\ln 2$$

where we have used the fact that for the solutions of Eq. (10.33):

$$\beta qJm^* + H = \frac{1}{2}\ln\left(\frac{1+m^*}{1-m^*}\right)$$

which can easily be proved from the definition of $\tanh(x) = (e^x - e^{-x})/(e^x + e^{-x})$. From the expression of the free energy we then find that it is minimized for $m^* = \pm m_0$ when Eq. (10.33) admits three roots, with $T < T_c$. This can be shown by the following considerations: with the help of Eqs (10.33) and (10.34), we can express the free energy per particle, in units of k_BT_c, as:

---

[7] This is an argument often invoked in dealing with partition functions; it is based on the fact that in the thermodynamic limit $N \to \infty$ the most favorable configuration of the system has a large weight, which falls off extremely rapidly for other configurations.

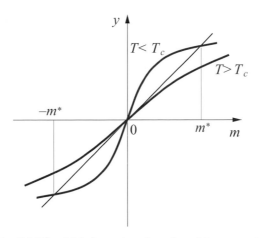

Fig. 10.5   Graphical solution of Eq. (10.33), which determines the value of the magnetization m^* for $T > T_c$ (red line) and $T < T_c$ (blue line).

$$\frac{F_N}{Nk_BT_c} = \frac{m^2}{2} - \left(\frac{T}{T_c}\right) \ln\left[\cosh\left(T_c/Tm\right)\right] - \left(\frac{T}{T_c}\right) \ln 2 \qquad (10.35)$$

where we have used m for the average magnetization per particle for simplicity. A plot of $F_N/(Nk_BT_c)$ as a function of m for various values of T/T_c is shown in Fig. 10.6. For $T > T_c$ the free energy has only one minimum at $m^* = 0$. For $T < T_c$ the free energy has two minima at $m^* = \pm m_0$, as expected. With decreasing T the value of m_0 approaches 1, and it becomes exactly one for $T - 0$. At $T - T_c$ the curve is very flat near $m = 0$, signifying the onset of the phase transition. The magnetization m plays the role of the *order parameter*, which determines the nature of the phases above and below the critical temperature. This type of behavior of the free energy as a function of the order parameter is common in second-order phase transitions, as illustrated in Fig. 10.6. The point at which the free energy changes behavior from having a single minimum to having two minima can be used to determine the critical temperature. An analysis of the phase transition in terms of the behavior of the free energy as a function of the order parameter is referred to as the Ginzburg–Landau theory. The presence of the external magnetic field H would break the symmetry of the free energy with respect to $m \rightarrow -m$, and would make one of the two minima at $\pm m_0$ lower than the other. Thus, the weak external magnetic field $H \rightarrow 0$ serves to pick one of the two equivalent ground states of the system.

It remains to elucidate the character of the transition. This is usually done through a set of *critical exponents* which describe the behavior of various physical quantities close to the critical temperature. These quantities are expressed as powers of $|T - T_c|$ for $T \rightarrow T_c$. Some examples of critical exponents are:

$$\text{order parameter} : m \sim |T_c - T|^\alpha$$

$$\text{susceptibility} : \chi = \frac{\partial m}{\partial H} \sim |T_c - T|^{-\gamma}$$

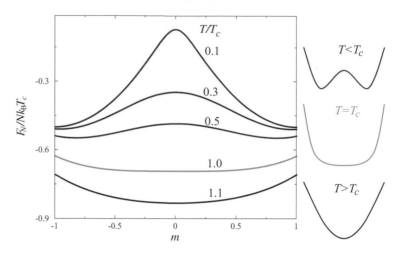

The free energy per particle F_N/N in units of $k_B T_c$, as a function of the average magnetization per particle m, for various values of T/T_c, from Eq. (10.35). For $T > T_c$ there is only one minimum at $m = 0$, while for $T < T_c$ there are two symmetric minima at $\pm m_0$. The value of m_0 approaches 1 as T approaches 0. Notice the very flat behavior of the free energy for $T = T_c$ near $m = 0$, signifying the onset of the phase transition. The diagram on the right illustrates the general behavior of the free energy as a function of the order parameter for $T < T_c$, $T = T_c$, and $T > T_c$.

We calculate the values of α and γ for the Ising model in the mean-field approximation. From Eqs (10.33) and (10.34) we obtain the following expression for the value of the average magnetization at temperature T below T_c, in the limit $H \to 0$:

$$m_0 = \frac{\exp\left[(T_c/T)\, m_0\right] - \exp\left[-(T_c/T)\, m_0\right]}{\exp\left[(T_c/T)\, m_0\right] + \exp\left[-(T_c/T)\, m_0\right]}$$

Near $T = T_c$ the value of m_0 is small, so we can use the lowest terms in the Taylor expansions of the exponentials to obtain:

$$m_0 = \frac{(T_c/T)\, m_0 + \frac{1}{6}\, (T_c/T)^3\, m_0^3}{1 + \frac{1}{2}\, (T_c/T)^2\, m_0^2}$$

which, with the definition $t = (T - T_c)/T$ gives:

$$m_0^2 = 2t\left(-\frac{t}{3} - \frac{2}{3}\right)^{-1}(t + 1)^2$$

Now we are only interested in the behavior of m_0 very close to the critical point, so we keep only the lowest-order term of t in the above expression to obtain:

$$m_0^2 \sim -3t = 3\frac{T_c - T}{T} \Rightarrow \alpha = \frac{1}{2}$$

For the calculation of γ we keep the external magnetic field, which is also a small quantity, in Eq. (10.33) to find:

$$m_0 = \frac{((T_c/T)m_0 + (H/k_B T)) + \frac{1}{6}((T_c/T)m_0 + (H/k_B T))^3}{1 + \frac{1}{2}((T_c/T)m_0 + (H/k_B T))^2}$$

Differentiating this expression with respect to H and taking the limit $H \to 0$, we find:

$$\chi = \frac{dm_0}{dH} = \left(\frac{T_c}{T}\chi + \frac{1}{k_B T}\right)\left[1 - \left(\frac{T_c}{T}m_0\right) + \frac{1}{2}\left(\frac{T_c}{T}m_0\right)^2\right]$$

where we have also kept only lowest-order terms in m_0. We can use $m_0 \sim t^{1/2}$ from our previous result to substitute in the above equation, which is then solved for χ to produce to lowest order in $T - T_c$:

$$\chi \sim \frac{1}{k_B(T - T_c)} \Rightarrow \gamma = 1.$$

The study of critical exponents for various models is an interesting subfield of statistical mechanics. Models that are at first sight different but have the same set of critical exponents are referred to as belonging to the same *universality class*, which implies that their physics is essentially the same.

10.4 Quantum Spins: Heisenberg Model

A model that not only captures the physics of quantum-mechanical interacting spins but is also of broader use in describing many-particle physics is the Heisenberg spin model. In this model, the spins are represented as $\mathbf{S}(\mathbf{R})$, with $\mathbf{S}$ denoting a quantum-mechanical spin variable and $\mathbf{R}$ the lattice site where it resides. A pair of spins at lattice sites $\mathbf{R}$ and $\mathbf{R}'$ interact by the "exchange integral," denoted by J, which depends on the relative distance $\mathbf{R} - \mathbf{R}'$ between the spins. The hamiltonian of the Heisenberg spin model is:

$$\hat{\mathcal{H}}_H = -\frac{1}{2}\sum_{\mathbf{R}\neq\mathbf{R}'} J(\mathbf{R} - \mathbf{R}')\ \mathbf{S}(\mathbf{R})\cdot\mathbf{S}(\mathbf{R}'), \quad J(\mathbf{R} - \mathbf{R}') = J(\mathbf{R}' - \mathbf{R}) \qquad (10.36)$$

If the exchange integral J is positive, then the model describes ferromagnetic order because the spins will tend to be oriented in the same direction to give a positive value for $\langle\mathbf{S}(\mathbf{R})\cdot\mathbf{S}(\mathbf{R}')\rangle$ as this minimizes the energy. In the opposite case, a negative value for J will lead to antiferromagnetic order, where nearest-neighbor spins will tend to be oriented in opposite directions. We first provide some motivation for defining this model hamiltonian, and then discuss the physics of two important cases, the ferromagnetic case, $J(\mathbf{R} - \mathbf{R}') > 0$, and the antiferromagnetic case, $J(\mathbf{R} - \mathbf{R}') < 0$.

10.4.1 Motivation of the Heisenberg Model

The simplest justification for considering a model like the Heisenberg spin model to represent the physics of a system of electrons, including explicitly their spin degrees of freedom, comes from the model of the hydrogen molecule discussed in Section 4.2. A different way to present the same ideas, in the context of a solid, is based on the Hubbard model. The two approaches are evidently related.

From a Two-Particle Problem: Hydrogen Molecule

In solving the hydrogen molecule, we had found that the ground state of the two-particle system is a spin singlet with energy

$$E_0^{(s)} = 2\epsilon + 2t\left[\lambda - \sqrt{\lambda^2 + 1}\right] = 2\left(\epsilon - t\sqrt{\lambda^2 + 1}\right) + \frac{U}{2}$$

and the first excited state of the two-particle system is a spin triplet with energy

$$E^{(t)} = 2\epsilon$$

where ϵ is the on-site energy of the single-particle states $|\phi_1\rangle, |\phi_2\rangle$, t is the hopping matrix element between different sites:

$$-\langle \phi_1|\hat{\mathcal{H}}^{sp}|\phi_2\rangle = -\langle \phi_2|\hat{\mathcal{H}}^{sp}|\phi_1\rangle$$

and U is the on-site Coulomb repulsion energy between two electrons:

$$U \equiv \langle \phi_i(\mathbf{r}_1)\phi_i(\mathbf{r}_2)| \frac{e^2}{|\mathbf{r}_1 - \mathbf{r}_2|} |\phi_i(\mathbf{r}_1)\phi_i(\mathbf{r}_2)\rangle, \quad i = 1, 2$$

that is, the energy cost of placing both electrons in the same single-particle state. We had found that there are also two more two-particle states, both spin singlets, with higher energies, given by:

$$E_1^{(s)} = 2\epsilon + U \quad \text{and} \quad E_2^{(s)} = 2\epsilon + 2t\left[\lambda + \sqrt{\lambda^2 + 1}\right] = 2\left(\epsilon + t\sqrt{\lambda^2 + 1}\right) + \frac{U}{2}$$

In the above expressions, $\lambda = U/4t$. We will work in the limit $U \gg t$ or equivalently $\lambda \gg 1$, which is the limit of strongly correlated electronic states. In this limit we obtain:

$$E_0^{(s)} \approx 2\epsilon - \frac{4t^2}{U}, \quad E_2^{(s)} \approx 2\epsilon + \frac{4t^2}{U} + 2U$$

and the sequence of energy levels for the two-particle states is:

$$E_0^{(s)} \approx \left(2\epsilon - \frac{4t^2}{U}\right) < E^{(t)} = (2\epsilon) \ll E_1^{(s)} = (2\epsilon + 2U) < E_2^{(s)} \approx \left(2\epsilon + 2U + \frac{4t^2}{U}\right)$$

Therefore, we can keep only the two lowest-energy two-particle states, the ground state being the spin singlet and the first excited state being the spin triplet, with energies $E_0^{(s)}, E^{(t)}$, and ignore the other two spin-singlet states whose energies are much higher. The goal is to use these two low-energy states to describe the system, including excitations associated with the spin degrees of freedom.

For the two-particle system, the total wavefunctions corresponding to the two lowest-energy states are, for the spin singlet, the unique ground state

$$\Psi_0^{(s)}(\mathbf{r}_1, s_1; \mathbf{r}_2, s_2) = \Psi_0^{(s)}(\mathbf{r}_1, \mathbf{r}_2) \frac{1}{\sqrt{2}} (\uparrow_1 \downarrow_2 - \downarrow_1 \uparrow_2)$$

and for the spin triplet, the threefold degenerate excited state

$$\Psi^{(t)}(\mathbf{r}_1, s_1; \mathbf{r}_2, s_2) = \Phi^{(t)}(\mathbf{r}_1, \mathbf{r}_2) \frac{1}{\sqrt{2}} (\uparrow_1 \downarrow_2 + \downarrow_1 \uparrow_2), \, \Phi^{(t)}(\mathbf{r}_1, \mathbf{r}_2)(\uparrow_1 \uparrow_2), \, \Phi^{(t)}(\mathbf{r}_1, \mathbf{r}_2)(\downarrow_1 \downarrow_2)$$

where the space-dependent parts are the wavefunctions derived in Eqs (4.32) and (4.28), respectively. Now, using the following identity for the spin operators:

$$2(\mathbf{S}_1 \cdot \mathbf{S}_2) = (\mathbf{S}_1 + \mathbf{S}_2)^2 - \mathbf{S}_1^2 - \mathbf{S}_2^2 = \mathbf{S}^2 - \mathbf{S}_1^2 - \mathbf{S}_2^2$$

where $\mathbf{S} = \mathbf{S}_1 + \mathbf{S}_2$ is the total spin, and the fact that the expectation values of the spin operators are:

$$\langle \mathbf{S}^2 \rangle = S(S+1), \quad \langle \mathbf{S}_i^2 \rangle = S_i(S_i + 1), \quad i = 1, 2$$

we find that for the singlet state, $S^{(s)} = 0$, the expectation value of the operator $(\mathbf{S}_1 \cdot \mathbf{S}_2)$ becomes:

$$\text{singlet}: \quad \langle \mathbf{S}_1 \cdot \mathbf{S}_2 \rangle = \frac{1}{2} \left(0 - \frac{3}{4} - \frac{3}{4} \right) = -\frac{3}{4}$$

while for the triplet state, $S^{(t)} = 1$, it becomes:

$$\text{triplet}: \quad \langle \mathbf{S}_1 \cdot \mathbf{S}_2 \rangle = \frac{1}{2} \left(2 - \frac{3}{4} - \frac{3}{4} \right) = \frac{1}{4}$$

where we have also used the fact that for each individual spin state $S_i = 1/2$, $i = 1, 2$. Taking advantage of these results, we can introduce a new hamiltonian for this system, defined as:

$$\hat{\mathcal{H}}_{\mathrm{H}} = \epsilon_0 - \frac{J}{2} \mathbf{S}_1 \cdot \mathbf{S}_2, \quad \epsilon_0 \equiv 2\epsilon - \frac{t^2}{U}, \quad J \equiv -\frac{8t^2}{U}$$

which involves the new parameters ϵ_0, J, given by the expressions above in terms of the parameters in the original hamiltonian, ϵ, t, U. This new hamiltonian has the same energy eigenvalues and spin eigenfunctions as the original model for the hydrogen molecule, for the space spanned by the two lowest-energy states identified above. Moreover, this hamiltonian is exactly of the Heisenberg form, Eq. (10.36), as desired. We already know that the lowest-energy state for this system is the spin singlet, that is, the spins are opposite and exchanging them flips the sign of the wavefunction, so this model with $J < 0$ describes the physics of an antiferromagnet, as we discuss in more detail below.

of, this special case; an example is the doped copper-oxide perovskites which are high-temperature superconductors (see Chapter 8).

10.4.2 Ground State of Heisenberg Ferromagnet

We study first the Heisenberg ferromagnet, defined by the hamiltonian of Eq. (10.36) with $J(\mathbf{R} - \mathbf{R}') > 0$. As a first step, we express the dot product of two spin operators with the help of the spin raising S_+ and lowering S_- operators (for details, see Appendix C):

$$S_+ \equiv S_x + iS_y, \quad S_- \equiv S_x - iS_y \Rightarrow S_x = \frac{1}{2}(S_+ + S_-), \quad S_y = \frac{1}{2i}(S_+ - S_-)$$

which give, for a pair of spins situated at $\mathbf{R}$ and $\mathbf{R}'$:

$$\mathbf{S}(\mathbf{R}) \cdot \mathbf{S}(\mathbf{R}') = S_x(\mathbf{R})S_x(\mathbf{R}') + S_y(\mathbf{R})S_y(\mathbf{R}') + S_z(\mathbf{R})S_z(\mathbf{R}')$$

$$= \frac{1}{2}\left[S_+(\mathbf{R})S_-(\mathbf{R}') + S_-(\mathbf{R})S_+(\mathbf{R}')\right] + S_z(\mathbf{R})S_z(\mathbf{R}')$$

When this result is substituted in the Heisenberg spin model, Eq. (10.36), taking into account that raising and lowering operators at different lattice sites commute with each other because they operate on different spins, we find:

$$\hat{\mathcal{H}}_{\mathrm{H}} = -\frac{1}{2} \sum_{\mathbf{R} \neq \mathbf{R}'} J(\mathbf{R} - \mathbf{R'})\left[S_-(\mathbf{R})S_+(\mathbf{R}') + S_z(\mathbf{R})S_z(\mathbf{R}')\right] \tag{10.37}$$

where the presence of a sum over all lattice vectors and the relation $J(\mathbf{R} - \mathbf{R}') = J(\mathbf{R}' - \mathbf{R})$ conspire to eliminate the two separate appearances of the lowering–raising operator product at sites $\mathbf{R}$ and $\mathbf{R}'$. If the lowering–raising operator product were neglected, and only the $S_z(\mathbf{R})S_z(\mathbf{R}')$ part were retained, the model would be equivalent to the classical Ising spin model discussed in the previous section.

Even though we developed this model with electrons in mind, we will allow the spin of the particles to be arbitrary. The z direction is determined by an external magnetic field which is taken to be vanishingly small, with its sole purpose being to break the symmetry and define a direction along which the spins tend to orient. If the particles were classical, the z component of the spin would be a continuous variable with values $S_z = S$ to $S_z = -S$. For quantum-mechanical particles the S_z values are quantized: $S_z = S, S - 1, \ldots, \quad S + 1, -S$. In the classical case, it is easy to guess the ground state of this system: it would be the state with all spins pointing in the same direction and S_z for each spin assuming the highest value it can take, $S_z = S$. It turns out that this is also the ground state of the quantum-mechanical system. We prove this statement in two stages. First, we will show that this state is a proper eigenfunction of the hamiltonian and calculate its energy; second, we will show that any other state we can construct cannot have lower energy.

We define the state with all $S_z(\mathbf{R}_i) = S, i = 1, \ldots, N$ as:

$$|S, \ldots, S\rangle \equiv |S^{(N)}\rangle$$

and apply the hamiltonian $\hat{\mathcal{H}}_{\mathrm{H}}$ in the form of Eq. (10.37) to it: the product of operators $S_-(\mathbf{R})S_+(\mathbf{R}')$ applied to $|S^{(N)}\rangle$ gives zero, because the raising operator applied to the maximum S_z value gives zero. Thus, the only terms in the hamiltonian that give non-vanishing contributions are the $S_z(\mathbf{R})S_z(\mathbf{R}')$ products, which give S^2 for each pair of spins and leave the spins unchanged. Therefore, the state $|S^{(N)}\rangle$ is indeed an eigenfunction of the hamiltonian with eigenvalue

$$E_0^{(N)} = -\frac{1}{2} \sum_{\mathbf{R} \neq \mathbf{R}'} J(\mathbf{R} - \mathbf{R}')S^2 \qquad (10.38)$$

If we consider a simplification of the model in which only spins at nearest-neighbor sites interact with a constant exchange-integral value J, the energy of the ground state becomes:

$$E_0^{(N)} = -\frac{J}{2} \sum_{\langle ij \rangle} S^2 = -\frac{J}{2}N\mathcal{Z}S^2 \qquad (10.39)$$

where $\mathcal{Z}$ is the number of nearest neighbors in the lattice.

Next we try to construct eigenstates of the hamiltonian that have different energy than $E_0^{(N)}$. Since the hamiltonian contains only pair interactions, we can focus on the possible configurations of a pair of spins situated at the given sites $\mathbf{R}, \mathbf{R}'$ and consider the spins at all other sites fixed. We are searching for the lowest-energy states, so it makes sense to consider only changes of one unit in the S_z value of the spins, starting from the configuration

$$|S_z(\mathbf{R}), S_z(\mathbf{R}')\rangle \equiv |S, S\rangle$$

There are two configurations that can be constructed with one S_z value lowered by one unit, namely $|S - 1, S\rangle$ and $|S, S - 1\rangle$. Since the hamiltonian contains a sum over all values of $\mathbf{R}, \mathbf{R}'$, both $S_-(\mathbf{R})S_+(\mathbf{R}')$ and $S_-(\mathbf{R}')S_+(\mathbf{R})$ will appear in it. When applied to the two configurations defined above, these operators will produce:

$$S_-(\mathbf{R}')S_+(\mathbf{R})|S - 1, S\rangle = 2S|S, S - 1\rangle$$
$$S_-(\mathbf{R})S_+(\mathbf{R}')|S, S - 1\rangle = 2S|S - 1, S\rangle$$

where we have used the general expression for the action of raising or lowering operators on a spin state of total spin S and z component S_z:

$$S_{\pm}|S_z\rangle = \sqrt{(S \mp S_z)(S + 1 \pm S_z)}|S_z \pm 1\rangle \qquad (10.40)$$

as discussed in Appendix C. Thus, we see that the raising–lowering operators turn $|S, S-1\rangle$ to $|S - 1, S\rangle$ and vice versa, which means that both of these configurations need to be included in the wavefunction in order to have an eigenstate of the hamiltonian. Moreover, the coefficients of the two configurations must have the same magnitude in order to produce an eigenfunction. Two simple choices are to take the coefficients of the two configurations to be equal or opposite:

$$|S^{(+)}\rangle = \frac{1}{\sqrt{2}}\left[|S, S - 1\rangle + |S - 1, S\rangle\right], \quad |S^{(-)}\rangle = \frac{1}{\sqrt{2}}\left[|S, S - 1\rangle - |S - 1, S\rangle\right]$$

Ignoring all other spins which are fixed to $S_z = S$, we can then apply the hamiltonian to the states $|S^{(+)}\rangle, |S^{(-)}\rangle$, to find that they are both eigenstates with energies

$$E^{(+)}(\mathbf{R}, \mathbf{R}') = -S^2J(\mathbf{R} - \mathbf{R}'), \quad E^{(-)}(\mathbf{R}, \mathbf{R}') = -(S^2 - 2S)J(\mathbf{R} - \mathbf{R}')$$

These energies should be compared to the contribution to the ground-state energy of the corresponding state $|S, S\rangle$, which from Eq. (10.38) is found to be:

$$E_0(\mathbf{R}, \mathbf{R}') = -S^2 J(\mathbf{R} - \mathbf{R}')$$

Thus, we conclude that of the two states we constructed, $|S^{(+)}\rangle$ has the same energy as the ground state, while $|S^{(-)}\rangle$ has higher energy by $2SJ(\mathbf{R} - \mathbf{R}')$ [recall that all $J(\mathbf{R} - \mathbf{R}')$ are positive in the ferromagnetic case]. This analysis shows that, when the S_z component of individual spins is reduced by one unit, only a special state has the same energy as the ground state; this state is characterized by equal coefficients for the spin pairs which involve one reduced spin. All other states which have unequal coefficients for pairs of spins with one of the two spins reduced, have higher energy. Notice that both states $|S^{(+)}\rangle$, $|S^{(-)}\rangle$ have only a single spin with $S_z = S - 1$ and all other spins with $S_z = S$, so they represent the smallest possible change from state $|S^{(N)}\rangle$. Evidently, lowering the z component of spins by more than one unit, or lowering the z component of several spins simultaneously, will produce states of even higher energy. These arguments lead to the conclusion that, except for the special state which is degenerate with the state $|S^{(N)}\rangle$, all other states have higher energy than $E_0^{(N)}$. It turns out that the state with the same energy as the ground state corresponds to the infinite wavelength, or zero wave-vector ($\mathbf{k} = 0$), spin-wave state, while the states with higher energy are spin-wave states with wave-vector $\mathbf{k} \neq 0$.

10.4.3 Spin Waves in Heisenberg Ferromagnet

The above discussion also leads us naturally to the low-lying excitations starting from the ground state $|S^{(N)}\rangle$. These excitations will consist of a linear superposition of configurations with the z component of only one spin reduced by one unit. In order to produce states that reflect the crystal periodicity, just as we did for the construction of states based on atomic orbitals in Chapter 2, we must multiply the spin configuration which has the reduced spin at site $\mathbf{R}$ by the phase factor $\exp[i\mathbf{k} \cdot \mathbf{R}]$. The resulting states are:

$$|S_{\mathbf{k}}^{(N-1)}\rangle \equiv \frac{1}{\sqrt{N}} \sum_{\mathbf{R}} e^{i\mathbf{k}\cdot\mathbf{R}} \frac{1}{\sqrt{2S}} S_-(\mathbf{R})|S^{(N)}\rangle \tag{10.41}$$

where the factors $1/\sqrt{N}$ and $1/\sqrt{2S}$ are needed to ensure proper normalization, assuming that $|S^{(N)}\rangle$ is a normalized state [recall the result of the action of S_- on the state with total spin S and z component $S_z = S$, Eq. (10.40)]. These states are referred to as "spin waves" or "magnons." We will show that they are eigenstates of the Heisenberg hamiltonian and calculate their energy and their properties.

To apply the Heisenberg ferromagnetic hamiltonian to the spin-wave state $|S_{\mathbf{k}}^{(N-1)}\rangle$, we start with the action of the operator $S_z(\mathbf{R}')S_z(\mathbf{R}'')$ on configuration $S_-(\mathbf{R})|S^{(N)}\rangle$:

$$S_z(\mathbf{R}')S_z(\mathbf{R}'')S_-(\mathbf{R})|S^{(N)}\rangle =$$
$$\left[S(S-1)(\delta_{\mathbf{R}'\mathbf{R}} + \delta_{\mathbf{R}''\mathbf{R}}) + S^2(1 - \delta_{\mathbf{R}'\mathbf{R}} - \delta_{\mathbf{R}''\mathbf{R}})\right]S_-(\mathbf{R})|S^{(N)}\rangle =$$
$$\left[S^2 - S\delta_{\mathbf{R}\mathbf{R}'} - S\delta_{\mathbf{R}\mathbf{R}''}\right]S_-(\mathbf{R})|S^{(N)}\rangle$$

so that the $S_z(\mathbf{R}')S_z(\mathbf{R}'')$ part of the hamiltonian when applied to state $|S_{\mathbf{k}}^{(N-1)}\rangle$ gives:

$$
\left[-\frac{1}{2} \sum_{\mathbf{R}' \neq \mathbf{R}''} J(\mathbf{R}' - \mathbf{R}'')S_z(\mathbf{R}')S_z(\mathbf{R}'') \right] |S_{\mathbf{k}}^{(N-1)}\rangle =
$$

$$
\left[-\frac{1}{2}S^2 \sum_{\mathbf{R}' \neq \mathbf{R}''} J(\mathbf{R}' - \mathbf{R}'') + S \sum_{\mathbf{R} \neq 0} J(\mathbf{R}) \right] |S_{\mathbf{k}}^{(N-1)}\rangle \qquad (10.42)
$$

We next consider the action of the operator $S_-(\mathbf{R}')S_+(\mathbf{R}'')$ on configuration $S_-(\mathbf{R})|S^{(N)}\rangle$:

$$
S_-(\mathbf{R}')S_+(\mathbf{R}'')S_-(\mathbf{R})|S^{(N)}\rangle = 2S\delta_{\mathbf{R}\mathbf{R}''}S_-(\mathbf{R}')|S^{(N)}\rangle \qquad (10.43)
$$

which gives, for the action of the $S_-(\mathbf{R}')S_+(\mathbf{R}'')$ part of the hamiltonian on state $|S_{\mathbf{k}}^{(N-1)}\rangle$:

$$
-\frac{1}{2} \sum_{\mathbf{R}' \neq \mathbf{R}''} J(\mathbf{R}' - \mathbf{R}'')S_-(\mathbf{R}')S_+(\mathbf{R}'')|S_{\mathbf{k}}^{(N-1)}\rangle = -S \sum_{\mathbf{R} \neq 0} J(\mathbf{R})e^{i\mathbf{k}\cdot\mathbf{R}}|S_{\mathbf{k}}^{(N-1)}\rangle \quad (10.44)
$$

Combining the two results, Eqs (10.42) and (10.44), we find that the state $|S_{\mathbf{k}}^{(N-1)}\rangle$ is an eigenfunction of the hamiltonian with energy

$$
E_{\mathbf{k}}^{(N-1)} - E_0^{(N)} + S \sum_{\mathbf{R} \neq 0} \left[1 - e^{i\mathbf{k}\cdot\mathbf{R}} \right] J(\mathbf{R}) \qquad (10.45)
$$

with the ground-state energy $E_0^{(N)}$ defined in Eq. (10.38). Since we have assumed the $J(\mathbf{R})$s to be positive, we see that the spin-wave energies are higher than the ground-state energy, except for $\mathbf{k} - 0$, which is degenerate with the ground state, as anticipated from our analysis based on a pair of spins. The excitation energy of the spin wave is defined as the amount of energy above the ground-state energy:

$$
\epsilon_{\mathbf{k}}^{SW} \equiv E_{\mathbf{k}}^{(N-1)} - E_0^{(N)} = S \sum_{\mathbf{R} \neq 0} \left[1 - e^{i\mathbf{k}\cdot\mathbf{R}} \right] J(\mathbf{R}) = 2S \sum_{\mathbf{R} \neq 0} J(\mathbf{R}) \sin^2 \left(\frac{1}{2}\mathbf{k}\cdot\mathbf{R} \right) \quad (10.46)
$$

where, to obtain the last expression, we have used the fact that $J(\mathbf{R}) = J(-\mathbf{R})$ to retain only the real part of the complex exponential that involves $\mathbf{k}$.

Specializing these results to the case of nearest-neighbor spin interactions only, we find that the energy is given by:

$$
\epsilon_{\mathbf{k}}^{SW} = JZS \left[1 - \gamma_{\mathbf{k}} \right], \quad \gamma_{\mathbf{k}} \equiv \frac{1}{Z} \sum_{i=1}^{Z} e^{i\mathbf{k}\cdot\mathbf{R}_i^{NN}} \qquad (10.47)
$$

where Z is the number of nearest neighbors in the lattice and $\mathbf{R}_i^{NN}, i = 1, \ldots, Z$, are the Z lattice vectors connecting one lattice point to its nearest neighbors. In the long-wavelength limit, $\mathbf{k} \to 0$, using a Taylor expansion for the exponential in $\gamma_{\mathbf{k}}$ we arrive at:

$$
\gamma_{\mathbf{k}} = \frac{1}{Z} \sum_{i=1}^{Z} \left[1 + i\mathbf{k}\cdot\mathbf{R}_i^{NN} - \frac{1}{2}(\mathbf{k}\cdot\mathbf{R}_i^{NN})^2 \right] = 1 - \frac{1}{2Z} \sum_{i=1}^{Z} (\mathbf{k}\cdot\mathbf{R}_i^{NN})^2
$$

since for any lattice we have:

$$\sum_{i=1}^{\mathcal{Z}} \mathbf{R}_i^{NN} = 0$$

Substituting this result in the above expression for the spin-wave excitation energy, we find:

$$\hbar\epsilon_{\mathbf{k}}^{SW} = \frac{JS}{2}\sum_{i=1}^{\mathcal{Z}}(\mathbf{k}\cdot\mathbf{R}_i^{NN})^2 = \frac{JS}{2}k^2\sum_{i=1}^{\mathcal{Z}}(\hat{\mathbf{k}}\cdot\mathbf{R}_i^{NN})^2$$

where we have expressed the wave-vector in terms of its magnitude k and its direction $\hat{\mathbf{k}}$, $\mathbf{k} = k\hat{\mathbf{k}}$. This result shows that the energy of these excitations, called "ferromagnetic magnons" or "ferromagnons," is quadratic in the wave-vector magnitude $k = |\mathbf{k}|$.

Next we wish to analyze the behavior of the system in a spin-wave state. To this end, we define the transverse spin–spin correlation operator which measures the correlation between the non-z components of two spins at sites $\mathbf{R}, \mathbf{R}'$. From its definition, this operator is:

$$S_\perp(\mathbf{R})S_\perp(\mathbf{R}') \equiv S_x(\mathbf{R})S_x(\mathbf{R}') + S_y(\mathbf{R})S_y(\mathbf{R}') = \frac{1}{2}\left[S_+(\mathbf{R})S_-(\mathbf{R}') + S_-(\mathbf{R})S_+(\mathbf{R}')\right]$$

The expectation value of this operator in state $|S_{\mathbf{k}}^{(N-1)}\rangle$ will involve the application of the lowering and raising operators $S_+(\mathbf{R})S_-(\mathbf{R}')$ or $S_-(\mathbf{R})S_+(\mathbf{R}')$ on configuration $S_-(\mathbf{R}'')|S^{(N)}\rangle$, which is similar to what we calculated above, Eq. (10.43), leading to:

$$\langle S_{\mathbf{k}}^{(N-1)}|S_\perp(\mathbf{R})S_\perp(\mathbf{R}')|S_{\mathbf{k}}^{(N-1)}\rangle = \frac{2S}{N}\cos[\mathbf{k}\cdot(\mathbf{R} - \mathbf{R}')] \tag{10.48}$$

This shows that spins which are apart by $\mathbf{R} - \mathbf{R}'$ have a difference in transverse orientation of $\cos[\mathbf{k}\cdot(\mathbf{R} - \mathbf{R}')]$, which provides a picture of what the spin-wave state means: spins are mostly oriented in the z-direction but have a small transverse component which changes orientation by an angle $\mathbf{k}\cdot(\mathbf{R} - \mathbf{R}')$ for spins separated by a distance $\mathbf{R} - \mathbf{R}'$. This angle obviously depends on the wave-vector $\mathbf{k}$ of the spin wave. An illustration for spins on a 2D square lattice is shown in Fig. 10.8.

Finally, we calculate the magnetization in the system with spin waves. First, we notice that in a spin-wave state only one out of N S_z spin components has been lowered by one unit in each configuration, defined as $|S_-(\mathbf{R})S^{(N)}\rangle$, which means that the z-component spin of each configuration and therefore of the entire $|S_{\mathbf{k}}^{(N-1)}\rangle$ state is $NS - 1$. Now suppose there are several spin-wave states excited at some temperature T. We will assume that we can treat the spin waves like phonons, so we can take a superposition of them to describe the state of the system, with boson occupation numbers

$$n_{\mathbf{k}}^{SW} = \frac{1}{\exp[\epsilon_{\mathbf{k}}^{SW}/k_B T] - 1}$$

Fig. 10.8 Illustration of a spin wave in the 2D square lattice with $\mathbf{k} = (\frac{1}{6}, \frac{1}{6})\frac{\pi}{a}$, where a is the lattice constant. Only the transverse component of the spins is shown, that is, the projection on the x, y plane.

with the excitation energies from Eq. (10.46). We must emphasize that treating the system as a superposition of spin waves is an approximation, because they cannot be described as independent bosons, as was the case for phonons (see Chapter 7). Within this approximation, the total z component of the spin, which we identify with the magnetization of the system, will be given by:

$$M(T) = NS - \sum_{\mathbf{k}} n_{\mathbf{k}}^{SW} = NS\left[1 - \frac{1}{NS}\sum_{\mathbf{k}}\frac{1}{\exp[\epsilon_{\mathbf{k}}^{SW}/k_B T] - 1}\right] \qquad (10.49)$$

since each state $|S_{\mathbf{k}}^{(N-1)}\rangle$ reduces the total z component of the ground-state spin, NS, by one unit. In the small $|\mathbf{k}|$ limit, turning the sum over $\mathbf{k}$ to an integral as usual, and defining the new variable $\mathbf{q} = \mathbf{k}/\sqrt{2k_B T/S}$, we obtain for the magnetization per particle $m(T) = M(T)/N$:

$$m(T) = S - \left(\frac{2k_B T}{n^{2/3}S}\right)^{3/2} \int \frac{d\mathbf{q}}{(2\pi)^3}\left[\exp\left(\sum_{\mathbf{R}\neq 0} J(\mathbf{R})(\mathbf{q}\cdot\mathbf{R})^2\right) - 1\right]^{-1} \qquad (10.50)$$

with n the density of particles. This expression gives the temperature dependence of the magnetization as being $\sim -T^{3/2}$, which is known as the Bloch $T^{3/2}$ law. This behavior is actually observed in experiments performed on isotropic ferromagnets, that is, systems in which $J(\mathbf{R} - \mathbf{R}')$ is the same for all operators appearing in the Heisenberg spin hamiltonian, as we have been assuming so far. There exist also anisotropic systems in which the value of $J_\perp(\mathbf{R} - \mathbf{R}')$ corresponding to the transverse operators, $S_\perp(\mathbf{R})S_\perp(\mathbf{R}') = S_x(\mathbf{R})S_x(\mathbf{R}') + S_y(\mathbf{R})S_y(\mathbf{R}')$, is different from the value $J_z(\mathbf{R} - \mathbf{R}')$ corresponding to the operator $S_z(\mathbf{R})S_z(\mathbf{R}')$, in which case the Bloch $T^{3/2}$ law does not hold.

One last thing worth mentioning, which is a direct consequence of the form for the magnetization given in Eq. (10.50), is that for small enough $|\mathbf{q}| = q$ we can expand the exponential to obtain:

$$\left[\exp\left(\sum_{\mathbf{R} \neq 0} J(\mathbf{R})(\mathbf{q} \cdot \mathbf{R})^2 \right) - 1 \right]^{-1} \approx \frac{1}{q^2} \left(\sum_{\mathbf{R} \neq 0} J(\mathbf{R})(\hat{\mathbf{q}} \cdot \mathbf{R})^2 \right)^{-1}$$

If we assume for the moment (we will analyze this assumption below) that only small values of $|\mathbf{q}|$ contribute to the integral, then the integrand contains the factor $1/q^2$, which in three dimensions is canceled by the factor q^2 in the infinitesimal $d\mathbf{q} = q^2 dq d\hat{\mathbf{q}}$, but in two and one dimensions is not canceled, because in those cases $d\mathbf{q} = qdqd\hat{\mathbf{q}}$ and $d\mathbf{q} = dqd\hat{\mathbf{q}}$, respectively. Since the integration includes the value $\mathbf{q} = 0$, we conclude that in three dimensions the integral gives finite value, but in two and one dimensions it gives infinite value. The interpretation of this result is that in two and one dimensions the presence of excitations above the ground state is so disruptive to the magnetic order in the system that the spin-wave picture itself breaks down at non-zero temperature. Another way to express this is that there can be no magnetic order at finite temperature in two and one dimensions within the isotropic Heisenberg model, a statement known as the Hohenberg–Mermin–Wagner theorem.[11] To complete the argument, we examine the validity of its basic assumption, namely the conditions under which only small values of $|\mathbf{k}|$ are relevant: from Eq. (10.49) we see that for large values of $\epsilon_{\mathbf{k}}^{SW}/k_B T$, the exponential in the denominator of the integrand leads to vanishing contributions to the integral. This implies that for large T, spin-wave states with large energies will contribute, but for small T, only states with very small energies can contribute to the integral. We have seen earlier that in the limit $\mathbf{k} \to 0$ the energy $\epsilon_{\mathbf{k}}^{SW} \to 0$, so that the states with the smallest $\mathbf{k}$ also have the lowest energy above the ground state. Therefore, at low T, only the states with lowest $|\mathbf{k}|$, and consequently lowest $|\mathbf{q}|$, will contribute to the magnetization. Thus, the Bloch $T^{3/2}$ law will hold for low temperatures. The scale over which the temperature can be considered low is set by the factor

$$\frac{S}{2} \sum_{\mathbf{R} \neq 0} J(\mathbf{R})(\mathbf{k} \cdot \mathbf{R})^2$$

which obviously depends on the value of the exchange integral $J(\mathbf{R})$.

10.4.4 Heisenberg Antiferromagnetic Spin Model

The Heisenberg spin model with negative exchange integral J can be used to study the physics of antiferromagnetism. The Heisenberg antiferromagnetic spin model is much more complicated than the ferromagnetic model. The basic problem is that it is considerably more difficult to find the ground state and the excitations of spins with

---

[11] N. D. Mermin and H. Wagner, *Phys. Rev. Lett.* **17**, 1133 (1966).

antiferromagnetic interactions. In fact, in this case the model is only studied in its nearest-neighbor interaction version, because it is obviously impossible to try to make all spins on a lattice point opposite from each other: two spins that are antiparallel to the same third spin must be parallel to each other. This is in stark contrast to the ferromagnetic case, where it is possible to make all spins on a lattice parallel, and this is actually the ground state of the model with $J > 0$. For nearest-neighbor interactions only, it is possible to create a state in which every spin is surrounded by nearest-neighbor spins pointing in the opposite direction, which is called the "Néel state." This is not true for every lattice, but only for lattices that can be split into two interpenetrating sublattices; these lattices are called *bipartite*. Some examples of bipartite lattices in three dimensions are the simple cubic lattice (with two interpenetrating cubic sublattices), the BCC lattice (also with two interpenetrating cubic sublattices), and the diamond lattice (with two interpenetrating FCC sublattices). In two dimensions, examples of bipartite lattices are the square lattice (with two interpenetrating square sublattices) and the graphitic lattice (with two interpenetrating hexagonal sublattices). The lattices that cannot support a Néel state are called "frustrated lattices," some examples being the FCC lattice in three dimensions and the hexagonal lattice in two dimensions.

The difficulty with solving the Heisenberg antiferromagnet is that, even in cases where the lattice supports such a state, the Néel state is not an eigenstate of the hamiltonian. This can easily be seen from the fact that the operators $S_+(i)S_-(j)$ or $S_-(i)S_+(j)$ which appear in the hamiltonian, when applied to the pair of antiparallel spins at nearest-neighbor sites $\langle ij \rangle$, flip both spins and as a result destroy the alternating-spin pattern and create a different configuration. This is illustrated in Fig. 10.9. It turns out that the energy of the ground state of the Heisenberg antiferromagnet, for a model with $\mathcal{Z}$ nearest-neighbor interactions of strength J between spins of magnitude S, is given by:

$$E_0^{\mathrm{AF}} = -2N\mathcal{Z}JS(S+1) + 2J\mathcal{Z}S \sum_{\mathbf{k}} \left(1 - \gamma_{\mathbf{k}}^2 \right)^{1/2} \qquad (10.51)$$

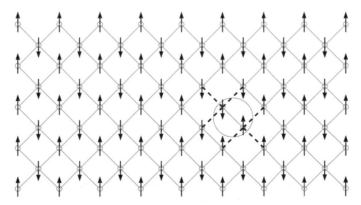

Fig. 10.9   Illustration of the Néel state on a 2D square lattice. Two spins, shown by a circle, have been flipped from their ideal orientation, destroying locally the antiferromagnetic order: neighboring pairs of spins with wrong orientations are highlighted by thicker dashed lines.

where $\gamma_{\mathbf{k}}$ is the same quantity defined in Eq. (10.47). The second term in this expression is referred to as the "zero-point energy" contribution. We can rewrite the expression for the ground-state energy as:

$$E_0^{AF} = -2N\mathcal{Z}JS^2 - 2J\mathcal{Z}S\left[N - \sum_{\mathbf{k}}\left(1 - \gamma_{\mathbf{k}}^2\right)^{1/2}\right]$$

In this expression, the first term represents the contribution to the energy of the strictly antiferromagnetically ordered lattice. Indeed, if $\gamma_{\mathbf{k}} = 0$, the energy is equal to this term. Since $\gamma_{\mathbf{k}} \neq 0$, the second term indicates that the actual ground state of the system is *not* the perfectly ordered antiferromagnetic configuration but includes fluctuations away from this order, with each sublattice having a small amount of disorder in the spin alignment. We define the excitation energy as the zero-point energy contribution:

$$\epsilon_{\mathbf{k}}^{AF} \equiv 2J\mathcal{Z}S\left(1 - \gamma_{\mathbf{k}}^2\right)^{1/2}$$

as this is the part that depends on the wave-vector $\mathbf{k}$ that describes the excitations above the ground state. In the long-wavelength limit, following the same steps as in the derivation for the spin-wave excitations of the ferromagnetic state, we then that find the excitation energy is proportional to the magnitude of the wave-vector, k, for simple lattices.

Much theoretical work has been directed toward understanding the physics of the Heisenberg antiferromagnet, especially after it was suggested by Anderson that this model may be relevant to the copper-oxide high-temperature superconductors.[12] Interestingly, spin waves may also be used to describe the physics of this model, at least in the limit of large spin S, assuming that deviations from the Néel state are small.[13]

10.5 Magnetic Domains

Having developed the theoretical background for describing magnetic order in ferromagnetic and antiferromagnetic ideal solids, we discuss next how this order is actually manifest in real materials. Firstly, a basic consideration is the temperature dependence of magnetic order. As we noted already in the introduction to this chapter, magnetic order is relatively weak and can easily be destroyed by thermal fluctuations. The critical temperature above which there is no magnetic order is called the Curie temperature (T_C) for ferromagnets and the Néel temperature (T_N) for antiferromagnets. Table 10.2 provides examples of elementary and compound magnetic solids and their Curie and Néel temperatures. In certain solids, these critical temperatures can be rather large, exceeding 1000 K, like in Fe and Co. The Néel temperatures are generally much lower than the Curie temperatures, the antiferromagnetic state being more delicate than the ferromagnetic one. Another interesting characteristic feature of ferromagnets is the number of Bohr magnetons per atom, n_B. This quantity is a measure of the difference in occupation of spin-up and spin-down states in

[12] P. W. Anderson, *Science* **235**, 1196 (1987).
[13] P. W. Anderson, *Phys. Rev.* **86**, 694 (1952); R. Kubo, *Phys. Rev.* **87**, 568 (1952).

Table 10.2 Examples of elemental and compound ferromagnets and antiferromagnets. The Curie temperature T_C and Néel temperature T_N are given in Kelvin; n_B is the number of Bohr magnetons per atom; M_s is the saturation magnetization in gauss. *Source*: D. R. Lide (Ed.), *CRC Handbook of Chemistry and Physics* (CRC Press, Boca Raton, FL, 1999–2000).

Elemental							
Ferromagnets				Antiferromagnets			
Solid	T_C	n_B	M_s	Solid	T_N	Solid	T_N
Fe	1043	2.2	1752	Cr	311	Sm	106
Co	1388	1.7	1446	Mn	100	Eu	91
Ni	627	0.6	510	Ce	13	Dy	176
Gd	293	7.0	1980	Nd	19	Ho	133

Compound						
Ferromagnets				Antiferromagnets		
Solid	T_C	Solid	T_C	Solid	T_N	Solid	T_N
MnB	152	Fe_3C	483	MnO	122	$FeKF_3$	115
MnAs	670	FeP	215	FeO	198	$CoKF_3$	125
MnBi	620	CrTe	339	CoO	291	MnF_2	67
MnSb	710	$CrBr_3$	37	NiO	600	$MnCl_2$	2
FeB	598	CrI_3	68	$MnRbF_3$	54	FeF_2	78
Fe_2B	1043	CrO_2	386	$MnKF_3$	88	CoF_2	38

the solid; it is determined by the band structure of a spin-polarized system of electrons in the periodic solid (values of n_B can readily be obtained from band-structure calculations of the type described in Chapter 5). The higher the value of n_B, the more pronounced is the magnetic behavior of the solid. However, there is no direct correlation between n_B and the Curie temperature.

Finally, the behavior of ferromagnets is characterized by the saturation magnetization M_s, that is, the highest value that the magnetization can attain when the system is subject to an external field. This value corresponds to all the spins being aligned in the same direction; values for this quantity for elemental ferromagnets are give in Table 10.2. The saturation magnetization is not attained spontaneously in a real ferromagnet but requires the application of an external field. This is related to interesting physics: in a real ferromagnet the dominant magnetic interactions at the atomic scale tend to orient the magnetic moments at neighboring sites parallel to each other, but this does not correspond to a globally optimal state from the energetic point of view. Other dipoles placed in the field of a given magnetic dipole would be oriented at various angles relative to the original dipole in order to minimize the magnetic dipole–dipole interactions. The dipole–dipole interactions are much weaker than the short-range interactions responsible for magnetic order. However, over macroscopically large distances the shear number of dipole–dipole interactions dictates that their contribution to the energy should also be optimized. The

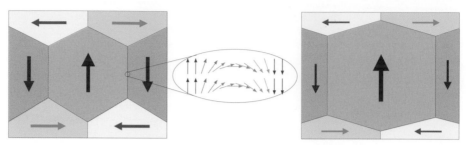

Fig. 10.10 Illustration of magnetic domains in a ferromagnet. The large arrows indicate the magnetization in each domain. The pattern of domain magnetizations follows the field of a magnetic dipole situated at the center of each figure. The figure on the left corresponds to a zero external magnetic field, with the total magnetization averaging zero. The figure on the right corresponds to a magnetic field in the same direction as the magnetization of the domain at the center, which has grown in size at the expense of the other domains. The inset illustrates the domain wall structure at microscopic scale in terms of individual spins.

system accommodates this by breaking into regions of different magnetization called domains. Within each domain the spins are oriented parallel to each other, as dictated by the atomic-scale interactions (the magnetization of a domain is the sum of all the spins in it). The magnetizations of different domains are oriented in a pattern that tries to minimize the magnetic dipole–dipole interactions over large distances. An illustration of this effect in a simple 2D case is shown in Fig. 10.10. The domains of different magnetization are separated by boundaries called domain walls. In a domain wall, the orientation of spins is gradually changed from that which dominates on one side of the wall to that of the other side. This change in spin orientation takes place over many interatomic distances in order to minimize the energy cost due to the disruption of order at the microscopic scale induced by the domain wall. If the extent of the domain wall were of order an interatomic distance, the change in spin orientation would be drastic across the wall, leading to high energy cost. Spreading the spin-orientation change over many interatomic distances minimizes this energy cost, very much like in the case of spin waves.

In an ideal solid the domain sizes and distribution would be such that the total magnetization is zero. In real materials the presence of a large number of defects introduces limitations in the creation and placement of magnetic domain walls. The magnetization M can be increased by the application of an external magnetic field H. The external field will favor energetically the domains of magnetization parallel to it and the domain walls will move to enlarge the size of these domains at the expense of domains with magnetization different from the external field. The more the magnetization of a certain domain deviates from the external field, the higher its energy will be and consequently the more this domain will shrink, as illustrated in Fig. 10.10. For relatively small fields this process is reversible, since the domain walls move only by a small amount and they can revert to their original configuration upon removal of the field. For large fields this process becomes irreversible, because if the walls have to move far, eventually they will cross regions with defects which will make their return to the original configuration impossible, after the field is removed. Thus, when the field is reduced to zero after saturation, the magnetization does not return to its original zero value but has a positive value because of limitations in the mobility of

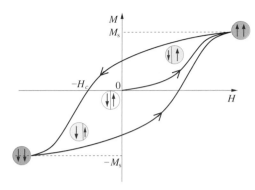

Fig. 10.11   Hysteresis curve of the magnetization M in a ferromagnet upon application of an external field H.
The original magnetization curve starts at $H = 0, M = 0$ and extends to the saturation point,
$M = M_s$. When the field is reduced back to zero, the magnetization has a finite value; zero
magnetization is obtained again for the reversed field value $-H_c$. The circular insets indicate
schematically the changes in magnetization due to domain wall motion, with the size of the arrows
indicating the dominant up or down domains.

domain walls due to defects in the solid. It is therefore necessary to apply a large field in
the opposite direction, denoted as $-H_c$ in Fig. 10.11, to reduce the magnetization M back
to zero; this behavior is referred to as "hysteresis." Continuing to increase the field in the
negative direction will again lead to a state with saturated magnetization, as indicated in
Fig. 10.11. When the field is removed, the magnetization now has a non-zero value in the
opposite direction to before, which requires the application of a positive field to reduce it
to zero. This response of the system to the external field is called a "hysteresis loop."

Further Reading

1. For a thorough review of the Heisenberg antiferromagnetic model, we refer the reader
   to: E. Manousakis, *Rev. Mod. Phys.* **63**, 1 (1991).
2. An extensive treatment of theoretical models for magnetic behavior can be found in:
   Theoretical Solid State Physics, Vol. 1, W. Jones and N. H. March (Wiley, London,
   1973).
3. A useful treatment of magnetic properties of materials and their applications is
   given in: *Magnetic Materials': Fundamentals and Device Applications*, N. Spaldin
   (Cambridge University Press, Cambridge, 2003).

Problems

1.   Verify the assignment of the total spin state S, angular momentum state L, and total
      angular momentum state J of the various configurations for the d and f electronic
      shells, according to Hund's rules, given in Table 10.1.

2.   Prove the expressions for the magnetization Eq. (10.14) and band energy Eq. (10.16) using the spin-up and spin-down filling of the two spin states defined in Eq. (10.13).

3.   In the Stoner theory of spontaneous magnetization of a system consisting of N electrons, the energy ϵ of an electron with magnetic moment $\pm \mu_B$ is modified by the term $\mp \mu_B(\mu_B M)$, where M is the magnetization number (the difference in spin-up and spin-down number of electrons). We introduce a characteristic temperature Θ_S for the onset of spontaneous magnetization by defining the magnetic contribution to the energy as:

$$E_{mag} = \bar{s} k_B \Theta_S$$

with k_B the Boltzmann constant and $\bar{s} = M/N$ the dimensionless magnetic moment per particle. At non-zero temperature T, the magnetization number M is given by:

$$M = V \int_{-\infty}^{\infty} [n(\epsilon - \bar{s} k_B \Theta_S) - n(\epsilon + \bar{s} k_B \Theta_S)] \, g(\epsilon) d\epsilon$$

while the total number of electrons is given by:

$$N = V \int_{-\infty}^{\infty} [n(\epsilon - \bar{s} k_B \Theta_S) + n(\epsilon + \bar{s} k_B \Theta_S)] \, g(\epsilon) d\epsilon$$

where V is the system volume, $n(\epsilon)$ is the Fermi occupation number

$$n(\epsilon) = \frac{1}{e^{(\epsilon - \mu)/k_B T} + 1}$$

$g(\epsilon)$ is the density of states, and μ is the chemical potential. For simplicity, we will consider the free-electron model or, equivalently, a single band with parabolic energy dependence, $\epsilon_\mathbf{k} = \hbar^2 \mathbf{k}^2 / 2\bar{m}$, with $\bar{m}$ the effective mass of electrons. We also define the Fermi integral function

$$F_{\frac{1}{2}}(y) = \int_0^{\infty} \frac{x^{\frac{1}{2}}}{e^{x-y} + 1} dx$$

Show that the number of electrons per unit volume, $n = N/V$, and the magnetic moment per unit volume, $m = M/V$, are given by:

$$n = \frac{2\pi}{\lambda^3} \left[F_{\frac{1}{2}}(y+z) + F_{\frac{1}{2}}(y-z) \right], \quad m = \frac{2\pi}{\lambda^3} \left[F_{\frac{1}{2}}(y+z) - F_{\frac{1}{2}}(y-z) \right]$$

where the variables y, z, λ are defined as:

$$y = \frac{\mu}{k_B T}, \quad z = \frac{\Theta_S}{T} \bar{s}, \quad \lambda = \left(\frac{4\pi^2 \hbar^2}{2\bar{m} k_B T} \right)^{1/2}$$

Next, show that the equations derived above lead to:

$$\bar{s} k_B \Theta_S = \frac{\epsilon_F}{2} \left[(1 + \bar{s})^{2/3} - (1 - \bar{s})^{2/3} \right]$$

with ϵ_F the Fermi energy of the unpolarized electrons. From this result, derive the condition for the existence of spontaneous magnetization:

$$k_B \Theta_S > \frac{2}{3}\epsilon_F$$

Find the upper limit of Θ_S. [*Hint:* Show that the Fermi integral function for large values of its argument can be approximated by:

$$F_{\frac{1}{2}}(y) = \frac{2}{3}y^{3/2} + \frac{\pi^2}{12}y^{-1/2} + \cdots$$

The results above are from the leading-order term in this expansion.]

4. In one dimension we can solve the nearest-neighbor spin-1/2 Ising model analytically by the "transfer matrix" method; the hamiltonian of the system is defined in Eq. (10.31). Consider a chain of N spins with periodic boundary conditions, that is, the right neighbor of the last spin is the first spin, and take for simplicity the spins to have values $s_i = \pm 1$.

(a) For $N = 2$, show that the matrix with entries

$$T_{ij}(s_1, s_2) = \exp\left[-\beta \hat{\mathcal{H}}(s_1, s_2)\right]$$

with $\hat{\mathcal{H}}(s_1, s_2)$ the hamiltonian of the two-spin system, is given by:

$$\underline{T} = \begin{pmatrix} \exp[\beta(J + H)] & \exp[-\beta J] \\ \exp[-\beta J] & \exp[\beta(J - H)] \end{pmatrix}$$

where the columns correspond to $s_1 = \pm 1$ and the rows correspond to $s_2 = \pm 1$. From this, show that the partition function of the two-spin system is:

$$Z_2 = \sum_{s_1, s_2} \exp\left[-\beta \hat{\mathcal{H}}(s_1, s_2)\right] = \text{Tr}[\underline{T}^2]$$

(b) Show that the hamiltonian of the N-spin system with periodic boundary conditions can be written as:

$$\hat{\mathcal{H}}(s_1, \ldots, s_N) = -\left(\frac{H}{2}s_1 + Js_1 s_2 + \frac{H}{2}s_2\right) - \left(\frac{H}{2}s_2 + Js_2 s_3 + \frac{H}{2}s_3\right) - \cdots$$
$$-\left(\frac{H}{2}s_{N-1} + Js_{N-1}s_N + \frac{H}{2}s_N\right) - \left(\frac{H}{2}s_N + Js_N s_1 + \frac{H}{2}s_1\right)$$

and therefore the partition function of the system can be written as:

$$Z_N = \sum_{s_1, \ldots, s_N} \exp\left[-\beta \hat{\mathcal{H}}(s_1, \ldots, s_N)\right] = \text{Tr}[\underline{T}^N] \qquad (10.52)$$

(c) The trace of a matrix $\underline{T}$ is unchanged if $\underline{T}$ is multiplied by another matrix $\underline{S}$ from the right and its inverse $\underline{S}^{-1}$ from the left:

$$\text{Tr}[\underline{T}] = \text{Tr}[\underline{S}^{-1} \underline{T} \underline{S}]$$

Show that this general statement allows us to multiply each power of $\underline{T}$ in Eq. (10.52) by $\underline{S}$ on the right and $\underline{S}^{-1}$ on the left without changing the value of the trace, to obtain:

$$Z_N = \text{Tr}\left[\left(\underline{S}^{-1}\,\underline{T}\,\underline{S}\right)^N\right]$$

We can then choose $\underline{S}$ to be the matrix of eigenvectors of $\underline{T}$, so that the product $\underline{S}^{-1}\,\underline{T}\,\underline{S}$ becomes a diagonal matrix with entries equal to the two eigenvalues of $\underline{T}$, which we denote as τ_1, τ_2. Thus, show that the partition function takes the form

$$Z_N = \text{Tr}\left[\begin{pmatrix} \tau_1 & 0 \\ 0 & \tau_2 \end{pmatrix}^N\right] = \tau_1^N + \tau_2^N$$

(d) Using the partition function, show that the free energy of the system is equal to:

$$F_N = -k_B T N \ln \tau_1 - k_B T \ln\left[1 + \left(\frac{\tau_2}{\tau_1}\right)^N\right]$$

Since $\underline{T}$ is a real symmetric matrix, it has two real eigenvalues, which in general are not equal. Assuming that $|\tau_2| < |\tau_1|$, show that in the thermodynamic limit $N \to \infty$ the free energy per particle takes the form

$$\frac{F_N}{N} = -k_B T \ln \tau_1 = -J - k_B T \ln\left[\cosh(\beta H) + \sqrt{\sinh^2(\beta H) + e^{-4\beta J}}\right]$$

(e) Show that the average magnetization per particle is given by:

$$m(H, T) = -\beta \frac{\partial}{\partial H}\left(\frac{F_N}{N}\right)_T = \frac{\sinh(\beta H)}{\sqrt{\sinh^2(\beta H) + e^{-4\beta J}}}$$

From this expression argue that the average magnetization per particle for $H \to 0$ goes to zero as long as the temperature is finite, and therefore there is no phase transition at any non-zero temperature.

APPENDICES

Appendix A **Mathematical Tools**

In this appendix we collect a number of useful mathematical results that are employed frequently throughout the text. The emphasis is not on careful derivations but on the motivation behind the results. Our purpose is to provide a reminder and a handy reference for the reader, whom we assume to have been exposed to the underlying mathematical concepts, but do not expect to recall the relevant formulae each instant the author needs to employ them.

A.1 **Differential Operators**

The differential operators encountered often in the description of the physical properties of solids are the gradient of a scalar field $\nabla_{\mathbf{r}}\Phi(\mathbf{r})$, the divergence of a vector field $\nabla_{\mathbf{r}} \cdot \mathbf{F}(\mathbf{r})$, the curl of a vector field $\nabla_{\mathbf{r}} \times \mathbf{F}(\mathbf{r})$, and the laplacian of a scalar field $\nabla_{\mathbf{r}}^2\Phi(\mathbf{r})$ (the laplacian of a vector field is simply the vector addition of the laplacians of its components, $\nabla_{\mathbf{r}}^2\mathbf{F} = \nabla_{\mathbf{r}}^2 F_x\hat{\mathbf{x}} + \nabla_{\mathbf{r}}^2 F_y\hat{\mathbf{y}} + \nabla_{\mathbf{r}}^2 F_z\hat{\mathbf{z}}$). These operators in three dimensions are expressed in the different sets of coordinates as follows:

In cartesian coordinates (x, y, z):

$$\mathbf{r} = x\hat{\mathbf{x}} + y\hat{\mathbf{y}} + z\hat{\mathbf{z}} \tag{A.1}$$

$$\mathbf{F}(\mathbf{r}) = F_x(\mathbf{r})\hat{\mathbf{x}} + F_y(\mathbf{r})\hat{\mathbf{y}} + F_z(\mathbf{r})\hat{\mathbf{z}} \tag{A.2}$$

$$\nabla_{\mathbf{r}}\Phi(\mathbf{r}) = \frac{\partial\Phi}{\partial x}\hat{\mathbf{x}} + \frac{\partial\Phi}{\partial y}\hat{\mathbf{y}} + \frac{\partial\Phi}{\partial z}\hat{\mathbf{z}} \tag{A.3}$$

$$\nabla_{\mathbf{r}} \cdot \mathbf{F}(\mathbf{r}) = \frac{\partial F_x}{\partial x} + \frac{\partial F_y}{\partial y} + \frac{\partial F_z}{\partial z} \tag{A.4}$$

$$\nabla_{\mathbf{r}} \times \mathbf{F}(\mathbf{r}) = \left(\frac{\partial F_z}{\partial y} - \frac{\partial F_y}{\partial z}\right)\hat{\mathbf{x}} + \left(\frac{\partial F_x}{\partial z} - \frac{\partial F_z}{\partial x}\right)\hat{\mathbf{y}} + \left(\frac{\partial F_y}{\partial x} - \frac{\partial F_x}{\partial y}\right)\hat{\mathbf{z}} \tag{A.5}$$

$$\nabla_{\mathbf{r}}^2\Phi(\mathbf{r}) = \frac{\partial^2\Phi}{\partial x^2} + \frac{\partial^2\Phi}{\partial y^2} + \frac{\partial^2\Phi}{\partial z^2} \tag{A.6}$$

In polar coordinates (r, θ, z):

$$r = \sqrt{x^2 + y^2}, \ \theta = \tan^{-1}(y/x), z \longleftrightarrow x = r\cos\theta, \ y = r\sin\theta, z \tag{A.7}$$

$$\mathbf{F}(\mathbf{r}) = F_r(\mathbf{r})\hat{\mathbf{r}} + F_\theta(\mathbf{r})\hat{\theta} + F_z(\mathbf{r})\hat{\mathbf{z}} \tag{A.8}$$

$$\nabla_\mathbf{r}\Phi(\mathbf{r}) = \frac{\partial\Phi}{\partial r}\hat{\mathbf{r}} + \frac{1}{r}\frac{\partial\Phi}{\partial\theta}\hat{\theta} + \frac{\partial\Phi}{\partial z}\hat{\mathbf{z}} \tag{A.9}$$

$$\nabla_\mathbf{r}\cdot\mathbf{F}(\mathbf{r}) = \frac{1}{r}\frac{\partial}{\partial r}(rF_r) + \frac{1}{r}\frac{\partial F_\theta}{\partial\theta} + \frac{\partial F_z}{\partial z} \tag{A.10}$$

$$\nabla_\mathbf{r}\times\mathbf{F}(\mathbf{r}) = \left(\frac{1}{r}\frac{\partial F_z}{\partial\theta} - \frac{\partial F_\theta}{\partial z}\right)\hat{\mathbf{r}} + \left(\frac{\partial F_r}{\partial z} - \frac{\partial F_z}{\partial r}\right)\hat{\theta} +$$
$$\left(\frac{\partial F_\theta}{\partial r} - \frac{1}{r}\frac{\partial F_r}{\partial\theta} + \frac{F_\theta}{r}\right)\hat{\mathbf{z}} \tag{A.11}$$

$$\nabla_\mathbf{r}^2\Phi(\mathbf{r}) = \frac{1}{r}\frac{\partial}{\partial r}\left(r\frac{\partial\Phi}{\partial r}\right) + \frac{1}{r^2}\frac{\partial^2\Phi}{\partial\theta^2} + \frac{\partial^2\Phi}{\partial z^2} \tag{A.12}$$

In spherical coordinates (r,θ,ϕ):

$$r = \sqrt{x^2 + y^2 + z^2},\ \theta = \cos^{-1}(z/r),\ \phi = \cos^{-1}(x/\sqrt{x^2 + y^2}) \longleftrightarrow$$
$$x = r\sin\theta\cos\phi,\ y = r\sin\theta\sin\phi,\ z = r\cos\theta \tag{A.13}$$

$$\mathbf{F}(\mathbf{r}) = F_r(\mathbf{r})\hat{\mathbf{r}} + F_\theta(\mathbf{r})\hat{\theta} + F_\phi(\mathbf{r})\hat{\phi} \tag{A.14}$$

$$\nabla_\mathbf{r}\Phi(\mathbf{r}) = \frac{\partial\Phi}{\partial r}\hat{\mathbf{r}} + \frac{1}{r}\frac{\partial\Phi}{\partial\theta}\hat{\theta} + \frac{1}{r\sin\theta}\frac{\partial\Phi}{\partial\phi}\hat{\phi} \tag{A.15}$$

$$\nabla_\mathbf{r}\cdot\mathbf{F}(\mathbf{r}) = \frac{1}{r^2}\frac{\partial}{\partial r}(r^2 F_r) + \frac{1}{r\sin\theta}\frac{\partial}{\partial\theta}(\sin\theta F_\theta) + \frac{1}{r\sin\theta}\frac{\partial F_\phi}{\partial\phi} \tag{A.16}$$

$$\nabla_\mathbf{r}\times\mathbf{F}(\mathbf{r}) = \frac{1}{r\sin\theta}\left[\frac{\partial}{\partial\theta}(\sin\theta F_\phi) - \frac{\partial F_\theta}{\partial\phi}\right]\hat{\mathbf{r}} + \left[\frac{1}{r\sin\theta}\frac{\partial F_r}{\partial\phi} - \frac{1}{r}\frac{\partial(rF_\phi)}{\partial r}\right]\hat{\theta}$$
$$+ \frac{1}{r}\left[\frac{\partial(rF_\theta)}{\partial r} - \frac{\partial F_r}{\partial\theta}\right]\hat{\phi} \tag{A.17}$$

$$\nabla_\mathbf{r}^2\Phi(\mathbf{r}) = \frac{1}{r^2}\frac{\partial}{\partial r}\left(r^2\frac{\partial\Phi}{\partial r}\right) + \frac{1}{r^2\sin\theta}\frac{\partial}{\partial\theta}\left(\sin\theta\frac{\partial\Phi}{\partial\theta}\right) + \frac{1}{r^2\sin^2\theta}\frac{\partial^2\Phi}{\partial\phi^2} \tag{A.18}$$

A useful relation that can easily be proved in any set of coordinates is that the curl of the gradient of a scalar field is identically zero:

$$\nabla_\mathbf{r}\times\nabla_\mathbf{r}\Phi(\mathbf{r}) = 0 \tag{A.19}$$

For example, if we define $\mathbf{F}(\mathbf{r}) = \nabla_\mathbf{r}\Phi(\mathbf{r})$, in cartesian coordinates we will have:

$$F_x = \frac{\partial\Phi}{\partial x},\quad F_y = \frac{\partial\Phi}{\partial y},\quad F_z = \frac{\partial\Phi}{\partial z}$$

in terms of which the x component of the curl will be:

$$\left(\frac{\partial F_z}{\partial y} - \frac{\partial F_y}{\partial z}\right) = \left(\frac{\partial^2\Phi}{\partial z\partial y} - \frac{\partial\Phi}{\partial y\partial z}\right) = 0$$

and similarly for the y and z components. Another relation of the same type is that the divergence of the curl of a vector field vanishes identically:

$$\nabla_{\mathbf{r}} \cdot \nabla_{\mathbf{r}} \times \mathbf{F} = \frac{\partial}{\partial x}\left(\frac{\partial F_z}{\partial y} - \frac{\partial F_y}{\partial z}\right) + \frac{\partial}{\partial y}\left(\frac{\partial F_x}{\partial z} - \frac{\partial F_z}{\partial x}\right) + \frac{\partial}{\partial z}\left(\frac{\partial F_y}{\partial x} - \frac{\partial F_x}{\partial y}\right)$$

$$\Rightarrow \nabla_{\mathbf{r}} \cdot (\nabla_{\mathbf{r}} \times \mathbf{F}(\mathbf{r})) = 0 \tag{A.20}$$

A third useful relation allows us to express the curl of the curl of a vector field in terms of its divergence and its laplacian:

$$\nabla_{\mathbf{r}} \times (\nabla_{\mathbf{r}} \times \mathbf{F}(\mathbf{r})) = \nabla_{\mathbf{r}}(\nabla_{\mathbf{r}} \cdot \mathbf{F}(\mathbf{r})) - \nabla_{\mathbf{r}}^2 \mathbf{F}(\mathbf{r}) \tag{A.21}$$

the proof of which is left as an exercise.

Lastly, we mention the two fundamental theorems that are often invoked in relation to integrals of the divergence or the curl of a vector field. The first, known as the divergence or **Gauss' theorem**, relates the integral of the divergence of a vector field over a volume V to the integral of the field over the surface S which encloses the volume of integration:

$$\int_V \nabla_{\mathbf{r}} \cdot \mathbf{F}(\mathbf{r})d\mathbf{r} = \oint_S \mathbf{F}(\mathbf{r}) \cdot \hat{\mathbf{n}}_S \, dS \tag{A.22}$$

where $\hat{\mathbf{n}}_S$ is the surface-normal unit vector on element dS. The second, known as the curl or **Stokes' theorem**, relates the integral of the curl of a vector field over a surface S to the line integral of the field over the contour C which encloses the surface of integration:

$$\int_S (\nabla_{\mathbf{r}} \times \mathbf{F}(\mathbf{r})) \cdot \hat{\mathbf{n}}_S \, dS = \oint_C \mathbf{F}(\mathbf{r}) \cdot d\mathbf{l} \tag{A.23}$$

where $\hat{\mathbf{n}}_S$ is the surface-normal unit vector on element dS and $d\mathbf{l}$ is the differential vector along the contour C.

A.2 Power Series Expansions

The Taylor series expansion of a continuous and infinitely differentiable function of x around x_0 in powers of $(x - x_0)$ is:

$$f(x) = f(x_0) + \frac{1}{1!}\left[\frac{\partial f}{\partial x}\right]_{x_0}(x - x_0) + \frac{1}{2!}\left[\frac{\partial^2 f}{\partial x^2}\right]_{x_0}(x - x_0)^2 + \frac{1}{3!}\left[\frac{\partial^3 f}{\partial x^3}\right]_{x_0}(x - x_0)^3 + \cdots$$

$$\tag{A.24}$$

with the first, second, third, ... derivatives evaluated at $x = x_0$. We have written the derivatives as partials to allow for the possibility that f depends on other variables as well; the above expression is easily generalized to multivariable functions. For x close to x_0, the expansion can be truncated to the first few terms, giving a very good approximation for $f(x)$ in terms of the function and its lowest few derivatives evaluated at $x = x_0$.

Using the general expression for the Taylor series, we obtain for the common exponential, logarithmic, trigonometric, and hyperbolic functions:

$$e^x = 1 + x + \frac{1}{2}x^2 + \frac{1}{6}x^3 + \cdots, \quad x_0 = 0 \tag{A.25}$$

$$\log(1 + x) = x - \frac{1}{2}x^2 + \frac{1}{3}x^3 + \cdots, \quad x_0 = 1 \tag{A.26}$$

$$\cos x = 1 - \frac{1}{2}x^2 + \frac{1}{24}x^4 + \cdots, \quad x_0 = 0 \tag{A.27}$$

$$\cos(\pi + x) = -1 + \frac{1}{2}x^2 - \frac{1}{24}x^4 + \cdots, \quad x_0 = \pi \tag{A.28}$$

$$\sin x = x - \frac{1}{6}x^3 + \frac{1}{120}x^5 + \cdots, \quad x_0 = 0 \tag{A.29}$$

$$\sin(\pi + x) = -x + \frac{1}{6}x^3 - \frac{1}{120}x^5 + \cdots, \quad x_0 = \pi \tag{A.30}$$

$$\tan x \equiv \frac{\sin x}{\cos x} = x + \frac{1}{3}x^3 + \frac{2}{15}x^5 + \cdots, \quad x_0 = 0 \tag{A.31}$$

$$\cot x \equiv \frac{\cos x}{\sin x} = \frac{1}{x} - \frac{1}{3}x - \frac{1}{45}x^3 + \cdots, \quad x_0 = 0 \tag{A.32}$$

$$\cosh x \equiv \frac{e^x + e^{-x}}{2} = 1 + \frac{1}{2}x^2 + \frac{1}{24}x^4 + \cdots, \quad x_0 = 0 \tag{A.33}$$

$$\sinh x \equiv \frac{e^x - e^{-x}}{2} = x + \frac{1}{6}x^3 + \frac{1}{120}x^5 + \cdots, \quad x_0 = 0 \tag{A.34}$$

$$\tanh x \equiv \frac{\sinh x}{\cosh x} = x - \frac{1}{3}x^3 + \frac{2}{15}x^5 + \cdots, \quad x_0 = 0 \tag{A.35}$$

$$\coth x \equiv \frac{\cosh x}{\sinh x} = \frac{1}{x} + \frac{1}{3}x - \frac{1}{45}x^3 + \cdots, \quad x_0 = 0 \tag{A.36}$$

$$\frac{1}{1 \pm x} = 1 \mp x + x^2 \mp x^3 + \cdots, \quad |x| < 1 \tag{A.37}$$

$$\frac{1}{\sqrt{1 \pm x}} = 1 \mp \frac{1}{2}x + \frac{3}{8}x^2 \mp \frac{5}{16}x^3 + \cdots, \quad |x| < 1 \tag{A.38}$$

The last expansion is useful for the so-called multipole expansion of $|\mathbf{r} - \mathbf{r}'|^{-1}$ for $r' = |\mathbf{r}'| \ll r = |\mathbf{r}|$:

$$\frac{1}{|\mathbf{r} - \mathbf{r}'|} = \frac{1}{r\sqrt{1 + r'^2/r^2 - 2\mathbf{r} \cdot \mathbf{r}'/r^2}}$$

$$= \frac{1}{r} + \frac{\mathbf{r} \cdot \mathbf{r}'}{r^3} + \frac{3(\mathbf{r} \cdot \mathbf{r}')^2 - r'^2 r^2}{2r^5} + \frac{5(\mathbf{r} \cdot \mathbf{r}')^3 - 3(\mathbf{r} \cdot \mathbf{r}')r'^2 r^2}{2r^7} + \cdots$$

with the first term referred to as the monopole term, the second the dipole term, the third the quadrupole term, and the fourth the octupole term.

Another very useful power-series expansion is the binomial expansion:

$$(a + b)^p = a^p + \frac{p}{1!}a^{p-1}b + \frac{p(p-1)}{2!}a^{p-2}b^2 + \frac{p(p-1)(p-2)}{3!}a^{p-3}b^3 + \cdots$$

$$+ \frac{p(p-1)(p-2)\cdots(p-(n-1))}{n!}a^{p-n}b^n + \cdots \tag{A.39}$$

with $a > 0, b, p$ real. This can easily be obtained as the Taylor series expansion of the function $f(x) = a^p(1 + x)^p$, with $x = (b/a)$, expanded around $x_0 = 0$. It is evident from this expression that for $p = n$: an integer, the binomial expansion is finite and has a total number of $(n + 1)$ terms; if p is not an integer, it has infinite terms.

A.3 Functional Derivatives

The derivative of a definite integral over a variable x with respect to a variable y that appears either in the limits of integration or in the integrand is given by:

$$\frac{\partial}{\partial y} \int_{a(y)}^{b(y)} f(x, y)dx = \int_{a(y)}^{b(y)} \frac{\partial f}{\partial y}(x, y)dx$$

$$+ \frac{db(y)}{dy}f(x = b, y) - \frac{da(y)}{dy}f(x = a, y) \qquad (A.40)$$

where $\partial f/\partial y$ is a function of both x and y. This expression can be used as the basis for performing functional derivatives: suppose that we have a functional $F[y]$ of a function $y(x)$ which involves the integration of another function of x and y over the variable x:

$$F[y] = \int_a^b f(x, y)dx \qquad (A.41)$$

where the limits of integration are constant. We want to find the condition that will give an extremum (maximum or minimum) of $F[y]$ with respect to changes in $y(x)$ in the interval $x \in [a, b]$. Suppose we change y to y' by adding the function $\delta y(x) = \epsilon \eta(x)$:

$$y'(x) = y(x) + \delta y(x) = y(x) + \epsilon \eta(x)$$

where ϵ is an infinitesimal quantity ($\epsilon \to 0$) and $\eta(x)$ an arbitrary function that satisfies the conditions $\eta(a) = \eta(b) = 0$ so that we do not introduce changes in the limits of integration. The corresponding F will be given by:

$$F[y'] = \int_a^b f(x, y')dx = \int_a^b f(x, y + \epsilon \eta)dx$$

Taking the derivative of this expression with respect to ϵ and setting it equal to zero to obtain an extremum, we find:

$$\frac{dF[y']}{d\epsilon} = \int_a^b \frac{\partial f}{\partial y'} \frac{dy'}{d\epsilon}dx = \int_a^b \frac{\partial f}{\partial y'}\eta(x)dx = 0$$

and since we want this to be true for $\epsilon \to 0$ and arbitrary $\eta(x)$, we conclude that we must have:

$$\frac{\delta F}{\delta y} \equiv \lim_{\epsilon \to 0} \frac{\partial f}{\partial y'}(x, y) = \frac{\partial f}{\partial y}(x, y) = 0$$

Thus, the necessary condition for finding the extremum of $F[y]$ with respect to variations in $y(x)$, when $F[y]$ has the form of Eq. (A.41), which we define as the variational functional derivative $\delta F / \delta y$, is the vanishing of the partial derivative with respect to y of the *integrand* appearing in F.

A.4 Fourier and Inverse Fourier Transforms

We derive the Fourier transform relations for 1D functions of a single variable x and the corresponding Fourier-space (also referred to as "reciprocal"-space) functions with variable k. We start with the definition of the Fourier expansion of a function $f(x)$ with period $2L$ in terms of complex exponentials:

$$f(x) = \sum_{n=-\infty}^{\infty} \hat{f}_n e^{in\pi x/L} \tag{A.42}$$

with n taking all integer values, which obviously satisfies the periodic relation $f(x + 2L) = f(x)$. Using the orthogonality of the complex exponentials:

$$\int_{-L}^{L} e^{-im\pi x/L} e^{in\pi x/L} dx = (2L)\delta_{nm} \tag{A.43}$$

with δ_{nm} the Kronecker delta:

$$\delta_{nm} = \begin{cases} 1 & \text{for } n = m \\ 0 & \text{for } n \neq m \end{cases} \tag{A.44}$$

we can multiply both sides of Eq. (A.43) by $\exp[-im\pi x/L]$ and integrate over x to obtain:

$$\hat{f}_n = \frac{1}{2L} \int_{-L}^{L} f(x) e^{-in\pi x/L} dx \tag{A.45}$$

We next define the reciprocal-space variable k through:

$$k = \frac{n\pi}{L} \Rightarrow dk = \frac{\pi}{L} \tag{A.46}$$

and take the limit $L \to \infty$, in which case k becomes a continuous variable, to obtain the Fourier transform $\hat{f}(k)$:

$$\hat{f}(k) \equiv \lim_{L \to \infty} \left[\hat{f}_n 2L \right] = \int_{-\infty}^{\infty} f(x) e^{-ikx} dx \tag{A.47}$$

Using the relation (A.46) in Eq. (A.42), in the limit $L \to \infty$, we obtain the inverse Fourier transform:

$$f(x) = \lim_{L \to \infty} \sum_{n=-\infty}^{\infty} (\hat{f}_n 2L) \frac{1}{2L} e^{in\pi x/L} = \frac{1}{2\pi} \int_{-\infty}^{\infty} \hat{f}(k) e^{ikx} dk \tag{A.48}$$

The expressions for the Fourier and inverse Fourier transforms in three dimensions are straightforward generalizations of Eqs (A.47) and (A.48) with x replaced by a vector $\mathbf{r}$ and k replaced by a vector $\mathbf{k}$:

$$\hat{f}(\mathbf{k}) = \int f(\mathbf{r}) e^{-i\mathbf{k}\cdot\mathbf{r}} d\mathbf{r} \Rightarrow f(\mathbf{r}) = \frac{1}{(2\pi)^3} \int \hat{f}(\mathbf{k}) e^{i\mathbf{k}\cdot\mathbf{r}} d\mathbf{k} \tag{A.49}$$

A.4.1 Fourier Transform of the Convolution

An important concept which applies in many physical situations is the convolution $h(\mathbf{r})$ of two functions $f(\mathbf{r})$ and $g(\mathbf{r})$, defined as:

$$h(\mathbf{r}) = \int f(\mathbf{r} - \mathbf{r}')g(\mathbf{r}')d\mathbf{r}' \tag{A.50}$$

The Fourier transform of the convolution gives:

$$\hat{h}(\mathbf{k}) = \int h(\mathbf{r})e^{-i\mathbf{k}\cdot\mathbf{r}}d\mathbf{r} = \int \left[\int f(\mathbf{r} - \mathbf{r}')g(\mathbf{r}')d\mathbf{r}'\right]e^{-i\mathbf{k}\cdot\mathbf{r}}d\mathbf{r}$$

$$= \int \left[\int f(\mathbf{r} - \mathbf{r}')e^{-i\mathbf{k}\cdot(\mathbf{r}-\mathbf{r}')}d\mathbf{r}\right]g(\mathbf{r}')e^{-i\mathbf{k}\cdot\mathbf{r}'}d\mathbf{r}' = \hat{f}(\mathbf{k})\hat{g}(\mathbf{k}) \tag{A.51}$$

that is, the Fourier transform of the convolution is the product of the Fourier transforms $\hat{f}(\mathbf{k})$ and $\hat{g}(\mathbf{k})$ of the functions $f(\mathbf{r})$ and $g(\mathbf{r})$. This is a very useful result, referred to as the "convolution theorem."

A.5 The δ-Function and its Fourier Transform

A.5.1 The δ-Function and the θ-Function

The standard definition of the δ-function, also known as the Dirac function, is:

$$\delta(0) \to \infty, \quad \delta(x \neq 0) = 0, \quad \int_{-\infty}^{\infty} \delta(x - x')dx' = 1 \tag{A.52}$$

that is, it is a function with an infinite peak at the zero of its argument, it is zero everywhere else, and it integrates to unity. From this definition, it follows that the product of the δ-function with an arbitrary function $f(x)$ integrated over all values of x must satisfy:

$$\int_{-\infty}^{\infty} f(x')\delta(x - x')dx' = f(x) \tag{A.53}$$

The δ-function is not a function in the usual sense, it is a function represented by a limit of usual functions. For example, a simple generalization of the Kronecker δ is:

$$w(a; x) = \begin{cases} \frac{1}{2a} & \text{for } |x| \leq a \\ 0 & \text{for } |x| > a \end{cases} \tag{A.54}$$

We will refer to this as the "window" function, since its product with any other function picks out the values of that function in the range $-a \leq x \leq a$, that is, in a window of width $2a$ in x. Taking the limit of $w(a; x)$ for $a \to 0$ produces a function with the desired behavior to represent the δ-function. Typically, we are interested in smooth, continuous functions whose appropriate limit can represent a δ-function. We give below a few examples of such functions.

A useful representation of the δ-function is a gaussian:

$$\delta(x - x') = \lim_{\beta \to 0} \left(\frac{1}{\beta \sqrt{\pi}} e^{-(x-x')^2/\beta^2} \right) \tag{A.55}$$

To prove that this function has the proper behavior, we note that around $x = x'$ it has width β which vanishes for $\beta \to 0$, its value at $x = x'$ is $1/\beta\sqrt{\pi}$ which is infinite for $\beta \to 0$, and its integral over all values of x' is 1 for any value of β (see discussion on normalized gaussians in Section A.7).

Another useful representation of the δ-function is a lorentzian:

$$\delta(x - x') = \lim_{\beta \to 0} \left(\frac{1}{\pi} \frac{\beta}{(x - x')^2 + \beta^2} \right) \tag{A.56}$$

To prove that this function has the proper behavior, we note that around $x = x'$ it has width β which vanishes for $\beta \to 0$, its value at $x = x'$ is $1/\pi\beta$ which is infinite for $\beta \to 0$, and its integral over all values of x' is 1 for any value of β, which is easily shown by contour integration (the function has simple poles at $x = x' \pm i\beta$).

Yet another useful representation of the δ-function is the following:

$$\delta(x - x') = \lim_{\beta \to 0} \left[\frac{1}{\pi\beta} \sin\left(\frac{x - x'}{\beta} \right) \frac{\beta}{x - x'} \right] \tag{A.57}$$

To prove that this function has the proper behavior, we note that the function $\sin(y)/y \to 1$ for $y \to 0$ and its width around $y = 0$ is determined by half the interval between y values at which it first becomes zero, which occurs at $y = \pm\pi$, that is, its width in the variable y is π. With $y = (x - x')/\beta$, we conclude that around $x = x'$ the function on the right-hand side of Eq. (A.57) has width $\beta\pi$ in the variable x, which vanishes for $\beta \to 0$, while its value at $x = x'$ is $1/\pi\beta$, which is infinite for $\beta \to 0$. Finally, its integral over all values of x' is 1 for any value of β, which is easily shown by expressing

$$\sin\left(\frac{x - x'}{\beta} \right) \frac{1}{x - x'} = \frac{e^{i(x-x')/\beta}}{2i(x - x')} - \frac{e^{-i(x-x')/\beta}}{2i(x - x')}$$

and using contour integration (each term in the integrand has a simple pole at $x = x'$).

In order to construct δ-functions in more than one dimension, we simply multiply δ-functions in each of the independent dimensions. For example, in cartesian coordinates we would have:

$$\delta(\mathbf{r} - \mathbf{r}') = \delta(x - x')\delta(y - y')\delta(z - z')$$

where $\mathbf{r} = x\hat{\mathbf{x}} + y\hat{\mathbf{y}} + z\hat{\mathbf{z}}$ and $\mathbf{r}' = x'\hat{\mathbf{x}} + y'\hat{\mathbf{y}} + z'\hat{\mathbf{z}}$ are two 3D vectors. A useful representation of the δ-function in three dimensions is given by:

$$\delta(\mathbf{r} - \mathbf{r}') = \frac{1}{4\pi} \nabla_\mathbf{r} \cdot \left(\frac{\mathbf{r} - \mathbf{r}'}{|\mathbf{r} - \mathbf{r}'|^3} \right) \tag{A.58}$$

This can be proved as follows: the expression on the right-hand side of this equation is identically zero everywhere except at $x = x', y = y', z = z'$, at which point it becomes infinite [both claims can easily be shown by direct application of the divergence expression,

Eq. (A.4)]; these are the two essential features of the δ-function. Moreover, using the divergence theorem, we have:

$$\int \nabla_{\mathbf{r}} \cdot \left(\frac{\mathbf{r} - \mathbf{r}'}{|\mathbf{r} - \mathbf{r}'|^3} \right) d\mathbf{r}' = \oint_{S'} \left(\frac{\mathbf{r} - \mathbf{r}'}{|\mathbf{r} - \mathbf{r}'|^3} \right) \cdot \hat{\mathbf{n}}_{S'} dS'$$

where S' is a surface enclosing the volume of integration and $\hat{\mathbf{n}}_{S'}$ is the surface-normal unit vector. Choosing the volume of integration to be a sphere centered at $\mathbf{r}$, in which case the surface-normal unit vector takes the form

$$\hat{\mathbf{n}}_{S'} = \frac{\mathbf{r} - \mathbf{r}'}{|\mathbf{r} - \mathbf{r}'|}$$

we obtain for the surface integral:

$$\oint_{S'} \left(\frac{\mathbf{r} - \mathbf{r}'}{|\mathbf{r} - \mathbf{r}'|^3} \right) \cdot \hat{\mathbf{n}}_{S'} dS' = \int_0^{2\pi} \int_0^\pi \frac{1}{|\mathbf{r} - \mathbf{r}'|^2} |\mathbf{r} - \mathbf{r}'|^2 \sin\theta d\theta d\phi = 4\pi$$

This completes the proof that the right-hand side of Eq. (A.58) is a properly normalized δ-function in three dimensions.

A function closely related to the δ-function is the socalled θ-function or step-function, also known as the Heavyside function:

$$\theta(x - x') = 0, \text{ for } x < x', \qquad \theta(x - x') = 1, \text{ for } x > x' \tag{A.59}$$

A useful representation of the θ-function is:

$$\theta(x - x') = \lim_{\beta \to 0} \left(\frac{1}{e^{-(x-x')/\beta} + 1} \right) \tag{A.60}$$

The δ-function and the θ-function are functions of their argument only in the limiting sense given in expressions (A.55) and (A.60). The behavior of these functions for several values of β is shown in Fig. A.1. From the definition of the θ-function, we can see that its derivative must be a δ-function:

$$\frac{d}{dx}\theta(x - x') = \delta(x - x')$$

It is easy to show that the derivative of the representation of the θ-function given in Eq. (A.60):

$$\lim_{\beta \to 0} \left(\frac{e^{-(x-x')/\beta}}{\beta(e^{-(x-x')/\beta} + 1)^2} \right)$$

is indeed another representation of the δ-function, as expected.

An expression often arises in which the δ-function has as its argument another function, $f(x)$, and the integral of its product with yet another function, $g(x)$, must be evaluated:

$$I = \int_{-\infty}^{\infty} g(x)\delta(f(x))dx$$

We consider the result of this evaluation, assuming that $f(x)$ has only simple roots, that is, $f(x_0) = 0, f'(x_0) \neq 0$, with $f'(x)$ the derivative of $f(x)$. Since the δ-function is zero when its argument is not zero, we work first only in the neighborhood of one simple root x_0 of $f(x)$,

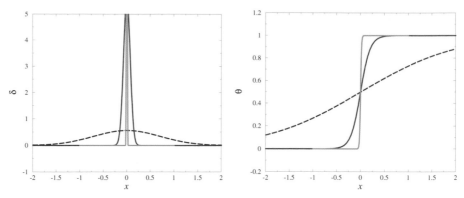

Left: The δ-function as represented by a normalized gaussian, Eq. (A.55), for $\beta = 1$ (dashed red line), 0.1 (blue dotted line), and 0.01 (green line). **Right**: The θ-function as represented by Eq. (A.60), for $\beta = 1$ (dashed red line), 0.1 (blue dotted line), and 0.01 (green line).

that is, $x \in [x_1, x_2]$, $x_1 < x_0 < x_2$. We define the variable $u = f(x)$, which at $x = x_0, x_1, x_2$ takes the values $0, u_1, u_2$, respectively. From this definition we have:

$$du = f'(x)dx \Rightarrow dx = \left(f'(x)\right)^{-1} du, \quad x = f^{-1}(u)$$

These expressions, when inserted in the integral under consideration with x in the neighborhood of x_0, give:

$$I_0 = \int_{u_1}^{u_2} g(f^{-1}(u))\delta(u) \left(f'(f^{-1}(u))\right)^{-1} du$$

In order to evaluate this integral we only need to consider the order of the limits. With x_1, x_2 close to x_0, we have:

$$f'(x_0) > 0 \Rightarrow f(x_1) = u_1 < f(x_0) = 0 < f(x_2) = u_2$$

$$f'(x_0) < 0 \Rightarrow f(x_1) = u_1 > f(x_0) = 0 > f(x_2) = u_2$$

which gives in turn:

$$f'(x_0) > 0 \Rightarrow I_0 = \int_{u_1}^{u_2} g(f^{-1}(u))\delta(u) \left(f'(f^{-1}(u))\right)^{-1} du = g(x_0) \left|f'(x_0)\right|^{-1}$$

$$f'(x_0) < 0 \Rightarrow I_0 = -\int_{u_2}^{u_1} g(f^{-1}(u))\delta(u) \left[-\left|f'(f^{-1}(u))\right|^{-1}\right] du = g(x_0) \left|f'(x_0)\right|^{-1}$$

Thus, in both cases we find that the integral is equal to the product of $g(x)$ with $\left|f'(x)\right|^{-1}$ evaluated at $x = x_0$. Since the argument of $\delta(f(x))$ vanishes at all the roots $x_0^{(i)}$ of $f(x)$, we will get similar contributions from each root, which gives as a final result:

$$\int_{-\infty}^{\infty} g(x)\delta(f(x))dx = \sum_i g(x_0^{(i)}) \left|f'(x_0^{(i)})\right|^{-1}, \quad \text{where } f(x_0^{(i)}) = 0 \tag{A.61}$$

with the summation on i running over all the simple roots of $f(x)$. These results imply directly the following relations:

$$\delta(a(x - x_0)) = \frac{1}{|a|}\delta(x - x_0)$$

$$\delta(f(x)) = \sum_i \left| f'(x_0^{(i)}) \right|^{-1} \delta(x - x_0^{(i)}), \quad \text{where } f(x_0^{(i)}) = 0$$

A.5.2 Fourier Transform of the δ-Function

To derive the Fourier transform of the δ-function we start with its Fourier expansion representation as in Eq. (A.42):

$$\delta(x) = \sum_{n=-\infty}^{\infty} d_n e^{i\frac{n\pi}{L}x} \Rightarrow d_n = \frac{1}{2L} \int_{-L}^{L} \delta(x) e^{-i\frac{n\pi}{L}x} dx = \frac{1}{2L} \tag{A.62}$$

which in the limit $L \to \infty$ produces:

$$\hat{\delta}(k) = \lim_{L \to \infty} (2L d_n) = 1 \tag{A.63}$$

Therefore, the inverse Fourier transform of the δ-function gives:

$$\delta(x - x') = \frac{1}{2\pi} \int_{-\infty}^{\infty} \hat{\delta}(k) e^{ik(x-x')} dk = \frac{1}{2\pi} \int_{-\infty}^{\infty} e^{ik(x-x')} dk \tag{A.64}$$

These relations for a function of the 1D variable x can be generalized straightforwardly to functions of 3D variables $\mathbf{r} = x\hat{\mathbf{x}} + y\hat{\mathbf{y}} + z\hat{\mathbf{z}}$ and $\mathbf{k} = k_x\hat{\mathbf{x}} + k_y\hat{\mathbf{y}} + k_z\hat{\mathbf{z}}$:

$$\delta(\mathbf{r} - \mathbf{r}') = \frac{1}{(2\pi)^3} \int e^{i(\mathbf{r}-\mathbf{r}')\cdot\mathbf{k}} d\mathbf{k} \tag{A.65}$$

or equivalently, by inverting the roles of $\mathbf{k}$ and $\mathbf{r}$, we obtain:

$$\delta(\mathbf{k} - \mathbf{k}') = \frac{1}{(2\pi)^3} \int e^{-i(\mathbf{k}-\mathbf{k}')\cdot\mathbf{r}} d\mathbf{r} \tag{A.66}$$

Notice that both δ-functions are symmetric functions of their argument:

$$\delta(\mathbf{r} - \mathbf{r}') = \delta(\mathbf{r}' - \mathbf{r}), \quad \delta(\mathbf{k} - \mathbf{k}') = \delta(\mathbf{k}' - \mathbf{k}).$$

A.5.3 The δ-Function Sums for Crystals

Consider the sum over all points in the first BZ: $\sum_{\mathbf{k}\in BZ} \exp[i\mathbf{k} \cdot \mathbf{R}]$. Let us assume that all $\mathbf{k}$ vectors lie within the first BZ. If we shift $\mathbf{k}$ by an arbitrary vector $\mathbf{k}_0 \in BZ$, this sum does not change, because for a vector $\mathbf{k} + \mathbf{k}_0$ which lies outside the first BZ there is a vector within the first BZ that differs from it by $\mathbf{G}$ and $\exp[i\mathbf{G} \cdot \mathbf{R}] = 1$. Moreover, all points translated outside the first BZ by the addition of $\mathbf{k}_0$ have exactly one image by translation inside the first BZ through $\mathbf{G}$, since all BZs have the same volume. Therefore, we have:

$$\sum_{\mathbf{k} \in BZ} e^{i\mathbf{k}\cdot\mathbf{R}} = \sum_{\mathbf{k} \in BZ} e^{i(\mathbf{k}+\mathbf{k}_0)\cdot\mathbf{R}} = e^{i\mathbf{k}_0\cdot\mathbf{R}} \left[\sum_{\mathbf{k} \in BZ} e^{i\mathbf{k}\cdot\mathbf{R}} \right] \Rightarrow e^{i\mathbf{k}_0\cdot\mathbf{R}} = 1 \Rightarrow \mathbf{R} = 0 \qquad (A.67)$$

where the last equality must hold because the relation $\exp[i\mathbf{k}_0 \cdot \mathbf{R}] = 1$ must hold for any $\mathbf{k}_0 \in BZ$, assuming that the sum itself is not zero. Thus we have proved that the sum $\sum_{\mathbf{k} \in BZ} \exp[i\mathbf{k} \cdot \mathbf{R}]$, if it is not equal to zero, necessitates that $\mathbf{R} = 0$, in which case all terms in the sum are 1 and the sum itself becomes equal to N, the total number of PUCs in the crystal. If $\mathbf{R} \neq 0$ then $\exp[i\mathbf{k}_0 \cdot \mathbf{R}] \neq 1$ for an arbitrary value of $\mathbf{k}_0$, and therefore the sum itself must be zero in order for the original equality in Eq. (A.67) to hold. These results put together imply:

$$\frac{1}{N} \sum_{\mathbf{k} \in BZ} e^{i\mathbf{k}\cdot\mathbf{R}} = \delta(\mathbf{R}) \qquad (A.68)$$

where $\delta(\mathbf{R})$ represents a Kronecker δ, defined in Eq. (A.44): $\delta(\mathbf{R}) = 1$ for $\mathbf{R} = 0$ and $\delta(\mathbf{R}) = 0$ for $\mathbf{R} \neq 0$. This is consistent with the fact that, strictly speaking, the summation over $\mathbf{k}$ vectors is a discrete sum over the N values that $\mathbf{k}$ takes in the BZ. In the limit of an infinite crystal, $N \to \infty$, it is customary to think of $\mathbf{k}$ as a continuous variable and to treat $\delta(\mathbf{k})$ as a δ-function rather than a Kronecker δ. The definition of $\delta(\mathbf{R})$ implies that for an arbitrary function $f(\mathbf{R})$ we will have:

$$\frac{1}{N} \sum_{\mathbf{R}} \sum_{\mathbf{k} \in BZ} f(\mathbf{R}) e^{i\mathbf{k}\cdot\mathbf{R}} = \sum_{\mathbf{R}} f(\mathbf{R})\delta(\mathbf{R}) = f(0) \qquad (A.69)$$

By a similar line of argument, we can show:

$$\frac{1}{N} \sum_{\mathbf{R}} e^{i\mathbf{k}\cdot\mathbf{R}} = \delta(\mathbf{k} - \mathbf{G}) \qquad (A.70)$$

where we treat the summation over $\mathbf{R}$ vectors as a discrete sum, in which case the right-hand side is a Kronecker δ. To prove this, we change every vector $\mathbf{R}$ in the argument of the exponential of the sum by $\mathbf{R}_0$; but each new vector $\mathbf{R} + \mathbf{R}_0$ is also a translational vector, so that the sum does not change because the summation extends already over all possible translations, that is:

$$\sum_{\mathbf{R}} e^{i\mathbf{k}\cdot\mathbf{R}} = \sum_{\mathbf{R}} e^{i\mathbf{k}\cdot(\mathbf{R}+\mathbf{R}_0)} = e^{i\mathbf{k}\cdot\mathbf{R}_0} \left[\sum_{\mathbf{R}} e^{i\mathbf{k}\cdot\mathbf{R}} \right] \Rightarrow e^{i\mathbf{k}\cdot\mathbf{R}_0} = 1 \Rightarrow \mathbf{k} = \mathbf{G} \qquad (A.71)$$

where the last equation is a consequence of the fact that the relation $\exp[i\mathbf{k} \cdot \mathbf{R}_0] = 1$ must hold for an arbitrary translational vector $\mathbf{R}_0$, assuming that the sum itself is not zero. If $\mathbf{k} = \mathbf{G}$, a reciprocal-space vector, then all terms in the above sum are 1, and the sum itself is equal to N (there are N terms in the sum, N being the total number of PUCs in the crystal). If $\mathbf{k} \neq \mathbf{G}$, then the sum itself must be 0, so that the original equality in Eq. (A.71) can still hold with $\mathbf{R}_0$ an arbitrary translational vector. The two results together imply the relation Eq. (A.70), where $\delta(\mathbf{k})$ represents a Kronecker δ, defined in Eq. (A.44): $\delta(\mathbf{k} - \mathbf{G}) = 1$ for $\mathbf{k} = \mathbf{G}$ and $\delta(\mathbf{k} - \mathbf{G}) = 0$ for $\mathbf{k} \neq \mathbf{G}$. This is consistent with the fact that the summation over $\mathbf{R}$ vectors is a discrete sum over the N values that $\mathbf{R}$ takes for the entire crystal. The definition of $\delta(\mathbf{k} - \mathbf{G})$ implies that for an arbitrary function $g(\mathbf{k})$, if $\mathbf{k}$ is restricted to the first BZ, we will have:

$$\frac{1}{N} \sum_{\mathbf{k} \in BZ} \sum_{\mathbf{R}} g(\mathbf{k}) e^{i\mathbf{k} \cdot \mathbf{R}} = \sum_{\mathbf{k} \in BZ} g(\mathbf{k}) \delta(\mathbf{k}) = g(0) \tag{A.72}$$

because the only reciprocal lattice vector contained in the first BZ is $\mathbf{G} = 0$.

A.6 The Coulomb Integral

Having developed the tools to handle δ-functions and their Fourier transforms, we discuss an example of an integral involving such transformations that we encountered in Chapter 4, Eq. (4.55). We refer to this as the "Coulomb integral":

$$\int_{k' < k_{\mathrm{F}}} \frac{\mathrm{d}\mathbf{k}'}{(2\pi)^3} \int \frac{e^{-i(\mathbf{k} - \mathbf{k}') \cdot (\mathbf{r} - \mathbf{r}')}}{|\mathbf{r} - \mathbf{r}'|} \mathrm{d}\mathbf{r}' \tag{A.73}$$

We first work out the Fourier and inverse Fourier transforms of the function $|\mathbf{r} - \mathbf{r}'|^{-1}$, which are given by:

$$\int e^{-i\mathbf{k} \cdot \mathbf{r}} \frac{1}{|\mathbf{r} - \mathbf{r}'|} \mathrm{d}\mathbf{r} = \frac{4\pi}{k^2} e^{-i\mathbf{k} \cdot \mathbf{r}'} \Longleftrightarrow \int \frac{\mathrm{d}\mathbf{k}}{(2\pi)^3} \frac{4\pi}{k^2} e^{i\mathbf{k} \cdot (\mathbf{r} - \mathbf{r}')} = \frac{1}{|\mathbf{r} - \mathbf{r}'|} \tag{A.74}$$

To prove these expressions, notice that we can shift the origin by $\mathbf{r}'$, so that, from the definition of the Fourier transform in 3D space, Eq. (A.49), we need to calculate the following integral:

$$\int e^{-i\mathbf{k} \cdot \mathbf{r}} \frac{1}{|\mathbf{r} - \mathbf{r}'|} \mathrm{d}\mathbf{r} = e^{-i\mathbf{k} \cdot \mathbf{r}'} \int e^{-i\mathbf{k} \cdot (\mathbf{r} - \mathbf{r}')} \frac{1}{|\mathbf{r} - \mathbf{r}'|} \mathrm{d}\mathbf{r} = e^{-i\mathbf{k} \cdot \mathbf{r}'} \int e^{-i\mathbf{k} \cdot \mathbf{r}} \frac{1}{r} \mathrm{d}\mathbf{r}$$

The Fourier transform can now be calculated with a useful general trick: we multiply the function $1/r$ by $\exp[-\eta r]$ where $\eta > 0$ and in the end we take the limit $\eta \to 0$. Using spherical coordinates $\mathrm{d}\mathbf{r} = r^2 \mathrm{d}r \sin\theta \mathrm{d}\theta \mathrm{d}\phi = \mathrm{d}x\mathrm{d}y\mathrm{d}z$, with the convenient choice of the z-axis along the $\mathbf{k}$ vector (so that $\mathbf{k} \cdot \mathbf{r} = kr \cos\theta$), we obtain:

$$\int \frac{e^{-\eta r}}{r} e^{-i\mathbf{k} \cdot \mathbf{r}} \mathrm{d}\mathbf{r} = \int_0^{2\pi} \int_0^{\pi} \int_0^{\infty} \left[r^2 \sin\theta \frac{e^{-\eta r}}{r} e^{-ikr\cos\theta} \right] \mathrm{d}r \mathrm{d}\theta \mathrm{d}\phi$$

$$= 2\pi \int_0^{\infty} \int_{-1}^{1} \left[r e^{-(ikw + \eta)r} \right] \mathrm{d}w \mathrm{d}r$$

$$= \frac{2\pi}{-ik} \left(\left[\frac{1}{-ik - \eta} e^{(-ik - \eta)r} \right]_0^{\infty} - \left[\frac{1}{ik - \eta} e^{(ik - \eta)r} \right]_0^{\infty} \right) = \frac{4\pi}{k^2 + \eta^2}$$

which, in the limit $\eta \to 0$, gives the desired result, the first of the Fourier relations in Eq. (A.74). Identifying the Fourier transform of $f(\mathbf{r}) = |\mathbf{r} - \mathbf{r}'|^{-1}$ as:

$$\hat{f}(\mathbf{k}) = \frac{4\pi}{k^2} e^{-i\mathbf{k} \cdot \mathbf{r}'}$$

and inserting it in the expression for the inverse Fourier transform, Eq. (A.49), produces the second Fourier relation in Eq. (A.74).

Using the inverse Fourier transform of $1/|\mathbf{r} - \mathbf{r}'|$ in the following form:

$$\frac{1}{|\mathbf{r} - \mathbf{r}'|} = \int \frac{d\mathbf{q}}{(2\pi)^3} \frac{4\pi}{q^2} e^{i\mathbf{q}\cdot(\mathbf{r}-\mathbf{r}')} \tag{A.75}$$

and substituting this expression in the Coulomb integral, Eq. (A.73), we obtain:

$$\int \frac{d\mathbf{k}'}{(2\pi)^3} \int e^{-i(\mathbf{k}-\mathbf{k}')\cdot(\mathbf{r}-\mathbf{r}')} \int \frac{d\mathbf{q}}{(2\pi)^3} \frac{4\pi}{q^2} e^{i\mathbf{q}\cdot(\mathbf{r}-\mathbf{r}')} d\mathbf{r}'$$

$$= 4\pi \int_{k'<k_F} \frac{d\mathbf{k}'}{(2\pi)^3} \int d\mathbf{q} \frac{1}{q^2} \left[\frac{1}{(2\pi)^3} \int e^{-i(\mathbf{k}-\mathbf{k}'-\mathbf{q})\cdot(\mathbf{r}-\mathbf{r}')} d\mathbf{r}' \right] \tag{A.76}$$

At this point it is useful to employ the Fourier transform representation of the δ-function: This allows us to identify the quantity in square brackets in the last expression with a δ-function in momentum space:

$$\frac{1}{(2\pi)^3} \int e^{-i(\mathbf{k}-\mathbf{k}'-\mathbf{q})\cdot(\mathbf{r}-\mathbf{r}')} d\mathbf{r}' = \delta(\mathbf{q} - (\mathbf{k} - \mathbf{k}')) \tag{A.77}$$

which, upon integration over $\mathbf{q}$, gives:

$$4\pi \left[\int_{k'<k_F} \frac{d\mathbf{k}'}{(2\pi)^3} \frac{1}{|\mathbf{k} - \mathbf{k}'|^2} \right] \tag{A.78}$$

which reduces to elementary integrals to give as a final result:

$$\int_{k'<k_F} \frac{d\mathbf{k}'}{(2\pi)^3} \int \frac{e^{-i(\mathbf{k}-\mathbf{k}')\cdot(\mathbf{r}-\mathbf{r}')}}{|\mathbf{r} - \mathbf{r}'|} d\mathbf{r}' = \frac{k_F}{\pi} F(k/k_F) \tag{A.79}$$

with $k = |\mathbf{k}|$, where the function $F(x)$ is defined as:

$$F(x) = 1 + \frac{1 - x^2}{2x} \ln \left| \frac{1 + x}{1 - x} \right| \tag{A.80}$$

A.7 Normalized Gaussians

In order to find the constant factor which will ensure normalization of the gaussian function, we must calculate its integral over all values of the variable. This integral is most easily performed by using Gauss' trick of taking the square root of the square of the integral and turning the integration over cartesian coordinates $dxdy$ into one over polar coordinates $rdrd\theta$:

$$\int_{-\infty}^{\infty} e^{-\alpha^2 x^2} dx = \left[\int_{-\infty}^{\infty} e^{-\alpha^2 x^2} dx \int_{-\infty}^{\infty} e^{-\alpha^2 y^2} dy \right]^{\frac{1}{2}}$$

$$= \left[\int_{-\infty}^{\infty} \int_{-\infty}^{\infty} e^{-\alpha^2 (x^2 + y^2)} dx dy \right]^{\frac{1}{2}} = \left[\int_{0}^{2\pi} \int_{0}^{\infty} e^{-\alpha^2 r^2} r dr d\theta \right]^{\frac{1}{2}}$$

$$= \left[2\pi \int_{0}^{\infty} e^{-\alpha^2 r^2} \frac{1}{2} dr^2 \right]^{\frac{1}{2}} = \frac{\sqrt{\pi}}{\alpha} \tag{A.81}$$

so that in three dimensions the normalized gaussian centered ar $\mathbf{r}_0$ is:

$$\left(\frac{\alpha}{\sqrt{\pi}} \right)^3 e^{-\alpha^2 |\mathbf{r} - \mathbf{r}_0|^2} \Rightarrow \left(\frac{\alpha}{\sqrt{\pi}} \right)^3 \int e^{-\alpha^2 |\mathbf{r} - \mathbf{r}_0|^2} d\mathbf{r} = 1 \tag{A.82}$$

In the limit of $\alpha \to \infty$ this function tends to an infinitely sharp spike of infinite height which integrates to unity, that is, a properly defined δ-function [compare to Eq. (A.55) with $\beta = 1/\alpha$].

The above derivation provides an easy way of calculating the even moments of the normalized gaussian, by simply taking derivatives with respect to α^2 of both sides of Eq. (A.81):

$$\int_{-\infty}^{\infty} x^2 e^{-\alpha^2 x^2} dx = -\frac{\partial}{\partial(\alpha^2)} \int_{-\infty}^{\infty} e^{-\alpha^2 x^2} dx = \frac{1}{2} \frac{\sqrt{\pi}}{\alpha^3}$$

$$\int_{-\infty}^{\infty} x^4 e^{-\alpha^2 x^2} dx = -\frac{\partial}{\partial(\alpha^2)} \int_{-\infty}^{\infty} x^2 e^{-\alpha^2 x^2} dx = \frac{3}{4} \frac{\sqrt{\pi}}{\alpha^5}$$

$$\int_{-\infty}^{\infty} x^{2n} e^{-\alpha^2 x^2} dx = -\frac{\partial^n}{\partial(\alpha^2)^n} \int_{-\infty}^{\infty} e^{-\alpha^2 x^2} dx = \frac{(2n-1)!!}{2^n} \frac{\sqrt{\pi}}{\alpha^{2n+1}}$$

where the symbol $(2n-1)!!$ denotes the product of all odd integers up to $(2n-1)$.

A.8 Potential of a Gaussian Function

We calculate the electrostatic potential of a charge distribution described by a normalized gaussian function, that is, a gaussian function which integrates to unity. This result is a central piece of the calculation of the Madelung energy through the Ewald method. We consider the charge distribution of a normalized gaussian centered at the origin of the coordinate axes:

$$\rho_g(\mathbf{r}) = \left(\frac{\alpha}{\sqrt{\pi}} \right)^3 e^{-\alpha^2 |\mathbf{r}|^2} \tag{A.83}$$

as determined in Eq. (A.82). The potential generated by this charge distribution is:

$$\Phi_g(\mathbf{r}) = \left(\frac{\alpha}{\sqrt{\pi}} \right)^3 \int \frac{e^{-\alpha^2 |\mathbf{r}'|^2}}{|\mathbf{r}' - \mathbf{r}|} d\mathbf{r}' \tag{A.84}$$

Changing variables to $\mathbf{t} = \mathbf{r}' - \mathbf{r} \Rightarrow \mathbf{r}' = \mathbf{t} + \mathbf{r}$ in the above equation, we obtain:

$$\Phi_g(\mathbf{r}) = \left(\frac{\alpha}{\sqrt{\pi}}\right)^3 \int \frac{e^{-\alpha^2(t^2+r^2+2\mathbf{t}\cdot\mathbf{r})}}{t}\, d\mathbf{t} \tag{A.85}$$

We use spherical coordinates to perform the integration $d\mathbf{t} = t^2 dt \sin\theta\, d\theta\, d\phi$, with the convenient choice of $\mathbf{r}$ defining the z-axis of the integration variable $\mathbf{t}$, which gives $\mathbf{t}\cdot\mathbf{r} = tr\cos\theta$, to obtain:

$$
\begin{aligned}
\Phi_g(\mathbf{r}) &= \left(\frac{\alpha}{\sqrt{\pi}}\right)^3 \int_0^{2\pi}\int_0^{\pi}\int_0^{\infty} \frac{e^{-\alpha^2(t^2+r^2+2tr\cos\theta)}}{t}\, t^2 dt \sin\theta\, d\theta\, d\phi \\
&= \left(\frac{\alpha}{\sqrt{\pi}}\right)^3 2\pi e^{-\alpha^2 r^2} \int_0^{\infty} e^{-\alpha^2 t^2}\left[\int_{-1}^{1} e^{-2\alpha^2 trw}\, dw\right] t\, dt \\
&= -\frac{\alpha}{r\sqrt{\pi}}\left[\int_0^{\infty} e^{-\alpha^2(t+r)^2}\, dt - \int_0^{\infty} e^{-\alpha^2(t-r)^2}\, dt\right] \\
&= -\frac{\alpha}{r\sqrt{\pi}}\left[\int_r^{\infty} e^{-\alpha^2 t^2}\, dt - \int_{-r}^{\infty} e^{-\alpha^2 t^2}\, dt\right] = \frac{\alpha}{r\sqrt{\pi}}\int_{-r}^{r} e^{-\alpha^2 t^2}\, dt \\
&= \frac{2\alpha}{r\sqrt{\pi}}\int_0^{r} e^{-\alpha^2 t^2}\, dt
\end{aligned}
$$

The last expression is related to the error function, the definition of which is:

$$\text{erf}(r) \equiv \frac{2}{\sqrt{\pi}}\int_0^{r} e^{-t^2}\, dt \tag{A.86}$$

Notice that for large values of r, the error function becomes

$$\text{erf}(r \to \infty) \to \frac{2}{\sqrt{\pi}}\int_0^{\infty} e^{-t^2}\, dt = 1$$

Thus, we can express the potential of the gaussian using the error function as:

$$\Phi_g(\mathbf{r}) = \frac{\text{erf}(\alpha r)}{r} \tag{A.87}$$

A.9 Green's Function Method

Green's function is a general mathematical tool for solving differential equations. Suppose that we are given the following equation to solve for the unknown function $f(\mathbf{r}, t)$:

$$\hat{O}(\mathbf{r}, t)f(\mathbf{r}, t) = F(\mathbf{r}, t) \tag{A.88}$$

where $\hat{O}(\mathbf{r}, t)$ is a linear operator in the variables $\mathbf{r}, t$ (it includes functions of $\mathbf{r}, t$ and derivatives), and $F(\mathbf{r}, t)$ is a given function. We define the Green's function $G(\mathbf{r}, \mathbf{r}'; t, t')$ such that

$$\hat{O}(\mathbf{r}, t)G(\mathbf{r}, \mathbf{r}'; t, t') = \delta(\mathbf{r} - \mathbf{r}')\delta(t - t') \tag{A.89}$$

then, multiplying both sides of this equation with $F(\mathbf{r}', t')$ and integrating over the variables $\mathbf{r}', t'$, we obtain:

$$\int\int \hat{\mathcal{O}}(\mathbf{r}, t)G(\mathbf{r}, \mathbf{r}'; t, t')F(\mathbf{r}', t')d\mathbf{r}'dt' = \int\int F(\mathbf{r}', t')\delta(\mathbf{r} - \mathbf{r}')\delta(t - t')d\mathbf{r}'dt' \Rightarrow$$

$$\hat{\mathcal{O}}(\mathbf{r}, t)\int\int G(\mathbf{r}, \mathbf{r}'; t, t')F(\mathbf{r}', t')d\mathbf{r}'dt' = F(\mathbf{r}, t)$$

where we have moved the operator $\hat{\mathcal{O}}(\mathbf{r}, t)$ outside the integrals over $\mathbf{r}', t'$ since it does not contain these variables. By comparing the last expression to the original equation we conclude:

$$f(\mathbf{r}, t) = \int\int G(\mathbf{r}, \mathbf{r}'; t, t')F(\mathbf{r}', t')d\mathbf{r}'dt' \tag{A.90}$$

that is, if we know the Green's function $G(\mathbf{r}, \mathbf{r}'; t, t')$ we can obtain the unknown function $f(\mathbf{r}, t)$ by simply integrating Green's function with the known function $F(\mathbf{r}', t')$ over all values of the variables $\mathbf{r}', t'$. In this sense, we can think of Green's function as the inverse of the operator $\hat{\mathcal{O}}(\mathbf{r}, t)$, that is:

$$G(\mathbf{r}, \mathbf{r}'; t, t') = [\hat{\mathcal{O}}(\mathbf{r}, t)]^{-1} \delta(\mathbf{r} - \mathbf{r}')\delta(t - t') \tag{A.91}$$

when this expression (with an operator in the denominator) can be assigned proper mathematical meaning. Finding the exact Green's function may not be easy, but finding good approximations to it, and therefore good approximations to the desired solution of the original equation, may be quite feasible.

We can now use this method in the case of the single-particle time-dependent Schrödinger equation:

$$\left[i\frac{\partial}{\partial t} - \hat{\mathcal{H}}^{\text{sp}}(\mathbf{r})\right]\psi(\mathbf{r}, t) = 0 \tag{A.92}$$

where $\hat{\mathcal{H}}^{\text{sp}}(\mathbf{r})$ is the time-independent single-particle hamiltonian. We introduce the time dependence of the solution as the exponential $\exp(-i\epsilon t/\hbar)$, which leads to the equation for the time-independent part of the solution $\psi(\mathbf{r})$:

$$\left[\epsilon - \hat{\mathcal{H}}^{\text{sp}}(\mathbf{r})\right]\psi(\mathbf{r}) = 0 \tag{A.93}$$

According to what we discussed above, Green's function for this case is then given by:

$$G(\mathbf{r}, \mathbf{r}'; \epsilon) = [\epsilon - \hat{\mathcal{H}}^{\text{sp}}(\mathbf{r})]^{-1} \delta(\mathbf{r} - \mathbf{r}') \tag{A.94}$$

which depends on the value of the energy ϵ. This is a mathematically meaningful expression because we can use the identity

$$[1 - x]^{-1} = 1 + x + x^2 + x^3 + \cdots$$

to write the inverse of the operator $[\epsilon - \hat{\mathcal{H}}^{\text{sp}}(\mathbf{r})]$ in terms of powers of the single-particle hamiltonian which are well-defined linear operators. Now suppose that the eigenfunctions of this hamiltonian are $\psi_i(\mathbf{r})$ with corresponding eigenvalues ϵ_i, then Green's function takes the form:

$$G(\mathbf{r}, \mathbf{r}'; \epsilon) = \sum_i \frac{\delta(\mathbf{r} - \mathbf{r}')}{\epsilon - \epsilon_i} = \sum_i \frac{\psi_i(\mathbf{r})\psi_i^*(\mathbf{r}')}{\epsilon - \epsilon_i} \tag{A.95}$$

This can easily be proved as follows. Since the eigenfunctions of the hamiltonian form a complete orthonormal set, any function $f(\mathbf{r})$ can be expressed as:

$$f(\mathbf{r}) = \sum_i c_i \psi_i(\mathbf{r})$$

But the definition of the δ-function is:

$$\int f(\mathbf{r}) \delta(\mathbf{r} - \mathbf{r}') d\mathbf{r} = f(\mathbf{r}')$$

and we can see that the expression

$$\sum_j \psi_j^*(\mathbf{r})\psi_j(\mathbf{r}') \tag{A.96}$$

satisfies exactly the above definition, since:

$$\int \sum_i c_i \psi_i(\mathbf{r}) \left(\sum_j \psi_j^*(\mathbf{r})\psi_j(\mathbf{r}') \right) d\mathbf{r} = \sum_{i,j} c_i \psi_j(\mathbf{r}') \delta_{ij} = \sum_i c_i \psi_i(\mathbf{r}') = f(\mathbf{r}')$$

where we have used the orthonormality of the wavefunctions:

$$\int \psi_j^*(\mathbf{r})\psi_i(\mathbf{r}) d\mathbf{r} = \delta_{ij}$$

Therefore, we can take $\delta(\mathbf{r} - \mathbf{r}')$ to be given by the expression of Eq. (A.96), insert it in the general expression for Green's function, Eq. (A.94), and obtain the result of Eq. (A.95).

Appendix B Classical Electrodynamics

Electrodynamics is the theory of fields and forces associated with stationary or moving electric charges. The classical theory is fully described by Maxwell's equations, the crowning achievement of nineteenth-century physics. There is also a quantum version of the theory which reconciles quantum mechanics with special relativity, but the scales of phenomena associated with electromagnetic fields in solids, that is, the energy, length, and time scale, are such that it is not necessary to invoke quantum electrodynamics. For instance, the scale of electron velocities in solids, set by the Fermi velocity $v_F = \hbar k_F/m_e$, is well below the speed of light, so electrons behave as non-relativistic point particles. We certainly have to take into account the quantized nature of electrons in a solid, embodied in the wavefunctions and energy eigenvalues that characterize the electronic states, but we can treat the electromagnetic fields as classical variables. It is often convenient to incorporate the effects of electromagnetic fields on solids using perturbation theory; this is treated explicitly in Appendix C. Accordingly, we provide here a brief account of the basic concepts and equations of classical electrodynamics. For detailed discussions, proofs, and applications, we refer the reader to standard textbooks on the subject, a couple of which are mentioned in Further Reading.

B.1 Electrostatics and Magnetostatics

The force on a charge q at $\mathbf{r}$ due to the presence of a point charge q' at $\mathbf{r}'$ is given by:

$$\mathbf{F} = \frac{qq'}{|\mathbf{r} - \mathbf{r}'|^3}(\mathbf{r} - \mathbf{r}') \tag{B.1}$$

which is known as Coulomb's force law. The corresponding electric field is defined as the force on a unit charge at $\mathbf{r}$, or, taking q' to be at the origin:

$$\mathbf{E}(\mathbf{r}) = \frac{q'}{|\mathbf{r}|^2}\hat{\mathbf{r}} \tag{B.2}$$

Forces in electrostatics are additive, so that the total electric field at $\mathbf{r}$ due to a continuous charge distribution is given by:

$$\mathbf{E}(\mathbf{r}) = \int \frac{\rho(\mathbf{r}')}{|\mathbf{r} - \mathbf{r}'|^3}(\mathbf{r} - \mathbf{r}')d\mathbf{r}' \tag{B.3}$$

with $\rho(\mathbf{r})$ the charge density (which has dimensions of electric charge per unit volume). Taking the divergence of both sides of this equation and using Gauss's theorem [see Eq. (A.22) in Appendix A], we find:

$$\nabla_{\mathbf{r}} \cdot \mathbf{E}(\mathbf{r}) = 4\pi\rho(\mathbf{r}) \tag{B.4}$$

where we have used the fact that the divergence of $(\mathbf{r} - \mathbf{r}')/|\mathbf{r} - \mathbf{r}'|^3$ is equal to $4\pi\,\delta(\mathbf{r} - \mathbf{r}')$ (see Appendix A). Integrating both sides of the above equation over the volume enclosed by a surface S, we obtain

$$\int \nabla_{\mathbf{r}} \cdot \mathbf{E}(\mathbf{r})\mathrm{d}\mathbf{r} = \oint_S \mathbf{E}(\mathbf{r}) \cdot \hat{\mathbf{n}}_s \,\mathrm{d}S = 4\pi Q \tag{B.5}$$

where Q is the total electric charge enclosed by the surface S and $\hat{\mathbf{n}}_s$ is the unit vector normal to the surface element $\mathrm{d}S$; this expression is known as Gauss's law. The electrostatic potential $\Phi(\mathbf{r})$ is defined through:

$$\mathbf{E}(\mathbf{r}) = -\nabla_{\mathbf{r}}\Phi(\mathbf{r}) \tag{B.6}$$

The potential is defined up to a constant, since the gradient of a constant is always zero; we can choose this constant to be zero, which is referred to as the Coulomb gauge. In terms of the potential, Eq. (B.4) becomes:

$$\nabla_{\mathbf{r}}^2\Phi(\mathbf{r}) = -4\pi\rho(\mathbf{r}) \tag{B.7}$$

a relation known as Poisson's equation. We note that the definition of the electrostatic potential, Eq. (B.6), implies that the curl of $\mathbf{E}$ must be zero, since

$$\nabla_{\mathbf{r}} \times \mathbf{E} = -\nabla_{\mathbf{r}} \times \nabla_{\mathbf{r}}\Phi$$

and the curl of a gradient is identically zero (see Appendix A). This is indeed true for the electric field defined in Eq. (B.2), as can be proved straightforwardly by calculating the line integral of $\mathbf{E}(\mathbf{r})$ around any closed loop and invoking Stokes' theorem [see Eq. (A.23) in Appendix A] to relate this integral to $\nabla_{\mathbf{r}} \times \mathbf{E}$. Because of the additive nature of electrostatic fields, this result applies to any field deriving from an arbitrary distribution of charges, hence it is always possible to express such a field in terms of the potential, as indicated by Eq. (B.6). In particular, the potential of a continuous charge distribution $\rho(\mathbf{r})$ turns out to be:

$$\Phi(\mathbf{r}) = \int \frac{\rho(\mathbf{r}')}{|\mathbf{r} - \mathbf{r}'|}\mathrm{d}\mathbf{r}' \tag{B.8}$$

which immediately leads to Eq. (B.3) for the corresponding electrostatic field.

From the above definitions it is simple to show that the energy W required to assemble a set of point charges q_i at positions $\mathbf{r}_i$ is given by:

$$W = \frac{1}{2}\sum_i q_i\Phi(\mathbf{r}_i)$$

where $\Phi(\mathbf{r})$ is the total potential due to the charges. If we now generalize this expression to a continuous distribution of charge $\rho(\mathbf{r})$, use Eq. (B.4) to relate $\rho(\mathbf{r})$ to $\mathbf{E}(\mathbf{r})$, integrate by parts and assume that the field dies at infinity, we find that the electrostatic energy W associated with an electric field $\mathbf{E}$ is given by:

$$W_e = \frac{1}{8\pi} \int |\mathbf{E}(\mathbf{r})|^2 d\mathbf{r} \tag{B.9}$$

The force on a charge q moving with velocity $\mathbf{v}$ in a magnetic field $\mathbf{B}$ is given by:

$$\mathbf{F} = q\left(\frac{\mathbf{v}}{c} \times \mathbf{B}\right) \tag{B.10}$$

where c is the speed of light; this is known as the Lorentz force law. The motion of a charge is associated with a current, whose density per unit area perpendicular to its flow is defined as $\mathbf{J}(\mathbf{r})$ (the dimensions of $\mathbf{J}$ are electric charge per unit area per unit time). The current density and the charge density $\rho(\mathbf{r})$ are related by:

$$\nabla_{\mathbf{r}} \cdot \mathbf{J}(\mathbf{r}) = -\frac{\partial \rho(\mathbf{r})}{\partial t} \tag{B.11}$$

which is known as the continuity relation. This relation is a consequence of the conservation of electric charge (a simple application of the divergence theorem produces the above expression). The magnetic field due to a current density $\mathbf{J}(\mathbf{r})$ is given by:

$$\mathbf{B}(\mathbf{r}) = \frac{1}{c} \int \frac{1}{|\mathbf{r}-\mathbf{r}'|^3} \mathbf{J}(\mathbf{r}') \times (\mathbf{r}-\mathbf{r}') d\mathbf{r}' \tag{B.12}$$

which is known as the Biot–Savart law. From this equation, it is a straightforward exercise in differential calculus to show that the divergence of $\mathbf{B}$ vanishes identically:

$$\nabla_{\mathbf{r}} \cdot \mathbf{B}(\mathbf{r}) = 0$$

while its curl is related to the current density:

$$\nabla_{\mathbf{r}} \times \mathbf{B}(\mathbf{r}) = \frac{4\pi}{c} \mathbf{J}(\mathbf{r}) \tag{B.13}$$

Integrating the second equation over an area S which is bounded by a contour C and using Stokes' theorem leads to:

$$\int (\nabla_{\mathbf{r}} \times \mathbf{B}(\mathbf{r})) \cdot \hat{\mathbf{n}}_s \, dS = \frac{4\pi}{c} \int \mathbf{J}(\mathbf{r}) \cdot \hat{\mathbf{n}}_s \, dS \Rightarrow \oint_C \mathbf{B}(\mathbf{r}) \cdot d\mathbf{l} = \frac{4\pi}{c} I \tag{B.14}$$

where I is the total current passing through the area enclosed by the loop C; the last expression is known as Ampère's law.

By analogy to the electrostatic case, we define a vector potential $\mathbf{A}(\mathbf{r})$ through which we obtain the magnetic field as:

$$\mathbf{B}(\mathbf{r}) = \nabla_{\mathbf{r}} \times \mathbf{A}(\mathbf{r}) \tag{B.15}$$

The vector form of the magnetic potential is dictated by the fact that $\nabla_{\mathbf{r}} \times \mathbf{B}$ does not vanish, which is in contrast to what we had found for the electrostatic potential $\mathbf{E}$. Similar to that situation, however, the magnetic potential $\mathbf{A}$ is defined up to a function whose curl vanishes; we exploit this ambiguity by choosing the vector field so that

$$\nabla_{\mathbf{r}} \cdot \mathbf{A}(\mathbf{r}) = 0 \tag{B.16}$$

which is known as the Coulomb gauge. With this choice, and using the relations between **B** and **J** that we derived above, we find that the laplacian of the vector potential is given by:

$$\nabla_{\mathbf{r}}^2 \mathbf{A}(\mathbf{r}) = -\frac{4\pi}{c}\mathbf{J}(\mathbf{r}) \tag{B.17}$$

which is formally the same as the Poisson equation, Eq. (B.7), only applied to vector rather than scalar quantities. From this relation, the vector potential itself can be expressed in terms of the current density as:

$$\mathbf{A}(\mathbf{r}) = \frac{1}{c}\int \frac{\mathbf{J}(\mathbf{r}')}{|\mathbf{r} - \mathbf{r}'|}\mathrm{d}\mathbf{r}'$$

Finally, by analogy to the derivation of the electrostatic energy associated with an electric field **E**, it can be shown that the magnetostatic energy associated with a magnetic field **B** is:

$$W_m = \frac{1}{8\pi}\int |\mathbf{B}(\mathbf{r})|^2 \mathrm{d}\mathbf{r} \tag{B.18}$$

Having obtained the expressions that relate the electric and magnetic potentials to the distributions of charges and currents, we can then calculate all the physically relevant quantities (such as the fields and from those the forces) which completely determine the behavior of the system. The only thing missing is the boundary conditions that are required to identify uniquely the solution to the differential equations involved; these are determined by the nature and the geometry of the physical system under consideration. For example, for conductors we have the following boundary conditions:

(a) The electric field $\mathbf{E}(\mathbf{r})$ must vanish inside the conductor.
(b) The electrostatic potential $\Phi(\mathbf{r})$ must be a constant inside the conductor.
(c) The charge density $\rho(\mathbf{r})$ must vanish inside the conductor and any net charge must reside on the surface.
(d) The only non-vanishing electric field $\mathbf{E}(\mathbf{r})$ must be perpendicular to the surface just outside the conductor.

These conditions are a consequence of the presence of free electric charges which can move within the conductor to shield any non-vanishing electric fields.

B.2   Fields in Polarizable Matter

In many situations the application of an external electric or magnetic field on a substance can instigate a response which is referred to as "polarization." Usually, the response is proportional to the applied field, which is called linear response. We refer to the polarization of electric nature as simply the polarization, $\mathbf{P}(\mathbf{r})$, and to the polarization of magnetic nature as the magnetization, $\mathbf{M}(\mathbf{r})$. We will also refer to the polarization or magnetization of the unit volume (or elementary unit, such as the unit cell of a crystal) in terms of the induced dipole moment $\mathbf{p}(\mathbf{r})$ or the induced magnetic moment $\mathbf{m}(\mathbf{r})$, respectively.

To conform to historical conventions (which are actually motivated by physical considerations), we define the total, induced, and net electric fields as:

$$\mathbf{E}(\mathbf{r}) \;:\; \text{total electric field}$$

$$-4\pi\,\mathbf{p}(\mathbf{r}) = \mathbf{P}(\mathbf{r}) \;:\; \text{polarization (induced electric field)}$$

$$\mathbf{D}(\mathbf{r}) \;:\; \text{net electric field}$$

where the net field is defined as the total minus the induced field ($\mathbf{D}$ is also called the electric displacement). These conventions lead to the following expressions:

$$\mathbf{D}(\mathbf{r}) = \mathbf{E}(\mathbf{r}) + 4\pi\,\mathbf{p}(\mathbf{r}) \Rightarrow \mathbf{E}(\mathbf{r}) = \mathbf{D}(\mathbf{r}) - 4\pi\,\mathbf{p}(\mathbf{r})$$

For linear response, we define the electric susceptibility χ_e through the relation[1]

$$\mathbf{p}(\mathbf{r}) = \chi_e\,\mathbf{E}(\mathbf{r})$$

It is also customary to define the dielectric constant ε as the factor that relates the total field to the net field through

$$\mathbf{D} = \varepsilon\mathbf{E} \Rightarrow \varepsilon = 1 + 4\pi\,\chi_e$$

where the last expression between the dielectric constant and the electric susceptibility holds for linear response. The historical and physical reason for this set of definitions is that, typically, an external field produces an electric dipole moment that tends to shield the applied field (the dipole moment produces an opposite field), hence it is natural to define the induced field with a negative sign.

The corresponding definitions for the magnetic field are as follows:

$$\mathbf{B}(\mathbf{r}) \;:\; \text{total magnetic field}$$

$$4\pi\,\mathbf{m}(\mathbf{r}) = \mathbf{M}(\mathbf{r}) \;:\; \text{magnetization (induced magnetic field)}$$

$$\mathbf{H}(\mathbf{r}) \;:\; \text{net magnetic field}$$

where, as before, the net field is defined as the total minus the induced field. These conventions lead to the following expressions:

$$\mathbf{H}(\mathbf{r}) = \mathbf{B}(\mathbf{r}) - 4\pi\,\mathbf{m}(\mathbf{r}) \Rightarrow \mathbf{B}(\mathbf{r}) = \mathbf{H}(\mathbf{r}) + 4\pi\,\mathbf{m}(\mathbf{r})$$

For linear response, we define the magnetic susceptibility χ_m through the relation

$$\mathbf{m}(\mathbf{r}) = \chi_m\mathbf{H}(\mathbf{r})$$

It is also customary to define the magnetic permeability μ as the factor that relates the total field to the net field through

$$\mathbf{B} = \mu\mathbf{H} \Rightarrow \mu = 1 + 4\pi\,\chi_m$$

where the last expression between the magnetic permeability and the magnetic susceptibility holds for linear response. This set of definitions is not exactly analogous to the

---

[1] Here we treat the susceptibility as a scalar quantity, which is appropriate for isotropic solids. In anisotropic solids, such as crystals, the susceptibility must be generalized to a second-rank tensor $\chi_{\alpha\beta}$, with α, β taking three independent values each in three dimensions.

definitions concerning electric polarization; the reason is that in a substance which exhibits magnetic polarization the induced field, typically, tends to be aligned with the applied field (it enhances it), hence the natural definition of the induced field has the same sign as the applied field.

The net fields are associated with the presence of charges and currents which are called "free," while the electric and magnetic polarization are associated with induced charges and currents that are called "bound." Thus, in the case of electric polarization which produces an electric dipole moment $\mathbf{p}(\mathbf{r}) = \mathbf{P}(\mathbf{r})/4\pi$, the bound charges are given by:

$$\sigma_b(\mathbf{r}) = \mathbf{P}(\mathbf{r}) \cdot \hat{\mathbf{n}}_s(\mathbf{r}), \quad \rho_b(\mathbf{r}) = -\nabla_{\mathbf{r}} \cdot \mathbf{P}(\mathbf{r})$$

where σ_b is a surface charge density, $\hat{\mathbf{n}}_s$ is the surface-normal unit vector and ρ_b is a bulk charge density. The potential due to the induced dipole moment can then be expressed in terms of these charges as:

$$\Phi^{\mathrm{ind}}(\mathbf{r}) = \int \frac{\mathbf{P}(\mathbf{r}') \cdot (\mathbf{r} - \mathbf{r}')}{|\mathbf{r} - \mathbf{r}'|^3} d\mathbf{r}' = \oint_{S'} \frac{\sigma_b(\mathbf{r}')}{|\mathbf{r} - \mathbf{r}'|} dS' + \int_{V'} \frac{\rho_b(\mathbf{r}')}{|\mathbf{r} - \mathbf{r}'|} d\mathbf{r}'$$

where V' is the volume of the polarized substance and S' is the surface that encloses it. Similarly, in the case of magnetic polarization which produces a magnetic dipole moment $\mathbf{m}(\mathbf{r}) = \mathbf{M}(\mathbf{r})/4\pi$, the bound currents are given by:

$$\mathbf{K}_b(\mathbf{r}) = \mathbf{M}(\mathbf{r}) \times \hat{\mathbf{n}}_s(\mathbf{r}), \quad \mathbf{J}_b(\mathbf{r}) = \nabla_{\mathbf{r}} \times \mathbf{M}(\mathbf{r}),$$

where $\mathbf{K}_b$ is a surface current density and $\hat{\mathbf{n}}_s$ is the surface-normal unit vector and $\mathbf{J}_b$ is a bulk current density. The potential due to the induced dipole moment can then be expressed in terms of these currents as:

$$\mathbf{A}^{\mathrm{ind}}(\mathbf{r}) = \int \frac{\mathbf{M}(\mathbf{r}') \times (\mathbf{r} - \mathbf{r}')}{|\mathbf{r} - \mathbf{r}'|^3} d\mathbf{r}' = \oint_{S'} \frac{\mathbf{K}_b(\mathbf{r}')}{|\mathbf{r} - \mathbf{r}'|} dS' + \int_{V'} \frac{\mathbf{J}_b(\mathbf{r}')}{|\mathbf{r} - \mathbf{r}'|} d\mathbf{r}'$$

with the meaning of V', S' the same as in the electric case. The values of the bound charges and currents are determined by the nature of the physical system and its geometrical features. Having identified the induced fields with the bound charges or currents, we can now obtain the net fields in terms of the free charges described by the charge density $\rho_f(\mathbf{r})$ or free currents described by the current density $\mathbf{J}_f(\mathbf{r})$:

$$\nabla_{\mathbf{r}} \cdot \mathbf{D}(\mathbf{r}) = 4\pi \rho_f(\mathbf{r}), \quad \nabla_{\mathbf{r}} \times \mathbf{H}(\mathbf{r}) = \frac{4\pi}{c} \mathbf{J}_f(\mathbf{r})$$

These are identical expressions to those relating the total fields $\mathbf{E}, \mathbf{B}$ to the charge or current densities in free space, Eqs (B.4) and (B.13), respectively.

B.3   Electrodynamics

The combined presence of electric charges and currents gives rise to electric and magnetic fields which are related to each other. Our discussion of electrostatics and magnetostatics leads to the following expression for the total force on a charge q moving with velocity $\mathbf{v}$ in the presence of external electric and magnetic fields $\mathbf{E}$ and $\mathbf{B}$, respectively:

$$\mathbf{F} = q \left(\mathbf{E} + \frac{\mathbf{v}}{c} \times \mathbf{B} \right) \tag{B.19}$$

Since this charge is moving, it corresponds to a current density $\mathbf{J}$. If the current density is proportional to the force per unit charge:

$$\mathbf{J} = \sigma \frac{\mathbf{F}}{q} = \sigma \left(\mathbf{E} + \frac{\mathbf{v}}{c} \times \mathbf{B} \right)$$

we call the behavior ohmic, with the constant of proportionality σ called the conductivity (not to be confused with the surface charge density). The inverse of the conductivity is called the resistivity $\rho = 1/\sigma$ (not to be confused with the bulk charge density). When the velocity is much smaller than the speed of light or the magnetic field much weaker than the electric field, the above relation reduces to:

$$\mathbf{J} = \sigma \mathbf{E} \tag{B.20}$$

which is known as Ohm's law. The conductivity is a characteristic material property and can be calculated if we have detailed knowledge of the electronic states (their eigenfunctions and corresponding energies; see Chapter 6).

The expressions we derived relating the fields to static charge and current density distributions are not adequate to describe the physics if we allow the densities and the fields to have a time dependence. These have to be augmented by two additional relations, known as Faraday's law and Maxwell's extension of Ampère's law. To motivate Faraday's law we introduce first the notion of the magnetic flux ϕ: it is the projection of the magnetic field $\mathbf{B}$ onto a surface element $\hat{\mathbf{n}}_s \, dS$, integrated over some finite surface area

$$\phi = \int \mathbf{B}(\mathbf{r}) \cdot \hat{\mathbf{n}}_s \, dS$$

where $\hat{\mathbf{n}}_s$ is the surface-normal unit vector corresponding to the surface element dS; the magnetic flux is a measure of the amount of magnetic field passing through the surface. Now let us consider the change with respect to time of the magnetic flux through a fixed surface, when the magnetic field is a time-dependent quantity:

$$\frac{d\phi}{dt} = \int \frac{\partial \mathbf{B}(\mathbf{r}, t)}{\partial t} \cdot \hat{\mathbf{n}}_s \, dS$$

The electromagnetic fields propagate with the speed of light, so if we divide both sides of the above equation by c we will obtain the change with respect to time of the total magnetic flux passing through the surface, normalized by the speed of propagation of the field:

$$\frac{1}{c} \frac{d\phi}{dt} = \frac{1}{c} \int \frac{\partial \mathbf{B}(\mathbf{r}, t)}{\partial t} \cdot \hat{\mathbf{n}}_s \, dS$$

From the definition of the force due to a magnetic field, Eq. (B.10), which shows that the magnetic field has the dimensions of force per unit charge, we conclude that the expression on the right-hand side of the last equation has the dimensions of energy per unit charge. Therefore, the time derivative of the magnetic flux divided by c is a measure of the energy per unit charge passing through the surface due to changes in the magnetic field with respect to time. To counter this expense of energy when the magnetic field changes, a current can be set up along the boundary C of the surface S. In order to induce such a current, an electric field $\mathbf{E}$ must be introduced which will move charges along C. The definition of the electric field as the force per unit charge implies that $\mathbf{E} \cdot \mathrm{d}\mathbf{l}$ is the elementary energy per unit charge moving along the contour C, with $\mathrm{d}\mathbf{l}$ denoting the length element along C. Requiring that the energy needed to move a unit charge around C under the influence of $\mathbf{E}$ exactly counters the energy per unit charge passing through the surface S due to changes in the magnetic field leads to:

$$\oint_C \mathbf{E}(\mathbf{r}, t) \cdot \mathrm{d}\mathbf{l} + \frac{1}{c} \frac{\mathrm{d}\phi}{\mathrm{d}t} = 0 \Rightarrow \oint_C \mathbf{E}(\mathbf{r}, t) \cdot \mathrm{d}\mathbf{l} = -\frac{1}{c} \int \frac{\partial \mathbf{B}(\mathbf{r}, t)}{\partial t} \cdot \hat{\mathbf{n}}_s \, \mathrm{d}S$$

Turning this last relation into differential form using Stokes' theorem, we find:

$$\nabla_{\mathbf{r}} \times \mathbf{E}(\mathbf{r}, t) = -\frac{1}{c} \frac{\partial \mathbf{B}(\mathbf{r}, t)}{\partial t}$$

which is Faraday's law.

To motivate Maxwell's extension of Ampère's law, we start from the continuity relation, Eq. (B.11), and allow for both spatial and temporal variations of the charge and current densities. Relating the charge density $\rho(\mathbf{r}, t)$ to the electric field $\mathbf{E}(\mathbf{r}, t)$ through Eq. (B.4), we find:

$$\nabla_{\mathbf{r}} \cdot \mathbf{J}(\mathbf{r}, t) = -\frac{\partial \rho(\mathbf{r}, t)}{\partial t} = -\frac{\partial}{\partial t} \left(\frac{1}{4\pi} \nabla_{\mathbf{r}} \cdot \mathbf{E}(\mathbf{r}, t) \right) = -\nabla_{\mathbf{r}} \cdot \left(\frac{1}{4\pi} \frac{\partial \mathbf{E}(\mathbf{r}, t)}{\partial t} \right)$$

A comparison of the first and last expressions in this set of equalities shows that the quantity in parentheses on the right side is equivalent to a current density and as such should be included in Eq. (B.13). In particular, this current density is generated by the temporal variation of the fields induced by the presence of external current densities, so its role will be to counteract those currents and therefore it must be subtracted from the external current density, which leads to:

$$\nabla_{\mathbf{r}} \times \mathbf{B}(\mathbf{r}, t) = \frac{4\pi}{c} \mathbf{J}(\mathbf{r}) + \frac{1}{c} \frac{\partial \mathbf{E}(\mathbf{r}, t)}{\partial t}$$

This is Ampère's law augmented by the second term, as argued by Maxwell.

The set of equations which relate the spatially and temporally varying charge and current densities to the corresponding electric and magnetic fields are known as Maxwell's equations:

$$\nabla_{\mathbf{r}} \cdot \mathbf{E}(\mathbf{r}, t) = 4\pi \rho(\mathbf{r}, t) \tag{B.21}$$

$$\nabla_{\mathbf{r}} \cdot \mathbf{B}(\mathbf{r}, t) = 0 \tag{B.22}$$

$$\nabla_{\mathbf{r}} \times \mathbf{E}(\mathbf{r}, t) = -\frac{1}{c} \frac{\partial \mathbf{B}(\mathbf{r}, t)}{\partial t} \tag{B.23}$$

$$\nabla_{\mathbf{r}} \times \mathbf{B}(\mathbf{r}, t) = \frac{4\pi}{c} \mathbf{J}(\mathbf{r}) + \frac{1}{c} \frac{\partial \mathbf{E}(\mathbf{r}, t)}{\partial t} \tag{B.24}$$

The second and third of these equations are not changed in polarizable matter, but the first and fourth can be expressed in terms of the net fields $\mathbf{D}, \mathbf{H}$ and the corresponding free charge and current densities in an exactly analogous manner:

$$\nabla_{\mathbf{r}} \cdot \mathbf{D}(\mathbf{r}, t) = 4\pi \rho_f(\mathbf{r}, t) \tag{B.25}$$

$$\nabla_{\mathbf{r}} \times \mathbf{H}(\mathbf{r}, t) = \frac{4\pi}{c} \mathbf{J}_f(\mathbf{r}) + \frac{1}{c} \frac{\partial \mathbf{D}(\mathbf{r}, t)}{\partial t} \tag{B.26}$$

Maxwell's equations can also be put in integral form. Specifically, integrating both sides of each equation, over a volume enclosed by a surface S for the first two equations and over a surface enclosed by a curve C for the last two equations, and using Gauss's theorem or Stokes' theorem (see Appendix A) as appropriate, we find that the equations in polarizable matter take the form

$$\oint_S \mathbf{D}(\mathbf{r}, t) \cdot \hat{\mathbf{n}}_s \, dS = 4\pi Q_f \tag{B.27}$$

$$\oint_S \mathbf{B}(\mathbf{r}, t) \cdot \hat{\mathbf{n}}_s \, dS = 0 \tag{B.28}$$

$$\oint_C \mathbf{E}(\mathbf{r}, t) \cdot d\mathbf{l} = -\frac{1}{c} \frac{\partial}{\partial t} \int \mathbf{B}(\mathbf{r}, t) \cdot \hat{\mathbf{n}}_s \, dS \tag{B.29}$$

$$\oint_C \mathbf{H}(\mathbf{r}, t) \cdot d\mathbf{l} = \frac{4\pi}{c} I_f + \frac{1}{c} \frac{\partial}{\partial t} \int \mathbf{D}(\mathbf{r}, t) \cdot \hat{\mathbf{n}}_s \, dS \tag{B.30}$$

where $\hat{\mathbf{n}}_s$ is the surface-normal unit vector associated with the surface element of S and $d\mathbf{l}$ is the length element along the curve C; Q_f is the total free charge in the volume of integration and I_f is the total free current passing through the surface of integration.

A direct consequence of Maxwell's equations is that the electric and magnetic fields can be expressed in terms of the scalar and vector potentials $\Phi, \mathbf{A}$, which now include both spatial and temporal dependence. From the second Maxwell equation, Eq. (B.22), we conclude that we can express the magnetic field as the curl of the vector potential $\mathbf{A}(\mathbf{r}, t)$, since the divergence of the curl of a vector potential vanishes identically (see Appendix A):

$$\mathbf{B}(\mathbf{r}, t) = \nabla_{\mathbf{r}} \times \mathbf{A}(\mathbf{r}, t) \tag{B.31}$$

Substituting this in the third Maxwell equation, Eq. (B.23), we find:

$$\nabla_{\mathbf{r}} \times \mathbf{E}(\mathbf{r}, t) = -\frac{1}{c} \frac{\partial}{\partial t} (\nabla_{\mathbf{r}} \times \mathbf{A}(\mathbf{r}, t)) \Rightarrow \nabla_{\mathbf{r}} \times \left(\mathbf{E}(\mathbf{r}, t) + \frac{1}{c} \frac{\partial \mathbf{A}(\mathbf{r}, t)}{\partial t} \right) = 0$$

which allows us to define the expression in the last parentheses as the gradient of a scalar field, since its curl vanishes identically (see Appendix A):

$$\mathbf{E}(\mathbf{r}, t) + \frac{1}{c} \frac{\partial \mathbf{A}(\mathbf{r}, t)}{\partial t} = -\nabla_{\mathbf{r}} \Phi(\mathbf{r}, t) \Rightarrow \mathbf{E}(\mathbf{r}, t) = -\nabla_{\mathbf{r}} \Phi(\mathbf{r}, t) - \frac{1}{c} \frac{\partial \mathbf{A}(\mathbf{r}, t)}{\partial t} \tag{B.32}$$

Inserting this expression in the first Maxwell equation, Eq. (B.21), we obtain:

$$\nabla_{\mathbf{r}}^2 \Phi(\mathbf{r}, t) + \frac{1}{c} \frac{\partial}{\partial t} (\nabla_{\mathbf{r}} \cdot \mathbf{A}(\mathbf{r}, t)) = -4\pi \rho(\mathbf{r}, t) \tag{B.33}$$

which is the generalization of Poisson's equation to spatially and temporally varying potentials and charge densities. Inserting the expressions for $\mathbf{B}(\mathbf{r}, t)$ and $\mathbf{E}(\mathbf{r}, t)$ in terms of the vector and scalar potentials, Eqs (B.31) and (B.32), in the last Maxwell equation, Eq. (B.24), and using the identity which relates the curl of the curl of a vector potential to its divergence and its laplacian, Eq. (A.21), we obtain:

$$\left(\nabla_\mathbf{r}^2 \mathbf{A}(\mathbf{r},t) - \frac{1}{c^2} \frac{\partial^2 \mathbf{A}(\mathbf{r},t)}{\partial t^2} \right) - \nabla_\mathbf{r} \left(\nabla_\mathbf{r} \cdot \mathbf{A}(\mathbf{r},t) + \frac{1}{c} \frac{\partial \Phi(\mathbf{r},t)}{\partial t} \right) = -\frac{4\pi}{c} \mathbf{J}(\mathbf{r},t) \quad \text{(B.34)}$$

The last two equations, Eqs (B.33) and (B.34), can be used to obtain the vector and scalar potentials $\mathbf{A}(\mathbf{r},t), \Phi(\mathbf{r},t)$ when the charge and current density distributions $\rho(\mathbf{r},t), \mathbf{J}(\mathbf{r},t)$ are known. In the Coulomb gauge, Eq. (B.16), the first of these equations reduces to the familiar Poisson equation, which, as in the electrostatic case, leads to:

$$\Phi(\mathbf{r},t) = \int \frac{\rho(\mathbf{r}',t)}{|\mathbf{r}-\mathbf{r}'|} \mathrm{d}\mathbf{r}' \quad \text{(B.35)}$$

and the second equation then takes the form

$$\nabla_\mathbf{r}^2 \mathbf{A}(\mathbf{r},t) - \frac{1}{c^2} \frac{\partial^2 \mathbf{A}(\mathbf{r},t)}{\partial t^2} = \frac{1}{c} \nabla_\mathbf{r} \left(\frac{\partial \Phi(\mathbf{r},t)}{\partial t} \right) - \frac{4\pi}{c} \mathbf{J}(\mathbf{r},t) \quad \text{(B.36)}$$

As a final point, note that the total energy of the electromagnetic field including spatial and temporal variations is given by:

$$E^{em} = \frac{1}{8\pi} \int \left(|\mathbf{E}(\mathbf{r},t)|^2 + |\mathbf{B}(\mathbf{r},t)|^2 \right) \mathrm{d}\mathbf{r} \quad \text{(B.37)}$$

which is a simple generalization of the expressions given for the static electric and magnetic fields.

B.4 Electromagnetic Radiation

The propagation of electromagnetic fields in space and time is referred to as electromagnetic radiation. In vacuum, Maxwell's equations become:

$$\nabla_\mathbf{r} \cdot \mathbf{E}(\mathbf{r},t) = 0, \quad \nabla_\mathbf{r} \cdot \mathbf{B}(\mathbf{r},t) = 0$$

$$\nabla_\mathbf{r} \times \mathbf{E}(\mathbf{r},t) = -\frac{1}{c} \frac{\partial \mathbf{B}(\mathbf{r},t)}{\partial t}, \quad \nabla_\mathbf{r} \times \mathbf{B}(\mathbf{r},t) = \frac{1}{c} \frac{\partial \mathbf{E}(\mathbf{r},t)}{\partial t}$$

Taking the curl of both sides of the third and fourth Maxwell equations and using the identity that relates the curl of the curl of a vector field to its divergence and its laplacian, Eq. (A.21), we find:

$$\nabla_\mathbf{r}^2 \mathbf{E}(\mathbf{r},t) = \frac{1}{c^2} \frac{\partial^2 \mathbf{E}(\mathbf{r},t)}{\partial t^2}, \quad \nabla_\mathbf{r}^2 \mathbf{B}(\mathbf{r},t) = \frac{1}{c^2} \frac{\partial^2 \mathbf{B}(\mathbf{r},t)}{\partial t^2}$$

where we have also used the first and second Maxwell equations to eliminate the divergence of the fields. Thus, both the electric and the magnetic field obey the wave equation with speed c. The plane-wave solution to these equations is:

$$\mathbf{E}(\mathbf{r},t) = \mathbf{E}_0 \mathrm{e}^{\mathrm{i}(\mathbf{k}\cdot\mathbf{r}-\omega t)}, \quad \mathbf{B}(\mathbf{r},t) = \mathbf{B}_0 \mathrm{e}^{\mathrm{i}(\mathbf{k}\cdot\mathbf{r}-\omega t)}$$

with $\mathbf{k}$ the wave-vector and ω the frequency of the radiation, which are related by:

$$|\mathbf{k}| = \frac{\omega}{c} \quad \text{(free space)}$$

With these expressions for the fields, from the first and second Maxwell equations we deduce that:

$$\mathbf{k} \cdot \mathbf{E}_0 = \mathbf{k} \cdot \mathbf{B}_0 = 0 \quad \text{(free space)} \tag{B.38}$$

that is, the vectors $\mathbf{E}_0, \mathbf{B}_0$ are perpendicular to the direction of propagation of radiation which is determined by $\mathbf{k}$; in other words, the fields have only transverse and no longitudinal components. Moreover, from the third Maxwell equation, we obtain the relation

$$\mathbf{k} \times \mathbf{E}_0 = \frac{\omega}{c}\mathbf{B}_0 \quad \text{(free space)} \tag{B.39}$$

which implies that the vectors $\mathbf{E}_0$ and $\mathbf{B}_0$ are also perpendicular to each other and have the same magnitude, since $|\mathbf{k}| = \omega/c$. The fourth Maxwell equation leads to the same result.

Inside a material with dielectric constant ε and magnetic permeability μ, in the absence of any free charges or currents, Maxwell's equations become:

$$\nabla_{\mathbf{r}} \cdot \mathbf{D}(\mathbf{r}, t) = 0, \quad \nabla_{\mathbf{r}} \cdot \mathbf{B}(\mathbf{r}, t) = 0$$

$$\nabla_{\mathbf{r}} \times \mathbf{E}(\mathbf{r}, t) = -\frac{1}{c}\frac{\partial \mathbf{B}(\mathbf{r}, t)}{\partial t}, \quad \nabla_{\mathbf{r}} \times \mathbf{H}(\mathbf{r}, t) = \frac{1}{c}\frac{\partial \mathbf{D}(\mathbf{r}, t)}{\partial t}$$

Using the relations $\mathbf{D} = \varepsilon \mathbf{E}, \mathbf{B} = \mu \mathbf{H}$, the above equations lead to the same wave equations for $\mathbf{E}$ and $\mathbf{B}$ as in free space, except for a factor $\varepsilon\mu$:

$$\nabla_{\mathbf{r}}^2 \mathbf{E}(\mathbf{r}, t) = \frac{\varepsilon\mu}{c^2}\frac{\partial^2 \mathbf{E}(\mathbf{r}, t)}{\partial t^2}, \quad \nabla_{\mathbf{r}}^2 \mathbf{B}(\mathbf{r}, t) = \frac{\varepsilon\mu}{c^2}\frac{\partial^2 \mathbf{B}(\mathbf{r}, t)}{\partial t^2}$$

which implies that the speed of the electromagnetic radiation in the solid is reduced by a factor $\sqrt{\varepsilon\mu}$. This has important consequences. In particular, assuming as before plane-wave solutions for $\mathbf{E}$ and $\mathbf{B}$ and using the equations which relate $\mathbf{E}$ to $\mathbf{B}$ and $\mathbf{H}$ to $\mathbf{D}$, we arrive at the following relations between the electric and magnetic field vectors and the wave-vector $\mathbf{k}$:

$$\mathbf{k} \times \mathbf{E}_0 = \frac{\omega}{c}\mathbf{B}_0, \quad \mathbf{k} \times \mathbf{B}_0 = -\varepsilon\mu\frac{\omega}{c}\mathbf{E}_0$$

which, in order to be compatible, require

$$|\mathbf{k}| = \frac{\omega}{c}\sqrt{\varepsilon\mu}, \quad |\mathbf{B}_0| = \sqrt{\varepsilon\mu}|\mathbf{E}_0| \tag{B.40}$$

As an application relevant to the optical properties of solids, we consider a situation where electromagnetic radiation is incident on a solid from the vacuum, with the wave-vector of the radiation at $90°$ angle to the surface plane (this is called normal incidence). First, we review the relevant boundary conditions. We denote the vacuum side by index 1 and the solid side by index 2. For the first two Maxwell equations, Eqs (B.27) and (B.28), we take the volume of integration to consist of an infinitesimal volume element with two surfaces parallel to the interface and negligible extent in the perpendicular direction, which gives:

$$D_\perp^{(1)} - D_\perp^{(2)} = 4\pi\sigma_f, \quad B_\perp^{(1)} - B_\perp^{(2)} = 0$$

where σ_f is the free charge per unit area at the interface. Similarly, for the last two Maxwell equations, Eqs (B.29) and (B.30), we take the surface of integration to consist of an infinitesimal surface element with two sides parallel to the interface and negligible extent in the perpendicular direction, which gives:

$$E_\parallel^{(1)} - E_\parallel^{(2)} = 0, \quad H_\parallel^{(1)} - H_\parallel^{(2)} = \frac{4\pi}{c}\mathbf{K}_f \times \hat{\mathbf{n}}_s$$

where $\mathbf{K}_f$ is the free current per unit area at the interface and $\hat{\mathbf{n}}_s$ the unit vector perpendicular to the surface element. We can also express the net fields $\mathbf{D}$ and $\mathbf{H}$ in terms of the total fields $\mathbf{E}$ and $\mathbf{B}$, using the dielectric constant and the magnetic permeability. Only the first and last equations change, giving:

$$\varepsilon^{(1)}E_\perp^{(1)} - \varepsilon^{(2)}E_\perp^{(2)} = 4\pi\sigma_f$$

$$\frac{1}{\mu^{(1)}}B_\parallel^{(1)} - \frac{1}{\mu^{(2)}}B_\parallel^{(2)} = \frac{4\pi}{c}\mathbf{K}_f \times \hat{\mathbf{n}}_s$$

We will specify next the physical situation to side 1 of the interface being the vacuum region, with $\varepsilon^{(1)} = 1$ and $\mu^{(1)} = 1$, and side 2 of the interface being the solid, with $\varepsilon^{(2)} = \varepsilon$ and $\mu^{(2)} \approx 1$ (most solids show negligible magnetic response). The direction of propagation of the radiation $\mathbf{k}$ will be taken perpendicular to the interface for normal incidence. We also assume that there are no free charges or free currents at the interface. This assumption is reasonable for metals where the presence of free carriers eliminates any charge accumulation. It also makes sense for semiconductors whose passivated, reconstructed surfaces correspond to filled bands which cannot carry current, hence there can only be bound charges. It is convenient to define the direction of propagation of the radiation as the z-axis and the interface as the x, y plane. Moreover, since the electric and magnetic field vectors are perpendicular to the direction of propagation and perpendicular to each other, we can choose them to define the x-axis and y-axis. The incident radiation will then be described by the fields

$$\mathbf{E}^{(I)}(\mathbf{r}, t) = E_0^{(I)}e^{i(kz-\omega t)}\hat{\mathbf{x}}, \quad \mathbf{B}^{(I)}(\mathbf{r}, t) = E_0^{(I)}e^{i(kz-\omega t)}\hat{\mathbf{y}}$$

The reflected radiation will propagate in the opposite direction with the same wave-vector and frequency:

$$\mathbf{E}^{(R)}(\mathbf{r}, t) = E_0^{(R)}e^{i(-kz-\omega t)}\hat{\mathbf{x}}, \quad \mathbf{B}^{(R)}(\mathbf{r}, t) = -E_0^{(R)}e^{i(-kz-\omega t)}\hat{\mathbf{y}}$$

where the negative sign in the expression for $\mathbf{B}^{(R)}$ is dictated by Eq. (B.39). Finally, the transmitted radiation will propagate in the same direction as the incident radiation and will have the same frequency but a different wave-vector given by $k' = k\sqrt{\varepsilon}$ because the speed of propagation has been reduced by a factor $\sqrt{\varepsilon}$, as argued above. Therefore, the transmitted fields will be given by:

$$\mathbf{E}^{(T)}(\mathbf{r}, t) = E_0^{(T)}e^{i(k'z-\omega t)}\hat{\mathbf{x}}, \quad \mathbf{B}^{(T)}(\mathbf{r}, t) = \sqrt{\varepsilon}E_0^{(T)}e^{i(k'z-\omega t)}\hat{\mathbf{y}}$$

where we have taken advantage of the result derived above, Eq. (B.40), to express the magnitude of the magnetic field as $\sqrt{\varepsilon}$ times the magnitude of the electric field in the solid. The general boundary conditions we derived above applied to the situation at hand give:

$$E_0^{(I)} + E_0^{(R)} = E_0^{(T)}, \quad E_0^{(I)} - E_0^{(R)} = \sqrt{\varepsilon}E_0^{(T)}$$

where we have used only the equations for the components parallel to the interface, since there are no components perpendicular to the interface because those would correspond to longitudinal components in the electromagnetic waves. The equations we obtained can easily be solved for the amplitude of the transmitted and reflected radiation in terms of the amplitude of the incident radiation, leading to:

$$\left|\frac{E_0^{(T)}}{E_0^{(I)}}\right| = \left|\frac{2\sqrt{\varepsilon}}{\sqrt{\varepsilon}+1}\right|, \quad \left|\frac{E_0^{(R)}}{E_0^{(I)}}\right| = \left|\frac{\sqrt{\varepsilon}-1}{\sqrt{\varepsilon}+1}\right| \tag{B.41}$$

These ratios of the amplitudes are referred to as the transmission and reflection coefficients; their squares give the power of transmitted and reflected radiation.

As a final exercise, we consider the electromagnetic fields inside a solid in the presence of free charges and currents. We will assume again that the fields, as well as the free charge and current densities, can be described by plane waves:

$$\rho_f(\mathbf{r}, t) = \rho_0 e^{i(\mathbf{k}\cdot\mathbf{r}-\omega t)}, \quad \mathbf{J}_f(\mathbf{r}, t) = \mathbf{J}_0 e^{i(\mathbf{k}\cdot\mathbf{r}-\omega t)}$$
$$\mathbf{E}(\mathbf{r}, t) = \mathbf{E}_0 e^{i(\mathbf{k}\cdot\mathbf{r}-\omega t)}, \quad \mathbf{B}(\mathbf{r}, t) = \mathbf{B}_0 e^{i(\mathbf{k}\cdot\mathbf{r}-\omega t)} \tag{B.42}$$

where all the quantities with subscript zero are functions of $\mathbf{k}$ and ω. We will also separate the fields in longitudinal (parallel to the wave-vector $\mathbf{k}$) and transverse (perpendicular to the wave-vector $\mathbf{k}$) components: $\mathbf{E}_0 = \mathbf{E}_{0,l} + \mathbf{E}_{0,t}$, $\mathbf{B}_0 = \mathbf{B}_{0,l} + \mathbf{B}_{0,t}$. From the general relations we derived earlier, we expect the transverse components of the fields to be perpendicular to each other. Accordingly, it is convenient to choose the direction of $\mathbf{k}$ as the z-axis, the direction of $\mathbf{E}_{0,t}$ as the x-axis, and the direction of $\mathbf{B}_{0,t}$ as the y-axis. We will also separate the current density into longitudinal and transverse components, $\mathbf{J}_0 = \mathbf{J}_{0,l} + \mathbf{J}_{0,t}$, the direction of the latter component to be determined by Maxwell's equations. In the following, all quantities that are not in boldface represent the magnitude of the corresponding vectors, for instance, $k = |\mathbf{k}|$.

From the first Maxwell equation we obtain:

$$\nabla_\mathbf{r} \cdot \mathbf{D}(\mathbf{r}, t) = 4\pi\rho_f(\mathbf{r}, t) \Rightarrow i\mathbf{k} \cdot \varepsilon(\mathbf{E}_{0,l} + \mathbf{E}_{0,t}) = 4\pi\rho_0 \Rightarrow E_{0,l} = -i\frac{4\pi}{\varepsilon}\frac{1}{k}\rho_0 \tag{B.43}$$

From the second Maxwell equation we obtain:

$$\nabla_\mathbf{r} \cdot \mathbf{B}(\mathbf{r}, t) = 0 \Rightarrow i\mathbf{k} \cdot (\mathbf{B}_{0,l} + \mathbf{B}_{0,t}) = 0 \Rightarrow B_{0,l} = 0 \tag{B.44}$$

From the third Maxwell equation we obtain:

$$\nabla_\mathbf{r} \times \mathbf{E}(\mathbf{r}, t) = -\frac{1}{c}\frac{\partial \mathbf{B}(\mathbf{r}, t)}{\partial t} \Rightarrow i\mathbf{k} \times (\mathbf{E}_{0,l} + \mathbf{E}_{0,t}) = i\frac{\omega}{c}\mathbf{B}_{0,t} \Rightarrow B_{0,t} = \frac{ck}{\omega}E_{0,t} \tag{B.45}$$

Finally, from the fourth Maxwell equation we obtain:

$$\nabla_{\mathbf{r}} \times \mathbf{H}(\mathbf{r}, t) = \frac{4\pi}{c} \mathbf{J}_f(\mathbf{r}, t) + \frac{1}{c}\frac{\partial \mathbf{D}(\mathbf{r}, t)}{\partial t} \Rightarrow$$

$$i\mathbf{k} \times \frac{1}{\mu}\mathbf{B}_{0,t} = \frac{4\pi}{c}(\mathbf{J}_{0,l} + \mathbf{J}_{0,t}) - i\frac{\omega}{c}\varepsilon(\mathbf{E}_{0,l} + \mathbf{E}_{0,t}) \tag{B.46}$$

Separating components in the last equation, we find that $\mathbf{J}_{0,t}$ is only in the $\hat{\mathbf{x}}$ direction and that

$$E_{0,l} = -i\frac{4\pi}{\varepsilon}\frac{1}{\omega}J_{0,l} \tag{B.47}$$

$$B_{0,t} = i\frac{4\pi\mu}{kc}J_{0,t} + \frac{\omega}{kc}\varepsilon\mu E_{0,t} \tag{B.48}$$

In the last expression we use the result of Eq. (B.45) to obtain:

$$E_{0,t} = -i\frac{4\pi\omega}{c^2}\frac{\mu}{\omega^2\varepsilon\mu/c^2 - k^2}J_{0,t} \tag{B.49}$$

With this last result we have managed to determine all the field components in terms of the charge and current densities.

It is instructive to explore the consequences of this solution. First, we note that we have obtained $E_{0,l}$ as two different expressions, Eqs (B.43) and (B.47), which must be compatible, requiring that:

$$\frac{1}{k}\rho_0 = \frac{1}{\omega}J_{0,l} \Rightarrow J_{0,l}k = \omega\rho_0$$

which is, of course, true due to the charge conservation equation:

$$\nabla_{\mathbf{r}}\mathbf{J}_f(\mathbf{r}, t) = -\frac{\partial \rho_f(\mathbf{r}, t)}{\partial t} \Rightarrow i\mathbf{k} \cdot (\mathbf{J}_{0,l} + \mathbf{J}_{0,t}) = i\omega\rho_0$$

the last equation implying a relation between $J_{0,l}$ and ρ_0 identical to the previous one. Another interesting aspect of the solution we obtained is that the denominator appearing in the expression for $E_{0,t}$ in Eq. (B.49) cannot vanish for a physically meaningful solution. Thus, for these fields $k \neq (\omega/c)\sqrt{\varepsilon\mu}$, in contrast to what we found for the case of zero charge and current densities. We will define this denominator to be equal to $-\kappa^2$:

$$-\kappa^2 \equiv \frac{\omega^2}{c^2}\varepsilon\mu - k^2 \tag{B.50}$$

We also want to relate the electric field to the current density through Ohm's law, Eq. (B.20). We will use this requirement, as it applies to the transverse components, to relate the conductivity σ to κ through Eq. (B.49):

$$\sigma^{-1} = i\frac{4\pi\omega}{c^2}\frac{\mu}{\kappa^2} \Rightarrow \kappa^2 = i\sigma\frac{4\pi}{c^2}\mu\omega$$

The last expression for κ when substituted in its definition, Eq. (B.50), yields:

$$k^2 = \frac{\omega^2}{c^2}\mu\varepsilon + i\mu\omega\frac{4\pi}{c^2}\sigma$$

which is an interesting result, revealing that the wave-vector has now acquired an imaginary component. Indeed, expressing the wave-vector in terms of its real and imaginary parts, $k = k_R + ik_I$, and using the above equation, we find:

$$k_R = \frac{\omega}{c}\sqrt{\varepsilon\mu}\left[\frac{1}{2}\left(1 + \left(\frac{4\pi\sigma}{\varepsilon\omega}\right)^2\right)^{1/2} + \frac{1}{2}\right]^{1/2}$$

$$k_I = \frac{\omega}{c}\sqrt{\varepsilon\mu}\left[\frac{1}{2}\left(1 + \left(\frac{4\pi\sigma}{\varepsilon\omega}\right)^2\right)^{1/2} - \frac{1}{2}\right]^{1/2}$$

These expressions show that, for finite conductivity σ, when $\omega \to \infty$ the wave-vector reverts to a real quantity only ($k_I = 0$). The presence of an imaginary component in the wave-vector has important physical implications: it means that the fields decay exponentially inside the solid as $\sim \exp[-k_I z]$. For large enough frequency the imaginary component is negligible, that is, the solid is transparent to such radiation. We note incidentally that with the definition of κ given above, the longitudinal components of the electric field and the current density obey Ohm's law but with an extra factor multiplying the conductivity σ:

$$-i\frac{4\pi}{\varepsilon}\frac{1}{\omega} = -\frac{1}{\sigma}\frac{\kappa^2 c^2}{\omega^2\mu\varepsilon} = \frac{1}{\sigma}\left(1 - \frac{k^2 c^2}{\omega^2\mu\varepsilon}\right) \Rightarrow J_{0,l} = \sigma\left(1 - \frac{k^2 c^2}{\omega^2\mu\varepsilon}\right)^{-1}E_{0,l}$$

Finally, we will determine the vector and scalar potentials which can describe the electric and magnetic fields. We define the potentials in plane-wave form as:

$$\mathbf{A}(\mathbf{r}, t) = \mathbf{A}_0 e^{i(\mathbf{k}\cdot\mathbf{r}-\omega t)}, \quad \Phi(\mathbf{r}, t) = \Phi_0 e^{i(\mathbf{k}\cdot\mathbf{r}-\omega t)} \tag{B.51}$$

From the standard definition of the magnetic field in terms of the vector potential, Eq. (B.31) and using the Coulomb gauge, Eq. (B.16), we obtain that the vector potential must have only a transverse component in the same direction as the transverse magnetic field:

$$\mathbf{A}_{0,t} = -i\frac{1}{k}\mathbf{B}_{0,t} \tag{B.52}$$

From the standard definition of the electric field in terms of the scalar and vector potentials, Eq. (B.32), we then deduce that the transverse components must obey:

$$\mathbf{E}_{0,t} = \frac{i\omega}{c}\mathbf{A}_{0,t} = \frac{\omega}{kc}\mathbf{B}_{0,t}$$

which is automatically satisfied because of Eq. (B.45), while the longitudinal components must obey:

$$\mathbf{E}_{0,l} = -i\mathbf{k}\Phi_0 \Rightarrow \Phi_0 = \frac{4\pi}{\varepsilon}\frac{1}{k^2}\rho_0 \tag{B.53}$$

where for the last step we have used Eq. (B.43). It is evident that the last expression for the magnitude of the scalar potential is also compatible with Poisson's equation:

$$\nabla_{\mathbf{r}}^2\Phi(\mathbf{r}, t) = -\frac{4\pi}{\varepsilon}\rho_f(\mathbf{r}, t) \Rightarrow \Phi_0 = \frac{4\pi}{\varepsilon}\frac{1}{|\mathbf{k}|^2}\rho_0$$

These results are useful in making the connection between the dielectric function and the conductivity from microscopic considerations, as discussed in Chapter 6.

Further Reading

1. *Introduction to Electrodynamics*, D. J. Griffiths (3rd edn, Prentice-Hall, New Jersey, 1999).
2. *Modern Electrodynamics*, A. Zangwill (Cambridge University Press, Cambridge, 2013).

Appendix C Quantum Mechanics

Quantum mechanics is the theory that captures the particle–wave duality of matter. Quantum mechanics applies in the microscopic realm, that is, at length scales and at time scales relevant to subatomic particles like electrons and nuclei. It is the most successful physical theory: it has been verified by every experiment performed to check its validity. It is also the most counter-intuitive physical theory, since its premises are at variance with our everyday experience, which is based on macroscopic observations that obey the laws of classical physics. When the properties of physical objects (like solids, clusters, and molecules) are studied at a resolution at which the atomic degrees of freedom are explicitly involved, the use of quantum mechanics becomes necessary.

In this appendix we attempt to give the basic concepts of quantum mechanics relevant to the study of solids, clusters, and molecules, in a reasonably self-contained form but avoiding detailed discussions. We refer the reader to standard texts of quantum mechanics for more extensive discussion and proper justification of the statements that we present here, some of which we mention in Further Reading.

C.1 The Schrödinger Equation

There are different ways to formulate the theory of quantum mechanics. In the following we will discuss the Schrödinger wave mechanics picture. The starting point is the form of a free traveling wave

$$\psi(\mathbf{r}, t) = e^{i(\mathbf{k}\cdot\mathbf{r} - \omega t)} \tag{C.1}$$

of wave-vector $\mathbf{k}$ and frequency ω. The free traveling wave satisfies the equation

$$i\hbar \frac{\partial \psi(\mathbf{r}, t)}{\partial t} = -\frac{\hbar^2}{2m} \nabla_{\mathbf{r}}^2 \psi(\mathbf{r}, t) \tag{C.2}$$

if the wave-vector and the frequency are related by:

$$\hbar\omega = \frac{\hbar^2 \mathbf{k}^2}{2m}$$

where $\hbar, m$ are constants (for the definition of the operator $\nabla_{\mathbf{r}}^2$, see Appendix A). Schrödinger postulated that $\psi(\mathbf{r}, t)$ can also be considered to describe the motion of a free particle with mass m and momentum $\mathbf{p} = \hbar\mathbf{k}$, with $\hbar\omega = \mathbf{p}^2/2m$ the energy of the free particle. Thus, identifying

$$i\hbar\frac{\partial}{\partial t} \rightarrow \epsilon : \text{energy}, \quad -i\hbar\nabla_{\mathbf{r}} \rightarrow \mathbf{p} : \text{momentum}$$

introduces the quantum-mechanical operators for the energy and the momentum for a free particle of mass m. Here, $\hbar$ is related to Planck's constant h by:

$$\hbar = \frac{h}{2\pi}$$

$\psi(\mathbf{r},t)$ is called the *wavefunction*, the absolute value squared, $|\psi(\mathbf{r},t)|^2$, of which is interpreted as the probability of finding the particle at position $\mathbf{r}$ and time t.

When the particle is not free, we add to the wave equation the potential energy term $\mathcal{V}(\mathbf{r},t)\psi(\mathbf{r},t)$, so that the equation obeyed by the wavefunction $\psi(\mathbf{r},t)$ reads:

$$\left[-\frac{\hbar^2\nabla_{\mathbf{r}}^2}{2m} + \mathcal{V}(\mathbf{r},t) \right] \psi(\mathbf{r},t) = i\hbar\frac{\partial\psi(\mathbf{r},t)}{\partial t} \tag{C.3}$$

which is known as the time-dependent Schrödinger equation. If the absolute value squared of the wavefunction $\psi(\mathbf{r},t)$ is to represent a probability, it must be properly normalized, that is:

$$\int |\psi(\mathbf{r},t)|^2 d\mathbf{r} = 1$$

so that the probability of finding the particle anywhere in space is unity. The wavefunction and its gradient must also be continuous and finite everywhere for this interpretation to have physical meaning. One other requirement on the wavefunction is that it decays to zero at infinity. If the external potential is independent of time, we can write the wavefunction as:

$$\psi(\mathbf{r},t) = e^{-i\epsilon t/\hbar}\psi(\mathbf{r}) \tag{C.4}$$

which, when substituted in the time-dependent Schrödinger equation, gives:

$$\left[-\frac{\hbar^2\nabla_{\mathbf{r}}^2}{2m} + \mathcal{V}(\mathbf{r}) \right] \psi(\mathbf{r}) = \epsilon\psi(\mathbf{r}) \tag{C.5}$$

This is known as the time-independent Schrödinger equation (TISE). The quantity inside the square brackets is called the hamiltonian, $\hat{\mathcal{H}}$:

$$\hat{\mathcal{H}} = -\frac{\hbar^2\nabla_{\mathbf{r}}^2}{2m} + \mathcal{V}(\mathbf{r})$$

In the TISE, the hamiltonian corresponds to the energy operator. Notice that the energy ϵ has now become the eigenvalue of the wavefunction $\psi(\mathbf{r})$ in the second-order differential equation represented by the TISE.

In most situations of interest we are faced with the problem of solving the TISE once the potential $\mathcal{V}(\mathbf{r})$ has been specified. The solution gives the eigenvalues ϵ and eigenfunctions $\psi(\mathbf{r})$, which together provide a complete description of the physical system. There are usually many (often infinite) solutions to the TISE, which are identified by their eigenvalues labeled by some index or set of indices, denoted here collectively by the subscript i:

$$\hat{\mathcal{H}}\psi_i(\mathbf{r}) = \epsilon_i\psi_i(\mathbf{r})$$

It is convenient to choose the wavefunctions that correspond to different eigenvalues of the energy to be orthonormal:

$$\int \psi_i^*(\mathbf{r})\psi_j(\mathbf{r})d\mathbf{r} = \delta_{ij} \qquad (C.6)$$

Such a set of eigenfunctions is referred to as a complete basis set, spanning the Hilbert space of the hamiltonian $\hat{\mathcal{H}}$. We can then use this set to express a general state of the system $\chi(\mathbf{r})$ as:

$$\chi(\mathbf{r}) = \sum_i c_i\psi_i(\mathbf{r})$$

where the coefficients c_i, due to the orthonormality of the wavefunctions, Eq. (C.6), are given by:

$$c_i = \int \psi_i^*(\mathbf{r})\chi(\mathbf{r})d\mathbf{r}$$

The notion of the Hilbert space of the hamiltonian is a very useful one: we imagine the eigenfunctions of the hamiltonian as the axes in a multidimensional space (the Hilbert space), in which the state of the system is a point. The position of this point is given by its projection on the axes, just like the position of a point in 3D space is given by its cartesian coordinates, which are the projections of the point on the x, y, z axes. In this sense, the coefficients c_i defined above are the projections of the state of the system on the basis set comprising the eigenfunctions of the hamiltonian.

A general feature of the wavefunction is that it has oscillating wave-like character when the potential energy is lower than the total energy (in which case the kinetic energy is positive) and decaying exponential behavior when the potential energy is higher than the total energy (in which case the kinetic energy is negative). This is most easily seen in a 1D example, where the TISE can be written as:

$$-\frac{\hbar^2}{2m}\frac{d^2\psi(x)}{dx^2} = [\epsilon - \mathcal{V}(x)]\psi(x) \Rightarrow \frac{d^2\psi(x)}{dx^2} = -[k(x)]^2\psi(x)$$

where we have defined the function

$$k(x) = \sqrt{\frac{2m}{\hbar^2}[\epsilon - \mathcal{V}(x)]}$$

The above expression then shows that, if we treat k as constant:

$$\text{for } \epsilon > \mathcal{V}(x) \Rightarrow k^2 > 0 \longrightarrow \psi(x) \sim e^{\pm i|k|x}$$
$$\text{for } \epsilon < \mathcal{V}(x) \Rightarrow k^2 < 0 \longrightarrow \psi(x) \sim e^{\pm |k|x}$$

and in the last expression we choose the sign that makes the wavefunction vanish for $x \to \pm\infty$ as the only physically plausible choice. This is illustrated in Fig. C.1 for a square barrier and a square well, so that in both cases the function $[k(x)]^2$ is a positive or

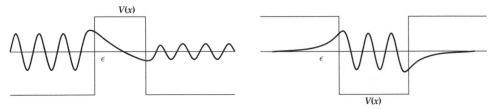

Fig. C.1 Illustration of the oscillatory and decaying exponential nature of the wavefunction (shown in blue) in regions where the potential energy $V(x)$ (shown in red) is lower than or higher than the total energy ϵ (indicated by a horizontal black line): on the left is a square barrier, on the right a square well.

negative constant everywhere. Notice that in the square barrier, the wavefunction before and after the barrier has the same wave-vector (k takes the same value before and after the barrier), but the amplitude of the oscillation has decreased, because only part of the wave is transmitted while another part is reflected from the barrier. The points at which $[k(x)]^2$ changes sign are called the "turning points," because they correspond to the positions where a classical particle would be reflected at the walls of the barrier or the well. The quantum-mechanical nature of the particle allows it to penetrate the walls of the barrier or leak out of the walls of the well, as a wave would, in sharp contrast to classical behavior; this phenomenon is referred to as "quantum tunneling." In terms of the wavefunction, a turning point corresponds to a value of x at which the curvature changes sign, that is, it is an inflection point.

For a specific application of these concepts, we consider a 1D square barrier with height $V_0 > 0$ in the range $0 < x < L$, and an incident plane wave of energy ϵ. We can write the wavefunction as a plane wave with wave-vector q in the region where the potential is zero, and an exponential with decay constant κ for $\epsilon < V_0$ or a plane wave with wave-vector k for $\epsilon > V_0$. We then use the following expressions to represent the wavefunction in various regions:

$$\text{incident}: \ e^{iqx}, \ x < 0; \quad \text{reflected}: \ Re^{-iqx}, \ x < 0; \quad \text{transmitted}: \ Te^{iqx}, \ x > L$$

$$Ae^{\kappa x} + Be^{-\kappa x}, \ 0 < x < L, \ \text{for}: \ \epsilon < V_0$$

$$Ce^{ikx} + De^{-ikx}, \ 0 < x < L, \ \text{for}: \ \epsilon > V_0$$

We next employ the conditions on the continuity of the wavefunction and its derivative at $x = 0, x = L$, from which we obtain that the transmission coefficient defined as $|T|^2$ is given by:

$$|T|^2 = \frac{1}{1 + \sinh^2(\kappa L)/4\lambda(1 - \lambda)}, \quad \kappa = \left(\frac{2m(V_0 - \epsilon)}{\hbar^2}\right)^{1/2} \quad \text{for} \ \epsilon < V_0$$

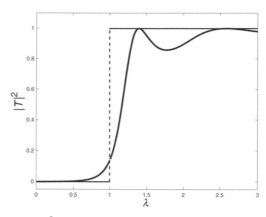

Fig. C.2   The transmission coefficient $|T|^2$ (blue line) of a square barrier of height $\mathcal{V}_0$ and width L, for an incident quantum-mechanical particle with energy ϵ, as a function of $\lambda = \epsilon/\mathcal{V}_0$. The red line shows the classical value, which is 0 for $\epsilon < \mathcal{V}_0$ ($\lambda < 1$) and 1 for $\epsilon > \mathcal{V}_0$ ($\lambda > 1$).

$$|T|^2 = \frac{1}{1 + \sin^2(kL)/4\lambda(\lambda - 1)}, \quad k = \left(\frac{2m(\epsilon - \mathcal{V}_0)}{\hbar^2}\right)^{1/2} \quad \text{for} \quad \epsilon > \mathcal{V}_0$$

with $\lambda = \epsilon/\mathcal{V}_0$, and

$$|T|^2 = \frac{1}{1 + m\mathcal{V}_0 L^2/2\hbar^2}, \quad \text{for} \quad \epsilon = \mathcal{V}_0$$

while the reflection coefficient is given by $|R|^2 = 1 - |T|^2$. The resulting value for the transmission coefficient is shown in Fig. C.2, as a function of $\lambda = \epsilon/\mathcal{V}_0$. This result is in sharp contrast to the classical value in two ways: first, there is non-zero transmission even for $\epsilon < \mathcal{V}_0$, while for a classical system the transmission would be strictly zero in this case; second, for $\epsilon > \mathcal{V}_0$, the transmission can be *lower* than unity, while for the classical system the transmission would always be equal to 1 in this regime.

C.2 Bras, Kets, and Operators

Once the wavefunction of a state has been determined, the value of any physical observable can be calculated by taking the expectation value of the corresponding operator between the wavefunction and its complex conjugate. The expectation value of an operator $\hat{O}(\mathbf{r})$ in a state described by wavefunction $\psi(\mathbf{r})$ is defined as:

$$\int \psi^*(\mathbf{r})\hat{O}(\mathbf{r})\psi(\mathbf{r})d\mathbf{r}$$

An example of an operator is $-i\hbar\nabla_{\mathbf{r}}$ for the momentum $\mathbf{p}$, as we saw earlier. As an application of these concepts, we consider the following operators in one dimension:

$$\hat{X} = x - \bar{x}, \quad \hat{P} = p - \bar{p}, \quad p = -i\hbar\frac{\partial}{\partial x}$$

with $\bar{x}$ and $\bar{p}$ denoting the expectation values of x and p, respectively:

$$\bar{x} \equiv \int \psi^*(x)x\psi(x)\mathrm{d}x, \quad \bar{p} \equiv \int \psi^*(x)p\psi(x)\mathrm{d}x$$

First, notice that the expectation values of $\hat{X}$ and $\hat{P}$ vanish identically. With the assumption that the wavefunction vanishes at infinity, it is possible to show that the expectation values of the squares of the operators $\hat{X}$ and $\hat{P}$ obey the following relation:[1]

$$\int \psi^*(x)\hat{X}^2\psi(x)\mathrm{d}x \int \psi^*(x)\hat{P}^2\psi(x)\mathrm{d}x \geq \frac{\hbar^2}{4} \tag{C.7}$$

We can interpret the expectation value as the average, in which case the expectation value of $\hat{X}^2$ is the variance of x, also denoted as $(\Delta x)^2$, and similarly for $\hat{P}^2$, with $(\Delta p)^2$ the variance of p. Then, Eq. (C.7) becomes:

$$(\Delta x)(\Delta p) \geq \frac{\hbar}{2}$$

which is known as the **Heisenberg uncertainty relation**. This seemingly abstract relation between the position and momentum variables can be used to yield very practical results. For instance, electrons associated with an atom are confined by the Coulomb potential of the nucleus to a region of $\sim 1\,\text{Å}$, that is, $\Delta x \sim 1\,\text{Å}$, which means that their typical momentum will be of order $p = \hbar/2\Delta x$. This gives a direct estimate of the energy scale for these electronic states:

$$\frac{p^2}{2m_e} = \frac{\hbar^2}{2m_e(2\Delta x)^2} \sim 1\,\text{eV}$$

which is very close to the binding energy scale for valence electrons (the core electrons are more tightly bound to the nucleus and therefore their binding energy is higher). The two variables linked by the Heisenberg uncertainty relation are referred to as conjugate variables. There exist other pairs of conjugate variables linked by the same relation, such as the energy ϵ and the time t:

$$(\Delta\epsilon)(\Delta t) \geq \frac{\hbar}{2}$$

In the calculation of expectation values it is convenient to introduce the so-called "bra" $\langle\psi|$ and "ket" $|\psi\rangle$ notation, with the first representing the wavefunction and the second its complex conjugate. In the bra and ket expressions the spatial coordinate $\mathbf{r}$ is left deliberately unspecified so that they can be considered as wavefunctions independent of the representation; when the coordinate $\mathbf{r}$ is specified, the wavefunctions are considered to

---

[1] A proof of this relation is given in the book of L. I. Schiff, mentioned in Further Reading, where references to the original work of Heisenberg are also provided.

be expressed in the "position representation." Thus, the expectation value of an operator $\hat{O}$ in state $|\psi\rangle$ is:

$$\langle\psi|\hat{O}|\psi\rangle \equiv \int \psi^*(\mathbf{r})\hat{O}(\mathbf{r})\psi(\mathbf{r})d\mathbf{r}$$

where the left-hand-side expression is independent of representation and the right-hand-side expression is in the position representation. In terms of the bra and ket notation, the orthonormality of the energy eigenfunctions can be expressed as:

$$\langle\psi_i|\psi_j\rangle = \delta_{ij}$$

and the general state of the system χ can be expressed as:

$$|\chi\rangle = \sum_i \langle\psi_i|\chi\rangle|\psi_i\rangle$$

from which we can deduce that the expression

$$\sum_i |\psi_i\rangle\langle\psi_i| = 1$$

is the identity operator. The usefulness of the above expressions is their general form, which is independent of representation.

This representation-independent notation can be extended to the time-dependent wavefunction $\psi(\mathbf{r},t)$, leading to an elegant expression. We take advantage of the series expansion of the exponential to define the following operator:

$$e^{-i\hat{\mathcal{H}}t/\hbar} = \sum_{n=0}^{\infty} \frac{(-i\hat{\mathcal{H}}t/\hbar)^n}{n!} \tag{C.8}$$

where $\hat{\mathcal{H}}$ is the hamiltonian, which we assume to contain a time-independent potential. When this operator is applied to the time-independent part of the wavefunction, it gives:

$$e^{-i\hat{\mathcal{H}}t/\hbar}\psi(\mathbf{r}) = \sum_{n=0}^{\infty} \frac{(-i\hat{\mathcal{H}}t/\hbar)^n}{n!}\psi(\mathbf{r}) = \sum_{n=0}^{\infty} \frac{(-i\epsilon t/\hbar)^n}{n!}\psi(\mathbf{r})$$

$$= e^{-i\epsilon t/\hbar}\psi(\mathbf{r}) = \psi(\mathbf{r},t) \tag{C.9}$$

where we have used Eq. (C.5) and the definition of the time-dependent wavefunction, Eq. (C.4). This shows that in general we can write the time-dependent wavefunction as the operator $\exp[-i\hat{\mathcal{H}}t/\hbar]$ applied to the wavefunction at $t = 0$. In a representation-independent expression, this statement gives:

$$|\psi(t)\rangle = e^{-i\hat{\mathcal{H}}t/\hbar}|\psi(0)\rangle \rightarrow \langle\psi(t)| = \langle\psi(0)|e^{i\hat{\mathcal{H}}t/\hbar} \tag{C.10}$$

with the convention that for a bra, the operator next to it acts to the left.

Now consider a general operator $\hat{O}$ corresponding to a physical observable; we assume that $\hat{O}$ itself is a time-independent operator, but the value of the observable changes with time because of changes in the wavefunction:

$$\langle\hat{O}\rangle(t) \equiv \langle\psi(t)|\hat{O}|\psi(t)\rangle = \langle\psi(0)|e^{i\hat{\mathcal{H}}t/\hbar}\hat{O}e^{-i\hat{\mathcal{H}}t/\hbar}|\psi(0)\rangle \tag{C.11}$$

We define a new operator

$$\hat{O}(t) \equiv e^{i\hat{\mathcal{H}}t/\hbar}\hat{O}e^{-i\hat{\mathcal{H}}t/\hbar}$$

which includes explicitly the time dependence, and whose expectation value in the state $|\psi(0)\rangle$ is exactly the same as the expectation value of the original time-independent operator in the state $|\psi(t)\rangle$. Working with the operator $\hat{O}(t)$ and the state $|\psi(0)\rangle$ is called the "Heisenberg picture," while working with the operator $\hat{O}$ and the state $|\psi(t)\rangle$ is called the "Schrödinger picture." The two pictures give identical results as far as the values of physical observables are concerned, as Eq. (C.11) shows, so the choice of one over the other is a matter of convenience. In the Heisenberg picture the basis is fixed and the operator evolves in time, whereas in the Schrödinger picture the basis evolves and the operator is independent of time. We can also determine the evolution of the time-dependent operator from its definition, as follows:

$$\frac{d}{dt}\hat{O}(t) = \frac{i\hat{\mathcal{H}}}{\hbar}e^{i\hat{\mathcal{H}}t/\hbar}\hat{O}e^{-i\hat{\mathcal{H}}t/\hbar} - \frac{i}{\hbar}e^{i\hat{\mathcal{H}}t/\hbar}\hat{O}\hat{\mathcal{H}}e^{-i\hat{\mathcal{H}}t/\hbar} = \frac{i}{\hbar}\hat{\mathcal{H}}\hat{O}(t) - \frac{i}{\hbar}\hat{O}(t)\hat{\mathcal{H}}$$

$$\Rightarrow \frac{d}{dt}\hat{O}(t) = \frac{i}{\hbar}[\hat{\mathcal{H}},\hat{O}(t)] \tag{C.12}$$

The last expression is defined as the "commutator" of the hamiltonian with the time-dependent operator $\hat{O}(t)$. The commutator is a general concept that applies to any pair of operators $\hat{O}_1,\hat{O}_2$:

$$\text{commutator}: [\hat{O}_1,\hat{O}_2] \equiv \hat{O}_1\hat{O}_2 - \hat{O}_1\hat{O}_2$$

The bra and ket notation can be extended to situations that involve more than one particle, as in the many-body wavefunction relevant to electrons in a solid. For example, a many-body wavefunction may be denoted by $|\Psi\rangle$ and when expressed in the position representation it takes the form

$$\langle \mathbf{r}_1,\ldots,\mathbf{r}_N|\Psi\rangle \equiv \Psi(\mathbf{r}_1,\ldots,\mathbf{r}_N)$$

where N is the total number of particles in the system. When a many-body wavefunction refers to a system of indistinguishable particles, it must have certain symmetries: in the case of fermions it is antisymmetric (it changes sign upon interchange of all the coordinates of any two of the particles), while for bosons it is symmetric (it is the same upon interchange of all the coordinates of any two of the particles). We can define operators relevant to a many-body wavefunction in the usual way. A useful example is the density operator: it represents the probability of finding any of the particles involved in the wavefunction at a certain position in space. For one particle by itself in state $\psi(\mathbf{r})$ the meaning we assigned to the wavefunction already gives this probability as:

$$|\psi(\mathbf{r})|^2 = n(\mathbf{r})$$

which can also be thought of as the density at $\mathbf{r}$, since the integral over all space gives the total number of particles (in this case 1):

$$\int n(\mathbf{r})d\mathbf{r} = \int \psi^*(\mathbf{r})\psi(\mathbf{r})d\mathbf{r} = 1$$

The corresponding operator must be defined as $\delta(\mathbf{r} - \mathbf{r}')$, with the second variable an arbitrary position in space. This choice of the density operator, when we take its matrix elements in the state $|\psi\rangle$ by inserting a complete set of states in the position representation, gives:

$$\langle\psi|\delta(\mathbf{r} - \mathbf{r}')|\psi\rangle = \int \langle\psi|\mathbf{r}'\rangle\delta(\mathbf{r} - \mathbf{r}')\langle\mathbf{r}'|\psi\rangle d\mathbf{r}'$$

$$= \int \psi^*(\mathbf{r}')\delta(\mathbf{r} - \mathbf{r}')\psi(\mathbf{r}')d\mathbf{r}' = |\psi(\mathbf{r})|^2 = n(\mathbf{r})$$

as desired. Generalizing this result to the N-particle system, we define the density operator as:

$$\mathcal{N}(\mathbf{r}) = \sum_{i=1}^{N} \delta(\mathbf{r} - \mathbf{r}_i)$$

with $\mathbf{r}_i, i = 1, \ldots, N$, the variables describing the positions of the particles. The expectation value of this operator in the many-body wavefunction gives the particle density at $\mathbf{r}$:

$$n(\mathbf{r}) = \langle\Psi|\mathcal{N}(\mathbf{r})|\Psi\rangle$$

In the position representation this takes the form

$$n(\mathbf{r}) = \int \langle\Psi|\mathbf{r}_1, \ldots, \mathbf{r}_N\rangle\mathcal{N}(\mathbf{r})\langle\mathbf{r}_1, \ldots, \mathbf{r}_N|\Psi\rangle d\mathbf{r}_1 \cdots d\mathbf{r}_N$$

$$= \int \Psi^*(\mathbf{r}_1, \ldots, \mathbf{r}_N) \sum_{i=1}^{N} \delta(\mathbf{r} - \mathbf{r}_i)\Psi(\mathbf{r}_1, \ldots, \mathbf{r}_N)d\mathbf{r}_1 \cdots d\mathbf{r}_N$$

$$= N \int \Psi^*(\mathbf{r}, \mathbf{r}_2, \ldots, \mathbf{r}_N)\Psi(\mathbf{r}, \mathbf{r}_2, \ldots, \mathbf{r}_N)d\mathbf{r}_2 \cdots d\mathbf{r}_N \qquad (C.13)$$

where the last equation applies to a system of N indistinguishable particles. By analogy to the expression for the density of a system of indistinguishable particles, we can define a function of two independent variables $\mathbf{r}$ and $\mathbf{r}'$, the so-called one-particle density matrix $\gamma(\mathbf{r}, \mathbf{r}')$:

$$\gamma(\mathbf{r}, \mathbf{r}') \equiv N \int \Psi^*(\mathbf{r}, \mathbf{r}_2, \ldots, \mathbf{r}_N)\Psi(\mathbf{r}', \mathbf{r}_2, \ldots, \mathbf{r}_N)d\mathbf{r}_2 \cdots d\mathbf{r}_N \qquad (C.14)$$

whose diagonal components are equal to the density: $\gamma(\mathbf{r}, \mathbf{r}) = n(\mathbf{r})$. An extension of these concepts is the pair correlation function, which describes the probability of finding two particles simultaneously at positions $\mathbf{r}$ and $\mathbf{r}'$. The operator for the pair correlation function is:

$$\mathcal{G}(\mathbf{r}, \mathbf{r}') = \frac{1}{2} \sum_{i\neq j; i, j=1}^{N} \delta(\mathbf{r} - \mathbf{r}_i)\delta(\mathbf{r}' - \mathbf{r}_j)$$

and its expectation value in the many-body wavefunction in the position representation gives $g(\mathbf{r}, \mathbf{r}')$:

$$g(\mathbf{r}, \mathbf{r}') = \int \langle \Psi | \mathbf{r}_1, \ldots, \mathbf{r}_N \rangle \mathcal{G}(\mathbf{r}, \mathbf{r}') \langle \mathbf{r}_1, \ldots, \mathbf{r}_N | \Psi \rangle d\mathbf{r}_1 \cdots d\mathbf{r}_N$$

$$= \int \Psi^*(\mathbf{r}_1, \ldots, \mathbf{r}_N) \frac{1}{2} \sum_{i \neq j; i, j = 1}^{N} \delta(\mathbf{r} - \mathbf{r}_i) \delta(\mathbf{r}' - \mathbf{r}_j) \Psi(\mathbf{r}_1, \ldots, \mathbf{r}_N) d\mathbf{r}_1 \cdots d\mathbf{r}_N$$

$$= \frac{N(N-1)}{2} \int \Psi^*(\mathbf{r}, \mathbf{r}', \mathbf{r}_3, \ldots, \mathbf{r}_N) \Psi(\mathbf{r}, \mathbf{r}', \mathbf{r}_3, \ldots, \mathbf{r}_N) d\mathbf{r}_3 \cdots d\mathbf{r}_N$$

where the last equation applies to a system of N indistinguishable particles. By analogy to the expression for the pair-correlation function of a system of indistinguishable particles, we can define a function of four independent variables $\mathbf{r}_1, \mathbf{r}_2, \mathbf{r}_1', \mathbf{r}_2'$, the so-called two-particle density matrix $\Gamma(\mathbf{r}_1, \mathbf{r}_2 | \mathbf{r}_1', \mathbf{r}_2')$:

$$\Gamma(\mathbf{r}_1, \mathbf{r}_2 | \mathbf{r}_1', \mathbf{r}_2') \equiv \frac{N!}{2! (N-2)!} \times \tag{C.15}$$

$$\int \Psi^*(\mathbf{r}_1, \mathbf{r}_2, \mathbf{r}_3, \ldots, \mathbf{r}_N) \Psi(\mathbf{r}_1', \mathbf{r}_2', \mathbf{r}_3, \ldots, \mathbf{r}_N) d\mathbf{r}_3 \cdots d\mathbf{r}_N$$

whose diagonal components are equal to the pair-correlation function: $\Gamma(\mathbf{r}, \mathbf{r}' | \mathbf{r}, \mathbf{r}') = g(\mathbf{r}, \mathbf{r}')$. These functions are useful when dealing with one-body and two-body operators in the hamiltonian of the many-body system. An example of this use is given below, after we have defined the many-body wavefunction in terms of single-particle states. The density matrix concept can readily be generalized to n particles with $j \leq N$.

The many-body wavefunction is often expressed as a product of single-particle wavefunctions, giving rise to expressions that include many single-particle states in the bra and the ket, as in the Hartree and Hartree–Fock theories discussed in Chapter 4:

$$|\Psi\rangle = |\psi_1 \cdots \psi_N\rangle$$

In such cases we adopt the convention that the order of the single-particle states in the bra or the ket is meaningful, that is, when expressed in a certain representation the jth independent variable of the representation is associated with the jth single-particle state in the order it appears in the many-body wavefunction. For example, in the position representation we will have:

$$\langle \mathbf{r}_1, \ldots, \mathbf{r}_N | \Psi \rangle = \langle \mathbf{r}_1, \ldots, \mathbf{r}_N | \psi_1 \cdots \psi_N \rangle \equiv \psi_1(\mathbf{r}_1) \cdots \psi_N(\mathbf{r}_N)$$

Thus, when expressing matrix elements of the many-body wavefunction in the position representation, the set of variables appearing as arguments in the single-particle states of the bra and the ket must be in exactly the same order. For example, in the Hartree theory, Eq. (4.37), the Coulomb repulsion term is represented by:

$$\langle \psi_i \psi_j | \frac{1}{|\mathbf{r} - \mathbf{r}'|} | \psi_i \psi_j \rangle \equiv \int \psi_i^*(\mathbf{r}) \psi_j^*(\mathbf{r}') \frac{1}{|\mathbf{r} - \mathbf{r}'|} \psi_i(\mathbf{r}) \psi_j(\mathbf{r}') d\mathbf{r} d\mathbf{r}'$$

Similarly, in the Hartree–Fock theory, the exchange term is given by:

$$\langle \psi_i \psi_j \mid \frac{1}{\mid \mathbf{r} - \mathbf{r}' \mid} \mid \psi_j \psi_i \rangle \equiv \int \psi_i^*(\mathbf{r}) \psi_j^*(\mathbf{r}') \frac{1}{\mid \mathbf{r} - \mathbf{r}' \mid} \psi_j(\mathbf{r}) \psi_i(\mathbf{r}') d\mathbf{r} d\mathbf{r}'$$

When only one single-particle state is involved in the bra and the ket but more than one variable appears in the bracketed operator, the variable of integration is evident from the implied remaining free variable. For example, in Eq. (4.37) all terms in the square brackets are functions of $\mathbf{r}$, therefore the term involving the operator $1/\mid \mathbf{r} - \mathbf{r}' \mid$, the so-called Hartree potential $\mathcal{V}^{\mathrm{H}}(\mathbf{r})$, must be:

$$\mathcal{V}^{\mathrm{H}}(\mathbf{r}) = e^2 \sum_{j \neq i} \langle \psi_j \mid \frac{1}{\mid \mathbf{r} - \mathbf{r}' \mid} \mid \psi_j \rangle = e^2 \sum_{j \neq i} \int \psi_j^*(\mathbf{r}') \frac{1}{\mid \mathbf{r} - \mathbf{r}' \mid} \psi_j(\mathbf{r}') d\mathbf{r}'$$

An expression for the many-body wavefunction in terms of products of single-particle states, which is by construction totally antisymmetric, that is, it changes sign upon interchange of the coordinates of any two particles, is the so-called Slater determinant:

$$\Psi(\{\mathbf{r}_i\}) = \frac{1}{\sqrt{N!}} \begin{vmatrix} \psi_1(\mathbf{r}_1) & \psi_1(\mathbf{r}_2) & \cdots & \psi_1(\mathbf{r}_N) \\ \psi_2(\mathbf{r}_1) & \psi_2(\mathbf{r}_2) & \cdots & \psi_2(\mathbf{r}_N) \\ \cdot & \cdot & & \cdot \\ \cdot & \cdot & & \cdot \\ \cdot & \cdot & & \cdot \\ \psi_N(\mathbf{r}_1) & \psi_N(\mathbf{r}_2) & \cdots & \psi_N(\mathbf{r}_N) \end{vmatrix} \tag{C.16}$$

The antisymmetric nature of this expression comes from the fact that if two rows or two columns of a determinant are interchanged, which corresponds to interchanging the coordinates of two particles, then the determinant changes sign. This expression is particularly useful when dealing with systems of fermions.

As an example of the various concepts introduced above, we describe how the expectation value of an operator $\hat{\mathcal{O}}$ of a many-body system that can be expressed as a sum of single-particle operators:

$$\hat{\mathcal{O}}(\{\mathbf{r}_i\}) = \sum_{i=1}^{N} o(\mathbf{r}_i)$$

can be obtained from the matrix elements of the single-particle operators in the single-particle states used to express the many-body wavefunction. We will assume that we are dealing with a system of fermions described by a many-body wavefunction which has the form of a Slater determinant. In this case, the many-body wavefunction can be expanded as:

$$\Psi^N(\mathbf{r}_1, ..., \mathbf{r}_N) = \frac{1}{\sqrt{N}} \left[\psi_1(\mathbf{r}_i) \Psi_{1,i}^{N-1} - \psi_2(\mathbf{r}_i) \Psi_{2,i}^{N-1} + \cdots \right] \tag{C.17}$$

where the $\Psi_{j,i}^{N-1}$ are determinants of size $N - 1$ from which the row and column corresponding to states $\psi_j(\mathbf{r}_i)$ are missing. With this, the expectation value of $\hat{\mathcal{O}}$ takes the form

$$\langle \hat{O} \rangle \equiv \langle \Psi^N | \hat{O} | \Psi^N \rangle$$

$$= \frac{1}{N} \sum_i \int \left[\psi_1^*(\mathbf{r}_i) o(\mathbf{r}_i) \psi_1(\mathbf{r}_i) + \psi_2^*(\mathbf{r}_i) o(\mathbf{r}_i) \psi_2(\mathbf{r}_i) + \cdots \right] d\mathbf{r}_i$$

where the integration over all variables other than $\mathbf{r}_i$, which is involved in $o(\mathbf{r}_i)$, gives unity for properly normalized single-particle states. The one-particle density matrix in the single-particle basis is expressed as:

$$\gamma(\mathbf{r}, \mathbf{r}') = \sum_{i=1}^{N} \psi_i(\mathbf{r}) \psi_i^*(\mathbf{r}') = \sum_{i=1}^{N} \langle \mathbf{r} | \psi_i \rangle \langle \psi_i | \mathbf{r}' \rangle \tag{C.18}$$

which gives for the expectation value of $\hat{O}$:

$$\langle \hat{O} \rangle = \int \left[o(\mathbf{r}) \gamma(\mathbf{r}, \mathbf{r}') \right]_{\mathbf{r}'=\mathbf{r}} d\mathbf{r} \tag{C.19}$$

With the following definitions of $\gamma_{i,j}$, $o_{i,j}$:

$$\gamma_{j,i} = \langle \psi_j | \gamma(\mathbf{r}', \mathbf{r}) | \psi_i \rangle, \quad o_{j,i} = \langle \psi_j | o(\mathbf{r}) | \psi_i \rangle \tag{C.20}$$

where the brackets imply integration over all real-space variables that appear in the operators. We obtain the general expression for the expectation value of $\hat{O}$:

$$\langle \hat{O} \rangle = \sum_{i,j} o_{i,j} \gamma_{j,i} \tag{C.21}$$

This expression involves exclusively matrix elements of the single-particle operators $o(\mathbf{r})$ and the single-particle density matrix $\gamma(\mathbf{r}, \mathbf{r}')$ in the single-particle states $\psi_i(\mathbf{r})$, which is very convenient for actual calculations of physical properties (see, for example, the discussion of the dielectric function in Chapter 6).

C.3   Solution of TISE

We discuss here some representative examples of how the TISE is solved to determine the eigenvalues and eigenfunctions of some potentials that appear frequently in relation to the physics of solids. These include free particles and particles in a harmonic oscillator potential or a Coulomb potential.

C.3.1   Free Particles

For free particles the external potential is zero everywhere. We have already seen that in this case the time-independent part of the wavefunction is:

$$\psi(\mathbf{r}) = C e^{i\mathbf{k} \cdot \mathbf{r}}$$

which describes the spatial variation of a plane wave; the constant C is the normalization. The energy eigenvalue corresponding to such a wavefunction is simply

$$\epsilon_{\mathbf{k}} = \frac{\hbar^2 \mathbf{k}^2}{2m}$$

which is obtained directly by substituting $\psi(\mathbf{r})$ in the TISE. All that remains is to determine the constant of normalization. To this end we assume that the particle is inside a box of dimensions (L_x, L_y, L_z) in cartesian coordinates, that is, the values of x range between $-L_x/2, L_x/2$ and similarly for the other two coordinates. The wavefunction must vanish at the boundaries of the box, or equivalently it must have the same value at the two edges in each direction, which implies that:

$$\mathbf{k} = k_x \hat{\mathbf{x}} + k_y \hat{\mathbf{y}} + k_z \hat{\mathbf{z}} \text{ with}: \quad k_x = \frac{2\pi n_x}{L_x}, \quad k_y = \frac{2\pi n_y}{L_y}, \quad k_z = \frac{2\pi n_z}{L_z}$$

with n_x, n_y, n_z integers. From the form of the wavefunction we find that:

$$\int |\psi(\mathbf{r})|^2 d\mathbf{r} = V|C|^2$$

where $V = L_x L_y L_z$ is the volume of the box. This shows that we can choose

$$C = \frac{1}{\sqrt{V}}$$

for the normalization, which completes the description of wavefunctions for free particles in a box. For $L_x, L_y, L_z \to \infty$ the spacing of values of k_x, k_y, k_z, that is, $dk_x = 2\pi/L_x$, $dk_y = 2\pi/L_y$, $dk_z = 2\pi/L_z$, becomes infinitesimal, that is, $\mathbf{k}$ becomes a continuous variable. Since the value of $\mathbf{k}$ specifies the wavefunction, we can use it as the only index to identify the wavefunctions of free particles:

$$\psi_{\mathbf{k}}(\mathbf{r}) = \frac{1}{\sqrt{V}} e^{i\mathbf{k}\cdot\mathbf{r}}$$

These results are related to the Fourier and inverse Fourier transforms, which are discussed in Appendix A. We note that:

$$\langle \psi_{\mathbf{k}'} | \psi_{\mathbf{k}} \rangle = \frac{1}{V} \int e^{-i\mathbf{k}'\cdot\mathbf{r}} e^{i\mathbf{k}\cdot\mathbf{r}} d\mathbf{r} = 0, \quad \text{unless} \quad \mathbf{k} = \mathbf{k}'$$

which we express by the statement that wavefunctions corresponding to different wave-vectors are *orthogonal* (see Appendix A, the discussion of the δ-function and its Fourier representation). The wavefunctions we found above are also eigenfunctions of the momentum operator with momentum eigenvalues $\mathbf{p} = \hbar\mathbf{k}$:

$$-i\hbar\nabla_{\mathbf{r}}\psi_{\mathbf{k}}(\mathbf{r}) = -i\hbar\nabla_{\mathbf{r}} \frac{1}{\sqrt{V}} e^{i\mathbf{k}\cdot\mathbf{r}} = (\hbar\mathbf{k})\psi_{\mathbf{k}}(\mathbf{r})$$

Thus, the free-particle eigenfunctions are an example where the energy eigenfunctions are simultaneously eigenfunctions of some other operator. In such cases, the hamiltonian and this other operator commute, that is:

$$[\hat{\mathcal{H}}, \hat{O}] = \hat{\mathcal{H}}\hat{O} - \hat{O}\hat{\mathcal{H}} = 0$$

C.3.2 Harmonic Oscillator Potential

We consider a particle of mass m in a harmonic oscillator potential in one dimension:

$$\left[-\frac{\hbar^2}{2m}\frac{d^2}{dx^2} + \frac{1}{2}\kappa x^2 \right] \psi(x) = \epsilon \psi(x) \tag{C.22}$$

where κ is the spring constant. This is a potential that arises frequently in realistic applications because near the minimum of the potential energy at $\mathbf{r} = \mathbf{r}_0$, the behavior is typically quadratic for small deviations from a Taylor expansion:

$$\mathcal{V}(\mathbf{r}) = \mathcal{V}(\mathbf{r}_0) + \frac{1}{2}\left[\nabla_{\mathbf{r}}^2 \mathcal{V}(\mathbf{r}) \right]_{\mathbf{r}=\mathbf{r}_0} (\mathbf{r} - \mathbf{r}_0)^2$$

with the first derivative of the potential vanishing by definition at the minimum. We can take the position of the minimum $\mathbf{r}_0$ to define the origin of the coordinate system, and use a separable form of the wavefunction $\psi(\mathbf{r}) = \psi_1(x)\psi_2(y)\psi_3(z)$ to arrive at Eq. (C.22) for each spatial coordinate separately with $\left[\nabla_{\mathbf{r}}^2 \mathcal{V}(\mathbf{r}) \right]_0 = \kappa$. We will find it convenient to introduce the frequency of the oscillator:

$$\omega = \sqrt{\frac{\kappa}{m}}$$

The following change of variables:

$$r_c = \sqrt{\frac{\hbar}{m\omega}}, \quad \gamma = \frac{2\epsilon}{\hbar\omega}, \quad u = x/r_c, \quad \psi(x) = CH(u)e^{-u^2/2}$$

produces a differential equation of the form

$$\frac{d^2 H(u)}{du^2} - 2u\frac{dH(u)}{du} + (\gamma - 1)H(u) = 0$$

which, with $\gamma = 2n + 1$ and n an integer, is solved by the so-called Hermite polynomials, defined recursively as:

$$H_{n+1}(u) = 2uH_n(u) - 2nH_{n-1}(u)$$
$$H_0(u) = 1, \quad H_1(u) = 2u$$

With these polynomials, the wavefunction of the harmonic oscillator becomes:

$$\psi_n(x) = C_n H_n(x/r_c)e^{-x^2/2r_c^2} \tag{C.23}$$

The Hermite polynomials satisfy the relation

$$\int_{-\infty}^{\infty} H_n(u)H_m(u)e^{-u^2}du = \delta_{nm}\sqrt{\pi}2^n n! \tag{C.24}$$

that is, they are orthogonal with a weight function $\exp[-u^2]$, which makes the wavefunctions $\psi_n(x)$ orthogonal. The above relation also allows us to determine the normalization C_n:

$$C_n = \sqrt{\frac{1}{(\sqrt{\pi}2^n n!)r_c}} \tag{C.25}$$

Table C.1 The lowest six eigenfunctions ($n = 0$–5) of the 1D harmonic oscillator potential. ϵ_n is the energy, $H_n(u)$ the Hermite polynomial, and $\psi_n(x)$ the full wavefunction, including the normalization.

n	ϵ_n	$H_n(u)$	$\psi_n(x)$
0	$\frac{1}{2}\hbar\omega$	1	$(1/\sqrt{\pi}r_c)^{1/2}\,e^{-x^2/2r_c^2}$
1	$\frac{3}{2}\hbar\omega$	$2u$	$(2/\sqrt{\pi}r_c)^{1/2}(x/r_c)e^{-x^2/2r_c^2}$
2	$\frac{5}{2}\hbar\omega$	$4(u^2 - \frac{1}{2})$	$(2/\sqrt{\pi}r_c)^{1/2}\left[(x/r_c)^2 - \frac{1}{2}\right]e^{-x^2/2r_c^2}$
3	$\frac{7}{2}\hbar\omega$	$8(u^3 - \frac{3}{2}u)$	$(4/3\sqrt{\pi}r_c)^{1/2}\left[(x/r_c)^3 - \frac{3}{2}(x/r_c)\right]e^{-x^2/2r_c^2}$
4	$\frac{9}{2}\hbar\omega$	$16(u^4 - 3u^2 + \frac{3}{4})$	$(2/3\sqrt{\pi}r_c)^{1/2}\left[(x/r_c)^4 - 3(x/r_c)^2 + \frac{3}{4}\right]e^{-x^2/2r_c^2}$
5	$\frac{11}{2}\hbar\omega$	$32(u^5 - 5u^3 + \frac{15}{4}u)$	$(4/15\sqrt{\pi}r_c)^{1/2}\left[(x/r_c)^5 - 5(x/r_c)^3 + \frac{15}{4}(x/r_c)\right]e^{-x^2/2r_c^2}$

which completely specifies the wavefunction with index n. The lowest six wavefunctions ($n = 0$–5) are given explicitly in Table C.1 and shown in Fig. C.3. We note that the original equation then takes the form

$$\left[-\frac{\hbar^2}{2m}\frac{d^2}{dx^2} + \frac{1}{2}\kappa x^2\right]\psi_n(x) = \left(n + \frac{1}{2}\right)\hbar\omega\psi_n(x) \tag{C.26}$$

that is, the eigenvalues that correspond to the wavefunctions we have calculated are quantized:

$$\epsilon_n = \left(n + \frac{1}{2}\right)\hbar\omega \tag{C.27}$$

C.3.3 Coulomb Potential

Another very common case is a potential that behaves as $1/|\mathbf{r}|$, known as the Coulomb potential, from the electrostatic interaction between particles with electrical charges at distance $|\mathbf{r}|$. We will discuss this case for the simplest physical system where it applies, the hydrogen atom. For simplicity, we take the proton fixed at the origin of the coordinate system and the electron at $\mathbf{r}$. The hamiltonian for this system takes the form

$$\hat{\mathcal{H}} = -\frac{\hbar^2}{2m_e}\nabla_{\mathbf{r}}^2 - \frac{e^2}{|\mathbf{r}|} \tag{C.28}$$

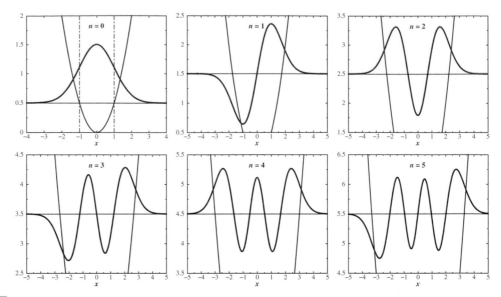

Fig. C.3 The lowest six eigenfunctions ($n = 0$–5) of the 1D harmonic oscillator potential (units used are such that all the constants appearing in the wavefunctions and the potential are equal to unity). The wavefunctions, in blue lines, have been shifted up by their energy in each case, which is shown as a horizontal black line. The harmonic oscillator potential is also shown as a red line. Notice the inflection points in the wavefunctions at the values of x where the energy eigenvalue becomes equal to the value of the harmonic potential (shown explicitly as two vertical dashed lines for the lowest eigenfunction).

with m_e the mass of the electron and $+e$, $-e$ the proton and electron charges, respectively. For this problem, we change coordinates from the cartesian x, y, z to the spherical r, θ, ϕ, which are related by:

$$x = r \sin \theta \cos \phi, \quad y = r \sin \theta \sin \phi, \quad z = r \cos \theta$$

and write the wavefunction as a product of two terms, one that depends only on r, called $R(r)$, and one that depends only on θ and ϕ, called $Y(\theta, \phi)$. Then, the hamiltonian becomes

$$-\frac{\hbar^2}{2m_e} \left[\frac{1}{r^2} \frac{\partial}{\partial r} \left(r^2 \frac{\partial}{\partial r} \right) + \frac{1}{r^2 \sin \theta} \frac{\partial}{\partial \theta} \left(\sin \theta \frac{\partial}{\partial \theta} \right) + \frac{1}{r^2 \sin^2 \theta} \frac{\partial^2}{\partial \psi^2} \right] - \frac{e^2}{r}$$

and the TISE equation, with ϵ the energy eigenvalue, takes the form

$$\frac{1}{R} \frac{d}{dr} \left(r^2 \frac{dR}{dr} \right) + \frac{2m_e r^2}{\hbar^2} \left[\epsilon + \frac{e^2}{r} \right] = -\frac{1}{Y} \left[\frac{1}{\sin \theta} \frac{\partial}{\partial \theta} \left(\sin \theta \frac{\partial Y}{\partial \theta} \right) + \frac{1}{\sin^2 \theta} \frac{\partial^2 Y}{\partial \psi^2} \right]$$

Since in the above equation the left-hand side is exclusively a function of r while the right-hand side is exclusively a function of θ, ϕ, they must each be equal to a constant which we denote by λ, giving rise to the following two differential equations:

$$\frac{1}{r^2}\frac{d}{dr}\left(r^2\frac{dR}{dr}\right)+\frac{2m_e}{\hbar^2}\left[\epsilon+\frac{e^2}{r}\right]R=\frac{\lambda}{r^2}R \tag{C.29}$$

$$\left[\frac{1}{\sin\theta}\frac{\partial}{\partial\theta}\left(\sin\theta\frac{\partial Y}{\partial\theta}\right)+\frac{1}{\sin^2\theta}\frac{\partial^2 Y}{\partial\psi^2}\right]=-\lambda Y \tag{C.30}$$

We consider the equation for $Y(\theta,\phi)$ first. This equation is solved by the functions

$$Y(\theta,\phi)=(-1)^{(|m|+m)/2}\left[\frac{(2l+1)(l-|m|)!}{4\pi(l+|m|)!}\right]^{1/2}P_l^m(\cos\theta)e^{im\phi} \tag{C.31}$$

where l and m are integers with the following range:

$$l\geq 0,\quad m=-l,-l+1,\ldots,0,\ldots,l-1,+l$$

and $P_l^m(w)$ are the functions

$$P_l^m(w)=(1-w^2)^{|m|/2}\frac{d^{|m|}}{dw^{|m|}}P_l(w) \tag{C.32}$$

with $P_l(w)$ the Legendre polynomials defined recursively as:

$$P_{l+1}(w)=\frac{2l+1}{l+1}wP_l(w)-\frac{l}{l+1}P_{l-1}(w)$$
$$P_0(w)=1,\quad P_1(w)=w$$

The $Y_{lm}(\theta,\phi)$ functions are called "spherical harmonics."

The spherical harmonics are the functions that give the anisotropic character of the eigenfunctions of the Coulomb potential, since the remaining part $R(r)$ is spherically symmetric. Taking into account the correspondence between the (x,y,z) cartesian coordinates and the (r,θ,ϕ) spherical coordinates, we can relate the spherical harmonics to functions of $x/r, y/r$ and z/r. Y_{00} is a constant which has spherical symmetry and is referred to as an s state; it represents a state of zero angular momentum. Linear combinations of higher spherical harmonics that correspond to $l=1$ are referred to as p states, those for $l=2$ as d states, and those for $l=3$ as f states. Still higher angular momentum states are labeled $g,h,j,k,\ldots$, for $l=4,5,6,7,\ldots$ The linear combinations of spherical harmonics for $l=1$ that correspond to the usual p states as expressed in terms of $x/r, y/r, z/r$ (up to constant factors that ensure proper normalization) are given in Table C.2. Similar combinations of spherical harmonics for higher values of l correspond to the usual d,f states defined as functions of $x/r, y/r, z/r$. The character of selected p,d and f states is shown in Fig. C.4. The connection between the spherical harmonics and angular momentum is explained next.

The eigenvalue of the spherical harmonic Y_{lm} in Eq. (C.30) is $\lambda=-l(l+1)$. When this is substituted in Eq. (C.29) and $R(r)$ is expressed as $R(r)=Q(r)/r$, this equation takes the form

$$\left[-\frac{\hbar^2}{2m_e}\frac{d^2}{dr^2}+\frac{\hbar^2l(l+1)}{2m_er^2}-\frac{e^2}{r}\right]Q(r)=\epsilon Q(r) \tag{C.33}$$

Table C.2 The spherical harmonics $Y_{lm}(\theta, \phi)$ for $l = 0, 1, 2, 3$ along with the x, y, z representation of the linear combinations for given l and $|m|$ and the identification of those representations as s, p, d, f orbitals.

l	m	$Y_{lm}(\theta,\phi)$	x, y, z representation	s, p, d, f orbitals
0	0	$(1/4\pi)^{1/2}$	1	s
1	0	$(3/4\pi)^{1/2}\cos\theta$	z/r	p_z
1	± 1	$\mp(3/8\pi)^{1/2}\sin\theta e^{\pm i\phi}$	$x/r, y/r$	p_x, p_y
2	0	$(5/16\pi)^{1/2}(3\cos^2\theta - 1)$	$(3z^2 - r^2)/r^2$	$d_{z^2 - r^2}$
2	± 1	$\mp(15/8\pi)^{1/2}\sin\theta\cos\theta e^{\pm i\phi}$	$xz/r^2, yz/r^2$	d_{xz}, d_{yz}
2	± 2	$(15/32\pi)^{1/2}\sin^2\theta e^{\pm 2i\phi}$	$xy/r^2, (x^2 - y^2)/r^2$	$d_{xy}, d_{x^2-y^2}$
3	0	$(7/16\pi)^{1/2}(5\cos^3\theta - 3\cos\theta)$	$(5z^2 - r^2)z/r^3$	$f_{z^3-zr^2}$
3	± 1	$\mp(21/64\pi)^{1/2}(5\cos^2\theta - 1)\sin\theta e^{\pm i\phi}$	$(5z^2 - r^2)x/r^3$	$f_{xz^2-xr^2}$
			$(5z^2 - r^2)y/r^3$	$f_{yz^2-yr^2}$
3	± 2	$(105/32\pi)^{1/2}\sin^2\theta\cos\theta e^{\pm i2\phi}$	$(x^2 - y^2)z/r^3$	$f_{zx^2-zy^2}$
			xyz/r^3	f_{xyz}
3	± 3	$\mp(35/64\pi)^{1/2}\sin^3\theta e^{\pm i3\phi}$	$(3x^2 - y^2)y/r^3$	$f_{yx^2-y^3}$
			$(3y^2 - x^2)x/r^3$	$f_{xy^2-x^3}$

which has the form of a 1D TISE. We notice that the original potential between the particles has been changed by the factor $\hbar^2 l(l+1)/2m_e r^2$, which corresponds to a centrifugal term if we take $\hbar^2 l(l+1)$ to be the square of the angular momentum. By analogy to classical mechanics, we define the quantum-mechanical angular momentum operator as $\mathbf{L} = \mathbf{r} \times \mathbf{p}$ which, using the quantum-mechanical operator for the momentum $\mathbf{p} = -i\hbar\nabla_{\mathbf{r}}$, has the following cartesian components (expressed both in cartesian and spherical coordinates):

$$L_x = yp_z - zp_y = -i\hbar\left(y\frac{\partial}{\partial z} - z\frac{\partial}{\partial y}\right) = i\hbar\left(\sin\phi\frac{\partial}{\partial\theta} + \cot\theta\cos\phi\frac{\partial}{\partial\phi}\right)$$

$$L_y = zp_x - xp_z = -i\hbar\left(z\frac{\partial}{\partial x} - x\frac{\partial}{\partial z}\right) = -i\hbar\left(\cos\phi\frac{\partial}{\partial\theta} - \cot\theta\sin\phi\frac{\partial}{\partial\phi}\right)$$

$$L_z = xp_y - yp_x = -i\hbar\left(x\frac{\partial}{\partial z} - z\frac{\partial}{\partial x}\right) = -i\hbar\frac{\partial}{\partial\phi} \tag{C.34}$$

It is a straightforward exercise to show from these expressions that:

$$\mathbf{L}^2 = L_x^2 + L_y^2 + L_z^2 = -\hbar^2\left[\frac{1}{\sin\theta}\frac{\partial}{\partial\theta}\left(\sin\theta\frac{\partial}{\partial\theta}\right) + \frac{1}{\sin^2\theta}\frac{\partial^2}{\partial\psi^2}\right] \tag{C.35}$$

and that the spherical harmonics are eigenfunctions of the operators $\mathbf{L}^2$ and L_z, with eigenvalues

$$\mathbf{L}^2 Y_{lm}(\theta,\phi) = l(l+1)\hbar^2 Y_{lm}(\theta,\phi), \quad L_z Y_{lm}(\theta,\phi) = m\hbar Y_{lm}(\theta,\phi) \tag{C.36}$$

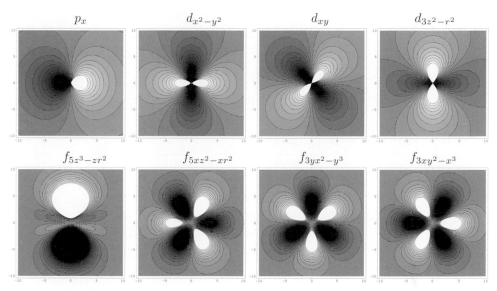

Fig. C.4 Contours of constant value for representative p, d, and f orbitals, with the positive values shown in white and the negative values in black. **Top row**: p_x on the x, y plane; $d_{x^2-y^2}$ on the x, y plane; d_{xy} on the x, y plane; $d_{3z^2-r^2}$ on the x, z plane. **Bottom row**: $f_{5z^3-zr^2}$ on the x, z plane; $f_{5xz^2-xr^2}$ on the x, z plane; $f_{3yx^2-y^3}$ on the x, y plane; $f_{3xy^2-x^3}$ on the xy, plane.

as might have been expected from our earlier identification of the quantity $\hbar^2 l(l+1)$ with the square of the angular momentum. This is another example of simultaneous eigenfunctions of two operators, which according to our earlier discussion must commute: $[\mathbf{L}^2, L_z] = 0$. Thus, the spherical harmonics determine the angular momentum l of a state and its z component, which is equal to m.

In addition to L_x, L_y, L_z, there are two more interesting operators, defined as:

$$L_\pm = I_x \pm iL_y = \hbar e^{\pm i\phi}\left[\pm\frac{\partial}{\partial\theta} + i\cot\theta\frac{\partial}{\partial\phi}\right] \tag{C.37}$$

which when applied to the spherical harmonics give the following result:

$$L_\pm Y_{lm}(\theta, \phi) = C_{lm}^{(\pm)} Y_{lm\pm1}(\theta, \phi) \tag{C.38}$$

that is, they raise or lower the value of the z component by one unit. For this reason they are called the raising and lowering operators. The value of the constants $C_{lm}^{(\pm)}$ can be obtained from the following considerations. First, we notice from the definition of L_+, L_- that:

$$L_+L_- = (L_x + iL_y)(L_x - iL_y) = L_x^2 + L_y^2 - iL_xL_y + iL_yL_x = L_x^2 + L_y^2 + i[L_x, L_y]$$

where in the last expression we have introduced the commutator of L_x, L_y. Next, we can use the definition of L_x, L_y to show that their commutator is equal to $i\hbar L_z$:

$$[L_x, L_y] = L_xL_y - L_yL_x = i\hbar L_z$$

which gives the following relation:

$$\mathbf{L}^2 = L_z^2 + L_+L_- - \hbar L_z$$

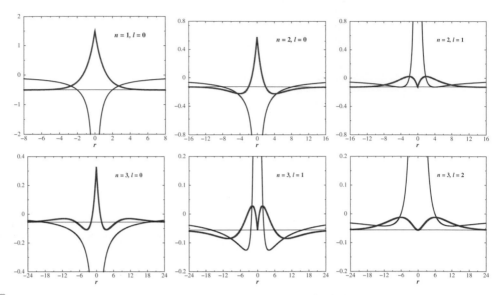

Fig. C.5 The lowest six radial eigenfunctions of the Coulomb potential for the hydrogen atom $R_{nl}(r)$ [$(n, l) =$ (1,0), (2,0), (2,1), (3,0), (3,1), (3,2)] (units used are such that all the constants appearing in the wavefunctions and the potential are equal to unity). The horizontal axis is extended to negative values to indicate the spherically symmetric nature of the potential and wavefunctions; this range corresponds to $\theta = \pi$. The wavefunctions (shown in blue) have been shifted by their energy in each case, which is shown as a horizontal black line. The total radial potential, including the Coulomb part and the angular momentum part, is also shown in red line: notice the large repulsive potential near the origin for $l = 1$ and $l = 2$, arising from the angular momentum part.

Using this last relation, we can take the expectation value of $L_{\pm}Y_{lm}$ with itself to obtain:

$$
\begin{aligned}
\langle L_{\pm}Y_{lm}|L_{\pm}Y_{lm}\rangle &= |C^{(\pm)}|^2\langle Y_{lm\pm1}|Y_{lm\pm1}\rangle = \langle Y_{lm}|L_{\mp}L_{\pm}Y_{lm}\rangle \\
&= \langle Y_{lm}|(\mathbf{L}^2 - L_z^2 \mp \hbar L_z)Y_{lm}\rangle = \hbar^2\left[l(l+1) - m(m\pm1)\right] \\
&\Rightarrow C^{(\pm)} = \hbar\left[l(l+1) - m(m\pm1)\right]^{1/2}
\end{aligned}
\tag{C.39}
$$

This is convenient because it provides an explicit expression for the result of applying the raising or lowering operators to spherical harmonics; it can be used to generate all spherical harmonics of a given l starting with one of them.

Finally, we consider the solution to the radial equation, Eq. (C.29). This equation is solved by the functions

$$
R_{nl}(r) = -\left[\left(\frac{2}{n(n+l)!\,a_0}\right)^3 \frac{(n-l-1)!}{2n}\right]^{1/2} e^{-\rho/2}\rho^l L_{n+l}^{2l+1}(\rho)
\tag{C.40}
$$

where we have defined two new variables:

$$
a_0 \equiv \frac{\hbar^2}{e^2 m_e}, \qquad \rho \equiv \frac{2r}{n a_0}
$$

Table C.3 The radial wavefunctions for $n = 1, 2, 3$ and the associated Laguerre polynomials $L_{n+l}^{2l+1}(r)$ used in their definition.

n	l	$L_{n+l}^{2l+1}(r)$	$a_0^{3/2} R_{nl}(r)$
1	0	-1	$2e^{-r/a_0}$
2	0	$-2(2-r)$	$\frac{1}{\sqrt{2}} e^{-r/2a_0} \left(1 - \frac{r}{2a_0}\right)$
2	1	$-3!$	$\frac{1}{\sqrt{6}} e^{-r/2a_0} \left(\frac{r}{2a_0}\right)$
3	0	$-3(6 - 6r + r^2)$	$\frac{2}{9\sqrt{3}} e^{-r/3a_0} \left(3 - 6\left(\frac{r}{3a_0}\right) + 2\left(\frac{r}{3a_0}\right)^2\right)$
3	1	$-24(4-r)$	$\frac{2\sqrt{2}}{9\sqrt{3}} e^{-r/3a_0} \left(\frac{r}{3a_0}\right)\left(2 - \frac{r}{3a_0}\right)$
3	2	$-5!$	$\frac{2\sqrt{2}}{9\sqrt{15}} e^{-r/3a_0} \left(\frac{r}{3a_0}\right)^2$

and the functions $L_n^l(r)$ are given by

$$L_n^l(r) = \frac{d^l}{dr^l} L_n(r) \tag{C.41}$$

with $L_n(r)$ the Laguerre polynomials defined recursively as:

$$L_{n+1}(r) = (2n + 1 - r)L_n(r) - n^2 L_{n-1}(r)$$

$$L_0(r) = 1, \quad L_1(r) = 1 - r \tag{C.42}$$

The index n is an integer that takes the values $n = 1, 2, \ldots$, while for a given n the index l is allowed to take the values $l = 0, \ldots, n - 1$, and the energy eigenvalues are given by:

$$\epsilon_n = -\frac{e^2}{2a_0 n^2}$$

The first few radial wavefunctions are given in Table C.3. The nature of these wavefunctions is illustrated in Fig. C.5. It is trivial to extend this description to a nucleus of charge Ze with a single electron around it: the factor a_0 gets replaced everywhere by a_0/Z while there is an extra factor of Z in the energy to account for the charge of the nucleus. In our treatment we have also considered the nucleus to be infinitely heavy and fixed at the origin of the coordinate system. If we wish to include the finite mass m_n of the nucleus then the mass of the electron in the above equations is replaced by the reduced mass of nucleus and electron, $\mu = m_n m_e/(m_n + m_e)$. Since nuclei are much heavier than electrons, $\mu \approx m_e$ is a good approximation.

C.4 Spin Angular Momentum

In addition to the usual terms of kinetic energy and potential energy, quantum-mechanical particles possess another property called spin, which has the dimensions of angular momentum. The values of spin are quantized to half-integer or integer multiples of $\hbar$. In the following we omit the factor of $\hbar$ when we discuss spin values, for brevity. If the total spin of a particle is s then there are $2s + 1$ states associated with it, because the projection of the spin onto a particular axis can have that many possible values, ranging from $+s$ to $-s$ in increments of 1. The axis of spin projection is usually labeled the z-axis, so a spin of $s = 1/2$ can have projections on the z-axis of $s_z = +1/2$ and $s_z = -1/2$; a spin of $s = 1$ can have $s_z = -1, 0, +1$, and so on.

The spin of quantum particles determines their statistics. Particles with half-integer spin are called fermions and obey the **Pauli exclusion principle**, that is, no two of them can be in exactly the same quantum-mechanical state as it is defined by all the relevant quantum numbers. A typical example is the electron, which is a fermion with spin $s = 1/2$. In the Coulomb potential there are three quantum numbers n, l, m associated with the spatial degrees of freedom as determined by the radial and angular parts of the wavefunction $R_{nl}(r)$ and $Y_{lm}(\theta, \phi)$. A particle with spin $s = 1/2$ can only have two states associated with a particular set of n, l, m values, corresponding to $s_z = \pm 1/2$. Thus, each state of the Coulomb potential characterized by a set of values n, l, m can accommodate two electrons with spin $\pm 1/2$, to which we refer as the "spin-up" and "spin-down" states. This rule basically explains the sequence of elements in the Periodic Table, as follows.

$n = 1$: This state can only have the angular momentum state with $l = 0, m = 0$ (the s state), which can be occupied by one or two electrons when spin is taken into consideration; the first corresponds to H, the second to He, with atomic numbers 1 and 2.

$n = 2$: This state can have angular momentum states with $l = 0, m = 0$ (an s state) or $l = 1, m = \pm 1, 0$ (a p state), which account for a total of eight possible states when the spin is taken into consideration, corresponding to the elements Li, Be, B, C, N, O, F, and Ne with atomic numbers 3–10.

$n = 3$: This state can have angular momentum states with $l = 0, m = 0$ (an s state), $l = 1, m = \pm 1, 0$ (a p state), or $l = 2, m = \pm 2, \pm 1, 0$ (a d state), which account for a total of 18 possible states when the spin is taken into consideration, corresponding to the elements Na, Mg, Al, Si, P, S, Cl, and Ar with atomic numbers 11–18 (in which the s and p states are gradually filled) and elements Sc, Ti, V, Cr, Mn, Fe, Co, Ni, Cu, and Zn with atomic numbers 21–30 (in which the d state is gradually filled).

There is a jump in the sequential occupation of states of the Coulomb potential, namely we pass from atomic number 18 with the $n = 3, l = 0$ and $l = 1$ states filled, to atomic number 21 in which the $n = 3, l = 2$ states begin to be filled. The reason for this jump has to do with the fact that each electron in an atom does not experience only the pure Coulomb potential of the nucleus, but a more complex potential which is also due to the presence of all the other electrons. For this reason the true states in the atoms are somewhat different from the states of the pure Coulomb potential, which is what makes the states with $n = 4$ and $l = 0$ start filling before the states with $n = 3$ and $l = 2$; the $n = 4, l = 0, m = 0$ states correspond to the elements K and Ca, with atomic numbers 19 and 20. Overall, however, the sequence of states in real atoms is remarkably close to what would be expected from the pure Coulomb potential. The same pattern is followed for states with higher n and l values.

The other type of particles, with integer spin values, do not obey the Pauli exclusion principle. They are called bosons and obey different statistics from fermions. For instance, under the proper conditions, all bosons can collapse into a single quantum-mechanical state, a phenomenon referred to as Bose–Einstein condensation that has been observed experimentally.

Since spin is a feature of the quantum-mechanical nature of particles, we need to introduce a way to represent it and include it in the hamiltonian as appropriate. Spins are represented by "spinors," which are 1D vectors of 0s and 1s. The spinors identify the exact state of the spin, including its magnitude s and its projection on the z-axis s_z: the magnitude is included in the length of the spinor, which is $2s + 1$, while the s_z value is given by the non-zero entry of the spinor in a sequential manner going through the values s to $-s$. The spinor of spin 0 is a vector of length 1, with a single entry [1]. The two spinors that identify the states corresponding to total spin $s = 1/2$ have length 2 and are:

$$\begin{bmatrix} 1 \\ 0 \end{bmatrix} \rightarrow \quad s_z = +\frac{1}{2} \ (\uparrow), \qquad \begin{bmatrix} 0 \\ 1 \end{bmatrix} \rightarrow \quad s_z = -\frac{1}{2} \ (\downarrow)$$

where we have also included the usual notation for spins $1/2$, as up and down arrows. The three spinors that identify the states corresponding to total spin $s = 1$ have length 3 and are:

$$\begin{bmatrix} 1 \\ 0 \\ 0 \end{bmatrix} \rightarrow \quad s_z = +1, \qquad \begin{bmatrix} 0 \\ 1 \\ 0 \end{bmatrix} \rightarrow \quad s_z = 0, \qquad \begin{bmatrix} 0 \\ 0 \\ 1 \end{bmatrix} \rightarrow \quad s_z = -1$$

The same pattern applies to higher spin values.

When spin needs to be included explicitly in the hamiltonian, the appropriate operators must be used that can act on spinors. These are square matrices of size $(2s + 1) \times (2s + 1)$ which multiply the spinors to produce other spinors or combinations of spinors. Since there are three components of angular momentum in 3D space, there must exist three matrices corresponding to each component of the spin. For spin $s = 1/2$ these are the following 2×2 matrices:

$$J_x = \frac{\hbar}{2} \begin{pmatrix} 0 & 1 \\ 1 & 0 \end{pmatrix}, \quad J_y = \frac{\hbar}{2} \begin{pmatrix} 0 & -i \\ i & 0 \end{pmatrix}, \quad J_z = \frac{\hbar}{2} \begin{pmatrix} 1 & 0 \\ 0 & -1 \end{pmatrix}$$

The matrices in square brackets without the constants in front are called the Pauli matrices and are denoted by $\sigma_x, \sigma_y, \sigma_z$:

$$\sigma_x = \begin{pmatrix} 0 & 1 \\ 1 & 0 \end{pmatrix}, \quad \sigma_y = \begin{pmatrix} 0 & -i \\ i & 0 \end{pmatrix}, \quad \sigma_z = \begin{pmatrix} 1 & 0 \\ 0 & -1 \end{pmatrix} \tag{C.43}$$

For spin $s = 1$ the spin operators are the following 3×3 matrices:

$$J_x = \frac{\hbar}{\sqrt{2}} \begin{pmatrix} 0 & 1 & 0 \\ 1 & 0 & 1 \\ 0 & 1 & 0 \end{pmatrix}, \quad J_y = \frac{\hbar}{\sqrt{2}} \begin{pmatrix} 0 & -i & 0 \\ i & 0 & -i \\ 0 & i & 0 \end{pmatrix}, \quad J_z = \hbar \begin{pmatrix} 1 & 0 & 0 \\ 0 & 0 & 0 \\ 0 & 0 & -1 \end{pmatrix}$$

It is easy to show that the matrix defined by $J^2 = J_x^2 + J_y^2 + J_z^2$ in each case is a diagonal matrix with all diagonal elements equal to 1 multiplied by the factor $s(s+1)\hbar^2$. Thus, the spinors are eigenvectors of the matrix J^2 with eigenvalue $s(s+1)\hbar^2$; for this reason, it is customary to attribute the value $s(s+1)\hbar^2$ to the square of the spin magnitude. It is also evident from the definitions given above that each spinor is an eigenvector of the matrix J_z, with an eigenvalue equal to the s_z value to which the spinor corresponds. It is easy to show that the linear combination of matrices $J_+ = (J_x + iJ_y)$ is a matrix which, when multiplied by a spinor corresponding to s_z, gives the spinor corresponding to $s_z + 1$, while the linear combination $J_- = (J_x - iJ_y)$ is a matrix, which when multiplied by a spinor corresponding to s_z, gives the spinor corresponding to $s_z - 1$. For example, for the $s = 1/2$ case:

$$J_+ = \hbar \begin{pmatrix} 0 & 1 \\ 0 & 0 \end{pmatrix}, \quad J_- = \hbar \begin{pmatrix} 0 & 0 \\ 1 & 0 \end{pmatrix}$$

$$J_+ \begin{bmatrix} 1 \\ 0 \end{bmatrix} = 0, \quad J_+ \begin{bmatrix} 0 \\ 1 \end{bmatrix} = \hbar \begin{bmatrix} 1 \\ 0 \end{bmatrix}, \quad J_- \begin{bmatrix} 1 \\ 0 \end{bmatrix} = \hbar \begin{bmatrix} 0 \\ 1 \end{bmatrix}, \quad J_- \begin{bmatrix} 0 \\ 1 \end{bmatrix} = 0$$

where the operator J_+ applied to the up spin gives zero because there is no state with higher s_z and similarly the operator J_- applied to the down spin gives zero because there is no state with lower s_z. These two matrices are therefore the raising and lowering operators, as might be expected by the close analogy to the definition of the corresponding operators for angular momentum, given in Eqs (C.37) and (C.38).

Finally, we consider the addition of spin angular momenta. We start with the case of two spin-1/2 particles as the most common situation encountered in the study of solids; the generalization to arbitrary values is straightforward and will also be discussed briefly. We denote the spin of the combined state and its z component by capital letters (S, S_z), in distinction to the spins of the constituent particles, denoted by $(s^{(1)}, s_z^{(1)})$ and $(s^{(2)}, s_z^{(2)})$. We can start with the state in which both spinors have $s_z = 1/2$, that is, they are the up spins. The combined spin will obviously have projection on the z-axis $S_z = 1$. We will show that this is one of the states of the total spin $S = 1$ manifold. We can apply the lowering operator, denoted by $S_- = s_-^{(1)} + s_-^{(2)}$, which is of course composed of the two individual lowering operators, to the state with $S_z = 1$ to obtain a state of lower S_z value. Notice that $s_-^{(i)}$ applies only to particle i:

$$S_- [\uparrow\uparrow] = \left[(s_-^{(1)} \uparrow) \uparrow + \uparrow (s_-^{(2)} \uparrow) \right] = \hbar\sqrt{2} \times \frac{1}{\sqrt{2}} [\downarrow\uparrow + \uparrow\downarrow]$$

where we have used up and down arrows as shorthand notation for spinors with $s_z = \pm 1/2$ and the convention that the first arrow corresponds to the spin of the first particle and the second arrow to the spin of the second particle. Now notice that the state in square brackets with the factor $1/\sqrt{2}$ in front is a properly normalized spinor state with $S_z = 0$, and its coefficient is $\hbar\sqrt{2}$, precisely what we would expect to get by applying the lowering operator on a state with angular momentum 1 and z projection 1, according to Eq. (C.39). If we apply the lowering operator once again to the new state with $S_z = 0$, we obtain

$$S_- \frac{1}{\sqrt{2}}[\downarrow\uparrow + \uparrow\downarrow] = \frac{1}{\sqrt{2}}\left[\downarrow (s_-^{(2)} \uparrow) + (s_-^{(1)} \uparrow) \downarrow\right] = \hbar\sqrt{2}[\downarrow\downarrow]$$

which is a state with $S_z = -1$ and coefficient $\hbar\sqrt{2}$, again consistent with what we expected from applying the lowering operator on a state with angular momentum 1 and z projection 0. Thus, we have generated the three states in the $S = 1$ manifold with $S_z = +1, 0, -1$. We can construct one more state, which is orthogonal to the $S_z = 0$ state found above and also has $S_z = 0$, by taking the linear combination of up–down spins with opposite relative sign:

$$\frac{1}{\sqrt{2}}[\uparrow\downarrow - \downarrow\uparrow].$$

Moreover, we can determine that the total spin value of this new state is $S = 0$, because applying the lowering or raising operator on it gives 0. This completes the analysis of the possible spin states obtained by combining two spin-$1/2$ particles which are the following:

$$S = 1 : \quad S_z = +1 \to [\uparrow\uparrow], \quad S_z = 0 \to \frac{1}{\sqrt{2}}[\uparrow\downarrow + \downarrow\uparrow], \quad S_z = -1 \to [\downarrow\downarrow]$$

$$S = 0 : \quad S_z = 0 \to \frac{1}{\sqrt{2}}[\uparrow\downarrow - \downarrow\uparrow]$$

The generalization of these results to addition of two angular momenta with arbitrary values is as follows. Consider the two angular momenta to be S and L; this notation usually refers to the situation where S is the spin and L the orbital angular momentum, a case for which we specify the following discussion as the most relevant to the physics of solids. The z-projection of the first component will range from $S_z = +S$ to $-S$, and of the second component from $L_z = +L$ to $-L$. The resulting total angular momentum J will take values ranging from $L + S$ to 0, because the possible z components will range from a maximum of $S_z + L_z = S + L$ to a minimum of $-(S + L)$, with all the intermediate values, each differing by a unit:

$$J = L + S : \quad J_z = L + S, L + S - 1, \ldots, -(L + S),$$
$$J = L + S - 1 : \quad J_z = L + S - 1, L + S - 2, \ldots, -(L + S - 1), \text{ etc.}$$

An important application of these rules is the calculation of expectation values of operators in the basis of states with definite J and J_z. Since these states are obtained by combinations of states with spin S and orbital angular momentum L, we denote them as $|JLSJ_z\rangle$, which form a complete set of $(2J + 1)$ states for each value of J, and a complete set for all possible states resulting from the addition of L and S when all allowed values of J are included:

$$\sum_{JJ_z} |JLSJ_z\rangle\langle JLSJ_z| = 1 \qquad (C.44)$$

A general theorem, known as the Wigner–Eckart theorem (see, for example, Schiff, p. 222), states that the expectation value of any vector operator in the space of the $|JLSJ_z\rangle$

states is proportional to the expectation value of $\hat{\mathbf{J}}$ itself and does not depend on J_z (in the following we use bold symbols with a hat, $\hat{\mathbf{O}}$, to denote vector operators and bold symbols, $\mathbf{O}$, to denote their expectation values). Let us consider as an example a vector operator which is a linear combination of $\hat{\mathbf{L}}$ and $\hat{\mathbf{S}}$, $(\hat{\mathbf{L}} + \lambda\hat{\mathbf{S}})$ with λ an arbitrary constant. From the Wigner–Eckart theorem we have:

$$\langle JLSJ_z'|(\hat{\mathbf{L}} + \lambda\hat{\mathbf{S}})|JLSJ_z\rangle = g_\lambda(JLS)\langle JLSJ_z'|\hat{\mathbf{J}}|JLSJ_z\rangle$$

where we have written the constants of proportionality as $g_\lambda(JLS)$. In order to obtain the values of these constants, note that the expectation value of the dot product $(\hat{\mathbf{L}} + \lambda\hat{\mathbf{S}}) \cdot \hat{\mathbf{J}}$ will be given by:

$$\langle JLSJ_z'|(\hat{\mathbf{L}} + \lambda\hat{\mathbf{S}}) \cdot \hat{\mathbf{J}}|JLSJ_z\rangle =$$
$$\langle JLSJ_z'|(\hat{\mathbf{L}} + \lambda\hat{\mathbf{S}}) \cdot \sum_{J''J_z''} |J''LSJ_z''\rangle\langle J''LSJ_z''|\hat{\mathbf{J}}|JLSJ_z\rangle =$$
$$\langle JLSJ_z'|(\hat{\mathbf{L}} + \lambda\hat{\mathbf{S}}) \cdot \sum_{J''J_z''} |J''LSJ_z''\rangle\mathbf{J}\delta_{JJ''}\delta_{J_zJ_z''} =$$
$$g_\lambda(JLS)\langle JLSJ_z'|\hat{\mathbf{J}}^2|JLSJ_z\rangle = g_\lambda(JLS)J(J+1) \tag{C.45}$$

where in the first step we inserted the unity operator of Eq. (C.44) between the two vectors of the dot product, in the second step we used the fact that the states $|JLSJ_z\rangle$ are eigenstates of the operator $\hat{\mathbf{J}}$ and therefore $\hat{\mathbf{J}}|JLSJ_z\rangle = |JLSJ_z\rangle\mathbf{J}$, in the third step we used the Wigner–Eckart theorem, and in the last step we used the fact that the expectation value of $\hat{\mathbf{J}}^2$ in the basis $|JLSJ_z\rangle$ is simply $J(J+1)$. Thus, for the expectation value of the original operator we obtain

$$\langle JLSJ_z'|\hat{\mathbf{L}} \cdot \hat{\mathbf{J}}|JLSJ_z\rangle + \lambda\langle JLSJ_z'|\hat{\mathbf{S}} \cdot \hat{\mathbf{J}}|JLSJ_z\rangle = g_\lambda(JLS)J(J+1)$$

From the vector addition of $\hat{\mathbf{L}}$ and $\hat{\mathbf{S}}$, namely $\hat{\mathbf{J}} = \hat{\mathbf{L}} + \hat{\mathbf{S}}$, we find:

$$\hat{\mathbf{S}} = \hat{\mathbf{J}} - \hat{\mathbf{L}} \Rightarrow \hat{\mathbf{L}} \cdot \hat{\mathbf{J}} = \frac{1}{2}\left[\hat{\mathbf{J}}^2 + \hat{\mathbf{L}}^2 - \hat{\mathbf{S}}^2\right]$$

$$\hat{\mathbf{L}} = \hat{\mathbf{J}} - \hat{\mathbf{S}} \Rightarrow \hat{\mathbf{S}} \cdot \hat{\mathbf{J}} = \frac{1}{2}\left[\hat{\mathbf{J}}^2 + \hat{\mathbf{S}}^2 - \hat{\mathbf{L}}^2\right]$$

while the expectation values of $\hat{\mathbf{L}}^2$ and $\hat{\mathbf{S}}^2$ in the basis $|JLSJ_z\rangle$ are $L(L+1)$ and $S(S+1)$. These results, employed in the above equation for $g_\lambda(JLS)$, lead to:

$$g_\lambda(JLS) = \frac{1}{2}(\lambda + 1) + \frac{1}{2}(\lambda - 1)\left[\frac{S(S+1) - L(L+1)}{J(J+1)}\right] \tag{C.46}$$

These expressions are known as the "Landé g-factors" in the context of the total angular momentum of atoms or ions with partially filled electronic shells, which are relevant to the magnetic behavior of insulators.

C.5 Stationary Perturbation Theory

Perturbation theory in general is a very useful method in quantum mechanics: it allows us to find approximate solutions to problems that do not have simple analytic solutions. In stationary perturbation theory (SPT), we assume that we can view the problem at hand as a slight change from another problem, called the unperturbed case, which we can solve exactly. The basic idea is that we wish to find the eigenvalues and eigenfunctions of a hamiltonian $\hat{\mathcal{H}}$ which can be written as two parts:

$$\hat{\mathcal{H}} = \hat{\mathcal{H}}_0 + \hat{\mathcal{H}}_1$$

where $\hat{\mathcal{H}}_0$ is a hamiltonian whose solutions we know analytically (as in one of the examples discussed above) and $\hat{\mathcal{H}}_1$ is a small perturbation of $\hat{\mathcal{H}}_0$. We can use the solutions of $\hat{\mathcal{H}}_0$ to express the unknown solutions of $\hat{\mathcal{H}}$:

$$\hat{\mathcal{H}}|\psi_i\rangle = \epsilon_i|\psi_i\rangle \tag{C.47}$$

$$|\psi_i\rangle = |\psi_i^{(0)}\rangle + |\delta\psi_i\rangle \tag{C.48}$$

$$\epsilon_i = \epsilon_i^{(0)} + \delta\epsilon_i \tag{C.49}$$

where quantities with superscript (0) identify the solutions of the hamiltonian $\hat{\mathcal{H}}_0$, which form a complete orthonormal set:

$$\hat{\mathcal{H}}_0|\psi_i^{(0)}\rangle = \epsilon_i|\psi_i^{(0)}\rangle, \quad \langle\psi_i^{(0)}|\psi_j^{(0)}\rangle = \delta_{ij} \tag{C.50}$$

and quantities without superscripts correspond to the solutions of the hamiltonian $\hat{\mathcal{H}}$. The problem reduces to finding the quantities $|\delta\psi_i\rangle$ and $\delta\epsilon_i$, assuming that they are small compared to the corresponding wavefunctions and eigenvalues. Notice in particular that the wavefunctions $|\psi_i\rangle$ are not normalized in the way we wrote them above, but their normalization is a trivial matter once we have calculated $|\delta\psi_i\rangle$. Moreover, each $|\delta\psi_i\rangle$ will only include the part which is orthogonal to $|\psi_i^{(0)}\rangle$, that is, it can include any wavefunction $|\psi_j^{(0)}\rangle$ with $j \neq i$. Our goal then is to express $|\delta\psi_i\rangle$ and $\delta\epsilon_i$ using the known wavefunctions and eigenvalues of $\hat{\mathcal{H}}_0$, which we will do as an expansion in successive terms:

$$|\delta\psi_i\rangle = |\delta\psi_i^{(1)}\rangle + |\delta\psi_i^{(2)}\rangle + \cdots$$

$$\delta\epsilon_i = \delta\epsilon_i^{(1)} + \delta\epsilon_i^{(2)} + \cdots$$

where the superscripts indicate the order of the approximation, that is, superscript (1) is the first-order approximation which includes the perturbation $\hat{\mathcal{H}}_1$ only to first power, etc.

C.5.1 Non-degenerate Perturbation Theory

We start with the simplest case, assuming that the eigenfunctions of the unperturbed hamiltonian are non-degenerate, that is, $\epsilon_i^{(0)} \neq \epsilon_j^{(0)}$ for $i \neq j$. When we substitute (C.48) and (C.49) in (C.47), we find:

$$\left(\hat{\mathcal{H}}_0 + \hat{\mathcal{H}}_1\right)\left(|\psi_i^{(0)}\rangle + |\delta\psi_i\rangle\right) = \left(\epsilon_i^{(0)} + \delta\epsilon_i\right)\left(|\psi_i^{(0)}\rangle + |\delta\psi_i\rangle\right)$$

Next, we multiply from the left both sides of the above equation by $\langle\psi_j^{(0)}|$ and use Eq. (C.50) to find:

$$\epsilon_i^{(0)}\delta_{ij} + \langle\psi_j^{(0)}|\hat{\mathcal{H}}_1|\psi_i^{(0)}\rangle + \epsilon_j^{(0)}\langle\psi_j^{(0)}|\delta\psi_i\rangle + \langle\psi_j^{(0)}|\hat{\mathcal{H}}_1|\delta\psi_i\rangle =$$
$$\epsilon_i^{(0)}\delta_{ij} + \delta\epsilon_i\delta_{ij} + \epsilon_i^{(0)}\langle\psi_j^{(0)}|\delta\psi_i\rangle + \delta\epsilon_i\langle\psi_j^{(0)}|\delta\psi_i\rangle \tag{C.51}$$

If, in Eq. (C.51), we take $j = i$ and keep only first-order terms, that is, terms that involve only one of the small quantities $\hat{\mathcal{H}}_1, |\delta\psi_i\rangle, \delta\epsilon_i$, we find:

$$\delta\epsilon_i^{(1)} = \langle\psi_i^{(0)}|\hat{\mathcal{H}}_1|\psi_i^{(0)}\rangle \tag{C.52}$$

where we have introduced the superscript (1) in the energy correction to indicate the order of the approximation. This is a simple and important result: the change in the energy of state i to first order in the perturbation $\hat{\mathcal{H}}_1$ is simply the expectation value of $\hat{\mathcal{H}}_1$ in the unperturbed wavefunction of that state $|\psi_i^{(0)}\rangle$. If, in Eq. (C.51), we take $j \neq i$ and keep only first-order terms, we obtain:

$$\langle\psi_j^{(0)}|\delta\psi_i\rangle = -\frac{\langle\psi_j^{(0)}|\hat{\mathcal{H}}_1|\psi_i^{(0)}\rangle}{\epsilon_j^{(0)} - \epsilon_i^{(0)}} \tag{C.53}$$

which is simply the projection of $|\delta\psi_i\rangle$ on the unperturbed state $\langle\psi_j^{(0)}|$. Therefore, $|\delta\psi_i\rangle$ will involve a summation over all such terms, each multiplied by the corresponding state $|\psi_j^{(0)}\rangle$, which gives for the first-order correction to the wavefunction of state i:

$$|\delta\psi_i^{(1)}\rangle = \sum_{j \neq i}\frac{\langle\psi_j^{(0)}|\hat{\mathcal{H}}_1|\psi_i^{(0)}\rangle}{\epsilon_i^{(0)} - \epsilon_j^{(0)}}|\psi_j^{(0)}\rangle \tag{C.54}$$

This is another important result, showing that the change in wavefunction of state i to first order in the perturbation $\hat{\mathcal{H}}_1$ involves the matrix elements of $\hat{\mathcal{H}}_1$ between the unperturbed state i and all the other unperturbed states j, divided by the unperturbed energy difference between state i and j.

If, in Eq. (C.51), we take $j = i$ and keep up to second-order terms, we obtain:

$$\delta\epsilon_i^{(2)} = \langle\psi_i^{(0)}|\hat{\mathcal{H}}_1|\delta\psi_i\rangle$$

In order to arrive at this expression we have used the fact that by construction $\langle\psi_i^{(0)}|\delta\psi_i\rangle = 0$, from the orthogonality of the $|\psi_i^{(0)}\rangle$s; we have also used the fact that we already know the first-order correction, given by Eq. (C.52). Substituting in the above equation the expression that we found for $|\delta\psi_i\rangle$ to first order, Eq. (C.54), we obtain for the change in the energy of state i to second order in $\hat{\mathcal{H}}_1$:

$$\delta\epsilon_i^{(2)} = \sum_{j \neq i}\frac{\langle\psi_j^{(0)}|\hat{\mathcal{H}}_1|\psi_i^{(0)}\rangle\langle\psi_i^{(0)}|\hat{\mathcal{H}}_1|\psi_j^{(0)}\rangle}{\epsilon_i^{(0)} - \epsilon_j^{(0)}} \tag{C.55}$$

It is a simple extension of these arguments to obtain $|\delta\psi_i\rangle$ to second order, which turns out to be:

$$|\delta\psi_i^{(2)}\rangle = -\sum_{j\neq i} \frac{\langle\psi_j^{(0)}|\hat{\mathcal{H}}_1|\psi_i^{(0)}\rangle\langle\psi_i^{(0)}|\hat{\mathcal{H}}_1|\psi_i^{(0)}\rangle}{(\epsilon_i^{(0)} - \epsilon_j^{(0)})^2}|\psi_j^{(0)}\rangle$$

$$+ \sum_{j,k\neq i} \frac{\langle\psi_j^{(0)}|\hat{\mathcal{H}}_1|\psi_k^{(0)}\rangle\langle\psi_k^{(0)}|\hat{\mathcal{H}}_1|\psi_i^{(0)}\rangle}{(\epsilon_i^{(0)} - \epsilon_j^{(0)})(\epsilon_i^{(0)} - \epsilon_k^{(0)})}|\psi_j^{(0)}\rangle \qquad \text{(C.56)}$$

It is also possible to go to even higher orders in both $\delta\epsilon_i$ and $|\delta\psi_i\rangle$. Usually the first-order approximation in the wavefunction and the second-order approximation in the energy are adequate.

C.5.2 Degenerate Perturbation Theory

The approach we developed above will not work if the original set of states involves degeneracies, that is, states whose unperturbed energies are the same, because then the factors that appear in the denominators in Eqs (C.54) and (C.55) will vanish. In such cases we need to apply degenerate perturbation theory. We outline here the simplest case of degenerate perturbation theory which involves two degenerate states, labeled i and j. By assumption, the two unperturbed states $|\psi_i^{(0)}\rangle$ and $|\psi_j^{(0)}\rangle$ have the same unperturbed energy $\epsilon_i^{(0)} = \epsilon_j^{(0)}$. We consider a linear combination of the two degenerate states and try to find the effect of the perturbation on the energy and the wavefunction of this state, denoted by $\delta\epsilon$ and $|\delta\phi\rangle$. We will then have for the unperturbed case:

$$\hat{\mathcal{H}}_0\left(a_i|\psi_i^{(0)}\rangle + a_j|\psi_j^{(0)}\rangle\right) = \epsilon_i^{(0)}\left(a_i|\psi_i^{(0)}\rangle + a_j|\psi_j^{(0)}\rangle\right)$$

and for the perturbed case

$$(\hat{\mathcal{H}}_0 + \hat{\mathcal{H}}_1)\left(a_i|\psi_i^{(0)}\rangle + a_j|\psi_j^{(0)}\rangle + |\delta\phi\rangle\right) = (\epsilon_i^{(0)} + \delta\epsilon)\left(a_i|\psi_i^{(0)}\rangle + a_j|\psi_j^{(0)}\rangle + |\delta\phi\rangle\right)$$

Closing both sides of the last equation first with $\langle\psi_i^{(0)}|$ and then with $\langle\psi_j^{(0)}|$, keeping first-order terms only in the small quantities $\hat{\mathcal{H}}_1, \delta\epsilon, |\delta\phi\rangle$ and taking into account that the original unperturbed states are orthonormal, we obtain the following two equations:

$$a_i\langle\psi_i^{(0)}|\hat{\mathcal{H}}_1|\psi_i^{(0)}\rangle + a_j\langle\psi_i^{(0)}|\hat{\mathcal{H}}_1|\psi_j^{(0)}\rangle = a_i\delta\epsilon$$

$$a_i\langle\psi_j^{(0)}|\hat{\mathcal{H}}_1|\psi_i^{(0)}\rangle + a_j\langle\psi_j^{(0)}|\hat{\mathcal{H}}_1|\psi_j^{(0)}\rangle = a_j\delta\epsilon$$

In order for this system of linear equations to have a non-trivial solution the determinant of the coefficients a_i, a_j must vanish, which gives:

$$\delta\epsilon = \frac{1}{2}\left[\langle\psi_i^{(0)}|\hat{\mathcal{H}}_1|\psi_i^{(0)}\rangle + \langle\psi_j^{(0)}|\hat{\mathcal{H}}_1|\psi_j^{(0)}\rangle\right]$$

$$\pm \frac{1}{2}\left[\left(\langle\psi_i^{(0)}|\hat{\mathcal{H}}_1|\psi_i^{(0)}\rangle + \langle\psi_j^{(0)}|\hat{\mathcal{H}}_1|\psi_j^{(0)}\rangle\right)^2 - 4\left|\langle\psi_j^{(0)}|\hat{\mathcal{H}}_1|\psi_i^{(0)}\rangle\right|^2\right]^{\frac{1}{2}} \qquad \text{(C.57)}$$

This last equation gives two possible values for the first-order correction to the energy of the two degenerate states. In general these two values are different, and we associate them with the change in energy of the two degenerate states; this is referred to as "splitting of the degeneracy" by the perturbation term $\hat{\mathcal{H}}_1$. If the two possible values for $\delta\epsilon$ are not different, which implies that in the above equation the expression under the square root vanishes, then we need to go to higher orders of perturbation theory to find how the degeneracy is split by the perturbation. A similar approach can also yield the splitting in the energy of states with higher degeneracy, in which case the subspace of states which must be included involves all the unperturbed degenerate states.

C.6 Time-Dependent Perturbation Theory

In time-dependent perturbation theory we begin with a hamiltonian $\hat{\mathcal{H}}_0$ whose eigenfunctions $|\psi_k^{(0)}\rangle$ and eigenvalues $\epsilon_k^{(0)}$ are known and form a complete orthonormal set. We then turn on a time-dependent perturbation $\hat{\mathcal{H}}_1(t)$ and express the new time-dependent wavefunction in terms of the $|\psi_k^{(0)}\rangle$s:

$$|\psi(t)\rangle = \sum_k c_k(t) e^{-i\epsilon_k^{(0)}t/\hbar} |\psi_k^{(0)}\rangle \tag{C.58}$$

where we have explicitly included a factor $\exp[-i\epsilon_k^{(0)}t/\hbar]$ in the time-dependent coefficient that accompanies the unperturbed wavefunction $|\psi_k^{(0)}\rangle$ in the sum. The wavefunction $|\psi(t)\rangle$ satisfies the time-dependent Schrödinger equation for the full hamiltonian:

$$\left[\hat{\mathcal{H}}_0 + \hat{\mathcal{H}}_1(t)\right]|\psi(t)\rangle = i\hbar\frac{\partial}{\partial t}|\psi(t)\rangle$$

Introducing the expression of Eq. (C.58) on the left-hand side of this equation, we obtain

$$\left[\hat{\mathcal{H}}_0 + \hat{\mathcal{H}}_1(t)\right]\sum_k c_k(t) e^{-i\epsilon_k^{(0)}t/\hbar} |\psi_k^{(0)}\rangle =$$

$$\sum_k c_k(t) e^{-i\epsilon_k^{(0)}t/\hbar} \epsilon_k^{(0)} |\psi_k^{(0)}\rangle + \sum_k c_k(t) e^{-i\epsilon_k^{(0)}t/\hbar} \hat{\mathcal{H}}_1(t)|\psi_k^{(0)}\rangle$$

while the same expression introduced on the right-hand side of the above equation gives

$$i\hbar\frac{\partial}{\partial t}\sum_k c_k(t) e^{-i\epsilon_k^{(0)}t/\hbar} |\psi_k^{(0)}\rangle =$$

$$\sum_k c_k(t) e^{-i\epsilon_k^{(0)}t/\hbar} \epsilon_k^{(0)} |\psi_k^{(0)}\rangle + i\hbar\sum_k \frac{dc_k(t)}{dt} e^{-i\epsilon_k^{(0)}t/\hbar} |\psi_k^{(0)}\rangle$$

By comparing the two results, we arrive at:

$$i\hbar\sum_k \frac{dc_k(t)}{dt} e^{-i\epsilon_k^{(0)}t/\hbar} |\psi_k^{(0)}\rangle = \sum_k c_k(t) e^{-i\epsilon_k^{(0)}t/\hbar} \hat{\mathcal{H}}_1(t)|\psi_k^{(0)}\rangle \tag{C.59}$$

Now, multiplying both sides of this last equation from the left by $\langle \psi_j^{(0)}|$ and using the completeness and orthonormality properties of the unperturbed wavefunctions, we obtain

$$i\hbar \frac{dc_j(t)}{dt} = \sum_k \langle \psi_j^{(0)}|\hat{\mathcal{H}}_1(t)|\psi_k^{(0)}\rangle e^{i(\epsilon_j^{(0)}-\epsilon_k^{(0)})t/\hbar} c_k(t) \tag{C.60}$$

This last expression is very useful for finding the transition probability between states of the system induced by the perturbation. Suppose that at $t = 0$ the system is in the eigenstate i of $\hat{\mathcal{H}}_0$, in which case $c_i(0) = 1$ and $c_j(0) = 0$ for $j \neq i$. Then, writing the above expression in differential form at time dt we obtain

$$dc_j(t) = -\frac{i}{\hbar}\langle \psi_j^{(0)}|\hat{\mathcal{H}}_1(t)|\psi_i^{(0)}\rangle e^{i(\epsilon_j^{(0)}-\epsilon_i^{(0)})t/\hbar} dt$$

which, with the definitions

$$\hbar\omega_{ji} \equiv \epsilon_j^{(0)} - \epsilon_i^{(0)}, \quad \mathcal{V}_{ji}(t) \equiv \langle \psi_j^{(0)}|\hat{\mathcal{H}}_1(t)|\psi_i^{(0)}\rangle$$

can be integrated over time to produce

$$c_j(t) = -\frac{i}{\hbar}\int_0^t \mathcal{V}_{ji}(t')e^{i\omega_{ji}t'}dt' \tag{C.61}$$

If we assume that the perturbation $\hat{\mathcal{H}}_1(t)$ has the simple time dependence

$$\hat{\mathcal{H}}_1(t) = \mathcal{V}_1 e^{-i\omega t}$$

then the integration over t' can be performed easily to give:

$$c_j(t) = \frac{\langle \psi_j^{(0)}|\mathcal{V}_1|\psi_i^{(0)}\rangle}{\hbar(\omega_{ji}-\omega)}\left(1 - e^{i(\omega_{ji}-\omega)t}\right)$$

Typically, the quantity of interest is the probability for the transition from the initial state of the system at time 0, in the present example identified with the state $|\psi_i^{(0)}\rangle$, to another state such as $|\psi_j^{(0)}\rangle$ at time t. This probability is precisely equal to $|c_j(t)|^2$, which from the above expression takes the form

$$|c_j(t)|^2 = 2\frac{\left|\langle \psi_j^{(0)}|\mathcal{V}_1|\psi_i^{(0)}\rangle\right|^2}{[\hbar(\omega_{ji}-\omega)]^2}[1 - \cos(\omega_{ji}-\omega)t] \tag{C.62}$$

In this discussion we have made the implicit assumption that the states $|\psi_j^{(0)}\rangle$ have a discrete spectrum. Generally, the spectrum of the unperturbed hamiltonian $\hat{\mathcal{H}}_0$ may be a continuum with density of states $g(\epsilon)$, in which case we need to include in the expression for the transition probability all possible final states with energy $\epsilon_j^{(0)}$; the number of such states in an interval $d\epsilon_j$ is $g(\epsilon_j)d\epsilon_j$. These considerations lead to the following expression for the *rate* of the transition between an initial state $|\psi_i^{(0)}\rangle$ with energy $\epsilon_i^{(0)}$ and a continuum of final states $|\psi_f^{(0)}\rangle$ with energy $\epsilon_f^{(0)}$ in an interval $d\epsilon_f$:

$$dP_{i\to f} = \frac{d}{dt}|c_f(t)|^2 g(\epsilon_f)d\epsilon_f = \frac{2}{\hbar^2}\left|\langle \psi_f^{(0)}|\mathcal{V}_1|\psi_i^{(0)}\rangle\right|^2\left[\frac{\sin(\omega_{fi}-\omega)t}{(\omega_{fi}-\omega)}\right]g(\epsilon_f)d\epsilon_f$$

If we now let the time of the transition be very long, $t \to \infty$, the function inside the large square brackets in the above expression becomes a δ-function [see Appendix A, Eq. (A.57)]:

$$\lim_{t \to \infty} \left[\frac{\sin(\omega_{fi} - \omega)t}{(\omega_{fi} - \omega)} \right] = \pi \delta(\omega_{fi} - \omega)$$

which leads to:

$$\frac{dP_{i \to f}}{d\epsilon_f} = \frac{2\pi}{\hbar} \left| \langle \psi_f^{(0)} | \mathcal{V}_1 | \psi_i^{(0)} \rangle \right|^2 \delta(\epsilon_i^{(0)} - \epsilon_f^{(0)} - \hbar\omega) g(\epsilon_f) \tag{C.63}$$

This last expression is known as **Fermi's golden rule**. For transitions from one single-particle state $|\psi_i^{(0)}\rangle$ to another single-particle state $|\psi_f^{(0)}\rangle$, in which case neither the density of states $g(\epsilon_f)$ nor the dependence of the transition probability on ϵ_f enter, the transition rate takes the form

$$P_{i \to f}(\omega) = \frac{2\pi}{\hbar} \left| \langle \psi_f^{(0)} | \mathcal{V}_1 | \psi_i^{(0)} \rangle \right|^2 \delta(\epsilon_i^{(0)} - \epsilon_f^{(0)} - \hbar\omega) \tag{C.64}$$

C.7 The Electromagnetic Field Term

An example of the application of perturbation theory is the motion of a particle of mass m and charge q in an external electromagnetic field. The electric field $\mathbf{E}$ is given in terms of the scalar and vector potentials $\Phi, \mathbf{A}$ as:

$$\mathbf{E}(\mathbf{r}, t) = -\frac{1}{c} \frac{\partial \mathbf{A}(\mathbf{r}, t)}{\partial t} - \nabla_{\mathbf{r}} \Phi(\mathbf{r}, t)$$

where c is the speed of light. In terms of these potentials, the classical hamiltonian which describes the motion of the particle is given by:

$$\hat{\mathcal{H}} = \frac{1}{2m} \left[\mathbf{p} - \frac{q}{c} \mathbf{A} \right]^2 + q\Phi + \mathcal{V}(\mathbf{r})$$

where $\mathcal{V}(\mathbf{r})$ is any other potential that the particles are subject to. We will adopt this expression as the quantum-mechanical hamiltonian of the particle, by analogy to what we did for the free particle. Moreover, we can choose the vector and scalar potentials of the electromagnetic fields such that $\nabla_{\mathbf{r}} \cdot \mathbf{A} = 0, \Phi = 0$, called the Coulomb gauge (see Appendix B).

Our goal now is to determine the interaction part of the hamiltonian, that is, the part that describes the interaction of the charged particle with the external electromagnetic field, assuming that the latter is a small perturbation in the motion of the particle, which is governed by the unperturbed hamiltonian

$$\hat{\mathcal{H}}_0 = \frac{\mathbf{p}^2}{2m} + \mathcal{V}(\mathbf{r})$$

We will also specialize the discussion to electrons, in which case $m = m_e, q = -e$. Expanding $[\mathbf{p} + (e/c)\mathbf{A}]^2$ and keeping only up to first-order terms in $\mathbf{A}$, we find:

$$\frac{1}{2m_e}\left[\mathbf{p} + \frac{e}{c}\mathbf{A}\right]^2 \psi(\mathbf{r}) = \frac{1}{2m_e}\left[\mathbf{p}^2 + \frac{e}{c}\mathbf{p}\cdot\mathbf{A} + \frac{e}{c}\mathbf{A}\cdot\mathbf{p}\right]\psi(\mathbf{r})$$

$$= \frac{1}{2m_e}\mathbf{p}^2\psi(\mathbf{r}) + \frac{e\hbar}{2im_ec}\left[\nabla_\mathbf{r}\cdot(\mathbf{A}\psi(\mathbf{r})) + \mathbf{A}\cdot(\nabla_\mathbf{r}\psi(\mathbf{r}))\right]$$

$$= \frac{1}{2m_e}\mathbf{p}^2\psi(\mathbf{r}) + \frac{e\hbar}{2im_ec}\left[2\mathbf{A}\cdot(\nabla_\mathbf{r}\psi(\mathbf{r})) + (\nabla_\mathbf{r}\cdot\mathbf{A})\psi(\mathbf{r})\right]$$

and taking into account that $\nabla_\mathbf{r}\cdot\mathbf{A} = 0$, due to our choice of the Coulomb gauge, gives for the interaction term due to the electromagnetic field the expression

$$\hat{\mathcal{H}}^{\text{int}}(\mathbf{r}, t) = \frac{e}{m_ec}\mathbf{A}(\mathbf{r}, t)\cdot\mathbf{p} \tag{C.65}$$

since $(\hbar/i)\nabla_\mathbf{r} = \mathbf{p}$. Typically, we are interested in calculating matrix elements of this interaction term between eigenstates of the unberturbed hamiltonian, to obtain the values of physical observables, like the optical response of solids. We take two such eigenstates and corresponding eigenvalues of the hamiltonian $\hat{\mathcal{H}}_0$ to be $|\psi_i\rangle, \epsilon_i$ and $|\psi_f\rangle, \epsilon_f$.[2] The matrix elements of the interaction hamiltonian will then be:

$$\langle\psi_f|\hat{\mathcal{H}}^{\text{int}}|\psi_i\rangle = \langle\psi_f|\frac{e}{m_ec}\mathbf{A}(\mathbf{r}, t)\cdot\mathbf{p}|\psi_i\rangle$$

In the Coulomb gauge, we express the vector potential as:

$$\mathbf{A}(\mathbf{r}, t) = \mathbf{A}_0[e^{i(\mathbf{q}\cdot\mathbf{r}-\omega t)} + e^{-i(\mathbf{q}\cdot\mathbf{r}-\omega t)}] \text{ and } \mathbf{q}\cdot\mathbf{A}_0 = 0 \tag{C.66}$$

where $\mathbf{A}_0$ is a real, constant vector. We can use the commutation relation

$$[\hat{\mathcal{H}}_0, \mathbf{r}] = -\frac{i\hbar}{m_e}\mathbf{p} \Rightarrow \frac{\mathbf{p}}{m_e} = \frac{i}{\hbar}(\hat{\mathcal{H}}_0\mathbf{r} - \mathbf{r}\hat{\mathcal{H}}_0)$$

to rewrite the matrix elements of the interaction hamiltonian as:

$$\langle\psi_f|\hat{\mathcal{H}}^{\text{int}}|\psi_i\rangle = \frac{ie}{\hbar c}\left[e^{-i\omega t}\mathbf{A}_0\cdot\langle\psi_f|e^{i\mathbf{q}\cdot\mathbf{r}}(\hat{\mathcal{H}}_0\mathbf{r} - \mathbf{r}\hat{\mathcal{H}}_0)|\psi_i\rangle + \text{c.c.}\right]$$

with c.c. denoting the complex conjugate. In the long-wavelength limit $|\mathbf{q}| \to 0$, using the fact that $|\psi_i\rangle, |\psi_f\rangle$ are eigenstates of the unperturbed hamiltonian $\hat{\mathcal{H}}_0$ with eigenvalues ϵ_i, ϵ_f, we obtain

$$\langle\psi_f|\hat{\mathcal{H}}^{\text{int}}|\psi_i\rangle = \frac{ie}{\hbar c}\left[e^{-i\omega t}\mathbf{A}_0\cdot\langle\psi_f|\mathbf{r}|\psi_i\rangle(\epsilon_f - \epsilon_i) + \text{c.c.}\right] \tag{C.67}$$

that is, the matrix elements of the interaction hamiltonian can be expressed as matrix elements of the position operator $\mathbf{r}$ between eigenstates of the unperturbed hamiltonian.

A different way to express this last result is the following. Since the energy must be conserved in any process involving transitions induced by the interaction hamiltonian, the

---

[2] Here we are using the conventional notation and denote the two states as "initial" (subscript i) and "final" (subscript f) states for a transition induced by the electromagnetic field.

difference in energy between initial and final states must be:

$$\epsilon_f - \epsilon_i = \hbar\omega$$

which leads to:

$$\langle\psi_f|\hat{\mathcal{H}}^{\mathrm{int}}|\psi_i\rangle = \frac{\omega}{ic}\left[e^{-i\omega t}\mathbf{A}_0 \cdot \langle\psi_f|\hat{\mathbf{d}}|\psi_i\rangle + \mathrm{c.c.}\right] \qquad (C.68)$$

where we have defined the dipole operator $\hat{\mathbf{d}} \equiv q\mathbf{r} = -e\mathbf{r}$ or, more generally:

$$\hat{\mathbf{d}} \equiv \rho(\mathbf{r})\mathbf{r}$$

where $\rho(\mathbf{r})$ is the charge density when we are interested in quantities per unit volume. This is a useful expression because it makes the connection between the external electric field, expressed in terms of the time-dependent part of the vector potential $e^{-i\omega t}\mathbf{A}_0$, and the dipole moment of the system $\mathbf{d}$, given by the expectation value:

$$\mathbf{d}_{fi} = \langle\psi_f|\hat{\mathbf{d}}|\psi_i\rangle$$

This expression is used to calculate optical absorption by solids or molecules.

C.8 Relativistic Quantum Mechanics

In this section we review the quantum-mechanical description of particles that can have any velocity, up to the speed of light; the description also includes explicitly the spin degrees of freedom. This is based on the Dirac equation. The Schrödinger equation that we studied in detail in the main text is the non-relativistic limit of the Dirac equation, that is, the zeroth-order expansion of the Dirac equation in powers of v/c, as we show below. We restrict the discussion to particles with spin $s = 1/2$, since our main concern is to be able to handle electrons in solids.

C.8.1 Dirac Equation

The Dirac equation for spin $s = 1/2$ particles is:

$$i\hbar\frac{\partial}{\partial t}\psi(\mathbf{r}, s, t) = \hat{\mathcal{H}}\psi(\mathbf{r}, s, t), \quad \hat{\mathcal{H}} = \left[c\,\boldsymbol{\alpha}\cdot\mathbf{p} + mc^2\,\beta + \mathcal{V}(\mathbf{r})\,I_4\right] \qquad (C.69)$$

where c is the speed of light, m the rest mass of the particle, $\mathbf{p} = -i\hbar\nabla_\mathbf{r}$ the momentum operator, $\mathcal{V}(\mathbf{r})$ the external potential, and $\boldsymbol{\alpha}, \beta$ are 4×4 matrices defined as:

$$\boldsymbol{\alpha} = \alpha_x\hat{x} + \alpha_y\hat{y} + \alpha_z\hat{z}, \quad \alpha_j = \begin{pmatrix} 0 & \sigma_j \\ \sigma_j & 0 \end{pmatrix}, j = x, y, z, \quad \beta = \begin{pmatrix} I_2 & 0 \\ 0 & -I_2 \end{pmatrix} \qquad (C.70)$$

where I_2 is the 2×2 identity matrix and I_4 the 4×4 identity matrix, and $\sigma_x, \sigma_y, \sigma_z$ are the 2×2 Pauli spin matrices defined earlier [see Section C.4, Eq. (C.43)]. The wavefunction

$\psi(\mathbf{r}, s, t)$ is a four-component spinor. For a stationary state with energy E which is a solution to the Dirac equation, we write the time-dependent state $|\psi\rangle$ as:

$$|\psi\rangle = \begin{bmatrix} |\phi\rangle \\ |\chi\rangle \end{bmatrix} e^{-iEt/\hbar}$$

where we have separated out the time-dependent part from the stationary part composed of $|\phi\rangle$ and $|\chi\rangle$ which are two-component spinors. Substituting this expression in the Dirac equation, Eq. (C.69), and defining the non-relativistic energy $\epsilon = E - mc^2$, gives:

$$(mc^2 + \epsilon) \begin{bmatrix} |\phi\rangle \\ |\chi\rangle \end{bmatrix} = c\boldsymbol{\sigma} \cdot \mathbf{p} \begin{bmatrix} |\chi\rangle \\ |\phi\rangle \end{bmatrix} + mc^2 \begin{bmatrix} |\phi\rangle \\ -|\chi\rangle \end{bmatrix} + \mathcal{V}(\mathbf{r}) \begin{bmatrix} |\phi\rangle \\ |\chi\rangle \end{bmatrix} \tag{C.71}$$

where we have defined the vector operator $\boldsymbol{\sigma} = \sigma_x \hat{x} + \sigma_y \hat{y} + \sigma_z \hat{z}$, with $\sigma_x, \sigma_y, \sigma_z$ the Pauli spin matrices. The lower and upper components lead to, respectively:

$$[2mc^2 + \epsilon - \mathcal{V}(\mathbf{r})]|\chi\rangle = [c\boldsymbol{\sigma} \cdot \mathbf{p}]|\phi\rangle, \quad [\epsilon - \mathcal{V}(\mathbf{r})]|\phi\rangle = [c\boldsymbol{\sigma} \cdot \mathbf{p}]|\chi\rangle \tag{C.72}$$

which shows that two spinors $|\phi\rangle$ and $|\chi\rangle$ are intimately linked. Below we work out some important limiting cases of the Dirac equation.

C.8.2 Free Relativistic Particles

For free particles, we have $\mathcal{V}(\mathbf{r}) = 0$. The equations, Eq. (C.72), for the two parts $|\phi\rangle, |\chi\rangle$ of the four-component spinor $|\psi\rangle$ become

$$(2mc^2 + \epsilon)|\chi\rangle = [c\boldsymbol{\sigma} \cdot \mathbf{p}]|\phi\rangle, \quad \epsilon|\phi\rangle = [c\boldsymbol{\sigma} \cdot \mathbf{p}]|\chi\rangle \tag{C.73}$$

with the energy eigenvalue E given by $E = \epsilon + mc^2$. Solving for the $|\phi\rangle$ component from the second equation and substituting it in the first gives:

$$\frac{\epsilon}{c^2}(2mc^2 + \epsilon)|\chi\rangle = [\boldsymbol{\sigma} \cdot \mathbf{p}][\boldsymbol{\sigma} \cdot \mathbf{p}]|\chi\rangle$$

We then employ the following vector identity:

$$(\boldsymbol{\sigma} \cdot \mathbf{a})(\boldsymbol{\sigma} \cdot \mathbf{b}) = \mathbf{a} \cdot \mathbf{b} + i\boldsymbol{\sigma} \cdot (\mathbf{a} \times \mathbf{b}) \tag{C.74}$$

which is valid for any two vectors $\mathbf{a}, \mathbf{b}$ and the spin operator $\boldsymbol{\sigma}$; this relation can easily be derived from the definition of the spin operator in terms of the Pauli matrices. This vector identity, with $\mathbf{a} = \mathbf{p}$ and $\mathbf{b} = \mathbf{p}$, gives:

$$\frac{\epsilon}{c^2}(2mc^2 + \epsilon)|\chi\rangle = \mathbf{p}^2|\chi\rangle$$

since $\mathbf{p} \times \mathbf{p} = 0$. A solution to this equation is a plane wave with wave-vector $\mathbf{k}$, of the form

$$|\chi\rangle = |s\rangle e^{i\mathbf{k}\cdot\mathbf{r}}, \quad E_{\mathbf{k}}^{(\pm)} = \pm\sqrt{|\hbar c\mathbf{k}|^2 + (mc^2)^2}$$

as can easily be shown, using the facts that:

$$\mathbf{p}e^{i\mathbf{k}\cdot\mathbf{r}} = -i\nabla_{\mathbf{r}}e^{i\mathbf{k}\cdot\mathbf{r}} = \hbar\mathbf{k} \quad \text{and} \quad \epsilon(2mc^2 + \epsilon) = \left(E_{\mathbf{k}}^{(\pm)}\right)^2 - \left(mc^2\right)^2 = |\hbar c\mathbf{k}|^2$$

Here $|s\rangle$ represents only the spin degree of freedom, since the rest have been explicitly accounted for in the wavefunction. Of the two energy eigenvalues, $E_{\mathbf{k}}^{(+)}$ is positive; to find the corresponding wavefunctions we start with the two possible choices for the spin part of $|\phi\rangle$:

$$|\uparrow\rangle = \begin{bmatrix} 1 \\ 0 \end{bmatrix}, \quad |\downarrow\rangle = \begin{bmatrix} 0 \\ 1 \end{bmatrix}$$

which lead to the solutions:

$$|\psi_{\mathbf{k},\uparrow}^{(+)}\rangle = \mathcal{N}^{(+)} \begin{bmatrix} 1 \\ 0 \\ \hbar c k_z/\kappa^{(+)} \\ \hbar c(k_x + ik_y)/\kappa^{(+)} \end{bmatrix}, \quad |\psi_{\mathbf{k},\downarrow}^{(+)}\rangle = \mathcal{N}^{(+)} \begin{bmatrix} 0 \\ 1 \\ \hbar c(k_x - ik_y)/\kappa^{(+)} \\ -\hbar c k_z/\kappa^{(+)} \end{bmatrix}$$

with $\kappa^{(+)} = E_{\mathbf{k}}^{(+)} + mc^2$ and $\mathcal{N}^{(+)}$ a normalization constant, which is easily found to be:

$$\mathcal{N}^{(+)} = \left(\frac{\kappa^{(+)}}{2E_{\mathbf{k}}^{(+)}} \right)^{1/2}$$

For the negative-energy eigenvalue $E_{\mathbf{k}}^{(-)}$, we start with the two possible choices for the spin part of $|\chi\rangle$, which lead to the solutions:

$$|\psi_{\mathbf{k},\uparrow}^{(-)}\rangle = \mathcal{N}^{(-)} \begin{bmatrix} -\hbar c k_z/\kappa^{(-)} \\ -\hbar c(k_x + ik_y)/\kappa^{(-)} \\ 1 \\ 0 \end{bmatrix}, \quad |\psi_{\mathbf{k},\downarrow}^{(-)}\rangle = \mathcal{N}^{(-)} \begin{bmatrix} \hbar c(-k_x + ik_y)/\kappa^{(-)} \\ \hbar c k_z/\kappa^{(-)} \\ 0 \\ 1 \end{bmatrix}$$

with $\kappa^{(-)} = mc^2 + E_{\mathbf{k}}^{(-)} = -(|E_{\mathbf{k}}^{(-)}| - mc^2)$ and $\mathcal{N}^{(-)}$ a normalization constant, which is easily found to be:

$$\mathcal{N}^{(-)} = \left(\frac{\kappa^{(-)}}{2E_{\mathbf{k}}^{(-)}} \right)^{1/2}$$

From the definition of the velocity operator we obtain:

$$i\frac{d\mathbf{r}}{dt} = [\mathbf{r}, \hat{\mathcal{H}}] \Rightarrow \frac{d\mathbf{r}}{dt} = c\,\boldsymbol{\alpha}$$

by using Eq. (C.69) with $\mathcal{V}(\mathbf{r}) = 0$ (we leave the proof of this relation to the reader). Then, the current operator $\mathbf{j} = q\langle\mathbf{v}\rangle$, where q is the particle charge, for the plane-wave states we obtained above, with the choice of wave-vector $\mathbf{k} = k\hat{z}$, gives:

$$\mathbf{j}_{\mathbf{k},\uparrow}^{(+)} = q\langle\psi_{\mathbf{k},\uparrow}^{(+)}|c\,\boldsymbol{\alpha}|\psi_{\mathbf{k},\uparrow}^{(+)}\rangle$$

from which we find for the three components of the current:

$$\left[\mathbf{j}_{\mathbf{k},\uparrow}^{(+)}\right]_x = qc\left(\frac{\kappa^{(+)}}{2E_{\mathbf{k}}^{(+)}}\right)\begin{bmatrix} 1 & 0 & \dfrac{\hbar c k\hat{z}}{\kappa^{(+)}} & 0 \end{bmatrix}\begin{pmatrix} 0 & 0 & 0 & 1 \\ 0 & 0 & 1 & 0 \\ 0 & 1 & 0 & 0 \\ 1 & 0 & 0 & 0 \end{pmatrix}\begin{bmatrix} 1 \\ 0 \\ \hbar c k\hat{z}/\kappa^{(+)} \\ 0 \end{bmatrix} = 0$$

and similarly the y component also vanishes, while

$$\left[j_{\mathbf{k},\uparrow}^{(+)}\right]_z = qc\left(\frac{\kappa^{(+)}}{2E_{\mathbf{k}}^{(+)}}\right)\begin{bmatrix} 1 & 0 & \frac{\hbar ck\hat{z}}{\kappa^{(+)}} & 0 \end{bmatrix}\begin{pmatrix} 0 & 0 & 1 & 0 \\ 0 & 0 & -1 & 0 \\ 1 & 0 & 0 & 0 \\ 0 & -1 & 0 & 0 \end{pmatrix}\begin{bmatrix} 1 \\ 0 \\ \hbar ck\hat{z}/\kappa^{(+)} \\ 0 \end{bmatrix} = \frac{qc^2\hbar k}{E_{\mathbf{k}}^{(+)}}$$

The current for the other states can be calculated similarly.

Stationary Particles

For stationary particles, $\mathbf{k} = 0$ and $E_0^{(\pm)} = \pm mc^2$, the states $|\psi_{0,s}^{(+)}\rangle$ become the solutions describing spin-up or spin-down particles with mass $+m$, namely:

$$\psi_{0,\uparrow}^{(+)}(\mathbf{r}, t) = \begin{bmatrix} 1 \\ 0 \\ 0 \\ 0 \end{bmatrix} e^{-imc^2t/\hbar}, \quad \psi_{0,\downarrow}^{(+)}(\mathbf{r}, t) = \begin{bmatrix} 0 \\ 1 \\ 0 \\ 0 \end{bmatrix} e^{-imc^2t/\hbar}$$

while the states $|\psi_{0,s}^{(-)}\rangle$ become the solutions describing spin-up or spin-down anti-particles with mass $-m$, namely:

$$\psi_{0,\uparrow}^{(-)}(\mathbf{r}, t) = \begin{bmatrix} 0 \\ 0 \\ 1 \\ 0 \end{bmatrix} e^{imc^2t/\hbar}, \quad \psi_{0,\downarrow}^{(-)}(\mathbf{r}, t) = \begin{bmatrix} 0 \\ 0 \\ 0 \\ 1 \end{bmatrix} e^{imc^2t/\hbar}$$

Massless Particles

For massless particles, $m = 0$, the equations, Eq. (C.73), for the two components $|\phi\rangle$, $|\chi\rangle$ of the four-component spinor become:

$$E|\chi\rangle = [c\boldsymbol{\sigma} \cdot \mathbf{p}]|\phi\rangle, \quad E|\phi\rangle = [c\boldsymbol{\sigma} \cdot \mathbf{p}]|\chi\rangle$$

We can define the following expressions:

$$|\psi^{(R)}\rangle \equiv \frac{1}{\sqrt{2}}[|\phi\rangle + |\chi\rangle], \quad |\psi^{(L)}\rangle \equiv \frac{1}{\sqrt{2}}[|\phi\rangle - |\chi\rangle]$$

and by adding and subtracting the two equations for the spinors $|\phi\rangle$, $|\chi\rangle$ we find that the newly defined spinors obey:

$$E|\psi^{(R)}\rangle = [c\boldsymbol{\sigma} \cdot \mathbf{p}]|\psi^{(R)}\rangle, \quad E|\psi^{(L)}\rangle = -[c\boldsymbol{\sigma} \cdot \mathbf{p}]|\psi^{(L)}\rangle \tag{C.75}$$

that is, the two-component spinors $|\psi^{(R)}\rangle$ and $|\psi^{(L)}\rangle$ obey the same equation with opposite energy eigenvalues; this equation is referred to as the "Weyl equation." The Weyl equation admits as a solution a plane wave of wave-vector $\mathbf{k}$ and energy eigenvalues $E_{\mathbf{k}}^{(\text{R})} = \hbar|\mathbf{k}|c$, $E_{\mathbf{k}}^{(\text{L})} = -\hbar|\mathbf{k}|c$. For the eigenvalue $E_{\mathbf{k}}^{(\text{R})}$, the corresponding eigenfunctions are:

$$\psi_{\mathbf{k},\uparrow}^{(\text{R})}(\mathbf{r}) = \left(\begin{bmatrix} 1 \\ 0 \end{bmatrix} + \frac{1}{|\mathbf{k}|} \begin{bmatrix} k_z \\ k_x + ik_y \end{bmatrix} \right) \frac{e^{i\mathbf{k}\cdot\mathbf{r}}}{\sqrt{2}} \tag{C.76}$$

$$\psi_{\mathbf{k},\downarrow}^{(\text{R})}(\mathbf{r}) = \left(\begin{bmatrix} 0 \\ 1 \end{bmatrix} + \frac{1}{|\mathbf{k}|} \begin{bmatrix} k_x - ik_y \\ -k_z \end{bmatrix} \right) \frac{e^{i\mathbf{k}\cdot\mathbf{r}}}{\sqrt{2}} \tag{C.77}$$

and they describe massless particles of positive helicity (right-handed). For the eigenvalue $E_{\mathbf{k}}^{(\text{L})}$, the corresponding eigenfunctions are:

$$\psi_{\mathbf{k},\uparrow}^{(\text{L})}(\mathbf{r}) = \left(\begin{bmatrix} 1 \\ 0 \end{bmatrix} - \frac{1}{|\mathbf{k}|} \begin{bmatrix} k_z \\ k_x + ik_y \end{bmatrix} \right) \frac{e^{i\mathbf{k}\cdot\mathbf{r}}}{\sqrt{2}} \tag{C.78}$$

$$\psi_{\mathbf{k},\downarrow}^{(\text{L})}(\mathbf{r}) = \left(\begin{bmatrix} 0 \\ 1 \end{bmatrix} - \frac{1}{|\mathbf{k}|} \begin{bmatrix} k_x - ik_y \\ -k_z \end{bmatrix} \right) \frac{e^{i\mathbf{k}\cdot\mathbf{r}}}{\sqrt{2}} \tag{C.79}$$

and they describe massless particles of negative helicity (left-handed).

C.8.3 Weak Relativistic Corrections: Spin–Orbit Coupling

The weak relativistic limit of the Dirac equation is relevant to many situations of interest in the physics of solids. By "weak relativistic limit" we mean that we can ignore terms of order $(v/c)^n$ for $n = 2$ or higher, with v the typical velocity of the particles, that is, the particle velocity is much smaller than the speed of light. An important reason for doing this is to examine what terms may be missing from the non-relativistic limit, that is, the Schrödinger equation. It turns out that at least one of these terms plays a role in the physics of solids, even in the non-relativistic limit.

   The weak relativistic limit consists of taking $\epsilon, \mathcal{V} \ll mc^2$, expanding in Taylor series, and neglecting terms that are of higher order in the small quantity (v/c). In this limit, $|\chi\rangle \sim (\mathbf{p}c)/2mc^2|\phi\rangle \sim (v/c)|\phi\rangle$, so that corrections of order $(v/c)^2$ in $|\phi\rangle$ will be of order $(v/c)^3$ in the equation for $|\chi\rangle$, which is already of order $(v/c)^2$ for a positive-mass particle like the electron. These considerations indicate that we can neglect the equation for $|\chi\rangle$ altogether and just substitute for $|\chi\rangle$, obtained from the lower-component equation, in the upper-component equation for $|\phi\rangle$:

$$|\chi\rangle = [2mc^2 + \epsilon - \mathcal{V}(\mathbf{r})]^{-1}[c\boldsymbol{\sigma}\cdot\mathbf{p}]|\phi\rangle \Rightarrow [\epsilon - \mathcal{V}(\mathbf{r})]|\phi\rangle = [c\boldsymbol{\sigma}\cdot\mathbf{p}][2mc^2 + \epsilon - \mathcal{V}(\mathbf{r})]^{-1}[c\boldsymbol{\sigma}\cdot\mathbf{p}]|\phi\rangle$$

The expansion to the two lowest orders in $(v/c)^2$ leads to:

$$\left[\frac{\mathbf{p}^2}{2m} + \mathcal{V}(\mathbf{r}) - \frac{\mathbf{p}^4}{8m^3c^2} - \frac{i\hbar\nabla_{\mathbf{r}}\mathcal{V}(\mathbf{r})\cdot\mathbf{p}}{4m^2c^2} + \frac{\hbar}{4m^2c^2}\boldsymbol{\sigma}\cdot(\nabla_{\mathbf{r}}\mathcal{V}(\mathbf{r})\times\mathbf{p}) \right]|\phi\rangle = \epsilon|\phi\rangle \tag{C.80}$$

This result is obtained by the following steps. Starting from the full relativistic equation, Eq. (C.71), after eliminating the lower component $|\chi\rangle$ and expanding to first order in the small quantity $(\epsilon - \mathcal{V}(\mathbf{r}))/mc^2$, we obtain the equation for the upper component $|\phi\rangle$:

$$\left[\boldsymbol{\sigma}\cdot\mathbf{p}\frac{1}{2m}\left(1 - \frac{\epsilon - \mathcal{V}(\mathbf{r})}{2mc^2} \right)\boldsymbol{\sigma}\cdot\mathbf{p} + \mathcal{V}(\mathbf{r}) \right]|\phi\rangle = \epsilon|\phi\rangle.$$

We employ Eq. (C.74) again, with $\mathbf{a} = \mathbf{p}$ and $\mathbf{b} = \mathbf{p}$, which leads to the zeroth-order terms of the hamiltonian:

$$\left[\frac{\mathbf{p}^2}{2m} + \mathcal{V}(\mathbf{r}) \right]$$

from the unity factor in the parentheses of the previous equation. Next, we employ the following vector identity:

$$(\boldsymbol{\sigma} \cdot \mathbf{p})\mathcal{V}(\mathbf{r}) = \boldsymbol{\sigma} \cdot (\mathcal{V}(\mathbf{r})\mathbf{p}) + \boldsymbol{\sigma} \cdot \left[\mathbf{p}, \mathcal{V}(\mathbf{r}) \right]$$

which is also derived from the definitions of the spin and momentum operators, in conjunction with the earlier vector identity and the commutator of the momentum with potential

$$\left[\mathbf{p}, \mathcal{V}(\mathbf{r}) \right] = -i\hbar\nabla_{\mathbf{r}}\mathcal{V}(\mathbf{r})$$

to show that the first-order corrections are given by:

$$\left[-\frac{\mathbf{p}^2}{4m^2c^2}(\epsilon - \mathcal{V}(\mathbf{r})) - \frac{i\hbar}{4m^2c^2}\nabla_{\mathbf{r}}\mathcal{V}(\mathbf{r}) \cdot \mathbf{p} + \frac{\hbar}{4m^2c^2}\boldsymbol{\sigma} \cdot (\nabla_{\mathbf{r}}\mathcal{V}(\mathbf{r}) \times \mathbf{p}) \right]$$

Combining the zeroth-order and first-order terms, and using the approximation

$$(\epsilon - \mathcal{V}(\mathbf{r})) \approx \frac{\mathbf{p}^2}{2m}$$

we obtain the hamiltonian of Eq. (C.80).

We comment on the physical meaning of the terms that make up the new hamiltonian. In the first two terms on the left-hand side we recognize the familiar terms of the non-relativistic Schrödinger equation. The third term is the first correction to the non-relativistic kinetic energy:

$$\mathcal{K} = \left[(mc^2)^2 + (c\mathbf{p})^2 \right]^{1/2} - mc^2 = (mc^2)\left[1 + \frac{(c\mathbf{p})^2}{(mc^2)^2} \right]^{1/2} - mc^2 \approx \frac{\mathbf{p}^2}{2m} - \frac{\mathbf{p}^4}{8m^3c^2} + \cdots$$

The fourth term, called the Darwin term, is related to relativistic effects (Lorentz contraction) of the potential term. Thus, the first four terms are the familiar non-relativistic kinetic and potential energy terms and their lowest-order corrections. The last term is a new one and contains explicitly the spin operator $\boldsymbol{\sigma}$ which couples to both the gradient of the potential and the momentum operator. If we assume a spherically symmetric potential, $\mathcal{V}(\mathbf{r}) = \mathcal{V}(r)$, then we have (see Appendix A):

$$\nabla_{\mathbf{r}}\mathcal{V}(\mathbf{r}) = \frac{1}{r}\frac{\partial\mathcal{V}}{\partial r}\mathbf{r}$$

which, when inserted in the expression for the last term, gives:

$$\Delta\hat{\mathcal{H}}_{SO} = \frac{\hbar}{4m^2c^2r}\frac{\partial\mathcal{V}}{\partial r}\boldsymbol{\sigma} \cdot \mathbf{L}, \quad \text{where} \quad \mathbf{L} = \mathbf{r} \times \mathbf{p} \tag{C.81}$$

This is called the "spin–orbit coupling" term, since it includes the coupling between the spin operator $\boldsymbol{\sigma}$ and the orbital angular momentum $\mathbf{L}$; it is important in cases where there are heavy ions in the crystal, in which relativistic corrections become important (typically for atomic numbers larger than ~ 50).

C.9 Second Quantization

Finally, we discuss a formalism that has been developed to handle problems involving many-particle states in a more efficient way. This is often referred to as "second quantization," a term we use here to adhere to literature conventions, although it is a misnomer: it does not lead to a higher (second) quantization level, it merely changes the notation to a more convenient form. A more appropriate description is "occupation number representation," the term we used in Chapter 8 when discussing the BCS many-body wavefunction for the superconducting state (see Section 8.4.1). We will concentrate here on fermionic many-body wavefunctions, although a similar formalism can be derived for bosonic systems.

In the second quantization formalism, the goal is to express a many-body state as formed by creating single-particle states starting from a vacuum. We will use the notation $|0\rangle$ for the vacuum, defined as a state with no particles, and the symbol c_i^+ for the operator that creates the single-particle state $|\psi_i\rangle$; we refer to c_i^+ as the "creation" operator. Its adjoint, denoted as c_i, removes a single-particle state $|\psi_i\rangle$ from the system, and is hence called the "destruction" (or "annihilation") operator. We emphasize that the index i is used here to represent all the relevant degrees of freedom, including the spin of the particles. The many-body state of a system of electrons, containing the single-particle states $|\psi_i\rangle, i = 1, \ldots, N$, in a Slater-determinant form, will then be given as:

$$|\Psi\rangle = \frac{1}{\sqrt{N!}} \sum_P \mathrm{sgn}(P)P\left(|\psi_1\rangle \cdots |\psi_i\rangle \cdots |\psi_N\rangle\right) \equiv c_1^+ \cdots c_i^+ \cdots c_N^+ |0\rangle$$

where $P(\cdot)$ refers to a permutation of the objects in parentheses and $\mathrm{sgn}(P)$ is the sign of this permutation. Starting from the original order of the objects, an even permutation has positive sign and an odd one has negative sign. The advantage of the above expression is that all the symmetry properties of the Slater determinant are now embodied in the product of creation operators. For this to be true, the creation operators have to satisfy certain rules. The first rule is that the order in which the operators appear is important, because it signifies the order of the single-particle states in the product, which corresponds to specific permutations. By convention, we take the last operator of a product to act as the first one on a state to its right, that is, in the above expression the operator c_N^+ acts first on the vacuum state and the operator c_1^+ acts last on it, creating the single-particle states in the product $|\psi_1\rangle, \ldots, |\psi_N\rangle$ from left to right. Therefore, the following relation must hold:

$$c_i^+ c_j^+ = -c_j^+ c_i^+, \quad \text{for } i \neq j$$

since creating the states $|\psi_i\rangle$ and $|\psi_j\rangle$ in the reverse order introduces an overall minus sign in the product, as it is equivalent to a permutation; this is based on the assumption that the two states are different ($i \neq j$). For the same single-particle state, its creation twice should not be allowed at all, since we are dealing with fermions, therefore:

$$c_i^+ c_i^+ = 0$$

The adjoint (destruction) operators must satisfy similar relations:

$$c_i c_j = -c_j c_i, \quad \text{for } i \neq j, \quad \text{and} \quad c_i c_i = 0$$

On similar grounds, the products of creation and destruction operators corresponding to different states must involve a minus sign when their order is switched and the adjoints are taken:

$$c_i^+ c_j = -c_j^+ c_i, \quad \text{for } i \neq j$$

These operators move an electron from state i (if it is occupied) to state j (if it is empty) or the other way around, and since the order of application in the two products is reversed, corresponding to a permutation, they must come with opposite signs. However, something different happens when we consider the product of a creation and a destruction operator with the same index, namely $c_i^+ c_i$; this operation can have two outcomes.

 (i) If the single-particle state is already present in the many-body state, then this product first destroys it (the action of the operator on the right) and then creates it again (the action of the operator on the left), which leaves the many-body state unchanged.
(ii) If the single-particle state is *not* present in the many-body state, then this product must give zero, since the single-particle state cannot be destroyed if it does not exist.

Therefore, this product simply counts the number of particles in the state $|\psi_i\rangle$, which for fermions can be only one or zero, and accordingly we define the single-particle occupation number:[3]

$$n_i \equiv c_i^+ c_i \tag{C.82}$$

Conversely, the product $c_i c_i^+$ first creates and then destroys the single-particle state $|\psi_i\rangle$, so if this single-particle state does *not* exist already in the many-body state, it simply gives the same many-body state back, but if this state already exists, it gives zero. Therefore, we must have:

$$c_i c_i^+ = 1 - n_i = 1 - c_i^+ c_i \tag{C.83}$$

The rules we have stated so far for the creation–destruction operators are summarized in the following relations:

$$c_i^+ c_j + c_j^+ c_i = \delta_{ij}, \quad c_i^+ c_j^+ + c_j^+ c_i^+ = 0, \quad c_i c_j + c_j c_i = 0 \tag{C.84}$$

known as the "anticommutation" relations of the operators. From the definitions of these operators, we also conclude that when we act with the operator c_i^+ on a many-body state with N particles in which the single-particle state $|\psi_i\rangle$ does not exist, we create a state with $N + 1$ particles. Conversely, when we act with the operator c_i on a many-body state with N particles in which the single-particle state $|\psi_i\rangle$ exists, we create a state with $N - 1$ particles.

---

[3] This justifies our preferred name for this formalism.

Only when we act on a many-body state with pairs of creation–destruction operators do we leave the total number of particles N unchanged, if the result is not zero. The total number of particles in the many-body state of the system can then be calculated by applying the total number operator:

$$\hat{N} = \sum_i n_i = \sum_i c_i^+ c_i$$

An important notion in the use of the occupation number representation is a complete basis of single-particle states. This simply means that our set of single-particle states $|\psi_i\rangle$ covers all possible such states in which particles can exist. In this case, the following relation holds:

$$\sum_i |\psi_i\rangle\langle\psi_i| = 1, \quad \langle\psi_i|\psi_j\rangle = \delta_{ij} \tag{C.85}$$

where we have also assumed that the basis is orthonormal. Thus, the sum over all the ket–bra single-particle products is the identity operator. This is justified as follows. Suppose we have an arbitrary state $|\chi\rangle$ expressed in terms of the single-particle basis states as:

$$|\chi\rangle = \sum_k \alpha_k |\psi_k\rangle$$

with α_k numerical constants. Then a single ket–bra product projects out of this state the part that corresponds to it:

$$|\psi_i\rangle\langle\psi_i|\chi\rangle = \sum_k \alpha_k |\psi_i\rangle\langle\psi_i|\psi_k\rangle = \sum_k \alpha_k |\psi_i\rangle\delta_{ik} = \alpha_i|\psi_i\rangle$$

so the sum over all ket–bra products projects out of the arbitrary state $|\chi\rangle$ all its parts, and since the basis is complete, it gives back the entire state $|\chi\rangle$, in other words, this sum is equivalent to the identity. This is very useful, because we can insert this identity operator in many expressions to obtain more convenient results. We apply these ideas to the case of operators that can be written in terms of single-particle expressions, two-particle expressions, and so on.

Suppose we have an operator $\hat{O}$ that is expressed in terms of the one-particle operators $\hat{o}(\mathbf{r}_i)$ as follows:

$$\hat{O} = \sum_k \hat{o}(\mathbf{r}_k)$$

We want to apply this operator on a state represented as a product of single-particle states:

$$\hat{O}\{\psi_1(\mathbf{r}_1)\psi_2(\mathbf{r}_2)\cdots\psi_i(\mathbf{r}_i)\cdots\} = \sum_k \hat{o}(\mathbf{r}_k)\{\psi_1(\mathbf{r}_1)\psi_2(\mathbf{r}_2)\cdots\psi_i(\mathbf{r}_i)\cdots\}$$

which is a cumbersome expression involving many terms. We insert factors of identity in every term of the sum, the first before the first term in the product, the second before the second term in the product, and so on, which leads to:

$$\hat{O}\left\{\psi_1(\mathbf{r}_1)\psi_2(\mathbf{r}_2)\cdots\psi_i(\mathbf{r}_i)\cdots\right\} = \sum_{l_1}\langle\psi_{l_1}|\hat{o}|\psi_1\rangle\left\{\psi_{l_1}(\mathbf{r}_1)\psi_2(\mathbf{r}_2)\cdots\psi_i(\mathbf{r}_i)\cdots\right\}$$
$$+\sum_{l_2}\langle\psi_{l_2}|\hat{o}|\psi_2\rangle\left\{\psi_1(\mathbf{r}_1)\psi_{l_2}(\mathbf{r}_2)\cdots\psi_i(\mathbf{r}_i)\cdots\right\}+\cdots$$
$$+\sum_{l_i}\langle\psi_{l_i}|\hat{o}|\psi_i\rangle\left\{\psi_1(\mathbf{r}_1)\psi_2(\mathbf{r}_2)\cdots\psi_{l_i}(\mathbf{r}_i)\cdots\right\}+\cdots$$

In this last expression, each product of single-particle states is missing one state, which had been replaced by another state; we can therefore take advantage of the notation of the creation–destruction operators to write each such product as the result of applying the appropriate operators, which leads to the much simpler expression for the operator $\hat{O}$:

$$\hat{O} = \sum_{ij}\langle\psi_i|\hat{o}|\psi_j\rangle c_i^+c_j, \quad \langle\psi_i|\hat{o}|\psi_j\rangle = \int\psi_i^*(\mathbf{r})\hat{o}(\mathbf{r})\psi_j(\mathbf{r})d\mathbf{r}$$

that produces exactly the same result when it acts on the original product of single-particle states. The only cost involved in rewriting the operator $\hat{O}$ in this new form is the computation of the matrix elements of the one-particle operator $\hat{o}(\mathbf{r})$ between the single-particle states $\psi_i^*(\mathbf{r})$ and $\psi_j(\mathbf{r})$. By an analogous set of steps, we can derive an expression for an operator which is a sum of two-particle operators:

$$\hat{Q} = \frac{1}{2}\sum_{kl}\hat{q}(\mathbf{r}_k,\mathbf{r}_l) \rightarrow \frac{1}{2}\sum_{ijkl}\langle\psi_i\psi_j|\hat{q}|\psi_k\psi_l\rangle c_i^+c_j^+c_lc_k$$

that involves the matrix elements of the operator $\hat{q}(\mathbf{r},\mathbf{r}')$ between two pairs of states:

$$\langle\psi_i\psi_j|\hat{q}|\psi_k\psi_l\rangle = \int\psi_i^*(\mathbf{r})\psi_j^*(\mathbf{r}')\hat{q}(\mathbf{r},\mathbf{r}')\psi_k(\mathbf{r})\psi_l(\mathbf{r}')d\mathbf{r}d\mathbf{r}'$$

We note here an important detail: in the expression for the operator $\hat{Q}$, the order of the creation–destruction operator indices is i, j, l, k, whereas in the corresponding matrix element the order of the wavefunction indices is i, j, k, l; this is a consequence of the anti-commutation relations of the creation–destruction operators, Eq. (C.84). Operators that involve more particles can be defined in a similar fashion.

As an application of these concepts, we consider the hamiltonian for electrons in a solid, which is expressed as:

$$\hat{\mathcal{H}} = \sum_i\frac{-\hbar^2\nabla_{\mathbf{r}_i}^2}{2m_e} + \sum_i\mathcal{V}^{\mathrm{ion}}(\mathbf{r}_i) + \frac{1}{2}\sum_{i\neq j}\frac{e^2}{|\mathbf{r}_i-\mathbf{r}_j|}$$

containing, in the order they appear above, the kinetic energy, the interaction of electrons with ions, and the electron–electron Coulomb repulsion (for details, see Chapter 4). We can separate the terms into one-particle, $\hat{h}(\mathbf{r})$, and two-particle, $\hat{v}(\mathbf{r},\mathbf{r}')$, operators:

$$\hat{\mathcal{H}} = \sum_i\hat{h}(\mathbf{r}_i) + \frac{1}{2}\sum_{i\neq j}\hat{v}(\mathbf{r}_i,\mathbf{r}_j), \quad \hat{h}(\mathbf{r}) = \frac{-\hbar^2\nabla_{\mathbf{r}}^2}{2m_e} + \mathcal{V}^{\mathrm{ion}}(\mathbf{r}), \quad \hat{v}(\mathbf{r},\mathbf{r}') = \frac{e^2}{|\mathbf{r}-\mathbf{r}'|}$$

We can then use this last expression, in conjunction with the formalism of second quantization, to express the hamiltonian as:

$$\hat{\mathcal{H}} = \sum_{ij} \langle \psi_i | \hat{h} | \psi_j \rangle c_i^+ c_j + \frac{1}{2} \sum_{ijkl} \langle \psi_i \psi_j | \hat{v} | \psi_k \psi_l \rangle c_i^+ c_j^+ c_l c_k$$

where the single-particle states $|\psi_i\rangle$ are any convenient basis for calculating the matrix elements of the one-particle and two-particle operators. For instance, if we assume that these are atomic-like states localized at the sites of individual atoms in the solid, in the spirit of the tight-binding approximation (see Chapter 2), the matrix elements become integrals of these localized single-particle states and the one-particle and two-particle operators. For the first one, we have:

$$t_{ij} \equiv \langle \psi_i | \hat{h} | \psi_j \rangle = \int \psi_i^*(\mathbf{r}) \left[\frac{-\hbar^2 \nabla_{\mathbf{r}}^2}{2m_e} + \mathcal{V}^{\text{ion}}(\mathbf{r}) \right] \psi_j(\mathbf{r}) d\mathbf{r}$$

which, in our TBA example, represents the energy related to moving a particle from the localized state $|\psi_i\rangle$ to the localized state $|\psi_j\rangle$. If the atomic orbitals belong to different sites, then this is what we called the "hopping matrix element" for electron motion from the site labeled i to the site labeled j. If the orbitals are the same, this is the energy associated with this orbital, what we called the "on-site energy" in the TBA:

$$\epsilon_i \equiv t_{ii} = \langle \psi_i | \hat{h} | \psi_i \rangle$$

The two-particle matrix elements involve the Coulomb integrals:

$$\langle \psi_i \psi_j | \hat{v} | \psi_k \psi_l \rangle = \int \psi_i^*(\mathbf{r}) \psi_j^*(\mathbf{r}') \frac{e^2}{|\mathbf{r} - \mathbf{r}'|} \psi_k(\mathbf{r}) \psi_l(\mathbf{r}') d\mathbf{r} d\mathbf{r}'$$

From the nature of the two-particle operator, we realize that the largest contribution to this term comes from states that have a very large wavefunction overlap. To maximize the overlap, we can put the electrons in the same orbital at the same site, with the only difference in their single-particle states being the spin, so that we do not violate the Pauli exclusion principle. To represent the state $|\psi_i\rangle$, we use the explicit notation for the orbital character $|\phi_\nu\rangle$ and the spin state s: $|\psi_i\rangle \rightarrow |\phi_\nu, s\rangle$. Thus, to maximize the Coulomb interaction, we can take:

$$|\psi_i\rangle = |\psi_k\rangle = |\phi_\nu, \uparrow\rangle, \quad |\psi_j\rangle = |\psi_l\rangle = |\psi_\nu, \downarrow\rangle$$

but then from our earlier discussion, Eq. (C.82), we realize that:

$$c_i^+ c_j^+ c_l c_k = n_{\nu,\uparrow} n_{\nu,\downarrow}$$

where $n_{\nu,s}$ is the occupation number of orbital $|\phi_\nu\rangle$ by an electron with spin s. Defining the integral as:

$$U_\nu \equiv \int \phi_\nu^*(\mathbf{r}) \phi_\nu^*(\mathbf{r}') \frac{e^2}{|\mathbf{r} - \mathbf{r}'|} \phi_\nu(\mathbf{r}) \phi_\nu(\mathbf{r}') d\mathbf{r} d\mathbf{r}'$$

and limiting ourselves to this dominant contribution of the electron Coulomb repulsion, we find that the hamiltonian can be written as:

$$\hat{\mathcal{H}} = \sum_i \epsilon_i n_i + \sum_{i \neq j} t_{ij} c_i^+ c_j + \sum_\nu U_\nu n_{\nu,\uparrow} n_{\nu,\downarrow}$$

which is the Hubbard model we have encountered in several contexts (see Chapters 4 and 10).

Further Reading

1. *An Introduction to Quantum Physics: A First Course for Physicists, Chemists, Materials Scientists and Engineers*, S. Trachanas (Wiley-VCH, Weinheim, 2018). This is a modern and wonderfully pedagogical introduction to quantum physics with many examples relevant to the physics of materials.
2. *Quantum Mechanics*, L. I. Schiff (McGraw-Hill, New York, 1968). This is one of the standard references, containing an advanced and comprehensive treatment of quantum mechanics.
3. *Quantum Mechanics*, L. D. Landau and L. Lifshitz (3rd edn, Pergamon Press, Oxford, 1977). This is another standard reference with an advanced but somewhat terse treatment of the subject.
4. *Advanced Quantum Mechanics*, J. J. Sakurai (Addison-Wesley, Reading, MA, 1967). This book contains extensive and careful discussion of relativistic quantum mechanics, albeit with an emphasis on high-energy physics applications.

Appendix D Thermodynamics and Statistical Mechanics

D.1 The Laws of Thermodynamics

Thermodynamics is the empirical science that describes the state of macroscopic systems without reference to their microscopic structure. The laws of thermodynamics are based on experimental observations. The physical systems described by thermodynamics are considered to be composed of a very large number of microscopic particles (atoms or molecules). In the context of thermodynamics, a macroscopic system is described in terms of the external conditions that are imposed on it, determined by scalar or vector fields, and the values of the corresponding variables that specify the state of the system for given external conditions. The usual fields and corresponding thermodynamic variables are:

$$\text{temperature} : T \longleftrightarrow S : \text{entropy}$$
$$\text{pressure} : P \longleftrightarrow V : \text{volume}$$
$$\text{chemical potential} : \mu \longleftrightarrow N : \text{number of particles}$$
$$\text{magnetic field} : \mathbf{H} \longleftrightarrow \mathbf{M} : \text{magnetization}$$
$$\text{electric field} : \mathbf{E} \longleftrightarrow \mathbf{P} : \text{polarization}$$

with the last two variables referring to systems that possess internal magnetic or electric dipole moments, so they can respond to the application of external magnetic or electric fields. The temperature, pressure, [1] and chemical potential fields are determined by putting the system in contact with an appropriate reservoir; the values of these fields are set by the values they have in the reservoir. The fields are *intensive* quantities (they do not depend on the amount of substance), while the variables are *extensive* quantities (they are proportional to the amount of substance) of the system. Finally, each system is characterized by its internal energy E, which is a state function, that is, it depends on the state of the system and not on the path that the system takes to arrive at a certain state. Consequently, the quantity dE is an exact differential.

Thermodynamics is based on three laws.

(1) **The first law of thermodynamics** states that:

Heat is a form of energy

---

[1] If the pressure field is not homogeneous we can introduce a tensor to describe it, called the stress; in this case the corresponding variable is the strain tensor field. These notions are discussed in Chapter 7.

which can be put into a quantitative expression as follows: the sum of the change in the internal energy ΔE of a system and the work done by the system ΔW is equal to the heat ΔQ absorbed by the system:

$$\Delta Q = \Delta E + \Delta W$$

Heat is not a concept that can be related to the microscopic structure of the system, and therefore Q cannot be considered a state function and dQ is not an exact differential. From the first law, heat is equal to the increase in the internal energy of the system when its temperature goes up without the system having done any work.

(2) **The second law of thermodynamics** states that:

> *There can be no thermodynamic process whose sole effect is to transform heat entirely to work*

A direct consequence of the second law is that, for a cyclic process which brings the system back to its initial state:

$$\oint \frac{dQ}{T} \leq 0 \tag{D.1}$$

where equality holds for a *reversible* process. This must be true because otherwise the released heat could be used to produce work, this being the sole effect of the cyclic process since the system returns to its initial state, which would contradict the second law. For a reversible process, the heat absorbed during some infinitesimal part of the process must be released at some other infinitesimal part, the net heat exchange summing to zero, or the process would not be reversible. From this we draw the conclusion that for a reversible process we can define a state function S through:

$$\int_{\text{initial}}^{\text{final}} \frac{dQ}{T} = S_{\text{final}} - S_{\text{initial}} \quad : \text{reversible process} \tag{D.2}$$

which is called the entropy. Its differential is given by:

$$dS = \frac{dQ}{T} \tag{D.3}$$

Thus, even though heat is not a state function, and therefore dQ is not an exact differential, the entropy S defined through Eq. (D.2) *is* a state function and dS is an exact differential. Since the entropy is defined with reference to a reversible process, for an arbitrary process we will have:

$$\int_{\text{initial}}^{\text{final}} \frac{dQ}{T} \leq S_{\text{final}} - S_{\text{initial}} \quad : \text{arbitrary process} \tag{D.4}$$

This inequality is justified by the following argument: we can imagine the arbitrary process between initial and final states to be combined with a reversible process between final and initial states, which together form a cyclic process for which the inequality (D.1) holds, and $S_{\text{final}} - S_{\text{initial}}$ is defined by the reversible part. Hence, for the combined cyclic process:

$$\oint \frac{dQ}{T} = \int_{\text{initial}}^{\text{final}} \left(\frac{dQ}{T}\right)_{\text{arbitrary}} + \int_{\text{final}}^{\text{initial}} \left(\frac{dQ}{T}\right)_{\text{reversible}} =$$

$$\int_{\text{initial}}^{\text{final}} \left(\frac{dQ}{T} \right)_{\text{arbitrary}} - (S_{\text{final}} - S_{\text{initial}}) \leq 0$$

which leads to Eq. (D.4). Having defined the inequality (D.4) for an arbitrary process, the second law can now be cast in a more convenient form: in any thermodynamic process which takes an isolated system (for which $dQ = 0$) from an initial to a final state, the following inequality holds for the difference in the entropy between the two states:

$$\Delta S = S_{\text{final}} - S_{\text{initial}} \geq 0 \tag{D.5}$$

where the equality holds for reversible processes. Therefore, *the equilibrium state of an isolated system is a state of maximum entropy.*

(3) **The third law of thermodynamics** states that:

> *The entropy at the absolute zero of the temperature is a universal constant*

a statement which holds for all substances. We can choose this universal constant to be zero:

$$S(T = 0) = S_0 = 0 : \quad \textit{universal constant}$$

The three laws of thermodynamics must be supplemented by the equation of state (EOS) of a system, together providing a complete description of all the possible states in which the system can exist, as well as the possible transformations between such states. Since these states are characterized by macroscopic experimental measurements, it is often convenient to introduce and use quantities like:

$$C_V \equiv \left(\frac{dQ}{dT} \right)_V : \quad \text{constant-volume specific heat}$$

$$C_P \equiv \left(\frac{dQ}{dT} \right)_P : \quad \text{constant-pressure specific heat}$$

$$\alpha = \frac{1}{V} \left(\frac{\partial V}{\partial T} \right)_P : \quad \text{thermal expansion coefficient}$$

$$\kappa_T \equiv -\frac{1}{V} \left(\frac{\partial V}{\partial P} \right)_T : \quad \text{isothermal compressibility}$$

$$\kappa_S \equiv -\frac{1}{V} \left(\frac{\partial V}{\partial P} \right)_S : \quad \text{adiabatic compressibility}$$

A familiar EOS is that of an ideal gas:

$$PV = Nk_B T \tag{D.6}$$

where k_B is Boltzmann's constant. For such a system it is possible to show, simply by manipulating the above expressions and using the laws of thermodynamics, several other relations:

$$E = E(T) \tag{D.7}$$

$$C_P - C_V = \frac{TV\alpha^2}{\kappa_T} \tag{D.8}$$

$$C_V = \frac{TV\alpha^2\kappa_S}{(\kappa_T - \kappa_S)\kappa_T} \tag{D.9}$$

$$C_P = \frac{TV\alpha^2}{(\kappa_T - \kappa_S)} \tag{D.10}$$

We prove here the first of the above relations, Eq. (D.7), that is, for the ideal gas the internal energy E is a function of the temperature only; the proof of the others is left as an exercise for the reader. The first law of thermodynamics, for the case when the work done on the system is mechanical, becomes:

$$dQ = dE + dW = dE + PdV$$

Using as independent variables V and T, we obtain:

$$dQ = \left(\frac{\partial E}{\partial T}\right)_V dT + \left(\frac{\partial E}{\partial V}\right)_T dV + PdV$$

The definition of the entropy Eq. (D.2), allows us to write dQ as TdS, so that the above equation takes the form

$$dS = \frac{1}{T}\left(\frac{\partial E}{\partial T}\right)_V dT + \left[\frac{1}{T}\left(\frac{\partial E}{\partial V}\right)_T + \frac{P}{T}\right]dV \tag{D.11}$$

and the second law of thermodynamics tells us that the left-hand side is an exact differential, so that taking the cross derivatives of the two terms on the right-hand side we should have:

$$\left(\frac{\partial}{\partial V}\right)_T\left[\frac{1}{T}\left(\frac{\partial E}{\partial T}\right)_V\right] = \left(\frac{\partial}{\partial T}\right)_V\left[\frac{1}{T}\left(\frac{\partial E}{\partial V}\right)_T + \frac{P}{T}\right]$$

Carrying out the differentiations and using the EOS of the ideal gas, Eq. (D.6):

$$P = \frac{Nk_BT}{V} \Rightarrow \left(\frac{\partial P}{\partial T}\right)_V = \frac{Nk_B}{V}$$

we obtain as the final result:

$$\left(\frac{\partial E}{\partial V}\right)_T = 0 \Rightarrow E = E(T)$$

For a general system, the internal energy is a function of temperature and volume, $E = E(T, V)$. This leads to a useful expression for the entropy as a function of temperature and volume: starting from Eq. (D.11) and integrating both sides with respect to T at fixed volume V produces

$$S(T, V) = \int_0^T \frac{1}{T'}\left(\frac{\partial E}{\partial T'}\right)_V dT' \tag{D.12}$$

since the constant of integration is $S(T = 0, V) = S_0 = 0$ from the third law of thermodynamics.

D.2 Thermodynamic Potentials

Another very useful concept in thermodynamics is the definition of "thermodynamic potentials" or "free energies," appropriate for different situations. For example, the definition of the

$$\textbf{enthalpy}: \Theta = E + PV \tag{D.13}$$

is useful because in terms of this the specific heat at constant pressure takes the form

$$C_P = \left(\frac{dQ}{dT}\right)_P = \left(\frac{\partial E}{\partial T}\right)_P + P\left(\frac{\partial V}{\partial T}\right)_P = \left(\frac{\partial \Theta}{\partial T}\right)_P \tag{D.14}$$

that is, the enthalpy is the appropriate free energy which we need to differentiate by temperature in order to obtain the specific heat at constant pressure. In the case of the ideal gas, the enthalpy takes the form

$$\Theta = (C_V + Nk_B)T \tag{D.15}$$

where we have used the result derived above, namely $E = E(T)$, to express the specific heat at constant volume as:

$$C_V = \left(\frac{dQ}{dT}\right)_V = \left(\frac{\partial E}{\partial T}\right)_V = \frac{dE}{dT} \Rightarrow E(T) = C_V T \tag{D.16}$$

Now we can take advantage of these expressions to obtain the following relation between specific heats of the ideal gas at constant pressure and constant volume:

$$C_P - C_V = Nk_B$$

The statement of the second law in terms of the entropy motivates the definition of another thermodynamic potential, the

$$\textbf{Helmholtz free energy}: F = E - TS \tag{D.17}$$

Specifically, according to the second law as expressed in Eq. (D.4), for an arbitrary process at constant temperature we will have:

$$\int \frac{dQ}{T} \leq \Delta S \Rightarrow \frac{\Delta Q}{T} = \frac{\Delta E + \Delta W}{T} \leq \Delta S \tag{D.18}$$

For a system that is mechanically isolated, $\Delta W = 0$, and consequently

$$0 \leq -\Delta E + T\Delta S = -\Delta F \Rightarrow \Delta F \leq 0 \tag{D.19}$$

which proves that *the equilibrium state of a mechanically isolated system at constant temperature is one of minimum Helmholtz free energy*. More explicitly, the above relation tells us that changes in F due to changes in the state of the system (which of course must be consistent with the conditions of constant temperature and no mechanical work done on or by the system) can only decrease the value of F; therefore at equilibrium, when F does not change any longer, its value must be at a minimum.

Finally, if in addition to the temperature the pressure is also constant, then by similar arguments we will have the following relation:

$$\int \frac{dQ}{T} \leq \Delta S \Rightarrow \frac{\Delta Q}{T} = \frac{\Delta E + \Delta W}{T} \leq \Delta S \Rightarrow P\Delta V \leq -\Delta E + T\Delta S = -\Delta F \quad \text{(D.20)}$$

We can then define a new thermodynamic potential, called the

$$\textbf{Gibbs free energy}: G = F + PV \quad\quad\quad\quad \text{(D.21)}$$

for which the above relation implies that:

$$\Delta G = \Delta F + P\Delta V \leq 0 \quad\quad\quad\quad \text{(D.22)}$$

which proves that *the equilibrium state of a system at constant temperature and pressure is one of minimum Gibbs free energy*. The logical argument that leads to this conclusion is identical to the one invoked for the Helmholtz free energy.

The thermodynamic potentials are also useful because the various thermodynamic fields (or variables) can be expressed as partial derivatives of the potentials with respect to the corresponding variable (or field) under proper conditions. Specifically, it can be shown directly from the definition of the thermodynamic potentials that the following relations hold:

$$P = -\left(\frac{\partial F}{\partial V}\right)_T, \quad P = -\left(\frac{\partial E}{\partial V}\right)_S \quad\quad \text{(D.23)}$$

$$V = +\left(\frac{\partial G}{\partial P}\right)_T, \quad V = +\left(\frac{\partial \Theta}{\partial P}\right)_S \quad\quad \text{(D.24)}$$

$$S = -\left(\frac{\partial F}{\partial T}\right)_V, \quad S = -\left(\frac{\partial G}{\partial T}\right)_P \quad\quad \text{(D.25)}$$

$$T = +\left(\frac{\partial \Theta}{\partial S}\right)_P, \quad T = +\left(\frac{\partial E}{\partial S}\right)_V \quad\quad \text{(D.26)}$$

For example, from the first and second laws of thermodynamics we have:

$$dE = dQ - PdV = TdS - PdV$$

and E is a state function, so its differential using as free variables S and V must be expressed as:

$$dE = \left(\frac{\partial E}{\partial S}\right)_V dS + \left(\frac{\partial E}{\partial V}\right)_S dV$$

Comparing the last two equations we obtain the second of Eq. (D.23) and the second of Eq. (D.26). Notice that in all cases a variable and its corresponding field are connected by the equations (D.23)–(D.26), that is, V is given as a partial derivative of a potential with respect to P and vice versa and S is given as a partial derivative of a potential with respect to T and vice versa. The thermodynamic potentials and their derivatives which relate thermodynamic variables and fields are summarized below.

Thermodynamic potentials and their derivatives

In each case the potential, the variables or fields associated with it, and the relations connecting fields and variables through partial derivatives of the potential are given.

$$\text{Internal energy}: \; E \; (S, V) \rightarrow P = -\left(\frac{\partial E}{\partial V}\right)_S, \quad T = +\left(\frac{\partial E}{\partial S}\right)_V$$

$$\text{Helmholtz free energy}: \; F = E - TS \; (T, V) \rightarrow P = -\left(\frac{\partial F}{\partial V}\right)_T, \quad S = -\left(\frac{\partial F}{\partial T}\right)_V$$

$$\text{Gibbs free energy}: \; G = F + PV \; (T, P) \rightarrow V = +\left(\frac{\partial G}{\partial P}\right)_T, \quad S = -\left(\frac{\partial G}{\partial T}\right)_P$$

$$\text{Enthalpy}: \; \Theta = E + PV \; (S, P) \rightarrow V = +\left(\frac{\partial \Theta}{\partial P}\right)_S, \quad T = +\left(\frac{\partial \Theta}{\partial S}\right)_P$$

In cases where the work is of magnetic nature, the relevant thermodynamic field is the magnetic field **H** and the relevant thermodynamic variable is the magnetization **M**; similarly, for electrical work the relevant thermodynamic field is the electric field **E** and the relevant thermodynamic variable is the polarization **P**. Note, however, that the identification of the thermodynamic field and variable is not always straightforward; in certain cases the proper analogy is to identify the magnetization as the relevant thermodynamic *field* and the external magnetic field is the relevant thermodynamic *variable* (for instance, in a system of non-interacting spins on a lattice under the influence of an external magnetic field, as discussed in Chapter 10).

As an example, we consider the case of a magnetic system. For simplicity, we assume that the fields and the magnetic moments are homogeneous, so we can work with scalar rather than vector quantities. The magnetic moment is defined as $m(\mathbf{r})$, in terms of which the magnetization M is given by:

$$M = \int m(\mathbf{r}) \mathrm{d}\mathbf{r} = Vm \tag{D.27}$$

where the last expression applies to a system of volume V in which the magnetic moment is constant, $m(\mathbf{r}) = m$. The differential of the internal energy $\mathrm{d}E$ for such a system in an external field H is given by:

$$\mathrm{d}E = T\mathrm{d}S + H\mathrm{d}M$$

where now H plays the role of the pressure in a mechanical system, and $-M$ plays the role of the volume. The sign of the HM term in the magnetic system is opposite from that of the PV term in the mechanical system, because an increase in the magnetic field H increases the magnetization M whereas an increase in the pressure P decreases the volume V. By analogy to the mechanical system, we define the thermodynamic potentials of Helmholtz free energy F, Gibbs free energy G, and enthalpy Θ as:

$$F = E - TS \Rightarrow dF = dE - TdS - SdT = -SdT + HdM$$
$$G = F - HM \Rightarrow dG = dF - HdM - MdH = -SdT - MdH$$
$$\Theta = E - HM \Rightarrow d\Theta = dE - HdM - MdH = TdS - MdH$$

which give the following relations between the thermodynamic potentials, variables, and fields:

$$T = +\left(\frac{\partial E}{\partial S}\right)_M, \quad H = +\left(\frac{\partial E}{\partial M}\right)_S \Rightarrow \left(\frac{\partial T}{\partial M}\right)_S = \left(\frac{\partial H}{\partial S}\right)_M$$

$$S = -\left(\frac{\partial F}{\partial T}\right)_M, \quad H = +\left(\frac{\partial F}{\partial M}\right)_T \Rightarrow \left(\frac{\partial S}{\partial M}\right)_T = -\left(\frac{\partial H}{\partial T}\right)_M$$

$$S = -\left(\frac{\partial G}{\partial T}\right)_H, \quad M = -\left(\frac{\partial G}{\partial H}\right)_T \Rightarrow \left(\frac{\partial S}{\partial H}\right)_T = \left(\frac{\partial M}{\partial T}\right)_H$$

$$T = +\left(\frac{\partial \Theta}{\partial S}\right)_H, \quad M = -\left(\frac{\partial \Theta}{\partial H}\right)_S \Rightarrow \left(\frac{\partial T}{\partial H}\right)_S = -\left(\frac{\partial M}{\partial S}\right)_H$$

Similar expressions for thermodynamic potentials and relations between them, and the thermodynamic variables and fields, can be defined for a system with electric moment $\mathbf{p}(\mathbf{r})$ in an external electric field $\mathbf{E}$.

When there is significant interaction at the microscopic level among the elementary moments induced by the external field (electric or magnetic), then the thermodynamic field should be taken as the total field, including the external and induced contributions. Specifically, the total magnetic field is given by:

$$\mathbf{B}(\mathbf{r}) = \mathbf{H} + 4\pi \mathbf{m}(\mathbf{r})$$

and the total electric field is given by:

$$\mathbf{D}(\mathbf{r}) = \mathbf{E} + 4\pi \mathbf{p}(\mathbf{r})$$

with $\mathbf{m}(\mathbf{r})$, $\mathbf{p}(\mathbf{r})$ the local magnetic or electric moment and $\mathbf{H}, \mathbf{E}$ the external magnetic or electric fields, usually considered to be constant.

D.3 Application: Phase Transitions

As an application of the concepts of thermodynamics we consider the case of the van der Waals gas and show how the minimization of the free energy leads to the idea of a first-order phase transition. In the van der Waals gas the particles interact with an attractive potential at close range. The effect of the attractive interaction is to reduce the pressure that the kinetic energy P_{kin} of the particles would produce. The interaction between particles changes from attractive to repulsive if the particles get too close. Consistent with the macroscopic view of thermodynamics, we attempt to describe these effects not from a detailed microscopic perspective but through empirical parameters. Thus, we write for the actual pressure P exerted by the gas on the walls of the container:

$$P = P_{\text{kin}} - \frac{\eta}{V^2}$$

where η is a positive constant with dimensions energy $\times$ volume and V is the volume of the box. This equation describes the reduction in pressure due to the attractive interaction between particles; the term $1/V^2$ comes from the fact that collisions between *pairs* of particles are responsible for the reduction in pressure and the probability of finding two particles within the range of the attractive potential is proportional to the square of the probability of finding one particle at a certain point within an infinitesimal volume, the latter probability being proportional to $1/V$ for a homogeneous gas. The effective volume of the gas will be equal to:

$$V_{\text{eff}} = V - V_0$$

where V_0 is a constant equal to the excluded volume due to the repulsive interaction between particles at very small distances. The van der Waals equation of state asserts that the same relation exists between the effective volume and the pressure due to kinetic energy, as in the ideal gas between volume and pressure, that is:

$$P_{\text{kin}} V_{\text{eff}} = N k_B T$$

In this equation we substitute the values of P and V, which are the measurable thermodynamic field and variable, using the expressions we discussed above for P_{kin} and V_{eff}, to obtain

$$(P + \frac{\eta}{V^2})(V - V_0) = N k_B T$$

This equation can be written as a third-order polynomial in V:

$$V^3 - \left(V_0 + \frac{N k_B T}{P}\right) V^2 + \frac{\eta}{P} V - \frac{\eta V_0}{P} = 0 \qquad (D.28)$$

For a fixed temperature, this is an equation that relates pressure and volume, called an isotherm. In general it has three roots, as indicated graphically in Fig. D.1. There is a range of pressures in which a horizontal line corresponding to a given value of the pressure intersects the isotherm at three points.

From our general thermodynamic relations we have:

$$dF = dE - T dS = dE - dQ = -dW = -P dV$$

This shows that we can calculate the free energy as:

$$F = - \int P dV$$

which produces the second plot in Fig. D.1. Notice that since the pressure P and volume V are always positive, the free energy F is always negative and monotonically decreasing. The free energy must be minimized for a mechanically isolated system. As the free energy plot of Fig. D.1 shows, it is possible to reduce the free energy of the system between states

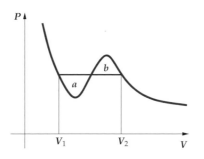 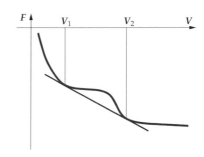

Fig. D.1 The Maxwell construction. **Left**: $P–V$ curve for the van der Waals gas at fixed temperature. **Right**: The corresponding $F–V$ curve. The common tangent construction on the free-energy curve determines the mixture of two phases corresponding to states 1 and 2 that will give lower free energy than the single-phase condition. The common tangent (red line) of the free energy corresponds to a horizontal line in the $P–V$ curve, which is determined by the requirement that the areas in regions labeled a and b are equal.

1 and 2 (with volumes V_1 and V_2) by taking a mixture of them rather than having a pure state at any volume between those two states. The mixture of the two states will have a free energy which lies along the common tangent between the two states. When the system follows the common tangent in the free energy plot, the pressure will be constant, since:

$$P = - \left(\frac{\partial F}{\partial V} \right)_T$$

The constant value of the pressure is determined by the common tangent and equal pressure conditions, which together imply:

$$\frac{F_2 - F_1}{V_2 - V_1} = \left(\frac{\partial F}{\partial V} \right)_{V=V_1} = \left(\frac{\partial F}{\partial V} \right)_{V=V_2} \Rightarrow - \left(\frac{\partial F}{\partial V} \right)_{V=V_1} (V_2 - V_1) = -(F_2 - F_1)$$

which in turn produces the equation

$$P_1(V_2 - V_1) = \int_{V_1}^{V_2} P dV$$

The graphical interpretation of this equation is that the areas a and b between the pressure–volume curve and the constant pressure line must be equal. This is known as the Maxwell construction.

   The meaning of this derivation is that the system can lower its free energy along an isotherm by forming a mixture at constant pressure between two phases along the isotherm, rather than being in a homogeneous single phase. This corresponds to the transition between the liquid phase on the left (smaller volume, higher pressure) and the gas phase on the right (higher volume, lower pressure). This particular transition, in which pressure and temperature remain constant while the volume changes significantly because of the different volumes of the two phases, is referred to as a *first-order phase transition*. The difference in volume between the two phases leads to a discontinuity in the first derivative of the Gibbs free energy as a function of pressure for fixed temperature. Related to this behavior is a discontinuity in the first derivative of the Gibbs free energy as a function of temperature for fixed pressure, which is due to a difference in the entropy of the two phases. If the first derivatives of the free energy are continuous, then the transition is referred to as *second-order*; second-order phase transitions usually have discontinuities in higher derivatives of the relevant free energy. A characteristic feature of a first-order phase transition is the existence of a *latent heat*, that is, an amount of heat which is released when going from one phase to the other at constant temperature and pressure.

The above statements can be proven by considering the Gibbs free energy of the system consisting of two phases and recalling that, as discussed above, it must be at a minimum for constant temperature and pressure. We denote the Gibbs free energy per unit mass as g_1 for the liquid and g_2 for the gas, and the corresponding mass in each phase as m_1 and m_2. At the phase transition the total Gibbs free energy G will be equal to the sum of the two parts:

$$G = g_1 m_1 + g_2 m_2$$

and it will be a minimum, therefore variations of it with respect to changes other than in temperature or pressure must vanish, $\delta G = 0$. The only relevant variation at the phase transition is transfer of mass from one phase to the other, but because of conservation of mass we must have $\delta m_1 = -\delta m_2$, therefore:

$$\delta G = g_1 \delta m_1 + g_2 \delta m_2 = (g_1 - g_2)\delta m_1 = 0 \Rightarrow g_1 = g_2$$

since the mass changes are arbitrary. But using the relations between thermodynamic potentials and variables listed earlier (p. 616) we can write:

$$\left(\frac{\partial (g_2 - g_1)}{\partial T} \right)_P = -(s_2 - s_1), \quad \left(\frac{\partial (g_2 - g_1)}{\partial P} \right)_T = (\omega_2 - \omega_1)$$

where s_1, s_2 are the entropies per unit mass of the two phases and ω_1, ω_2 the volumes per unit mass; these are the discontinuities in the first derivatives of the Gibbs free energy as a function of temperature for fixed pressure, or as a function of pressure for fixed temperature, that we mentioned above.

Denoting the differences in Gibbs free energy, entropy, and volume per unit mass between the two phases as $\Delta g, \Delta s, \Delta \omega$, we can write the above relationships as:

$$\left(\frac{\partial \Delta g}{\partial T} \right)_P \left(\frac{\partial \Delta g}{\partial P} \right)_T^{-1} = -\frac{\Delta s}{\Delta \omega}$$

Now the Gibbs free energy difference Δg between the two phases is a function of T and P, which can be inverted to produce expressions for T as a function of Δg and P or for P as a function of Δg and T. From these relations it is easy to show that the following relation holds between the partial derivatives:

$$\left(\frac{\partial \Delta g}{\partial T} \right)_P \left(\frac{\partial P}{\partial \Delta g} \right)_T = -\left(\frac{\partial P}{\partial T} \right)_{\Delta g} = \left(\frac{\partial \Delta g}{\partial T} \right)_P \left(\frac{\partial \Delta g}{\partial P} \right)_T^{-1}$$

which, when combined with the earlier equation, gives:

$$\left(\frac{\partial P}{\partial T} \right)_{\Delta g} = \frac{\Delta s}{\Delta \omega} \Rightarrow \frac{dP(T)}{dT} = \frac{\Delta s}{\Delta \omega}$$

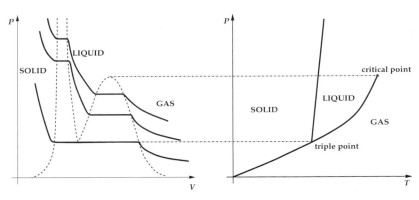

Fig. D.2 Isotherms on the P–V plane (left plot) and phase boundaries between the solid, liquid, and gas phases on the P–T plane (right plot). The dashed lines in the P–V plane identify the regions that correspond to the transitions between the phases which occur at constant pressure (horizontal lines). The triple point and critical point are identified on both diagrams.

where, in the last relation, we have used the fact that at the phase transition $\Delta g = g_2 - g_1 = 0$ is constant and therefore the pressure P is simply a function of T. We can define the latent heat L through the difference in entropy of the two phases to obtain an expression for the derivative of the pressure:

$$L = T\Delta s \Rightarrow \frac{\mathrm{d}P}{\mathrm{d}T} = \frac{L}{T\Delta\omega} \tag{D.29}$$

known as the Clausius–Clapeyron equation .[2] A similar analysis applies to the solid–liquid phase-transition boundary. The change in volume per unit mass in going from the liquid to the gas phase is positive and large, while that going from the solid to the liquid phase is much smaller and can even be negative (see discussion below).

A general diagram of typical isotherms on the P–V plane and the corresponding curves on the P–T plane that separate different phases is shown in Fig. D.2. Some interesting features in these plots are the so-called *triple point* and *critical point*. At the triple point, there is coexistence between all three phases and there is no liquid phase for lower temperatures. At the critical point, the distinction between liquid and gas phases has disappeared. This last situation corresponds to the case where all three roots of Eq. (D.28) have collapsed to one. At this point, the equation takes the form

$$(V - V_c)^3 = V^3 - 3V_c V + 3V_c^2 V - V_c^3 = 0 \tag{D.30}$$

where V_c is the volume at the critical point. We can express the volume V_c, pressure P_c, and temperature T_c at the critical point in terms of the constants that appear in the van der Waals equation of state:

$$V_c = 3V_0, \quad P_c = \frac{\eta}{27V_0^2}, \quad T_c = \frac{8\eta}{27Nk_B V_0}$$

---

[2] The latent heat and the change of volume are defined here per unit mass, so they can both be scaled by the same factor without changing the Clausius–Clapeyron equation

Moreover, if we use the reduced variables for volume, pressure, and temperature:

$$\omega = \frac{V}{V_c}, \quad p = \frac{P}{P_c}, \quad t = \frac{T}{T_c}$$

we obtain the following expression for the van der Waals equation of state:

$$\left(p + \frac{3}{\omega^3}\right)\left(\omega - \frac{1}{3}\right) = \frac{8}{3}t$$

which is the same for all substances since it does not involve any substance-specific constants. This equation is called the "law of corresponding states" and is obeyed rather accurately by a number of substances which behave like a van der Waals gas.

In Fig. D.2 we have shown the boundary between the solid and liquid phases with a large positive slope, as one might expect for a substance that expands when it melts: a higher melting temperature is needed to compensate for the additional constraint imposed on the volume by the increased pressure. The slope of curves that separate two phases is equal to the latent heat divided by the temperature and the change in volume, as determined by the Clausius–Clapeyron relation, Eq. (D.29). For solids that expand when they melt, the change in volume is positive, and of course so are the latent heat and temperature, leading to a curve with large positive slope, since the change in volume going from solid to liquid is usually small. Interestingly, there are several solids that contract upon melting, in which case the corresponding phase boundary on the P–T plane has *negative* slope. Examples of such solids are ice and silicon. The reason for this unusual behavior is that the bonds in the solid are of a nature that actually makes the structure of the solid rather open, in contrast to most common solids where the atoms are closely packed. In the latter case, when the solid melts the atoms have larger average distance from their neighbors in the liquid than in the solid due to the increased kinetic energy, leading to an increase in the volume. In the case of solids with open structures, melting actually reduces the average distance between atoms when the special bonds that keep them at fixed distances from each other in the solid disintegrate due to the increased kinetic energy in the liquid, and the volume decreases.

D.4 The Principles of Statistical Mechanics

Statistical mechanics is the theory that describes the behavior of macroscopic systems in terms of thermodynamic variables (like the entropy, volume, average number of particles, etc.), using as a starting point the microscopic structure of the physical system of interest. The difference between thermodynamics and statistical mechanics is that the first theory is based on empirical observations while the second theory is based on knowledge (true or assumed) of the microscopic constituents and their interactions. The similarity between the two theories is that they both address the macroscopic behavior of the system: thermodynamics does it by dealing exclusively with macroscopic quantities and using empirical laws, while statistical mechanics does it by constructing averages over all states consistent with the external conditions imposed on the system (like temperature, pressure, chemical potential, etc.). Thus, the central principle of statistical mechanics is to identify all possible states of the system in terms of their microscopic structure, and take an average of the physical quantities of interest over those states that are consistent with the external conditions. The average must involve the proper weight for each state, which is related to the likelihood of this state occurring, given the external conditions.

As in thermodynamics, statistical mechanics assumes that we are dealing with systems composed of a very large number of microscopic particles (typically atoms or molecules). The variables that determine the state of the system are the position, momentum, electric charge, and magnetic moment of the particles. All these are microscopic variables and their values determine the state of individual particles. A natural thing to consider within statistical mechanics is the average occupation number of microscopic states by particles. The space that corresponds to all the allowed values of these microscopic variables is called the *phase space* of the system. A central postulate of statistical mechanics is that all relevant portions of phase must be sampled properly in the average. This is formalized through the notion of *ergodicity*: a sampling procedure is called ergodic if it does not exclude in principle any state of the system that is consistent with the imposed external conditions. There exists a theorem, called Poincaré's theorem, which says that given enough time a system will come arbitrarily close to any of the states consistent with the external conditions. Sampling the phase space of a system composed of many particles is exceedingly difficult; while Poincaré's theorem assures us that a system will visit all the states that are consistent with the imposed external conditions, it will take a huge amount of time to sample all the relevant states by evolving the system in a causal manner between states. To circumvent this difficulty, the idea of *ensembles* was developed, which makes calculations feasible in the context of statistical mechanics. In the next section we develop the notions of average occupation numbers and of different types of ensembles, and give some elementary examples of how these notions are applied to simple systems.

D.5 Average Occupation Numbers

The average occupation numbers can be obtained in the simplest manner by making the assumption that the system exists in its most probable state, that is, the state with the highest probability of occurring given the external conditions. We define our physical system as consisting of particles which can exist in a (possibly infinite) number of microscopic states labeled by their energy, ϵ_i. The energy of the microscopic states is bounded from below but not necessarily bounded from above. If there are n_i particles in the microscopic state i, then the total number N of particles in the system and its total energy E will be given by:

$$N = \sum_i n_i, \quad E = \sum_i n_i \epsilon_i \tag{D.31}$$

We will denote by $\{n_i\}$ the distribution of particles that consists of a particular set of values n_i that determine the occupation of the levels ϵ_i. The number of states of the entire system corresponding to a particular distribution $\{n_i\}$ will be denoted as $W(\{n_i\})$. This number is proportional to the volume in phase space occupied by this particular distribution $\{n_i\}$. The most probable distribution must correspond to the largest volume in phase space. That is, if we denote the most probable distribution of particles in microscopic states by $\{\bar{f}_i\}$, then $W(\{\bar{f}_i\})$ is the maximum value of W. Let us suppose that the degeneracy or multiplicity of the microscopic state labeled i is g_i, that is, there are g_i individual microscopic states with the same energy ϵ_i. With these definitions, and with the restriction of constant N and E, we can now derive the average occupation of level i, which will be the same as $\bar{f}_i$. There are three possibilities, which we examine separately.

D.5.1 Classical Maxwell–Boltzmann Statistics

In the case of classical distinguishable particles, there are $g_i^{n_i}$ ways to put n_i particles in the same level i, and there are

$$W(\{n_i\}) = \frac{N!}{n_1!\,n_2!\cdots n_i!\cdots}n_1^{n_1}n_2^{n_2}\cdots g_i^{n_i}\cdots$$

ways of arranging the particles in the levels with energy ϵ_i. The ratio of factorials gives the number of permutations for putting n_i particles in level i, since the particles are distinguishable (it does not matter which n_i of the N particles were put in level i). Since we are dealing with large numbers of particles, we will use the following approximation:

$$\textbf{Stirling's formula}: \quad \ln(N!) = N\ln N - N \tag{D.32}$$

which is very accurate for large N. With the help of this, we can now try to find the maximum of $W(\{n_i\})$ by considering variations in n_i, under the constraints of Eq. (D.31). It is actually more convenient to find the maximum of $\ln W$, which will give the maximum of W since W is a positive quantity ≥ 1. With Stirling's formula, $\ln W$ takes the form

$$\ln W = N\ln N - N + \sum_i n_i\ln g_i - \sum_i (n_i\ln n_i - n_i) = N\ln N + \sum_i n_i(\ln g_i - \ln n_i)$$

We include the constraints through the Lagrange multipliers α and β, and perform the variation of W with respect to n_i to obtain:

$$0 = \delta\ln W - \alpha\delta N - \beta\delta E \Rightarrow$$

$$0 = \delta\left[N\ln N + \sum_i n_i(\ln g_i - \ln n_i)\right] - \alpha\delta\sum_i n_i - \beta\delta\sum_i n_i\epsilon_i \Rightarrow$$

$$0 = \sum_i \delta n_i\left[\ln g_i - \ln n_i - \alpha - \beta\epsilon_i + \ln N\right] \tag{D.33}$$

where we have used $N = \sum_i n_i \Rightarrow \delta N = \sum_i \delta n_i$. Since Eq. (D.33) must hold for arbitrary variations δn_i, and it applies to the maximum value of W which is obtained for the distribution $\{\bar{f}_i\}$, we conclude that:

$$0 = \left[\ln g_i - \ln \bar{f}_i - \gamma - \beta\epsilon_i\right] \Rightarrow \bar{f}_i^{MB} = g_i e^{-\gamma}e^{-\beta\epsilon_i} \tag{D.34}$$

where we have defined $\gamma = \alpha - \ln N$. Thus, we have derived the Maxwell–Boltzmann distribution, in which the average occupation number of level i with energy ϵ_i is proportional to $\exp[-\beta\epsilon_i]$. All that remains to do in order to have the exact distribution is to determine the values of the constants that appear in Eq. (D.34). Recall that these constants were introduced as the Lagrange multipliers that take care of the constraints of constant number of particles and constant energy. The constant β must have the dimensions of inverse energy. The only other energy scale in the system is the temperature, so we conclude that $\beta = 1/k_B T$. It is actually possible to show that this must be the value of β through a much more elaborate argument based on the Botzmann transport equation (see, for example, the book by Huang, mentioned in Further Reading). The other constant, γ, is obtained by normalization, that is, by requiring that when we sum $\{f_i\}$ over all the values of the index i we obtain the total number of particles in the system. This can only be done explicitly for specific systems where we can evaluate g_i and ϵ_i.

Example: Consider the case of the classical ideal gas which consists of particles of mass m, with the only interaction between the particles being binary hard-sphere collisions. The particles are contained in a volume V and the gas has density $n = N/V$. In this case, the energy ϵ_i of a particle with momentum $\mathbf{p}$ and position $\mathbf{r}$, and its multiplicity g_i, are given by:

$$\epsilon_i = \frac{\mathbf{p}^2}{2m}, \quad g_i = \mathrm{d}\mathbf{r}\,\mathrm{d}\mathbf{p}$$

Then, summation over all the values of the index i gives:

$$N = \sum_i n_i = \int \mathrm{d}\mathbf{r} \int e^{-\gamma} e^{-\beta\mathbf{p}^2/2m} \mathrm{d}\mathbf{p} = V e^{-\gamma} (2\pi m/\beta)^{\frac{3}{2}}$$

$$\Rightarrow e^{-\gamma} = \frac{n}{(2\pi m/\beta)^{\frac{3}{2}}} \tag{D.35}$$

which completely specifies the average occupation number for the classical ideal gas. With this we can calculate the total energy of the system as:

$$E = \sum_i n_i \epsilon_i = \frac{n}{(2\pi m/\beta)^{\frac{3}{2}}} \int \mathrm{d}\mathbf{r} \int \frac{\mathbf{p}^2}{2m} e^{-\beta\mathbf{p}^2/2m} \mathrm{d}\mathbf{p}$$

$$= V \frac{n}{(2\pi m/\beta)^{\frac{3}{2}}} \left[-\frac{\partial}{\partial\beta} \int e^{-\beta\mathbf{p}^2/2m} \mathrm{d}\mathbf{p} \right] = \frac{3}{2}\frac{N}{\beta} \tag{D.36}$$

D.5.2 Quantum Fermi–Dirac Statistics

In the case of quantum-mechanical particles that obey Fermi–Dirac statistics, each level i has occupation 0 or 1, so the number of particles that can be accommodated in a level of energy ϵ_i is $n_i \leq g_i$, where g_i is the degeneracy of the level due to the existence of additional good quantum numbers. Since the particles are indistinguishable, there are

$$\frac{g_i!}{n_i!\,(g_i - n_i)!}$$

ways of distributing n_i particles in the g_i states of level i. The total number of ways of distributing N particles in all the levels is then given by:

$$W(\{n_i\}) = \prod_i \frac{g_i!}{n_i! \, (g_i - n_i)!} \Rightarrow$$

$$\ln W(\{n_i\}) = \sum_i \left[g_i(\ln g_i - 1) - n_i(\ln n_i - 1) - (g_i - n_i)(\ln(g_i - n_i) - 1) \right]$$

where we have used Stirling's formula for $\ln(g_i!)$, $\ln(n_i!)$, $\ln(g_i - n_i!)$. Using the same variational argument as before, in terms of δn_i, we obtain for the most probable distribution $\{\bar{f_i}\}$, which corresponds to the maximum of W:

$$\bar{f}_i^{FD} = g_i \frac{1}{e^{\beta(\epsilon_i - \mu)} + 1} \tag{D.37}$$

where $\mu = -\alpha/\beta$. The constants are fixed by normalization, that is, summation of n_i over all values of i gives N and summation of $n_i \epsilon_i$ gives E. This is the Fermi–Dirac distribution, with μ the highest energy level which is occupied by particles at zero temperature, called the Fermi energy; μ is also referred to as the chemical potential, since its value is related to the total number of particles in the system. In Chapter 1 we calculated the Fermi energy, which is the same quantity as μ, for a system of N non-interacting fermions in volume V, with $g_i = 2$ (spin 1/2), in terms of the density $n = N/V$, Eq. (1.7).

D.5.3 Quantum Bose–Einstein Statistics

In the case of quantum-mechanical particles that obey Bose statistics, any number of particles can be at each of the g_i states associated with level i. We can think of the g_i states as identical boxes in which we place a total of n_i particles, in which case the system is equivalent to one consisting of $n_i + g_i$ objects which can be arranged in a total of

$$\frac{(n_i + g_i)!}{n_i! \, g_i!}$$

ways, since both boxes and particles can be interchanged among themselves arbitrarily. This gives for the total number of states $W(\{n_i\})$ associated with a particular distribution $\{n_i\}$ in the levels with energies ϵ_i:

$$W(\{n_i\}) = \prod_i \frac{n_i + g_i!}{n_i! \, (g_i)!} \Rightarrow$$

$$\ln W(\{n_i\}) = \sum_i \left[(n_i + g_i)(\ln(n_i + g_i) - 1) - n_i(\ln n_i - 1) - g_i(\ln g_i - 1) \right]$$

where again we have used Stirling's formula for $\ln(g_i!)$, $\ln(n_i!)$, $\ln(n_i + g_i!)$. Through a similar variational argument as in the previous two cases, we obtain for the most probable distribution $\{\bar{f_i}\}$ which corresponds to the maximum of W:

$$\bar{f}_i^{BE} = g_i \frac{1}{e^{\beta(\epsilon_i - \mu)} - 1} \tag{D.38}$$

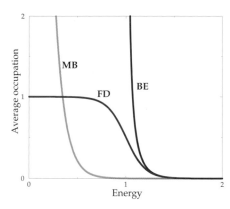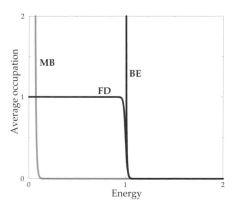

Fig. D.3   Average occupation numbers in the Maxwell–Boltzmann (MB), Fermi–Dirac (FD), and Bose–Einstein (BE) distributions, for $\beta = 10$ (left) and $\beta = 100$ (right), on an energy scale in which $\mu = 1$ for the FD and BE distributions. The factor $n/(2\pi m)^{3/2}$ in the MB distribution of the ideal gas is set to be $= 1$.

where $\mu = -\alpha/\beta$. The constants are once more fixed by the usual normalization conditions, as in the previous two cases. In this case, μ is the value of energy just below the lowest occupied level: the values of ϵ_i of occupied states must be above μ at finite temperature.

The three distributions are compared in Fig. D.3 for two different values of the temperature ($\beta = 10$ and 100), on an energy scale in which $\mu = 1$. The multiplicities g_i for the Fermi–Dirac (FD) and Bose–Einstein (BE) distributions are taken to be 1. For the Maxwell–Boltzmann (MB) distribution we use the expression derived for the ideal gas, with the factor $n/(2\pi m)^{3/2} = 1$. As these plots show, the FD distribution approaches a step function, that is, a function equal to 1 for $\epsilon < \mu$ and 0 for $\epsilon > \mu$, when the temperature is very low (β very large). For finite temperature T, the FD distribution has a width of order $1/\beta = k_{\mathrm{B}}T$ around μ over which it decays from 1 to 0. The BE distribution also shows interesting behavior close to μ: as the temperature decreases, f changes very sharply as ϵ approaches μ from above, becoming infinite near μ. This is suggestive of a phase transition, in which all particles occupy the lowest allowed energy level $\epsilon = \mu$ for sufficiently low temperature. Indeed, in systems composed of particles obeying Bose statistics, there is a phase transition at low temperature in which all particles collapse to the lowest energy level. This is known as the Bose–Einstein condensation and has received considerable attention recently because it has been demonstrated in a system consisting of trapped atoms.

D.6 Ensemble Theory

As mentioned earlier, constructing an average of a physical quantity (like the energy of the system) is very difficult if the system has many particles, and we insist on evolving the system between states until we have included enough states to obtain an accurate estimate.

For this reason, the idea of an ensemble of states has been developed: an ensemble is a set of states of the system which are consistent with the imposed external conditions. These states do not have to be causally connected but can be at very different regions of phase space; as long as they satisfy the external conditions they are admissible as members of the ensemble. Taking an average of a physical quantity over the members of the ensemble is a more efficient way of obtaining accurate averages: with a sample consisting of N images in an ensemble it is possible to sample disparate regions of phase space, while a sample based on the same number N of states obtained by causal evolution from a given state is likely to be restricted to a much smaller region of the relevant phase space. A key notion of ensemble theory is that the selection of the images of the system included in the ensemble is not biased in any way, known formally as the *postulate of equal a priori probabilities*.

D.6.1　Definition of Ensembles

We consider systems which consist of a large number N of particles and whose states are described in terms of the microscopic variables ($\{s\}$) and the total volume V that the system occupies. These variables can, for example, be the $3N$ positions ($\{q\}$) and the $3N$ momenta ($\{p\}$) of the particles in 3D space, or the N values of the spin or the dipole moment of a system of particles in an external magnetic field. The energy E of a state of the system is the value that the hamiltonian $\hat{\mathcal{H}}_N(\{s\})$ of the N-particle system takes for the values of ($\{s\}$) that correspond to that particular state.

An ensemble is defined by the density of states ρ included in it. In principle, all states that satisfy the imposed external conditions are considered to be members of the ensemble. There are three types of ensembles, the microcanonical, canonical, and grand-canonical ensemble. The precise definition of the density ρ for each ensemble depends on the classical or quantum-mechanical nature of the system.

Classical Case

Microcanonical ensemble. In the microcanonical ensemble, the density of states is defined as:

$$\rho(\{s\}) = \frac{1}{Q} \quad \text{for } E \leq \hat{\mathcal{H}}_N(\{s\}) \leq E + \delta E$$
$$= 0 \quad \text{otherwise} \tag{D.39}$$

with δE an infinitesimal quantity on the scale of E. Thus, in the microcanonical ensemble the energy of the system is fixed within an infinitesimal range. We assume that in this ensemble the total number of particles N and the total volume V of the system are also fixed. The value of Q is chosen so that summation over all the values of $\rho(\{s\})$ for all the allowed states $\{s\}$ gives 1. For example, if the system consists of N indistinguishable particles there are $N!$ equivalent ways of arranging them in a given configuration $\{s\}$, so the weight of a particular configuration must be $1/Q = 1/N!$. This counting of states has its origin in quantum-mechanical considerations: the wavefunction of the system contains a factor of $1/\sqrt{N!}$, and any average involves expectation values of an operator within the system wavefunction, which produces a factor $1/N!$ for every configuration. Without this factor we would be led to inconsistencies. Similarly, if the system is composed of n types of indistinguishable particles, then the relevant factor would be $1/(N_1! N_2! \cdots N_n!)$, where N_i is the number of indistinguishable particles of type i ($i = 1, 2, \ldots, n$). If the particles are described by their positions and momenta in 3D space, that is, the state of the system is specified by $3N$ position values ($\{q\}$) and $3N$ momentum values ($\{p\}$), an additional factor is needed to cancel the dimensions of the integrals over positions and momenta when averages are taken. This factor is taken to be h^{3N}, where h is Planck's constant. This choice is a consequence of the fact that the elementary volume in position–momentum space is h, by Heisenberg's uncertainty relation (see Appendix C). Again, the proper normalization comes from quantum-mechanical considerations.

Canonical ensemble. In the canonical ensemble the total volume V and total number of particles N in the system are fixed, but the energy E is allowed to vary. The system is considered to be in contact with an external reservoir with temperature T. The density of states in the canonical ensemble is:

$$\rho(\{s\}) = \frac{1}{Q} \exp\left[-\beta \hat{\mathcal{H}}_N(\{s\})\right] \qquad (D.40)$$

where $\beta = 1/k_B T$ with T the externally imposed temperature. The factor $1/Q$ serves the purpose of normalization, as in the case of the microcanonical ensemble. For a system of N identical particles $Q = N!$, as in the microcanonical ensemble.

Grand-canonical ensemble. In the grand-canonical ensemble we allow for fluctuations in the volume and in the total number of particles in the system, as well as in the energy. The system is considered to be in contact with an external reservoir with temperature T and chemical potential μ, and under external pressure P. The density of states in the grand-canonical ensemble is:

$$\rho(\{s\}, N, V) = \frac{1}{Q} \exp\left[-\beta(\hat{\mathcal{H}}_N(\{s\}) - \mu N + PV)\right] \qquad (D.41)$$

with Q the same factor as in the canonical ensemble.

It is a straightforward exercise to show that even though the energy varies in the canonical ensemble, its variations around the average are extremely small for large enough systems. Specifically, if $\langle\hat{\mathcal{H}}\rangle$ is the average energy and $\langle\hat{\mathcal{H}}^2\rangle$ is the average of the energy square, then:

$$\frac{\langle\hat{\mathcal{H}}\rangle^2 - \langle\hat{\mathcal{H}}^2\rangle}{\langle\hat{\mathcal{H}}\rangle^2} \sim \frac{1}{N} \qquad (D.42)$$

which shows that for large systems the deviation of the energy from its average value is negligible. Similarly, in the grand-canonical ensemble the deviation of the number of particles from its average is negligible. Thus, in the limit $N \rightarrow \infty$ the three types of ensembles are equivalent.

Quantum-Mechanical Case

Microcanonical ensemble. We consider a system with N particles and volume V, both fixed. For this system, the quantum-mechanical microcanonical ensemble is defined through the wavefunctions of the states of the system whose energies lie in a narrow range δE which is much smaller than the energy E, by analogy to the classical case:

$$\rho(\{s\}) = \sum_{K=1}^{N} |\Psi_K(\{s\})\rangle\langle\Psi_K(\{s\})| \quad \text{for} \ \ E \leq \langle\Psi_K|\hat{\mathcal{H}}|\Psi_K\rangle \leq E + \delta E \tag{D.43}$$

where the wavefunctions $|\Psi_K(\{s\})\rangle$ represent the quantum-mechanical states of the system with energy $\langle\psi_K(\{s\})|\hat{\mathcal{H}}(\{s\})|\psi_K(\{s\})\rangle$ in the range $[E, E + \delta E]$; the set of variables $\{s\}$ includes all the relevant degrees of freedom in the system which enter in the description of the hamiltonian $\hat{\mathcal{H}}$ and the wavefunctions (to simplify the notation we will not include these variables in association with the hamiltonian and the wavefunctions in most expressions of what follows).

The thermodynamic average of an operator $\hat{\mathcal{O}}$ is given by the average of its expectation value over all the states in the ensemble, properly normalized by the sum of the norms of these states:

$$\overline{\hat{\mathcal{O}}} = \sum_{K=1}^{N} \langle\Psi_K|\hat{\mathcal{O}}|\Psi_K\rangle \left[\sum_{K=1}^{N} \langle\Psi_K|\Psi_K\rangle \right]^{-1} \tag{D.44}$$

If we assume that the wavefunctions $|\Psi_K\rangle$ are properly normalized, $\langle\Psi_K|\Psi_K\rangle = 1$, then the normalization factor that enters into the average becomes:

$$\left[\sum_{K=1}^{N} \langle\Psi_K|\Psi_K\rangle \right]^{-1} = \frac{1}{N}$$

The form of the normalization factor used in Eq. (D.44) is more general. In the definition of the quantum ensembles we will not need to introduce factors of $1/Q$ as we did for the classical ensembles, because these factors come out naturally when we include in the averages the normalization factor mentioned above.

Inserting complete sets of states $\sum_I |\Phi_I\rangle\langle\Phi_I|$ between the operator and the wavefunctions $\langle\Psi_K|$ and $|\Psi_K\rangle$, we obtain:

$$\overline{\hat{\mathcal{O}}} = \sum_{K=1}^{N} \sum_{IJ} \langle\Psi_K|\Phi_I\rangle\langle\Phi_I|\hat{\mathcal{O}}|\Phi_J\rangle\langle\Phi_J|\Psi_K\rangle \left[\sum_{K=1}^{N} \langle\Psi_K|\Psi_K\rangle \right]^{-1}$$

which can be rewritten in terms of matrix notation, with the following definitions of matrix elements:

$$\hat{\mathcal{O}}_{IJ} = \langle\Phi_I|\hat{\mathcal{O}}|\Phi_J\rangle$$

$$\rho_{JI} = \langle\Phi_J| \left[\sum_{K=1}^{N} |\Psi_K\rangle\langle\Psi_K| \right] |\Phi_I\rangle \tag{D.45}$$

In terms of these matrix elements the sum of the norms of the states $|\Psi_K\rangle$ becomes:

$$\sum_{K=1}^{N}\langle\Psi_K|\Psi_K\rangle = \sum_{K=1}^{N}\langle\Psi_K|\left[\sum_I|\Phi_I\rangle\langle\Phi_I|\right]|\Psi_K\rangle$$

$$= \sum_I\langle\Phi_I|\left[\sum_{K=1}^{N}|\Psi_K\rangle\langle\Psi_K|\right]|\Phi_I\rangle = \sum_I\rho_{II} = \mathrm{Tr}[\rho]\qquad(D.46)$$

where Tr denotes the trace, that is, the sum of diagonal matrix elements. With this expression the thermodynamic average of the operator $\hat{O}$ takes the form

$$\overline{\hat{O}} = \frac{1}{\mathrm{Tr}[\rho]}\sum_{K=1}^{N}\sum_{IJ}\langle\Psi_K|\Phi_I\rangle\hat{O}_{IJ}\langle\Phi_J|\Psi_K\rangle$$

$$= \frac{1}{\mathrm{Tr}[\rho]}\sum_{IJ}\hat{O}_{IJ}\langle\Phi_J|\left[\sum_{K=1}^{N}|\Psi_K\rangle\langle\Psi_K|\right]|\Phi_I\rangle$$

$$= \frac{1}{\mathrm{Tr}[\rho]}\sum_J\left[\sum_I\rho_{JI}\hat{O}_{IJ}\right] = \frac{1}{\mathrm{Tr}[\rho]}\sum_J[\rho\hat{O}]_{JJ} = \frac{\mathrm{Tr}[\rho\hat{O}]}{\mathrm{Tr}[\rho]}\qquad(D.47)$$

Notice that this last expression is general, and depends on the states included in the ensemble only through the definition of the density matrix, Eq. (D.45). The new element that was introduced in the derivation of this general expression was the complete state of states $|\Phi_I\rangle$, which can always be found for a given quantum-mechanical system. In most situations it is convenient to choose these states to be the energy eigenstates of the system. Moreover, assuming that the energy interval δE is smaller than the spacing between energy eigenvalues, the states $|\Psi_K\rangle$ are themselves energy eigenstates. In this case we can identify the set of states $|\Phi_I\rangle$ with the set of states $|\Psi_K\rangle$, since the two sets span the same Hilbert space and are therefore related to each other by at most a rotation. Having made this identification, we see that the matrix elements of the density operator become diagonal:

$$\rho_{JI} = \langle\Psi_J|\left[\sum_{K=1}^{N}|\Psi_K\rangle\langle\Psi_K|\right]|\Psi_I\rangle = \sum_{K=1}^{N}\delta_{JK}\delta_{KI} = \delta_{JI}\qquad(D.48)$$

if the set of states $|\Psi_K\rangle$ is orthonormal. This choice of a complete set of states also allows straightforward definitions of the canonical and grand-canonical ensembles, by analogy to the classical case.

Canonical ensemble. We consider a system with N particles and volume V, both fixed, at a temperature T. The quantum canonical ensemble for this system in the energy representation, that is, using as basis a complete set of states $|\Phi_I\rangle$ which are eigenfunctions of the hamiltonian $\hat{\mathcal{H}}$, is defined through the density:

$$\rho(\{s\}) = \sum_I|\Phi_I\rangle e^{-\beta E_I(\{s\})}\langle\Phi_I| = \sum_I|\Phi_I\rangle e^{-\beta\hat{\mathcal{H}}(\{s\})}\langle\Phi_I| = e^{-\beta\hat{\mathcal{H}}(\{s\})}\qquad(D.49)$$

where we have taken advantage of the relations

$$\hat{\mathcal{H}}|\Phi_I\rangle = E_I|\Phi_I\rangle,\qquad \sum_I|\Phi_I\rangle\langle\Phi_I| = 1$$

that hold for this complete set of energy eigenstates. Then, the average of a quantum-mechanical operator $\hat{O}$ in this ensemble is calculated as:

$$\bar{\hat{O}} = \frac{\mathrm{Tr}[\rho\hat{O}]}{\mathrm{Tr}[\rho]} = \frac{\mathrm{Tr}[e^{-\beta\hat{\mathcal{H}}}\hat{O}]}{\mathrm{Tr}[e^{-\beta\hat{\mathcal{H}}}]} \tag{D.50}$$

Grand-canonical ensemble. Finally, for the quantum grand-canonical ensemble we define the density by analogy to the classical case as:

$$\rho(\{s\}, N, V) = e^{-\beta(\hat{\mathcal{H}}(\{s\}) - \mu N + PV)} \tag{D.51}$$

where μ is the chemical potential and P the pressure, both determined by the reservoir with which the system under consideration is in contact. In this case the number of particles in the system N and its volume V are allowed to fluctuate, and their average values are determined by the chemical potential and pressure imposed by the reservoir; for simplicity, we consider only fluctuations in the number of particles, the extension to volume fluctuations being straightforward. In principle we should define an operator that counts the number of particles in each quantum-mechanical state and use it in the expression for the density $\rho(\{s\}, N, V)$. However, matrix elements of either the hamiltonian or the particle number operators between states with different numbers of particles vanish identically, so in effect we can take N in the above expression to be the number of particles in the system and then sum over all values of N when we take averages. Therefore, the proper definition of the average of the operator $\hat{O}$ in this ensemble is given by:

$$\bar{\hat{O}} = \frac{\mathrm{Tr}[\rho\hat{O}]}{\mathrm{Tr}[\rho]} = \frac{\sum_{N=0}^{\infty} \mathrm{Tr}[e^{-\beta\hat{\mathcal{H}}}\hat{O}]_N e^{\beta\mu N}}{\sum_{N=0}^{\infty} \mathrm{Tr}[e^{-\beta\hat{\mathcal{H}}}]_N e^{\beta\mu N}} \tag{D.52}$$

where the traces inside the summations over N are taken for a fixed value of N, as indicated by the subscripts.

D.6.2 Derivation of Thermodynamics

The average values of physical quantities calculated through ensemble theory should obey the laws of thermodynamics. In order to derive thermodynamics from ensemble theory we have to make the proper identification between the usual thermodynamic variables and quantities that can be obtained directly from the ensemble. In the microcanonical ensemble the basic equation that relates thermodynamic variables to ensemble quantities is:

$$S(E, V) = k_B \ln \Omega(E, V) \tag{D.53}$$

where S is the entropy and $\Omega(E, V)$ is the number of microscopic states with energy E and volume V. This definition of the entropy is the only possible one that satisfies all the requirements, such as the extensive nature of S with E and V and the second and third laws of thermodynamics. Having defined the entropy, we can further define the temperature T and the pressure P as:

$$\frac{1}{T} = \left(\frac{\partial S(E, V)}{\partial E}\right)_V, \quad P = T\left(\frac{\partial S(E, V)}{\partial V}\right)_E$$

and through those we can calculate all the other quantities of interest. Although it is simple, the microcanonical ensemble is not very useful because it is rather difficult to actually calculate $\Omega(E, V)$, as Eq. (D.53) requires. The other two ensembles are much more convenient for calculations, so we concentrate our attention on those.

For calculations within the canonical ensemble, it is convenient to introduce the partition function

$$Z_N(V, T) = \sum_{\{s\}} \rho(\{s\}) = \sum_{\{s\}} \frac{1}{Q} e^{-\beta \hat{\mathcal{H}}_N(\{s\})} \tag{D.54}$$

which is simply a sum over all values of the density $\rho(\{s\})$. We have used the subscript N, the total number of particles in the system (which is fixed), to denote that we are in the canonical ensemble. The partition function in the case of a system of N indistinguishable particles with momenta $\{p\}$ and coordinates $\{q\}$ takes the form

$$Z_N(V, T) = \frac{1}{N! \, h^{3N}} \int d\{q\} \int d\{p\} e^{-\beta \hat{\mathcal{H}}_N(\{p\},\{q\})} \tag{D.55}$$

Using the partition function, which embodies the density of states relevant to the canonical ensemble, we define the free energy

$$F_N(V, T) = -k_B T \ln Z_N(V, T) \tag{D.56}$$

and through it the entropy and pressure:

$$S_N = -\left(\frac{\partial F_N}{\partial T}\right)_V, \quad P_N = -\left(\frac{\partial F_N}{\partial V}\right)_T$$

from which we can obtain all other thermodynamic quantities of interest.

For the grand-canonical ensemble we define the grand partition function as:

$$Z(\mu, V, T) = \sum_{\{s\},N} \frac{1}{Q} \rho(\{s\}, N, V) e^{\beta PV} = \sum_{N=0}^{\infty} e^{\beta \mu N} \frac{1}{Q} \sum_{\{s\}} e^{-\beta \hat{\mathcal{H}}_N(\{s\})} \tag{D.57}$$

A more convenient representation of the grand partition function is based on the introduction of the variable z, called the fugacity:

$$z = e^{\beta \mu}$$

With this, the grand partition becomes the form

$$Z(z, V, T) = \sum_{N=0}^{\infty} z^N Z_N(V, T) \tag{D.58}$$

where we have also used the partition function in the canonical ensemble $Z_N(V, T)$ that we defined earlier. In the case of a system consisting of N indistinguishable particles with positions $\{q\}$ and momenta $\{p\}$, the grand partition function takes the form

$$Z(\mu, V, T) = \sum_{N=0}^{\infty} e^{\beta \mu N} \frac{1}{N! \, h^{3N}} \int d\{q\} \int d\{p\} e^{-\beta \hat{\mathcal{H}}_N(\{p\},\{q\})} \tag{D.59}$$

In terms of the grand partition function we can calculate the average energy $\bar{E}$ and the average number of particles in the system $\bar{N}$:

$$\bar{E}(z, V, T) = -\left(\frac{\partial \ln Z(z, V, T)}{\partial \beta}\right)_{z,V}, \quad \bar{N}(z, V, T) = z\left(\frac{\partial \ln Z(z, V, T)}{\partial z}\right)_{V,T} \tag{D.60}$$

We can then define the free energy

$$F(z, V, T) = -k_B T \ln\left[\frac{Z(z, V, T)}{z^{\bar{N}}}\right] \tag{D.61}$$

which can be considered either as a function of z or a function of $\bar{N}$, since these two variables are related by Eq. (D.60). From the free energy we can obtain all other thermodynamic quantities. For example, the pressure P, entropy S, and chemical potential μ are given as:

$$P = -\left(\frac{\partial F}{\partial V}\right)_{\bar{N},T}, \quad S = -\left(\frac{\partial F}{\partial T}\right)_{\bar{N},V}, \quad \mu = \left(\frac{\partial F}{\partial \bar{N}}\right)_{V,T}$$

Finally, we find from the definition of the grand partition function and the fact that $\rho(\{s\}, N, V)$ is normalized to unity that:

$$\frac{PV}{k_B T} = \ln[Z(z, V, T)] \tag{D.62}$$

which determines the equation of state of the system.

The above expressions hold also for the quantum-mechanical ensembles. The only difference is that the partition function in this case is defined as the trace of the density matrix. For example, in the quantum canonical ensemble the partition function becomes

$$Z_N(V, T) = \mathrm{Tr}[\rho(\{s\})] = \sum_I \rho_{II} = \sum_I \langle \Phi_I | e^{-\beta\hat{\mathcal{H}}(\{s\})} | \Phi_I \rangle = \sum_I e^{-\beta E_I(\{s\})}$$

where we have assumed a complete set of energy eigenstates $|\Phi_I\rangle$ as the basis for calculating the density matrix elements. This form of the partition function is essentially the same as in the classical case, except for the normalization factors which will enter automatically in the averages from the proper definition of the normalized wavefunctions $|\Phi_I\rangle$. Similarly, the partition function for the quantum grand-canonical ensemble will have the same form as in Eq. (D.58), with $Z_N(V, T)$ the partition function of the quantum canonical system with N particles as defined above, and $z = \exp[\beta\mu]$ the fugacity.

D.7 Applications of Ensemble Theory

D.7.1 Equipartition and the Virial

A general result that is a direct consequence of ensemble theory is the equipartition theorem, which we state here without proof (for a proof see the book by Huang, mentioned in Further Reading). For a system consisting of particles with coordinates $q_i, i = 1, 2, \ldots, 3N$ and momenta $p_i, i = 1, 2, \ldots, 3N$, the equipartition theorem is:

$$\langle p_i \frac{\partial \hat{\mathcal{H}}}{\partial p_i} \rangle = \langle q_i \frac{\partial \hat{\mathcal{H}}}{\partial q_i} \rangle = k_B T \tag{D.63}$$

A direct consequence of the equipartition theorem is the following. Using the hamiltonian equations of motion for such a system:

$$\frac{\partial p_i}{\partial t} = -\frac{\partial \hat{\mathcal{H}}}{\partial q_i}$$

we obtain the expression

$$\mathcal{V} = \langle \sum_{i=1}^{3N} q_i \frac{\partial p_i}{\partial t} \rangle = -\langle \sum_{i=1}^{3N} q_i \frac{\partial \hat{\mathcal{H}}}{\partial q_i} \rangle = -3N k_B T \tag{D.64}$$

where $\mathcal{V}$ is known as the "virial." These general relations are useful in checking the behavior of complex systems in a simple way. For example, for a hamiltonian which is quadratic in the degrees of freedom, like that of harmonic oscillators with spring constants κ:

$$\hat{\mathcal{H}}_{HO} = \sum_{i=1}^{3N} \frac{p_i^2}{2m} + \sum_{i=1}^{3N} \frac{1}{2} \kappa q_i^2$$

there is a thermal energy equal to $k_B T/2$ associated with each harmonic degree of freedom, because from the equipartition theorem we have:

$$\langle q_i \frac{\partial \hat{\mathcal{H}}}{\partial q_i} \rangle = \langle \kappa q_i^2 \rangle = k_B T \Rightarrow \langle \frac{1}{2} \kappa q_i^2 \rangle = \frac{1}{2} k_B T$$

and similarly for the momentum variables.

D.7.2 Ideal Gases

The ideal gas is defined as consisting of N particles confined in a volume V, with the only interaction between particles being binary elastic collisions, that is, collisions which preserve energy and momentum and take place at a single point in space when two particles meet. This model is a reasonable approximation for dilute systems of atoms or molecules which interact very weakly. Depending on the nature of the particles that compose the ideal gas, we distinguish the classical and quantum-mechanical cases. Of course, all atoms or molecules should ultimately be treated quantum mechanically, but at sufficiently high temperatures that their quantum-mechanical nature is not apparent. Therefore, the classical ideal gas is really the limit of quantum-mechanical ideal gases at high temperature.

Classical Ideal Gas

We begin with the classical ideal gas, in which any quantum-mechanical features of the constituent particles are neglected. This is actually one of the few systems that can be treated in the microcanonical ensemble. In order to obtain the thermodynamics of the classical ideal gas in the microcanonical ensemble, we need to calculate the entropy from the total number of states $\Omega(E, V)$ with energy E and volume V, according to Eq. (D.53). The quantity $\Omega(E, V)$ is given by:

$$\Omega(E, V) = \frac{1}{N!\, h^{3N}} \int_{E \leq \hat{\mathcal{H}}(\{p\},\{q\}) \leq E+\delta E} \mathrm{d}\{p\}\mathrm{d}\{q\}$$

However, it is more convenient to use the quantity

$$\mathcal{N}(E, V) = \frac{1}{N!\, h^{3N}} \int_{0 \leq \hat{\mathcal{H}}(\{p\},\{q\}) \leq E} \mathrm{d}\{p\}\mathrm{d}\{q\}$$

in the calculation of the entropy. It can be shown that this gives a difference in the value of the entropy of order $\ln N$, a quantity negligible relative to N, to which the entropy is proportional as an extensive variable. In the above expression we have taken the value of 0 to be the lower bound of the energy spectrum.

$\mathcal{N}(E, V)$ is much easier to calculate: each integral over a position variable $\mathbf{q}$ gives a factor of V, while each variable $\mathbf{p}$ ranges in magnitude from a minimum value of 0 to a maximum value of $\sqrt{2mE}$, with E the total energy and m the mass of the particles. Thus, the integration over $\{q\}$ gives a factor V^N, while the integration over $\{p\}$ gives the volume of a sphere in $3N$ dimensions with radius $\sqrt{2mE}$. This volume in momentum space is given by:

$$\Pi_{3N} = \frac{\pi^{3N/2}}{(\frac{3N}{2} + 1)!}(2mE)^{3N/2}$$

so that the total number of states with energy up to E becomes:

$$\mathcal{N}(E, V) = \frac{1}{N!\, h^{3N}} \Pi_{3N} V^N$$

and using Stirling's formula we obtain, for the entropy:

$$S(E, V) = Nk_\mathrm{B} \ln\left[\frac{V}{N}\left(\frac{E}{N}\right)^{3/2}\right] + \frac{3}{2}Nk_\mathrm{B}\left[\frac{5}{3} + \ln\left(\frac{4\pi m}{3h^2}\right)\right] \tag{D.65}$$

From this expression we can calculate the temperature and pressure as discussed earlier, obtaining:

$$\frac{1}{T} = \frac{\partial S(E, V)}{\partial E} \Rightarrow T = \frac{2}{3}\frac{E}{Nk_\mathrm{B}} \tag{D.66}$$

$$P = T\frac{\partial S(E, V)}{\partial V} \Rightarrow P = \frac{Nk_\mathrm{B}T}{V} \tag{D.67}$$

with the last expression being the familiar equation of state of the ideal gas.

An interesting application of the above results is the calculation of the entropy of mixing of two ideal gases of the same temperature and density. We first rewrite the entropy of the ideal gas in terms of the density n, the energy per particle u, and the constant entropy per particle s_0, defined through:

$$n = \frac{N}{V}, \quad u = \frac{E}{N}, \quad s_0 = \frac{3}{2}k_\mathrm{B}\left[\frac{5}{3} + \ln\left(\frac{4\pi m}{3h^2}\right)\right]$$

With these definitions the entropy becomes:

$$S = \frac{3}{2}Nk_B \ln(u) - Nk_B \ln(n) + Ns_0$$

Now consider two gases at the same density $n_1 = n_2 = n$ and temperature, which implies $u_1 = u_2 = u$, from the result we obtain above, Eq. (D.66). The first gas occupies volume V_1 and the second gas volume V_2. We assume that the particles in the two gases have the same mass m. If the gases are different[3] and we allow them to mix and occupy the total volume $V = V_1 + V_2$, the new density of each gas will be $n_1' = N_1/V \neq n, n_2' = N_2/V \neq n$, while their energy will not have changed after mixing, since the temperature remains constant (no work is done on or by either gas). Then the entropy of mixing ΔS, defined as the entropy of the system after mixing minus the entropy before, will be:

$$\Delta S = N_1 k_B \ln\left(1 + \frac{V_2}{V_1}\right) + N_2 k_B \ln\left(1 + \frac{V_1}{V_2}\right) > 0$$

If the gases were the same, then the density of the gas after mixing will be the same because:

$$n = \frac{N_1}{V_1} = \frac{N_2}{V_2} = \frac{N_1 + N_2}{V_1 + V_2}$$

and the temperature will remain the same as before, which implies that $\Delta S = 0$, that is, there is no entropy of mixing. This is the expected result since the state of the gas before and after mixing is exactly the same. We note that if we had not included the $1/N!$ factor in the definition of the ensemble density, this last result would be different, that is, the entropy of mixing two parts of the same gas at the same temperature and density would be positive, an unphysical result. This is known as the "Gibbs paradox," which led Gibbs to hypothesize the existence of the $1/N!$ factor in the ensemble density long before its quantum-mechanical origin was understood.

Finally, we calculate for future reference the canonical partition function of the N-particle classical ideal gas:

$$
\begin{aligned}
Z_N(\beta, V) &= \frac{1}{N! h^{3N}} \int d\mathbf{r}_1 \cdots d\mathbf{r}_N \int d\mathbf{p}_1 \cdots d\mathbf{p}_N \exp\left[-\beta \sum_i \frac{\mathbf{p}_i^2}{2m}\right] \\
&= \frac{1}{N! h^{3N}} V^N \prod_{i=1}^{N} \int d\mathbf{p}_i \exp\left[-\beta \frac{\mathbf{p}_i^2}{2m}\right] \\
&= \frac{1}{N!} V^N \left(\frac{m k_B T}{2\pi \hbar^2}\right)^{3N/2}
\end{aligned}
\tag{D.68}
$$

where we have used the relations $\mathbf{p} = \hbar \mathbf{k}$, with $\hbar = h/2\pi$, for the momentum and $\beta = 1/k_B T$ for the temperature.

---

[3] This can be achieved experimentally, for instance by the right-hand and left-hand isomers of the same molecule.

Quantum Ideal Gases

For the quantum-mechanical ideal gases we begin with the construction of a properly normalized wavefunction for N indistinguishable free particles of mass m. In this case the hamiltonian of the system consists of only the kinetic energy term:

$$\hat{\mathcal{H}}_N = -\sum_{i=1}^{N} \frac{\hbar^2 \nabla_{\mathbf{r}_i}^2}{2m}$$

The individual free particles are in plane-wave states with momenta $\mathbf{k}_1, \mathbf{k}_2, \ldots, \mathbf{k}_N$:

$$|\mathbf{k}_i\rangle = \frac{1}{\sqrt{V}} e^{i\mathbf{k}_i \cdot \mathbf{r}_i}$$

with V the volume of the system. A product over all such states is an eigenfunction of the hamiltonian

$$\langle \mathbf{r}_1, \ldots, \mathbf{r}_N | \mathbf{k}_1, \ldots, \mathbf{k}_N \rangle = \frac{1}{\sqrt{V^N}} e^{i(\mathbf{k}_1 \cdot \mathbf{r}_1 + \cdots + \mathbf{k}_N \cdot \mathbf{r}_N)}$$

with total energy given by:

$$\sum_{i=1}^{N} \frac{\hbar^2 \mathbf{k}_i^2}{2m} = \frac{\hbar^2 K^2}{2m} = E_K$$

where K is taken to be a positive number. However, a single such product is not a proper wavefunction for the system of indistinguishable particles, because it does not embody the requisite symmetries. The N-body wavefunction $|\Psi_K^{(N)}\rangle$ must include all possible permutations of the individual momenta, which are $N!$ in number:

$$|\Psi_K^{(N)}\rangle = \frac{1}{\sqrt{N!}} \sum_P s_P |P[\mathbf{k}_1, \ldots, \mathbf{k}_N]\rangle = \frac{1}{\sqrt{N!}} \sum_P s_P |P\mathbf{k}_1, \ldots, P\mathbf{k}_N\rangle \qquad \text{(D.69)}$$

where P denotes a particular permutation of the individual momenta and s_P the sign associated with it; $P\mathbf{k}_i$ is the value of the momentum of the particle i under permutation P, which must be equal to one of the other values of the individual momenta. For bosons, all signs $s_P = 1$, so that the wavefunction is symmetric with respect to interchange of any two particles. For fermions, the signs of odd permutations are -1, while those for even permutations are $+1$, so that the exchange of any two particles will produce an overall $-$ sign for the entire wavefunction.

As a first step in the analysis of the thermodynamic behavior of this system, we calculate the partition function in the canonical ensemble and in the position representation. A matrix element of the canonical density in the position representation is given by:

$$\langle \mathbf{r}_1, \ldots, \mathbf{r}_N | \rho | \mathbf{r}_1', \ldots, \mathbf{r}_N' \rangle = \langle \mathbf{r}_1, \ldots, \mathbf{r}_N | \sum_K |\Psi_K^{(N)}\rangle e^{-\beta E_K} \langle \Psi_K^{(N)} | \mathbf{r}_1', \ldots, \mathbf{r}_N' \rangle \qquad \text{(D.70)}$$

We will manipulate this matrix element to obtain a useful expression by performing the following steps, which we explain below:

$$[1] \quad \langle\{\mathbf{r}_i\}| \sum_K |\Psi_K^{(N)}\rangle e^{-\beta E_K} \langle\Psi_K^{(N)}|\{\mathbf{r}_i'\}\rangle =$$

$$[2] \quad \frac{1}{N!} \sum_K \sum_{PP'} s_P s_{P'} \langle\{\mathbf{r}_i\}|\{P\mathbf{k}_i\}\rangle e^{-\beta E_K} \langle\{P'\mathbf{k}_i\}|\{\mathbf{r}_i'\}\rangle =$$

$$[3] \quad \frac{1}{N!} \sum_P s_P \langle\{P\mathbf{r}_i\}| \sum_K \sum_{P'} s_{P'} |\{\mathbf{k}_i\}\rangle e^{-\beta E_K} \langle\{P'\mathbf{k}_i\}|\{\mathbf{r}_i'\}\rangle =$$

$$[4] \quad \frac{1}{(N!)^2} \sum_P s_P \langle\{P\mathbf{r}_i\}| \sum_{\mathbf{k}_1,\dots,\mathbf{k}_N} \sum_{P'} s_{P'} |\{\mathbf{k}_i\}\rangle e^{-\beta E_K} \langle\{P'\mathbf{k}_i\}|\{\mathbf{r}_i'\}\rangle =$$

$$[5] \quad \frac{1}{N!} \sum_P s_P \langle\{P\mathbf{r}_i\}| \sum_{\mathbf{k}_1,\dots,\mathbf{k}_N} |\{\mathbf{k}_i\}\rangle e^{-\beta E_K} \langle\{\mathbf{k}_i\}|\{\mathbf{r}_i'\}\rangle =$$

$$[6] \quad \frac{1}{N!} \sum_P s_P \sum_{\mathbf{k}_1,\dots,\mathbf{k}_N} \langle\{\mathbf{r}_i\}|\{P\mathbf{k}_i\}\rangle e^{-\beta E_K} \langle\{\mathbf{k}_i\}|\{\mathbf{r}_i'\}\rangle =$$

$$[7] \quad \frac{1}{N!} \sum_P s_P \sum_{\mathbf{k}_1,\dots,\mathbf{k}_N} e^{-\beta E_K} \prod_{n=1}^{N} \langle\mathbf{r}_n|P\mathbf{k}_n\rangle \langle\mathbf{k}_n|\mathbf{r}_n'\rangle =$$

$$[8] \quad \frac{1}{N!} \sum_P s_P \sum_{\mathbf{k}_1,\dots,\mathbf{k}_N} \frac{1}{V^N} \prod_{n=1}^{N} \exp\left[i(P\mathbf{k}_n \cdot \mathbf{r}_n - \mathbf{k}_n \cdot \mathbf{r}_n') - \frac{\beta\hbar^2}{2m}\mathbf{k}_n^2 \right]$$

In step [1] we rewrite Eq. (D.70) with slightly different notation, denoting as $\{\mathbf{r}_i\}$, $\{\mathbf{r}_i'\}$ the unprimed and primed coordinate sets. In step [2] we use the expression of the wavefunction Ψ_N from Eq. (D.69), which introduces the set of wave-vectors $\{\mathbf{k}_i\}$ for the bra and ket, and the associated sums over permutations P and P' with signs s_P and s_P', respectively. In step [3] we make use of the fact that in the position representation the individual single-particle states are plane waves $\exp[iP\mathbf{k}_i \cdot \mathbf{r}_i]$ which involve the dot products between the *permuted* $\mathbf{k}_i$ vectors and the $\mathbf{r}_i$ vectors, but the sum over all these dot products is equal to the sum over dot products between the $\mathbf{k}_i$ vectors and the *permuted* $\mathbf{r}_i$ vectors, with the same signs for each permutation; we can therefore transfer the permutation from the $\{\mathbf{k}_i\}$ vectors to the $\{\mathbf{r}_i\}$ vectors. In step [4] we use the fact that we can obtain the value of K^2 as the sum over any of the $N!$ permutations of the $\{\mathbf{k}_i\}$ vectors, therefore if we replace the sum over values of K by the N sums over values of the individual $\mathbf{k}_i$ vectors we are overcounting by a factor of $N!$; accordingly, we introduce an extra factor of $N!$ in the denominator to compensate for the overcounting. In step [5] we use the fact that all permutations P' of the vectors $\mathbf{k}_i$ give the same result, because now the $\mathbf{k}_i$s have become dummy summation variables since the sums are taken over all of them independently; therefore, we can replace the $N!$ terms from the permutations P' by a single term multiplied by $N!$, and since this term can be any of the possible permutations we pick the first permutation (with unchanged order of $\mathbf{k}_j$s) for convenience. In this way we get rid of the summation over the permutations P' and restore the factor $1/N!$ in front of the entire sum, rather than the $1/(N!)^2$ we had in step [4]. In step [6] we move the permutations P from the $\mathbf{r}_i$s to the $\mathbf{k}_i$s, by the same arguments used above for the reverse move. In step [7] we write explicitly the bra and ket of the many-body free-particle wavefunction as products of single-particle states and in step [8] we use the plane-wave expression for these single-particle states.

Let us consider how this general result applies to $N = 1$ and $N = 2$ particles for illustration. For $N = 1$ it takes the form

$$
\begin{aligned}
\langle \mathbf{r} | e^{-\beta \hat{\mathcal{H}}_1} | \mathbf{r}' \rangle &= \int \frac{d\mathbf{k}}{(2\pi)^3} \exp \left[-\frac{\beta \hbar^2}{2m} k^2 + i\mathbf{k} \cdot (\mathbf{r} - \mathbf{r}') \right] \\
&= \left(\frac{m}{2\pi \beta \hbar^2} \right)^{3/2} e^{-m(\mathbf{r}-\mathbf{r}')^2/2\beta \hbar^2}
\end{aligned} \tag{D.71}
$$

where we have used the usual replacement of $(1/V) \sum_{\mathbf{k}} \to \int d\mathbf{k}/(2\pi)^3$ and obtained the last expression by completing the square in the integral. We define the thermal wavelength λ as:

$$
\lambda = \sqrt{\frac{2\pi \hbar^2}{mk_{\mathrm{B}}T}} \tag{D.72}
$$

in terms of which the matrix element of $\exp[-\beta \hat{\mathcal{H}}_1]$ in the position representation takes the form

$$
\langle \mathbf{r} | e^{-\beta \hat{\mathcal{H}}_1} | \mathbf{r}' \rangle = \frac{1}{\lambda^3} e^{-\pi (\mathbf{r}-\mathbf{r}')^2/\lambda^2} \tag{D.73}
$$

The physical interpretation of this expression is that it represents the probability of finding the particle at positions $\mathbf{r}$ and $\mathbf{r}'$, or the spread of the particle's wavefunction in space; the thermal wavelength gives the scale over which this probability is non-vanishing. When $\lambda \to 0$ this expression tends to a δ-function, which is the expected limit since it corresponds to $T \to \infty$, at which point the behavior becomes classical and the particle is localized in space. Thus, λ gives the extent of the particle's wavefunction.

The normalization factor required for thermodynamic averages becomes:

$$
\mathrm{Tr}[e^{-\beta \hat{\mathcal{H}}_1}] = \int \langle \mathbf{r} | e^{-\beta \hat{\mathcal{H}}_1} | \mathbf{r} \rangle d\mathbf{r} = V \left(\frac{m}{2\pi \beta \hbar^2} \right)^{3/2} = \frac{V}{\lambda^3}
$$

With these expressions we can calculate thermodynamic averages, for instance the energy of the quantum-mechanical free particle, which involves matrix elements of the hamiltonian in the position representation given by:

$$
\langle \mathbf{r} | \hat{\mathcal{H}}_1 | \mathbf{r}' \rangle = \left[-\frac{\hbar^2}{2m} \nabla_{\mathbf{r}'}^2 \right]_{\mathbf{r}'=\mathbf{r}}
$$

giving, for the average energy:

$$
E_1 = \overline{\hat{\mathcal{H}}_1} = \frac{\mathrm{Tr}[\hat{\mathcal{H}}_1 e^{-\beta \hat{\mathcal{H}}_1}]}{\mathrm{Tr}[e^{-\beta \hat{\mathcal{H}}_1}]} = \frac{3}{2} k_{\mathrm{B}}T \tag{D.74}
$$

as expected for a particle with only kinetic energy at temperature T.

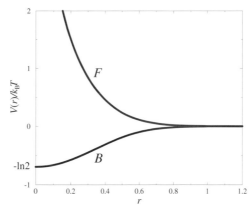

Effective interaction potential for a pair of bosons or fermions: the inter-particle distance r is
measured in units of the thermal wavelength λ.

For the case of two quantum-mechanical particles, $N = 2$, the density matrix becomes:

$$\langle \mathbf{r}_1, \mathbf{r}_2 | e^{-\beta \hat{\mathcal{H}}_2} | \mathbf{r}_1', \mathbf{r}_2' \rangle = \int \frac{d\mathbf{k}_1 d\mathbf{k}_2}{2(2\pi)^6} \exp\left[-\frac{\beta \hbar^2}{2m} (\mathbf{k}_1^2 + \mathbf{k}_2^2) \right] \times$$

$$\left[e^{i(\mathbf{k}_1 \cdot \mathbf{r}_1 - \mathbf{k}_1 \cdot \mathbf{r}_1')} e^{i(\mathbf{k}_2 \cdot \mathbf{r}_2 - \mathbf{k}_2 \cdot \mathbf{r}_2')} \pm e^{i(\mathbf{k}_2 \cdot \mathbf{r}_1 - \mathbf{k}_1 \cdot \mathbf{r}_1')} e^{i(\mathbf{k}_1 \cdot \mathbf{r}_2 - \mathbf{k}_2 \cdot \mathbf{r}_2')} \right]$$

where the $+$ sign corresponds to bosons and the $-$ sign to fermions. With this expression,
the normalization factor becomes:

$$\text{Tr}[e^{-\beta \hat{\mathcal{H}}_2}] = \int \langle \mathbf{r}_1, \mathbf{r}_2 | e^{-\beta \hat{\mathcal{H}}_2} | \mathbf{r}_1, \mathbf{r}_2 \rangle d\mathbf{r}_1 d\mathbf{r}_2 = \frac{1}{2} \left(\frac{V}{\lambda^3} \right)^2 \left[1 \pm \frac{\lambda^3}{2^{3/2} V} \right]$$

so that the normalized diagonal matrix element of the density takes the form

$$\frac{\langle \mathbf{r}_1, \mathbf{r}_2 | e^{-\beta \hat{\mathcal{H}}_2} | \mathbf{r}_1, \mathbf{r}_2 \rangle}{\text{Tr}[e^{-\beta \hat{\mathcal{H}}_2}]} = \frac{1}{V^2} \frac{1 \pm \exp\left[-2\pi (\mathbf{r}_1 - \mathbf{r}_2)^2 / \lambda^2 \right]}{1 \pm \lambda^3 / 2^{3/2} V} \tag{D.75}$$

with the $+$ signs for bosons and the $-$ signs for fermions. This matrix element represents
the probability of finding the two particles situated at positions $\mathbf{r}_1$ and $\mathbf{r}_2$ at temperature
T, which is equivalent to the finite-temperature pair correlation function. The expression
we obtained for this probability is interesting, because it reveals that, even though we have
assumed particles to be non-interacting, there is an effective interaction between them: if
there were no interaction at all, this probability should simply be $1/V^2$, the product of
the independent probabilities of finding each particle at position $\mathbf{r}$ in space. Therefore, the
remaining term represents the effective interaction which is due to the quantum-mechanical
nature of the particles. To explore this feature further, we first note that we can neglect the
denominator of the remaining term, since $\lambda^3 \ll V$: this is justified on the grounds that λ^3
is the volume over which the single-particle wavefunction is non-vanishing, as we found
above, which we assume to be much smaller than the volume of the system V. We can then

define an effective interaction potential $V(r)$ through the numerator of the remaining term in Eq. (D.75):

$$e^{-\beta V(r)} = 1 \pm \exp\left(-\frac{2\pi r^2}{\lambda^2}\right) \Rightarrow V(r) = -k_B T \ln\left[1 \pm \exp\left(-\frac{2\pi r^2}{\lambda^2}\right)\right]$$

where we have expressed the potential in terms of the relative distance between the particles $r = |\mathbf{r}_1 - \mathbf{r}_2|$. A plot of $V(r)/k_B T$ as a function of r (in units of λ) is shown in Fig. D.4. From this plot it is evident that the effective interaction is *repulsive* for fermions and *attractive* for bosons. The effective interaction is negligible for distances $r > \lambda$, which indicates that quantum-mechanical effects become noticeable only for inter-particle distances within the thermal wavelength. As the temperature increases, λ decreases, making the range over which quantum effects are important vanishingly small in the large-T limit, thus reaching the classical regime as expected.

Generalizing the above results, it is easy to show that for an N-particle system the diagonal matrix element of the density takes the form

$$\langle \mathbf{r}_1, \ldots, \mathbf{r}_N | e^{-\beta \hat{\mathcal{H}}_N} | \mathbf{r}_1, \ldots, \mathbf{r}_N \rangle = \frac{1}{N!\, \lambda^{3N}}\left[1 \pm \sum_{i<j} f_{ij} f_{ji} + \sum_{i<j<k} f_{ij} f_{jk} f_{ki} \pm \cdots\right] \quad (D.76)$$

with $+$ signs for bosons and $-$ signs for fermions, and the factors f_{ij} defined as:

$$f_{ij} = \exp\left[-\frac{\pi}{\lambda^2}(\mathbf{r}_i - \mathbf{r}_j)^2\right]$$

The average inter-particle distance in this system is $|\mathbf{r}_i - \mathbf{r}_j| \sim (V/N)^{1/3}$, while the extent of single-particle wavefunctions is λ. For

$$\lambda \ll (V/N)^{1/3} \sim |\mathbf{r}_i - \mathbf{r}_j|$$

which corresponds to the classical limit, the exponentials f_{ij} are vanishingly small and the normalizing factor becomes:

$$\mathrm{Tr}[e^{-\beta \hat{\mathcal{H}}_N}] = \int \langle \mathbf{r}_1, \ldots, \mathbf{r}_N | e^{-\beta \hat{\mathcal{H}}_N} | \mathbf{r}_1, \ldots, \mathbf{r}_N \rangle d\mathbf{r}_1 \cdots d\mathbf{r}_N = \frac{1}{N!}\left(\frac{V}{\lambda^3}\right)^N \quad (D.77)$$

which is exactly the same as the partition function $Z_N(\beta, V)$ that we calculated for the classical free-particle gas, Eq. (D.68), only now the factors $N!$ and h^{3N} needed in the denominator have appeared naturally from the underlying quantum-mechanical formulation of the problem.

As a last application of ensemble theory to the quantum-mechanical ideal gas, we obtain the equation of state for fermions and bosons. To this end, the most convenient approach is to use the grand-canonical partition function and the momentum representation, in which the matrix elements of the density are diagonal. This gives, for the canonical partition function:

$$Z_N(\beta, V) = \sum_{\mathbf{k}_1,\ldots,\mathbf{k}_N} \langle \mathbf{k}_1,\ldots,\mathbf{k}_N | e^{-\beta \hat{\mathcal{H}}_N} | \mathbf{k}_1,\ldots,\mathbf{k}_N \rangle$$

$$= \sum_{\mathbf{k}_1,\ldots,\mathbf{k}_N} \exp\left[-\beta E(n_{\mathbf{k}_1},\ldots,n_{\mathbf{k}_N}) \right]$$

$$= \sum_{\{n_{\mathbf{k}}\}} \exp\left[-\beta E(\{n_{\mathbf{k}}\}) \right] \tag{D.78}$$

where we have used the occupation numbers $n_{\mathbf{k}}$ of states $\mathbf{k}$ to express the energy of the system as:

$$E(\{n_{\mathbf{k}}\}) = \sum_{i=1}^{N} n_{\mathbf{k}_i} \epsilon_{\mathbf{k}_i}, \qquad \epsilon_{\mathbf{k}_i} = \frac{\hbar^2 \mathbf{k}_i^2}{2m}, \qquad \sum_{i=1}^{N} n_{\mathbf{k}_i} = N$$

With this result for the canonical partition function, the grand-canonical partition function becomes:

$$Z(z,\beta,V) = \sum_{N=0}^{\infty} z^N Z_N(\beta,V) = \sum_{N=0}^{\infty} z^N \sideset{}{'}\sum_{\{n_{\mathbf{k}}\}} \exp\left[-\beta \sum_{\mathbf{k}} n_{\mathbf{k}} \epsilon_{\mathbf{k}} \right]$$

$$= \sum_{N=0}^{\infty} \sideset{}{'}\sum_{\{n_{\mathbf{k}}\}} \prod_{\mathbf{k}} \left(z e^{-\beta \epsilon_{\mathbf{k}}} \right)^{n_{\mathbf{k}}} \tag{D.79}$$

where the symbol $\sum'_{\{n_{\mathbf{k}}\}}$ is used to denote that the summation over the occupation numbers $\{n_{\mathbf{k}}\}$ must be done with the restriction $\sum_{\mathbf{k}} n_{\mathbf{k}} = N$. However, since we also have an additional summation over all the values of N, we can simply substitute the restricted summation over $\{n_{\mathbf{k}}\}$ with an unrestricted summation and omit the summation over the values of N, thus obtaining:

$$Z(z,\beta,V) = \sum_{\{n_{\mathbf{k}}\}} \prod_{\mathbf{k}} (z e^{-\beta \epsilon_{\mathbf{k}}})^{n_{\mathbf{k}}} = \sum_{n_{\mathbf{k}_1}} \sum_{n_{\mathbf{k}_2}} \cdots (z e^{-\beta \epsilon_{\mathbf{k}_1}})^{n_{\mathbf{k}_1}} (z e^{-\beta \epsilon_{\mathbf{k}_2}})^{n_{\mathbf{k}_2}} \cdots$$

$$= \prod_{\mathbf{k}} \sum_{n_{\mathbf{k}}} (z e^{-\beta \epsilon_{\mathbf{k}}})^{n_{\mathbf{k}}} \tag{D.80}$$

The individual occupation numbers of indistinguishable particles can be:

$$n_{\mathbf{k}} = 0, 1, 2, \ldots \quad \text{for bosons}$$

$$n_{\mathbf{k}} = 0, 1, \ldots, (2s+1) \quad \text{for fermions with spin } s$$

For simplicity we only discuss below the case of spin $s = 0$ particles, the extension to the general case being straightforward.

Ideal Bose gas. The case of the ideal Bose gas is straightforward. The grand partition function contains a geometric sum which can be summed to give:

$$Z^{(B)}(z,\beta,V) = \prod_{\mathbf{k}} \frac{1}{1 - z e^{-\beta \epsilon_{\mathbf{k}}}} \tag{D.81}$$

and the logarithm of this expression gives the equation of state:

$$\frac{PV}{k_B T} = \ln Z^{(B)}(z, \beta, V) = -\sum_{\mathbf{k}} \ln(1 - ze^{-\beta\epsilon_{\mathbf{k}}}) \tag{D.82}$$

while the average occupation number of state $\mathbf{k}$ becomes:

$$\bar{n}_{\mathbf{k}}^{(B)} = -\frac{1}{\beta}\frac{\partial}{\partial\epsilon_{\mathbf{k}}} \ln Z^{(B)}(z, \beta, V) = \frac{1}{z^{-1}e^{\beta\epsilon_{\mathbf{k}}} - 1} \tag{D.83}$$

which, with $z = \exp[\beta\mu]$, is identical to the expression derived earlier, Eq. (D.38). The average total number of particles in the system is:

$$\bar{N}^{(B)} = z\frac{\partial}{\partial z} \ln Z^{(B)}(z, \beta, V) = \sum_{\mathbf{k}} \frac{ze^{-\beta\epsilon_{\mathbf{k}}}}{1 - ze^{-\beta\epsilon_{\mathbf{k}}}} = \sum_{\mathbf{k}} \bar{n}_{\mathbf{k}}^{(B)} \tag{D.84}$$

Ideal Fermi gas. For spinless fermions we have $n_{\mathbf{k}} = 0, 1$ and the grand partition function becomes:

$$Z^{(F)}(z, \beta, V) = \prod_{\mathbf{k}}(1 + ze^{-\beta\epsilon_{\mathbf{k}}}) \tag{D.85}$$

and the logarithm of this expression gives the equation of state:

$$\frac{PV}{k_B T} = \ln Z^{(F)}(z, \beta, V) = \sum_{\mathbf{k}} \ln(1 + ze^{-\beta\epsilon_{\mathbf{k}}}) \tag{D.86}$$

while the average occupation number of state $\mathbf{k}$ becomes:

$$\bar{n}_{\mathbf{k}}^{(F)} = -\frac{1}{\beta}\frac{\partial}{\partial\epsilon_{\mathbf{k}}} \ln Z^{(F)}(z, \beta, V) = \frac{1}{z^{-1}e^{\beta\epsilon_{\mathbf{k}}} + 1} \tag{D.87}$$

which, with $z = \exp[\beta\mu]$, is identical to the expression derived earlier, Eq. (D.37), through the argument based on the most probable distribution. The average total number of particles in the system, since we are dealing with the grand-canonical ensemble which allows fluctuations in this number, is given by:

$$\bar{N}^{(F)} = z\frac{\partial}{\partial z} \ln Z^{(F)}(z, \beta, V) = \sum_{\mathbf{k}} \frac{ze^{-\beta\epsilon_{\mathbf{k}}}}{1 + ze^{-\beta\epsilon_{\mathbf{k}}}} = \sum_{\mathbf{k}} \bar{n}_{\mathbf{k}}^{(F)} \tag{D.88}$$

Using the expression for the grand partition function of the spinless Fermi gas, Eq. (D.85), the definition of the corresponding free energy, Eq. (D.61), and the general relation between the free energy and entropy, Eq. (D.62), and noting that from the definition of $\bar{n}_{\mathbf{k}}^{(F)}$, Eq. (D.87), we have:

$$\bar{n}_{\mathbf{k}}^{(F)} = \frac{ze^{-\beta\epsilon_{\mathbf{k}}}}{1 + ze^{-\beta\epsilon_{\mathbf{k}}}}, \quad 1 - \bar{n}_{\mathbf{k}}^{(F)} = \frac{1}{1 + ze^{-\beta\epsilon_{\mathbf{k}}}}$$

we can show straightforwardly that the entropy of a spinless Fermi gas is given by:

$$S^{(F)} = k_B \sum_{\mathbf{k}} \left[\bar{n}_{\mathbf{k}}^{(F)} \ln \bar{n}_{\mathbf{k}}^{(F)} + \left(1 - \bar{n}_{\mathbf{k}}^{(F)}\right) \ln \left(1 - \bar{n}_{\mathbf{k}}^{(F)}\right) \right] \tag{D.89}$$

We derive next some expressions for the equation of state of the ideal Fermi gas that are useful in understanding material properties (see, for example, applications in magnetic properties, Chapter 10). We use the usual substitution $(1/V)\sum_{\mathbf{k}} \rightarrow \int d\mathbf{k}/(2\pi)^3$ and define the volume per particle $v = V/\bar{N}$ to obtain, for the fermion gas from Eqs (D.86) and (D.88):

$$\frac{P}{k_B T} = \int \ln\left(z\exp\left[-\beta\frac{\hbar^2 \mathbf{k}^2}{2m}\right]+1\right)\frac{d\mathbf{k}}{(2\pi)^3} = \frac{1}{\lambda^3}h_5(z) \tag{D.90}$$

$$h_5(z) \equiv \frac{4}{\sqrt{\pi}}\int_0^\infty u^2 \ln(1+ze^{-u^2})du$$

$$\frac{1}{v} = \int\left(z^{-1}\exp\left[\beta\frac{\hbar^2 \mathbf{k}^2}{2m}\right]+1\right)^{-1}\frac{d\mathbf{k}}{(2\pi)^3} = \frac{1}{\lambda^3}h_3(z) \tag{D.91}$$

$$h_3(z) \equiv \frac{4}{\sqrt{\pi}}\int_0^\infty u^2 \frac{ze^{-u^2}}{1+ze^{-u^2}}du$$

where we have introduced the variable

$$u = \left(\frac{\beta\hbar^2}{2m}\right)^{1/2}|\mathbf{k}| = \frac{\lambda}{\sqrt{4\pi}}|\mathbf{k}|$$

with λ the thermal wavelength. We have also defined the integrals over $\mathbf{k}$ in the previous equations as the functions h_5 and h_3 of z, which is the only remaining parameter. The subscripts of these two functions come from their series expansions in z:

$$h_n(z) = \sum_{m=1}^\infty \frac{(-1)^{m+1}z^m}{m^{n/2}}, \quad n = 3, 5 \tag{D.92}$$

Further Reading

1. *Statistical Mechanics*, K. Huang (2nd edn, Wiley, New York, 1987). This is a standard treatment of the subject at an advanced level, on which most of the discussion in this appendix is based.

Index